ILLUSTRATED FLORA

Part of Western Uttar Pradesh and Delhi NCR, India

Wild growth of *Equisetum ramosissimum*

Shiv Nadar University

Shiv Nadar University is a comprehensive, multidisciplinary, research-focused and student-centric institution that is bringing a paradigm shift in higher education in India through its innovative curriculum, interdisciplinary focus, and cross-disciplinary thinking across a wide range of disciplines. The University is building an eco-system of knowledge to promote recognition of the inter-connectedness of ideas, systems, and environments in the world inside the campus, and those outside it. The University has 5 Schools, 18 Departments and 6 Research Centers engaged in teaching, practice, and research in disciplines as diverse as Engineering, Humanities & Social Sciences, Management, Natural Sciences, Art, Design, Performing Arts, Communication, and Extended Education & Professional Development. The schools offer Bachelor, Masters, and Doctoral degrees along with multidisciplinary curriculum to enable students to explore subjects and disciplines that may be widely different from their chosen Majors.

ILLUSTRATED FLORA

Part of Western Uttar Pradesh and Delhi NCR, India

Jyoti K Sharma
Amit K Tripathi
Mohd Ahmad
Center for Environmental Sciences and Engineering
School of Natural Sciences
Shiv Nadar University
Greater Noida, Uttar Pradesh 201314

2025
Daya Publishing House®
A Division of
Astral International Pvt. Ltd.
New Delhi – 110 002

Reprinted, 2025

ISBN: 978-93-88982-05-4 (HB)

Photo Credit: Dr. Amit K Tripathi
Front Cover: Date Palm (Phoenix sylvestris) Grove

SNU Library Cataloguing-in-Publication Data

Names: Sharma, Jyoti K. | Tripathi, Amit K. | Ahmad, Mohd
Title: Illustrated Flora: Part of Western Uttar Pradesh and Delhi NCR, India
Description: Delhi: Astral International (P) Ltd.; Greater Noida: Shiv Nadar University | Includes Bibliographic references and index | Pages: 606 | Illustrations: 1061 | Size: 22 x 29 cm
Identifiers: ISBN: 978-93-88982-05-4 (Print)
Subjects: Botany | Plant Taxonomy | Illustrated Flora | Taxonomic description | Medicinal value and uses | Chemical constituents
Classification: DDC 581.9542 SHA

Citation of the Book

Sharma, Jyoti K., Tripathi, Amit K. and Ahmad, Mohd 2018. Illustrated Flora: Part of Western Uttar Pradesh and Delhi NCR, India. Astral International (P) Ltd and Shiv Nadar University, India, p. 606.

This book forms part of the research project "Biodiversity documentation of Chithara Village Panchayat, Dadri, Gautam Buddha Nagar, Uttar Pradesh, India" funded by Shiv Nadar University, India.

Published by : **Daya Publishing House®**
A Division of
Astral International Pvt. Ltd.
– ISO 9001:2015 Certified Company –
4736/23, Ansari Road, Darya Ganj
New Delhi-110 002
Ph. 011-43549197, 23278134
E-mail: info@astralint.com
Website: www.astralint.com &

Shiv Nadar University,
Greater Noida, Tehsil- Dadri,
District- Gautam Buddha Nagar
UP 201314 (India)

Digitally Printed at : **Replika Press Pvt. Ltd.**

Foreword

SHIV NADAR UNIVERSITY
NH-91, Tehsil Dadri, District Gautam Buddha Nagar
UP 201314, India

Professor Rupamanjari Ghosh, Ph.D.
Founding Director, School of Natural Sciences, and
Vice-Chancellor

During my very first visit to the Shiv Nadar University when it was just about starting its operations, I was struck by the abundant natural wealth of the area. There were also visible signs of the pressures of development or economic growth, increasing population and growing consumerism, taking a rapid toll on the health of this rich environment. I started searching for an answer as to what was really 'worth' conserving in the face of these pressures, and the first step towards the answer was clearly a proper **scientific documentation** of the local biodiversity which would then help formulate a much-needed **strategy** for conservation. An urgent focused research on documentation of plant and animal diversity was to be carried out so that the inevitable human intervention could be informed, integrated and holistic.

It was heartening for me to learn that the Biological Diversity Act (2002) directs every *Panchayat* in the country to prepare its own People's Biodiversity Registers, but no such Register existed for this region. Fortunately, when Prof. Jyoti Kumar Sharma with his vast experience and expertise in this field joined the University in early 2013, this was the first project that he readily agreed to lead. The study of flora is the first major step in documenting and conserving the local plant diversity. The current documentation of the flora of the Chithara Village Panchayat in Block Dadri, District Gautam Buddha Nagar, Uttar Pradesh, India, which should serve as a model for the local Biodiversity Register, describes 272 plant species under 203 genera belonging to 69 plant families. I understand that this is the first time such a comprehensive floristic study has been carried out in a Village Panchayat in the country. This pioneering initiative is in fact in line with the overall vision and commitment of the University to creating knowledge and solutions that would serve national and global needs. I do hope that the book will play a crucial role in enhancing awareness about our local plant diversity and give impetus to other researchers to undertake such floristic studies in their regions so that we sustainably utilize the wild plant diversity and conserve it for posterity.

The book, besides giving valuable taxonomic information about plants of the region, provides details on their chemical and medicinal properties and ethno-medicinal uses. I am sure that the pharmacological information on medicinal properties of plants for treating various diseases/ailments and their chemical composition will motivate researchers to isolate effective biomolecules for drug development for important diseases. Excellent photographs of all the plants recorded will aid their easy identification in the field for further study and research.

I congratulate Prof. Jyoti Kumar Sharma and his team members, Dr. Amit K Tripathi and Mr. Mohd. Ahmad, for bringing out an excellent reference book on Plant Taxonomy, which will go a long way in fulfilling a critical need felt by students, teachers and researchers in India.

R. Ghosh

30 September 2017

Rupamanjari Ghosh

Preface

Floristic studies of a particular area/region are vital for conservation of native and endemic species of flora and for bringing awareness about the value of wild plants and their uses. The study area Chithara Village Panchayat in Gautam Buddha Nagar district in western Uttar Pradesh, also a part of the National Capital Region (NCR) of Delhi, is one of the 7 districts of Uttar Pradesh included in NCR. While NCR is arid and semi-arid, area of western Uttar Pradesh, situated in the Gangetic plains is one of the most fertile regions of the country and equally rich in wild plant diversity. One of earliest and pioneering detailed floristic studies in India "Flora of Upper Gangetic Plains, and of the adjacent Siwalik and sub-Himalayan tracts" carried out by Duthie during 1903-1929, is still considered as the classical work on plant taxonomy not in India but whole of the world. However, in later years detailed floristic diversity studies, except for a few districts in NCR and also in western Uttar Pradesh have not got due attention in northern India. The prominent flora in NCR are of Delhi, Ghaziabad, Baghpat, Jhajjar and Alwar districts and Bulandshahr, Muzaffarnagar, Saharanpur, Pilibhit, Agra, Aligarh and Mathura districts in western Uttar Pradesh. Except a handful of them none of these flora are exhaustive and provide illustrations of all the plant species for authentic field identification purpose. Since floristic diversity keep on changing in time and space due to various anthropogenic and natural factors studies need to be carried out periodically. The present book on "Illustrated Flora of Part of Western Uttar Pradesh and Delhi NCR, India", though deals with a small geographical area, provides for the first time an account of a detailed floristic study carried out in a local body in the region. The book also provides the first of its kind of floristic treatment of an educational institution in the country. In the past a few studies have been undertaken on the Campus flora in the country. While some have just the checklist of plants (Flora of Amar Singh College, Jammu) others provide details of the floristic diversity (Flora of BHU Campus, Flora of Pondicherry University Campus, Flora of Bharathiar University Campus, Biodiversity of IIT Bombay Campus) but without illustrations or providing photographs of only a few plant species.

The documentation of plant diversity covering all the seasons undertaken during 2014-2017 is unique as it provides taxonomic details of wild plant species recorded along with illustrations and medicinal properties as most of the wild plants recorded have been used in traditional/folklore medicines. Though during the course of this investigation many of the wild areas were brought under developmental activities in the wetlands, it is envisaged that at least some of the wild growth towards the wetland will be left as such without further disturbance so that unique plant diversity, especially rare species of *Equisetum ramosissimum* and *Equisetum ramosissimum* subsp. *debile* discovered in the SNU Campus and surrounding wetland areas and reported for the first time from northern India, are well preserved and conserved.

Authentic identification/recognition of plants in the field is a major challenge since there are a large number of vernacular and common names for the same species as well as different plant species have the same common name. Hence, besides the taxonomic description, to facilitate field identification for the first time the book provides original photographs of all the plants recorded in the flora illustrating morphology of leaf, flower and fruits besides their common names, vernacular names in Hindi and Sanskrit, habit, habitat and distribution. All the photographs in the book are taken by Amit K. Tripathi.

Also provided for each species are chemical properties, medicinal properties and folklore and traditional medicinal uses which have been compiled from extensive literature search. The book also provides a Glossary of medical and taxonomic terms used as well as an appendix giving the properties of medicinal plant species used for various diseases and ailments in traditional and folk systems of medicines.

It is hoped that the book will be equally useful to undergraduate and postgraduate students, teachers, and researchers of Botany/Plant taxonomy/Ethnobotany/Ayurveda/Biochemistry in Colleges and Universities in India and elsewhere and it will generate interest in local plants and their conservation.

We are thankful to Dr. Sunil Srivastava, Additional Director, Western Zone BSI, Dehradun for their generous help in the identification of plant specimens. We also record our appreciation for the initial work carried out by Dr. Ganesh Datt Bhatt, Research Associate in the collection of plants in the SNU Campus and land use land cover studies.

Last but not the least we owe a great deal to Prof. Rupamanjari Ghosh, Vice Chancellor, SNU for her keen interest in preserving and studying the biodiversity of the region and especially of the SNU Campus and supporting to undertake this important study.

Dr Jyoti K Sharma

Dr Amit K Tripathi

Mohd Ahmad

30 September 2017

Center for Environmental Sciences & Engineering
School of Natural Sciences
Shiv Nadar University

Contents

1

Introduction

The dynamics and patterns of biodiversity, which are measured in terms of species richness, distribution of organisms, and their ecological roles in a defined, physical environment, are influenced by the characteristics of both the immediate physical environment and the surrounding landscape. Many biodiversity rich areas in the tropics are undergoing rapid, wide ranging changes in land cover due to deforestation on account of anthropogenic pressure and diversion of land for agriculture, infrastructure development, etc. As a consequence the tropical biodiversity is getting depleted at a faster rate. Conservation efforts of these species are often hampered by the absence of basic information upon which to build conservation strategies and reliable alternatives to get rid of uncontrolled and probably dangerous developmental activities (Sharma 2016).

Biodiversity represents the variability among living organisms from all sources including, inter alia, terrestrial, marine and other aquatic ecosystems and the ecological complexes of which they are part; this includes diversity within species, between species and of ecosystems. The Convention on Biological Diversity (CBD), is the most significant of the pertinent international agreements. India signed the CBD on 5th June 1992, ratified it on 18 February 1994 and brought it into force on 19 May 1994. As a follow up to this Convention, India has enacted the Biological Diversity Act, 2002, which received the assent of the President on the 5th February, 2003. Subsequently, the Biological Diversity Rules were enacted during 2004.

The Biological Diversity Act extends to the whole of India and reaffirms the sovereign rights of the state over its biological resources. The three main tenets of the Act are: "Conservation of biological diversity, sustainable use of its components and fair and equitable sharing of the benefits arising out of the use of biological resources, knowledge and for matters connected therewith." Section – 41(1) of the Act states that "Every local body shall constitute a Biodiversity Management Committee (BMC) within its area for the purpose of promoting conservation, sustainable use and documentation of biological diversity including preservation of habitats, conservation of land races, folk varieties and cultivars, domesticated stocks and breeds of animals and microorganisms and chronicling of knowledge relating to biological diversity." Subsequently, the Government of India published Biological Diversity Rules, 2004 and Section 22 (6) of which states –"The main function of the BMC is to prepare People's Biodiversity Registers in consultation with local people. The Register shall contain comprehensive information on availability and knowledge of local biological resources, their medicinal or any other use or any other traditional knowledge associated with them."

Since, biodiversity and associated knowledge found in different ecosystems, is being eroded day by day information on the biological resources present in various ecological communities and the species present therein is of paramount importance in conservation planning and natural resource management. However, in the absence of knowledge about the species present and taxonomic identification of individual species make it difficult to assess the richness of plant diversity and whether particular communities are present on particular sites and how they may be affected by anthropogenic factors or proposed development. In this respect floristic surveys play an important role in biodiversity conservation and they form an important element of the documentation. "Flora" refers to the actual plant species growing within an ecological community of a particular region or period, listed as species and considered as a group. Floristic surveys are carried out usually by exploring a defined area or region many times during one or more growing seasons to identify plants and collecting specimens for preparing herbarium for generating baseline data which plays a crucial role in floristic studies. A Herbarium of all the plant specimens collected is a standard practice of flora work; herbarium is a collection of plants that usually have been dried, pressed, preserved on sheets and arranged according to any accepted system of classification for future reference and study.

As a first step towards conservation and sustainable utilization of biodiversity, work on biodiversity documentation was undertaken in Chithara Village Panchayat, Dadri Tehsil, Gautam Buddha Nagar, Uttar Pradesh where so far no efforts have been made to prepare the People's Biodiversity Register as mandated in the Biological Diversity Act (2002). The biodiversity documentation of Chithara Village Panchayat (Population: 7,656; Area: 770.78 ha; Coordinates: latitude 28°32'13.48"N, longitude 77°34'09.13"E; Altitude 210 msl) gave us an opportunity to carry out the detailed floristic study of the area, including the rich flora of SNU Campus of 115.74 ha also located in the Panchayat surrounded by wetlands and agricultural fields. The campus is rich in biodiversity due to adjacent wetlands, where more than 200 avi fauna have been sighted in the past and is considered to be the home for one of the endangered birds, Bristled grass bird (*Chaetornis striatus*).

Table 1. Floristic details of studies undertaken in the districts under Western Uttar Pradesh and Delhi NCR

SN	District (Area km^2)	Floristic study			
		Total plant species recorded	No. of plant Families	No. of Genera	Reference
1	Agra (4027)	609	103	384	Sharma and Dhakre (1995)
2	Aligarh (3700.4)	751*	134	525	Shazia (2013)
3	Amroha (2,321)	-	-	-	-
4	Badaun (4,234)	-	-	-	-
5	Baghpat (NCR)	566	102	371	Ahamed and Gupta (2010)
6	Bijnor (4,049)	-	-	-	-
7	Bulandshahr (NCR)	376	-	-	Singh (1969)
8	Bareilly (4120)	151	45	91	Misra (2015)#
9	Etah (4,446)	-	-	-	-
10	Etawah (2,311)	560	102	340	Singh and Misra (2014)$
11	Firozabad (2,362)	-	-	-	-
12	Gautam Buddha Nagar (NCR)	257	65	214	Ansari *et al.* (2016)
13	Ghaziabad (NCR)	1,654*	171	868	Vardhana (2007)
14	Hapur (NCR)	-	-	-	-
15	Kasganj (1,993)	-	-	-	-
16	Hathras (1,840)	-	-	-	-
17	Mainpuri (2,745)	-	-	-	-
18	Mathura (3,329.4)	507	97	321	Sharma *et al.* (2005b)
19	Meerut (NCR)	862	118	521	Shishodia (2013)
20	Moradabad (3493)	235	46	154	Bhati *et al.* (2014)
21	Muzaffarnagar (NCR)	577	77	321	Malik (2015a)
22	Pilibhit (3,504)	723*	128	471	Malhotra (1979)
23	Rampur (2,367)	-	-	-	-
24	Sambhal	-	-	-	-
25	Saharanpur (3,860)	142	-	74	Malik (2015b)
26	Shahjahanpur (4,575)	151	45	91	Misra (2015)#
27	Shamli (1,063)	-	-	-	-

**Wild and cultivated; # - Shahjahanpur, study also included Bareilly, Lakhimpur-Kheri and Pilibhit districts; $ - Study also included Auraiya district of Uttar Pradesh*

The study area in Gautam Buddha Nagar district in western Uttar Pradesh, situated in National Capital Region (NCR), is one of the 22 districts from western Uttar Pradesh, Haryana and Rajasthan included in NCR. While most of the southern NCR encompassing Haryana and Rajasthan districts is arid and semi-arid, area of western Uttar Pradesh, situated in the Gangetic plains is one of the most fertile regions of the country and equally rich in wild plant diversity. Even though there has been tremendous land use and land cover changes during the past two decades as a result of urbanization, however, due to the presence of extensive wetlands (Surajpur and Bil Akbarpur), agricultural fields and wastelands with wild plant growth the area is still rich in wild plant diversity. After one of the earliest and pioneering detailed floristic studies in India "Flora of the upper Gangetic Plains, and of the adjacent Siwalik and sub-Himalayan tracts" by Duthie (1903-29) at regional level and Flora of Delhi by Maheshwari (1963) subsequently, no such detailed studies have been carried out on regional basis and most of the studies were confined to administrative boundaries of districts.

Since, the study area in Gautam Buddha Nagar district is part of western Uttar Pradesh and which is also included in NCR, floristic studies conducted both in western Uttar Pradesh and NCR were searched in the literature. In Western Uttar Pradesh, the most fertile region of the country, of the 27 districts, floristic studies have been done only in 15 districts (Table 1; Appendix 1). Except a few, most of the floristic studies are sketchy describing a few hundreds of plant species. Ghaziabad district recorded the highest number of plant species (1,654) and genera (858), indicating exhaustiveness of the study. This was followed by flora of Meerut, Aligarh, Pilibhit, and Agra districts. None of these flora are illustrated and do not provide any information on uses of plants. In eleven districts of western Uttar Pradesh no floristic studies have been undertaken so far.

In NCR, which consists of a total of 22 districts including those of Uttar Pradesh (7), Haryana (13) and Rajasthan (2), in addition to NCT of Delhi, floristic studies have been done only in 14 districts, including NCT (Table 2; Appendix 2). The highest number of species (1,654) were recorded from Ghaziabad district; NCT of Delhi which is higher in geographical area recorded only 531 plant species. This difference in floristic diversity could be due to presence of ridge forests and also more urbanization in Delhi as compared to Ghaziabad and also presence of varied habitats in Ghaziabad than in Delhi. Ghaziabad district was followed by Meerut and Jhajjar districts recording higher floristic diversity than Delhi; of course both have higher geographic area than Delhi. Out of seven districts of UP in NCR floristic studies have been done in six; of 13 districts of Haryana floristic studies have been done only in five districts; of two district of Rajasthan both have their floristic studies done. Except Maheshwari's Flora of Delhi, which provided excellent line drawings of plants, none of the other flora included any illustrations of plant species which makes difficult for the students to identify uncommon or rare plant species in the field.

Further, in some of the districts the floristic studies were restricted only to sanctuary areas. In Gautam Buddha Nagar district the floristic study has been carried out only in Surajpur Wetlands Birds Sanctuary (area 3.08 km^2). Similarly, the floristic study in Bharatpur and Gurgaon were restricted only to Keoladeo Birds Sanctuary, Bharatpur (28.7 km^2) and Sultanpur National Birds Sanctuary, Gurgaon (1.42 km^2), respectively. Since, except a few all the studies are not exhaustive from the point of view of conservation of plant diversity in this region of the country, it is essential that detailed floristic studies are conducted in all the remaining districts of western Uttar Pradesh and NCR where either the information is sketchy or no information on plant diversity is available.

Besides the detailed illustrated floristic study of Chithara, another unique feature of the book is detailed floristic diversity study of Shiv Nadar University Campus (115.74 ha). In the past a few studies have been undertaken on the Campus flora in the country. While some have just the checklist of plants (Flora of Amar Singh College, Jammu) others provide details of the floristic diversity (Flora of BHU Campus, Flora of Pondicherry University Campus, Flora of Bharathiar University Campus, Biodiversity of IIT Bombay Campus) but without illustrations or providing photographs of only a few plant species. Among a few, one of the well documented campus flora is of BHU Campus which provides details of the floristic diversity, including wild, cultivated and domesticated horticultural species. However, in the absence of any illustrations the usefulness of the well documented flora could not be achieved by the residents of the campus. The details of horticultural flora of SNU Campus, including that of Botanic Garden are provided in a separate Chapter.

Table 2. Floristic details of studies undertaken in NCT and districts of Uttar Pradesh, Haryana and Rajasthan included in Delhi NCR

SN	District (Area km² and State)	Floristic study			
		Total plant species recorded	No. of plant Families	No. of Genera	Reference
National Capital Region (NCR)-58,332 km²					
1	National Capital Territory (NCT) of Delhi (1,483)	531	-	-	Maheshwari (1963)
2	Baghpat (1,321) – UP	566	102	371	Ahamed and Gupta (2010)
3	Bulandshahr (4,353) – UP	376	-	-	Singh (1969)
4	Gautam Buddha Nagar (1442) – UP	257	65	214	Ansari *et al.* (2016)
5	Ghaziabad (1273) – UP	1,654*	171	868	Vardhana (2007)
6	Hapur (660) – UP	-	-	-	-
7	Meerut (2,522) – UP	862	118	521	Shishodia (2013)
8	Muzaffarnagar (3,008) – UP	577	77	321	Malik (2015a)
9	Bhiwani (5,140) – HR	-	-	-	-
10	Faridabad (2151) – HR	-	-	-	-
11	Gurgaon (1,253) – HR	151	54	129	Debta and Srivastava (2015)
12	Jind (2,702) – HR	282	76	-	Singh and Kumar (2013)
13	Jhajjar (1,834) – HR	731	114	459	Singh (2009)
14	Karnal (2520) – HR	345	77	245	Kaur *et al.* (2016)
15	Mahendragarh, (1,899) – HR	50	-	-	Yadav *et al.* (2010)
16	Nuh, (1,860) – HR	-	-	-	-
17	Palwal (1,359) – HR	-	-	-	-
18	Panipat (1268) – HR	-	-	-	-
19	Riwari (1,594) – HR	-	-	-	-
20	Rohtak (1,668) – HR	-	-	-	-
21	Sonipat (2,260) – HR	-	-	-	-
22	Alwar (8,380) – RJ	403	86	271	Dular (2015)
23	Bharatpur (5,066) – RJ	323	77	237	Kumar and Chauhan (2011)

**Wild and cultivated; UP - Uttar Pradesh; HR – Haryana; RJ - Rajasthan*

In the present floristic study, carried out in the NCR region and part of western Uttar Pradesh, besides the taxonomic description of each plant species, to facilitate field identification for the first time the book provides illustrations of all the plants recorded in the flora along with vernacular/common names, habit, habitat and distribution.

Most of the weeds are known to possess medicinal properties and have been used all over the world in folk medicines, traditional medicines and even modern system of medicines. Chemical constituents/composition, medicinal properties and medicinal uses for all the species were collected through extensive literature search and have been included under each species. Also, the medicinal properties and medicinal uses of all the recorded species for various diseases and ailments were analyzed and details are included in a separate Chapter. To make the book useful to reader of any category a glossary of medical terms used and an appendix listing various diseases and ailments for which various plant species have been used in traditional and folk system of medicines are also provided.

As the Indian flora consists of a large number of alien species (Nayar 1997) nativity of all the plant species recorded in the flora was ascertained from the published literature and their status in the flora analyzed critically and details are included in a separate Chapter.

2

Methodology

The Study Area

Western Uttar Pradesh, is a region in India that comprises the western districts of Uttar Pradesh state, including the areas of Rohilkhand and Braj. The region has some demographic, economic and cultural patterns that are distinct from other parts of Uttar Pradesh, and more closely resemble those of Haryana and Rajasthan states. Western Uttar Pradesh has experienced rapid economic growth, due to intensive agriculture under Green Revolution and infrastructure development similar to Haryana and Punjab. A major part of western Uttar Pradesh is a part of National Capital Region of India. Western Uttar Pradesh, is a region in India that comprises of 27 western districts of Uttar Pradesh: Meerut, Bulandshahr, Gautam Buddha Nagar, Ghaziabad, Hapur, Saharanpur, Muzaffarnagar, Shamli, Moradabad, Bijnor, Rampur, Amroha, Sambhal, Etawah, Badaun, Bareilly, Pilibhit, Shahjahanpur, Agra, Firozabad, Mainpuri, Mathura, Aligarh, Etah, Hathras, Kasganj, Baghpat (Fig. 1).

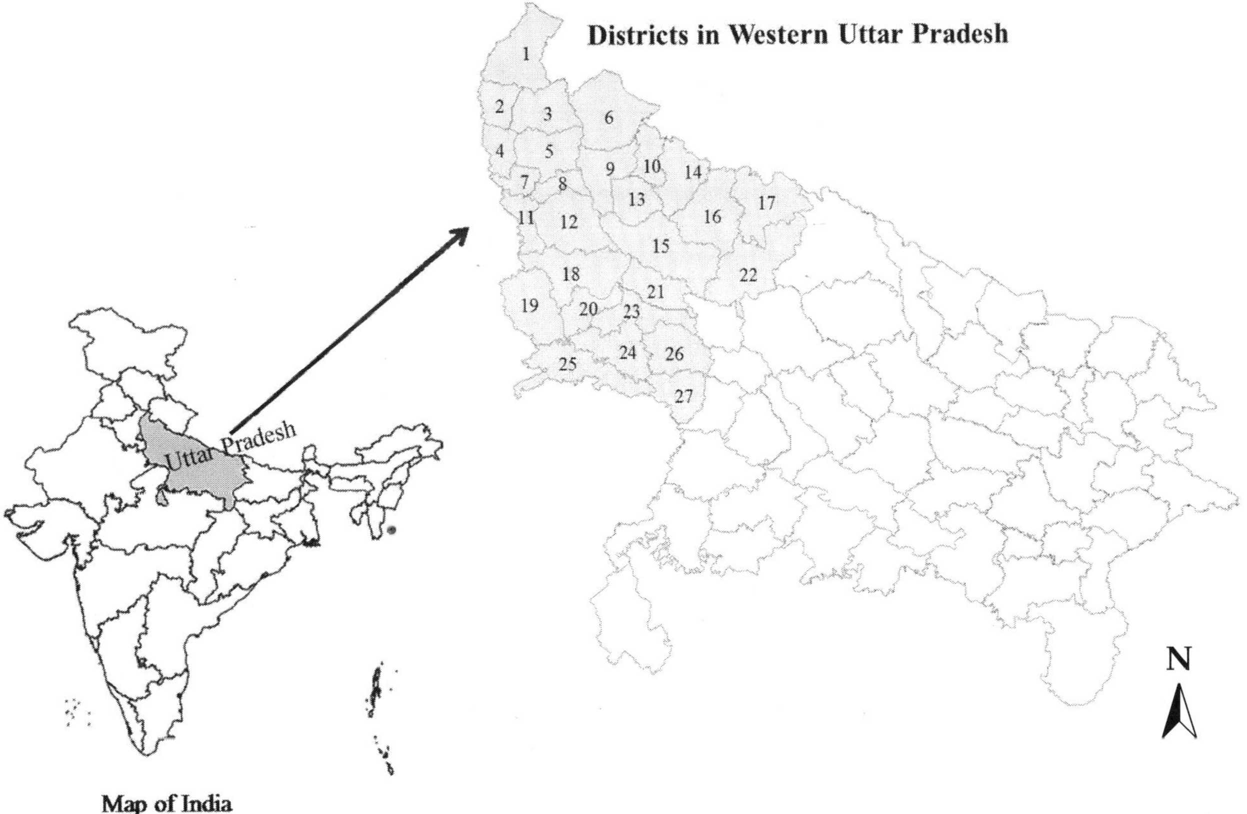

Fig. 1: Districts in Western Uttar Pradesh

(1. Saharanpur; 2. Shamli; 3. Muzaffarnagar; 4. Baghpat; 5. Meerut; 6. Bijnor; 7. Ghaziabad; 8. Hapur; 9. Amroha; 10. Moradabad; 11. **Gautam Buddha Nagar**; 12. Bulandshahr; 13. Sambhal; 14. Rampur; 15. Badaun; 16. Bareilly; 17. Pilibhit; 18. Aligarh; 19. Mathura; 20. Harhras; 21. Kasganj; 22. Shahjahanpur; 23. Etah; 24. Firozabad; 25. Agra; 26. Mainpuri; 27. Etawah)

The National Capital Region (NCR) is a coordinated planning region centred upon the National Capital Territory of Delhi in India. It encompasses the entire NCT of Delhi and several districts surrounding it from the states of Haryana, Uttar Pradesh and Rajasthan. The NCR is a *rural-urban* region, with a population of over 46,069,000 and an urbanization level of 62.6%. As well as the cities and towns the NCR contains ecologically sensitive areas like the Aravalli ridge, forests, wildlife and bird sanctuaries.

A total of 22 districts in three neighbouring states of Haryana, Uttar Pradesh and Rajasthan along with whole of the National Capital Territory of Delhi (total area: 1,483 km^2) constitute the National Capital Region (NCR) of India (total area: 58,332 km^2) (Fig. 2). The districts included from the neighbouring states in NCR are: Uttar Pradesh (Meerut, Muzaffarnagar, Ghaziabad, Gautam Budh Nagar, Bulandshahr, Baghpat, Hapur – total area:14,858 km^2); Haryana (Faridabad, Gurgaon, Mahendragarh, Bhiwani, Nuh, Rohtak, Sonipat, Rewari, Jhajjar, Panipat, Palwal, Jind, Karnal - total area: 28,545 km^2) and Rajasthan (Alwar, Bharatpur - total area: 13,447 km^2).

Fig. 2: National Capital Region showing the study area district (Gautam Buddha Nagar)

Gautam Buddha Nagar district and Chithara Village Panchayat

The Gautam Buddha Nagar, having a total area of 1,442 km^2, is situated at an altitude of 200 m latitude: 28°24′14.73″ N and longitude: 77°32′33.87″ E. The district situated in western Uttar Pradesh is adjacent to NCT. Dadri is one of the Tehsils of the district in which Chithara Village Panchayat is located. The district comes under Western Plain Zone of western Uttar Pradesh with Sandy loam and loamy soil texture, canal and tube well irrigation, medium rainfall, sub-tropical climate, rice-wheat crop rotation and dairy based farming system. Major farming systems/ enterprises include Mixed Farming system, Agri-pastural (Dairy), Dairy, Agri-horti (Vege), Pastural Farming system, and Agri-horti (Vege) Pastural Farming system. Major crops & Enterprises include Paddy, Jowar, Wheat, Mustard, Lentil, and Vegetables & Dairy.

The climatic parameters taken from Ground Water Department, Uttar Pradesh for Gautam Buddha Nagar District are given below (Anon. 2008-2009)

Climate

The climate of the district is sub-humid and characterized by hot summer and bracing cold season. After February there is continuous increase in temperature till June which is generally the hottest month with average mean temperature of 32.85°C followed by May with 31.9°C. The coldest month is January with average mean temperature of 14.2°C followed by December with 15.4°C.

During the southwest monsoon season the relative humidity is high and after the withdrawal of the monsoon humidity decreases. The mean monthly maximum relative humidity in the morning of August month is 84% and mean monthly minimum relative humidity measured in the morning of May is 41 percent. The normal annual mean wind speed in the district is 6.7 km/h the highest normal wind speed is 9.2 km/h in the month of June followed by 8.3 km/h in May. The annual normal potential evapotranspiration (PET) of the district is 1545.3 mm. The maximum PET occurs in the month of June followed by May with 225.3 mm and 222.2 mm respectively.

Rainfall

The annual normal rainfall (1901-1970) of the district is 700.6 mm as observed in the nearest rain gauge station at Sikandrabad. The maximum rainfall occurs during the monsoon period i.e., June to September having the normal value of 600 mm which is 85.7 % of annual rainfall. August is the wettest month having the normal rainfall of 205.8 mm followed by July when normal rainfall received about 194.4 mm.

Soil Types

The soil ranges from pure sand to stiff clays and including all combination of the two extreme litho units. The pure sand is called *Bhur* and clay is called *Matiar*. The mixture of sand and clay in equal proportion forms *Dumat* or loam, a good agriculture soil. Several subcategories of *Dumat* are possible depending upon the contents of clay and sand. The bad land patches (*Kallor*) which are ingested with *Reh* at places do not support any vegetation growth. Alluvial soils occurring in flood plain of rivers is called Kemp which yield good crops. *Kankars* invariably associated with clay and at times retards ground water movement.

Table 3. Range of various chemical parameters recorded in water from the district

Chemical Parameter	Range
pH	7.95 - 8.15
Specific Conductance (SC/EC) (μs/cm at 25°C)	419 - 2270
Bicarbonate (mg/l)	152 - 750
Chloride (mg/l)	14 - 347
Nitrate (mg/l)	3.3 - 141
Sulphate (mg/l)	9.6 - 125
Fluoride {mg/l)	0.27 - 1.7
Calcium (mg/l)	8 - 136
Magnesium (mg/l)	4.8 - 50
Total Hardness as CaCO3 (mg/l)	150 - 450
Sodium (mg/l)	35 - 506
Potassium (mg/l)	4.8 - 18
Arsenic (mg/l)	0.0007 - 0.001

Water Quality

In general ground water quality in the district is good and all the constituents are well within the permissible limits as prescribed by ICMR (1975). The general range of various important chemical constituents in the ground water samples are given in Table 3. Ground water in shallow aquifers in general is good and safe for drinking and irrigation purposes. The ground water is generally alkaline type (pH 7.95 - 8.15) and salinity limits within the permissible limits except in few samples collected from Jahangirpur, Luharli, Bilaspur, Dadri, Dankaur and Jewar where it is more than 1000 μs/cm at 25°C. The ground water quality tends to deteriorate in deeper zones where it gets brackish to saline.

Digitization of Chithara Village boundary

The study area, Chithara Village Panchayat, where Shiv Nadar University is also located, comes under

Fig. 3: Location of the study area, Chithara Village Panchayat in Dadri Tehsil

Dadri Block/ Tehsil of Gautam Buddha Nagar District in western Uttar Pradesh in NCR. Chithara Village Panchayat, one of the 56 Panchayats under Dadri Block Panchayat/Tehsil, is 770.78 ha in area. Delhi is about 40 km and Noida 20 km from Chithara Village. A digitized map of Chithara Village Panchayat was prepared (Fig. 3), including the village boundary showing all the land holdings.

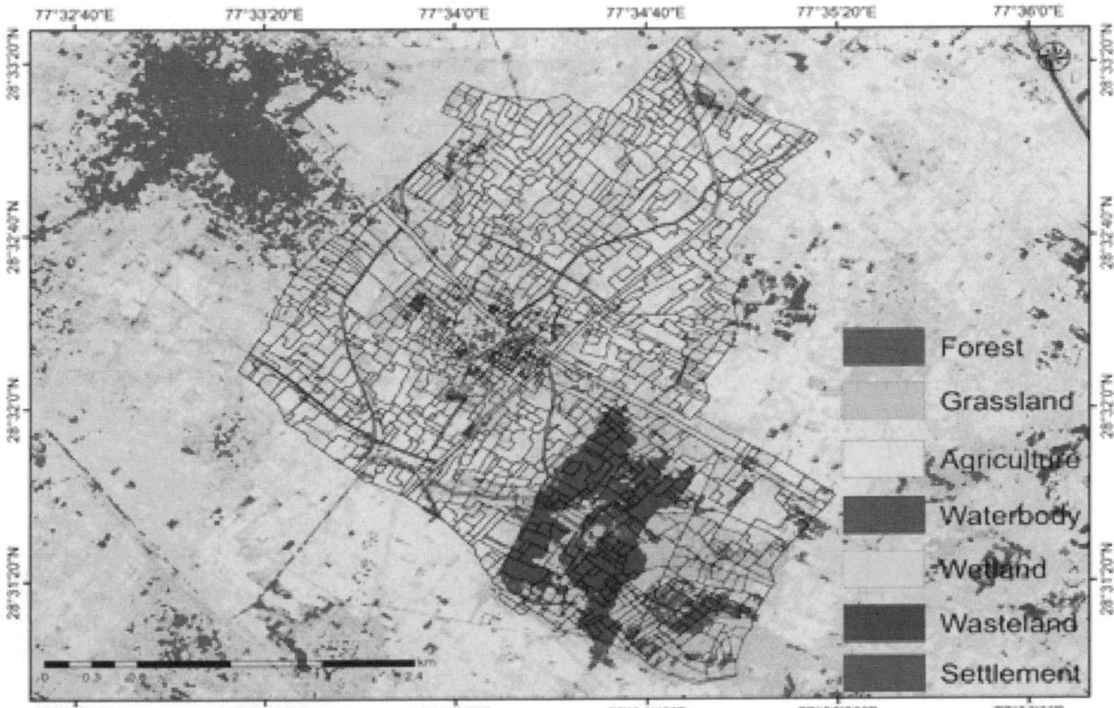

Fig. 4: Land use and land cover map of Chithara (IRS P6 LISS-IV imagery of 04 April 2010) showing SNU Campus as shaded area

Map of SNU Campus

As the SNU Campus comes under Chithara Panchayat, a detailed floristic study was also conducted in the campus of Shiv Nadar University (28°31′36.88″ N latitude and 77°34′30.77″E longitude), located on National Highway No. 91 in Village Chithara (Fig. 4). The University has a campus of 115.74 ha. The mapping of the Campus of SNU was done using a Trimble Juno Handheld GPS. The whole area was covered by foot with GPS. The area was covered in the polygon mode. The GPS (Vector data) data, processed and analyzed, was overlaid on the Google Earth image and final map was generated (Fig. 5).

Fig. 5: Location of SNU Campus in Google Earth image (2017)

Floristic Survey

For documenting the flora of Chithara Village Panchayat inventory of wild plants (herbs, shrubs, climbers and trees) uninhabited areas such as wastelands, swamps, grasslands, area along the canal, drains and agricultural field boundaries were visited frequently in all the three seasons during 2014-2017 and plant collections made. For studying the flora of the SNU Campus all the developed and undeveloped areas and wetlands were surveyed frequently throughout the year. All the horticultural plants introduced in the SNU Campus and Thematic Botanic Garden were surveyed and documented separately and list prepared. Besides the wild herbaceous flora, the original tree flora of the Campus, including the date palms (*Phoenix sylvestris* (L.) Roxb.) was surveyed and their GPS observations recorded for ease in locating them in the campus. The Campus is dotted with numerous smaller or larger date palm groves which have specialized ecological niche due to shade, moisture and rich humus; the largest grove has more than 300 palms. These groves were also surveyed for associated tree, herb and shrub species. The north-eastern side of the SNU Campus has a part of the wetland area with perennial water as well as low lying marshy area. The aquatic vegetation of this area was also enumerated and included in the flora. For collecting information on horticultural diversity of the Campus, all the landscaped areas were also surveyed for wild as well as cultivated and domesticated plants introduced in the lawns, gardens, avenues, hedges, etc.

Collection of plant specimens

The specimens were collected in the vasculum or polythene bags and carried to the laboratory for processing. In the field, the tools usually used were trowel (digger) for digging roots (if required in the case of herbs only), scissors/ secateurs and knife for cutting twigs, a stick with a hook for collection of parts of tall trees, a field note book, vasculum, polythene bags, camera, GPS (Fig. 6). The collected specimens were tagged with a field number and necessary information recorded in a field note book and transported in a vasculum/poly bags to prevent willing to the laboratory. To avoid damage during transportation and preservation at least 5-6 specimens of a plant species were collected and processed without any delay.

Fig. 6: Implements and equipments used in the flora studies

Illustrations

Frequent field visits were also required to take photographs of the plants and their flowers and fruits, etc. Photographs of plants showing their habit, and details of flower, fruits, etc. were taken at different times of the year to capture different stages of growth of the plants.

Processing of Plant Specimens- Pressing and Drying

The collected plant specimens were sorted and trimmed appropriately, spread out then pressed in between the folds of old newspapers avoiding overlapping of plant parts and kept in the wooden plant press (Fig. 7).

Fig. 7: Plant specimens in the plant press

Fig. 8: A pressed dried plant spread on the newspaper

Often a heavy object (such as a piece of stone slab) was kept over the plant press/newspapers to apply force. After 24 hours press was opened and newspapers changed. The specimens were transferred to fresh newspapers frequently, to reduce discoloration of the foliage and flowers as well as fungal growth (Fig. 8).

Preservation

Fully dried and well preserved specimens were treated with 2% mercuric chloride alcoholic solution (Absolute Ethanol) by immersing in the solution in large plastic trays (45 cm x 60 cm). The specimens were handled with the help of forceps while dipping in the solution. As a precautionary measure plastic gloves and mask were used while processing the specimens. The treated specimens were transferred to fresh newspapers for drying and kept in the plant press.

Mounting of plant on a herbarium sheet

Mounting is the process of affixing a dried, pressed and treated plant specimen and its label to a herbarium sheet made of handmade acid-free heavy paper obtained from Pune. This provides physical support that allows the specimen to be handled and stored with a minimum of damage. The treated plant specimens were sorted and trimmed appropriately and spread out and mounted on the herbarium sheet of standard size (42 x 29 cm; 300-500 gsm) (Maharashtra State Khadi and Village Industries Board, Handmade Paper Institute, M. S. Publishers and Printers, Pune; hmpipune@yahoo.in). If the specimen did not fit the dimensions of the sheet it was folded or cut into sections. Mounting was done with the help of Fevicol adhesive; Fevicol was used sparingly using a paint brush to attach the larger portions of the plant, such as stems, large leaves, tendril, and fruits (Fig. 9).

The bulky plant parts like dry fruits, seeds, etc. were dried without pressing and treated with 2% mercuric chloride and were put in a small butter paper envelop. The mounted specimens were again sprayed with 2% alcoholic solution of mercuric chloride.Prior to attachment, the specimen and its label were laid out on the paper to allow maximum observation of diagnostic (usually reproductive) features as well as the range of variation in vegetative structures, including both sides of the leaves. Plants were generally positioned in a life-like arrangement (that is, with roots or lower stem toward the bottom of the sheet and flowers toward the top). When laying out the plant it was ensured to leave space on the sheet for the specimen label on the right hand bottom corner.

Fig. 9: A dried and treated specimen is being mounted on the herbarium sheet

Labelling

A printed label (Fig. 10) was pasted on the lower right hand corner of the herbarium sheet. The label indicated the information about the family, botanical name, vernacular/common name, habit, date of collection, locality, collectors name, etc. Fig. 11 shows two herbarium sheets with mounted plant specimens.

Catalogue-collection number: A sequential numbering system starting from 00001 was used.

Herbarium of Shiv Nadar University
School of Natural Sciences
Center for Environmental Studies

Catalogue No............................
Family..
Botanical name..
Vernacular/Common name...
Habit..
Date of Collection...
Locality..
....
Collected by..
Identity determined by ....
Remarks ....
....

Fig. 10: A specimen of SNU Herbarium label

Habit: It describes the form of the plant such as tree, shrub, and herb (climber and creeper)

Date of Collection: A format with the month spelled out or abbreviated and 4 digit year to prevent confusion was used consistently: e.g., 3 May 2015, not 3/5/15 or 5/3/15

Locality: Description of the area/locality from where the plant was collected.

Collector: The full name(s) of person(s) collected the specimen in the field.

Identity determiner of the scientific name: The name of the person who identified the plant.

Remarks: Some additional information about the plant.

Storage

The mounted herbarium sheets were placed in between fold of a brown paper sheet. A similar label with all the plant information was also affixed on this brown sheet. All the herbarium specimens belonging to a family were placed in the designated family drawer of the herbarium cupboard; the family position was marked according to and finally they are incorporated into the herbarium cupboards in their proper position according to Bentham and Hooker's system of classification.

Fig. 11: Herbarium sheets with mounted plant specimens

Identification of plant specimens

The identification of plant specimens requires a considerable amount of time and effort. After a thorough literature review all the relevant literature pertaining to local or regional floras of Delhi, Uttar Pradesh were collected with the assistance of the University Library for the identification process. Several publications, especially the following were consulted for confirming the taxonomic identity the plant specimens:

1. Flora of Ghaziabad District by Rashtra Vardhana (2007)
2. Flora of the Upper Gangetic Plain and of the Adjacent Siwalik and Sub-Himalayan Tracts by J.F. Duthie (Reprint 1960)
3. Flora of Agra District by A.K. Sharma and J.S. Dhakre (1995)
4. The Flora of Delhi by J. K. Maheshwari (1963)
5. Illustrations of the Flora of Delhi by J. K. Maheshwari (1966)
6. The Pteridophyte Flora of the Upper Gangetic Plains by Chowdhury (1973)

Illustrations of some of the important morphological characteristics of the plants used in the identification of species such as Leaf Margin (Fig. 12), Leaf Apex (Fig. 13), Leaf Base (Fig. 14) and Leaf Shape (Fig. 15) are provided.

The accepted name of the species, family, and number of genera and species were verified from 'The Plant List, Version 1.1' (http://www.theplantlist.org). Wherever it was not possible to identify the plants, especially in the case of grasses, help was taken from the Northern Regional Office of Botanical Survey of India (BSI), Ministry of Environment, Forests & Climate Change, Dehradun and Forest Research Institute, Dehradun. A specimen of each identified plant of SNU Herbarium was deposited in BSI Herbarium, Dehradun as part of the record and contribution of Shiv Nadar University Herbarium.

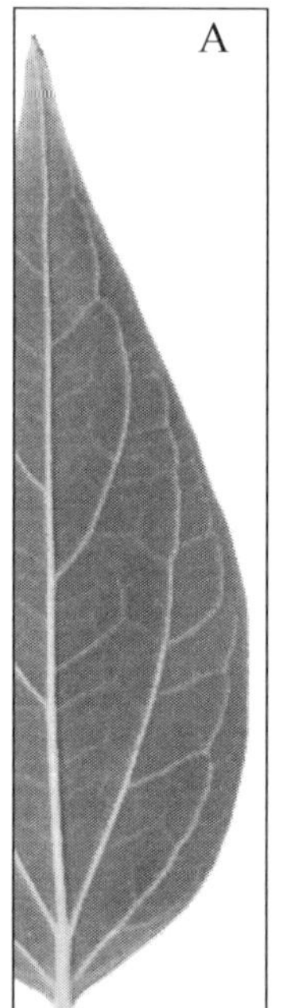

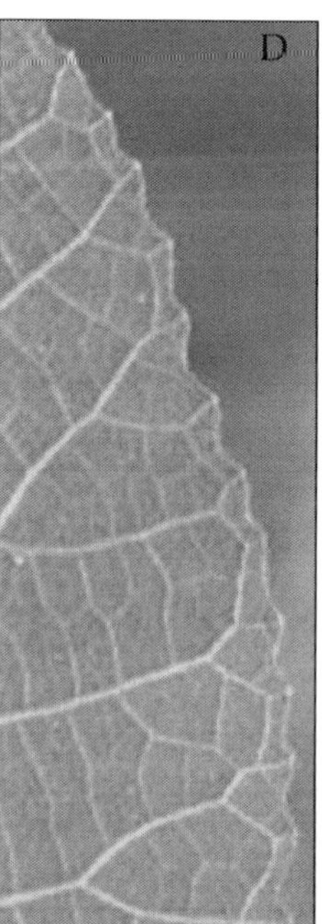

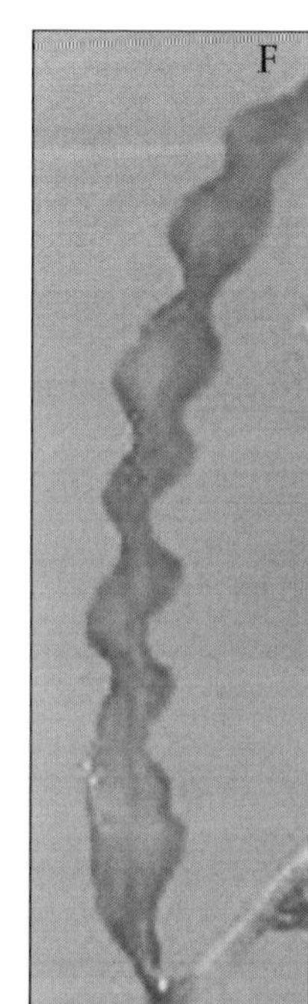

Fig. 12: Types of leaf margin: A-Entire; B-Serrate; C-Serrulate; D-Dentate; E-Crenate; F-Undulate

Taxonomic Classification followed

For arranging the plants/families, Bentham & Hooker's natural system of classification for seed plants, published in *Genera plantarum ad exemplaria imprimis in herbariis kewensibus servata definita (three volumes)* (G. Bentham & J.D. Hooker (1862–1883) was followed. This system of classification shares similarity with the classificatory systems of Bessey and Hutchinson. Bentham and Hooker described families and genera of seed plants known based upon their overall similarity. It is the most popular and widely used system of classification due to its ease to use. It has a great practical application and is followed in curricula in various educational institutes. Their classification system is now used in various commonwealth countries including India. Bentham and Hooker grouped advanced, seed bearing plants into a major division called Phanerogamia, which was divided into the following three classes: Dicotyledonae, Gymnospermae and Monocotyledoneae. Genera Plantarum classified 97,205 species of flowering plants which were grouped into 202 orders (now called as Families).

In the description of the flora the plants have been arranged family-wise. For each species the following details were given: Scientific Name, Common Name and Vernacular Name (Hindi, Sanskrit, wherever available), Distribution, Habit, Habitat, Taxonomic Description, Chemical Constituents, Medicinal Properties and Medicinal Uses (Traditional/Folklore). The season-wise occurrence of all the plants is tabulated in Appendix 4 and glossary of taxonomic terms used in the description of a species is given in Appendix 10.

Fig. 13: Types of leaf apex: A-Acute; B-Acuminate; C-Caudate; D-Obtuse; E-Rounded; F-Retuse; G-Emarginate; H-Obcordate; I-Truncate; J-Mucronate; K-Apiculate

Fig. 14: Types of leaf base: A-Amplexicaul; B-Attenuate; C-Auriculate; D-Cordate; E-Oblique; F-Cuneate; G-Rounded; H-Truncate

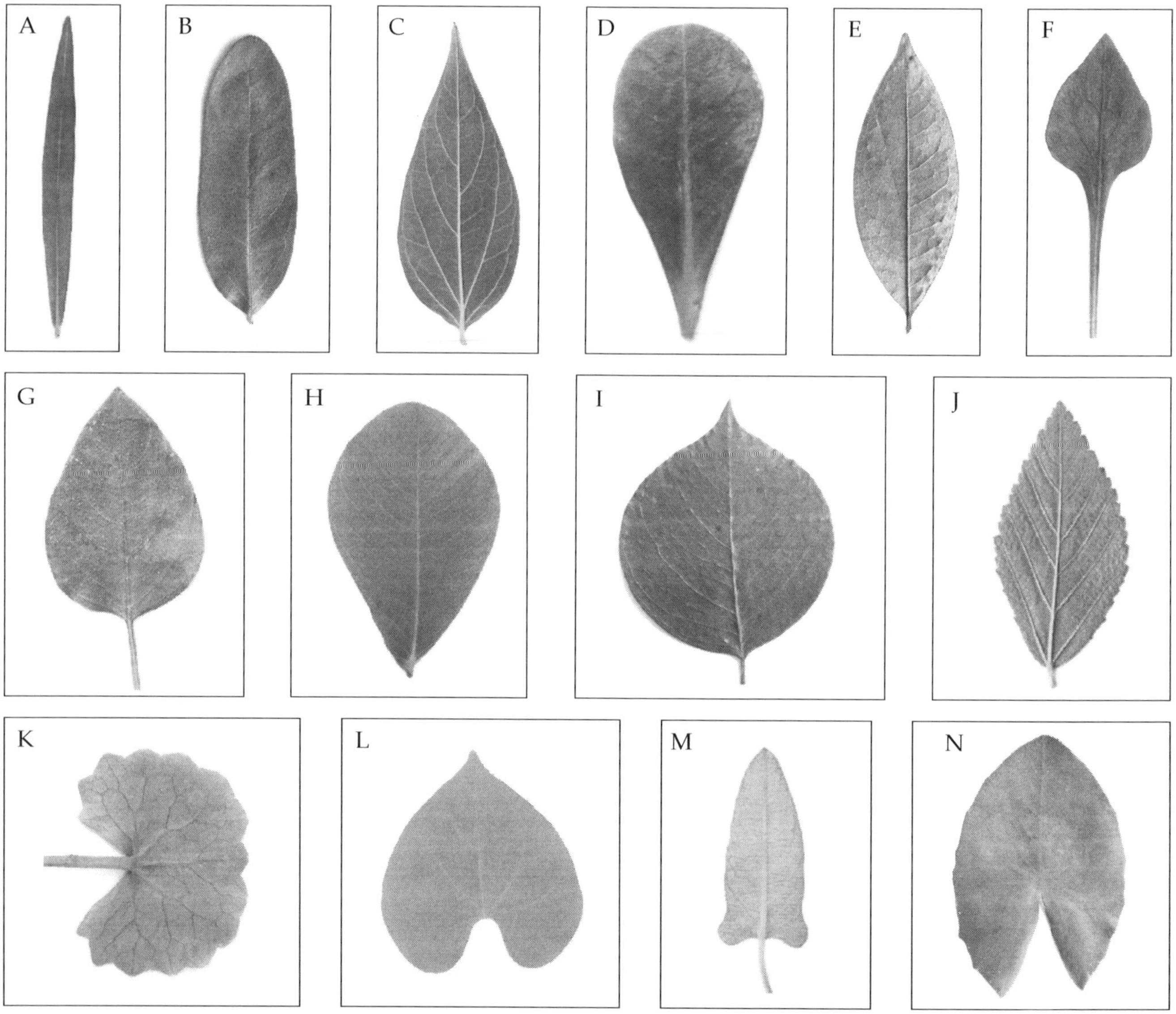

A B C D E F

G H I J

K L M N

Fig. 15: Types of leaf shape: A-Linear; B-Oblong; C-Lanceolate; D- Oblanceolate; E-Elliptic; F-Spathulate; G-Ovate; H-Obovate; I-Orbicular; J-Rhomboid; K-Reniform; L-Cordate; M-Hastate; N-Sagittate

Alien species in the Flora

The recorded plant species were also enumerated alphabetically in a tabular form, followed by author's citation, name of the family and nativity. The information on alien status of each species was searched in the available literature and nativity confirmed (Maheshwari and Paul 1975, Sharma and Panday 1984, Saxena 1991, Srivastava 1993, Pandey and Parmar 1994, Singh and Srivastav 1999, Weber 2003, Khuroo *et al.* 2007, Naithani *et al.* 2007, Reddy *et al.* 2008, Dogra *et al.* 2010, Singh *et al.* 2010c, Mandal 2011, Sood *et al.* 2011, Sekar 2012, Srivastava *et al.* 2014b, Mishra *et al.* 2015d, ENVIS (BSI) 2016); CABI (Invasive Species Compendium).

Medicinal properties and uses of each species were searched in the literature and documented. Based on this information each plant species was ranked in 1-5 scale taking in to account their medicinal properties and uses in folklore and traditional medicines in India and other countries.

Medicinal properties of plants

Since, about 95 percent of the plants recorded in the flora were known to have medicinal value, information on chemical constituents, medicinal properties and medicinal uses (folklore and traditional) of each plant species was collected from the literature and has been presented along with the plant description. Extensive bibliography given at the end includes books, journals, pamphlets, reports, relevant websites, etc. from which this information was collected; no citations have been given in the text due to the enormity of the published information consulted. Information cited in the book on various diseases/ailments and plants used to cure them has been compiled and given in at the end (Appendix 8) to give an idea about the efficacy of various plants in curing a disease/ailment. Also provided a detailed glossary of all the medical/technical terms used while describing the medicinal properties of plants (Appendix 9).

Horticultural flora of the Campus

The Campus of Shiv Nadar University having an area of (115.74 ha), has well laid out landscaped gardens, lawns, hedges and avenue tree plantations to merge with the majestic buildings. All the perennial plants in the gardens were recorded, identified scientifically and their checklist was prepared. The details of all the horticultural plants of the Campus with scientific name, family, common name and Hindi names are given in Appendix 5.

Thematic Botanic Garden

Brief Profile of Thematic Botanic Garden– Plant diversity

The Thematic Botanic Garden is being developed in an area of 9.54 acres (3.86 ha) adjacent to the wetland area of

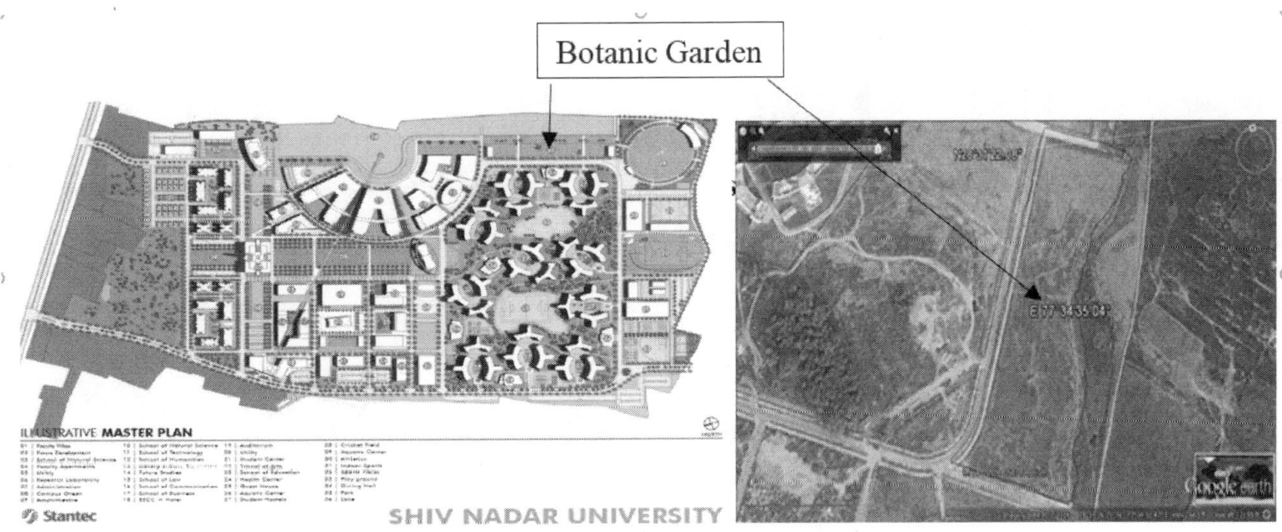

Fig. 16: Location of Botanic Garden in SNU Campus

the Campus, including 3.26 acres (1.32 ha) of low lying wetland area; the land area of the Garden is 6.28 acres (2.54 ha). The Garden includes a part of the perennial wetland of Bil Akbarpur spread out on the Eastern side of the

Campus adjoining the University, provides habitat for aquatic life – flora and fauna, especially the migratory and other aquatic birds, fishes and hydrophytes.

For the Garden development a detailed lay out plan was prepared in 1 cm: 5 meter scale for the designated area and plant species to be planted finalized.

For establishing the following 18 theme gardens in the Thematic Botanic Garden, taxonomically identified plants of each theme were collected from various sources within Delhi and outside such as research institutions and standard nurseries, wherever available and planted systematically in the thematic gene pool gardens as per the layout plan or maintained in the green house nursery until out planted in designated areas. Initially, six to seven themes will be established and gradually the number of themes will be increased to cover a wide variety of habit and habitat of plants.

Theme 1: Endemic and RET Species Garden

Theme 2: Orchidarium (Green House)

Theme 3: Fruit Plants Garden

Theme 4: Bamboo Garden (Bambusetum)

Theme 5: Palm Garden (Palmetum)

Theme 6: Ficus Garden

Theme 7: Medicinal Plants Garden

Theme 8: Gymnosperms Garden

Theme 9: Aromatic and Essential Oil Plant Garden

Theme 10: Anticancer Plants Garden

Theme 11: Antimalarial and Mosquito Repellant Plants Garden

Theme 12: Anti-Oxidant Plants Garden

Theme 13: Aquatic Plants Garden and Conservation Area

Theme 14: Xerophytes Garden (Arid & Semi-Arid Zone Plants)

Theme 15: Cacti and Succulents Garden (Green House)

Theme 16: Thallophyte & Bryophyte Garden (Cryptogams Garden)

Theme 17: Fern and Allies House (Green House)

Theme 18: Butterfly Garden

3

Flora – Family-wise Systematic Description of Plants

Floristic analysis

The floristic survey of Chithara Village Panchayat including, SNU Campus, conducted during all the three seasons, winter, summer and rainy throughout the year during 2014-17, covering different types of habitat such as agricultural fields, canal bunds, roadsides, village streets, wastelands, grasslands, etc., recorded rich plant diversity. There were a total of 272 vascular plant species belonging to 203 genera under 69 plant families recorded in the flora. Even though the SNU Campus (115.74 ha) is much smaller in area as compared to Chithara Village (770.78 ha) it recorded proportionately higher species diversity possibly due to its proximity to wetlands and agriculture fields. Very unique and prominent feature of Chithara flora is the presence of numerous date palm trees, *Phoenix sylvestris* (L.) Roxb., dotted in agricultural fields, wetlands and roadsides, either individually or as a grove usually, associated with some other tree species, herbs and shrubs. The largest date palm grove was found in the SNU Campus and second largest in Chithara Village in the middle of agricultural fields. Further details about date palms are given under SNU flora.

There were 45 plant species which were recorded exclusively in Chithara Village and not recorded in the SNU Campus (Table 4). Except plant families *viz.,* Lythraceae (*Ammannia baccifera*), Pontederiaceae (*Eichhornia crassipes*), Salicaceae (*Salix tetrasperma),* Sphenocleaceae *(Sphenoclea zeylanica),* Scrophulariaceae (*Verbascum chinense, V. thapsus*), Thelypteridaceae *(Ampelopteris prolifera)* and Urticaceae (*Gonostegia pentandra*) which occurred in Chithara Village, 62 plant families were recorded in the Campus of Shiv Nadar University. A total of 16 families (Acanthaceae, Amaranthaceae, Compositae, Convolvulaceae, Cyperaceae, Euphorbiaceae, Hydrocharitaceae, Lamiaceae, Leguminosae, Meliaceae, Moraceae, Onagraceae, Phyllanthaceae, Poaceae, Solanaceae, Verbenaceae) were common to both Chithara and SNU Campus.

Table 4. Plant species recorded outside SNU Campus in Chithara

SN	Family	Name of plant species	Types	Habit
1	Acanthaceae	*Barleria prionitis* L.	Dicot	US
2	Acanthaceae	*Rungia pectinata* (L.) Nees	Dicot	H
3	Amaranthaceae	*Alternanthera sessilis* (L.) R.Br. ex DC.	Dicot	H
4	Compositae	*Acmella paniculata* (Wall. ex DC.) R.K.Jansen	Dicot	H
5	Compositae	*Blumea laciniata* (Wall. ex Roxb.) DC.	Dicot	H
6	Compositae	*Caesulia axillaris* Roxb.	Dicot	H
7	Compositae	*Carthamus oxyacantha* M.Bieb.	Dicot	H
8	Convolvulaceae	*Cuscuta reflexa* Roxb.	Dicot	CL/CR
9	Convolvulaceae	*Merremia hederacea* (Burm. f.) Hallier f.	Dicot	CL/CR
10	Convolvulaceae	*Operculina turpethum* (L.) Silva Manso	Dicot	CL/CR
11	Cyperaceae	*Cyperus iria* L.	Monocot	SE
12	Cyperaceae	*Fimbristylis ferruginea* (L.) Vahl	Monocot	SE
13	Euphorbiaceae	*Jatropha gossypiifolia* L.	Dicot	SH
14	Hydrocharitaceae	*Vallisneria spiralis* L.	Monocot	H

Table 4. continued...

SN	Family	Name of plant species	Types	Habit
15	Lamiaceae	*Nepeta hindostana* (B.Heyne ex Roth) Haines	Dicot	H
16	Leguminosae	*Acacia nilotica* (L.) Delile	Dicot	T
17	Leguminosae	*Albizia lebbeck* (L.) Benth.	Dicot	T
18	Leguminosae	*Alhagi maurorum* Medik.	Dicot	US
19	Leguminosae	*Indigofera linnaei* Ali	Dicot	H
20	Leguminosae	*Medicago lupulina* L.	Dicot	H
21	Leguminosae	*Mimosa rubicaulis* Lam.	Dicot	SH
22	Leguminosae	*Senna obtusifolia* (L.) H.S.Irwin & Barneby	Dicot	H
23	Leguminosae	*Teramnus labialis* (L.f.) Spreng.	Dicot	CL/CR
24	Lythraceae	*Ammannia baccifera* L.	Dicot	H
25	Meliaceae	*Melia azedarach* L.	Dicot	T
26	Moraceae	*Ficus palmata* Forssk.	Dicot	SH
27	Onagraceae	*Ludwigia hyssopifolia* (G.Don) Exell	Dicot	H
28	Phyllanthaceae	*Phyllanthus tenellus* Roxb.	Dicot	H
29	Poaceae	*Brachiaria ramosa* (L.) Stapf	Monocot	GR
30	Poaceae	*Chloris barbata* Sw.	Monocot	GR
31	Poaceae	*Echinochloa crus-galli* (L.) P.Beauv.	Monocot	GR
32	Poaceae	*Eleusine indica* (L.) Gaertn.	Monocot	GR
33	Poaceae	*Eragrostis amabilis* (L.) Wight & Arn.	Monocot	GR
34	Poaceae	*Eragrostis japonica* (Thunb.) Trin.	Monocot	GR
35	Poaceae	*Leptochloa panicea* (Retz.) Ohwi	Monocot	GR
36	Poaceae	*Paspalum distichum* L.	Monocot	GR
37	Pontederiaceae	*Eichhornia crassipes* (Mart.) Solms	Monocot	H
38	Salicaceae	*Salix tetrasperma* Roxb.	Dicot	T
39	Scrophulariaceae	*Verbascum chinense* (L.) Santapau	Dicot	H
40	Scrophulariaceae	*Verbascum thapsus* L.	Dicot	H
41	Solanaceae	*Solanum torvum* Sw.	Dicot	SH
42	Sphenocleaceae	*Sphenoclea zeylanica* Gaertn.	Dicot	H
43	Thelypteridaceae	*Ampelopteris prolifera* (Retz.) Copel.	Pteridophyte	PH
44	Urticaceae	*Gonostegia pentandra* (Roxb.) Miq.	Dicot	H
45	Verbenaceae	*Clerodendrum phlomidis* L.f.	Dicot	SH

(H-Herb; SH-Shrub; US-Undershrub; T-Tree; CL/CR-Climber/Creeper; GR-Grass; SE-Sedge; PH-Pteridophytic herb)

The flora was enriched mainly by plant species belonging to Angiosperms. Of the 272 vascular plants recorded, 268 species (approx. 99%) belonged to Angiosperms (flowering plants), whereas 4 species were that of Pteridophytes. Approximately, 79 percent of these plants belonged to Dicotyledons (215 species), 20 percent to Monocotyledons (53 species) and 1 percent to Pteridophytes (4 species) (Fig. 17).

All the plant species of Chithara Village Panchayat including the campus of Shiv Nadar University are tabulated in Appendix 3.

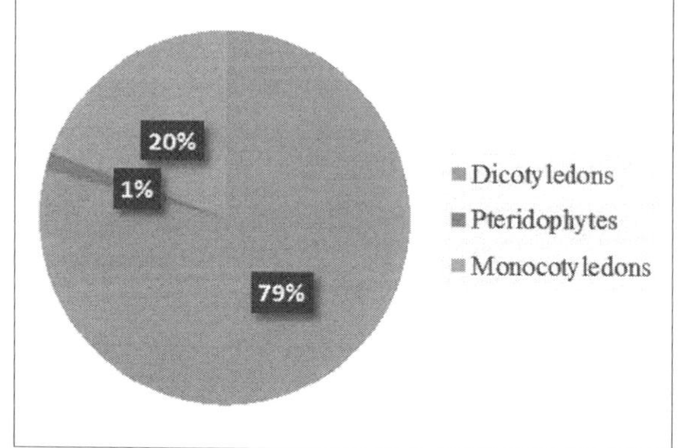

Fig. 17: Percentage of species recorded in various categories of plants in Chithara Village

Habit of recorded plant species

The flora of Chithara Village including SNU Campus was dominated by herbs which accounted for 147 species followed by sedges and grasses (38 species), climber and creepers (24 species), trees (22 species), undershrubs (20 species), shrubs (17 species) and Pteridophytes (4 species) (Fig. 18).

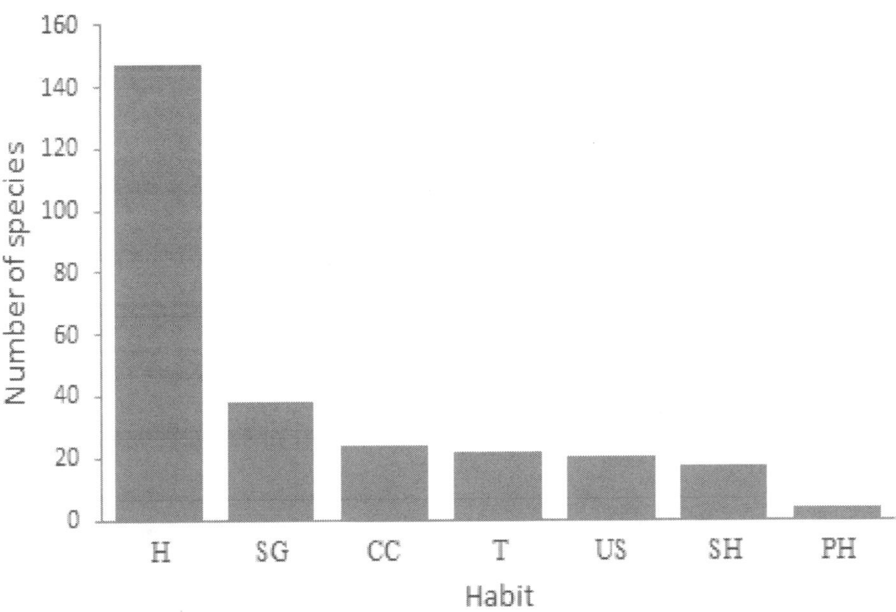

Fig. 18: Number of species recorded in various habit classes in Chithara Village

(H-Herbs; SG-Sedges and Grasses; CL/CR-Climber and Creepers; T-Trees; US-Undershrubs; SH-Shrubs; PH-Pteridophytic Herbs)

Prominent Families

Floristic analysis of wild flora of Chithara Village showed that of the 69 plant families Luguminosae with 39 plant species was the most dominant family followed by Poaceae (31 species), Compositae (24 species), Convolvulaceae (14 species) and Amaranthaceae and Malvaceae (12 species each) (Fig. 19).

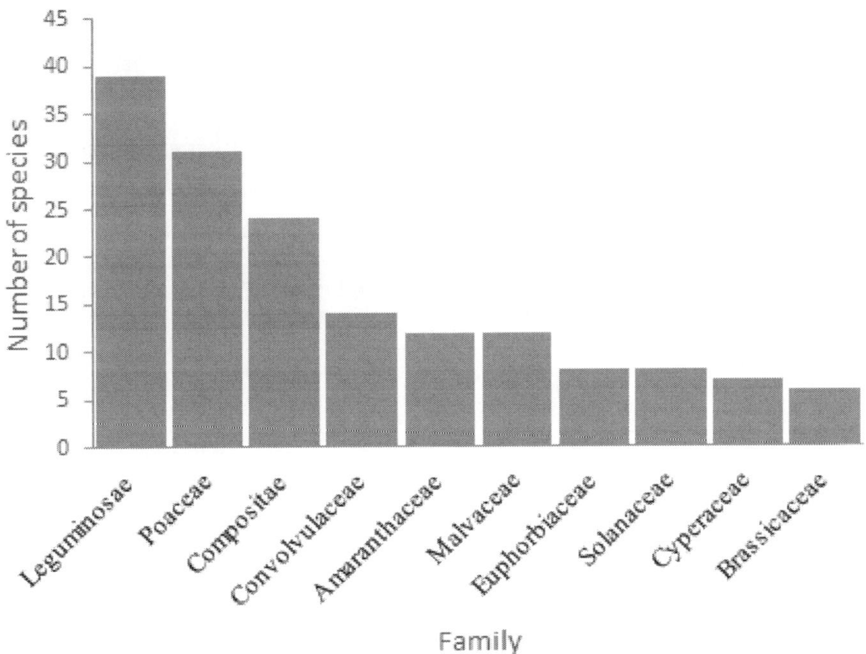

Fig. 19: Ten prominent plant families with the number of species in Chithara Village Panchayat

Out of 272 vascular plant species, Dicotyledons were represented by 56 families, 159 genera, 215 species; Monocotyledons by 10 families, 41 genera, 53 species and Pteridophytes by 3 families, 3 genera and 4 species (Fig. 20).Among the Dicots, the most dominant family was Leguminosae represented by 24 genera and 39 species followed by Compositae with 20 genera and 24 species. Out of 56 families in Dicots, these two families (Leguminosae and Compositae) contributed about 30 percent of species diversity.

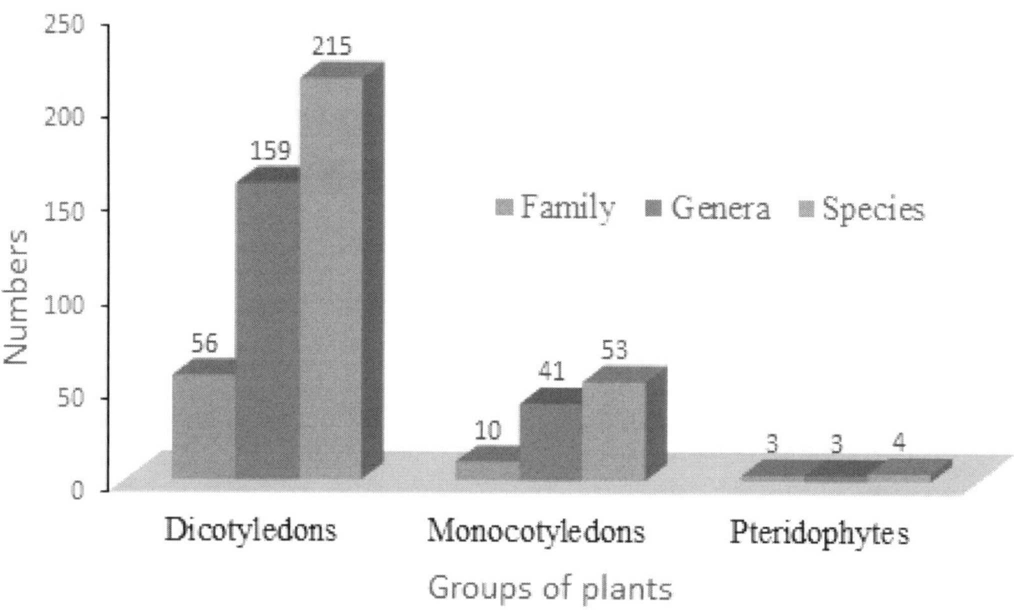

Fig. 20: Vascular plant diversity in Chithara flora

Of the 10 families represented in Monocotyledons, Poaceae was the most versatile family with highest number of species belonging to 25 genera and 31 species and contributed more than 50 percent species of total Monocots (Appendix 3).

In some of the families there were a few genera which had numerous species. Under this category Leguminosae had maximum number of 11 genera having 2 or more than 2 species (Table 5). However, in Convolvulaceae, which was represented by 14 species and 6 genera in the flora, genus *Ipomoea* had the highest number of 7 species. This was followed by *Euphorbia* (Euphorbiaceae) and *Cyperus* (Cyperaceae) having 5 species each. There were a total of 46 genera under 23 families which had 2 or more than 2 species. Details of genera with ≥2 species in 23 prominent families are given in Table 5. Most of these plants are known to possess medicinal properties and used in traditional and folk medicines in India and other countries.

Table 5. Genera represented by two or more number of species in plant families

SN	Family	Number of Genera	Number of species	Genera with ≥2 species	
				Genus	Number of species
Dicots					
1.	Leguminosae	24	39	*Indigofera*	4
				Medicago	3
				Tephrosia	3
				Alysicarpus	2
				Crotalaria	2
				Desmodium	2
				Melilotus	2
				Prosopis	2
				Rhynchosia	2
				Senna	2
				Sesbania	2
2.	Compositae	20	24	*Blumea*	3
				Ageratum	2

Table 5. continued...

SN	Family	Number of Genera	Number of species	Genera with ≥2 species	
				Genus	Number of species
				Sonchus	2
3.	Amaranthaceae	8	12	*Alternanthera*	4
				Chenopodium	2
4.	Malvaceae	8	12	*Corchorus*	3
				Sida	3
5.	Convolvulaceae	6	14	*Ipomoea*	7
				Convolvulus	2
				Cuscuta	2
6.	Solanaceae	5	8	*Solanum*	3
				Physalis	2
7.	Euphorbiaceae	4	8	*Euphorbia*	5
8.	Acanthaceae	4	5	*Justicia*	2
9.	Plantaginaceae	4	5	*Veronica*	2
10.	Polygonaceae	4	5	*Persicaria*	2
11.	Moraceae	3	4	*Ficus*	2
12.	Apocynaceae	2	3	*Calotropis*	2
13.	Phyllanthaceae	1	4	*Phyllanthus*	4
14.	Linderniaceae	1	2	*Lindernia*	2
15.	Onagraceae	1	2	*Ludwigia*	2
16.	Oxalidaceae	1	2	*Oxalis*	2
17.	Portulacaceae	1	2	*Portulaca*	2
18.	Rhamnaceae	1	2	*Ziziphus*	2
19.	Scrophulariaceae	1	2	*Verbascum*	2
Monocots					
20.	Poaceae	24	31	*Brachiaria*	2
				Echinochloa	2
				Eragrostis	2
				Paspalum	2
				Saccharum	2
				Setaria	2
21.	Commelinaceae	3	4	*Commelina*	2
22.	Cyperaceae	2	7	*Cyperus*	5
				Fimbristylis	2
Pteridophytes					
23.	Equisetaceae	1	2	*Equisetum*	2

Seasonality in occurrence of species

The annual herbs, grasses, sedges, undershrubs, etc. showed seasonality in their occurrence. Most of the grasses and large number of other weeds start germinating towards the end of June and early July when the monsoon starts. However, some plants were found growing in summer and winter seasons while a few species were observed growing throughout the year.

Numerous plants showed seasonality in their occurrence and were recorded in a specific season. A checklist of wild plant species recorded in different seasons from Chithara Village is given in Appendix 4. Out of 272 plant species, 39 species from 35 genera and 22 families were present as the permanent vegetation in Chithara Village including SNU Campus, being shrubs, trees or perennial climber; Leguminosae was the most diverse family having 10 species. The number of families, genera and species represented in different seasons are given in Fig. 21.

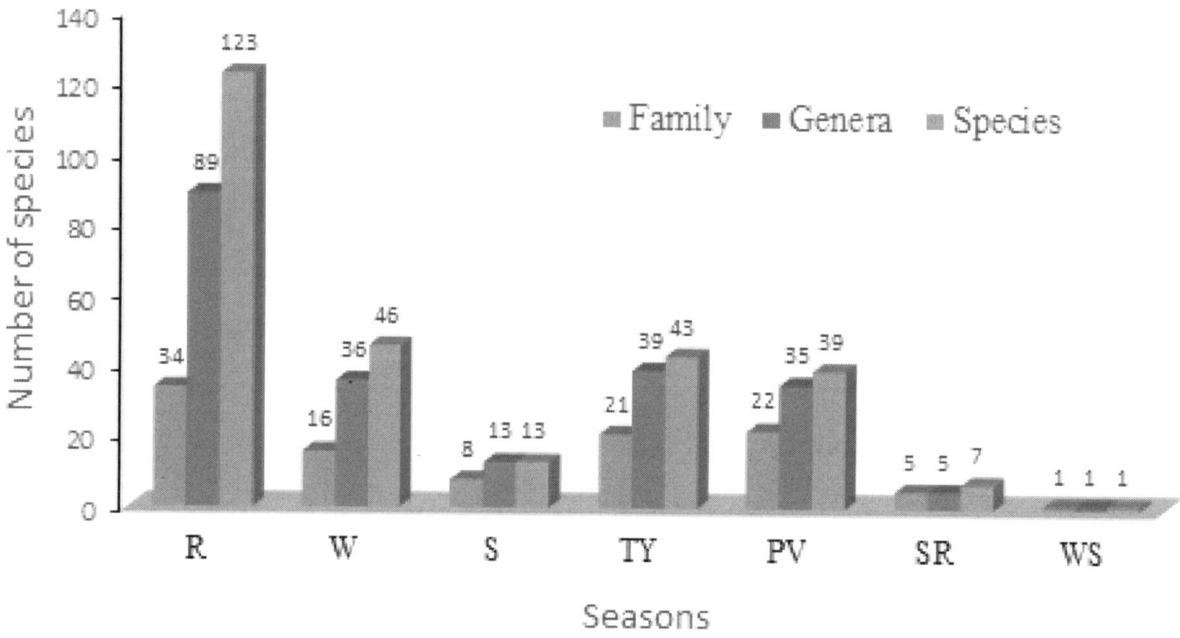

Fig. 21: Plant species recorded in different seasons
(R-Rainy; W-Winter; S-Summer; TY-Throughout Year; PV-Permanent Vegetation; SR-Summer and Rainy; WS-Winter and Summer)

Maximum number of 123 species was recorded during the rainy season followed by winter (46 species) and summer (13 species). A total of 43 plant species were found to be growing throughout the year. Interestingly, a few species were represented during two seasons. *Emex australis* was recorded both during winter and summer seasons whereas seven species such as *Artemisia scoparia* Waldst. & Kitam., *Nepeta hindostana* (B.Heyne ex Roth) Haines, *Pistia stratiotes* L., *Portulaca oleracea* L., *Portulaca quadrifida* L., *Verbascum chinense* (L.) Santapau, and *Verbascum thapsus* L. continued to grow during summer and rainy seasons.

The 123 plant species which were recorded during rainy season belonged to 89 genera and 34 families. Poaceae was the most speciose family with 23 species followed by Leguminosae (17 species), Convolvulaceae and Malvaceae (9 species each), Cyperaceae (7 species) and Amaranthaceae (6 species). In winter season, 46 plant species were represented by 36 genera and 16 families; Compositae (10 species) was the most dominant family followed by Leguminosae (8 species) and Brassicaceae (6 species). As compared to other seasons, least number of species (13 species) belonging to 13 genera and 8 families were recorded during the summer season; Compositae was the most dominant family with 4 species followed by Leguminosae (3 species). The 43 species which were recorded throughout the year, were represented by 39 genera under 21 families; Compositae and Poaceae (5 species each) were the most dominant family followed by Solanaceae (4 species).

New record of occurrence of plant species

The survey documented a few unique and rare species as they were recorded for the first time in this region and their occurrence was very restricted. These species were *Emex australis* Steinh, *Phyllanthus tenellus* Roxb. and *Ludwigia hyssopifolia* (G.Don). *Emex australis,* belonging to family Polygonaceae, was found growing gregariously at a few places in the SNU Campus and now spreading in the surrounding areas. *E. australis,* a native of South Africa, is commonly called as Spiny Emex or Doublegee; it is a prostrate, annual, monoecious, winter herb. *E. australis* is widespread serious weed of cereal crops in Western Australia. In India, the genus *Emex* is distributed very scarcely. Since the only reported occurrence of *E. australis* is Jammu (Sharma and Jamwal, 1989) this is the new record for Uttar Pradesh and Upper Gangetic Plains.

Ludwigia, belonging to family Onagraceae, is a widely distributed genus in the world and many parts of India. *L. hyssopifolia,* commonly called water primrose, was found growing along the canal side in marshy areas in Chithara Village. Possibly native of America, it is a new addition to the regional flora of Western Uttar Pradesh. It has become an extremely aggressive weed of rice and of marshy areas in several countries such as in Malaysia, Indonesia, Borneo, Thailand, Nigeria, Sri Lanka, etc. and probably been introduced in India. In India, earlier it has been recorded from Assam, Bihar, Madhya Pradesh, Orissa, Tamilnadu, West Bengal, Uttarakhand, Rajasthan, Haryana, Tamil Nadu and Kerala.

Phyllanthus tenellus, an annual herb, commonly known as long-stalk phyllanthus, is a member of family Phyllanthaceae. It is native to Mascarine Islands and reported to occur in Australia; in India the plant is found distributed in Mallapuram, Alappuzha and Thiruvanathapuram districts of Kerala state. The plant was found growing wild in several places in the SNU Campus and this is the new record for Uttar Pradesh and Upper Gangetic Plains.

Apart from this, two species of *Equisetum* i.e., *Equisetum ramosissimum* Desf. and *Equisetum ramosissimum* subsp. *debile* (Roxb. ex Vaucher) Hauke were also recorded from the SNU Campus. This is the first report of the occurrence of both the sub species from western Uttar Pradesh and NCR Delhi. The details of both the sub species are given in another subsection of this chapter.

A few other species like, *Capparis sepiaria, Crateva adansonii* subsp. *odora, Holoptelea integrifolia, Mitragyna parviflora, Phyllanthus reticulatus, Streblus asper*, etc. which were also documented from the study area had very rare occurrence in the wild.

Plant diversity in SNU Campus

Floristic analysis

The floristic survey of SNU Campus, conducted in some of the relatively undisturbed areas, part of wetlands and developed areas, recorded a very rich plant diversity. Of the total 272 plant species recorded in Chithara, a total of 227 plant species belonging to 175 genera under 62 plant families were found in the SNU Campus. More than 80 percent (approx. 83%) of total diversity of Chithara Village Panchayat was recorded in SNU Campus which is just 15 percent of the total area of Chithara.

Habit of recorded plant species

Like Chithara Village, the rich campus flora of Shiv Nadar University was also dominated by herbs with 128 species followed by sedges and grasses (28 species), climbers and creepers (20 species), trees (18 species), undershrubs (18 species), shrubs (12 species) and Pteridophytes (3 species) (Fig. 22).

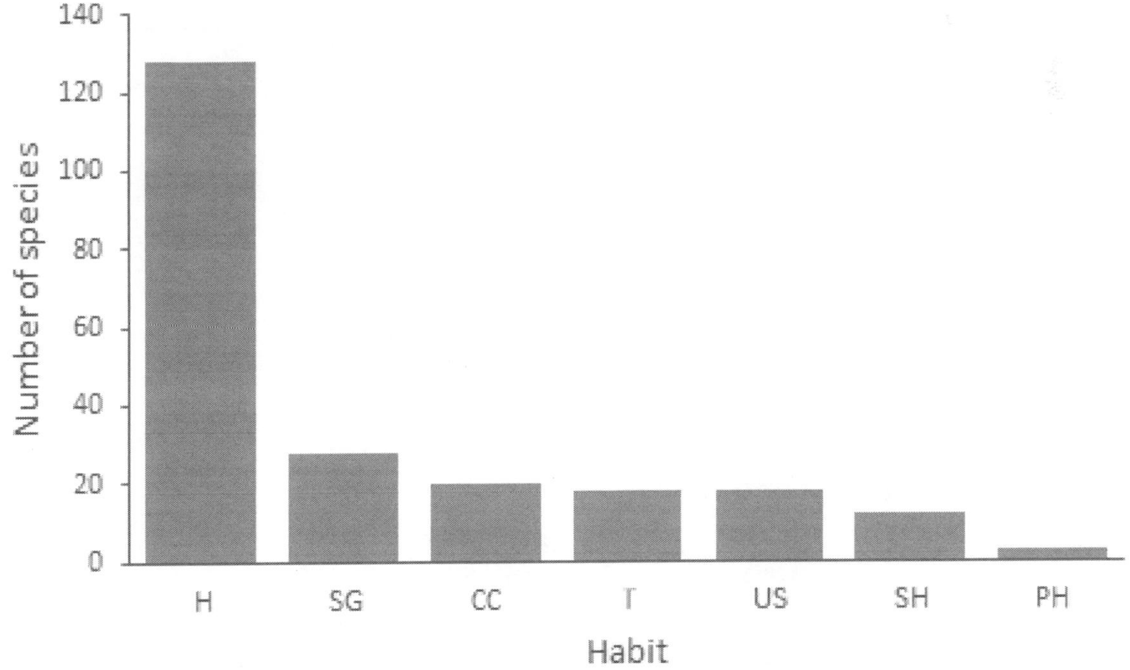

Fig. 22: Number of species recorded in various habit classes in SNU Campus

(H-Herbs; SG-Sedges and Grasses; CL/CR-Climber and Creepers; T-Trees; US-Undershrubs; SH-Shrubs; PH-Pteridophytic Herbs)

Prominent Families

Floristic analysis of SNU Campus revealed that the Luguminosae was the most speciose family with 31 plant species followed by Poaceae (23 species), Compositae (20 species), Malvaceae (12 species), Amaranthaceae and Convolvulaceae (11 species each) (Fig. 23). Euphorbiaceae, Solanaceae and Brassicaceae were represented by 7, 7 and 6 species, respectively. Family Cucurbitaceae, Cyperaceae, Plantaginaceae and Polygonaceae had 5 species. A total of 30 families were represented by only one species; 10 families by 2 species each; seven families by 3 species each; 2 families by 4 species each.

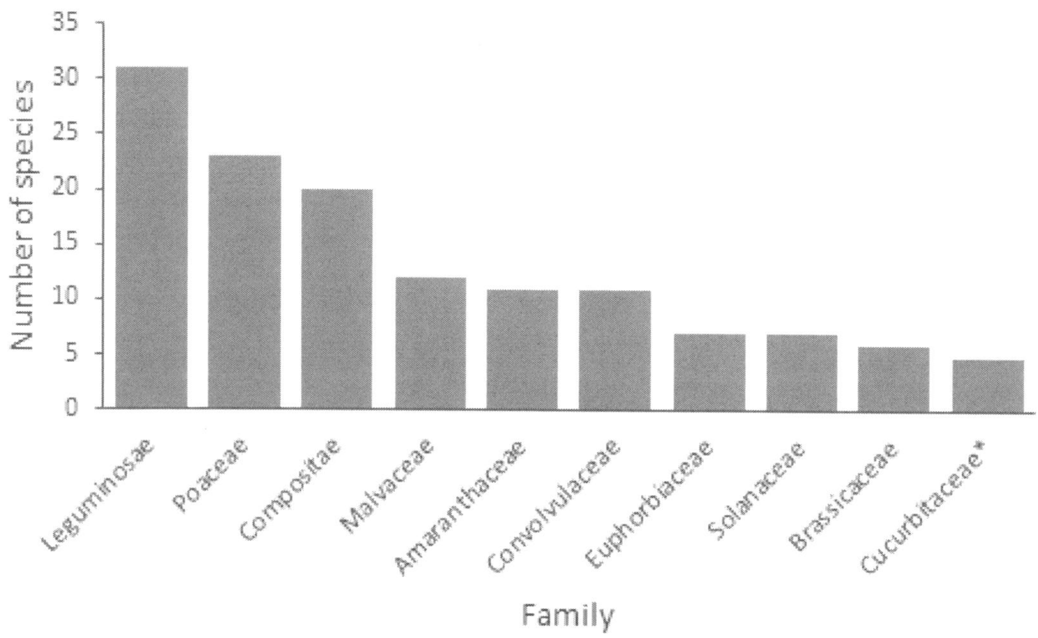

Fig. 23: Ten prominent plant families with the number of species in SNU Campus

(Cyperaceae, Plantaginaceae and Polygonaceae also had 5 species along with Cucurbitaceae)*

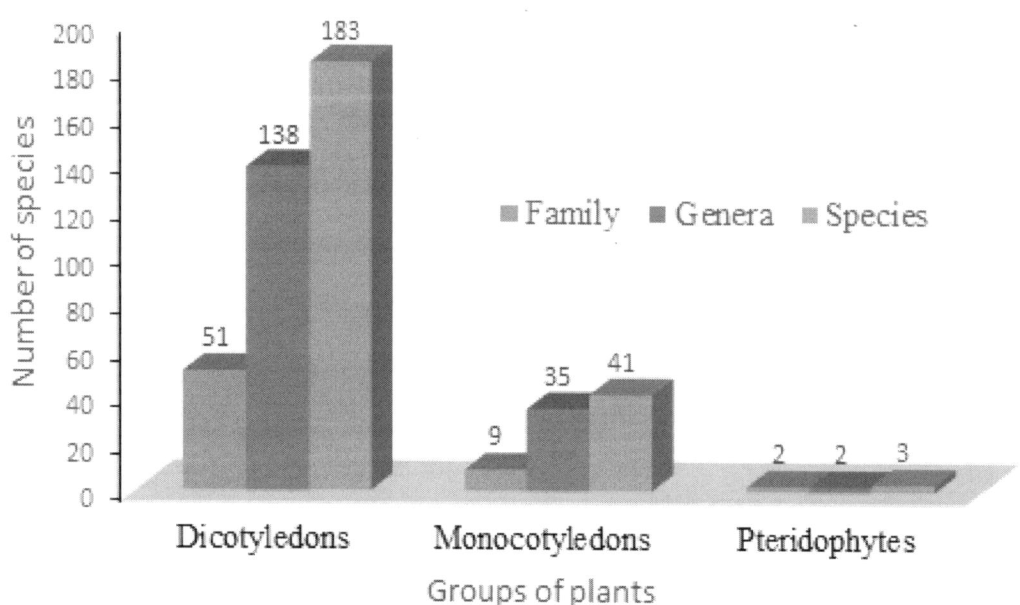

Fig. 24: Vascular plant diversity in SNU Campus

The vascular plant diversity analysis of SNU Campus showed that out of 227 plant species, Dicotyledons were represented by 51 families, 138 genera, 183 species followed by Monocotyledons (9 families, 35 genera, 41 species). The Pteridophytes were least represented (2 families, 2 genera and 3 species) (Fig. 24).

In terms of genera, the genus *Ipomoea* had the highest number of 7 species in SNU Campus, followed by Euphorbia (5 species) and *Cyperus* (4 species). A total of 6 genera such as *Alternanthera, Corchorus, Indigofera, Phyllanthus, Sida* and *Tephrosia* had 3 species each in SNU Campus. Altogether 27 genera were represented by 2 species each whereas a large number of genera (139) were found to have only one species (Appendix 3). Aapproximately 81 percent of the species belonged to Dicotyledons (183 species), 18 percent to Monocotyledons (41 species) and 1 percent to Pteridophytes (3 species).

Unique wild flora

The SNU flora has a few unique and prominent features such as occurrence of natural trees, date palm groves, preponderance of hydrophytes due to nearby wetlands and occurrence of *Equisetum,* which is very rare in this part of the region, besides recording of a few rare and unrecorded species. The campus has prominent feature of groves of date palm, *Phoenix sylvestris* (L.) Roxb. associated with some other tree species, herbs and shrubs.

Tree flora

The list of original wild trees of Shiv Nadar University is given in Table 6. The wild tree flora of the Campus recorded a total of 199 individuals belonging to 18 species with *Phoenix sylvestris* (Date palm) as the most dominant species scattered as an individual tree, in small groups and in a large grove; the total number does not include the date palms present in a large grove which alone had more than 300 palms of all age groups.

Table 6. Wild tree flora of Shiv Nadar University

SN	Name of the plant species	Family	No. of individuals
1	*Azadirachta indica* A.Juss.	Meliaceae	9
2	*Butea monosperma* (Lam.) Taub.	Leguminosae	1
3	*Cordia dichotoma* G.Forst.	Boraginaceae	3
4	*Crateva adansonii* subsp. *odora* (Buch.-Ham.) Jacobs	Capparaceae	3
5	*Dalbergia sissoo* DC.	Leguminosae	7
6	*Diospyros montana* Roxb.	Ebenaceae	4
7	*Ficus religiosa* L.	Moraceae	26
8	*Holoptelea integrifolia* Planch.	Ulmaceae	1
9	*Leucaena leucocephala* (Lam.) de Wit	Leguminosae	2
10	*Mitragyna parvifolia* (Roxb.) Korth.	Rubiaceae	1
11	*Morus alba* L.	Moraceae	11
12	*Phoenix sylvestris* (L.) Roxb.*	Arecaceae	110
13	*Prosopis cineraria* (L.) Druce	Leguminosae	3
14	*Prosopis juliflora* (Sw.) DC.	Leguminosae	12
15	*Sesbania sesban* (L.) Merr.	Leguminosae	1
16	*Streblus asper* Lour.	Moraceae	1
17	*Syzygium cumini* (L.) Skeels	Myrtaceae	3
18	*Ziziphus jujuba* Mill.	Rhamnaceae	1

** Phoenix sylvestris (Date palm) scattered as an individual tree or in small groups (>300 individual palms of the largest date palm grove of SNU Campus are not included in this table)*

The 18 tree species recorded in the Campus belonged to 17 genera representing 11 plant families, Leguminosae being the most dominant with 6 species. In terms of number of individuals, *Phoenix sylvestris,* which was recorded as an individual tree or in small groups was most dominant with 110 individuals followed by *Ficus religiosa* (26 individuals), *Prosopis juliflora* (12 individuals), *Morus alba* (11 individuals) and *Azadirachta indica* (9 individuals); *Dalbergia sissoo,* the most common tree of this region had 7 individuals (Table 6). For all the individual trees of all the species their GPS coordinates were recorded for future reference.

Date palms

Date palm is native to southern Pakistan, most of India, Sri Lanka, Nepal, Bhutan, Myanmar, Indo-China, Andaman Islands and Bangladesh. Mostly growing in plains and scrubland up to 1,300 m above sea level, *Phoenix sylvestris* ranges from 4 to 15 m in height and 40-50 cm in diameter. More taxonomic details of the species are given under flora. A total of 110 individuals of *P. sylvestris* were found scattered or grouped in very small groves whereas more than 300 individuals of date palm were recorded in one large date palm grove (Lat. 28°31′17.27″ N, Long. 77°34′24.59″ E).

Fig. 25: Largest date palm (*Phoenix sylvestris*) grove in SNU Campus

The large date palm grove (Fig. 25), which has very dense growth of palms of varying maturities and consequently poor light intensity within the grove provided specific ecological niche for several other plants recorded only in the grove such as *Abrus precatorius* L., *Butea monosperma* (Lam.) Taub., *Capparis sepiaria* L., *Cissampelos pareira* L., *Crateva adansonii* subsp. *odora* (Buch.-Ham.) Jacobs, *Diospyros montana* Roxb., *Justicia adhatoda* L., *Phyllanthus reticulatus* Poir., *Streblus asper* Lour. etc. Interestingly, these species were not recorded elsewhere in the Campus. In fact some of these species do not occur in wild any more in this region.

Hydrophytes

Due to the proximity to the wetlands, the Campus flora recorded rich hydrophyte diversity having species

Fig. 26: *Pistia stratiotes*: A free floating hydrophyte

such as *Ceratophyllum demersum* L., *Hydrilla verticillata* (L.f.) Royle, *Lemna perpusilla* Torr., *Nymphaea pubescens* Willd., *Nymphoides cristata* (Roxb.) Kuntze, *Pistia stratiotes* L. (Fig. 26), *Potamogeton crispus* L., *Sagittaria guayanensis* Kunth, *Spirodela polyrrhiza* (L.) Schleid., *Stuckenia pectinata* (L.) Börner, *Typha domingensis* Pers., *Utricularia stellaris* L.f., etc. Specific taxonomic details of these species are given under flora.

Equisetum

While documenting the campus flora, the most exciting find was the occurrence of two species of *Equisetum* i.e., *Equisetum ramosissimum* Desf. (*E. ramosissimum* subsp. *ramosissimum*) and *Equisetum ramosissimum* subsp. *debile* (Roxb. ex Vaucher) Hauke (Fig. 27). Curiously, *E. ramosissimum* was recorded alone at one location (latitude 28°31′52.96″ N and longitude 77°34′38.77″ E) whereas at another location adjacent to wetlands (latitude 28°31′16.91″ N and longitude 77°34′37.42″ E), both *E. ramosissimum* and *E. ramosissimum* subsp. *debile* were recorded growing in close proximity. At one location there was intermixing of the populations of both the species; however, at nearby eastern boundary wall of the campus only *E. ramosissimum* was found growing intermixed with *Saccharum benghalense* and *Saccharum spontaneum*. Though there are several records of *E. ramosissimum* from India, this is the first record of occurrence of these two species of *Equisetum* from Upper Gangetic Plains and that too in the same locality. *E. ramosissimum* has been recorded earlier from Jammu & Kashmir (Ladakh, Leh, Pahlgam), Assam and Meghalaya whereas *E. ramosissimum* subsp. *debile* has been recorded earlier from Arunachal Pradesh, Meghalaya, Manipur, West Bengal and Assam. Populations of both the species were visited regularly to record their growth as well as formation of strobilus and their maturation/dehiscence. Detailed description of both the species is given under flora.

Fig. 27: A; *Equisetum ramosissimum* (*E. ramosissimum* subsp. *ramosissimum*); B; *E. ramosissimum* subsp. *debile*

FAMILY-WISE SYSTEMATIC DESCRIPTION OF PLANTS

Taxonomical details, including botanical name, common name, vernacular name, habit, habitat, distribution, flowering and fruiting time, chemical constituents, medicinal properties and medicinal uses and description of 272 plants recorded in the Chithara Village are given below. Plants are arranged family-wise as per Bentham and Hooker's system of classification. The plant species are arranged in alphabetical order in each family.

I. DICOTYLEDONS

Fig. 28: A mesophytic (A; *Tribulus terrestris*) and a hydrophytic (B; *Nymphoides cristata*) dicot

The Dicotyledons, also known as dicots, which may be mesophytes or hydrophytes (Fig. 28) or xerophytes, are one of the two traditional major groups into which all the flowering plants or angiosperms were formerly divided; whose members typically have two cotyledons, or embryonic leaves, in their seeds, leaves with reticulate venation. The flowers generally have parts in fours or fives (Fig. 29 A, B), or multiplies thereof. The Compositae is the largest family under Dicotyledons; the flower of this family looks like a single flower but actually it is a cluster of many petal like rayflorets and central disk floret flowers (Fig. 29 C).

Fig. 29: Dicot flowers (A - Family Brassicaceae with 4 petals; B - Family Oxalidaceae with 5 petals; C – Family Compositae with cluster of many smaller flowers)

There are around 200,000 species within this group making it the larger of the two traditional angiosperm groups. However, the dicotyledons no longer are regarded as a group. That is because the current systems of biological classification (*Chase MW & Reveal JL 2009. A phylogenetic classification of the land plants to accompany APG III. Botanical Journal of the Linnean Society 161: 122–127.*) considers evolutionary relatedness as one of the important criterion of classification. Since, the dicotyledons are not a group made up of all the descendants of a common ancestor (i.e. they are not amonophyletic group), the traditional dicots are thus aparaphyletic group. The largest clade of the dicotyledons are known as the eudicots. These may be distinguished from all other flowering plants by the structure of their pollen. Eudicots have tricolpate pollen, or forms derived from it, involving the pollen having three or more pores set in furrows called colpi; other dicotyledons and monocotyledons have monosulcate pollen, or derived forms.

1. RANUNCULACEAE

Ranunculaceae – Buttercup or Crowfoot family

The name of the family is derived from Latin *ranunculus*, "little frog", from *rāna* "frog". The family having over 2,377 known species of flowering plants in 65 genera is distributed worldwide. It includes herbs and some climbers, mostly perennial. Plants have primitive flowers with many separate parts spirally arranged and divided leaves. The family was represented by one genus and a single species in Chithara Village.

Habit

1.1 *Ranunculus sceleratus* L.

(Syn. *Ranunculus indicus* Roxb.)

Common Name: Cursed Buttercup, Poisonous Buttercup, Celery-Leaved Buttercup, Blister Buttercup.

Vernacular Name: *Hindi:* Jaldhaniya (जलधनिया); *Sanskrit:* Kandakatuka (कंदकतुका), Kandira (कंदिरा).

Habit: Annual herb.

Habitat: Moist places, generally near water.

Distribution: It has a circumpolar distribution in the northern hemisphere; native to temperate and boreal North America and Eurasia, where it grows in wet and moist habitats. It occurs in wet plains of Northern India and warm valley of the Himalayas, from Kashmir to Assam, at banks of rivers, up to 2,000 m above msl.

Flowering and Fruiting Time: February-April.

Taxonomic Description: An erect, annual herb, up to 60 cm high, glabrous or slightly pubescent, stem grooved. Leaves clustered at base; radical leaves deeply 3-lobed with long petioles, each lobe further dissected; cauline leaves shortly petiolate. Flowers actinomorphic, bisexual, yellow. Calyx slightly reflexed; sepals-5, caducous, pubescent. Corolla oblanceolate; petals-5, nearly as long as sepals, yellow. Stamens numerous. Fruit a head of small achenes, oblong; achenes several, compressed, obliquely obovate.

Radical leaf

Flowering twig

Flower

Chemical Constituents: Plant contains anemonin, ranunculine, serotonin and other triptamine derivatives such as 5-hydroxy tryptamine, apigenin, apigenin 4′-o-α-rhamnopyranoside, apigenin 7-O-β-glucopyranosyl-4′-O-α-rhamnopyranoside, tricin 7-O-β-glucopyranoside, isoscopoletin, tricin, protocatechuyl aldehyde, and protoanemonin. (Bhutya 2011, Aslam *et al.* 2012).

Medicinal Properties: Fresh plant is highly acrid, rubefacient, vesicant and toxic; causes inflammation of digestive tract. The dry plant has antiseptic, digestive, appetizer, deobstruent, anthelmintic, and emmenagogue and galactagogue properties. Seretonin present in the plant is potent vasco constrictor. Protoanemonin has strong antibacterial, antiviral, cytopathogenic and vermicidal properties. It is effective against both gram-positive and gram-negative bacteria, similar to Penicillin. It inhibits the growth of *Escherichia coli, Staphylococcus aureus* and *Candida albicans.*

Whole plant, used as an astringent, is slightly bitter herb that is specifically antihaemorrhoidal internally and externally used for perineal damage after child birth. A strong fungicidal, anodyne, used in skin disorders; leaves ground with Brassica oil and applied externally to cure eczema. This medicinal herb is capable of promoting blood circulation by removing blood stasis, expelling cold, relieving swelling, and removing excessive heat from the liver and the gall bladder. It can also cure internal abscess, malaria, scrofula, snake or scorpion venom, and acute icteric hepatitis. Fresh *R. sceleratus* is used to relieve hydrops articulation of the knee, with an 80 percent cure rate. (Bhutya 2011, Mei *et al.* 2012; http://medplants.blogspot.in).

Medicinal Uses (Traditional/Folklore): *R. sceleratus* is a common traditional Chinese medicinal herb with excellent therapeutic effects, which was recorded in the earliest Chinese literature. It has been used to treat blisters and as counter-irritant since time immemorial. In China, it is used to treat colds, arthritis and rheumatisms and to invigorate health. In Laos, Cambodia and Vietnam, the seeds are eaten to promote appetite and digestion, treat kidney diseases and heal abscesses. Plant juice is given in bronchitis and cough. Whole plant is crushed and tied on ulcers on feet. Paste of roots with mustard oil is used on swellings; paste is applied regularly on male sex organ for increasing erectile power and for treating impotence.

Decoction made from the whole plant is also used as an insecticide. Powdered plant mixed with flour and sugar is used for rat poisoning. Leaves are vesicant and used as poison for arrow points. All types of livestock can become ill upon ingestion of the plant, but cattle are most commonly affected; horses and goats are also affected and get poisoned. (Mei *et al.* 2012; http://medplants.blogspot.in).

2. MENISPERMACEAE

Menispermaceae – Moonseed family

It contains 68 genera and 448 species, most of which are woody climbers in tropical forests, although some genera extend into temperate regions in North America and Japan. One species belonging to this family was recorded in the study area.

2.1 *Cissampelos pareira* L.

(Syn. *Cissampelos argentea* Kunth; *Cissampelos hirsuta* Buch.-Ham. ex DC.)

Common Name: Velvet Leaf, False Pareira Brava, Abuta.

Vernacular Name: *Hindi:* Bhatvel (भटवेल), Padha (पढ़ा), Parhi (पाढ़ी); *Sanskrit:* Laghu Patha (लघु पाठा).

Habit: Climber.

Habitat: Generally climbing on trees and bushes along roadsides, canal sides, and railway lines.

Distribution: First described from Latin America. The plant occurs throughout the tropics. Found in tropical and subtropical parts of India.

Flowering and Fruiting Time: July-September.

Taxonomic Description: An extensive climber, branchlets densely tomentose. Leaves alternate, petiolate, nearly as long as broad, 3-11 cm across, palmately 5-7 nerved, broad ovate, orbicular, apex usually emarginate, base cordate, margin entire. Flowers unisexual. Male flowers minute, axillary, bracteate, greenish yellow; sepals-4, hairy outside; petals-4, hispid; stamens-4, filament connate. Female flowers axillary or terminal, bracts foliaceous. Carpel solitary, pilose. Fruits drupes, small, obovoid, pubescent, shining red when ripe.

Habit

Flowering twig

Flowers

Chemical Constituents: The plant contains alkaloids *viz.,* hayatine (+ curine), hayatinine, hayatidine and other bisbenzylisoquinoline alkaloids, including some non-nitrogenous. Hayatine (dl-berberine) is the principal alkaloid of root. Its derivatives, methiodide and methochloride are potent neuro-muscular blocking agent. Other chemicals present are hayatinine, hayatinine, hayatidine, quercitol and sterol. (Bhutya 2011, Arora *et al.* 2012, Jhuma *et al.* 2012).

Medicinal Properties: The plants have a bitter taste and possess diuretic, purgative and antiperiodic and antimalarial properties. Furthermore, they are judged to be good against dyspepsia, diarrhoea, dropsy and cough, urinary difficulties like cystitis, dysentery, asthma and heart diseases; mainly extracts of the roots are used. The roots are bitter, astringent, antispasmodic, analgesic, antilithic, emmenagogue, rubefacient, anthelmintic, carminative, stomachic, depurative, digestive, anti-inflammatory, diuretic, expectorant, febrifuge and galactagogue. In the simplest cases leaves are good as an antiseptic against inflammation and applied on wounds in order to heal sores. (Bhutya 2011, Arora *et al.* 2012).

Medicinal Uses (Traditional/Folklore): Velvet Leaf is known as Midwife's herb as mainly used in women aliments. It is used for menstrual problems, painful menstruation, and hormonal imbalance, ease in childbirth, postpartum pain, to prevent threatened miscarriage, and control uterine haemorrhages, hormonal acne and premenstrual syndrome. It is very effective herb in vaginal discharge and good for overall health of women reproductive system. Chemical compound methalonic obtained from extract of leaves is used as antifertility agent because it alters the secretion of gonadotropins and estradiol. It is also used for heart problems, kidney stones, kidney infections and pains, asthma, dysuria, arthritis, muscle cramps, stomach pains, non-healing ulcers, fistula, leucorrhoea, gonorrhoea, hypogalactia (agalactia or agalactorrhea), sexually transmitted diseases, snake bites and conjunctivitis.

In the traditional folk medicine the roots are used against a number of ailments. Roots are used in dyspepsia, indigestion, flatulence, abdominal pains, diarrhoea, dysentery, blood disorders, cardiac disorders, colic, edema, leprosy, sensation, cough, coryza, asthma, bronchitis, cystitis and dysuria and lactation disorders. Root paste is applied topically on scabies and eruptions. It is also used in non-healing ulcers, skin disorders, scabies, leprosy, migraine, leucorrhoea and gonorrhoea. Leaves are used in eye trouble, skin ailments and burns, burning sensation, wounds, fever and cold. The plant is also used for the treatment of malaria and dengue.

In Ayrveda, *Patha* is useful in colicky pains, fever, vomiting, and skin conditions. The therapeutic uses of *Patha* according to the Ayurvedic Pharmacopoeia of India are in the treatment of abdominal pain, cardiac pains, burning feeling, pruritus, poisons, breathing difficulties and worms; also effective in treating diarrhoea, skin diseases, and fever. It is very effective antidote and wound healer. It is also very effective in kushtha roga (leprosy). Due to these properties external paste of leaves and roots are used to cure various skin disorders, and snake bite poisoning. *Patha* is very effective appetizer and used to cure anorexia, indigestion and abdominal pain. It is used to stimulate digestive fire and also used to cure diarrhoea and dysentery. Being a powerful expectorant it is used to cure cough and dyspnoea. It has anti-inflammatory properties and used as blood purifier. It is very effective for heart and blood related disorders. Paste of roots and leaves are used to reduce inflammation. It is good for female reproductive system and it is used to purify breast milk secretion and various other disorders associated with breast. Being a diuretic, it is very effective in dysuria and haematuria. The plant has anthelminthic action and used in worm infestation. Leaves are used to prepare poultice to treat scabies, abscess, itching, acne and wounds. This herb is also used to cure poisoning and dog bites. (Arora *et al.* 2012, Jhuma *et al.* 2012).

3. NYMPHAEACEAE

Nymphaeaceae – Water lily family

The plants of this family having 8 genera and 70 known species are commonly called as water lilies. They are found as rhizomatous aquatic herbs in temperate and tropical climates around the world. The family is further characterized by scattered vascular bundles in the stems, and frequent presence of latex, usually with distinct, stellate-branched sclereids projecting into the air canals. One species belonging to this family was recorded in Chithara Village.

3.1 *Nymphaea pubescens* Willd.

(Syn. *Nymphaea lotus* var. *pubescens* (Willd.) Hook. f. & Thomson)

Common Name: White Water Lily, White Lotus.

Vernacular Name: *Hindi:* Ko Kaa (को का); *Sanskrit:* Kumud (कुमुद).

Habit: Aquatic herb.

Habitat: Ponds and lakes.

Distribution: The plant is common in shallow lakes and ponds throughout the temperate and tropical Asia: Bangladesh, India, Sri Lanka, Yunnan, Taiwan, Philippines, Cambodia, Laos, Myanmar, Thailand, Vietnam, Indonesia and Malaysia.

Flowering and Fruiting Time: August-November.

Habit

Flowers with long petiole

Flower

Taxonomic Description: An aquatic, perennial herb with laticiferous root. Leaves sub-sagittate when young, orbicular at maturity, about 10-30 cm across, sinuate toothed, deeply cordate at base, obtuse to retuse at apex, densely pubescent beneath, palmately nerved, petiole up to 1 m long. Flowers solitary, complete, bisexual, white, floating on the surface of water, generally 8-10 cm across, long peduncled. Sepals-4, elliptic-lanceolate, greenish-pink, minutely pubescent, 5-10 nerved. Petals numerous, oblong, elliptic-lanceolate, white; outer petals tinged with purple, up to 7 cm long. Stamens numerous; filaments flat, longer than the anthers. Ovary superior, several-loculed. Berries broadly globose, 3-4 cm in diameter. Seeds numerous, ovoid, nearly 1 mm across.

Chemical Constituents: The plant contains alkaloids, flavonoids, glycosides, terpenoids, tannins, phenols, saponins and steroids. (Shajeela *et al.* 2012).

Medicinal Properties: This plant is mentioned in Siddha system of medicine, for the treatment of bleeding piles, diabetes and as cardiotonic of the heart. Used as blood purifier and in the treatment of jaundice. The seeds are sweet, cooling, constipating, aphrodisiac, stomachic and restorative. They are useful in vitiated conditions of pita, dipsia, diarrhoea and dermatopathy. The rhizome is cooling, sweet, bitter and tonic and is useful in diarrhoea, dysentery, dipsia and general debility. The flowers are astringent and cardiotonic. The seeds are sweet, cooling, constipating, aphrodisiac, stomachic and restorative. The root stock and leaves are cooling, bitter, sweet, and aromatic, used as alterant, diuretic and tonic. (Angadi *et al.* 2013, Muthulingam 2010).

Medicinal Uses (Traditional/Folklore): *N. pubescens* plants have been used as medicine since time immemorial and the uses of plant are mentioned in Charaka Samhita and Sushrutha Samhitha which are the oldest books in India. Plant is an aphrodisiac for both men and women and a general remedy for all kind of illnesses. Leaves are used in cutaneous and subcutaneous parasitic infection, eye treatments and pregnancy. The whole plant or the parts of the plant used in siddha and folk medicine for treating diabetes, bleeding piles, dyspepsia, jaundice and eye disorders. Rhizome ground with seeds of *Piper nigrum* and paste is applied externally on neck against Goitre. Dried rhizome powder (20 g) mixed with honey is given orally in the early morning to cure jaundice. (Prasad and Savithramma 2015).

4. PAPAVERACEAE

Papaveraceae – Poppy family

An economically important family of about 41 genera and approximately 920 known species, the family has plants that are ready colonizers of vacant land. The best example is the Mexican poppy, a native of America that has spread throughout the tropical countries of the world. This group has both annuals as well as herbaceous perennials, rarely trees; some of the plants are poisonous. The plants have latex which may be white or yellowish in colour. Two species belonging to two genera of this family were recorded in Chithara Village.

4.1 *Argemone mexicana* L.

(Syn. *Papaver mexicanum* (L.) E.H.L.Krause)

Habit

Common Name: Mexican Prickly Poppy.

Vernacular Name: *Hindi:* Satyanashi (सत्यानाशी), Kataila (कटैला).

Habit: Annual herb.

Habitat: Roadsides, open and waste places.

Distribution: Native of America; found throughout the warmer parts of India, in areas up to 1,500 m elevation on roadsides and wastelands.

Flowering and Fruiting Time: January to May.

Taxonomic Description: An erect, prickly, greenish-white, annual herb, 40-90 cm high. Stem and branches with yellow latex. Leaves variable in size and shape, sinuate-pinnatified, up to 18 cm long, glaucous, semi-amplexicaul, spiny margin, upper surface white blotched. Flowers terminal, yellow, sessile or sub-sessile, solitary subtended by foliaceous bracts. Sepals-2 or 3, calyptrate, horn at the apex. Petals twice the number of sepals, bright yellow, present in two series, deciduous. Stamens numerous, anther and filament yellow. Stigma dark red, 4-6 lobed. Capsules elliptic, oblong, prickly. Seeds numerous, spherical, black.

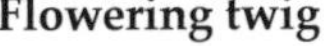

Flowering twig

Flower

Chemical Constituents: Most of the isolated compounds belong to the class of alkaloids; besides, terpenoids, flavonoids, phenolics, long-chain aliphatic compounds, and few aromatic compounds are found to be other constituents of this plant. Four quaternary isoquinoline alkaloids, dehydrocorydalmine, jatrorrhizine, columbamine, and oxyberberine, have been isolated from the whole plant of *A. mexicana.* The seed pods secrete a pale yellow latex when cut open; Berberine (0.74%), protopine (0.36%) and free aminoacids are found in latex. Seeds contain 22–36%

of a pale yellow non-edible oil, called *argemone oil* or *katkar oil*, which contains the toxic alkaloids sanguinarine, allocryptonine and dihydrosanguinarine. Flowers contain isorhamentin 3 – glucoside and 3, 7-diglucoside. From roots and stems protopine, berberine, resin and a toxic principle, and from aerial parts protopine and allocryptopine have been isolated. The other chemical compounds isolated are: norsanguinarine, norchelerthrine, cryptopine, berberine, benzophenanthridine, coptisine, sanguinarine and chelerythrine. (Shrivastava *et al.* 2013, Pandya 2009).

Medicinal Properties: The plant is acrid, anodyne, anthelmintic, anti-inflammatory, antipyretic, antidiarrhoeal, antifungal, antileprotic, antiviral, bitter, cooling, diuretic, expectorant, aphrodisiac, emetic, depurative, detersive, diuretic, diaphoretic, hypnotic, ophthalmic, purgative, resolutive, stomachic, sedative, toxic and vulnerary. Berberine exhibits significant myocardial depressant activity and total alkaloidal fraction reported to cause fall in blood pressure. (Warrier *et al.* 2010).

Medicinal Uses (Traditional/Folklore): *A. mexicana* is considered as an important medicinal plant in India; the yellow juice, which exudes when the plant is injured, has long been used in India as traditional medicine for dropsy, jaundice, ophthalmia, scabies and cutaneous affections. Different parts of this plant are used in chronic skin diseases, and also as emetic, expectorant, demulcent and diuretic; the plant possesses antivenom property as well. The seeds and seed oil are useful as a remedy in vitiated conditions of cough, asthma, pertussis, skin diseases, leprosy, ulcers, wounds, dental caries, constipation, dysentery, intestinal affections, rheumatalgia and colic. Leaves and seeds are reported to find application in maintaining normal blood circulation and cholesterol level in human body, cough, wounds, ulcers and skin diseases. The roots are useful in guinea-worm infestation, skin diseases, pruritus, blennorrhagia, inflammations, all types of poisoning, constipation, flatulence, colic, malarial fever and vesicular calculus. Flowers are expectorant and have been used in the treatment of chronic coughs. The latex is useful in dropsy, jaundice, skin diseases, leprosy, ulcers, conjunctivitis, inflammations and malaria fever.

In Rajasthan, the plant is used for female sex disorders, eye diseases, mental disorders, skin diseases, tooth ache and wound healing. In Brazil, the plant is commonly known as 'cardo-santo' and used traditionally in the treatment of a number of diseases. Seeds of the plant are used as purgative, laxative and digestive while its latex is used against conjunctivitis. Besides, its infusion finds application against hypertension in Brazil. The Seri of Sonora, Mexico use the entire plant both fresh and dried for various ailments. An infusion is made to relieve kidney pain, to help expel a torn placenta, and in general to help cleanse the body after parturition. The seeds are taken as a laxative. An *Argemone mexicana* tea is used by traditional healers in Mali to treat malaria. This use has been studied clinically for the treatment of uncomplicated malaria.

In traditional/folklore medicines, latex, leaf juice extract and seed oil of the plant are applied externally to heal wounds faster, and in skin diseases. Its juice extract, latex, seed oil is applied externally on wounds and skin diseases. The root paste is applied externally as an anti-inflammatory agent on scorpion and insect bites. Seed paste and seed oil is applied ov er joints to relieve arthritis pain as an analgesic. For constipation, bloating and abdominal colic pain, root powder is administered in a dose of 1-2 grams per day. Root paste is administered (up to 3 grams dose) to relieve intestinal worm infestation. (Priya and Rao 2012, Brahmachari *et al.* 2013; https://easyayurveda.com).

4.2 *Fumaria indica* (Hausskn.) Pugsley

(Syn. *Fumaria parviflora* var. *indica* (Hausskn.) Parsa)

Common Name: Indian Fumitory.

Vernacular Name: *Hindi:* Papara (पापड़ा), Pittpapra (पित्तपापड़ा); *Sanskrit:* Parpata (परपता).

Habit: Annual herb.

Habitat: Moist places in gardens, cultivated and non-cultivated fields.

Distribution: The plant is distributed in Europe, Africa and many Asian countries, including India. Found throughout India, particularly on the banks of the Ganges and in the Himalayas as a weed in cultivated areas and roadsides up to an altitude of 2,700 m. It is also distributed in the higher elevations of the Mysore plateau and Nilgiris.

Habit

Flowering and Fruiting Time: January-February.

Taxonomic Description: An annual, diffuse, much branched herb. Leaves glaucous, finely dissected into narrow flat segments, linear-oblanceolate, margin entire, tip acute or mucronate. Flowers terminal or in leaf opposed racemes, bracts lanceolate. Sepals-2, much smaller than petals, lanceolate. Petals 2+2, purplish-pink, one outer posterior petal produced short spur at the base. Fruits globose, 2-2.5 mm in diameter, rugose, one seeded.

Leaf

Inflorescence

Fruits

Chemical Constituents: The herb contains isoquinoline alkaloids including protopine, sang-uinarine, cryptopin, d-bicuculine, fumaridine. The leaves contain kaempferol and quercetin glycosides. The chemical constituents reported to occur in the plants are protopine, cryptopine, dl-bicuculline, aldumine, fumaridine, fumarizine, spiroisoquinoline and d-hydrastine as chief alkaloids. Other minor alkaloids include parfumidine, parfumine, coptisine, tetrahydrocoptisine, fumariline, fumarilicine, narceimine, 8-methoxydihydrosanguinarine, oxysanguinarine, fumariflorine, lastourvilline, N-methyl corydaldine, oxycoptisine, raddeanine, N-methyl-hydrasteine, dehydrocheilanthifoline, narlumidine, papraine and paprazine. Tannins and fumaric acid and other non-alkaloidal constituents such as nonacosanol and sitosterol, 19-methyloctacosanol and 3-methyloctacosanol have also been isolated from the plant. (Bhutya 2011, Ravikanth *et al.* 2014).

Medicinal Properties: Plant is astringent, laxative, febrifuge, diaphoretic, antipyretic, diuretic, anthelmintic and blood purifier. It is useful in curing dyspepsia, scrofulous skin, liver diseases, pyrexia, influenza, and scabies, burning sensation, vertigo, epistaxis, sprue, anaemia and diarrhoea. (Bhutya 2011).

Medicinal Uses (Traditional/Folklore): Indian Fumitory is used in aches and pains, diarrhoea, fever, influenza and liver complaints. The herb mixed with honey is taken internally to prevent vomiting. A cold infusion of the plant is used to treat wasting diseases of children and to help cooling during fever and in the treatment of constipation and dyspepsia. It is used as a blood purifier for skin diseases and applied externally in leucoderma and as a fomentation for swollen joints. The dried plant is also used as an anthelmintic, diuretic and diaphoretic and, in combination with black pepper, for jaundice. The herb mixed with honey may be taken internally to prevent vomiting.

It is a very common plant in veterinary medicine and is used to overcome constipation and urinary problems in cattle. It is used as a fodder in many parts of India, particularly in Assam. The leaf is used in udder infection of ruminants.

The plant is sold under the name *Pitpapra* in Ayurveda. In the Unani system of medicine the plant is incorporated into *Trifala Shahtara*. (Ravikanth *et al.* 2014; http://www.flowersofindia.net).

5. BRASSICACEAE

Brassicaceae – Mustard/Cabbage family

The family has 372 genera and 4,060 accepted species containing some of the important economically important cruciferous vegetables, including broccoli, cabbage, cauliflower, turnip, lettuce, rapeseed, common radish, etc. The flowers are cruciform (e.g., in the form of a cross) with four petals and four sepals; the plants are commonly referred to as "crucifers" or "cruciferous" for that reason. The flowers usually have four long and two short stamens. The family was represented by six genera and six species in the flora.

5.1 *Arabidopsis thaliana* (L.) Heynh

(Syn. *Sisymbrium thalianum* (L.) J.Gay)

Common Name: Thale Cress, Mouse-Ear Cress.

Vernacular Name: None known.

Habit: Annual herb.

Habitat: Moist and shady areas, along water banks.

Distribution: Native to Eurasia and north-western Africa; considered as a weed

Flowering and Fruiting Time: January- April.

Basal leaves

Flowers and fruits

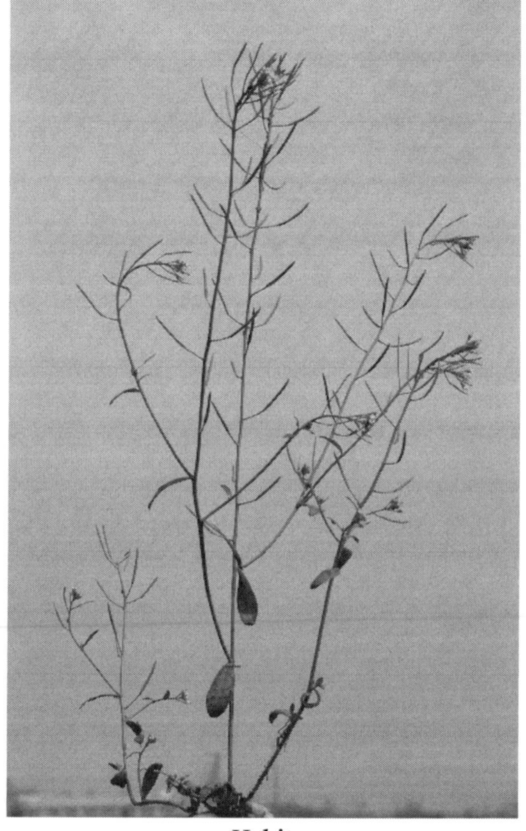

Habit

Taxonomic Description: An erect, pubescent, annual herb, 10-25 cm high. Basal leaves form a rosette, 2-7 cm long, shortly petioled, oblong, oblanceolate, entire or slightly dentate. Upper leaves smaller, sessile, oblong, linear, entire at margin, cuneate at the base. Inflorescence raceme, slender, terminal. Flowers complete, without bract, bisexual, pedicellate. Sepals-4, green, ovate. Petals-4, white. Fruit siliquae, cylindrical, linear, 1-1.8 cm long, glabrous, compressed; valves 1-nerved. Seeds numerous, brown.

Chemical Constituents: In total, 23 glucosinolates have been isolated which include four series of homologs and analogs. Fifteen possess aliphatic side chains, of which six contain methylthioalkyl and six contain methylsulphinylalkyl side chains; eight possess aromatic side chains, of which four constitute a homologous series of benzoic acid esters and three possess 3-indolylmethyl-based structures. 16 of the glucosinolates are detected for the first time and three of these, 4-hydroxybutyl glucosinolate, 5-benzoyloxypentyl glucosinolate, and 6-benzoyloxyhexyl glucosinolate, represent novel plant constituents. (Hogge *et al.* 1988).

Medicinal Properties: None known.

Medicinal Uses (Traditional/Folklore): Plant has minor medicinal value and is used to cure mouth sores. (http://pfaf.org).

5.2 *Capsella bursa-pastoris* (L.) Medik
(Syn. *Thlaspi bursa-pastoris* L.)

Common Name: Shepherd's Purse, Blind weed.

Vernacular Name: *Hindi:* Mumiri (मुमीरी).

Habit: Annual herb.

Habitat: Moist places, along roadsides, garden, cultivated fields.

Distribution: Native in different parts of the world, including Cyprus, Europe, Saudi Arabia, Turkey, Pakistan, India, Iran, Iraq, Azarbayjan, China and many other places in Asia. It is also found in North Africa, Central America and Eastern Europe. In India, found commonly in northern India from 2,800-3,000 m in waste places, crop fields and along roadsides.

Flowering and Fruiting Time: January-April.

Radical leaves

Flower and fruits

Habit

Taxonomic Description: Annual herb, slightly or sometimes densely pubescent, up to 40 cm high. Radical leaves rosulate, shortly stalked, oblong-lanceolate or oblanceolate, pinnatified, dentate-serrate or entire; cauline leaves linear-lanceolate, auricled, entire or dentate at margin, acute at apex. Inflorescence racemose, slender, up to 30 cm long in fruit. Flowers small, complete, pedicellate. Sepals-4, oblong, green. Petals-4, oblanceolate, white. Fruit silicles, nearly 5-6 mm across, triangular or obcordate. Seeds many, oblong, minute, brown.

Chemical Constituents: The plants contain a wide range of chemicals including flavonoids, polypeptides, choline, acetylcholine, histamine, tyramine, fatty acids, sterols, organic acids, amino acids, sulforaphane, many trace elements, vitamins and many other compounds. Specific compounds isolated are: Thiamin, choline, inositol, fumaric acid, ascorbic acid, riboflavin, calcium, potassium, phosphorus, beta carotene, vitamin K, niacin, iron, and rutin. Analysis of the isolated essential oils revealed the presence of 19 compounds accounting for 88.24 percent of total oil content. The main compounds in essential oils were 1, 1-dimethylcyclopentane, ethyl linoleate, palmitic acid and phytane. (Al-Snafi 2015d, Kamali *et al.* 2015).

Medicinal Properties: *C. bursa-pastoris* has antimicrobial, anti-inflammatory, antioxidant, cardiovascular, reproductive, anticancer, hepato-protective and sedative properties. The plants also have other pharmacological medicinal properties such as antibleeding, anticancer, antithrombin, antioxidant, antidiabetic, wound healing, and antipyretic (fever treatment). A tea made from the whole plant is antiscorbutic, astringent, diuretic, emmenagogue, haemostatic, hypotensive, oxytocic, stimulant, vasoconstrictor and vulnerary. (Dhiman 2004, Al-Snafi 2015d).

Medicinal Uses (Traditional/Folklore): Shepherd's Purse is one of the most important medicinal plants of the family Brassicaceae. The plant is used for heart and circulatory problems including mild heart failure, low blood pressure, and nervous heart complaints. It is also used for headache, vomiting blood, blood in the urine, diarrhoea, and bladder infections. Women use shepherd's purse for premenstrual problems, long periods, and menstrual cramps. The whole plant is used as astringent and antiscorbutic in haematuria and diarrhoea and as diuretic in dropsy. It is stated that the plant is used in Chinese medicines for ailments for the eyes and for dysentery. Its seeds are also used as astringent. The plant is considered as a good substitute for ergot of rye to arrest bleeding from the lungs, kidney and uterus. Its haemostyptic properties have long been known and are said to equal those of ergot and *Hydrastis* sp. (a natural antibiotic with anticancer properties). During the World Wars, when these were no longer obtainable in German commerce, a liquid extract of *C. bursa-pastoris* was used as a substitute made by exhausting the drug with boiling water. When dried and infused, it yields a tea which is still considered by herbalists one of the best specifics for stopping haemorrhages of all kinds - of the stomach, the lungs, or the uterus, and more especially bleeding from the kidneys.

The plant has been used medicinally as haemostatic, diuretic and antipyretic in China and Japan for many centuries. The whole plant was used to treat edema caused by nephritis, odynuria, hemafecia, menorrhagia, chyluria and hypertension. The leaves and roots of the plant were used as an edible vegetable, eaten raw or cooked in some countries. The plant has been ranked 7th amongst 250 potential antifertility plants in China. A homeopathic remedy is made from the fresh plant is used in the treatment of nose bleeds and urinary calculus.

The plant has been used in English domestic practice from early times as an astringent in the treatment of diarrhoea. It has been employed as fresh decoction in haematuria, haemorrhoids, chronic diarrhoea and dysentery, and locally as a vulnerary in nose-bleeding, which is checked by inserting the juice on cotton-wool. It is also used as an application in rheumatic affections, and has been found curative in various uterine haemorrhages, especially those with which uterine cramp and colic are associated, and also in various passive haemorrhages from mucous surfaces. It is a remedy of the first importance in catarrhal conditions of the bladder and ureters, also in ulcerated conditions and abscess of the bladder. It increases the flow of urine. Its use is specially indicated when there is white mucous matter voided with the urine; relief in these cases following at once.

The plant has antimosquito properties. The seed, when placed in water, attracts mosquitoes. It has a gummy substance that binds the insect's mouth to the seed. The seed also releases a substance toxic to the larvae. About ½ kilo of seed is said to be able to kill 10 million larvae.

As a veterinary medicine, the plant's decoction with milk was used to check active purging in calves in Britain. (Dhiman 2004, Al-Snafi 2015d; http://healingherbinfo.com, http://tropical.theferns.info, https://www.botanical.com).

5.3 *Cardamine hirsuta* L.

(Syn. *Cardamine multicaulis* Hoppe ex Schur)

Common Name: Hairy Bitter Cress, Flick Weed.

Vernacular Name: None Known.

Habit: Annual herb.

Habitat: Shady and moist places.

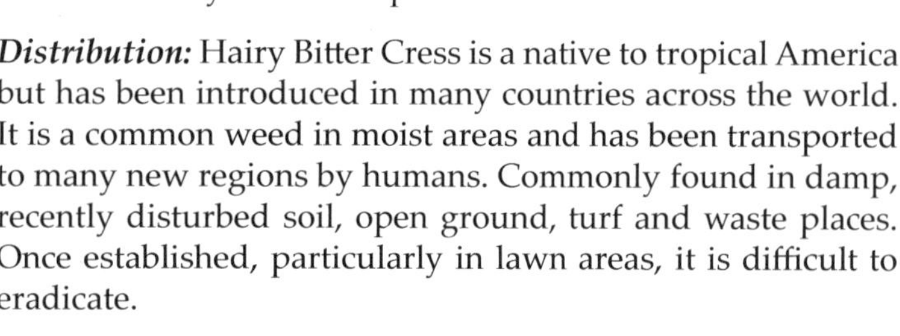

Distribution: Hairy Bitter Cress is a native to tropical America but has been introduced in many countries across the world. It is a common weed in moist areas and has been transported to many new regions by humans. Commonly found in damp, recently disturbed soil, open ground, turf and waste places. Once established, particularly in lawn areas, it is difficult to eradicate.

Habit

***Flowering and Fruiting Time*:** January-March.

Taxonomic Description: An erect to sub-erect, annual diffuse herb, up to 30 cm high. Stem much-branched from the base. Leaves variable, glabrous or slightly pubescent; radical leaves form a rosette, petiolate, imparipinnate, ovate or orbicular, sometimes toothed or irregularly lobed. Cauline or upper leaves narrow, sessile or shortly petiolate, oblong, lanceolate. Inflorescence raceme or subcorymbose. Flowers white, small. Calyx lobes-4, oblong. Corolla lobes-4, white, obovate, twice the length of sepals. Stamens 4-6. Fruit dehiscent siliquae, cylindical, slightly compressed, up to 3 cm long. Seeds many, brown.

Flowering twig

Flowers

Chemical Constituents: Vitamin C, calcium, magnesium, beta carotene, antioxidants and sulfur-containing compounds that boost immunity and help in cancer prevention. (https://naturesaccess.com, http://healingweeds.blogspot.in).

Medicinal Properties: Plant is tonic, stimulant and antioxidant. (https://naturesaccess.com).

Medicinal Uses (Traditional/Folklore): In Laos, Vietnam and Cambodia it is used to stop dysentery and to treat eye troubles; leaves are eaten as salads. Apatani tribe of Arunachal Pradesh use leaves for treating indigestion. Plant is used as a common diuretic in Manipur. (Kala 2005, Wiart 2006b, Yumnam and Tripathi 2012).

5.4 *Lepidium didymum* L.

(Syn. *Coronopus didymus* (L.) Sm.; *Senebiera pinnatifida* DC.)

Common Name: Bitter Cress, Lesser Swine-cress, Wart Cress.

Vernacular Name: *Hindi:* Jangli Hala (जंगली हाला).

Habit: Annual herb.

Habitat: Moist and waste places.

Distribution: *L. didymum* is native to tropical America. It is a cosmopolitan weed recorded from Ethiopia, Kenya, Tanzania, Namibia, Zimbabwe, South Africa and India. Found growing on cultivated and waste ground, in gardens and lawns, paths and roadsides.

Habit

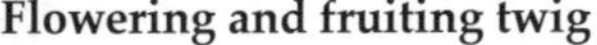

Flowering and fruiting twig

Fruits

Flowering and Fruiting Time: January-March.

Taxonomic Description: A prostrate, foetid, annual herb with long tap root. Stem spreading, angular, branched from the base; branches sparsely or densely covered with simple or 2-fid trichomes. Leaves pinnately compound, lobed; lobes obovate, apiculate, and entire. Inflorescence racemose, terminal or axillary. Flowers ebracteate, bisexual, greenish white or yellowish green. Sepals-4, ovate-oblong. Petals-4, linear. Fertile stamens 2. Fruit silicles, notched at apex, emarginated at base, didymous. Seeds 1 per locule, winged, reniform, brown, and slightly compressed.

Chemical Constituents: Phytochemical screening of *L. didymum* showed the presence of flavonoids, saponins and tannins. The bioactive flavonoid chrysoeriol and its glycoside chrysoeriol-6-O-acetyl-4'-β-D-glucoside have been isolated from the plant. (https://www.prota4u.org).

Medicinal Properties: It has antiscorbutic, digestive, expectorant, febrifuge, stimulant and tonic properties. Plants are used in the treatment for cancer, gangrene, haemorrhoids, and allergies and wound healing. (https://www.prota4u.org).

Medicinal Uses (Traditional/Folklore): A decoction of the whole plant is drunk to treat headache and fevers. Whole plant paste is applied to prevent hair fall. In India, the plant is valued in traditional medicine as a treatment for allergies and wounds. In Argentina, *L. didymum* is used as an expectorant, digestive and febrifuge, and for treating gangrene and haemorrhoids. (https://www.prota4u.org).

5.5 *Rorippa palustris* (L.) Besser

(Syn. *Cardamine palustris* Bubani; *Nasturtium palustre* (L.) DC.)

Common Name: Yellow Marsh Cress, Bog Yellow-cress.

Vernacular Name: *Hindi:* Chamsuru (चम्सूरु).

Habit: Annual herb.

Habitat: Marshy areas, generally near water bodies.

Distribution: It is widespread and native to parts of Africa and much of Asia, Europe, Eurasia, North America and the Caribbean. It is also found in other parts of the world as an introduced species and a common weed. The plant grows in many types of damp, wet, and aquatic habitat. In India, it is known to occur in humid marshy places, near streams and lakes, up to an altitude of 2,000-3,300 m in the states of Assam, Bihar, Himachal Pradesh, Jammu and Kashmir, and West Bengal.

Upper portion of plant

Flowering and Fruiting Time: January-March.

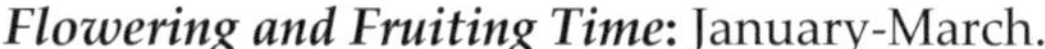

Flowering and fruiting twig

Fruits

Taxonomic Description: An erect, branched, annual herb, up to 80 cm high. Leaves variable, up to 20 cm long; radical leaves stalked, pinnately compound, lobes toothed; cauline leaves smaller, base semi-amplexicaul, margin irregularly crenate serrate. Flowers small, bright yellow in many flowered, 10-30 cm long racemes. Sepals-4, about 2-3 mm long. Petals-4, obovate, nearly equal to the sepals. Stamens-6. Pods oblong, 0.5-12 mm long, beaked, slightly curved. Seeds minute, numerous, orbicular, reddish-brown.

Chemical Constituents: From the plant, sixteen chemical compounds belonging to flavonoids (kaempferol and quercetin derivatives), phenolic acids (Gallic, coumaric and ferulic acids derivatives) and an anthocyanin (delphinine derivative) have been identified and characterized, of which ten compounds were detected for the first time from *R. palustris*. (Marzouk *et al.* 2016).

Medicinal Properties: It has diuretic and stimulant properties. The leaves are antiscorbutic. (Naskar 1993, Lassak and McCarthy 2001)

Medicinal Uses (Traditional/Folklore): Used as an eye medicine, especially in ceremonial eye washes. (Austin 2004).

5.6 *Sisymbrium irio* L.

(Syn. *Sisymbrium irioides* Boiss.)

Common Name: London Rocket.

Vernacular Name: *Hindi:* Khubkalan (खूबकलाँ), Asalio (असलियो); *Sanskrit:* Khakasi (खाकसी), Khubakalan (खूबकलाँ).

Habit: Annual herb.

Habitat: Moist waste places and gardens.

Distribution: Native to southern Europe, North Africa and temperate Asia but has been carried by migrants to North America, Australasia and South Africa, where it has naturalized and become a well-known, yellow-flowered weed of waste and neglected areas. In India, it is reported to occur in Jammu & Kashmir, Himachal Pradesh, Uttarakhand, and Uttar Pradesh.

Flowering and Fruiting Time: January-March.

Habit

Taxonomic Description: A small, erect, 15-80 cm high, annual herb. Leaves petiolate, pinnately compound or partite; lobes toothed, rarely hastate. Inflorescence racemose. Flowers bisexual, small, numerous, yellow in lax racemes; pedicels slender. Sepals-4. Petals-4. Stamens-6; filaments yellowish. Ovary superior, stigma capitate, sessile. Siliquae narrow, 3-5 cm long, sub-torulose; valve 3-nerved, Seeds numerous, small, brown.

Raceme with flower and fruits

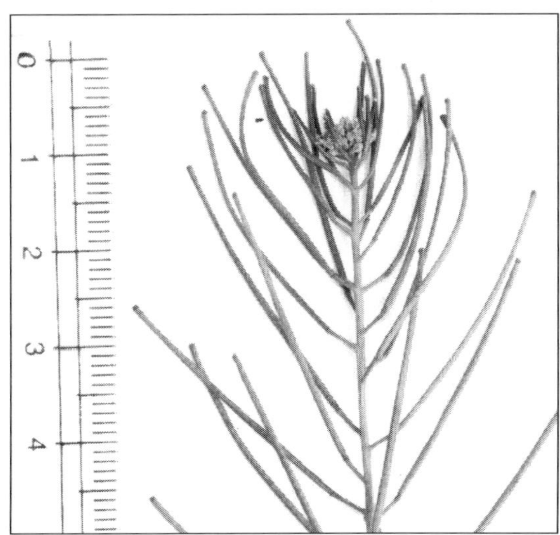

Fruit size with scale

Chemical Constituents: The plant is a rich source of flavoinoids and glucosinolates, phytoconstitents i.e., alkaloids, glycosides, carbohydrate, phenolics and tannins, phytosterols, fixed oils, fats, proteins and amino acids, flavonoids, saponins, gums and mucilage. It contains essential oils which has 53 components, representing 97.48% of the total oil. The main constituents of the oil were dioctyladipate (25.44%), N-(n-proyl) acetamide (14.77%), isopropyl isothiocyanat (11.55%), isobutyl isothiocyanate (6.75%), 3,7,11,15-tetramethyl-2-hexadecen-1-ol (6.52%), cis-8,11,14-eicosatrienoic acid (6.30%), heptacosane (3.89%), palmitic acid (3.45%), n-butyl isothiocyanate (2.85%) and dimethoxyacetophenone (2.54%). (Al-Qudah and Zarga 2010).

Medicinal properties: *S. irio* has antipyretic, analgesic, expectorant, febrifuge, antimicrobial and antioxidant properties. The seeds are expectorant, restorative and stimulant. They are used in the treatment of asthma. (Shah *et al.* 2013b).

Medicinal Uses (Traditional/Folklore): London Rocket is used in the Middle East to treat coughs and chest congestion, to relieve rheumatism, to detoxify the liver and spleen, and to reduce swelling and clean wounds. The seeds are placed under the lids of sore eyes in order to cause weeping and thereby wash foreign matter out of the eye. An infusion of the leaves is used in treating affections of the throat and chest. The plant is used in folk medicine as a febrifuge, a stimulating poultice, treating asthma and for infections of the throat and chest. Seeds are used as expectorant, febrifuge and used in treatment of voice disorders. Seed is also used for stomach problems and dysentery. (Al-Qudah and Zarga 2010, Al-saffar *et al.* 2016).

6. CLEOMACEAE

Cleomaceae

It is a small family of flowering plants comprising of 257 species under 12 genera. The plants are mostly herbs having palmately compound leaves and six stamens. Their inflorescence is more or less flat-topped, elongating only after the open flowers have faded. The petals are typically narrowed strongly at the base. One species belonging to this family was recorded in the flora of Chithara Village Panchayat.

6.1 *Cleome viscosa* L.

(Syn. *Polanisia viscosa* (L.) Blume)

Common Name: Asian Spider Flower, Dog Mustard.

Vernacular Name: *Hindi*: Bagra (बागरा), Hurhur (हुरहुर), Pili Hurhur (पीली हुरहुर).

Habit: Annual herb.

Habitat: Moist waste places and fallow fields.

Distribution: Native of tropical America. It is a pantropical weed, occurs throughout the tropical regions of world. It occurs in India, USA, Nigeria, China, Ceylon, Pakistan, and Africa. In India, it is found throughout the plains in woodland, grassland, and wasteland and often occurring on sandy soil and sometimes on rocky and calcareous soils.

Flowering and Fruiting Time: June-September.

Habit

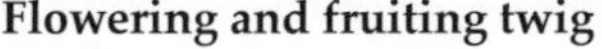

Flowering and fruiting twig

Flower

Taxonomic Description: An erect, branched, annual herb, up to 1 m high, glandular pubescent. Stem and branches viscid-pubescent, woody. Leaves 3-5 foliate, petiolate; leaflets sessile, sub-equal, middle largest, 2-5 x 1-2.5 cm, elliptic-oblong to oblanceolate, base cuneate, margin ciliate, tip obtuse. Flowers long pedicellate, bracteate, axillary or terminal, Sepals-4, lanceolate, glandular-pubescent outside. Petals-4, oblong-obovate, yellow. Stamens 12 to numerous; filaments variable in length. Capsules linear, cylindrical, up to 10 cm long, erect or curved, beaked, tapering to both ends, pubescent. Seeds reniform, numerous, reddish-brown to black.

Chemical Constituents: The phytochemistry of the plant shows the presence of tannins, saponins, flavonoids, steroids, alkaloids, phenols, terpenoids, carbohydrate, protein, fixed oil, flavonoids- quercetin, kaempferol. Seeds, found to be a rich source of various amino acids, contain umbeliferone derivative cleosandrin, coumarino-lignans (cleomiscosins). A glycoside eriodictyol-5-rhamnoside is isolated from the whole plant terpenes, flavonoids, phenol, carboxylic acid, and polyphenols. (Singh *et al.* 2015a).

Medicinal Properties: Traditionally, the plant possesses anthelmintic, carminative, anticonvulsant, antidiarhoeal, antimicrobial, and wound healing properties. Plants are acrid, thermogenic, antimicrobial, analgesic, antiemetic, antidiarrhoeal, hepatoprotective, antifibrotic, antitumour, anticonvulsant, sudorific and psychopharmacological properties. In traditional system of medicine, leaves, seeds and roots of the plant are widely used as an anthelmintic, antiscorbutic, antiseptic, cardiac stimulant, carminative, febrifuge and sudorific, anticonvulsant, antidiarrhoeal, and are also used to treat skin diseases. The roots are stimulant, antiscorbutic and vermifuge. The leaf juice is digestive and is good for otalgia. The seeds are anthelmintic, carminative, constipating, febrifuge and cardiac stimulant and are useful in fever, diarrhoea, worm infestations, cardiac disorders and dyspepsia. (Warrier *et al.* 2010, Singh *et al.* 2015a, Tripti *et al.* 2015).

Medicinal Uses (Traditional/Folklore): Plant and its parts (leaves, seeds, roots, etc.) are used traditionally to cure variety of diseases. In Ayurveda, it is used as an anthelmintic, and to treat pruritis and several other diseases like gastrointestinal disorders/infections. The whole herb is used in the treatment of ringworm, flatulence, colic, dyspepsia, cough, bronchitis, cardiac disorders. Leaves are used as the external application to get relief from inflammation of the middle ear and applied on wounds and ulcers; leaves have hepatoprotective activity. The leaves and seeds of the plant are used as rubefacient and to treat infection, rheumatism, fever and headache. Traditionally, this plant is used in various disorders such as diarrhoea, fever, inflammation, liver diseases, bronchitis, skin diseases, and malarial fever. A decoction is used as an expectorant and digestive stimulant and the vapor from a steaming decoction of the whole plant is inhaled to treat headache. The seeds of *C. viscosa* are reported to have nutritive value while the juice of leaves is applied to the skin as a counter-irritant. The root is a remedy for scurvy and rheumatism. All parts of the plant are used in liver diseases, chronic painful joints and mental disorders. (Singh *et al.* 2015a, Tripti *et al.* 2015).

7. CAPPARACEAE

Capparaceae – Caper family

The family contains 34 genera and about 449 species of herbs, shrubs and trees. Capers are known for their backward curving thorn. Butterflies belonging to white and yellow group prefer to lay eggs on plants of this group. The family was represented by two genera and two species in study area.

7.1 *Capparis sepiaria* L.

(Syn. *Capparis incanescens* DC.)

Habit

Common Name: Wild Caper Bush.

Vernacular Name: *Hindi:* Kanthari (कंठारि).

Habit: A large straggling shrub.

Habitat: Dry places and roadsides.

Distribution: Distributed in India, Sri Lanka, Bangladesh, Malesia, Indochina, China, Australia and tropical Africa. In India, common in the foothills, scrub jungles up to 900 m.

Flowering and Fruiting Time: May-July

Taxonomic Description: An evergreen, spreading, much-branched climbing shrub. Young stem and branches pubescent. Leaves 2-5.5 cm long, pubescent, shortly petiolate, ovate, elliptic-lanceolate, entire, emarginated. Stipular spines paired, recurved. Inflorescence terminal umbels. Sepals ovate, concave, ciliate along margin. Petals white, unequal, oblong. Stamens numerous. Fruits globose, glabrous, 1-seeded, deep purple when ripe.

Flowering twig Flower Fruits

Chemical Constituents: Plants are known to contain alkaloids, flavonoids and glycosides, proteins, reducing sugars, saponins, starch, steroids, tannins, gums and mucilage. (Rajesh *et al.* 2010).

Medicinal Properties: The plant is said to be alterative, febrifuge and tonic. The fruit is aphrodisiac and also used as stomachic, tonic and appetizer, removes *Kapha* and *Vata*, and cures fever, blood purifier, tumours, inflammation and disease of muscles. It possesses analgesic and hepatoprotective properties. (http://uses.plantnet-project.org/en/Capparis_sepiaria_(PROTA)).

Medicinal Uses (Traditional/Folklore): The bark and the roots are used as a wound dressing. In Tanzania, root powder is taken in porridge or as tea to treat anthrax and cancer; the root bark is used against fever and stomach complaints. In Kenya, root decoctions are inhaled to treat pain in the chest. Traditionally, the plant is used for the treatment of jaundice, inflammation and dysentery. Its flowers, leaves and roots are used in the treatment of cough and toxemia and root powder as a cure for snakebite. (http://tropical.theferns.info, http://uses.plantnet-project.org/en/Capparis_sepiaria_(PROTA)).

7.2 *Crateva adansonii* subsp. *odora* (Buch.-Ham.) Jacobs

(Syn. *Crateva roxburghii* R.Br.)

Common Name: Garlic Pear Tree, Caper Tree, Three-Leaf Caper.

Vernacular Name: *Hindi:* Barna (बर्ना), Barni (बर्नी).

Habit: Tree.

Habitat: Found growing wild and cultivated in the gardens and near religious places.

Distribution: It is widespread, from Mauritania, Senegal and Gambia eastward to Eritrea and Ethiopia, and southward to Tanzania and Zambia. In India, it is found in Peninsular India, Western India, Gangetic Plains, and Eastern India, up to Tripura and Manipur. It is also found in Sikkim and Andman and Nicobar Island. It is found mostly along the bank of the river and streams and near the temples.

A branch of tree

Flowering and Fruiting Time: March-May.

Taxonomic Description: A small, unarmed, deciduous tree. Leaves long-petioled, 3-foliate; leaflets glabrous, 5-11 cm long, ovate to elliptic-lanceolate, long acuminate at apex, entire at margin, tapering at base; lateral leaflets oblique. Flowers bisexual, in corymbose clusters. Sepals-4, ovate, petaloid, free. Petals-4, veined, ovate or oblong, greenish white to yellow. Stamens many, longer than the petals; filaments long, white to purple; anthers basifixed. Ovary ellipsoid. Stigma knob shaped, sessile. Fruits globose, orange-yellow when mature. Seeds numerous, reniform, brownish-black, smooth, buried in yellow pulp.

Flowering twig

Flower

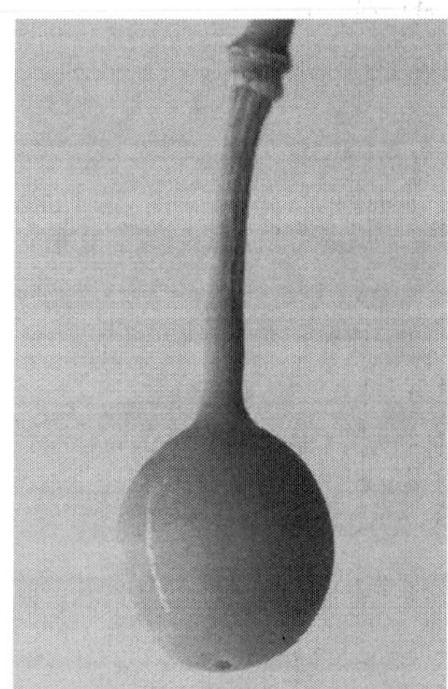

Fruit

Chemical Constituents: The main chemical constituents are saponin and tannin. An active principle lupeol, a triterpenoid which is mainly involved in the pharmacological activities of this plant has been isolated from the bark besides ceryl alcohol, friedelin, lupeol, betulinic acid and diosgenin. Lupeol is known to have anti-inflammatory activity and it reduces renal tubular damage. Leaf showed the presence of alkaloids tannins, flavonoids, saponins, cardiac glycoside and steroids. (Amos *et al.* 2015).

Medicinal Properties: The bark and roots are used in traditional and *folklore* medicines, as an appetizer and tonic having diuretic and lithotriptic/ lithontriptic, antilithiatic, rubefacient, diuretic, anti-inflammatory, laxative, antioxidant, hepatoprotectant, antirehumatic, antiperiodic, antimycotic, contraceptive, antipyretic, antihelminthic, demulcent and vasicant properties. Leaf extracts of *Crateva* spp. have showed antibacterial as well as antifungal activities. (Schmelzer and Gurib-Fakim 2013).

Medicinal Uses (Traditional/Folklore): Various plant parts of *C. adansonii* are used in Ayurvedic traditional medicines. Leaf decoctions are applied as a vapor bath against jaundice, yellow fever and eye complaints. They are taken to treat arterial hypertension, malaria, jaundice, oedema, epilepsy, measles, dysmenorrhoea, abscesses and skin diseases. Leaf powder is applied to cysts and to treat headache and rheumatism. In Burkina Faso, the leaves are commonly used to induce and stimulate milk production in breast-feeding women. It is useful in disorders of urinary organs, urinary tract infections, pain and burning micturition, renal and vesical calculi. The leaves are applied to the head as a mild counter-irritant for easing headaches. A steam bath of the leaves over the face is used as a remedy for all troubles due to poor vision. The leaves are used in fumigations for treating jaundice and yellow fever. The powdered leaves and bark are considered to be rubefacient and are used especially on cysts.

Powdered stem-bark is an appetizer, and used in renal and urinary troubles, gastro-intestinal and urinary affections; decoction of stem-bark compounded with root powder is given in gravel; bark-collyrium is used in eye affections; stem-bark and fresh leaf mixed together are used as rubefacient, in rheumatic pain; leaf-juice is given to treat rheumatism. Powdered or pounded stem bark is applied to cysts and swellings, whereas bark decoctions are taken to treat rheumatism and sterility, and as a tonic.

The bark removed from the base of the trunk has been used both internally and externally for treating sterility. It is used in combination with *Flacourtia flavescens* as a treatment for leprosy. The bark is said to be rubefacient and tonic. It is widely used as a remedy for stomach-troubles. It is used as a counter-irritant for treating headaches. Powdered and boiled in oil, it is used as an application for rheumatic condition. A bark paste is used as a poultice on swellings.

Dried ground roots are taken against fever and applied externally to treat headache and swellings of the body and taken internally to treat gastrointestinal complaints and rheumatism. Powdered root bark is administered as a remedy against carbuncles and anthrax. The root is used as a febrifuge. It is used in several treatments for syphilis. Extract of root-bark mixed with honey is given for treating scrofulous enlargement of glands under lower jaw. (Schmelzer and Gurib-Fakim 2013, Amos *et al.* 2015; http://tropical.theferns.info).

8. CARYOPHYLLACEAE

Caryophyllaceae – Pink family or Carnation family

It is a large family, with 91 genera and about 2,456 known species. This cosmopolitan family of mostly herbaceous plants is best represented in temperate climates, with a few species growing on tropical mountains. Many species are grown as ornamental plants, and some species are widespread weeds. The family was represented by four species belonging to four genera in Chithara Village.

8.1 *Arenaria serpyllifolia* L.

(Syn. *Arenaria viscida* Loisel.; *Stellaria serpyllifolia* (L.) Scop.)

Habit

Common Name: Thyme-Leaved Sandwort.

Vernacular Name: None known.

Habit: Annual herb.

Habitat: Cultivated and fallow fields, along water channels.

Distribution: Euro-Siberian Southern-temperate element; widely naturalized outside its native range. Generally, grows lowland, but found up to 610 m.

Flowering and Fruiting Time: February-March.

Taxonomic Description: An annual, pubescent, sub-erect or decumbent-ascending, slender herb, 10-20 cm long. Stem much branched from the base. Leaves small, up to 1 cm long, opposite-decussate, sessile, ovate or elliptic, apex acute, margin ciliate. Flowers bisexual, actinomorphic; bracts leafy, lanceolate. Sepals-5, free, pubescent, lanceolate, longer than the petals. Petals-5, white. Stamens usually 10, inserted on a disc; anthers dorsifixed. Ovary superior, ovoid; styles-3. Capsules ovoid, as long or little shorter than sepals. Seeds numerous, compressed, dark brown to black, tubercled.

Flowering twig

Leaves with ciliate margin

Flower

Chemical Constituents: The plant contains flavanoids and xanthones - quercetin, quercetin 7-O-α-L-rhamnoside, quercetin 3-*O*-α-L-rhamnoside, 3-O-methyl quercetin, 1,5-dihydroxyxanthone, 1,5-dihydroxyxanthone-6-O-β-D-glucoside, 5-hydroxy-1-methoxyxanthone and 6-deoxyisojacareubin. (Zhou *et al.* 2013).

Medicinal Properties: The plant is alexiteric, antitussive, depurative, diuretic and febrifuge, having refrigerant, pterygium, and antituberculosis properties. (http://temperate.theferns.info).

Medicinal Uses (Traditional/Folklore): Use of plant prevents or relieves cough, tends to purify and cleanse the blood, induces urination and has fever-reducing properties. A decoction of the leaves has been used in the treatment of dysentery. It has also been used in the treatment of urinary bladder complaints, calculus troubles (Kidney stone) and acute and chronic cystitis. (Chandra and Rawat 2015; http://temperate.theferns.info).

8.2 *Silene conoidea* L.

(Syn. *Conosilene conoidea* Fourr.)

Common Name: Cone Catchfly, Weed Campion, Weed Silene.

Vernacular Name: None Known.

Habit: Annual herb.

Habitat: Cultivated fields, gardens.

Distribution: It is native to Eurasia (France, Italy, Turkey, the Eastern Mediterranean, the Caucasus, Iran, Central Asia, the Himalaya, Xin-Jiang), and it is also known to occur in North Africa and North America.

Flower

Habit

Flowering and Fruiting Time: January- March.

Taxonomic Description: An erect, dichotomously branched, glandular-pubescent, annual herb, up to 40 cm high. Leaves ex-stipulate, 5-10 cm long, opposite-decussate, entire, acute, oblong, linear, lanceolate; radical leaves spathulate, cauline sessile. Flowers few, pink or purplish in leafy, dichotomous paniculate cyme. Calyx tubular, 2-2.5 cm long, narrow above, teeth linear. Corolla pink, obovate. Stamens 10. Ovary oblong, ovoid, styles-3. Capsules 1-1.5 cm long, ovoid, narrowed towards the apex, enclosed in inflated calyx. Seeds blackish-brown, tuberculate, kidney-shaped.

Chemical Constituents: It contains secondary metabolites such as α-spinasterol, α-spinasterolglucoiside, D-pinitol, sucrose and conoidene (2, 2′-(1, 3- butadiene-1, 4-diyl) bis 3-methoxy-5-(2- propen-1-yl) furan). The new constituents isolated from *S. conoidea* are 8 (4″-o-α-l-rhamnopranosyl)-C-β-dglucopyranosylidiosmetin, 8(4″-o-α-l-rhamnopyraosyl)-C-β-D-glucopyranosylapigenin. (Ahmad *et al.* 1998, Mamadalieva *et al.* 2014).

Medicinal Properties: Ophthalmia and emollient properties. (http://temperate.theferns.info).

Medicinal Uses (Traditional/Folklore): The plant is known as emollient and is used in bath or as a fumigant. A paste is prepared by grinding seeds and young leaves which is applied on pimples. This paste is also used for backache. The dry root is crushed and used as soap to wash wounds and hair. Flowers are used in the treatment of eye infections and sore eyes. (Shinwari *et al.* 2015; http://temperate.theferns.info).

8.3 *Spergula arvensis* L.

(Syn. *Spergularia arvensis* (L.) Cambess.)

Common Name: Corn Spurry.

Vernacular Name: *Hindi:* Ban Dhania (बन धनिया).

Habit: Annual herb.

Habitat: Cultivated and wastelands.

Distribution: It is native to Eurasia. It grows throughout Europe, Asia, Africa, North America, South America, Australia, and New Zealand.

Flowering and Fruiting Time: January-March.

Taxonomic Description: A small, annual herb, 5-30 cm high. Stem prostrate or ascending, branched from the base. Leaves 1-3 cm long in false whorls, linear, obtuse, entire; stipules small, ovate, acute. Flowers bracteate, pedicellate, in lax terminal dichasial cymes. Sepals-5, ovate-lanceolate. Petals-5, ovate, lanceolate, white. Stamens 6-7, inserted on a disk; filaments distinct. Ovary superior, styles-3. Capsules ovoid, about as long as sepals. Seeds numerous, black, winged.

Habit

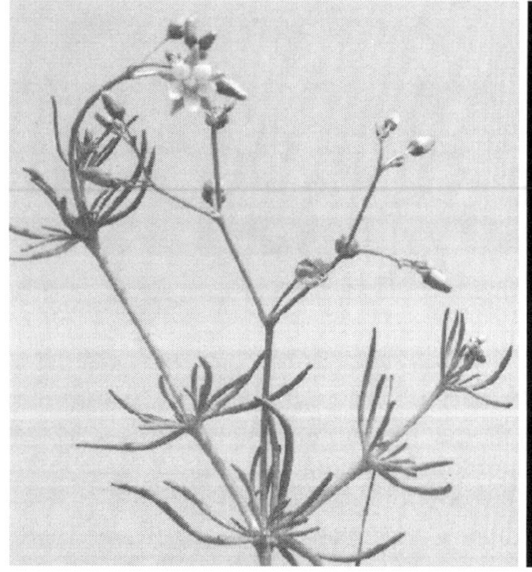

Flowering twig

Flower

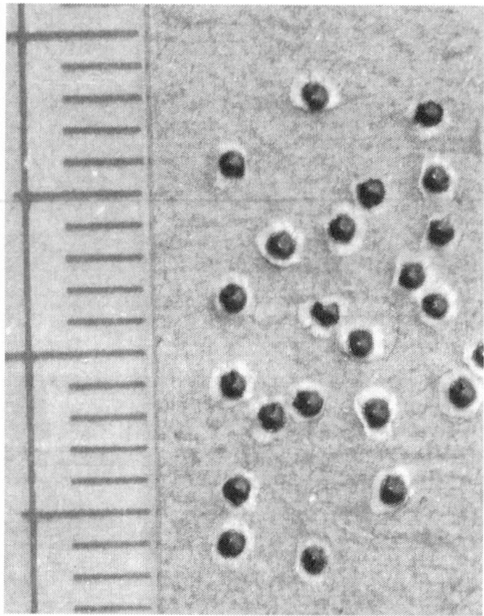

Winged seeds

Chemical Constituents: It contains phenol, tannins, proteins, carbohydrates, ß-carotene, lycopene and vitamin A and vitamin C. Leaves contain high concentration of potassium, calcium, sodium and moderate amount of zinc, cadmium and iron. Small amount of magnesium, manganese and chromium are also present. (Sundarapandian *et al.* 2016).

Medicinal Properties: The plant has antibacterial and antifungal activities, and laxative, diuretic, antiviral, antiparasitic, antioxidant and wound healing properties; it is also employed against pulmonary tuberculosis. (Najeeb *et al.* 2013, Sharmila *et al.* 2014, Sundarapandian *et al.* 2016).

Medicinal Uses (Traditional/Folklore): The plant paste is applied externally to mitigate shoulder and body pains. It is applied to wounds for its healing and antiseptic qualities. The seed and probably also the leaves, contain saponins. Saponins are much more toxic to some creatures, such as fish, and hunting tribes traditionally use it in large quantities in streams, lakes, etc. in order to stupefy or kill the fish as a fish poison. (Manikandan 2008, Sundarapandian *et al.* 2016; http://www.pfaf.org/user/Plant.aspx, http://temperate.theferns.info).

8.4 *Stellaria media* (L.) Vill.

(Syn. *Alsine media* L.)

Habit

Common Name: Chickweed, Starweed, Passerina.

Vernacular Name: Buch-Bucha (बुच बुचा).

Habit: Winter season annual herb.

Habitat: Moist places in gardens, wastelands, cultivated fields.

Distribution: It is native to Europe, but naturalized in many parts of North America and other parts of the world. In India, it is found extending from plains up to 3,800 m.

Flowering and Fruiting Time: January-March.

Taxonomic Description: A small, sub-erect, decumbent, branched, slender, annual herb. Leaves opposite, up to 3 cm long; ovate-oblong, elliptic, lanceolate, exstipulate, entire at margin, acute to acuminate at apex, cordate or cuneate at base; lower leaf petiolate, upper leaf sessile. Flowers axillary or terminal, in leafy cymes; pedicels long, pubescent. Sepals-5, ovate, lanceolate, hairy. Petals-5, bi-fid, white, deeply lobed, shorter than the sepals. Stamens 3-10, inserted on a disc. Styles-3. Fruits capsule, ovoid-oblong, longer than the sepals. Seeds brown, acutely tubercled.

Flowering twig

Flower

Chemical Constituents: Chickweed contains mucilage, saponins, glycosides, coumarins, hydroxycoumarins, flavonoids, carboxylic acids, triterpenoids, silica, minerals, vitamins A, B, C, and fatty acids; vitamin C is about 150-350 mg per l00 g. Some of the specific phytochemicals reported in the plant are: *Phenolic acids*: Vanillic acid, p-hydroxybenzoic acid, ferulic acid, caffeic acid, chlorogenic acid; *Flavonoids:* Apigenin, genistein, vicenin-2, c-glycosyl flavone, triterpenoid; *Saponin:* Gypsogenin; *Pentasaccharide:* Stellariose; *Lipids*: Hentriacontane, hexacosanyl palmitate, methyl stearate, pentacosanol, tricontanol, triacontanoic acid, ß-sitosterol and its ß-D-glucoside, 6,7-dimethyl heptacosane, 3-methyl- 6- hydroxy-heneicos-3-enyl acetate and 5-acetoxydotetracont-3 en-1-ol; *Others:* Emodin, physcion, questin, 1- hexacosanol, daucosterol, cyclo-(leu-lle), cyclo (val-tyr), aethyl-D pyrano-galactoside, kaempferol-3,7-ß-L-dirhamno side; *Amino acids:* Uracil, thymine, thymidine, guanosine, 2-chloroadenosine, aspartic acid, glutamic acid, Alanine, serine, leucine, glycine, threonine, tyrosine, lysine, histidine, proline, G-aminobutyric acid. The following constituents have been isolated from chickweed essential oil: borneol, menthol, linalol, 1, 8-cineole, and other terpenes such as epoxy-dehydro-caryophyllene, monoterpene alcohol-ester and caryophyllene. The carotenoid content was found to be 4.2 mg/100g.

Nutritional value of Chickweed: The plant contains ascorbic-acid, ß-carotene, calcium, coumarins, genistein, gamma-linolenic-acid, flavonoids, hentriacontanol, magnesium, niacin, oleic-acid, potassium, riboflavin, rutin, selenium, triterpenoid saponins, thiamin, and zinc. (Sharma and Arora 2012).

Medicinal Properties: It has a cooling and drying effect on wounds and skin eruptions. Minor burns respond quickly to chickweed's care. The whole plant is used in alternative medicine as an astringent, carminative, demulcent, diuretic, expectorant, laxative, vulnerary and refrigerant. A decoction of the whole plant of Chickweed is taken internally as a postpartum depurative, emmenagogue, galactagogue and circulatory tonic. It is also used to relieve constipation; an infusion of the dried herb is used in coughs and hoarseness, and is beneficial in the treatment of kidney complaints. New research indicates Chickweed's use as an effective antihistamine. The decoction is also used externally to treat rheumatic pains, wounds and ulcers. (Dhiman 2004).

Medicinal Uses (Traditional/Folklore): The plant is a traditional Chinese medicine that has been used in folk medicine for over 200 years, mainly for the treatment of dermatitis, itchy skin conditions and other skin diseases, pulmonary diseases, bronchitis, rheumatic pains, arthritis and period pain. The entire plant is described as a very useful in the intestinal, renal, reproductory and respiratory tract inflammations. The plant gives soothing action on dermoid surface and applied on skin inflammation and also used in erysipelas, scalds, burns, ulcers, haemorrhoids, eczema and eye inflammation. The powder of the plant is used for fomentation on swellings. Besides, the plant is also used in plaster for broken bones and swellings owing to be its cooling and binding nature. It has been used as a remedy to treat iron-deficiency, anaemia (for its high iron content), as well as for skin diseases. Chickweed is applied as a medicinal poultice to get relief from any kind of roseola; also effective in treating fragile superficial veins or itching skin conditions.

Common Chickweed has long been used in folkloric medicine as tonic to cleanse the kidney and liver. Chickweed water is an old wives' remedy for obesity. Among the many claimed health benefits from chickweed is its ability to heal skin wounds. It is used very effectively as a poultice to treat boils, cuts, burns, abscesses and ulcers. Chickweed is also used to treat itchiness associated with eczema and psoriasis. Another health benefit from Chickweed is its ability to act against inflammation and pain. It is used to treat rheumatism, arthritis and menstrual pain. Chickweed is also used to treat constipation, upset stomach and to promote digestion. Chickweed is also used to treat asthma and other respiratory problems such as cough and colds. It is believed that the Vitamin C in Chickweed can help boost body resistance and can even be used to treat scurvy.

The medicinal properties of this common garden weed combined with its nutritional properties make it a popular ingredient in salves and lotions. Chickweed is a nutritious diuretic. Chickweed can be eaten raw mixed with salad or prepared and taken as tea. A medicinal tea is prepared from the plant to be given for the relief from during cold or flu. For preparing the tea, one table spoon of dried herb or two table spoon full of fresh herb is added in 1 cup of boiling water and simmered for 10 min. Half a cup of this tea is taken 2 to 4 times daily for effective cure. (Dhiman 2004, Sharma and Arora 2012).

9. PORTULACACEAE

Portulacaceae – Purselane family

The family containing 10 plant genera and 258 species comprise of succulent, sun-loving, low growing herbs. Short lived flowers remain open during hottest part of the day. Many of the species are used as ornamentals. Two species belonging to a single genus of this family were recorded in Chithara Village.

9.1 *Portulaca oleracea* L.

(Syn. *Portulaca consanguinea* Schltdl.)

Habit

Common Name: Purslane.

Vernacular Name: *Hindi:* Lunia (लूनिया), Badi Noni (बड़ी नोनी); *Sanskrit:* Brihalloni (बृहल्लोनि), Lonamala (लोनमाला).

Habit: Annual herb.

Habitat: Moist waste places, roadsides.

Distribution: Common in the temperate and subtropical regions, although it extends into the tropics and higher latitudes. It is found throughout India as a weed up to 1,500 m; common in arable lands, drainages, wayside, generally in clayey soils.

Flowering and Fruiting Time: April-June.

Taxonomic Description: A sub-erect, succulent, annual herb. Stem glabrous, prostrate or ascending, branched at base, greenish-pink in colour. Leaves 1-4 cm long, entire, shortly petiolate, oblong, obovate, spathulate, apex truncate or retuse, base cuneate. Flowers in terminal heads, sessile, surrounded by a cluster of crowded leaves. Sepals-2, united at the base into a tube. Petals-5, obovate, yellow, notched. Stamens many; filaments filiform, yellow. Capsules ovoid, dehiscing above the base. Seeds numerous, black.

Flowering twig

Flower

Chemical Constituents: Diverse compounds have been isolated from *P. oleracea*, such as flavonoids, alkaloids, polysaccharides, fatty acids, terpenoids, sterols, proteins vitamins and minerals. This plant is also known to contain ascorbic acid, α-tocopherol, and B-complex vitamins, for example, niacin, pyridoxine, and riboflavin. Furthermore, it is rich in minerals like phosphorus, manganese, icon, calcium selenium, and the amino acids isoleucine, proline, leucine, lysine, phenylalanine, methionine, cystine, valine, threonine, and tyrosine. Other constituents isolated from this plant are β-carotene, glutathione, melatonin, portulacerebroside A, catechol, and bergapten. (Zhou *et al.* 2015).

Medicinal Properties: The stem and leaves are sour, bitter, thermogenic, stomachic, alexiteric, antibacterial, antiscorbutic, sudorific, aperient, alterant, diuretic, vulnerary and tonic. They are useful in asthma, diabetes, gasropathy, tumours, inflammation, nephropathy, cystalgia, cystitis, splenopathy, haemorrhoids, jaundice, diabetes, scalds and burns, dipsia, cephalalgia, otalgia, odontalgia, scurvy, burning sensation, dysentery, vomiting, skin diseases, cardiovascular diseases, dysuria, haematuria, gonorrhoea and ulcerative stomatitis. The seeds are cooling, diuretic and antidysenteric, antioxidant and are useful in burns and scalds, strangury and dysentery. The leaves are a rich source of omega-3 fatty acids, which is thought to be important in preventing heart attacks and strengthening the immune system. (Warrier *et al.* 2010, Gogoi and Zaman 2013, Zhou *et al.* 2015).

Medicinal Uses (Traditional/Folklore): Purslane has been used in folk medicine since ancient times and is included in the World Health Organization's list of most widely used medicinal plants. The fresh juice is used in the treatment of strangury, coughs, sores, etc. A tea made from the leaves is used in the treatment of stomach aches and headaches. The leaf juice is applied to earaches, it is also said to alleviate caterpillar stings. The leaves are poulticed and applied to burns. Plant is boiled and the water used for treating arthritis or muscle pains in Guyana. Juice from macerated plant is applied as an antiseptic to cuts and bruises to stop bleeding and to heal wounds. In Surinam, leaves are mixed with sugar or soap and applied onto mature abscesses. In French Guiana, an infusion of leaves is used as a cholagogue, to treat renal colic, and to prepare the body before taking a purgative. Leaves are crushed and mixed with coconut oil and vinegar, and placed on sprained limbs. Leaf infusion is antidiabetic and digestive; used as an emollient. Leaves are crushed and mixed with grease or oil and made into an unguent which is rubbed on sprained or torn muscles. The leaves also used to make a refreshing purgative drink, which also serves to diminish the leakages of albuminuria.

The decoction of the whole plant (9-15 g dried or 30-150 g fresh) is given daily in divided doses for bacterial dysentery, enteritis, urinary tract infections, leucorrhoea and mastitis. Several reports also cite its use for whooping cough. The crushed fresh plant can be applied externally to stop bleeding, including uterine bleeding following child delivery. It is also applied on bruises, skin ulcers, skin fungal infections, insect bites and strings. (Kaul and Dwivedi 2010, Gogoi and Zaman 2013, Zhou *et al.* 2015; http://medplants.blogspot.in, http://tropical.theferns.info).

9.2 *Portulaca quadrifida* L.

(Syn. *Portulaca meridiana* L.f.)

Habit

Common Name: Chicken Weed, Wild Purslane.

Vernacular Name: *Hindi*: Paviri (पविरी).

Habit: Annual herb.

Habitat: Dry areas, wastelands, roadsides.

Distribution: It is native of tropical America; found in waste places and along road-sides in the outer sub-Himalayan tracts.

Flowering and Fruiting Time: August-September.

Taxonomic Description: A much-branched, prostrate, annual herb. Stem slender, creeping, purplish red, often rooting at the nodes and each node with dense silvery white hairs. Leaves 5-6 mm long, succulent, opposite-decussate, sessile or shortly petiolate, oblong-elliptic, lanceolate, entire, acute at apex. Flowers yellow, solitary, terminal, subtended by long silky hairs and 4-leaved involucre. Sepals-2, basally connate into a short tube. Petals-4, yellow. Stamens usually 8. Capsules conical, 3-4 mm long. Seeds many, minute, reniform, black.

Flowering twig

Flower

Chemical Constituents: Phytochemical screening reported the presence of alkaloids, tannins, mucilage, steroidal compounds and carbohydrates. (Mulla and Swamy 2010).

Medicinal Properties: Seeds are used as antidysenteric, antifungal, diuretic, emetic, poultice, antiscorbutic, anthelmintic, antidiarrhoeal and demulcent. The fresh leaves of the plant are bruised and used as external application in abdominal complaints, haemorrhoids and erysipelas. The plant is sour, bitter, hot, alterative, and laxative; it causes biliousness and *"Kapha"*; cures fevers, asthma, cough, urinary disorders, inflammations, and good for eye diseases, abdominal diseases, debility, anaemia, skin diseases and ulcers. (Dhiman 2004, Mulla and Swamy 2010).

Medicinal Uses (Traditional/Folklore): It is a traditional medicinal plant, valued for its benefits in the management of urinary and inflammatory disorders. The leaf juice is applied to abscesses and decoction is given in dysentery. The decoction of the plant is used as anthelminthic and in the treatment of stomach complaints and gonorrhoea. In Indo-China the juice of leaves is applied to abscesses and used as a collyrium. In Nigeria, the leaves are used as a local application for treating swellings. A poultice of the plant is applied in abdominal complaints, erysipelas and haemorrhoids. (Vardhana 2008, Mulla and Swamy 2010).

10. TAMARICACEAE

Tamaricaceae

The family comprises of three plant genera and a total of 79 known species. It is native to drier areas of Europe, Asia and Africa. Many species grow on saline soils. The leaves are generally scale-like, measuring 1-5 mm long, overlap each other along the stem. The family was represented by a single species in the flora.

10.1 Tamarix indica Willd.

(Syn. *Tamarix gallica* var. *indica* (Willd.) Ehrenb.; *Tamarix troupii* Hole)

Common Name: Indian Tamarisk.

Vernacular Name: *Hindi:* Jhau (झाऊ), *Sanskrit:* Aphalah (अफल), Jhavukah (झावुक).

Habit: Shrub.

Habitat: Along the bank of rivers, wetlands, canals.

Distribution: Asia: Afghanistan, Bangladesh, India, Myanmar, Pakistan and Sri Lanka.

Flowering and Fruiting Time: May-October.

Habit

Taxonomic Description: An erect, large shrub or small tree, 2-8 m tall. Stem much-branched, bark reddish-brown. Leaves tiny, alternate, sessile, triangular-ovate, acute to acuminate at apex, partially sheathing. Flowers small, pink, bisexual, in large panicles or spike-like racemes. Bracts triangular, shorter than the flowers. Sepals-5, triangular, persistent. Petals-5, oblong, much longer than the sepals. Stamens-5; filaments filiform, twice the length of the petals. Stigma 3, capitate. Capsules conical, 5-6 mm long, triangular. Seeds many; coma (tuft of hairs) nearly 2.5 mm long.

Flowering twig

Flowers

Chemical Constituents: The phytochemical analysis of the plant shows that it is rich in antioxidants such as tannin (50%), tamarixin, tamarixetin, troupin, 4- methylcoumarin and 3, 3′-di-0-methylellagic acid. The other principle constituents are traces of aglycone, tamarixetin-3-glucoside, quercetin and kaempferol. The plant also contains a high level of tannin (ellagic and gallic acids) and quercetol (methyllic ester). Furthermore, the tamarix is found to be rich in polyphenolic compounds such as flavonoids, phenolic acids and coumarins. The quercetin glycosides are found in abundant in plants and their 30-o-methyl (isorhamnetin) derivatives are also not uncommon. Several types of polyphenols (anthocyanins, tannins, flavonones, isoflavonones, resveratrol and ellagic acid) have also been reported in addition to terpenoid antioxidant compounds like carotenoids and essential oils. (Drabu *et al.* 2012).

Medicinal Properties: Tamarix has been employed in traditional medicine as astringent, aperitif, stimulus of perspiration and diuretic. The bark is bitter and an astringent and tonic. Roots have antinociceptive, cytotoxic and diuretic properties. Several researches have proved antioxidant nature of the plant. Antimicrobial activity of bark has also been reported. (Drabu *et al.* 2012).

Medicinal Uses (Traditional/Folklore): The plant has been reported to be useful in leucoderma, spleen trouble and eye diseases. Traditionally, it is used as an anthelmintic, diarrhoea and gingivitis. In Algeria and surrounding areas, the plant has been used medicinally for rheumatism, diarrhoea, and other maladies. It is also used as prophylactic and therapeutic remedies to cure malaria as folk medicine. Fruit and roots are useful for treating dysentery and chronic diarrhoea. (Drabu *et al.* 2012).

11. MALVACEAE

Malvaceae – Cotton family

This family includes about 245 plant genera and more than 4,465 species. The plants are cosmopolitan in distribution, more abundant in tropical and subtropical regions. In India, Malvaceae is represented by 22 genera and 125 species. Plants may be annual herbs or perennial shrubs or trees. The members of this family have mucilaginous substance and stellate hairs occur on their young parts and flowers actinomorphic, hermaphrodite, hypogynous, pentamerous, with epicalyx, calyx free or united, corolla free; stamens indefinite, monadelphous; anthers monothecous, stamens united to form a tube. The family was represented by a total of 12 species under eight genera in the flora.

11.1 *Abutilon indicum* (L.) Sweet

(Syn. *Sida indica* L.)

Habit

Common Name: Indian Abutilon, Indian Mallow.

Vernacular Name: *Hindi:* Kanghi (कंघी), Kakahi (ककही); *Sanskrit:* Atibala (अतिबाला).

Habit: Shrub.

Habitat: Common along the roadsides, and in open and wastelands.

Distribution: Found throughout tropical and sub-tropical parts of India up to 1200 m.

Flowering and Fruiting Time: August-March.

Taxonomic Description: An erect, much-branched, perennial shrub up to 2 m tall. Leaves up to 15 cm long, alternate, petiolate, sub-orbicular, broadly cordate at base, irregular toothed at margin, acuminate at apex, slightly pubescent on both surface; petiole usually longer than the blade. Flowers axillary, solitary; peduncle longer than the petiole; pedicels jointed near the top. Calyx lobes-5, ovate, acute or apiculate, pubescent. Corolla orange yellow, lobes-5; petals separate, obovate. Stamens numerous, distinct; filaments filiform; anther ovoid. Ripe carpels 12-20, shortly-beaked. Seeds kidney-shaped, dark brown or black.

Flower

Fruit

Chemical Constituents: The plant contains mucilage, tannins, asparagines, gallic acid and sesquiterpens; hexoses, n-alkane mixtures, alkanols, β-sitasterol, vanilllic, p-coumaric, caffeic and fumaric acids and alantaolactone, isoalantolactone, alkaloids, leucoanthocynins, flavonoids, sterols, treterpenoids, saponins and glycosides are also reported from the plant. The main constituents of the roots are alantolactone, isoalantolactone and gallic acid. (Bhutya 2011, Mohite *et al.* 2012, Saini *et al.* 2015a).

Medicinal Properties: Seeds are laxative, expectorant, demulcent, nervine tonic, aphrodisiac and anodyne. Useful to cure cough, chronic cystitis, gleet, gonorrhoea; used for anti-ageing, promoting semen quantity and increasing physical strength. Psychosomatic medicinal use related to scorpion bite treatment is also reported. Roots are nervine tonic, anodyne; useful to cure nervous pains, paralysis, strangury, fevers, chest affection, and urethritis. Bark is astringent and diuretic. (Bhutya 2011, Mohite *et al.* 2012, Saini *et al.* 2015a).

Medicinal Uses (Traditional/Folklore): Leaves are cooked and eaten for bleeding piles, boils, ulcers, fomentation to painful parts of body and decoction used for toothache and tender gums. In ancient days, a spoonful of the powder made from the dried whole plant with a spoonful of honey, is given for six months until the day of marriage, for safe and quick pregnancy. (Bhutya 2011, Mohite *et al.* 2012, Saini *et al.* 2015a).

11.2 *Corchorus aestuans* L.

(Syn. *Corchorus acutangulus* Lam.)

Habit

Common Name: East Indian Mallow, Wild Jute, West African Mallow.

Vernacular Name: *Hindi*: Jangli Jute, Chonch (चोंच), Hade-Ka-Khet (हाडे का खेत); *Sanskrit*: Chunchuh (चुनचुह).

Habit: Annual herb.

Habitat: Roadsides, cultivated and un-cultivated fields.

Distribution: *C. aestuans* is a pantropical species.

Flowering and Fruiting Time: August-October.

Taxonomic Description: An erect, branched, annual herb, up to 60 cm tall. Leaves up to 10 cm long, stipulate, alternate, ovate-lanceolate, acute at apex, rounded at base, pubescent beneath, serrate, basal serrature prolonged into filiform appendages; petiole pilose. Flowers bracteate, 2-3, in leaf opposed cymes; pedicels short. Calyx teeth-5, linear-oblong, apiculate. Corolla lobes-5, obovate, yellow. Stamens numerous. Capsules angled, 2-3 cm long, apex divided into 3, bifid glabrous beak at top.

Flowering twig

Flower

Capsules

Chemical Constituents: Analysis of plant revealed the presence of fourteen different bioactive constituents namely 3,7,11,15-tetramethyl-2-hexadecen-1-ol, Trans-2-undecen-1-ol, E-7-Tetradecenol, n-Hexadecanoic acid, Phytol, 9,12,15-octadecatrienoic acid, methyl ester, (Z,Z,Z)-, Docosanoic acid, ethyl ester, 1-Eicosanol, 9,9-dimethoxybicyclo[3.3.1] nona-2,4-dione, Heptadecanoic acid, heptadecyl ester, Pentadecanoicacid,2,6,10,14-tetramethyl-, methylester, 3-Hexadecycloxycarbonyl-5-(2-hydroxyethyl)-4-methylimidazolium ion, Squalene, Vitamin E. (Al-Snafi 2016a).

Medicinal Properties: Anticancer, antilaeukemic, tonic, carminative and febrifuge. (Al-Snafi 2016a).

Medicinal Uses (Traditional/Folklore): In traditional African medicine, an extract of the roots or leaves is taken for the treatment of gonorrhoea, and an extract of the whole plant is used for making injections for the treatment of urethral discharges. Plant decoction is used in diarrhoea. In DR Congo, leaves are squeezed and the sap is sniffed for the treatment of headache. In India, the seeds are used for the treatment of stomachache and pneumonia. The medicinal uses in the Philippines are similar to those of *C. capsularis*: the leaves serve against headache, and the seeds, in the form of powder/decoction, as a tonic, carminative and febrifuge. An ethanolic extract of the entire plant was found to have anticancer activity against epidermal carcinoma of nasopharynx in tissue culture. Corchorusin D, one of various triterpenoid glycosides isolated from the aerial parts of the plant, has shown antileukaemic activity in U937 and HL-60 cell lines. (Shukla *et al.* 2010, Al-Snafi 2016a).

11.3 *Corchorus olitorius* L.

(Syn. *Corchorus catharticus* Blanco)

Habit

Common Name: Nalta Jute, Tossa Jute.

Vernacular Name: *Hindi:* Pat-Sag (पात-साग), Mitha-Pat (मीठापात); *Sanskrit*: Mahachanchu (महाचंचु).

Habit: Herb.

Habitat: Waste places, cultivated fields.

Distribution: *C. olitorius* is native to Africa; widely distributed in the wild in Asia.

Flowering and Fruiting Time: August-October.

Taxonomic Description: Annual, erect, branched herb, 60-120 cm high. Leaves up to 15 cm long, alternate, ovate-lanceolate, acute to acuminate, base rounded, margin serrated, basal two serratures prolonged into filiform tails. Petiole 2-3 cm long, pubescent. Stipules linear-lanceolate, filiform. Flowers solitary or 1 to 3, bracteate, sub-sessile, arranged in axillary or leaf opposed cymes. Calyx lobes-5, lobes linear-oblong, apiculate. Corolla lobes-5, yellow, obovate. Stamens many. Capsules sub-cylindrical, up to 7 cm long, linear, beaked, 3-6 valved, glabrous. Seeds trigonus, black.

Flowering twig

Flower

Capsules

Chemical Constituents: *C. olitorius* contains protein, fat, carbohydrate, fiber, ash, calcium, potassium, iron, sodium, phosphorous, beta-carotene, thiamine, riboflavin, niacin, ascorbic acid, etc. From the leaves of *Tossa Jute* flavonoid such as glycosides, astragalin (kaempferol 3-O-β-D-glucopyranoside), tolifolin (kaempferol 3-O-β-D-galactopyranoside), isoquercetin (quercetin 3-O-β-D-glucopyranoside) and juglanin (kaempferol 3-O-α-L-arabinopyranoside), oleanolic acid, glyceryl monopalmitate, β-sitosterol and β-sitosterol 3-O-β-D-glucopyranoside have been isolated. The seeds are poisonous to mammals and insects. They contain cardiac glycosides. (Kohda *et al.* 1994, Adegoke and Adebayo-Tayo 2009).

Medicinal Properties: It has been reported to be demulcent, deobstruent, lactagogue, purgative, diuretic, antipyretic, analgesic, antimicrobial and tonic. (Adegoke and Adebayo-Tayo 2009, Islam 2013).

Medicinal Uses (Traditional/Folklore): In Ayurveda and folk medicines, the leaves have been variously used for ascites, pain, piles, tumours, cystitis, dysuria, fever and gonorrhoea. The leaves are demulcent, diuretic, febrifuge and tonic and used in the treatment of chronic cystitis, gonorrhoea and dysuria. A cold infusion is said to restore the appetite and strength. The seeds are purgative. Injections of olitoriside, an extract from the plant, markedly improve cardiac insufficiencies and have no cumulative attributes; hence, it can serve as a substitute for strophanthin. (Adegoke and Adebayo-Tayo 2009, Islam 2013).

11.4 *Corchorus tridens* L.

(Syn. *Corchorus burmanni* DC.)

Habit

Common Name: Wild Jute, Jew's Mallow, Horn-Fruited Jute, African Jute.

Vernacular Name: *Hindi:* Kadvapat (कड़वापात), Hardikaket (हार्दिककेत).

Habit: Herb.

Habitat: Waste places, cultivated fields, along water channels.

Distribution: *C. tridens* is widespread over tropical Africa and is also found in Nepal, Pakistan, India and Northern Australia. In Africa, it occurs from Senegal in the west, east to Somalia and south to South Africa; it has been recorded from many countries.

Flowering and Fruiting Time: August-October.

Taxonomic Description: An erect or sub-erect, annual herb, up to 60 cm tall. Leaves 2-12 cm long, linear, oblong-lanceolate, shortly petiolate, apex acute, margin serrate, two basal serratures prolonged into filiform tails. Stipules linear, setaceous. Flowers bracteate, small, sub-sessile, 1-4 in leaf opposed cymes. Sepals-5, linear-oblong, apiculate. Petals-5, yellow, oblong, oblanceolate. Stamens usually 10. Ovary oblong, cylindrical. Capsules 2-4 cm long, cylindrical, often curved, ends usually into 3-spreading tips. Seeds black, truncate at both ends.

Flowering twig

Flower

Capsules

Chemical Constituents: Flavonoids, alkaloids, carbohydrates, quinines, steroids, tannins, and phenol. (Isaiah *et al.* 2016).

Medicinal Properties: Microbial activity and wound healing. (Isaiah *et al.* 2016).

Medicinal Uses (Traditional/Folklore): Roots extract is taken as backache medicine in Zimbabwe. In Namibia, plant is used for treating genital ulcers caused by syphilis or chancroid (a bacterial sexually transmitted disease (STD). Stem and leaves are crushed in water and applied on wounds for quick healing. Fruit juice is useful in jaundice and sexual problems. (Hedimbi and Chinsembu 2012, Maroyi 2013, Shaheen *et al.* 2014).

11.5 *Malva parviflora* L.

(Syn. *Althaea parviflora* (L.) Alef.)

Habit

Common Name: Cheeseweed Mallow, Small-Flowered Mallow.

Vernacular Name: *Hindi:* Panirak (पनिरक).

Habit: Annual herb.

Habitat: Along water channels, cultivated fields.

Distribution: Plant is native to Northern Africa, Europe and Asia and widely naturalized elsewhere. Distributed in fields and wastelands, grasslands, at forest edges, in woodland, on heavily grazed grassland, sometimes on stony hills, at the edges of ponds and lakes and easily becoming a weed in cultivation, elevations up to 1,500 m.

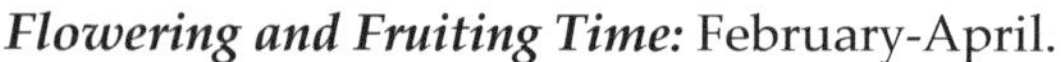

Flowering and Fruiting Time: February-April.

Taxonomic Description: A spreading, ascending or prostrate, annual herb. Stem many branched from the base. Leaves sub-orbicular, 5-10 cm across, base broadly cordate, apex rounded or obtuse, margin crenate-serrate, 5-7 lobed, petiolate; petioles 5 to 20 cm long. Stipules lanceolate, acuminate, deltoid. Flowers 2-6, in axillary or terminal clusters. Epicalyx linear, pubescent. Calyx lobes ovate, sepal enclosing fruit. Corolla bluish-white, hardly longer than the calyx; petals obovate. Schizocarps 2-3 mm long, mericarps 10. Seeds black to brown, kidney-shaped.

Flowering twig

Flower & Schizocarps

Chemical Constituents: Phytochemical screening of leaves, stems and roots of *M. parviflora* indicated the presence of active compounds including: saponin, flavonoids, alkaloids, vitamins, terpenoides and phenols in both wild and cultivated plants, while tannins were not detected in the former. (Farhan *et al.* 2012).

Medicinal Properties: *M. parviflora* leaf extracts possess anti-inflammatory, cytotoxic, antimicrobial and antioxidant properties. The whole plant is emollient and pectoral. Plants are very demulcent, with slight nervine tonic properties. The seeds are demulcent. (Akbar *et al.* 2014; http://pfaf.org, http://www.tropical.theferns.info).

Medicinal Uses (Traditional/Folklore): Traditional uses of this plant include the treatment of cough, inflammatory diseases of mucous membranes, nonspecific dermatitis, stomach ache, and sore throat. The most typical preparations are aqueous extractions. The whole plant is used as a poultice on swellings, running sores and boils. Seeds are used in the treatment of coughs and ulcers in the bladder. A decoction of the roots or leaves has been used as a hair rinse to remove dandruff and to soften the hair. In Lesotho, dried powder or an infusion made from leaves and roots of *M. parviflora* is used to clean wounds and sores. A hot poultice made from leaves is used to treat wounds and swelling and is incorporated into a lotion to treat bruised and broken limbs. The leaves are used by the Xhosa people of South Africa for treating swollen, inflamed purulent wounds. In Pakistan, the plants are used in the treatment of cough, throat infection and other bronchial problems as well as stomach and intestine irritations. The flowers and leaves are emollient and used for the softening of sensitive area of the skin. It is applied as poultice to reduce swelling and draw out toxins. The leaves help to reduce gut irritation and have laxative effects. Combine with *Eucalyptus*, it makes a good remedy for cough and other chest ailments. *M. parviflora* has also been used for the treatment of headache, fever, sores and various digestive complaints.

M. parviflora is reported to be highly toxic and may cause mortality in foraging livestock, such as sheep, horses and cattle. Sheep are the most often affected and develop clinical signs including staggering, trembling arched back and laboured breathing. This toxicity is due to the presence of malvalic acid, an unsaturated fatty acid previously referred to as halphen acid. (Farhan *et al.* 2012, Akbar *et al.* 2014; http://pfaf.org, http://www.tropical.theferns.info).

11.6 *Malvastrum coromandelianum* (L.) Garcke

(Syn. *Malva coromandeliana* L.)

Habit

Common Name: False Mallow, Broom Weed, Clock Plant, Prickly Malvastrum.

Vernacular Name: *Hindi:* Kharenti (खरैंटी).

Habit: Annual herb.

Habitat: Roadsides, railway lines, fallow fields.

Distribution: A native of America, now got naturalized in India; it is found along the roadsides, railway lines, in waste places and along forest edges up to 1,200 m.

Flowering and Fruiting Time: August-March.

Taxonomic Description: An erect or sub-erect, branched, annual herb, 20-60 cm high. Stem and branches covered with stellate hairs. Leaves up to 8 cm long and 3.5 cm wide, alternate, petiolate, ovate-oblong, lanceolate, coarsly serrate at margin, acute at apex and base, slightly pubescent on both surface. Stipules hairy, linear-lanceolate. Flowers axillary, solitary or 2-4 in terminal clusters. Bracteoles-3, linear. Calyx campanulate; lobes-5, ovate, acuminate, pubescent. Corolla yellow; petals obliquely obocordate. Schizocarps globular; mericarps 10-12, kidney-shaped. Seeds brown or black, glabrous.

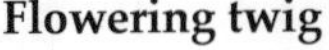

Flowering twig

Flower

Chemical Constituents: Phytochemical screening revealed the presence of alkaloids, tannins, amino acid, phenolic and flavonoids, proteins, and carbohydrates. The aerial parts contain β-phenylethylamine, dotriacontane, dotriacontanol, β-sitosterol, stigmasterol, campesterol, lutein, N-methyl-β-phenylethylamine and unidentified indole alkaloid. Also contains palmitic acid, palmitoleic acid, stearic acid, oleic acid, linoleic acid, malvalic acid and sterculic acid. (Gangrade *et al.* 2012, Sanghai *et al.* 2013).

Medicinal Properties: *M. coromandelianum* is used in traditional medicine as an anti-inflammatory, analgesic, antidysenteric, antipyretic, antibacterial and hypogylcemic and in the treatment of jaundice and ulcers. (Dhiman 2004, Gangrade *et al.* 2012, Sanghai *et al.* 2013).

Medicinal Uses (Traditional/Folklore): In India, Bhil tribes of Rajasthan use this plant in the form of decoction to cure jaundice. Whole plant is used to treat inflammation, hepatitis, liver infection, enteritis, diarrhoea, arthritis, sore throat and cough. Crushed leaves are applied to ringworm, wounds, sores; leaf decoction is helpful in dysentery. Roots are used as aphrodisiac; root bark is given with milk to increase strength. Dried seeds ground with sugar are given to cure premature ejaculation. It is also used for pains in the stomach.

The plant is mucilaginous and a refreshing; it is used in French Guiana as an emollient. In West Indies, plant is considered to be emollient, and resolvent. The leaves are applied to inflamed sores. Flowers are used in cough, chest and lung diseases. Decoction of flowers is used to reduce fever; decoction of whole plant is used to treat dysentery; poultice of leaves is used to heal wounds. In Mexico, leaf infusion of this plant is used to cure diabetes. Indigenous people take decoction of its leaves in dysentery and apply their paste on wounds. Leaves, roots together with leaves of *Sida acuta* are ground into a paste and applied externally for three times to cure blisters. (Balasubramanian *et al.* 1997, Dhiman 2004, Andrade-Cetto and Heinrich 2005, Gangrade *et al.* 2012, Sanghai *et al.* 2013; http://medplants.blogspot.in).

11.7 *Melochia corchorifolia* L.

(Syn. *Sida cuneifolia* Roxb.)

Habit

Common Name: Chocolate Weed.

Vernacular Name: *Hindi:* Chitrabeej (चित्रबीज).

Habit: Annual herb.

Habitat: Roadsides, cultivated fields.

Distribution: Native to tropical America.

Flowering and Fruiting Time: July-November.

Taxonomic Description: Herb, annual, erect, much branched, up to 80 cm tall. Leaves simple, alternate, petiolate, variable in size and shape, 3-8 cm long, ovate, lanceolate, cordate, rounded or truncate at base, serrated at margin, pointed at tip, slightly stellate pubescent beneath. Stipule lanceolate, ciliate. Flowers small, nearly sessile in dense terminal or axillary clusters, surrounded by bracteoles. Calyx lobes-5, sepals lanceolate. Corolla 5-lobed; petals obovate, white to pink, darker at centre. Capsules depressed-globose, loculicidally 5-valved, hispid. Seeds 1 in each locule, brown.

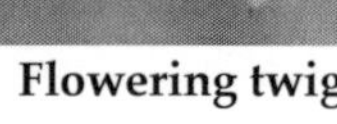

Flowering twig

Flowers

Capsules

Chemical Constituents: Phytochemical analysis of the leaves shows the presence of triterpenes (friedelin, friedelinol and β-amyrin), flavonol glycosides (hibifolin, triflin and melocorin), aliphatic compounds, flavonoids (vitexin and robunin), β-D-sitosterol and its stearate, β-D-glucoside and alkaloids. Mineral analysis of *M. corchorifolia* showed that the leaves contain a high level of potassium (7.25 mg/100 g DW), followed by calcium (750.37 mg/100 g DW) and then phosphorus (101.89 mg/100 g DW). Sodium content (94.00 mg/100 g DW) is the lowest among the macro elements determined. Other mineral composition in mg/100 g DW is: Cu (33.50), Fe (19.91), Mn (9.68) and Zn (6.73). (Pullaiah 2014).

Medicinal Properties: Plant extract showed hepatoprotective, antioxidant, antibacterial and anthelmintic activities. Fruit powder is used for ear problems, dysentery and abdominal swellings. (Pullaiah 2014).

Medicinal Uses (Traditional/Folklore): A decoction of the plant is applied in folk medicine in India as a cure for abdominal swelling, dysentery and snake bites. A leaf decoction is prescribed in a compound mixture of herbs against urinary disorders. A decoction of the leaves and roots is used internally to treat dysentery, and a decoction of the leaves to stop vomiting. The leaves are also used for poulticing sores and swellings of the abdomen, and the sap is applied as an antidote to wounds caused by arrows poisoned with *Antiaris toxicaria*. Leaves and roots are used externally for poulticing in case of smallpox. Stem and leaves, boiled in oil, are applied as a relief from water snake bite. It is also used to relieve gastralgia and headaches. (Chopra *et al.* 1956, Pullaiah 2014; http://medplants.blogspot.in).

11.8 *Sida acuta* Burm.f.

(Syn. *Malvastrum carpinifolium* (L. f.) A. Gray; *Sida carpinifolia* L.f.)

Common Name: Common Wireweed, Common Fanpetals.

Vernacular Name: *Hindi*: Baraira (बरैरा); *Sanskrit:* Bala (बाला).

Habit: Undershrub.

Habitat: Roadsides, railway lines, waste places.

Distribution: It is believed to have originated in Central America, but today has a pantropical distribution. It is common along road-sides, waste places and scrub jungles in warmer parts of Shiwaliks, Dehradun, Almora, Nainital and Haridwar Disricts including Terai-Bhabar up to 1,000 m.

Flowering and Fruiting Time: June-October.

Habit

Flowering twig

Flower

Taxonomic Description: An erect, much branched, undershrub, up to 60 cm tall. Branches rough with minute stellate hairs. Leaves up to 7 cm long, simple, alternate, petioled, ovate, lanceolate, obtuse at base, crenate serrate at margin, acute at apex, glabrous on both side. Stipules linear, filiform, subulate, longer than the petioles. Flowers solitary, axillary, pedicel short, peduncle jointed near the middle. Calyx lobes-5, campanulate, ovate-triangular. Corolla yellow; petals-5, minutely ciliate at margin, twice the length of calyx. Stamens numerous. Carpels 5-10, rugose, 2-awned.

Chemical Constituents: Phytoconstituents such as alkaloids, phytosterols, tannins, flavonoids, saponins, etc. are present in the plant. (Palaksha and Ravishankar 2012).

Medicinal Properties: The roots are used as a coolant, astringent, diaphoretic, antipyretic and a tonic useful in nervous and urinary diseases and in blood and bile disorders. It is bitter tonic used as a febrifuge and stomachic in chronic bowel complaints, and to expel worms. The leaves are a demulcent and diuretic. The whole plant is used for curing fever, bronchitis, ulcer, diarrhoea, and dysentery and skin diseases. (Mohideen *et al.* 2002, Dhiman 2004, Ajithabai *et al.* 2012).

Medicinal Uses (Traditional/Folklore): The aerial part of the plant is the most frequently used part. Whole plant boiled in gingelly oil is used on testicular swellings and elephantiasis and to hasten suppuration. The paste of leaves is mixed with coconut oil and applied on head regularly for the treatment of dandruff and also for strengthening hair. Decoction of the plant is taken orally to cure urinary diseases and impotency. Extracts of *Sida acuta* and *Sida cordifolia* are used to relieve hay fever and asthma. They also have antibacterial and antiprotozoal properties. In Central America, the plant is used to treat asthma, renal inflammation, cold, fever, headache, ulcers and worms. The juice of the leaves is mixed with vinegar to make an anti-inflammatory and digestive remedy. The plant is ground and mixed with soft grease and sugar to make a poultice that is applied to soften abscesses and release of pus from wounds. (Dhiman 2004, Karou *et al.* 2007, Ajithabai *et al.* 2012).

11.9 *Sida cordata* (Burm.f.) Borss.Waalk.

(Syn. *Melochia cordata* Burm.f.; *Sida veronicifolia* Lam.)

Common Name: Long-Stalk Sida.

Vernacular Name: *Hindi:* Bhuinii (भूइनी), Kharenti (खरैंटी); *Sanskrit:* Bhumibala (भूमिबाला), Nagabala (नागबाला).

Habit: Herb.

Habitat: Open and wastelands.

Distribution: Native to India, it is a weed that grows wild in wastelands and along roadsides. The plant is found throughout the tropical and sub-tropical plains all over India and Sri Lanka.

Flowering and Fruiting Time: August-November.

Habit

Taxonomic Description: A prostrate or ascending, 40-90 cm high, much-branched herb. Stem and branches covered with scattered, stellate hairs. Leaves 3-7.5 x 2-6 cm, broadly ovate to sub-orbicular, stellate-pubescent on both surfaces, margin crenate-serrate, apex acute to acuminate, base cordate, petiole pubescent, stipule filiform. Flowers nearly 1 cm across, yellow, axillary, generally solitary; pedicels stellate hairy, 1-2.5 cm long. Calyx campanulate, lobes-5, triangular. Corolla lobes-5, lobes obovate. Stamens numerous, anthers yellow. Mericarps-5. Seeds brown or black.

Flowering twig

Flower

Triangular calyx lobe

Chemical Constituents: *S. cordata* contains primary metabolites like carbohydrates, amino acids, proteins, etc. and secondary metabolites like the alkaloids, flavonoids, tannins, saponins, phenolics, terpenoids, glycosides, emodins, catechins, coumarins, anthraquinones, etc. (Gulnaz and Savitha 2013).

Medicinal Properties: Leaf and stem of *S. cordata* possess antimicrobial activity; leaves also have antidiarrhoeal activity. The flowers and ripe fruits are refrigerant and are useful in relieving burning sensation, hyperdiuresis, pectoral lesions and promoting strength. The roots are sweet, sour, astringent, bitter, acrid, thermogenic and tonic, and are useful in fever, uropathy and arthritis. Its roots are used as diuretic, astringent, stomachic, febrifuge and demulcent and seeds are applied as laxative, aphrodisiac and demulcent; recommended in cystitis, colic gonorrhoea, and piles. (Warrier *et al.* 2010).

Medicinal Uses (Traditional/Folklore): The plant has a potential for strengthening and glowing of the body. In folk medicine, females use it as soup in the last days of pregnancy to reduce pain of labour and reducing the period. It also leads to decrease the period of post parturition bleeding. The whole plant is used by the tribal people of Madekeri district, Karnataka, India to treat various aliments like hepatic disorder, dysentery, cholera, etc. It is also one of the components in herbal preparation in Tamil Nadu, India used on cuts and wounds, to relieve pain, etc. The bark of root is used for leucorrhoea, gonorrhoea and hyperdiuresis. (Warrier *et al.* 2010, Gulnaz and Savitha 2013)

11.10 *Sida cordifolia* L.

(Syn. *Sida herbacea* Cav.; *Sida rotundifolia* Lam.)

Common Name: Heart-Leaf Sida, Bala, Country Mallow, Flannel Weed.

Vernacular Name: *Hindi:* Kharinta (खरींटा), Kharenti (खरैंटी); *Sanskrit:* Balaa (बाला), Sumanganaa (सुमन गाना), Bhadrabalaa (भद्रबाला), Bhadraudani (भदरौदानी).

Habit: Undershrub.

Habitat: Along roadsides and wastelands.

Distribution: Native to India. It has naturalized throughout the world, and is considered an invasive weed in Africa, Australia, the southern United States, Hawaiian Islands, New Guinea, and French Polynesia. The plant is distributed throughout the tropical and sub-tropical regions of India, in moist places, up to 1800 m.

Flowering and Fruiting Time: April-November.

Flowering twig

Flower

Habit

Taxonomic Description: An erect herb, up to 70 cm high. Stem and branches appressed with long spreading hairs. Leaves alternate, petiolate, ovate, oblong, serrated, obtuse, cordate at base, covered with spreading hairs on both surface; petiole usually as long as the blade; stipules, linear, hairy. Flowers white to pale yellow. Calyx lobes-5, triangular, pointed. Corolla lobes-5, obliquely obovate. Peduncle jointed near the flower. Stamens many. Carpels 5, awned, awn longer than the calyx. Seeds dark brown.

Chemical Constituents: *S. cordifolia* contains ephedrine, which is an amphetamine-like stimulant that can cause harmful side effects. Besides ephedrine, the plants also contain pseudo-ephedrine, which is the major alkaloid in the aerial parts. The seeds contain fatty oil, steroids, proteins, ephedrine, pseudoephedrine, linoleic acid, malvalic acid, aterculic acid, coronaric acid, phytosterols, resine, resin acids, mucin and potacium nitrate. The root contains alkaloids, ephedrine, pseudoephedrine, β-phenethylamine, carboxilated triptamines and hypaphorine, quinazoline alkaloids- vasicinon, vasicinin, vasicimol, choline and betanine. (Bhutya 2011).

Medicinal Properties: The plant is vigorating, spermatopoietic, used for cure of spermatorrhea, fever, fits, leucorrhoea, micturition, gonorrhoea, colic, nervous disorders, general debility and heart irregularity. In addition, the plants have other wide range of properties such as antidiabetic, antihypercholesterolemic, anthelmintic, antioxidant, anti-inflammatory, wound healing properties; the plant also possesses adoptogenic, antipyretic, antiulcerogenic, antibacterial and antifungal, and diuretic properties. The plants are useful in cardiovascular problems, liver regeneration, and reproductive problems. Seeds are nervine tonic, aphrodisiac, used for sexual debility, and as a tonic. The leaves are demulcent and febrifuge, useful to cure dysentery, opthalmia and rheumatism. (Bhutya 2011, Jain *et al.* 2011; http://medplants.blogspot.in).

Medicinal Uses (Traditional/Folklore): Different parts of the plant are used in many diseases in Ayurvedic, Siddha and tribal systems. This herb is a very good cardiac tonic and reduces petechial haemorrhage. Ethnic communities of Ranchi, Hazaribag and Varanasi districts consider the plant useful in venereal diseases. Ethnic communities of Delhi area use seeds in spermatorrhoea and gonorrhoea. The herb has *rasayana* (rejuvenating) properties. It is very effective as medicated oil. The popular *Mahanarayana taila, Balati taila, Prabhanjana Vimardhana, Ksheera-bala taila* contain this herb. The oils are used topically to massage the sore muscles, sore joints in arthritis and rheumatism, in sciatica and neuritis of legs. Another oil called *Dhanvantari tailam* (21 and 101 times boiled) which contains Bala along with 47 other substances, and is prepared in milk and given for all disorders produced by the derangement of '*Vata*', such as emaciation, diseases of reproductive organs. For this reason the oil prepared using this herb is used to massage patients who suffer from paralysis, cervical spondylosis, facial paralysis, etc. The leaves are used as an infusion in treating fevers and delirium. Paste of the leaves relieves swelling of any parts of the body. Guava juice of one fruit, 1/4 liter Goat milk and root of *Sida cordifolia* are mixed together thoroughly. The preparation is administered orally and three doses are sufficient for the disappearance of symptoms like clear urine and removal of yellowness from eyes of the patients. Juice of the plant is invigorating, spermatopoietic, used in spermatorrhoea and also applied on cuts and wounds and to treat arthritis.

The seeds contain much larger quantities of alkaloids than the leaves and roots and are used as an aphrodisiac and a nervine tonic. Seed decoction is also useful in treating colic, tenesmus and gonorrhoea. Seed powder is given in dyspepsia. Seeds are useful in spermatorrhea, paralysis, sciatica, polyuria, dysuria, cystitis, strangury, hematuria, leucorrhoea, urinary tract infection, fevers and general debility. The root of this plant is astringent, diuretic, and tonic, and its infusion is useful in cystitis, strangury, haematuria, bleeding piles, chronic dysentery, leucorrhoea, and gonorrhoea. Root are used for the treatment of rheumatism, neurological disorders (hemiplegia, facial paralysis, and sciatica), polyuria, dysuria, cystitis, strangury and hematuria; leucorrhoea and other uterine disorders, fevers and general debility. Root infusion is given in suppressed urination. Root bark, pulverized and mixed with sesame oil and milk, has been said to be effective in cases of facial paralysis and sciatica.

Whole plant is boiled, and the water used as an herbal bath for washing the skin, as an antipyretic for chickenpox and measles, by the Guyana Patamona people; also leaves are boiled, and the water taken as an antipyretic. Leaves are cooked and eaten in cases of bleeding piles. The whole plant, pounded with a little water is given in doses of ¼ seer (app. 233 g) for spermatorrohea, rheumatism, and gonorrhoea. Made into paste with juice of palmyra palm tree, it is applied locally, in elephantiasis. (Singh *et al.* 2010d, Jain *et al.* 2011, Rama Shankar *et al.* 2012, Rameshkumar and Ramakritinan 2013, Basumatary *et al.* 2014; http://medplants.blogspot.in).

11.11 *Triumfetta rhomboidea* Jacq.

(Syn. *Bartramia indica* L.; *Triumfetta indica* Lam.)

Habit

Common Name: Diamond Burbark, Chinese Bur.

Vernacular Name: *Hindi:* Chiki (चिकी); *Sanskrit:* Jhinjharita (झिंझारिता), Gippit (गिपपित).

Habit: Undershrub.

Habitat: An abundant weed along roads, canal banks, waste places, thickets and fallow fields.

Distribution: A pantropic weed; native of tropical America. The plant is extensively naturalized in tropical regions.

Flowering and Fruiting Time: August-November.

Taxonomic Description: A much-branched, 30-80 cm high, undershrub with stellate-pubescent branchlets. Leaves stipulate, up to 8 cm long, rhomboid-ovate to broad-ovate, generally 3-lobed, hairy, irregularly serrate at margin. Flowers about 8 mm long, yellow in axillary or leaf-opposed cymes, crowded at the nodes, shortly pedicellate. Sepals-5, 6 mm long, apiculate, pubescent. Petals-5, obovate, spathulate, 5 mm long. Stamens 8-15, filaments yellow. Capsules nearly 5 mm in diameter, globular, clothed with hooked spines.

Flowering and fruiting twig

Flower

Chemical Constituents: The plant contains carbohydrates, glycosides, phytosterol, steroids, flavonoids, tannin, saponins, triterpenoids, alkaloids and phenolic compounds. (Devmurari *et al.* 2010).

Medicinal Properties: Leaves, flowers, fruits and bark are mucilaginous, antibacterial, demulcent, astringent, and also used in gonorrhoea and against leprosy. Root is tonic, styptic, galactagogue, aphrodisiac, cooling, useful in dysentery and as diuretic. Leaves have antihypertensive, astringent, diuretic, emollient, analgesic, antitumour, anticancer and anti-inflammatory properties. Leaves and bark are astringent, anticholera, demulcent and anti-inflammatory. (Devmurari *et al.* 2010, Sahoo *et al.* 2016).

Medicinal Uses (Traditional/Folklore): Various parts of *T. rhomboidea* such as fruits, flowers, leaves, bark and root have been used in ancient therapy. The plant also has been in use in Ayurveda and Sidha for the treatment of various ailments such as angina, pneumonia, inflammation, headache, snakebites, pregnancy/birth/puerperium disorders, diarrhoea, dysentery, gonorrhoea and hypertension. Roots are used for treating impotency and barrenness and root decoction is given to cure dysentery and treatment of cough; root paste is applied to treat pimples and intestinal worms. A decoction of the root is also used as a remedy for internal ulceration; pounded roots are given in the treatment of intestinal ulcers. A decoction of the whole plant prepared in rice water, or of the root and bark, is used to treat diarrhoea, dysentery, internal haemorrhages and gonorrhoea. The leaves and flowers are used in the treatment against leprosy. The fruits and pounded roots are known to promote childbirth; hot infusion of roots facilitates childbirth and hastens parturition. The crushed flowers are applied as a poultice on boils. Patients with severe colds are treated by giving them a daily sauna with the boiling leaves. Bark and fresh leaves are given for diarrhoea and dysentery. A decoction of leaves taken in cough and as blood purifier. A paste prepared from leaves and flowers crushed with water is applied on itching wounds and used against leprosy; leaves are mashed with rhizome of *Zingiber officinale*, lemon juice and coconut oil, fried, and taken as bechic (cough suppressant). Hill-tribes use *T. rhomboidea* to treat stomachache, indigestion and fever during menstruation. Leaves are used as a traditional cure for use in snake bite and treatment of cancer and tumour. (Shukla *et al.* 2010, Sahoo *et al.* 2016; http://medplants.blogspot.in).

11.12 *Urena lobata* L.

(Syn. *Urena trilobata* Vell.)

Habit

Common Name: Caesar weed.

Vernacular Name: *Hindi:* Bachita (बचीता), Lapetua (लपेटुआ).

Habit: Shrub.

Habitat: Along roadsides, railway lines, canal banks.

Distribution: This species has a pan-tropical distribution and it can be found growing throughout moist tropical and subtropical regions of the world, including Asia, tropical Africa, Australia, North, Central and South America, the West Indies and islands in the Pacific.

Flowering and Fruiting Time: May-November.

Taxonomic Description: An annual or perennial, branched shrub, slightly appressed with stellate hairs, 1 to 2 m tall. Leaves variable in shape and size, ovate, orbicular to oblong-lanceolate, 3-7 lobed, petiolate, up to 10 cm long, shallowly cordate or obtuse at base, crenate-serrate or sub entire at margin, mid rib glandular at base, softly hairy on both sides. Stipules linear, lanceolate. Flowers axillary, solitary to 2-3 in cluster, pink, darker in centre. Bracteoles-5, oblong, lanceolate, as long as the calyx. Schizocarps globose; mericarps-5, with several hooked bristles. Seeds kidney-shaped, brown.

Flowering twig

Flower

Schizocarps

Chemical Constituents: Phytochemicals detected in the leaves of *U. lobata* are alkaloids, flavonoids, cardiac glycoside, saponins, terpenoid and tannins. In branches and leaves ten flavonoid compounds, *viz.*, kaempferol, rutin, quercetin, afzelin, astragalin, tiliroside, kaempferol3oβdglycopyranoside7oαlrhamnoside, kaempferol7oαlrhamnoside, kaempferol7oαlrhamnoside4′oβdglycopyranoside, and crenuloside have been found. (Fagbohun *et al.* 2012, Babu *et al.* 2016).

Medicinal Properties: The plant is used traditionally as an antibacterial, anti-inflammatory and amoebicidal. The leaves are diuretic, antidiabetic, and emollient, emmenagogue, refrigerant, styptic and vulnerary. The roots of this plant are considered as diuretic and used externally as embrocation in lumbago and rheumatism and internally as a decoction of the same is used in cases of flatulent colic. Its flowers are considered expectorant and pectoral and are given in cough. (Dhiman 2004, Vardhana 2008, Purnomo *et al.* 2015, Babu *et al.* 2016).

Medicinal Uses (Traditional/Folklore): The plant is commonly used in traditional medicine, with the leaves and roots most commonly employed. The whole plant and its parts are used in bronchitis, gastritis, cough and cold, nephritis, diarrhoea, fever, pneumonia, gingivitis and menorhagia. It is also used for gonorrhoea, toothache, wounds, malaria, colds, dysentery, enteritis, goitre, indigestion, coughs, leucorrhoea, tonsillitis and respiratory diseases.

The juice of the leaves or roots is used widely to treat bowel complaints, especially colic, stomachache, diarrhoea and dysentery, and also to treat gonorrhoea and persistent fever from malaria. A decoction from the leaves and roots is taken to relieve pains all over the body due to excessive exertion. A decoction of a very old plant, boiled with eggs, is said to induce abortion. A lotion made from the plant is used to treat yaws and headache. The whole plant is macerated and used externally for treating fractures, wounds, mastitis and snake bites; also used to treat inflammation of the intestines and bladder. The leaves are used externally as a poultice on wounds and skin diseases. The roots are diuretic and used to treat rheumatism and lumbago. An infusion/decoction of roots is given to aid difficult childbirth and also used to treat colds, dysentery, enteritis, goitre, indigestion, leucorrhoea, malaria, rheumatism and tonsillitis. Roots are chewed and applied externally to swellings caused by filariasis. The twigs are chewed to treat toothache. The bark is used to heal cuts. The flowers are taken in decoction as a pectoral and expectorant in dry coughs. An infusion of the flowers is used as a gargle for aphthae and a sore throat. A decoction of the seeds is taken as a vermifuge.

Leaves, roots, and flowers are reportedly used in traditional medicine in Malaysia, Indo-China, Philippines, Papua New Guinea, Fiji, and India to treat such diverse ailments as colic, malaria, gonorrhoea, fever, wounds, toothache, and rheumatism. In Nigeria, people use *U. lobata* to treat diabetes mellitus. The whole plant is macerated and used externally for treating fractures, wounds, mastitis and snake bites. Whole plant boiled in sesame oil is applied externally on rheumatic pains by the tribal in western Maharashtra, India.

In South East Asia and Africa, *U. lobata* is considered a magic plant, and is used in healing rites, for protection, and in wedding and rice ceremonies. Seeds and parts of plants are used in Africa in stews and are eaten as famine food. In India, seeds are used to produce soap, while the charcoal of the whole plant is used for blackening teeth. (Purnomo *et al.* 2015, Babu *et al.* 2016; http://medplants.blogspot.in, http://tropical.theferns.info).

12. ZYGOPHYLLACEAE

Zygophyllaceae – Caltrop/Bean-Caper family

The family comprising of around 211 species in 26 genera contains the bean-caper and caltrop. Most of the species are trees, or shrubs, or herbs, often with jointed nodes; non-laticiferous; resinous, or not resinous. Flowers ebracteate, ebracteolate; corolla when present, 1 whorled, polypetalous. The family was represented by one species in the flora.

12.1 *Tribulus terrestris* L.

(Syn. *Tribulus lanuginosus* L.)

Common Name: Puncture Vine, Caltrop.

Vernacular Name: *Hindi:* Gokhru (गोखरू); *Sanskrit:* Gokshura (गोक्षुरा), Trikanta (त्रिकांता), Shvadamshtra (श्वदंष्ट्रां).

Habit: Prostrate annual herb.

Habitat: Roadsides, wastelands, fallow lands.

Distribution: *Gokhru,* a native of the tropical America, is widespread throughout the world from latitudes 35°S to 47°N. Distributed throughout India, up to 5,400 m. It grows as a weed along the roadsides, waste places and boundaries of agricultural fields.

Flowering and Fruiting Time: May to October.

Habit

Flowering and fruiting twig

Flower

Fruit

Taxonomic Description: A prostrate to decumbent annual herb, clothed with silky hairs. Stem branched from the base; branches up to 1.5 m long, silky-villous. Leaves abruptly paripinnate; leaflets small, 5-7 pairs, oblong, mucronate at apex, oblique at base, entire at margin, hairy. Stipules subulate. Flowers yellow, solitary, axillary or pseudoaxillary. Calyx lobes-5, lanceolate, pubescent, slightly ciliate at margin. Corolla lobes-5, free, oblanceolate. Stamens-10. Stigma 5 rayed. Fruits schizocarpic, spinous, villous, usually with 5 woody cocci, each cocci with 2 or 4 spines.

Chemical Constituents: The chemistry of *T. terrestris* has been extensively studied and the occurrence of saponins, flavonoids, glycosides, alkaloids, lignanamides and cinammic acid, amides and tannins has been reported. The fruit and root contain pharmacologically important metabolites such as phytosteroids, flavonoids, alkaloids and glycosides. A number of saponins and their different derivatives have also been identified in *T. terrestris*, of which diosgenin, gitogenin and chlorogenin are present in the leaf tissue. The presence of spirostanol and furostanol saponins is a characteristic feature of this plant, the latter being considered to be biogenetic precursors of their spiro analogs. Various derivatives of tigogenin, neotigogenin, gitogenin, neogitogenin, hecogenin, neohecogenin, diosgenin, ruscogenin, chlorogenin and sarsasapogenin have been reported in the plant. Leaves contain ascorbic acid, calcium, carbohydrate, fat, fiber, iron, oxalate, phosphorus, potassium, protein, and tribuloside. Fruits are rich in fat, amino and various fatty acids such as aspartic acid, glutamic acid, linoleic acid, oleic acid, palmitic acid and stearic acid. (Jameel *et al.* 2012, Chhatre *et al.* 2014, Hashim *et al.* 2014).

Medicinal Properties: The roots and fruits are sweet, cooling, diuretic, aphrodisiac, emollient, appetizer, digestive, anthelmintic, expectorant, anodyne, anti-inflammatory, alterant, laxative, cardiotonic, styptic, lithontriptic and tonic. They are useful in strangury, dysuria, renal and vesical calculi, anorexia, dyspepsia, helminthiasis, cough, asthma, consumption, inflammation, cardiopathy, haemoptysis, spermatorrhoea, anaemia, scabies, ophthalmia, ulocace and general weakness. The seeds are astringent, diuretic, abortifacient, aphrodisiac, strengthening and tonic, and are useful in epistaxis, haemorrhages and ulcerative stomatitis. Seeds are used for treating abscesses, anaemia, coughs, fluxes, heamorrhoids, and spermatorrhea. The leaves are astringent, diuretic, aphrodisiac, depurative, anthelmintic and tonic. They are useful in gonorrhoea, gleet, ulitis, inflammation, menorrhagia, strangury, leprosy, skin diseases, verminosis and general weakness and for treating scabious skin diseases and psoriasis. Fruit is given for headache, hepatitis, impotence, liver, ophthalmia and stomatitis, and vertigo; they are recommended for kidneys and liver ailments and vision improvement. The whole plant is recommended for the treatment of cancer and as aphrodisiac and in traditional medicine for treating male infertility. (Warrier *et al.* 2010, Jameel *et al.* 2012, Hashim *et al.* 2014, Shaheen *et al.* 2014)

Medicinal Uses (Traditional/Folklore): *T. terrestris* is mentioned in ancient Indian Ayurvedic medical texts dating back thousands of years for the treatment of sexual dysfunction and various urinary disorders. In India, *T. terrestris* is used in folk medicine as a tonic, aphrodisiac, palliative, astringent, stomachic, antihypertensive, lithotriptic, and diuretic, and for the treatment of painful urination, calculous affections and Bright's disease. *T. terrestris* has potential for the treatment of urinary stones, Parkinson's disease, liver and eye diseases, benign prostatic hyperplasia, can improve male sexual function and the performance of athletes. The dried fruit of the herb is very effective in most of the genitourinary tract disorders. It is a vital constituent of *Gokshuradi Guggulu*, a potent Ayurvedic medicine used to support proper functioning of the genitourinary tract. It has also been used for centuries in Ayurveda to treat impotence and venereal diseases. In Bulgaria, the plant is used as a folk medicine for treating impotence. In addition to all these applications, the Ayurvedic Pharmacopoeia of India attributes cardiotonic properties to the root and fruit.

In traditional Chinese medicine, the fruits were used for the treatment of eye trouble, edema, abdominal distension, emission, morbid leucorrhoea, and sexual dysfunction. The plant is described as a highly valuable drug in the Shern-Nong Pharmacopoeia (the oldest known pharmacological work in China) in restoring the depressed liver, for treatment of fullness in the chest, mastitis, flatulence, acute conjunctivitis, headache and vitiligo. In Unani medicine, the plant is used as diuretic, mild laxative and general tonic. The Greeks used it as a diuretic. The ash of the whole plant is good for external application in rheumatic-arthritis. Decoction of dried seeds is used for diuretic and antilithiatic activity. In China and Vietnam, it has been used in the treatment of postpartum haemorrhage, epistaxis and gastro intestinal bleeding. *T. terrestris* is being promoted as a testosterone booster for the purpose of building muscle and increasing libido. It does not work like DHEA and androstenedione 100, which are progenitors of testosterone. Instead, claims have been made that it enhances testosterone levels by increasing luteinizing hormone levels. (Jameel *et al.* 2012, Rahul 2013, Kor *et al.* 2013, Yadav 2013, Chhatre *et al.* 2014, Hashim *et al.* 2014, Shaheen *et al.* 2014).

13. OXALIDACEAE

Oxalidaceae – Wood-Sorrel family

It is a small family of eight genera and 601 accepted plant species of herbaceous plants, shrubs and small trees, with the great majority of the 376 species occurring in the genus *Oxalis* (wood sorrels). Members of this family typically have divided leaves, the leaflets showing "sleep movements", spreading open in light and closing in darkness with no stipules. Two species belonging to a single genus of this family were recorded in the flora.

13.1 *Oxalis corniculata* L.

(Syn. *Oxalis repens* Thunb.)

Habit

Common Name: Indian Sorrel, Creeping Wood Sorrel, Creeping Oxalis, Yellow Wood Sorrel.

Vernacular Name: *Hindi:* Amrul (अम्रुल), Khati-buti (खट्टी-बूटी); *Sanskrit:* Asmanthaka (असमन्थका) and Kushali (कुशाली).

Habit: Prostrate annual herb.

Habitat: Moist waste places, gardens, nurseries.

Distribution: A native of Europe; found throughout India, up to 3,000 m in all moist exposed localities.

Flowering and Fruiting Time: Most of the year.

Taxonomic Description: A small, prostrate or procumbent, pubescent, annual herb. Stem creeping, branched from the base, rooting at nodes. Leaves stipulate, long petioled, trifoliate; leaflets 1.5 cm long, broadly obcordate, ciliate at entire margin. Flowers axillary, solitary or few flowered in sub-umbellate clusters, peduncles long. Sepals-5, linear-lanceolate, pubescent. Petals-5, yellow, obcordate. Stamens 10. Ovary ellipsoid. Capsules 1.5-2 cm long, linear, sub-cylindrical, hairy. Seeds transversely ribbed, many, brown.

Leaf

Flower

Capsules

Chemical Constituents: The plant contains phytochemicals like flavanoids, tannins, phytosterols, phenol, glycoseides, fatty acids, galacto-glycerolipid and volatile oil. The leaves contain flavonoids, iso-vitexine and vitexine-2"-O-beta – D-glucopyrunoside. It is rich source of essential fatty acids like palmitic acid, oleic, linoleic, linolenic and stearic acids. They are good source of vitamin C (125 mg/100g), carotene (3.6 mg/100g), but contain a high content of oxalates (12% of dry material). The leaves and stem contain tartaric and citric acid while stem also contains malic acid. (Badwaik *et al.* 2011).

Medicinal Properties: The plant is sour, astringent, thermogenic, cooling to touch, anodyne, anthelmintic, anti-inflammatory, digestive, depurative, carminative, liver tonic, lithontripic, stomachic, styptic, diuretic, constipating, febrifuge, antibacterial, vermifuge, emmenagogue and antiseptic. It is useful in dyspepsia, haemorrhoids, anaemia, tympanitis, fever, urinary tract infections, dysentery, enteritis, diarrhoea, scurvy, corns and warts, excrescences of skin, burns and skin eruptions, inflamed ulcers, cephalalgia, ophthalmopathy, toxicity, cardiopathy, haemorrhages, dysmenorrhoea, amenorrhoea, strangury, hepatopathy and burning sensation. (Warrier *et al.* 2010, Badwaik *et al.* 2011, Rahul 2013; http://tropical.theferns.info).

Medicinal uses (Traditional/folklore): In Ayurvedic medicine, it is used for liver and digestive problems. The juice of the plant, mixed with butter, is applied to muscular swellings, boils and pimples. An infusion is used as a wash to rid children of hookworms. Fresh leaf extract is taken orally in the morning for 4- 5 days during indigestion, and liver bladder inflammation. Leaf paste is also applied topically on wounds to kill worms. The leaves are effective in certain skin diseases like warts, corns and other excrescences of the skin. They are applied locally in these conditions. The juice of whole plant mixed with onion is also applied to remove warts. A poultice of the leaves applied over an inflamed areas to relieve pain, and over boils, to ripen them. The leaf juice mixed with black pepper and ghee, gives relief from red spot and eruptions on the skin caused by biliousness. It is known to be very effective in controlling high fevers; the mild infusion of leaves when taken during high fever helps bring down the fever. The leaves being rich in vitamin C are used to relieve deficiency of vitamin C in the body (scurvy); the leaves are used by making an infusion for this purpose. Disorders like dysentery and enteritis in mild conditions can be relieved by boiling the leaves of the herb with buttermilk and having it twice a day; the juice of the leaves as well as fresh leaves are used to relieve many other stomach disorders. A few drops of the juice from the leaves cool the eyes and prevent the opacity of the cornea; leaf juice is used as eye drops for treating cataract and conjunctivitis. The leaf juice mixed with buttermilk is found to be very helpful in relieving jaundice and reduce the poly hydra condition brought on by diabetes. The leaves are also used in the treatment of traumatic injuries, sprains and poisonous snake bites, insect bites; leaves are used as an antidote to poisoning by the seeds of *Datura* spp., arsenic and mercury.

In Fiji, the plant is traditionally used as a remedy for convulsions in children and for healing fractured bones. In Nepal, it is used for stomach aches; the leaves are also used in rituals. (Abbasi *et al.* 2010, Senthil Kumar and Rajkapoor 2010, Badwaik *et al.* 2011, Rahul 2013; http://tropical.theferns.info).

13.2 *Oxalis debilis* var. *corymbosa* (DC.) Lourteig

(Syn. *Oxalis corymbosa* DC.; *Oxalis martiana* Zucc.)

Common Name: Large-Flowered Pink Sorrel, Lilac Oxalis, Pink Wood Sorrel.

Vernacular Name: *Hindi:* Khatti-mithi (खट्टी-मीठी).

Habit: Perennial herb.

Habitat: Moist waste places, roadsides.

Distribution: Native to South America; naturalized in Africa, Asia, Australasia, Europe and Northern America; widely naturalized in different subtropical and tropical parts of India. It occurs throughout India.

Flowering and Fruiting Time: November-April.

Habit

Taxonomic Description: A perennial diffused herb, bulbous at base; bulbils clustered, globose. Leaves 3-foliate, long petioled, stipulate; leaflets obcordate, 2-lobed, margin entire, base cuneate, slightly pubescent on both surface. Flowers bracteate, pedicellate; 5-7 flowers arranged on long axillary peduncles in sub-umbellate cymes. Calyx lobes-5; sepals oblong, lanceolate with 2-orange dots, pubescent. Corolla lobes-5; petals purple or violet with darker nerves. Stamens 10. Ovary 5-angled, oblong. Seeds numerous.

Trifoliate leaf with inflorescence

Flowers

Chemical Constituents: The plant contains carbohydrate, protein, crude fibre, ascorbic acid and phenol. (Sarma *et al.* 2015).

Medicinal Properties: Antioxidant and neutraceutical properties. (Sarma *et al.* 2015).

Medicinal Uses (Traditional/Folklore): It is used traditionally for treatment of various ailments such as diarrhoea, fevers, urinary problems and appetite loss. Plant juice is used for stomachache, scurvy, indigestion and piles; leaf paste is applied externally for treating skin diseases and boils. Decoction of leaves is used as a gargle to relieve angina pain. Rhizome juice is antidote. (Singh *et al.* 2010d, Saikia and Khan 2011).

14. MELIACEAE

Meliaceae – Mahogany family

The family includes about 52 genera and about 669 known species, with a pantropical distribution. Most of the species are trees and shrubs characterized by alternate, usually pinnate leaves without stipules, and by syncarpous, apparently bisexual (but actually mostly cryptically unisexual) flowers borne in panicles, cymes, spikes, or clusters. Most species are evergreen, but some are deciduous. The family was represented by two species belonging to two genera in Chithara Village.

14.1 *Azadirachta indica* A. Juss.

(Syn. *Melia azadirachta* L.)

Habit

Common Name: Neem, Indian Lilac, Margosa.

Vernacular Name: *Hindi:* Neem (नीम); *Sanskrit*: Arishtha (अरिष्ठा).

The meaning of Sanskrit name 'Arishtha' is 'reliever of sicknesses' and hence is considered as 'Sarbaroganibarini'.

Habit: Tree.

Habitat: Planted along roadsides as an avenue plantations as well as in villages; sometimes found wild.

Distribution: Native to India and the Indian subcontinent including Nepal, Pakistan, Bangladesh and Sri Lanka. It is typically grown in tropical and semi-tropical regions. Throughout India, in deciduous forests, also widely cultivated.

Flowering and Fruiting Time: March-July.

Taxonomic Description: A medium to large-sized tree with dark grey or blackish bark. Leaves alternate, exstipulate, imparipinnate; leaflets up to 15, shortly stalked, 3-7 cm long, oblong-lanceolate, opposite or subopposite, margin serrated, base oblique, apex acute to acuminate. Inflorescence axillary panicles. Flowers white, scented, about 6-8 mm across, pedicel about 5 mm long. Sepals-5. Petals-5, free, oblanceolate. Staminal tube cylindrical, nearly 4 mm long. Ovary 3-celled, stigma 3-lobed. Fruit a drupe, oblong-ellipsoid, 1.5-2 cm long, smooth, yellow when ripe. Seed-1.

Inflorescence

Fruits

Chemical Constituents: Investigations on the phytochemicals of the neem tree were carried out extensively in the middle of the twentieth century. Since the early report by Siddiqui (1942) on the isolation of nimbin, the first bitter compound isolated from neem oil, so far more than 135 compounds have been isolated from different parts of neem showing diverse chemistry. The compounds are divided into two major classes: isoprenoids and others. The isoprenoids include diterpenoids and triterpenoids containing protomeliacins, limonoids, azadirone and its derivatives, gedunin and its derivatives, vilasinin type of compounds and C-secomeliacins such as nimbin, salanin and azadirachtin. The nonisoprenoids include proteins (amino acids) and carbohydrates (polysaccharides), sulphur compounds, polyphenolics such as flavonoids and their glycosides, dihydrochalcone, coumarin and tannins, aliphatic compounds, etc. (Biswas *et al.* 2002).

Medicinal Properties: It has numerous medicinal properties such as anthelmintic, antifungal, antidiabetic, antibacterial, antiviral, contraceptive and sedative. The bark is bitter, astringent, acrid, refrigerant, depurative, antiperiodic, vulnerary, demulcent, insecticidal, liver tonic, expectorant, urinary astringent, anthelmintic and pectoral; it is useful in amenorrhoea, anorexia, bronchitis, burning sensation, cough, diabetes, dyspepsia, fatigue, haemorroids, hepatopathy, hyperdipsia, intermittent and malarial fevers, intestinal worms, inflammation, lumbago, otalgia, syphilis, skin diseases, leprosy, eczema, leucoderma, pruritus, tumour, tubercular glands, ulcers, urorrhea, vitiated conditions of pitta, vomiting and wounds.

Some of the potential medicinal properties of neem are described below.

Immunostimulant activity: Aqueous extract of bark and leaf have anticomplement and immunostimulant activity. Neem oil has been shown to possess activity by selectively activating the cell-mediated immune mechanisms to elicit an enhanced response to subsequent mutagenic or antigenic challenge.

Hypoglycaemic activity: Aqueous extract of neem leaves significantly decreases blood sugar level and prevents adrenaline as well as glucose-induced hyperglycemia.

Antiulcer effect: Neem leaf and bark aqueous extracts produce highly potent antacid, secretory and antiulcer activity.

Antimalarial activity: Neem seed and leaf extracts are effective against both chloroquine-resistant and sensitive strain malarial parasites.

Antifungal activity: Extracts of neem leaf, neem oil seed kernels are effective against certain fungi including *Trichophyton, Epidermophyton, Microspora, Trichosporum, Geotricum* and *Candida.*

Antibacterial activity: Oil from the leaves, seed and bark possesses a wide spectrum of antibacterial action against Gram-negative and Gram-positive microorganisms, including *Mycobacterium tuberculosis* and streptomycin resistant strains. In vitro, it inhibits *Vibrio cholera, Klebsiella pneumoniae, M. tuberculosis* and *M. pyogenes.* Antimicrobial effects of neem extract have been demonstrated against *Streptococcus mutans* and *S. faecalis.*

Antiviral activity: Aqueous leaf extract offers antiviral activity against Vaccinia virus, Chikungunya and measles virus.

Anticancer activity: Neem leaf aqueous extract effectively suppresses oral squamous cell carcinoma induced by 7, 12-dimethylbenz[a] anthracene (DMBA), as revealed by reduced incidence of neoplasm. Neem may exert its chemopreventive effect in the oral mucosa by modulation of glutathione and its metabolizing enzymes.

Antioxidant activity: The antioxidant activity of neem seed extract has been demonstrated *in vivo* during horse-grain germination. (Biswas *et al.* 2002, Warrier *et al.* 2010, Prashanth and Krishnaiah 2014).

Medicinal Uses (Traditional/Folklore): Neem is most important medicinal plant that has been declared worldwide as the "Tree of the 21st century" by the United Nations. In India, it is called "Divine Tree", "Life giving tree", "Nature's Drugstore", "Village Pharmacy" and "Panacea for all diseases". As a traditional practice, products made from neem trees have been used in India for over two millennia for their medicinal properties in Ayurveda, Siddha and Unani medicines, particularly prescribed for skin diseases. Neem leaves have also been used to treat skin diseases like eczema, psoriasis, etc. It is effective to cure ringworm, eczema and scabies. Lotion derived from neem leaf, when locally applied, can cure these dermatological diseases within 3-4 days in acute stage or a fortnight in chronic case. Neem leaf extract has been prescribed for oral use for the treatment of malaria by Ayurveda practitioners from time immemorial. Recently, a clinical trial has been carried out to ascertain the efficacy of neem extract to control hyperlipidemia in a group of malarial patients severely infected with *Plasmodium falciparum*. The lipid level, especially cholesterol, was found to be lower during therapy when compared to non-malaria patients.

Neem oil and the bark and leaf extracts have been therapeutically used as folk medicine to control leprosy, intestinal helminthiasis, respiratory disorders, and constipation and also as a general health promoter. Its use for the treatment of rheumatism, chronic syphilitic sores and indolent ulcer has also been evident. Neem oil finds use to control various skin infections. Bark, leaf, root, flower and fruit together cure blood morbidity, biliary afflictions, itching, skin ulcers, burning sensations and phthisis. In Rajasthan, neem is used to treat allergy, skin diseases, and snake bite. Neem oil is used for healthy hair, to improve liver function, detoxify the blood, and balance blood sugar levels. Use of neem to treat patients suffering from various forms of cancer are available in the literature. One patient with parotid tumour and another with epidermoid carcinoma responded successfully when treated with neem seed oi1. In a recent study, NIM- 76, a refined product from neem oil, prevented pregnancy with no adverse effects. (Biswas *et al.* 2002, Kaul and Dwivedi 2010, Singh *et al.* 2010d, Shukla *et al.* 2010, Pathak *et al.* 2013, Rahul 2013, Prashanth and Krishnaiah 2014, Shaheen *et al.* 2014; http://medplants.blogspot.in).

14.2 *Melia azedarach* L.

(Syn. *Azedarach sempervirens* Kuntze)

Common Name: Chinaberry Tree, Persian Lilac, Bead Tree.

Vernacular Name: *Hindi:* Bakain (बकैन), Mahaneem (महानीम); *Sanskrit:* Mahanimbah (महानिंब).

Habit: Small or medium-sized tree.

Habitat: Commonly planted along the roadsides and in gardens; also run wild in unused grounds.

Distribution: It is native to Pakistan, India, Indochina, Southeast Asia and Australia.

Flowering and Fruiting Time: March-June.

Habit

Taxonomic Description: A small to medium-sized branched tree, young branches slightly pubescent. Leaves 30-50 cm long or more, bipinnate, sometimes tripinnate; pinnae 5-9, imparipinnate; leaflets opposite, 3-9 on each pinnae, 3-6 x 1-2 cm, ovate-lanceolate, base oblique, tip acuminate, margin serrate to slightly lobed. Flowers lilac colour, fragrant, in long axillary lax panicles. Calyx 5-partite, pubescent. Corolla rotate; petals-5, slightly hairy. Staminal tube purple. Ovary sub-globose, 5-celled. Fruit a drupe, globose-ellipsoid, nearly 1.5 cm across, green when young, turning yellow and wrinkled at maturity.

Leafy twig

Unripe fruits

Chemical Constituents: Phytochemical screening of *M. azedarach*, showed the presence of number of organic molecules i.e., terpenoids, flavonoids, steroids, acids, anthraquinones, alkaloids, saponins, and tannins. Leaves, stem wood, fruits and roots contain bioactive terpenoids and limonoids. (Sharma and Paul 2013).

Medicinal Properties: The plants have antioxidant, antimicrobial, anti-inflammatory, cardio-protective, analgesic, anticancer, antiulcer, astringent, antiseptic, antipyretic, antiplasmodial and male contraceptive properties. Its various parts have antiviral, antimalarial, cathartic, emetic and emmenagogic properties. The leaf juice is anthelmintic, antilithic, diuretic, emmenagogue, expectorant, and vermifuge and their decoction is astringent, and stomachic. Flowers are astringent, refrigerant, stomachic, vermicidal, anodyne, diuretic, emmenagogue, resolvent, deobstruent and alexipharmic. Fruits are reputed to be powerfully vermifuge. Fruits are purgative, tonic and anthelmintic. Seeds are bitter, expectorant, anthelmintic and aphrodisiac. Roots are bitter, astringent, mildly thermogenic, anodyne, vulnerary, antiseptic, anthelmintic, constipating, expectorant, febrifuge, antiperiodic, urinary astringent, emmenagogue and bitter tonic in low doses. Root bark is deobstruent, resolvent, alexipharmic, astringent, tonic and antiperiodic, also emetic, anthelmintic, in large doses narcotic. (Azam *et al.* 2013, Sharma and Paul 2013).

Medicinal Uses (Traditional/Folklore): Traditionally, the plant is used as anthelmintic, antilithic diuretic, emmenagouge, astringent and stomachic. An aqueous extract of plant reduces the intensity of asthmatic attacks. Leaves are used in anaemia and measles, jaundice, treat malaria and to expel parasitic worms. The leaves are used externally to treat skin conditions. Leaves are applied in the form of poultice to relieve nervous headache and to cure the eruption on the scalp. Leaf juice is employed in hysteria, and used internally and externally in leprosy, scrofula and other skin diseases.

Flowers are applied as a poultice to relieve nervous headache. They are valuable in eruptive skin diseases and for killing lice. An herbal cream containing a methanolic extract of *M. azedarach* flowers was found potent against bacterial skin diseases like cellulitis, pustules, pyogenic infections, etc. in children. The results obtained are comparable to those with neomycin. In Manipur, leaves and flowers are used as poultice in nervous headache. Fruits are sweetish, and though said by some to be poisonous, is eaten by children without inconvenience. Paste of Fresh fruits and leaves is given to cattle twice a day for 2-3 days to treat gas trouble and indigestion. Ground the bark in rock sugar keep it for one day and take next day in the early morning to empty stomach against typhoid fever. Seeds are useful in helminthiasis, typhoid fever, pain in the pelvic region, uropathy, vitiated conditions of *vata* and scrofula. They are prescribed in rheumatism, urinary disorders and piles; oil obtained from seeds is applied locally in skin diseases. Seed-oil is used in rheumatism.

Roots are useful in sciatica, lumbago, headache, leprosy, leucoderma, skin diseases, wounds, ulcers, piles, worm infestation, cough, asthma, amenorrhoea, dysmenorrhoea, diabetes, alleviating sciatica, abnormal urethral discharge, chronic and intermittent fevers, vomiting, and post labour pain in uterus. (Azam *et al.* 2013, Rahul 2013, Sharma and Paul 2013, Shaheen *et al.* 2014; http://www.flowersofindia.net).

15. RHAMNACEAE

Rhamnaceae – Buckthorn family

The family includes 53 genera and more than 839 species mostly trees, shrubs and some vines. It has worldwide distribution, but are more common in the subtropical and tropical region and warm temperate region. The family was represented by two species belonging to a single genus in Chithara Village.

15.1 *Ziziphus jujuba* Mill.

(Syn. *Rhamnus jujuba* L.; *Ziziphus mauritiana* Lam.)

Common name: Indian Jujube, Indian Plum.

Vernacular name: *Hindi:* Ber (बेर).

Habit: A small tree.

Habitat: Cultivated for its edible fruits and also found wild.

Distribution: Its precise natural distribution is uncertain due to extensive cultivation, but is thought to be in southern Asia, between Lebanon, northern India, and southern and central China, and possibly also southeastern Europe though more likely introduced there. Distributed throughout India, in dry deciduous forests up to 1,500 m.

Fruiting branch of tree

Flowering and Fruiting Time: September-February.

Abaxial side of leaf

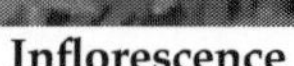

Inflorescence

Fruits

Taxonomic Description: A small, armed tree with spreading and drooping branches. Bark often dull-black or grey. Leaves alternate, petiolate, variable, 2-6 cm long, yellowish-green, oblong-elliptic or ovate-elliptic, basally 3-nerved, closely serrulate at margin, acute or obtuse at apex, rounded at base, shining glabrous above and white tomentose beneath, base usually oblique. Prickles solitary or in paired. Flowers shortly pedicellate, 4-5 mm across, greenish yellow, borne in short axillary, tomentose cymes or fascicles. Calyx lobes-5; sepals ovate, deltoid. Petals-5, obovate, concave. Stamens-5, nearly equal to the petals. Ovary 2-celled, styles 2-cleft. Fruits (drupes) oblong-ovoid, 1.5-2.5 cm long, reddish-purple when ripe.

Chemical Constituents: *Z. jujuba* contains flavonoids, saponins, tannins, vitamin A, vitamin B, sugars, mucilage, calcium, phosphate and iron. The pulp contains moisture, protein, fat, carbohydrate, calcium, phosphorus, iron, carotene, thiamine, riboflavin, Vitamin C. Ground seeds on extraction with petroleum ether gave 33% of bright yellowish oil. Fatty acid composition of the oil contains oleic acid- 71.7%, linoleic acid- 15%. *Z. jujuba* fruits are very rich in vitamins C, B_1 and B_2.

Several types of chemical components including triterpenic acids, flavonoids, cerebrosides, amino acids, phenolic acids, microelements, vitamins, total sugars and nucleosides have been isolated and identified during the past decades. (Preeti and Tripathi 2014, Preeti *et al.* 2014, Wang *et al.* 2014a).

Medicinal Properties: The roots are bitter, cooling, anodyne and tonic and are useful in vitiated condition of *pitta,* fever, wounds and ulcers and cephalalgia. The bark is astringent, constipating ant tonic, and is useful in dysentery, diarrhoea, gingivitis and boils. The leaves are bitter, cooling, astringent, anthelmintic, diaphoretic and antipyretic, and are useful in stomatitis, ulorrhea, wounds, syphilitic ulcers, asthma, typhoid fever, diarrhoea and obesity. The fruits are sweet, cooling, anodyne, purgative, mucilaginous, pectoral, styptic, aphrodisiac, invigorating, depurative, appetizer and tonic; dried fruits are anodyne, anticancer, pectoral, refrigerant, sedative, stomachache, styptic and tonic Fruits are useful in vitiated conditions of *pitta,* burning sensation, hyperdipsia, consumption, vomiting, constipation, flatulence, dyspepsia, nausea, leprosy, skin diseases, pruritus, wounds and ulcers, haemorrhages and general debility. The seeds are acrid, sweet, astringent, soporific, hypnotic, narcotic, sedative, stomachache and tonic. They are useful in ophthalmopathy, dipsia, cough, asthma, wounds, and vitiated conditions of *pitta,* burning sensation, diarrhoea, vomiting, leucorrhoea, general debility and insomnia.

Modern pharmacological studies have shown that the fruit is commonly used for tranquilizing mind, enhancing the immunologic functions, inhibiting tumour growth, resisting fatigue, beautifying and nourishing face, resisting oxidation and aging. (Warrier *et al.* 2010, Wang *et al.* 2014a).

Medicinal Uses (Traditional/Folklore): Jujube is both a delicious fruit and an effective herbal remedy. It aids weight gain, improves muscular strength and increases stamina. In Chinese medicine, it is prescribed as a tonic to strengthen liver function. It functions as antidote, diuretic, emollient and expectorant. The leaves are astringent and febrifuge; also, said to promote hair growth. The leaves are applied as poultices and are helpful in liver troubles, asthma and fever. Leaf powder mixed with powdered leaf of *Ficus glomerata* Roxb. is applied locally in scorpion sting. Leaf juice 10 ml mixed with little salt is taken orally (3 times daily) in the treatment of dysentery. The leaves are used for hypoglycemic effects, as diuretic, emollient, expectorant, to promote hair growth, anticancer, sedative, blood purifier and in treatment of diarrhoea.

Fruits are used as liver tonic, as an antioxidant, hepatoprotective, for weight gain, increasing stamina and reported to have anticancer properties. The dried fruits are considered to purify the blood and aid digestion. They are used internally in the treatment of chronic fatigue, loss of appetite, diarrhoea, anaemia, irritability and hysteria. The seed is used internally in the treatment of palpitations, insomnia, nervous exhaustion, night sweats and excessive perspiration. The root is used in the treatment of dyspepsia. A decoction of the root has been used in the treatment of fevers. The root is made into a powder and applied to old wounds and ulcers. The plant is a folk remedy for anaemia, hypertonia, nephritis and nervous diseases. The fruits are applied on cuts and ulcers and also employed in pulmonary ailments and fevers. (Dolui *et al.* 2004, Sharma *et al.* 2017; http://www.pfaf.org, http://medplants.blogspot.in).

15.2 *Ziziphus nummularia* (Burm.f.) Wight & Arn.

(Syn. *Rhamnus nummularia* Burm.f.; *Ziziphus rotundifolia* Lam.)

Common name: Jhar Beri, Chani Bor.

Vernacular name: *Hindi*: Jhar Beri (झाड-बेरी), Jhar Ber (झाड़ बेर); *Sanskrit:* Sukshmaphala (सूक्ष्मफल), Bhukartaka (भूकर्तक), Karkandhu (कर्कन्धुः).

Habit: Busy shrub.

Habitat: Common along roadsides, railway lines, canal banks and waste places.

Distribution: *Z. nummularia* is distributed in India, Pakistan, Iraq and Iran. In India, it occurs in dry and arid zones, mostly in Punjab, Rajasthan and parts of Uttar Pradesh.

Habit

Flowering and Fruiting Time: July-November.

Taxonomic Description: A much-branched, busy shrub, 2-5 m high. Branchlets bifarious, flexuose, tomentose with paired stipular spines. Leaves alternate, shortly petioled, 1-3 x 0.5-2 cm, ovate to orbicular, margin serrate, apex acute to obtuse, 3-nerved from base, base rounded or oblique, abaxial surface densely tomentose. Flowers about 5 mm across, pedicellate, pale yellow, borne in dense axillary cymes. Sepals-5, ovate-triangular. Petals-5, obovate. Disk 10-lobed. Stamens-5. Ovary bicarpellary, styles-2. Fruits (drupes) globose, 0.8-1.2 cm across, green when young and reddish black when ripe.

Flowering twig **Abaxial leaf surface** **Fruits**

Chemical Constituents: Leaves contain ascorbic acid and the root bark tannin and cyclopeptide alkaloids, numularines A and B and C, mucronine D and amphibine H. The stem bark contains numularine M, R and S, besides frangufoline. The other chemicals present are nummularogenin, zizynummin, lapachol and polysaccharides, pectin composed of l-rhamnose, d-galacturonic acid, dgalactose, l-arabinose, peptide alkaloids, cyclopeptides, and flavonoides, saponins, and triterpenoides, fatty acides, ziziphin N, O, P & Q and dodeca-acetylprodelphinidin B3. (Bhutya 2011, Kumar *et al.* 2011d).

Medicinal Properties: Leaves are cooling, used for washing hairs, treating sore throat, bleeding gums, scabies and skin diseases. Fruits are cooling, astringent, appetizer, and stomachic, used for bilious affections. The seeds are good for eyes, seminal weakness, pains, tonic, and as aphrodisiac; used for asthma, bronchitis and burning sensation. Numularine R and S, has antibacterial activity against Gram-negative and Gram-positive bacteria. The alkaloids fraction, obtained from the bark, has significant analgesic activity. (Bhutya 2011).

Medicinal Uses (Traditional/Folklore): Roots of *Z. nummularia* are crushed along with the roots of *Butea monosperma* in equal amount and filtered. One teaspoon of the filtrate is taken orally with water thrice a day to cure bowel movement (stool). Extract of fruit powder kept overnight is given to ladies in the morning to increase oogenesis. Fruits are used as blood purifier. In India, dried fruits are used medicinally as astringent in bilious affliction and leaves to treat scabies and other skin diseases. In Jaisalmer district of Rajasthan, the plants are used in various folk medicines. The leaves are boiled with tea leaves and decoction taken to cure cold and cough. Paste of leaves is applied on skin to cure various diseases, including scabies and boils. The root extract is given orally to cure diarrhoea. Leaves and fruits are used by tribal for treating cold, diarrhoea, dysentery, indigestion, and inflammation of gums. (Muhammad and Khan 2008, Singh *et al.* 2010b, Kumar *et al.* 2011d, Nigam *et al.* 2013, Shaheen *et al.* 2014, Singh 2015).

16. VITACEAE

Vitaceae – Vine family

The family has 16 genera and 985 known species, including the grapevine and Virginia creeper. The family name is derived from the genus *Vitis*. Plants in this group are mostly woody climbers and few shrubs. Typical vines, several of these climbers have tendrils for support. Fruits are typical grape-like fleshy berries. Several species of Hawkmoths lay eggs on some of these plants. The family was represented by one species in Chithara flora.

16.1 *Cayratia trifolia* (L.) Domin

(Syn. *Cayratia carnosa* (Lam.) Gagnep.; *Vitis trifolia* L.)

Common Name: Bush Grape, Fox-Grape, Three-Leaved Wild Vine, Threeleaf Cayratia.

Vernacular Name: *Hindi*: Amalbel (अमलबेल), Ramchana (रामचना); *Sanskrit*: Amlavetasah (अम्लवेतस), Atyamlaparni (अत्यमलपर्णी).

Habit: An extensive climber.

Habitat: Often found climbing on herbaceous or woody plant along roadsides, railway lines and canal banks.

Distribution: It is native to India, Asia and Australia. It is widely distributed in tropical and subtropical Asia, Africa, India, Australia and Pacific islands where its wide-ranging medicinal values are well-documented both in folk medicine and pharmacological studies.

Flowering and Fruiting Time: August-October.

Habit

Inflorescence

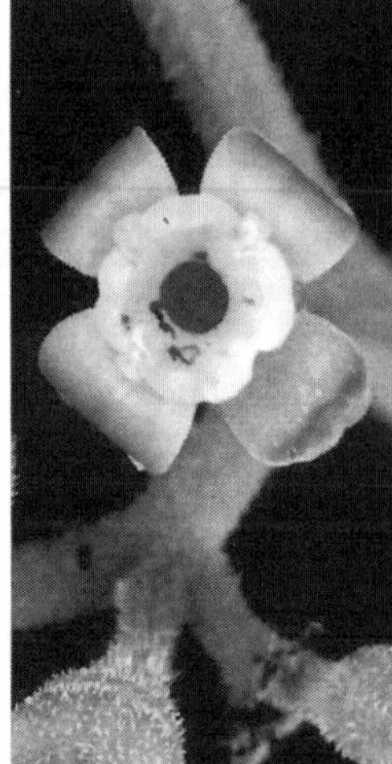

Flower

Fruits

Taxonomic Description: An extensive herbaceous climber, sometimes woody at base. Branches angular, young parts pubescent. Leaves 3-foliate, petioled; petiole up to 6 cm long; leaflets 3-9 x 2-7 cm long, elliptic, ovate or obovate, apex rounded or acute, margin crenate-serrate to sub-entire. Tendrils opposite the leaves, slender, wiry, 3-5 branched. Flowers greenish white, axillary in umbellate cymes, peduncle longer than the leaves. Calyx and corolla lobes-4, pubescent. Style distinct. Fruit a berry, globose, black when mature, 2-4 seeded. Seeds triangular.

Chemical Constituents: Whole plant of *C. trifolia* has been reported to contain yellow waxy oil, steroids/terpenoids, flavonoids, tannins upon preliminary phytochemical screening. Leaves contain stilbenes (piceid, reveratrol, viniferin, ampelopsin). Stem, leaves, roots are reported to possess hydrocyanic acid, delphinidin and several flavonoids such as cyanidin is reported in the leaves. The plant also contains kaempferol, myricetin, quercetin, triterpenes and epifriedelanol; seeds and fruits have cyanogenic compounds. Fruits contain calcium oxalate responsible for severe irritation in the mouth. (Kumar *et al.* 2011a, Kumar *et al.* 2012c).

Medicinal Properties: Whole plant is astringent, used as diuretic, in tumours, neuralgia and splenopathy. The bark extract shows the antiviral, antibacterial, antiprotozoal, hypoglycemic, anticancer and diuretic activity. Leaves are aphrodisiac, rubafacient, and anti-inflammatory. (Kumar *et al.* 2011a, Kumar *et al.* 2012c).

Medicinal Uses (Traditional/Folklore): All parts of *C. trifolia* are used as diuretic and astringent by tribes and is useful in treating tumours, neuralgia, splenopathy and leucorrhoea. Leaf, root and seeds are used as poultice to be applied on ulcers and boils. Hot decoction of leaves and root is used as diaphoretic and recommended in high fevers. Sap of stems and leaf juice are aphrodisiac. Leaves, which have rubifacient, property, are used to stop bleeding due to injuries. Leaf decoction or the juice of fresh leaves is used for the treatment of cure scurvy in the Philippines, to prevent head itch and dandruff in Java, to relieve inflammation and high fever in Thailand and Peninsular Malaysia while the young leaves are eaten as vegetable in Moluccas. Infusion of seeds mixed with the extract of tubers is taken orally to treat diabetes. Paste made from fresh leaves of *Gymnema sylvestris* and *C. trifolia* is applied locally for the treatment of eczema. The aqueous extract of *C. trifolia* leaf is reported to possess antimosquito and potent larvicidal properties against *Culex quinquefasciatus*. Paste of tubers is applied topically for treating snake bites and for early cure of wounds and edema. Extract of tuber along with infusion of seeds is taken orally to treat diabetes whereas powder made from tubers is taken orally with the milk for treating fractured bones. Root bark is found to be useful for muscular pain and as an antidote against snake bites; the stem is used as an aphrodisiac. Root is used in anemic conditions, stomachic diseases, and as an astringent; root paste topically applied as an antidote in snake bites, and also in complains of caruncles. The root, ground with black pepper, is applied to boils. Roots are grounded with black pepper and applied as poultice on boils; root paste is mixed with coconut oil and applied as decoction.

As a veterinary medicine, poultice of leaves is applied on the yoke sores of bullock and also used to cure swelling, injury and other infections. For controlling bullocks, climbers are wrapped around their neck and leaf poultice is applied to cure yoke sores. (Kumar *et al.* 2011a, Kumar *et al.* 2012c, Mishra *et al.* 2015e).

17. LEGUMINOSAE

Leguminosae – Legume, Pea, or Bean family

It is a large and economically important family of flowering plants including trees, shrubs, and perennial or annual herbaceous plants, which are easily recognized by their fruit (legume) and their compound, stipulated leaves. The family is widely distributed, and is the third-largest land plant family in terms of number of species, with about 946 genera and 24,505 known species, accounting to about 7 percent of flowering plant species. In study area, this family was represented by a large number of plant species (39) belonging to 24 genera.

17.1 *Abrus precatorius* L.

(Syn. *Abrus abrus* (L.) Wright; *Glycine abrus* L.)

Common Name: Licorice, Coral Bead Vine, Crab's Eye.

Vernacular Name: Ratti (रत्ती), Gunchi (गुंची).

Habit: Perennial climber.

Habitat: Generally found twining on the shrubs or trees along roadsides, forests and gardens.

Distribution: The plant is native to India and grows in tropical and subtropical areas of the world, including South Africa, China, Islands, West Indies, Brazil, *etc.* In India, the plant is found all throughout the plains from Himalaya down to Southern India and Sri Lanka.

Flowering and Fruiting Time: August-November.

Habit

Flowers in raceme

Open pods with seeds

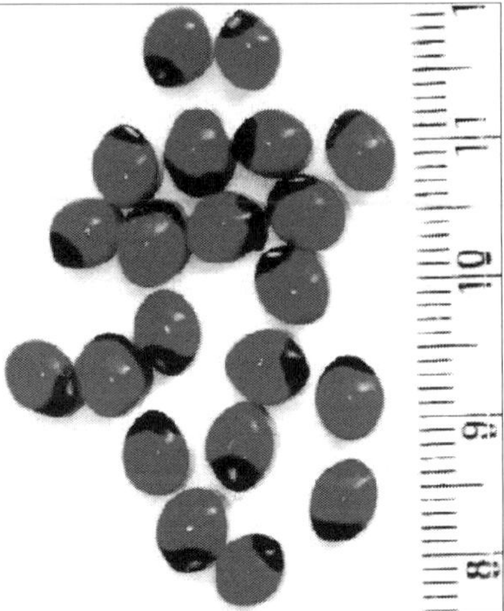

Seeds

Taxonomic Description: A perennial climber, deciduous. Leaves paripinnate; leaflets 10-18 pairs, opposite, 1-2.5 x 0.5-1 cm, oblong, membranous, base rounded, apex rounded-mucronate, margin entire, petiolule very short. Flowers whitish-pink, in dense axillary racemes. Calyx campanulate, 3-4 mm long, thinly silky. Corolla whitish-pink, nearly 1 cm long. Stamens-9, monadelphous. Pods oblong, 3-4 cm long, beaked, turgid and pubescent. Seeds 3-6, ovoid, bright red with a black spot at hilum.

Chemical Constituents: The plant contains several groups of secondary compounds, including alkaloids, steroids and other triterpenoids, isoflavanoquinones, anthocyanins, starch, tannin, protein, flavanoids, phenolic compound, fixed oil, amino acid and the flavones; also isolated are luteolin, abrectorin, orientin, isoorientin and desmethoxycentaureidin 7-0-rutinoside. Seeds are known to be poisonous and contain abrine. (Garaniya and Bapodra 2014).

Medicinal Properties: *A. precatorius* is potentially medicinally important having antidiabetic, antioxidative, neuroprotective, antiviral, neuromuscular, anticonvulsant, antiepileptic, immune-modulating, abortifacient, anti-implantation, anthelmintic, antidepression, memory enhancing, diuretic, antimicrobial, antiyeast, anti-inflammatory, antiarthritic, analgesic, anticancer, antifertility, antispermatogenic, antiestrogenic, antimalarial, antiallergic, antiasthmatic, anticataract, insecticidal, and antitoxicity properties. (Garaniya and Bapodra 2014; http://medplants.blogspot.in).

Medicinal Uses (Traditional/Folklore): The roots, leaves and seeds of *A. precatorius* are used for different medicinal purposes. The plant is traditionally used for the treatment of tetanus, scratches and sores and wounds caused by dogs, cats and mice and prevention of rabies; also used mixed with other ingredients to treat leucoderma. In Ayurveda, leaves of A. precatorius are used as laxative, expectorant and aphrodisiac and for treating urticaria, eczema, stomatitis, conjunctivitis, alopecia areata, migraine, lymphomas/leukaemia and dysmenorrhoea. The leaves of the plant are also used for curing fever, cough and cold and roots to treat jaundice and haemoglobinuria bile. For prevention of early graying of hair, a paste of leaves and seeds is regularly applied on the scalp. Paste of roots is used to cure abdominal pains, tumours and also used for effecting abortion. Root is chewed as a snake bite remedy by tribes. Hot water extract of fresh root possesses antimalarial and anticonvulsant properties. Decoction prepared from dried roots is used for treating bronchitis and hepatitis. Various African tribes use powdered seeds as oral contraceptives. Abrus seeds are purgative, emetic, tonic, antiphlogistic, aphrodisiac and antiophthalmic and also used for treating tuberculosis and painful swellings and for curing worm infection.

In veterinary medicine, the plant is used in the treatment of fractures. Seeds are antimicrobial and have the potential of good insecticide.

Seeds, very beautiful to look at, have uniform weight of 1/10th of a gram, hence used as weighing unit, called 'Ratti' in ancient India by the goldsmiths. (Amiya 2006, Kaul and Dwivedi 2010; Garaniya and Bapodra 2014; http://medplants.blogspot.in).

17.2 *Acacia nilotica* (L.) Delile

(Syn. *Acacia arabica* (Lam.) Willd.)

Common Name: Gum Arabic Tree, Babool.

Vernacular Name: *Hindi:* Babool (बबूल), Kikar (कीकर); *Sanskrit:* Babbula (बब्बूला).

Habit: A small tree.

Habitat: Generally planted along the roadsides and also found wild in wastelands and dry areas.

Distribution: Pantropical and subtropical, abundant throughout Asia, Australia, Africa and America. In India, it occurs naturally in traditional rural and agro-pastoral systems.

Flowering and Fruiting Time: July-January.

Branch of tree

Flowers

Pods

Taxonomic Description: A small, spinous tree. Bark rough, fissured and dark-brown. Spines pointed, straight. Leaves bipinnately compound; pinnae 3-6 pairs, gland present at lower pair of pinnae; leaflets 10-25 pairs, opposite, linear, 3-6 mm long, entire at margin obtuse at apex. Flowers yellow, fragrant, in axillary, fascicled, globose heads. Calyx campanulate. Corolla tubular, about double the length of the calyx. Stamens numerous, filaments filiform. Pods 8-20 cm long, stalked, moniliform, flat, white-tomentose. Seeds compressed, sub-orbicular, blackish-brown.

Chemical Constituents: The plant is acknowledged to be rich in alkaloids, volatile essential oils, phenols and phenolic glycosides, resins, oleosins, steroids, and tannins. (Ali *et al.* 2012, Verma 2016a).

Medicinal Properties: Different parts of the plant such as the leaves, roots, seeds, bark, fruits, flowers, gum and immature pods have anticancer, antimutagenic, spasmogenic, vasoconstrictor, antibacterial, antihypertensive, antispasmodic, antipyretic, antiasthmatic, antioxidant, cytotoxic, antidiabetic, antiplatelet agregatory, antiplasmodial, molluscicidal and antifungal properties; the plants have inhibitory activity against Hepatitis C Virus (HCV) and human immunodeficiency virus (HIV)-I. (Ali *et al.* 2012; http://medplants.blogspot.in).

Medicinal Uses (Traditional/Folklore): The leaves are antipyretic, astringent, tonic to the liver and the brain, enriches the blood. The tender leaves infusion used as an astringent and remedy for diarrhoea and dysentery. The leaves are consumed internally in regular doses of 12 g each to provide relief in the case of diarrhoea. For treating tonsillitis, mouthwashes made from leaves are used in gargling. In the case of conjunctivitis, the leaves of the *Kikar* tree are ground to prepare a paste which is applied over the eyes and left on overnight. For treating Epiphora, which is a disorder in which the eyes continuously keep watering and tears from them keep flowing, the leaves are boiled in water and this water applied regularly over the eyes. Paste made using foliar parts of this tree are also very effective in treating erectile dysfunction, night fall and other disorders that affect the male genital organs. For curing eczema, *Kikar* gum mixed with the pulp of mango is applied on the skin. The gum is said to be very useful in diabetes mellitus. The bark of the tree is used to make a preparation which provides a lot of relief in treating the genital disorder such as leucorrhoea. Decoction of bark is largely used as an astringent douche in gonorrhoea, cystitis, vaginitis, leucorrhoea, prolapse of the uterus and piles. Powder of the root is also useful in treating leucorrhoea, wound healing and burning sensation. Flowers are used in diarrhoea, dysentery, premature ejaculation and seminorrhoea. The pods are used for impotency, urino-genital disorder and in treating persistent dry cough. The seeds and leaf extracts are used for general body vigour. (Singh *et al.* 2010b, Ali *et al.* 2012, Rahul 2013, Verma 2016a; http://www.alwaysayurveda.com, http://medplants.blogspot.in).

17.3 *Albizia lebbeck* (L.) Benth.

(Syn. *Acacia lebbeck* (L.) Willd.; *Mimosa lebbeck* L.)

Habit

Common Name: Lebbeck Tree, Woman's Tongues Tree.

Vernacular Name: *Hindi*: Siris (सिरिस); *Sanskrit*: Sirisah (सिरसः).

Habit: Large Tree.

Habitat: Commonly planted in gardens, along the roadsides and also found wild.

Distribution: Native to Indo-Malaya, New Guinea and Northern Australia.

Flowering and Fruiting Time: April-September.

Taxonomic Description: A large deciduous tree with greyish bark. Leaves 2-pinnate with a large gland at the base and 1 or more interpinnular ones, pinnae 2-3 pairs; leaflets 5-9 pairs, opposite, shortly petioled, oblong or obovate-oblong, entire, oblique at base, retuse or obtuse at tip, glabrescent beneath. Flowers greenish-white or yellowish-white, fragrant, axillary, solitary or 2-4 peduncled heads; pedicels slender, pubescent. Calyx 5-lobed, teeth short, deltoid. Corolla campanulate, 5-lobed; lobes ovate-lanceolate, hairy outside. Stamens numerous; filaments exserted, united at the base. Pods straw-coloured, 10-25 x 2-4 cm, strap-shaped, straight, and compressed. Seeds 4-12, sub-orbicular or somewhat reniform, brownish, flattened.

Leaf

Flowers

Chemical Constituents: Phytochemical analysis reveals the presence of alkaloids, carbohydrates, protein, amino acid, glycosides, saponin, tannins, and flavanoids. Air-dried seeds of *A. lebbeck* contain albigenin (I), a triterpene and triterpene saponin, lebbekanin-A (II). The leaves contain three non-protein amino acids, which are uncommon to any other plant. Mature leaves accumulate a large amount of keto acids including phosphoenol-pyruvate, glyoxylate, oxaloacetate and α-oxoglutarate. The oil of *A. lebbeck* contains 3.5% of unsaponifiable fraction. The principal components of oil are sterols (38%), methyl sterols (3%), triterpene alcohol (18%), tocopherol (22%) and total hydrocarbons and carotenoids (19%). The main sterol of oil is β-sitosterol (78%). The other ingredients are cycloeucalenol, 24- ethylophenol, cycloartenol, β-amyrin, and α-tocopherol. (Une *et al.* 2001, Faisal *et al.* 2012; http://shodhganga.inflibnet.ac.in).

Medicinal Properties: Lebbeck, used as a tonic, has antiseptic, astringent, antibacterial, antiallergic, antidermatosis, antidysenteric and antiprotozoal properties. It is useful in vitiated conditions of *Pitta* and *Kapha*, cough and catarrh, asthma, enlarged cervical glands, opthalmopathy, skin eruption, leprosy, leucoderma, sprains, wounds, ulcers, and neuralgia. (Une *et al.* 2001, Warrier *et al.* 2010, Faisal *et al.* 2012).

Medicinal Uses (Traditional/Folklore): Ayurveda recommends all parts of the plant for the treatment of snakebite. In Unani system of medicine, the leaves are useful in night blindness. Bark is used in bronchitis and bark and seeds for treating piles. Flowers are effective in cough, bronchitis, tropical pulmonary eosinophilia, and asthma. The hot aqueous stem bark decoction and its butanolic fraction was found effective in the antiallergenic activity in various models like anti-PCA or mast cell stabilizing activity. The seeds are claimed to be aphrodisiac and tonic to the brain; the oil is applied topically in leucoderma. The leaves and seeds are used in the treatment of eye problems such as ophthalmia. It is taken internally to treat diarrhoea, dysentery and piles. The flowers are applied locally to mature boils and alleviate skin eruptions.

Lebbeck, an astringent, is also used to treat boils, cough, eye infections, flu, gingivitis, lung and pectoral problems, and abdominal tumours. The bark is used medicinally to treat inflammation. (Une *et al.* 2001, Singh *et al.* 2010d, Warrier *et al.* 2010, Faisal *et al.* 2012, Rahul 2013; http://tropical.theferns.info).

17.4 *Alhagi maurorum* Medik.

(Syn. *Alhagi camelorum* Fisch.; *Alhagi pseudalhagi* (M.Bieb.) Fisch.; *Hedysarum alhagi* L.)

Common Name: Camelthorn-Bush, Persian Manna Plant.

Vernacular Name: *Hindi*: Bharbharra (भरभर्रा); *Sanskrit:* Bhatuashak (भटुआशाक).

Habit: Undershrub.

Habitat: Abundantly found along the roadsides and in fallow fields.

Distribution: Native to the region extending from the Mediterranean (Cyprus and Egypt in the East) to Mongolia and Russia in the west, south to India and Saudi Arabia but has been introduced to many other areas of the world, including Australia, southern Africa, and the western United States. In India, it is distributed in Rajasthan, Delhi, Uttar Pradesh, Gujarat, Maharashtra and Punjab.

Flowering and Fruiting Time: April-June.

Habit

Taxonomic Description: An erect, spinous, undershrub, 30-100 cm high. Spines axillary, 3-5 cm long, green with yellow tips. Leaves alternate, sub-sessile, stipulate, 0.5-1.5 cm long, obovate-oblong, margin entire, base cuneate, apex obtuse-mucronate. Flowers red, shortly pedicelled, 1-6 on spine at the top of the plant. Calyx 2.5-3 mm long, campanulate; teeth very

Flowering twig

short, triangular. Corolla red to purple, 7-9 mm long. Stamens 9+1 (diadelphous). Pods constricted between the seeds, 2-4 cm long, somewhat falcate.

Chemical Constituents: Plants contain tannins, flavonoids, unsaturated sterols/triterpenes, carbohydrates, lactones, proteins/amino acids and saponins. Other phytochemicals include β-phenethylamine, N-methyl-β-phenethylamine, Nmethyltyramine, hordenine, 3,4-dihydroxy-β-phenethyltrimethylammonium hydroxide, 3-methoxy-4-hydroxy-β-phenethyltrimethylammonium hydroxide, N-methylmescaline and salsolidine. Flavonoid glycosides kaempferol, chrysoeriol, isorhamnetin, chrysoeriol-7-O-xylosoid, kaempferol-3-galactorhamnoside, and isorhamnetin 3-O-β-D-apio-furanosyl (1-2) β-D-galactopyranoside have been isolated from the ethanol extract of aerial parts. The plant extraction with benzene yield 24-alkyl sterols, and two new flavanone glycosides, alhagitin and alhagidin. The stem has seven β-phenethylamines and one tetrahydroisoquinoline alkaloid. Choline and traces of betaine were also obtained from the stem. (Srivastava *et al.* 2014a, Suthar *et al.* 2016).

Medicinal Properties: The plants are used as a medicinal herb for its gastroprotective, diaphoretic, diuretic, expectorant, laxative, antidiarrhoeal and antiseptic properties, and in the treatment of rheumatism and haemorrhoids. (Ahmad *et al.* 2015).

Medicinal Uses (Traditional/Folklore): Plants are used in folk medicine as a treatment for glandular tumours, nasal polyps, and ailments related to the bile ducts. It is used as laxative, diuretic and expectorant in Rajasthan. Leaves are smoked in the treatment of asthma in Mount Abu, Rajasthan. Crushed flowers along with sugar are taken orally to cure bleeding piles in the Shekhawati region of Rajasthan. In Palestine area of West Bank, Israel, the plant is used to cure problems of urinary systems and stones. In Golan Heights and the West Bank region of Israel, the decoction of roots is consumed orally to cure kidney stones and diarrhoea. People of Kalat and Khuzdar regions of Balochistan, Pakistan use powder of flowers ground in sugar for improvement of eyesight, powder of dry flowers in stomachache and water extract of roots for liver complaint. (Shaheen *et al.* 2014, Srivastava *et al.* 2014a, Ahmad *et al.* 2015, Suthar *et al.* 2016).

17.5 *Alysicarpus monilifer* (L.) DC.

(Syn. *Hedysarum moniliferum* L.)

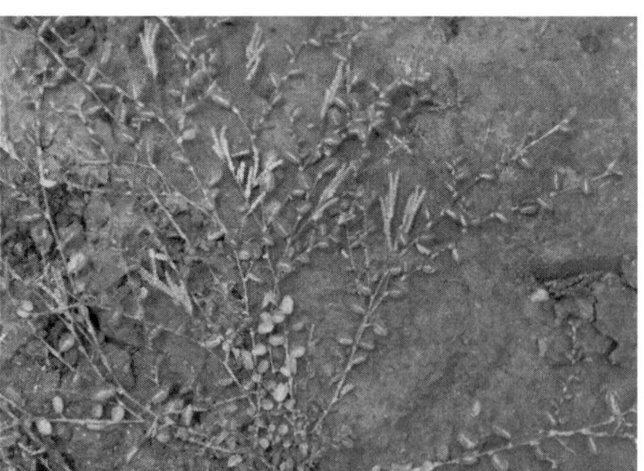
Habit

Common Name: Necklace-Pod Alyce Clover.

Vernacular Name: Jhuhi Ghas (झूहीं घास), Amera (अमेरा), Chatta Ki Ghas (छत्ता की घास).

Habit: Annual prostrate herb.

Habitat: Appears during the rains in grassy places, rocky hillocks and wastelands.

Distribution: Native to Africa and Asia. In India, it is distributed throughout the plains of Tamil Nadu, Jammu, Maharashtra, Punjab, Gujarat, except Kutch and Bulsar, Madhya Pradesh and Uttar Pradesh.

Flowering and Fruiting Time: August-October.

Taxonomic Description: A spreading or prostrate annual herb. Stems much branched, densely clothed with spreading hairs. Leaves alternate, petiolate, 0.6-2.5 cm long, oblong, obtuse, often cordate at base. Stipules lanceolate, 2-5 mm long. Flowers small, shortly stalked in close erect racemes. Calyx covered with scattered hairs; sepals-5, very narrow, rigid. Corolla pink to violet in colour; petals-5, unequal. Pods up to 2 cm long, moniliform in shape, distinctly 2-7 jointed, densely appressed with hooked hairs.

Flowering twig

Pods

Chemical Constituents: It contains carbohydrates, phytosterols, triterpenoidal saponins, fixed oils, phenolics and tannins; also present are alkaloids, proteins, free amino acids, glycosides and flavonoids. (Karthikeyan *et al.* 2015).

Medicinal Properties: Anti-inflammatory, diuretic, antidiarrhoeal and anti-snake venom properties. (Karthikeyan *et al.* 2014, Gholami 2015, Karthikeyan *et al.* 2015).

Medicinal Uses (Traditional/Folklore): *A. monilifer* has been used in indigenous system of medicine as anti-inflammatory and in stomachache, and as an antidote to snake bite. It is also used to treat skin diseases and as a diuretic. The leaves are used in fever and jaundice, leucoderma, diarrhoea, and kidney stones. Whole plant is used to treat diarrhoea in Gujarat. Paste of whole plant is warmed and applied on boils (Patel *et al.* 2013, Karthikeyan *et al.* 2014, Gholami 2015, Karthikeyan *et al.* 2015).

17.6 *Alysicarpus vaginalis* (L.) DC.

(Syn. *Hedysarum vaginale* L.)

Habit

Common Name: Buffalo Clover, Buffalo-Bur, One-Leaf Clover, Alyce Clover.

Vernacular Name: None known.

Habit: Annual herb.

Habitat: Common in gardens, moist waste places.

Distribution: It is native to parts of Africa and Asia; it has been introduced to other continents, such as Australia and the Americas. Found throughout India in moist places.

Flowering and Fruiting Time: August-October.

Taxonomic Description: A diffuse or ascending, branched, annual herb. Leaves simple, entire, much variable in shape, ovate-suborbicular, linear, oblong-lanceolate, entire, truncate or cordate at base. Inflorescence racemes, nearly 10 cm long. Flowers small, arranged in lax racemes at the end of branches. Calyx 5-lobed, as long as first joint of the pod. Corolla 5-lobed, tinged with pink or purple. Stamens diadelphous. Pods about 2 cm long, 6-8 jointed, slightly hairy.

Flowering twig

Flowers

Pods

Chemical Constituents: The phytochemical tests confirmed the presence of alkaloids, flavonoids, steroid, triterpenoids and phenolic substances in the different extracts of the plant. (Tapadia *et al.* 2017).

Medicinal Properties: Diuretic, purgative, anticancer, expectorant and antimicrobial. (Tapadia *et al.* 2017).

Medicinal Uses (Traditional/Folklore): In India, the roots of *A. vaginalis* are used traditionally by the tribal for diuretics, leprosy, pulmonary troubles and back pain. The leaf juice is used for the improvement of eye sight and earache. The entire plant is used for the treatment of renal calculi-kidney stone. In Uttarakhand, gujjars use the leaf paste as external application for treating skin allergy. The plant is also used in Southeast Asia as an Ayurvedic medicine. In folk medicine a decoction of roots has been reportedly used to treat cough in Java, while in Vietnam an infusion of powdered seeds is used against dysentery and colics. In China, the whole plant is used medicinally for treating wounds and bone fractures. In Nigeria, the plant has been used in indigenous system of medicine for the treatment of leprosy, cough, stomach pain, and also as a purgative and expectorant. (Tirkey 2006, Tapadia *et al.* 2017; https://www.cabi.org).

17.7 *Butea monosperma* (Lam.) Taub.

(Syn. *Butea frondosa* Willd.; *Erythrina monosperma* Lam.)

Common Name: Flame of the Forest, Bastard Teak.

Vernacular Name: *Hindi:* Palash (पलाश), Dhak (ढाक), Tesu (टेसू); *Sanskrit*: Palasha (पलाशा).

Habit: A deciduous tree.

Habitat: Deciduous forests, wild along roadsides, railway lines, canals as well as planted.

Distribution: Native to tropical and sub-tropical parts of the Indian Subcontinent and Southeast Asia, ranging across India, Bangladesh, Nepal, Sri Lanka, Myanmar, Thailand, Laos, Cambodia, Vietnam, Malaysia and western Indonesia. The plant is distributed throughout India, in deciduous forests up to 1,200 m elevation.

Upper branch of tree

Flowering and Fruiting Time: March-May.

Taxonomic Description: A small to medium-sized, deciduous tree with greyish-black bark. Branches irregular, crooked. Leaves alternate, tri-foliate, glabrous above and silky-tomentose beneath when mature, long-petiolate; stipules caducous; terminal leaflet obovate, rhomboid, 10-15 cm long, about as long as broad, margin entire, lateral leaflets oblong-ovate, base oblique, apex obtuse. Flowers bright orange-red, 5-8 cm long, borne in axillary and terminal, densely fascicled racemes. Bracts and bracteoles small. Calyx campanulate, about 1-1.3 cm long; lobes-5, deltoid, velvety. Stamens-10, diadelphous. Ovary shortly stalked, style curved, stigma capitate. Pods silky-tomentose, oblong, up to 20 cm long and 4 cm wide. Seeds orbicular, reniform, compressed, reddish-brown.

Tri-foliate leaf

Inflorescence

Chemical Constituents: Plants contain flavonoids, glucosides, butin, butrin, isobutrin, palastrin; flowers have flavonoids, butein, butrin, isobutrin, plastron, coreipsin and isocoreipsin5-6, monospermoside, sulphurein, chalcones. Seeds contain palasonin and seed oil which has d-Iactone of n-heneicosanoic acid, monospermine, new phytolectin. The compounds isolated from stem bark are stigmasterol, stigmasterol-βD-glucopyranoside, nonacosanoic acid, 3 α- hydxoxyeuph-25-ene and 2, 14-dihydroxy-11, 12-dimethyl-8-oxo-octadec-11- encyclohexane. (Chokchaisiri *et al.* 2009, Rajput *et al.* 2011, Sapkale *et al.* 2013).

Medicinal Properties: The bark is acrid, bitter, astringent, thermogenic, emollient, and aphrodisiac, appetizer, digestive, constipating, anthelmintic and tonic. The flowers have anticonvulsive, antihepatotoxic, anti-implantation, hypoglycaemic, astringent, diuretic, aphrodisiac properties and tonic. The stem bark possesses antifungal property. The gum is powerful astringent. The plant is useful in vitiated conditions of *Pitta and Kapha*, anorexia, dyspepsia, diarrhoea, dysentery, haemorrhoids, intestinal worms, bone fractures, rectal diseases, dysmenorrhoea, gonorrhoea, hepatopathy, ulcers, tumours, hydrocele and diabetes. (Kasture *et al.* 2000, Warrier *et al.* 2010).

Medicinal Uses (Traditional/Folklore): Almost all parts of the plant (flowers, leaves, gum obtained from the plant, seeds and the oil that can be extracted from the flowers and seeds) are used for medicinal purposes. The flowers of the tree are rich in glucosides, butin, neteroside and butrin whereas the seeds are a rich source of moodooga oil or kino- tree oil which is a fixed oil and is yellow in colour. This oil has various medicinal properties. The gum obtained from the tree is rich in gallic acid and tannic acid and is referred to as Bengal kino gum or simply Butea gum. The leaves are used as ingredients of tonics and aphrodisiacs and are also helpful in arresting bleeding or secretion. The flowers are widely used in the treatment of hepatic disorders, viral hepatitis, and diarrhoea. The roots are useful for treating night blindness, filariasis, piles, ulcers and tumours.

In Ayurveda, plant parts of *B. monosperma* are used for different ailments: *Bark*: useful against snake venom, wounds, indigestion, gastroenteritis, fever, tuberculosis; *Gum*: astringent, beneficial to children and women; *Leaf:* astringent, sex stimulant, useful in intestinal worms, dyspepsia, piles, menorrhagia, pimples, wounds in mouth/throat; *Flower*: diuretic, sex stimulant, helps menstruation, useful in gastroenteritis. In Siddha, flower-juice is used in preparation of the medicine Murukkam; *Seed*: useful against intestinal worms. In Unani, ingredient of the medicine called 'Dhak (tesu)' and 'Samaghke Dhak' are prepared from *B. monosperma.*

In folklore medicines different plant parts are used for various ailments: leaves are used for delirium (Khasi & Garo tribe), boils (Varanasi), skin diseases (Chandrapur, Maharashtra), Boil, inflammation (Garhwal, Uttarakhand) and Vermifuge (Sagar, Madhya Pradesh); flowers are used against diarrhoea, dysentery, pimples (Garhwal, Uttarakhand), eye complaints (Ratan Mahal Hills, Nilagiri, Tamil Nadu), aphrodisiac, astringent, diuretic (Delhi), dysentery (Korku), diabetes, diarrhoea, piles (Sagar) and antifertility (Kannanore, Kerala), in dog bite, urinary complaints (Khandla tribes) ; seeds are used for ascaris, contraceptive (Mayurbhanjan), anthelmintic (Delhi), vermifuge (Varanasi), abortifacient (Bhat tribes), anthelmintic, cooling agent (Garhwal); bark for dysentery (Mirzapur), bone fracture (Bhoxa tribes), diarrhoea (Dang tribes), as haemostatic, in wounds (Nilgiri); roots in the treatment of tuberculosis (Santhal Pargana), dog bite (Jhabua tribe) and gum for diarrhoea, dysentery, tonic (Shiwaliks), piles, urinary complaints (Bhoxa), diuretic (Tharu), astringent (Delhi), and diarrhoea (East Godawari).

Flowers yield a yellow dye of little permanency. Flowers are eaten as vegetables by tribes of Manbhum and Hazaribagh Districts of Bihar while fruits by Garhwalis. Tree is sacred to the Hindus and Buddhists. Flower is an essential item of Saraswati Puja. (Singh *et al.* 2010d, Rahul 2013, Dutta 2017; http://www.alwaysayurveda.com, http://www.bsienvis.nic.in, http://medplants.blogspot.in).

17.8 *Cajanus scarabaeoides* (L.) Thouars

(Syn. *Atylosia scarabaeoides* (L.) Benth.)

Habit

Common Name: Showy Pigeonpea, Peanut Grass, Tropical Clover.

Vernacular Name: Jangal Tor (जंगल तोर), Ban Kulthi (बन कुल्थी), Ban Kultha (बन कुलथा).

Habit: Trailing herb or climber; annual or perennial.

Habitat: Often twining on herbs and bushes along roadsides and canal banks.

Distribution: This is an extremely widely distributed species found naturally in both temperate and tropical zones and native to much of South-East Asia, China, Japan, India, Pakistan, Papua New Guinea and Australia. It is also introduced in coastal West Africa and is considered a weed of open grasslands in Pacific islands.

Flowering and Fruiting Time: August-October.

Taxonomic Description: A slender climber. Stem twining, densely brown-pubescent, angular. Leaves trifoliate, pubescent, gland dotted beneath; leaflets 4-6 cm long, elliptic, obovate, basally 3-nerved, entire, obtuse or sub-acute at apex, lateral leaflets oblique. Flowers 2-6, axillary, arranged on short peduncles. Calyx campanulate, silky-tomentose, lobes linear-lanceolate. Corolla yellow, 1-1.5 cm long. Stamens 9+1. Pods oblong, 2-3 cm long, moniliform, densely brown-tomentose. Seeds 2-6, brownish.

Trifoliate leaf

Pods

Chemical Constituents: The plant contains d-pinitol, vitexin, atyloside, alkaloids, glycosides, steroids, flavonoids, terpenoids, amino acids and peptides. (Poongothai and Sripathi 2013; https://www.bimbima.com/ayurveda/medicinal-plant-wild-kulthi/457/)

Medicinal Properties: The plants possess anthelmintic, antidiabetic, antidiarrhoel, fungicidal, anti-rheumatic, antipyretic properties. The essential oil obtained from the leaves possess fungicidal property. (Panda 2002, Poongothai and Sripathi 2013, Dey *et al.* 2017; https://www.bimbima.com/ayurveda/medicinal-plant-wild-kulthi/457/)

Medicinal Uses (Traditional/Folklore): The whole plant is used in Ayurveda for the cure of various ailments such as rheumatism, fever, pains sores, diabetes, anaemia, cholera and dysentery. Whole plant decoction mixed with honey is given orally to women after delivery as tonic. As a traditional medicine, in Orissa and Chota Nagpur, the plant decoction is given after delivery for many days as tonic and detoxification of the body. In addition, the leaves of the plant species have been used to improve indigestion in traditional medicines as well as limit the excessive production of urine. In case of swelling in legs and body, leaf paste is applied topically as well as taken orally. Fresh leaf paste is applied externally in rheumatism. Paste of fresh stem and leaves is applied on sores and venereal diseases. Powder of roots is taken internally for indigestion, stomachache and abdominal gas. Crushed root extract is used as ear drops for treating deafness. The Kulthi dal is also eaten for reducing pain. The paste of roots or whole plant mixed with coconut oil, is applied on scalp to cure hair fall. As a veterinary medicine the plant is given to treat diarrhoea and dysentery in cattle to be effective in reducing diarrhoea in cattle. A decoction of plant is given to cure foot and mouth disease as well as tongue and mouth sores in cattle. (Panda 2002, Poongothai and Sripathi 2013, Dey *et al.* 2017; https://www.bimbima.com/ayurveda/medicinal-plant-wild-kulthi/457/).

17.9 *Crotalaria medicaginea* Lam.

(Syn. *Crotalaria foliosa* Willd.)

Common Name: Trefoil Rattlepod.

Vernacular Name: Gulali (गुलाली), Adbaumethi (अदबउमेठी).

Habit: An annual or perennial herb.

Habitat: Common in wastelands, fallow fields, gardens.

Distribution: South Asia to Australia, US.

Flowering and Fruiting Time: June-October.

Taxonomic Description: An erect, much branched herb, 30-80 cm tall. Root tap-root, turmeric in colour. Leaves tri-foliate. Leaflets little longer than the petiole, up to 3 cm long, entire, oblanceolate, obovate, obtuse to emarginated at apex, glabrous above, obscurely silky beneath. Stipule minute, deciduous. Racemes terminal, many flowered and leaf opposed, peduncles exceeding the leaves; bract linear, minute. Flowers yellow. Calyx appressed silky, teeth linear. Corolla twice in length of calyx, yellow in colour. Pods obliquely sub-globose, clothed with appressed silky hairs. Seeds 2, kidney-shaped.

Habit

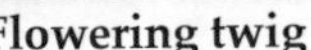

Flowering twig

Flowers

Fruits

Seeds inside pod

Chemical Constituents: Methyle ethers of supinidine and heliotridine have been isolated from the plant. (Sawhney and Atal 1970).

Medicinal Properties: Antidiabetic. (Pullaiah and Naidu 2003).

Medicinal Uses (Traditional/Folklore): For curing white discharge, one spoon full of leaf paste with one cup of milk is used in Chhattisgarh, India. Whole plant of *C. medicaginea* is used to treat diabetes. In Andhra Pradesh, leaf is used in the treatment of wounds and boils. (Pullaiah and Naidu 2003, Tirkey 2006, Maheswari *et al.* 2012).

17.10 *Crotalaria mysorensis* Roth

(Syn. *Crotalaria hirsuta* Roxb.)

Common Name: Rattlebox, Mysore Rattlepod.

Vernacular Name: None known.

Habit: Herb.

Habitat: Along the canal banks, grassy localities.

Distribution: Natural distribution is Indo-Malesia region. *C. mysorensis* is native to India, Nepal, Pakistan, Sri Lanka and the Malay Archipelago; found throughout India.

Flowering and Fruiting Time: September-October.

Taxonomic Description: A branched, erect herb, up to 60 cm high. Branches clothed with long spreading silky hairs. Leaves simple, shortly petiolate, entire, 3-8 cm long, linear, oblong, rounded at both ends, appressed with thinly silky hairs; stipules minute, linear, persistent. Racemes terminal, many flowered, up to 20 cm long; bracts leafy, lanceolate. Calyx lobes clothed with long silky hairs, linear, lanceolate. Corolla yellow, almost equal to the calyx, not exserted. Pods glabrous, 1.5-3.5 cm long, oblong. Seeds numerous.

Habit

Leaves

Pod

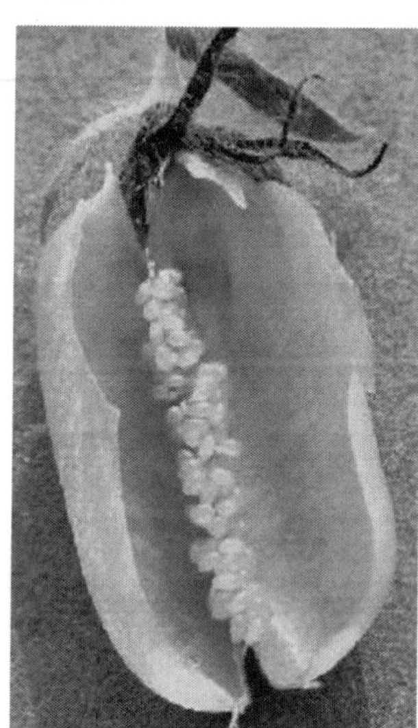

Seeds inside pod

Chemical Constituents: The plant contains alkaloides, benzopyrenoides-scopolatin, steroides, and β-sitosterol. (Sawhney and Atal 1968).

Medicinal Properties: Antiepileptic. (http://pankajinprintmedia.blogspot.in/2012/12/herbal-remedies-useful-with-crotalaria.html).

Medicinal Uses (Traditional/Folklore): Plants are used in epilepsy and associated troubles. (http://pankajinprintmedia.blogspot.in/2012/12/herbal-remedies-useful-with-crotalaria.html).

17.11 *Dalbergia sissoo* DC.

(Syn. *Amerimnon sissoo* (Roxb.) Kuntze)

Habit

Common Name: Indian Rosewood, North Indian Rosewood, Sisu.

Vernacular Name: Shisham (शीशम), Sisu (सिसु).

Habit: A medium-sized or large tree.

Habitat: Commonly planted along roadsides and also seen growing wild in waste places.

Distribution: *D. sissoo* is native to the Indian Subcontinent and southern Iran; it is state tree of Punjab (India). Found throughout the sub-Himalayan tracts up to 1,500 m, also cultivated in Punjab, Uttar Pradesh, Bengal and Assam. Also reported to occur in Afghanistan, Bangladesh, Bhutan, Malaysia, Pakistan, Cameroon, Cyprus, Ethiopia, Indonesia, Iraq, Israel, Kenya, Mauritius, Nigeria, Sudan, Tanzania, Thailand, Togo, US, and Zimbabwe.

Flowering and Fruiting Time: March-July.

Taxonomic Description: A medium-sized or large, deciduous tree with greyish-brown bark. Leaves alternate; leaflets 3-5, glabrescent, sub-orbicular, 3-6 cm long, base oblique, apex acuminate, margin entire, petiolules 5-10 mm long, stipules caducous. Inflorescence axillary, panicle racemes. Flowers small, yellowish-white. Calyx 5-lobed, 5-6 mm long. Corolla lobes-5, yellowish-white. Stamens-9, monadelphous. Pods oblong, flat, 3-8 cm long, strap-shaped, pale brown when dry. Seeds 1-3, brownish, nearly 8 mm long, flattened.

Fruiting twig

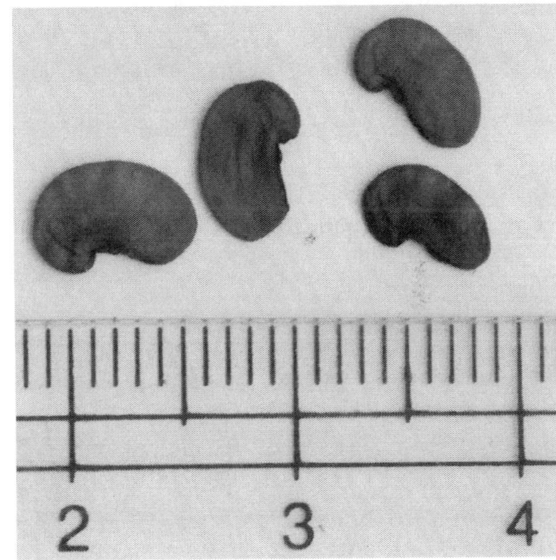

Seeds

Chemical constituents: Different parts of the plant contain various chemical constituents. *Leaves*: Isoflavone-O-glycoside. *Flowers*: Biochenin A, tectorigenin, 7, 4 dimethyle tectorigenin and 7-O- methyle tectorigenin. *Green pods*: Mesoinisitol, 7-O- methyle tectorigenin and 4'-rhamnoglucoside. *Mature pods*: Isocaviumin, tectorigenin, dalbergin, caviunin and tannins. *Stem bark*: Dalberginone, dalbergin, methyl dalbergin and dalbergichromene. *Heartwood:* Dalbergin, nordalberginones, dalbergichromene, fixid oil and essential oils. (Bharath *et al.* 2013, Bhattacharya *et al.* 2014).

Medicinal Properties: The roots are astringent and constipating, and are useful in diarrhoea and dysentery. The leaves are bitter, ophthalmic, styptic, digestive, constipating, anthelmintic, diuretic and stimulant, and useful in gonorrhoea, menorrhagia, excoriation, ophthalmopathy, dyspepsia, colic, diarrhoea, dysentery, vomiting, haemorrhoids and burning sensation. The bark and heart wood are astringent, acrid, bitter, thermogenic, anthelmintic, aphrodisiac, depurative, vulnerary, anti-inflammatory, emmenagogue, expectorant, antipyretic, skin diseases, leprosy, leucoderma, scabies, ulcers, dyspepsia, dysentery, syphilis, gastropathy, gout and intermittent fevers. (Warrier *et al.* 2010).

Medicinal Uses (Traditional/Folklore): In Ayurveda leaf juice is used for eye ailments, considering the wood and bark as abortifacient, anthelmintic, antipyretic, aphrodisiac, expectorant, and refrigerant. The wood and bark are used for anal disorders, blood diseases, burning sensations, dysentery, dyspepsia, leucoderma, and skin ailments. Different parts such as roots, bark, wood, leaves and seeds are being used as remedy in many diseases including skin diseases, blood diseases, syphilis, stomach problems, dysentery, nausea, eye and nose disorders, aphrodisiac, expectorant. Leaf extract has been used to treat sore throats, heart problems, dysentery, and syphilis. People use twigs of sissoo to clean their teeth. The plant is used in treatment of leprosy, jaundice, gonorrhoea and syphilis, etc. Decoction of leaves is useful in gonorrhoea. Root is astringent. Wood is alterative, useful in leprosy, boils, and eruptions and to allay vomiting. In Yunani medicines, the wood for blood disorders, burning sensations, eye and nose disorders, scabies, scalding urine, stomach problems, and syphilis. The alterative wood is used in India for boils, eruptions, leprosy and nausea.

The species has been associated with the life of traditional communities and a number of ethnomedicinal uses have been recorded. Many medicinal uses for its fresh leaves, dried bark, and wood rasping are reported from its native region. *D. sissoo* is reported to be a stimulant used in folk medicine and remedies. Various parts of *D. sissoo* are traditionally used in treating different diseases and are as follows. *Seeds*: Sissoo oil is used to treat blue itching, burning on the skin, and scabies. *Leaves*: Finely ground paste of 8-10 leaves of sissoo and 25g of palm candy taken in the morning alleviates profuse menstruation. 50-100 ml decoction of the leaves taken thrice in a day is useful in painful micturition and to cure boils and pimples. 10-15 ml juice (leaves) taken thrice in a day helps in eliminating pus in urine and in treating jaundice. The leaves warmed and tied on breast, and consuming the decoction of the leaves removes swelling of the breast. *Bark:* Powdered bark (3-6 g) or decoction of the leaves is helpful in gonorrhoea. Decoction of the bark and leaf is given in leprosy. Make a decoction of 10 g sissoo bark with 500 g of water and it should be boiled till the liquid reduces to half. Mix the juice of the bark and consume for forty days every morning which helps in leprosy. *Sissoo nectar*: Take 20 g of Sissoo nectar, 320 g water, and 160 g milk. Boil it till only milk remains and consume three times a day. This milk cures any type of fever.

As a veterinary medicine, In India and Nepal rural people use *D. sissoo* leaves to treat animals suffering from non-specific diarrhoea. Herbal preparation *of D. sissoo* and *Datura stramonium* with cow urine is used as a potent antiseptic preparation for prevention and treatment of chronic bacterial infections. (Singh *et al.* 2010d, Bharath *et al.* 2013, Bhattacharya *et al.* 2014).

17.12 *Desmodium gangeticum* (L.) DC.

(Syn. *Hedysarum gangeticum* L.)

Common Name: Sal Leaved Desmodium.

Vernacular Name: *Hindi:* Dirghamuli (दीर्घमूली), Shalaparni (शालपर्णी); *Sanskrit*: Vidarigandha, Shalaparni (शालपर्णी).

Habit: Perennial undershrub.

Habitat: Along roadsides, canal banks and railway lines.

Distribution: The plant is widely distributed in the tropics and subtropics from tropical Africa, East-Asia, China, Japan, Indian subcontinent, Myanmar, Thailand, Cambodia, Laos, Vietnam, Malaysia, Indonesia, Philippines up to Australia. The plant is distributed throughout India in dry forests up to 1,500 m elevation and as forest undergrowth in low elevations. It is also cultivated in the plains and in the lower Himalayas.

Flowering and Fruiting Time: August-October.

Taxonomic Description: A sub-erect, undershrub with angled and hairy stem. Leaves 5-13 x 2-6 cm, alternate, petioled, ovate-oblong, obtuse or sub-acute at apex, entire at margin, rounded at base, glabrous above and pubescent beneath; stipules scarious, linear-lanceolate. Inflorescence axillary or terminal racemes. Flowers violet or purplish-white; pedicels about 5 mm long, pubescent. Stamens 9+1. Pods oblong, slightly falcate, 1.5-2.5 cm long, 4-8 jointed, pubescent with hooked hairs.

Flowering and fruiting twig

Chemical Constituents: Plant contains alkaloids gangetin, gangetinin, desmodin. Leaflets contain pterocarpan, phytoalexin, and desmocarpin. Aerial parts contain 5-methoxy-N, N-dimethyltryptamine, N1 N-dimethyltryptamine, their N-oxides, N-rnethyltetrahydroharman, 6-rnethoxy-2-methyl-Beta-carbolinium cation; also known to contain indole-alkylamines and their derivatives and five tryptamine derivatives, Nb-Me-tetrahydroharman and 6-OMe-2-Me- ß-carbolinium cation. The root contains pterocarpanoids- gangetin, gangetinin, desmodin, N, N-dimethyl tryptamine, hypaphorine, hordenine, candicine, N-methyl tyramine and b-phenyl ethyl amine. Roots also contain several alkaloids, N, N-di-methyl-tryptamine and its N-oxide, N-methyltyramine, hypaphorine, candicine, hordenine and a ß-phenylethylamine base and several carboxylated and decarboxylated tryptamines and ß-alkylamine derivatives. Seeds contain ß-carboline alkaloid, indole-3-alkylamine, carbolines and five phospholipids. Seeds showed the presence of sugars, fatty oil and alkaloids. (Rastogi *et al.* 2011, Bhattacharjee *et al.* 2013, Ganjhu *et al.* 2014, Vedpal *et al.* 2016; http://medplants.blogspot.in).

Flowers

Pods

Medicinal Properties: The plant shows hypotensive activity. The root is bitter, anti-inflammatory, antipyretic, analgesic, aphrodisiac, thermogenic, nervine, cardio tonic, febrifuge, expectorant, constipating, diuretic, anthelmintic, laxative, cardiotonic, expectorant, astringent, anti-dysenteric, anti-diarrhoeal (used in post-natal care, chronic fever, cough, biliousness, vomiting), carminative, antiemetic, febrifuge, galactagogue and anti-catarrhal and debility; it is useful also in convulsions, fevers, oedema, kidney disorders and post-delivery complications. Root, bark and leaves are carminative and rejuvinative. Leaves are galactagogue. (Warrier *et al.* 2010, Rastogi *et al.* 2011, Ganjhu *et al.* 2014).

Medicinal Uses (Traditional/Folklore): *Desmodium* is one of the ten ingredients in the *Dasamula Kwatha* of Ayurvedic medicine. The herb is used in the preparation of Ayurvedic medicines such as: *Guru* (heavy to digest), *Chardijit* (useful in vomiting), *Jwarahara* (useful in fever), *Shwasahara* (useful in dyspnoea, Asthma, Bronchitis and Chronic respiratory disorders), *Atisarajit* (relieves dysentery and diarrhoea), *Doshatrayahara* (balances all the three Doshas), *Santapanashini* (relieves burning sensation). The Unani medicinal preparation "Arq dashmul" also contains the roots. It is considered a curative for leucorrhoea and for pains due to cold. In Siddha, it is used in *Curam, Cuvasa Noy, Kirumi, Mulam, Mekam, and Vali* medicinal preparations. A poultice of leaves is used for lumbago.

Roots are useful in vitiated conditions of *Vata,* anorexia, dyspepsia, haemorrhoids, dysentery, strangury, fever, gout, inflammations, cough, asthma, bronchitis, cardiopathy and debility. A decoction of the leaves is used against stones in the gall bladder, kidneys or bladder. The leaves are galactagogue; applied as a poultice to the head as a treatment for headache and for lumbago. A decoction of the root is employed to treat kidney problems, oedema, swellings, chronic fever, coughs, biliousness, diarrhoea and dysentery; or as a sedative in children. The roots are applied to the gums as a treatment for toothache. A decoction is used externally to clean wounds and ulcers. The whole plant is considered to be anthelmintic. Bark is used in diarrhoea and haemorrhages. (Rastogi *et al.* 2011, Bhattacharjee *et al.* 2013, Ganjhu *et al.* 2014, Vedpal *et al.* 2016; http://medplants.blogspot.in).

17.13 *Desmodium triflorum* (L.) DC.

(Syn. *Hedysarum triflorum* L.)

Common Name: Creeping Tick Trefoil, Tropical Trefoil, Three-Flower Beggarweed.

Vernacular Name: *Hindi:* Tipatiya (तिपतिया), Kudaliya (कुदालिया); *Sanskrit*: Hamsapadi (हंसपदी), Tripadi (त्रिपदी).

Habit: Annual herb.

Habitat: It grows in fields, open grasslands, along roadsides and in lawns.

Distribution: Throughout India as a weed up to 900 m.

Flowering and Fruiting Time: August-November.

Habit

Taxonomic Description: A small trailing herb, often rooting at nodes. Branches clothed with fine spreading hairs. Leaves trifoliate. Stipules minute, lance shaped, persistent. Leaflets obovate, truncate or emarginated at apex, glabrous above, pubescent beneath, more or less equal in length and width (less than 1 cm). Flowers axillary, pink or purple, 1-3 in the axils of the leaves. Calyx lobes lanceolate, covered with long whitish hairs. Corolla pink or purple. Pods about 1.5 cm long, joints 2-5, clothed with fine hairs. Seeds brown in colour.

Flowering twig

Flower

Pods

Chemical constituents: *D. triflorum* plant contains ursolic acid, vitexin, genistin, fucosterol and rare diholosylflavane, 2-0 Glucosylvitexin. Leaves contain alkanoid, β-phenethylamine (major alkaloid), indole-3-acetic acid, tyrumine, trigonelline, hypaphorine and choline. Root contains alkaloid including hypaphorine, which is the major alkaloid, and N, N-dimethyl tryptophan betaine and choline. (Gavalapu *et al.* 2013, Vedpal *et al.* 2016).

Medicinal Properties: It is acrid, cooling, sweet, expectorant and galactagogue, and is useful in vitiated conditions of *Pitta*, cough, bronchitis, wounds, abscess, sores, pruritus, erysipelas, dysentery, flatulence and burning sensation. Plant has astringent, laxative, diuretic, antipyretic, expectorant, antiseptic, expectorant, antispasmodic, sympathomimetic properties; used for central nervous system stimulation, curare-mimetic activity and on skin rashes, and for inducing sweat and promoting digestion. The roots are reputedly carminative, tonic and diuretic and used in bilious complaints. (Warrier *et al.* 2010).

Medicinal uses (Traditional/Folklore): Plant and its parts are used to cure various ailments in folklore and traditional medicines. A decoction is commonly used to treat diarrhoea and dysentery; and to quench thirst and as a mouthwash; the crushed plant, or a poultice of the leaves, is applied externally on wounds, ulcers, and for skin problems in general. The whole plant is used medicinally for inducing sweat and promoting digestion. The leaves are used in diarrhoea, convulsions and as a galactagogue. The fresh leaves of the plant are applied to wounds and abscesses that are usually difficult to heal. The paste is sometimes applied to sores and itch. The fresh juice of the plant is also recommended for use in dysentery and as a laxative. Dried powder of whole plant when taken on empty stomach is useful in curing bone fracture. Leaf paste in water is applied on forehead to bring down high fever. The leaves are ground with cow's milk; they are given daily in the morning. The main actions include antispasmodic, sympathomimetic, central nervous system stimulation, curare-mimetic activity and diuretic. The infusion of root is used to promote labour and treat vertigo. Roots are diuretic and used for cough, asthma, dysentery, hepatitis, red eye with inflammation, and lymph infection; roots are antitoxic anthelmintic action against *Ascaris lumbricoides*. Decoction of roots is used for colic and root juice dropped in eye to treat eye inflammation.

In Philippines, a decoction is used as mouth wash and as an expectorant. In Thailand, the whole plant is used as an antipyretic and to quench thirst. In Indonesia, Malaysia, Philippines, Laos and India, the plant in crushed form (or) a poultice of the leaves is externally applied on wounds, ulcers and for skin problems in general, apparently for its antiseptic properties. As an external use Bhil tribal people in India use the plant/ leaf paste to bring down unspecified high fever by applying once on the forehead. Used in Ayurveda and Sidha.; tender parts chewed to treat toothache; fresh juice of the plant given to children for coughs; plant juice used as ear drops to alleviate earache and also given as galactagogue; infusion of the leaves of *Hydrocotyle javanica*, mixed with plant of *D. triflorum*, leaves of *Kadsura scandens* and fruits of *Litsea cubeba*, drunk to treat measles, pox in small children. Leaves are used to treat diarrhoea and dysentery, convulsions, wounds and abscesses; fresh leaves used internally as a galactagogue, applied to wounds and abscesses; young leaves are chewed for toothache; a paste of bruised leaves applied to itches, nails and indolent sores; leaf juice applied for cuts, wounds and piles. (Gavalapu *et al.* 2013, Vedpal *et al.* 2016; http://medplants.blogspot.in).

17.14 *Indigofera astragalina* DC.

(Syn. None known)

Habit

Common Name: Silky Indigo.

Vernacular Name: Dagadia (दगडिया).

Habit: Annual herb.

Habitat: Moist deciduous forests, scrub jungles and in cultivated fields. Common in open waste places, along roadsides in the plains.

Distribution: Found in tropical America, tropical and South Africa, Bangladesh, Bhutan, India, Sri Lanka, Myanmar, China; naturalized in West Indies.

Flowering and Fruiting Time: September-November.

Taxonomic Description: A densely hairy undershrub with woody stem. Leaves imparipinnate; leaflets 5-11, opposite, 2-4 x 1.5-2.0 cm, obovate, oblong or elliptic, entire, mucronate at apex, pubescent on both the surfaces, petiolules 1.5-2 mm long, stipules lanceolate. Flowers crowded in long, erect racemes. Calyx densely brown-hirsute. Corolla pink. Pods about 2 cm long, hirsute. Seeds 3-6, black.

Leaf

Pods

Chemical Constituents: Plants are rich in organic and fatty acids, flavonoids such as rotenoids and coumarins, polyphenol, tannins, cardiac glycosides, saponins, alkaloids, resins and lignin (insoluble fibre). (Gafar and Itodo 2011).

Medicinal Properties: Plants are astringent, antidiarrhoeal and antimicrobial. *I. astragalina* extracts show activity against *Staphylococcus aureus* and *Escherichia coli.* (Gafar *et al.* 2010).

Medicinal Uses (Traditional/Folklore): Decoction of leaves is used in diarrhoea in Chhattisgarh by tribal. Roots are very effective for toothache and a decoction for cough; root powder is given for pain in the chest. (Gafar *et al.* 2010, Shukla *et al.* 2010; http://medplants.blogspot.in).

17.15 *Indigofera linifolia* (L.f.) Retz.

(Syn. *Hedysarum linifolium* L.f.)

Habit

Common Name: Narrowleaf Indigo.

Vernacular Name: *Hindi:* Ratnamala (रत्नमाला), Pandarphali (पाण्डरफली).

Habit: Annual herb.

Habitat: Common in open waste places, often mixed with grasses.

Distribution: It is found in America, Africa, South Asia and South-East Asia to China and Australia. It is found throughout India.

Flowering and Fruiting Time: May-October.

Taxonomic Description: A procumbent herb. Stem much branched, silvery-pubescent. Leaves up to 2 cm long, simple, alternate, linear, acute-mucronate at apex, sessile or shortly petioled, silvery above and beneath. Stipules minute, setaceous. Inflorescence raceme, about 3 cm long. Flowers 6-12 in axillary racemes. Calyx teeth setaceous, much longer than the tube. Corolla 2-3 times longer than the calyx, bright red. Pods minute, globose, apiculate, appressed with silvery white hairs. Seeds-1, globose, brown or black.

Flowering twig

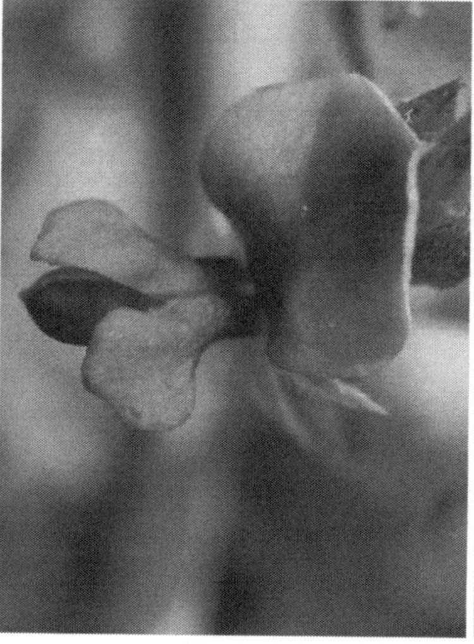

Flower

Pods

Chemical Constituents: The seeds of *I. linifolia* are rich in K, Ca, Mn and Cu and albumins and globulins constitute the major proportion of seed protein. Seed lipids contains a large proportion of unsaturated fatty acids with linoleic acid as the predominant one. Adequate levels of all the essential amino acid are also reported from seeds. (Siddhuraju *et al.* 1995)

Medicinal Properties: Antibacterial, antiscorbutic, diuretic, hepatoprotective, stimulant, vermifuge (Gritto *et al.* 2015; https://herbpathy.com/Uses-and-Benefits-of-Indigofera-Linifolia-Cid5796)

Medicinal Uses (Traditional/Folklore): The plant is useful in asthma and bronchitis. It is used to cure cuts and wounds in Delhi region. The plant is used for febrile eruptions and also as a vermifuge in Tamil Nadu. In Pakistan, the plant extract is useful in febrile eruptions and used in amenorrhea. (Vardhana 2008, Mahmood *et al.* 2011, Gritto *et al.* 2015).

17.16 *Indigofera linnaei* Ali

(Syn. *Hedysarum prostratum* L.; *Indigofera enneaphylla* L.)

Habit

Common Name: Birdsville Indigo, Bird's-foot Indigo.

Vernacular Name: *Hindi:* Leel (लील); *Sanskrit:* Vasuka (वासुका).

Habit: Annual or perennial, prostrate herb.

Habitat: Found growing in wastelands, roadsides, dry areas and gardens.

Distribution: Found in Africa, Malesia, Melanesia to New Guinea and Australia. Distributed throughout India.

Flowering and Fruiting Time: June-November.

Taxonomic Description: A much-branched, prostrate herb. Branchlets 20-30 cm long, densely hairy. Root-stock woody. Leaves 5-9 foliate, alternate, leaflets up to 1 cm long and 5 mm wide, sub-sessile, oblong-obovate or oblanceolate, margin entire, base cuneate, apex retuse-obtuse to mucronate, appressed hirsute on both the sides, stipules cuspidate. Flowers in dense axillary heads or small racemes, bracts membranous. Calyx campanulate, lobes-5, pubescent. Corolla lobes-5, bright red. Pods oblong, hairy, 4-5 mm long, usually with 2 seeds.

Chemical Constituents: Three nitropropanoyl esters of glucose namely 1,2,6-tri-*O*-(3-nitropropanoyl)-β-D-glucopyranose, 2,3,4,6-tetra-O-(3-nitropropanoyl)-α-D-glucopyranose and 3,4,6-tri-*O*-(3-nitropropanoyl)-α-D-glucopyranose have been isolated from the aerial parts of *I. linnaei.* (Kumar *et al.* 2011b).

Hirsute leaves

Fruiting twig

Medicinal Properties: The plants have antinociceptive, antitumour, cytotoxic, anti-inflammatory, antimicrobial, antioxidant, antidyslipidemic, and antiscorbutic activities. (Vardhana 2008, Kumar *et al.* 2011b, Sandhyavali *et al.* 2014).

Medicinal Uses (Traditional/Folklore): The plant is used in traditional system of medicine in treating ulcer, solid tumour, insect stings, snake bites and epilepsy. The juice of the plant is used as antiscorbutic and diuretic, for burns and epilepsy. It has long been used by the ethnic people and native medical practitioners to treat rheumatism, arthritis, inflammation, tumour and liver diseases. The leaf juice is given in doses of 10-12 ml along with honey twice daily in jaundice, inflammation of liver, etc. The leaves are useful in treating hydrophobia. The extract of plant is given for treating epilepsy and in neuropathy. In Tamil Nadu, equal quantities of leaves and roots are made in to a paste and one teaspoon of it is administered for treating asthma. (Kumar *et al.* 2011b, Sandhyavali *et al.* 2014).

17.17 *Indigofera tinctoria* L.

(Syn. *Indigofera indica* Lam.)

Common Name: True Indigo, Common Indigo.

Vernacular Name: *Hindi:* Neelini (नीलिनी), Neel (नील); *Sanskrit:* Neelika (नीलिका), Nilapushpa (नीलपुष्पा).

Habit: Undershrub.

Habitat: Common along the roadsides, railway lines and waste places.

Distribution: Widespread across tropical regions around the globe. Throughout India, usually between 0 and 300 m altitude; grows along forest margin, in secondary forests, grassy fields and sandy coasts and along roadsides and riverbanks.

Flowering and Fruiting Time: Almost throughout the year.

Habit

Taxonomic Description: An undershrub, up to 80 cm tall. Stem and branches appressed pubescent. Leaves imparipinnate; leaflets 5-13, entire, petiolate, elliptic, oblong to obovate, 1-4 x 0.5-1.5 cm, base cuneate, apex emarginate. Stipules minute. Inflorescence raceme. Flower pinkish red, pedicel 0.5 cm long. Calyx teeth triangular, 5-lobed, silvery pubescent. Corolla exerted. Stamens 10. Pods linear, pubescent, 3-4 cm long, slightly curved. Seeds 5-12, rectangular, black.

Flowering twig

Flowers

Pods

Chemical Constituents: The plant contains flavonoids, alkaloids, glycosides, and terpenoids. A galactomannan from seed and apigenin, indigotin, kaempferol, tutetion, rotenoids deguelin, dehydrodeguelin, rotenol, rotenone, tephrosin and sumatrol have been isolated from various parts of the plant. (Verma and Suresh 2002; https://easyayurveda.com).

Medicinal Properties: Roots, stem and leaves are bitter, thermogenic, laxative, antitoxic, expectorant, anthelmintic, tonic, diuretic, and are useful for promoting the growth of hair, and in gastropathy, splenomegaly, cephalalgia, cardiopathy, chronic bronchitis and asthma, ulcers and skin diseases. The juice expressed from leaves is useful in the treatment of hydrophobia. An extract of the plant is good for epilepsy and neuropathy. (Warrier *et al.* 2010).

Medicinal Uses (Traditional/Folklore): In Ayurveda, the plant, Neelini, is used for inducing mobility, diarrhoea, purgation, relieving constipation, improving hair strength and growth, as purgative and antitoxic agent. Paste of leaves is applied on hair, 20 minutes before bath, to relieve grey hairs and to promote hair growth. The plant is used for various diseases of eye, bronchitis, and skin diseases. The plant is also used as an ingredient in many hair oils such as *Neelibhringadi tailam*, used in treating hair fall and grey hair problems. Traditionally, leaf paste is applied externally to heal wound quickly, also useful in case of insect bites and skin disorders with itching; it is also used in treating scorpion bite. The other folk medicinal use include application of the leaf juice/paste as a prophylactic against rabies (hydrophobia), and as a decoction for blennorrhagia. For treating dog bite, 50 ml of juice extract is mixed with equal quantity of milk and administered in the morning to relieve hydrophobia. Plant extract is a good cure for epilepsy, nervous disorders, bronchitis, and as an ointment for sores, old ulcers, and haemorrhoids. Roots are used for hepatitis, scorpion bites, and urinary complaints. Chiefly a source of dye, indigo also has been used as a nematicide and treatment for a range of ills including scorpion bites and ovarian and stomach cancer. (Verma and Suresh 2002, Kumar *et al.* 2009, Laitonjam and Wangkheirakpam 2011; https://easyayurveda.com, https://www.drugs.com).

17.18 *Lathyrus aphaca* L.

(Syn. *Orobus aphaca* (L.) Doll)

Common Name: Yellow Pea, Yellow Vetchling.

Vernacular Name: *Hindi:* Jangli Mattar (जंगली मटर).

Habit: Annual herb.

Habitat: Dry places, such as sand, gravel and requires a well-drained habitat.

Distribution: It is native to Mediterranean region (southern Europe, parts of Asia, and North Africa).

Flowering and Fruiting Time: January-March.

Taxonomic Description: A sub-erect, glabrous, annual herb. Leaves reduced into tendrils; stipules in pairs, foliaceous, broadly ovate, acute, entire, hastate, 2-3 cm long. Flowers yellow to yellowish-white, axillary, peduncles longer than stipules. Calyx teeth-5, equal, lanceolate, tube bell shaped. Corolla lobed-5, unequal, standard broad. Stamens didelphous. Pods linear, oblong with persistent style. Seeds 4-6.

Habit

Flowering twig

Pod

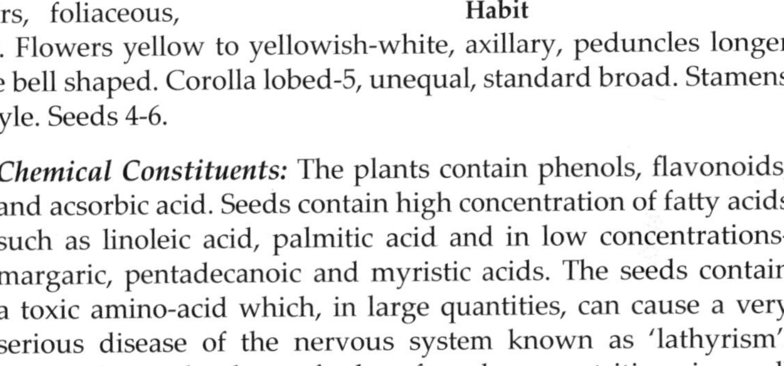

Chemical Constituents: The plants contain phenols, flavonoids, and acsorbic acid. Seeds contain high concentration of fatty acids such as linoleic acid, palmitic acid and in low concentrations-margaric, pentadecanoic and myristic acids. The seeds contain a toxic amino-acid which, in large quantities, can cause a very serious disease of the nervous system known as 'lathyrism'. The seed is said to be perfectly safe and very nutritious in small quantities. (Bagci *et al.* 2001; http://pfaf.org).

Medicinal Properties: The ripe seeds are narcotic and flowers are resolvent. (Vardhana 2008).

Medicinal Uses (Traditional/Folklore): Decoction of seed and flower is used for treating skin infection/skin related problems as a topical application. (Khan *et al.* 2014a).

17.19 *Leucaena leucocephala* (Lam.) de Wit

(Syn. *Acacia frondosa* Willd. *Leucaena glauca* Benth., *Mimosa leucocephala* Lam.)

Common Name: Leucaena, Ipil-Ipil, White Babool, Subabul.

Vernacular Name: *Hindi:* Subabul (सुबबूल).

Habit: Fast growing small tree or large shrub.

Habitat: *L. leucocephala* is a weed of open (often coastal or riverine) habitats, semi-natural, disturbed, degraded habitats and other ruderal sites. In India, it was introduced as a plantation species; also planted as hedge and boundary plants of gardens.

Distribution: It is native to tropial America but is now naturalized throughout the tropics.

Flowering and Fruiting Time: September-November.

Taxonomic Description: An unarmed, small tree or large shrub. Leaves bipinnately compound, pinnae 4-8 pairs;

Habit

leaflets 10-17 pairs, sessile, linear-oblong, base oblique, margin entire, apex obtuse to acute. Flowers white; peduncles long, axillary, solitary or in pairs, bearing a dense globular head of flowers; heads 1.5-2.0 cm in diameter. Calyx lobes-5. Corolla whitish. Stamens-10, exserted, twice as long as the corolla, anthers versatile, filaments filiform. Pods up to 16 cm long, strap shaped, clustered in umbels, linear, flat, reddish shining when mature. Seeds numerous, ovate, brown.

Twig with flowers and pods

Globular head of flowers

Pods

Chemical Constituents: The plant contains hydrocyanic acid, leucaenine, quercitrin and tannic acid. Leaves and seeds also contain glucoside quercetrin and amino acids arginine, cystine, histidine, isoleucine, leucine, lycine, methionine, methionine, cystine, phenylalanine, threonine, tryosine and valine. Seeds and young leaves yield 4 percent of mimosine, which causes loss of hair in non-ruminant animals, especially in horses, mules, donkeys and hogs. The seeds, rich in protein, also contain a dark green to brown oil containing saturated acids and unsaturated acids. The oil is reportedly rich in linoleic acid and contains significant quantities of arachidic and lignoceric acids. The oil also contains sterol (mainly β-sitosterol), methylsterols, triterpenoid alcohols, tocopherols (mainly α-tocopherol), glycolipids, hydrocarbons and carotenoids. The oil is the richest vegetable source of phosphatides. (Aderibigbe *et al.* 2011, Salem *et al.* 2011).

Medicinal Properties: Various parts of *L. leucocephala* have been reported to have medicinal properties ranging from control of stomach diseases to contraception and abortion and the seed gum has been reported to be useful as a binder in tablet formulation. Sulfated glycosylated form of polysaccharides from the seeds was reported to possess significant cancer chemopreventive and antiproliferative activities. Also, mimosine, an amino acid from the seeds was reported to possess anticancer activity and to inhibit the growth of hair. Other studies on the extracts of the seeds had shown varying activities including central nervous system depressant, anthelmintic and antidiabetic activities. The leaves and seeds have antiseptic and parasiticidal properties. (Aderibigbe *et al.* 2011, Hassan *et al.* 2014, Verma *et al.* 2016b).

Medicinal Uses (Traditional/Folklore): In Indonesia, an aqueous extract derived from boiling the seeds of *L. leucocephala* is taken orally to treat type-2 diabetes and to expel intestinal worms. Leaf paste is applied to poisonous stings and bites in Myanmar. In the Philippines, roasted seeds are used as emollient and to increase menstrual flow. Decoction of bark and roots are useful as contraceptive or for inducing abortion in Latin America. In Malaysia, the young leaves are pounded with rice and thin paste applied around the neck to treat coughs. It is also rubbed over the whole body to treat measles. Leaf ash mixed with coconut oil is applied on the body to treat scurf. Leaves are used as an herbal bath to cleanse the body. Shoots boiled with alum and water are used to clean the affected part of shingles. (Meena Devi *et al.* 2013, Verma 2016b, Raikwar and Chowbey 2017; http://medicinalplantsinsingapore.wikifoundry.com/page/Leucaena+leucocephala).

17.20 *Medicago lupulina* L.

(Syn. *Medicago cupaniana* Guss.)

Habit

Common Name: Black Medic, Nonesuch, Hop Clover.

Vernacular Name: *Hindi:* Lusan Ghas (लुसन घास); *Sanskrit:* Ashvabala (अश्वावाला).

Habit: Trailing herb.

Habitat: Common along moist places in gardens, cultivated and fallow fields.

Distribution: A native of the old world, Black Medick is found throughout Europe, North Africa, the near east, and most of Asia, including India, China, and Korea. It is naturalized in central Asia, Japan, South Africa, Australia, New Zealand, the United States, Canada, and much of South America. Black medick is found throughout the United States, including Hawaii and Alaska. In India, it is recorded from Haryana and Madhya Pradesh.

Flowering and Fruiting Time: January-March.

Flowering twig

Pods

Taxonomic Description: A trailing, much-branched herb with stem up to 40 cm long. Leaves 3-foliate, stipulate, appressed pubescent; leaflets 1-2 cm long, obovate, margin obscurely incise-crenate, base cuneate, apex retuse-apiculate. Flowers numerous, very small, yellow, borne in small, dense globose heads; peduncle exceeding the leaves. Calyx setaceous, 1.5-2 mm long. Corolla yellow, nearly as long as or somewhat longer than the calyx. Stamens diadelphous. Pods sickle-shaped, about 3 mm long, 1-seeded, hairy, turning black when ripe.

Chemical Constituents: From the roots of *M. lupulina*, four triterpene saponins like soyasaponin 1, hederagenin 3-O-β-D-glucopyranoside, medicagenic acid 3-O-β-D-glucopyranoside and medicagenic acid 3,28-di (O-β-D-glucopyranoside) have been isolated. (Oleszek *et al.* 1988).

Medicinal Properties: Laxatives, antibacterial and lenitive properties. (http://medplants.blogspot.in, http://medicinalplants.us).

Medicinal Uses (Traditional/Folklore): Decoction of leaves, stem, flowers and fruits are given orally to treat disorders of lungs, coughs and heal wounds. Extracts from the leaves *of M. lupulina* and *M. sativa* are used as aids to digestion, as laxatives, and have also been recommended for the treatment of peptic ulcers and for urinary and bowel problems.

As a veterinary medicine, the plant is galactagogue, lenitive and nutritive fodder. (http://medplants.blogspot.in, http://medicinalplants.us).

17.21 *Medicago monantha* (C.A.Mey.) Trautv.

(Syn. *Trigonella incisa* Benth.)

Habit

Common Name: Medick, Single-Flowered Medick.

Vernacular Name: None known.

Habit: Annual herb.

Habitat: Often found in cultivated fields, gardens.

Distribution: Asia: China, Mangolia, Afghanistan, Pakistan and India.

Flowering and Fruiting Time: January-April.

Taxonomic Description: A slender, diffuse, annual herb, clothed with appressed pubescent. Leaves trifoliate; leaflets up to 1.5 cm long, obovate, margin sharply incise-dentate towards the apex, base cuneate, stipule semi-sagitate with long points. Flowers axillary, 1-2 on very short peduncles. Calyx teeth setaceous, pubescent. Corolla yellow, slightly exserted. Stamens diadelphous. Pods 5-6 cm long, flat, slightly falcate, net-veined, clothed with soft hairs. Seeds numerous, brownish.

Flowering and fruiting twig

Pods

Chemical Constituents: photochemical screening reported the presence of alkaloids, phenol and saponins (Esmaeili *et al.* 2012, Rehman *et al.* 2017).

Medicinal Properties: Antioxidants, antibacterial, antifungal, anti-inflammatory and antiviral (Esmaeili *et al.* 2012, Rehman *et al.* 2017).

Medicinal Uses (Traditional/Folklore): *M. monantha* is therapeutically utilized as an appetite stimulant. (Esmaeili *et al.* 2012, Rehman *et al.* 2017).

17.22 *Medicago polymorpha* L.

(Syn. *Medicago denticulata* Willd.; *Medicago polymorpha* var. *polymorpha*)

Habit

Common Name: Bur Clover, Toothed Medic, California Bur Clover.

Vernacular Name: None known.

Habit: Annual herb.

Habitat: Common in cultivated fields, wastelands and along water channels.

Distribution: It is native to Europe, Central Asia, China, Japan, and North Africa; but found throughout the world.

Flowering and Fruiting Time: January-April.

Taxonomic Description: An annual, decumbent-ascending herb. Stem glabrous, branched at the base. Leaves 3-foliate; leaflets up to 2 cm long, obovate, toothed towards the top. Stipules auriculate at base, margin laciniate. Inflorescence usually umbel. Flowers 2-6 together, small. Calyx teeth lanceolate. Corolla yellow, twice the length of calyx. Pods sub-globose, 2-4 spiral, disk like, with double row of spines at margin.

Flowering twig

Flowers

Disc like pods

Chemical Constituents: The herb contains carotenoids, triterpene, saponins, isoflavonoids and coumarins. The young leaves contain protein, fat, carbohydrate and ash. They are rich in vitamins A, C, and E. (Rani *et al.* 2015; http://temperate.theferns.info, http://www.pfaf.org).

Medicinal Properties: Bur clover is used for medicinal purposes for skin plagues, dysentery, constipation, indigestion and against rheumatic pains. Leaves and young stem are used as digestive, laxative and tonic. (Khan *et al.* 2013b, Rani *et al.* 2015).

Medicinal Uses (Traditional/Folklore): The plant is effective against constipation and other digestive ailments. In Italy, bur clover leaf has been used for many centuries, such as medicinal purposes for treating rheumatic pains and wounds and is still used today. In India, Bur clover is used for medicinal purposes for skin plagues and dysentery. (Abbasi *et al.* 2013, Rani *et al.* 2015; https://plants.usda.gov/plantguide/pdf/pg_mepo3).

17.23 *Melilotus albus* Medik.

(Syn. *Melilotus leucanthus* DC.)

Habit

Common Name: White Sweet Clover, White Melilot.

Vernacular Name: *Hindi:* Safed Ban Methi (सफेद बन मेथी).

Habit: Annual herb.

Habitat: Cultivated fields, moist places, waste ground.

Distribution: It is native to Europe.

Flowering and Fruiting Time: January-April.

Taxonomic Description: An erect small, branched annual herb, up to 70 cm tall. Leaves 3-foliate; leaflets oblong-obovate, serrated towards upper part, entire at base, stipules lanceolate. Inflorescence raceme. Racemes long, terminal, axillary. Calyx teeth-5, lanceolate. Corolla white. Stamens diadelphous. Pods about 5 mm long, ovoid, elliptic, beaked with persistent calyx. Seeds usually 1 rarely 2.

Chemical Constituents: Active chemical constituents reported are coumarins, flavonoids, phenols and aromatic compounds. (Stefanović *et al.* 2015; http://pfaf.org).

Medicinal Properties: The plant is vermifuge, carminative, antimicrobial, antibiofilm, antioxidant, emollient and poultice. (Stefanović *et al.* 2015; http://pfaf.org).

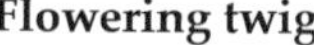

Flowering twig

Inflorescence

Pods

Medicinal Uses (Traditional/Folklore): The dried leaves of white Melilot contain coumarins, which is useful as an anticlotting agent for the blood. It is also utilizes as an ointments for external ulcers. Leaves grinded and paste is applied on forehead for headache, for stomach problem and indigestion. The extracts of leaves/juice are used as eye lotion for curing eye diseases. (Prakash 2014; http://pfaf.org, https://www.cabi.org).

17.24 *Melilotus indicus* (L.) All.

(Syn. *Melilotus indica* (L.) All.; *Melilotus parviflorus* Desf.; *Trifolium indicum* L.)

Habit

Common Name: Indian Sweet Clover, Yellow Sweet Clover.

Vernacular Name: *Hindi:* Ban Methi (बन मेथी), Morila (मोरिला).

Habit: Annual herb.

Habitat: A common weed of moist waste places, gardens, parks, cultivated fields, roadsides.

Distribution: It has a wide native distribution, ranging from Macaronesia and northern Africa, through Europe, and in temperate and tropical Asia.

Flowering and Fruiting Time: December-May.

Taxonomic Description: An erect, annual herb, up to 60 cm tall. Stem slender, branched. Leaves 3-foliate. Leaflets 1-2.5 cm long, glabrous, obovate or oblanceolate, retuse or emarginated, sharply dentate towards top. Inflorescence raceme. Raceme axillary or terminal, dense flowered, 3-5 cm long. Calyx teeth deltoid. Corolla yellow, almost twice the length of calyx. Stamens diadelphous (9+1). Pods 2.5 mm long, beaked, ovoid, obscurely reticulate with persistent calyx. Seeds usually 1, glabrous.

Flowering twig

Inflorescence

Pods

Chemical Constituents: Phytochemicals such as coumarin, herniarin, umbelliferene and scopoletin have been identified in the plant. The presence of β-sitosterol, a sterol or triterpene alcohol, choline and an aromatic compound have also been reported. (Sonju *et al.* 2017).

Medicinal Properties: The plant contains coumarin, which is an anticoagulant. Its characteristic sweet odor, intensified by drying, is derived from coumarin. It has been used in herbal medicines. The plant is known to have anticoagulant, antispasmodic, aromatic, antibacterial, discutient, purgative, strongly laxative, emollient, analgesic, insect repellent, tonic, astringent and narcotic properties. (Sonju *et al.* 2017; http://tropical.theferns.info).

Medicinal Uses (Traditional/Folklore): The plant having many medicinal values for the cure of several ailments is used in Ayurveda as well as traditional medicines. Seeds are taken to treat genital organ diseases. The infusion is made from aerial parts and used to cure bronchial disorder and abdominal pain. Aerial parts are cooked as potherb which is believed to be digestive and whole plant taken as a purgative, and as a very strong laxative.

The plant is also used for treatment of it is used for swellings, tumours, skin rash, wounds, gastrointestinal problems, colds, bowel complaints and infantile diarrhoea, improving blood circulation, and haemorrhoids, etc. The seed is made into a gruel and used in the treatment of bowel complaints and infantile diarrhoea. Infusion prepared from leaves and seeds is generally used for stomachache and as an emollient. Plant juice is applied to treat cuts and wounds and rubbed on skin to cure itching. Leaf extract along with lemon juice is applied over skin eruptions. The leaves and the aerial parts of the plant are known to as insect repellent. The whole plant is toxic to livestock. (Amjad *et al.* 2015, Sonju *et al.* 2017; http://tropical.theferns.info).

17.25 *Mimosa rubicaulis* Lam.

(Syn. *Mimosa octandra* Roxb.)

Flowering branch

Common Name: Himalayan Mimosa.

Vernacular Name: *Hindi:* Aila (अइला), Alral (अलरल), Shiahkanta (शियाहकांता), Arlu (अरलू); *Sanskrit:* Rala- Arlu (राला-अरलू), Sajjaka (सज्जका).

Habit: A large prickly shrub.

Habitat: Commonly found along the bank of canals and other aquatic bodies.

Distribution: Himalayan Mimosa is found in the Himalayas, from Afghanistan to Bhutan, at altitudes of 300-1,900 m. *M. rubicaulis* is found across India but prominently in Arunachal Pradesh, Assam, Meghalaya, Tripura; common in hilly forest, occasionally in forests in plains.

Flowering and Fruiting Time: June-October.

Taxonomic Description: A large, straggling, very prickly shrub. Branches pubescent, grooved with short hooked prickles. Leaves bi-pinnate with prickly rachis; pinnae 4-12 pairs, opposite; leaflets 8-14 pairs, 5-7 mm long and 1.5-2 mm wide, linear-oblong, entire, base oblique, apex obtuse-mucronate, appressed pubescent, stipules subulate. Flowers bracteate, pink, fading to white, in solitary or fascicled spherical heads, peduncles pilose, heads 1-1.5 cm across. Calyx very short, pubescent. Corolla tubular, lobes-4. Stamens 8-10, exserted. Pods oblong-linear, 8-12 cm long, thin, strap-shaped, curved, 5-10 seeded.

Flowers

Pods

Chemical Constituents: Plant contains phenols such as flavonoides (quercitin and luteolin), β-sitosterol and 4-ethylgallic acid. (Genest *et al.* 2008, Manda and Seru 2015).

Medicinal Properties: Plant has antipyretic, antihepatic, antiemetic, expectorant properties. (Genest *et al.* 2008; https://phytochem.nal.usda.gov/phytochem/search).

Medicinal Uses (Traditional/Folklore): The ethnobotanical uses of this species include its use in the treatment of bronchitis, cholera, cough, dyspepsia, fever, jaundice, smallpox, syphilis, dibetes and tuberculosis. Traditionally, the plant is used in snakebite, cholera, pleurisy, constipation, icterus, sunstroke, cachexia sores and measles. In India, leaves are used in the treatment of burns and the roots as an antiemetic agent. (Genest *et al.* 2008, Manda and Seru 2015; https://phytochem.nal.usda.gov/phytochem/search).

17.26 *Prosopis cineraria* (L.) Druce

(Syn. *Mimosa cineraria* L.; *Prosopis spicigera* L.)

Common Name: Khejri Tree.

Vernacular Name: *Hindi*: Khejri (खेजड़ी), Sami (सामी).

Habit: Small or medium sized tree.

Habitat: Open fields, roadsides.

Distribution: This species is possibly originated in North West India and eastern Pakistan. The tree grows in dry and arid regions of Arabia and in regions of India mainly Rajasthan, Haryana, Punjab, Gujarat, Western Uttar Pradesh and drier parts of Deccan and extends as far as South in Tuticorin. It is the state tree of Rajasthan and Telangana and also the national tree of the United Arab Emirates.

Flowering and Fruiting Time: April-June.

Taxonomic Description: A small to medium sized tree. Branches armed with conical prickles. Leaves bipinnate, pinnae usually 2-3 pairs; leaflets 8-12 pairs, dark green, 0.4-1.2 x 0.4 cm, sub-sessile, oblong, mucronate at apex, generally unequal sided. Inflorescence spike, axillary. Flowers small, yellow or creamy white. Calyx cup shaped, 5-toothed. Petals 5, tips recurved. Stamens 10. Pods 8-16 x 0.5-1 cm, hanging, torulose, cylindrical. Seeds many, oblong, brown.

Habit

Bipinnate leaves

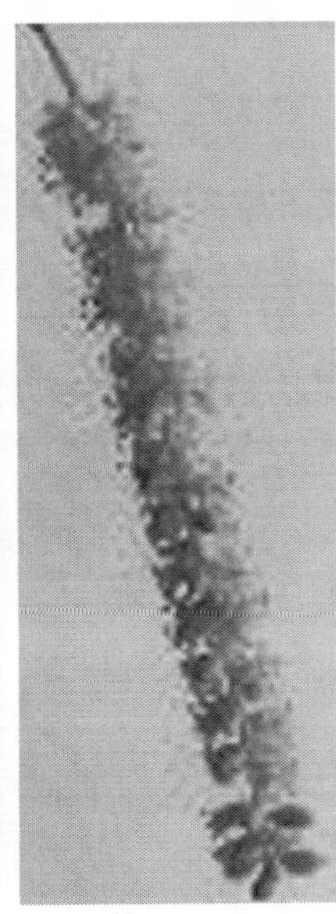

Inflorescence

Close up of flowers

Chemical Constituents: Various chemical constituents are present in plant parts. Flowers contain patuletin glycoside, patulitrin, luteolin, rutin, sitosterol, spicigerine, flavone derivatives, prosogerin A and prosogerin B. From leaves steroids like campesterol, cholestrol, sitosterol, stigmasterol, actacosanol, hentriacontane, methyl docosanoate, diisopropyl-10, 11-dihydroxyicosane-1, 20-dioate, tricosan-1-ol, and 7, 24-tirucalladien-3-one along with a piperidine and alkaloid spicigerine have been isolated. While seeds contain prosogerin C, prosogerin D, prosogerin E, gallic acid, patuletin, patulitrin, luteolin, and rutin. Phytoconstituents of bark are hexacosan-25-on-l –ol, a new keto alcohol alongwith ombuin and atriterpenoid glycoside, vitamin K1, n-octacosyl acetate, the long chain aliphatic acid. Presence of glucose, rhamnose, sucrose and starch is also reported. (Khatri *et al.* 2010, Prabha *et al.* 2014, Pareek *et al.* 2015).

Medicinal Properties: The plant is reported to be astringent, demulcent and pectoral. Leaves have antibacterial, antihyperglycemic, antihyperlipidemic and antioxidant activity. The bark of the tree is dry, acrid, bitter with a sharp taste, cooling, anthelmintic, tonic; it cures leprosy, dysentery, bronchitis, asthma, leucoderma and piles. The pod is said to be astringent. (Khatri *et al.* 2010, Prabha *et al.* 2014, Pareek *et al.* 2015).

Medicinal Uses (Traditional/Folklore): The whole plant is used in the Indigenous System of Medicine as a folk remedy for various ailments like leprosy, dysentery, bronchitis, asthma, leucoderma, piles, muscular tremor and wandering of the mind. Flowers are used for skin diseases, as a blood purifier and producing a cooling effect. It provides safeguard in pregnancy by preventing miscarriage; flowers are pounded, mixed with sugar and used during pregnancy as safeguard against miscarriage. The ashes are rubbed over the skin to remove hair. Smoke from the leaves is used for eye troubles. The pods, rich in protein, are used as vegetable in the dried and green form. The leaf is used for treating mouth ulcers and eye diseases. Stem bark is useful in the treatment of rheumatism, cough, common cold and anthelmintic disorders.

As a veterinary medicine leaf paste is applied on boils and blisters, including mouth ulcers in livestock and leaf infusion on open sores on the skin. The bark and leaf galls are used for tanning. (Khatri *et al.* 2010, Prabha *et al.* 2014, Pareek *et al.* 2015).

17.27 *Prosopis juliflora* (Sw.) DC.

(Syn. *Mimosa juliflora* Sw.)

Habit

Common Name: Algaroba, Mesquite.

Vernacular Name: *Hindi:* Junglee Kikar (जंगली कीकर), Vilayati Babul (विलायती बबूल).

Habit: A small tree with drooping branches.

Habitat: Grows wild in open, dry areas; planted along roadsides and in plantations in arid areas.

Distribution: It is native to Mexico, South America and the Caribbean. The plant is distributed throughout India, except Jammu & Kashmir, Himachal Pradesh, Sikkim, and Arunachal Pradesh.

Flowering and Fruiting Time: Almost throughout year.

Taxonomic Description: It is a small, drought resistant, evergreen tree, 10 to 15 m tall with drooping branches. Stem shallowly fissured, bark blackish-brown. Leaves 1-3 at each node, bipinnate, pinnae generally 4. Leaflets 17-20 pairs, sessile, unequal sided, 0.5-1 cm long, linear, oblong, entire, glabrous. Inflorescence spikes, axillary. Flowers yellow or cream coloured in hanging spikes. Pods 10-22 x 1-1.5 cm, pendulous, strongly compressed, slightly curved, torulose. Seeds numerous, brown.

Chemical Constituents: Phytochemical constituents present in the plant are alkaloids, carbohydrates, steroids, protein, tannins, phenols, flavonoids, gum and mucilage, glycosides, saponins and terpenes. Twelve of the flavonoids have been identified as 6 flavones--apigenin, luteolin, apigenin 6, 8-di-C-glycosides, chrysoeriol 7-O-glucoside, luteolin 7- O- glucoside, and 6 flavonols--kaempferol 3-O-methyl ether, quercetin 3-O-methyl ether, isorhamnetin 3-O-glucoside, isorhamnetin 3-orutinoside, quercetin 3-O-rutinoside, and quercetin 3-O-diglycoside (glucose, arabinose). (SivaKumar *et al.* 2009, Dave and Bhandari 2013, Prabha *et al.* 2014).

Bipinnate leaf

Spikes and single flower

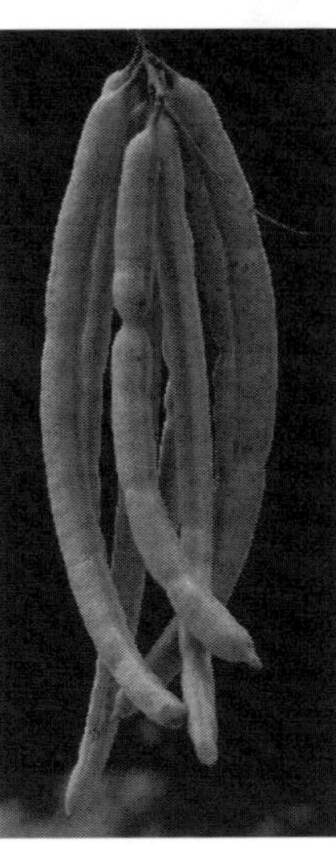

Pods

Medicinal Properties: The plant possesses antidiabetic, antiasthmatic, antiallergic, anti-inflammatory and antifungal properties. Mesquite has been used as a traditional medicine on different continents for curing catarrh, and sore throat and for the healing of wounds. (SivaKumar *et al.* 2009, Dave and Bhandari 2013, Prabha *et al.* 2014).

Medicinal Uses (Traditional/Folklore): Syrup prepared from the ground pods has various medicinal values. It is given to children showing weight deficiency. A tea made from *P. juliflora* is thought to be good for digestive disturbances and skin lesions. An astringent decoction is made from boiling wood chips, a bark extract is used as an antiseptic on wounds and gum is used to treat eye infections; stem extract is useful to treat fever. For treating sunburn decoction of the beans is very useful. Pods are used to make eyewashes and also prepared as a poultice that is applied to a sore throat or as a drink that is taken for animal stings. Gum is good for stomach ailments. It is taken for diarrhoea, stomach inflammation, system cleansing or to settle the intestines. Seed galactomannans, commonly known as seed gums, have widespread industrial applications in food paper, textile, petroleum, pharmaceuticals and cosmetics. (Sindhu *et al.* 2012, Dave and Bhandari 2013, Prabha *et al.* 2014).

17.28 *Rhynchosia capitata* (Roth) DC.

(Syn. *Glycine capitata* Roth)

Habit

Common Name: Papra.

Vernacular Name: *Hindi:* Kulata (कुलटा), Kulthi (कुलटी).

Habit: Trailing herb.

Habitat: Cultivated and fallow fields.

Distribution: India, Pakistan and Sri Lanka.

Flowering and Fruiting Time: June-October.

Taxonomic Description: An annual, branched, slender, trailing herb. Stem and branches hispid. Leaves alternate, 3-foliate; leaflets obovate, rhomboid-ovate, entire, cuneate at base, 2-5-4.0 cm in length, as long as broad. Inflorescence raceme. Racemes axillary, 4-6 flowered, ultimately becoming twisted and forming round heads of flowers, peduncles up to 4 cm long. Calyx segments long, subulate. Corolla yellow. Pods oblong, ovoid, softly pilose, slightly compressed. Seeds-2 with waxy aril, small, compressed.

Chemical Constituents: The plant contains polyphenolic compounds such as flavonoids, tannins and saponins. (Chaturvedi 2015).

Medicinal Properties: Aerial parts of *R. capitata* have shown potential as a good source of natural antioxidants. (Chaturvedi 2015).

Flowering twig

Flower

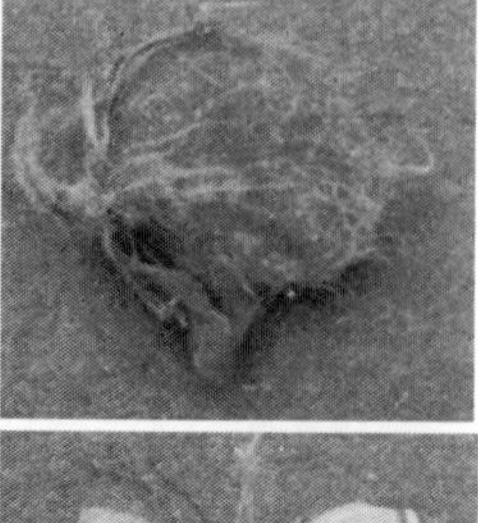
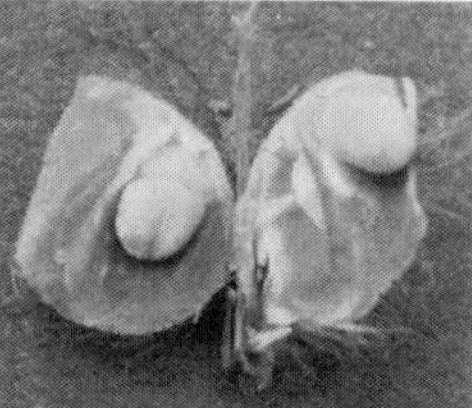
Closed and open pod

Medicinal Uses (Traditional/Folklore): Decoction of roots is helpful for stomach cleaning. (Tayade and Patil 2006).

17.29 *Rhynchosia minima* (L.) DC.

(Syn. *Dolicholus minimus* (L.) Medik.; *Rhynchosia medicaginea* (Lam.) DC.)

Habit

Common Name: Burn-Mouth-Vine, Least Snout-Bean, Rhynchosia.

Vernacular Name: *Hindi:* Kulata (कुलटा), Kulthi (कुलथी).

Habit: Climber.

Habitat: Twining on herbs, bushes, electric poles along roadsides, and railway lines.

Distribution: Tropical Africa, West Indies, U.S.A, Australia, Pakistan, India, Sri Lanka, Lower Myanmar.

Flowering and Fruiting Time: July-October.

Taxonomic Description: An annual, twining or climbing herb. Stem and branches nearly glabrous, slender. Leaves alternate, trifoliate; stipules minute, lanceolate; leaflets up to 5 cm long, deltoid, rhomboid-ovate, 3-nerved from the base, margin entire, nearly as broad as long, laterals oblique. Inflorescence raceme, slightly longer than the leaves, axillary. Flowers yellow. Stamens 10, diadelphous. Pods oblong, slightly recurved, 1-1.5 cm long, beaked at apex. Seeds-2 without aril, small, compressed.

Flowering and fruiting twig

Flowers

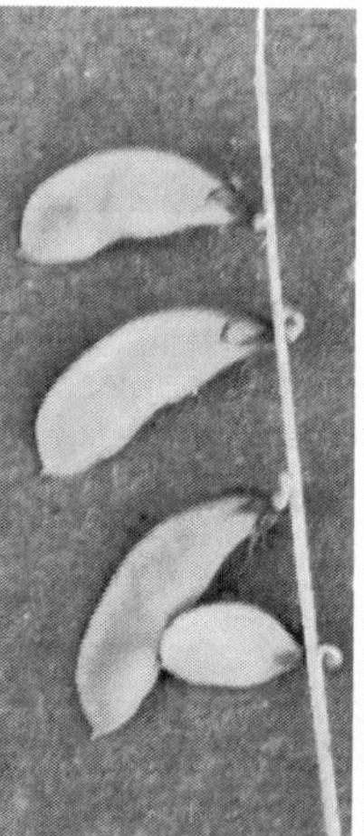
Pods

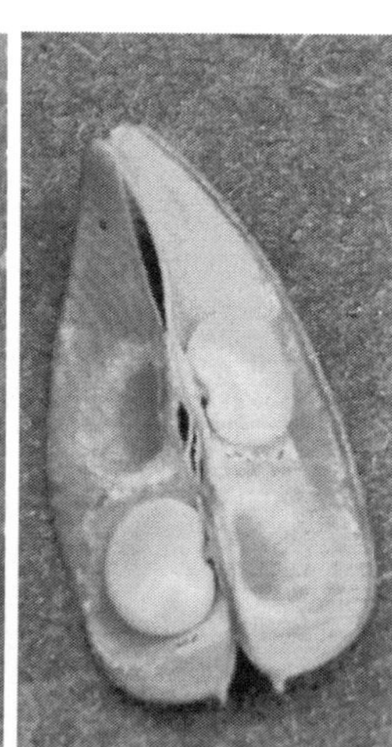
Open pod with seeds

Chemical Constituents: Seed coat and pericarp contain gallic acid, hydroquinone diacetate and other phenolics; flavonoids in the leaves are in the form of C-glycosyflavones. (Yellasubbaiah *et al.* 2015).

Medicinal Properties: *R. minima* is used as abortifacient, ecbolic and general healing medicine and to treat haemorrhoids, diarrhoea and dysentery. The seeds are bitter and poisonous and seed extract shows specific agglutinating action on human RBC (root-bark) and used for treating heart ailments. Root is healing, laxative, antimalarial and antipyretic and used for treating diarrhoea, and dysentery. Essential oils extracted show high antibacterial and antifungal activity against several bacterial and fungal species. The plant is also reported to be used as an antidote for snake bite. (Gundidza *et al.* 2009, Yellasubbaiah *et al.* 2015).

Medicinal Uses (Traditional/Folklore): The plant is used traditionally as a strong abortifacient, anthelmintic and for the treatment of wounds, asthma and piles. In Zimbabwe and states of Tamil Nadu, Andhra Pradesh and Uttar Pradesh (Vindhya), the plant extract is used as an effective abortifacient; in Zimbabwe it is also used to alleviate boils. In Satpura (Madhya Pradesh), root extract is used for general weakness and whole plant as a tonic. In Uttar Pradesh (Vindhya), the plant is used for treating asthma and piles. (Gundidza *et al.* 2009, Chaudhary 2010, Quattrocchi 2012, Singh *et al.* 2013a, Yellasubbaiah *et al.* 2015).

17.30 *Senna obtusifolia* (L.) H.S.Irwin & Barneby

(Syn. *Cassia obtusifolia* L.)

Common Name: Sicklepod, Chinese Senna.

Vernacular Name: *Hindi:* Chakwar **(चकवड़);** *Sanskrit:* Cakramarda (चक्रमर्दा), Prapunnata (प्रपुण्णता).

Habit: Annual herb.

Habitat: Occurring in open waste places, fallow fields along the roadsides and railway lines.

Distribution: Native to tropical America but has become widespread throughout the tropics and subtropics. However, the extent of its original distribution in the neotropics is unknown.

Flowering and Fruiting Time: July-November.

Taxonomic Description: An erect, branched, annual herb or rarely undershrub. Leaves alternate, paripinnate, up to 10 cm long; leaflets 3-pairs, 2-5.5 cm long, obovate or obovate-oblong, glabrous or somewhat pubescent, margin entire, base cuneate-rounded to oblique, apex mucronate; rachis grooved above with a gland between the lowest pair of leaflets, stipules linear. Flowers bright yellow, axillary. Calyx 5-partite, lobes green. Petals-5, obovate. Fertile stamens-7. Pods up to 20 cm long, linear, sub-cylindric. Seeds numerous, oblong, dark brown.

Habit

Chemical Constituents: Plants contain phytoconstituents like anthraquinones, phytosterols, triterpenoids, flavonoid, palmitic acid, stearic acid, quercetin and isoquercetin. Eight compounds isolated from the ethyl acetate fraction of 95% EtOH extract of the transformed roots *of S. obtusifolia* are betulinic acid, chrysophanol, physcion, stigmasterol, 1-hydroxy-7-methoxy-3-methyl-anthraquinone, 8-O-methylchrysophanol, 1-O-methylchrysophanol and aloe-emodin. (Yang *et al.* 2006, Anisuzzaman *et al.* 2014).

Medicinal Properties: The plants have anodyne, antifungal, antiviral, anti-inflammatory, antidysenteric, febrifuge, and stomachic properties. The leaves are used as a laxative, anthelminthic and as a poultice to treat skin infections, sores, ulcers and insect bites. (Anisuzzaman *et al.* 2014; https://www.prota4u.org, http://tropical.theferns.info).

Flower

Calyx

Medicinal Uses (Traditional/Folklore): *S. obtusifolia* has been traditionally used for the treatment of dizziness, dysentery asthma, blurred vision and eye inflammation. The plants are useful in fever, colic, eczema, stomachache, constipation, febrifuge, fatigue, blood purifier, and arthritis and herpes diseases. The leaves are used locally as a remedy for gout, sciatica, joints pain, stomachache and head-ache. A decoction of the leaves is used to treat eye complaints. The leaves are used as a laxative and as a poultice to treat skin infections, sores, ulcers and insect bites. The leaves are further used as an anthelminthic and against vomiting and stomachache. The leaf decoction is used as febrifuge and for treatment of gingivitis, urinary tract infections, diarrhoea, fever and cough. A decoction of the leaves is used to treat eye complaints in Senegal and Zanzibar. In DR Congo a fish poison is made from the crushed leaves. The seeds are eaten, combined with a leaf decoction, to treat conjunctivitis. In French Guiana, root tincture is rubbed on rheumatic areas and the leaf infusion is effective for renal calculi. It serves as laxative, which promotes the evacuation of bowels and soften the stool and cures constipation. It is beneficial in purifying and eliminating the impurities from the blood. The roots are used as a laxative and anthelminthic. (DeFilipps 2004, Anisuzzaman *et al.* 2014; http://tropical.theferns.info, https://www.prota4u.org).

17.31 *Senna occidentalis* (L.) Link

(Syn. *Cassia occidentalis* L.)

Common Name: Coffee Senna, Negro Coffee.

Vernacular Name: *Hindi:* Kasunda (कसुंडा), Bari Kasondi (बड़ी कसौंदी); *Sanskrit:* Kasamarda (कसमरदा), Vimarda (विमर्दा), Arimarda (अरिमर्दा).

Habit: Undershrub.

Habitat: Common along roadsides, open waste places.

Distribution: *S. occidentalis* has a pantropical distribution and probably originated in tropical south America. The plant occurs throughout India.

Flowering and Fruiting Time: July-December.

Habit

Taxonomic Description: An erect undershrub, up to 1.5 m tall. Leaves paripinnate, petiole with an ovoid gland at base; leaflets 3-5 pairs, 2-8 cm long, ovate, lanceolate, glabrous, apex acuminate, base oblique, margin entire. Stipules lanceolate, ovate, caducous. Flowers axillary, pedicellate, yellow, bracts linear-lanceolate. Calyx lobes-5, green, obtuse. Corolla yellow, 1-2 cm long. Stamens 10, fertile stamens-7, anthers oblong. Pods 8-13 cm long, slightly falcate, compressed, transversely septate. Seeds pale-brown, ovoid, numerous.

Apical portion of plant

Flower

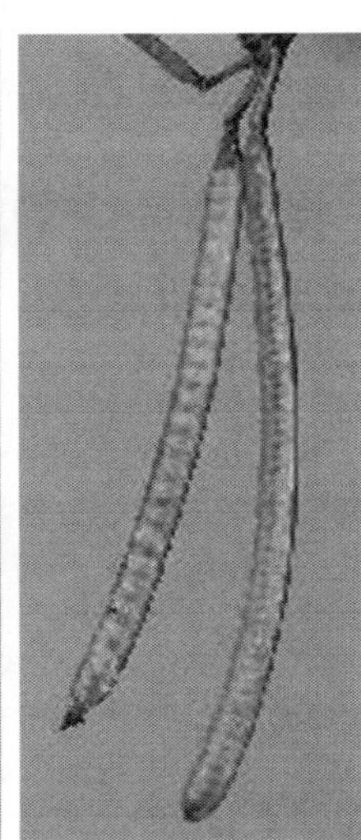

Pods

Chemical Constituents: The plant contains many chemical groups including alkaloids, anthocyanosides, phenolics, proteins, phlobatannins, steroids, tannins, flavonoids, anthroquinone, saponins, terpenes, resins, balsams, amino acids, carbohydrates, sugars and cardiac glycosides. (Odeja *et al.* 2014, Al-Snafi 2015a).

Medicinal Properties: The plant has numerous important medicinal properties such as antimicrobial, antifungal, insecticidal, insect antifeedant, antimalarial, anthelmintic, antioxidant, antianxiety, antidepressant, antimutogenic, purgative, laxative, expectorant, antidiabetic, wound healing, hepatoprotective, renoprotective, sun protective, smooth muscles relaxation, immune-modulating, anti-inflammatory, analgesic, antipyretic, vermifuge and febrifuge. (Yadav *et al.* 2010a, Odeja *et al.* 2014, Al-Snafi 2015a, Manikandaselvi *et al.* 2016).

Medicinal Uses (Traditional/Folklore): Most recorded uses of *S. occidentalis* are as traditional remedies for ailments such as typhoid, malaria, liver complaints and dog bites. The plant extract is used to cure eye inflammations in Ayurveda. The plant is a bitter tonic and mainly used to detoxify liver, cure internal bacterial and fungal disorders, kill parasites and viruses, enhances immunity and promote perspiration. It is also helpful in cough, convulsions, reduces blood pressure, reduces spasms and as cardiotonic. The seed is also bitter and has purgative properties and used as a diuretic, liver detoxifier, as a hepato-tonic (balances and strengthens the liver). Further, it is used in whooping cough and convulsion.

In Jamaican folk medicines, the plant is used for curing diarrhoea, dysentery, constipation, fever, cancer, eczema and venereal diseases. Leaves are used in the treatment of gonorrhoea, fevers, urinary tract disorders and edema. Leaves were also used externally and internally in skin diseases, itches, scabies and ringworm. Hot decoction preferred to quinine for its tonic properties. Paste of leaves and calcium hydroxide applied to abscesses for quick opening and clearing of pus. A tea made from roots and dried flowers is used as a treatment for colds and upset stomach. An ointment prepared from the leaves is applied as a remedy for ringworm and other affections of the skin.

Seeds are brewed into a coffee-like beverage for asthma, and a flower infusion is used for bronchitis. *S. occidentalis* is used as a coffee substitute in Egypt and elsewhere despite reports that the seeds are toxic to cattle. The seeds contain no caffeine or tannin. (Yadav *et al.* 2010a, Al-Snafi 2015a, Manikandaselvi *et al.* 2016; http://tropical.theferns.info).

17.32 *Sesbania bispinosa* (Jacq.) W.Wight

(Syn. *Aeschynomene bispinosa* Jacq.; *Sesbania aculeata* (Willd.) Pers.)

Common Name: Prickly Sesban.

Vernacular Name: Hindi: Dhaincha (ढैंचा), Ikad (इकड़), Dadon (दादों), Daden (ददेन).

Habit: Shrub.

Habitat: Cultivated as well as grows wild in fallow agricultural lands and field boundaries.

Distribution: Native to tropical America, it occurs throughout India.

Flowering and Fruiting Time: August-October.

Taxonomic Description: An erect annual shrub, up to 2 m tall. Branches and underside of leaf-rachis armed with small hooked prickles. Leaves up to 35 cm long, paripinnate; leaflets numerous pairs, glaucous beneath, 1-2 cm long, linear, oblong, obtuse, apiculate. Flowers yellow in axillary or terminal racemes. Racemes up to 10 cm long, 2-6 flowered, peduncle often with prickles. Stamens diadelphous. Pods up to 25 cm long, linear and slightly falcate, beaked, somewhat torulose, glabrous. Seeds numerous.

Habit

Chemical Constituents: Amino acids such as lysine, arginine, and histidine. (Anon. 2006).

Medicinal Properties: Plant decoction is antacid and febrifuge. Bark and seeds are astringent used in treating diarrhoea. The leaves and flowers are astringent. They are used in the treatment of inflammations, bacterial infections and tumours. It is a good source of trace and macro elements. (Parab and Vaidya 2016; http://tropical.theferns.info).

Medicinal Uses (Traditional /Folklore): Seed powder is given to induce hunger. Leaf paste is applied all over the body of small babies and also given bath against whooping cough. Flower smoke is used as mosquito and insect repellent. Roots have emetic properties; powdered root is given to a person bitten by a snake. The seed, mixed with flour, is used to treat ringworm and other skin diseases. They are prepared as poultices for external use or taken as a decoction for internal ailments.

Leaf **Flowers** **Pods**

As a veterinary medicine, leaf wash is given to cattle as a preventive measure to tsetse flies. (Vardhana 2008, Parab and Vaidya 2016; http://tropical.theferns.info, http://medplants.blogspot.in).

17.33 *Sesbania sesban* (L.) Merr.

(Syn. *Aeschynomene sesban* L.; *Sesban aegyptiaca* Poir.)

Habit

Common Name: Common Sesban, Egyptian Rattlepod, Egyptian Riverhemp.

Vernacular Name: *Hindi:* Jayanti (जंयती), Rawasan (रेवासन); *Sanskrit:* Jayantika (जयंतिका).

Habit: A fast growing small tree.

Habitat: Cultivated as a hedge or shade plant in villages and also grows wild.

Distribution: Throughout India, wild and cultivated, up to an altitude of 1,200 m.

Flowering and Fruiting Time: August-November.

Taxonomic Description: A soft-wooded, fast-growing, small tree or large shrub, 3-7 m high. Leaves up to 15 cm long, paripinnate, leaflets many pairs, opposite, glaucous beneath, 1.5-2.5 cm long, linear, oblong, shortly stalked, margin entire, apex obtuse-apiculate, petiolules very short; stipules linear, deciduous. Inflorescence raceme. Flowers 1-1.5 cm long, borne in 3-20 flowered, lax, drooping, axillary racemes. Calyx campanulate, lobes deltoid. Corolla yellow, sometimes tinged with purple. Pods 15-25 cm long, sub-cylindrical, torulose with numerous seeded.

Flowering twig

Flowers

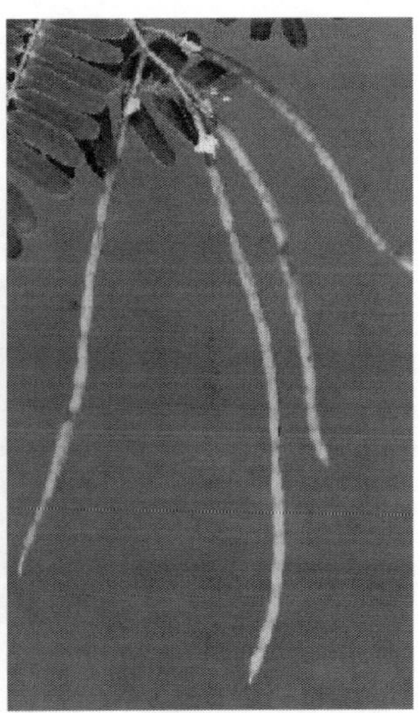

Pods

Chemical Constituents: The pods and leaves contain cholesterol, xampesterol and β- sitosterol. Flowers contain cyanidin, delfinidin, α-ketoglutaric, oxoloaxetic and pyruvic acids. Phytochemicals present in the seeds are oleanolic acid, stigmastane-5.24(28)-diene- 3β-O-β-D-galactopyranoside and galactomannan. The pods and leaves contain campesterol and betasitosterol. Pollen and pollen tubes are known to have alphaketoglutaric, oxaloacetic and pyruvic acids. The seeds are reported to contain crude protein and crude lipid. (Bhutya 2011, Nigussie and Alemayehu 2013).

Medicinal Properties: The leaves have antibiotic, anti-inflammatory, anthelmintic, antitumour and contraceptive properties known to be useful in curing dysuria. Seeds and bark are emmenagogue and astringent, useful to cure menorrhagia, spleen enlargement and diarrhoea. Seeds are cardiac depressant, antibacterial, antipoison, anti-ageing, anodyne and beneficial for hair and skin diseases. Bark is also useful in skin diseases. (Bhutya 2011; http://tropical.theferns.info).

Medicinal Uses (Traditional/Folklore): The leaves of *S. sesban* have traditionally been used as purgative, demulcent, maturant, anthelmintic and for all pains and inflammation. According to ethno medicinal claims the poultice of leaves promotes suppuration of boils and abscesses and absorption of inflammatory rheumatic swellings. Juice of fresh leaves is credited with anthelmintic properties. It is also used in veterinary medicine againts tsetse flies. Among the Haya people of Tanzania, it is used to treat sore throat, gonorrhoea, syphilis, spasmodic fits in children and jaundice during pregnancy. (Usman *et al.* 2013; http://tropical.theferns.info, http://www.worldagroforestry.org).

17.34 *Tephrosia pumila* (Lam.) Pers.

(Syn. *Tephrosia procumbens* (Buch.-Ham.) Gamble)

Common Name: Indigo Sauvage.

Vernacular Name: None known.

Habit: Procumbent herb.

Habitat: Usually hidden among grasses.

Distribution: Pantropical.

Flowering and Fruiting Time: July-September.

Taxonomic Description: A small, diffuse, procumbent or straggling herb. Stem branched, densely strigose. Leaves imparipinnate. Leaflets 5-9, oblanceolate, downy, 1-1.5 cm long, base cuneate, apex retuse or emarginated. Flowers white. Calyx appressed with silky hairs. Corolla white, rarely pink or red. Pods downy, linear, up to 4 cm long, shortly strigose with beaked apex. Seeds oblong, black to brown in colour.

A portion of plant

Flowering twig

Flower

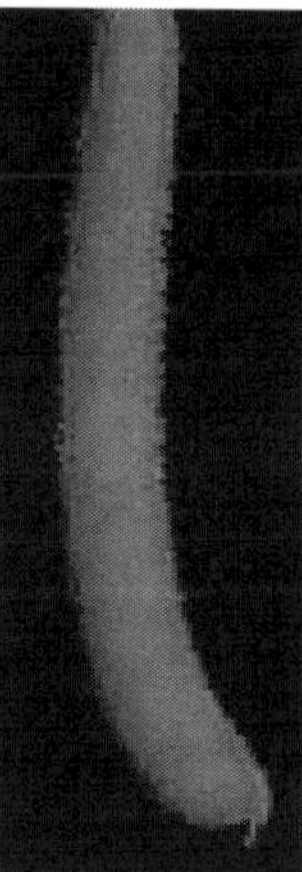

Pod

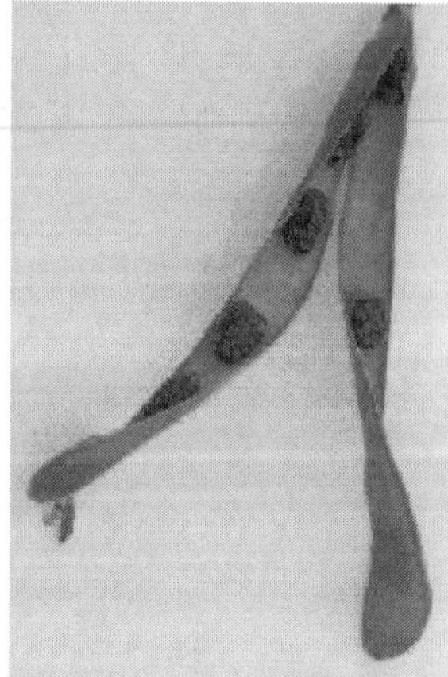

Mature pod with seeds

***Chemical Constituents*:** Plants contain flavonoids (pumilaisoflavones a, pumilaisoflavones b, pumilaisoflavones C, pumilaisoflavones D, pumilanol, tephrinone β-hydroxychalcone, praecansone-A), rotenoid (rotenone), triterpene (lupeol), and sterol (stigmasterol). Ursolic acid and calopogonium isoflavone B also have been reported from this plant. (Touqeer *et al.* 2013).

Medicinal Properties: The plants show antiprotozoal, antibacterial, antifungal and piscicidal activities. (Pethakamsetty *et al.* 2010, Touqeer *et al.* 2013).

Medicinal Uses (Traditional/Folklore): Roots are useful against rheumatism, fevers, pulmonary problems, bladder disorders, coughing, hair loss and reproductive disorders. Whole plant is burnt and ashes used for massaging after wetting with water against stroke. Juice of whole plant (without leaves) is given for the cure of diarrhoea in Nasik, Maharashtra. (Moshi *et al.* 2006, Bobbarala *et al.* 2009, Touqeer *et al.* 2013, Kshirsagar 2016).

17.35 *Tephrosia purpurea* (L.) Pers.

(Syn. *Cracca purpurea* L.)

Habit

Common Name: Wild Indigo, Fish Poison.

Vernacular Name: *Hindi:* Sarphonk (सरफोंक), Sharpunkha (शुखपुष्पा).

Habit: Perennial herb.

Habitat: Open waste places, fallow agricultural fields, along roadsides, and railway lines on poor soils.

Distribution: Pantropical distribution. Though native to eastern part of India, now the plant occurs throughout India.

Flowering and Fruiting Time: June-September.

Taxonomic Description: A small, erect to sub-erect, branched herb, 20-60 cm long. Stem glabrous or slightly pubescent. Leaves imparipinnate, stipulate, short petiolate with 7-19 leaflets. Leaflets oblong-elliptic, oblanceolate, 1.5 to 3.5 x 0.5-1.5 cm, glabrous above and obscurely silky beneath, apex obtuse-retuse or mucronate. Inflorescence racemose, 10-15 cm long, terminal at the end of branchlets; flowers pedicellate. Calyx densely silky, as long as the pedicel. Corolla red or deep purple. Style glabrous. Pods linear, glabrescent, 4-5 cm long, slightly recurved at the tip. Seeds 6-8 per pod, greyish brown, ellipsoid.

Flowers

Pods

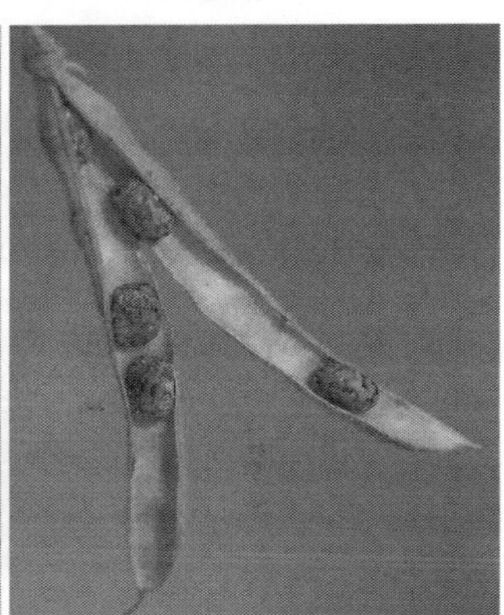
Mature pod with seeds

Chemical Constituents: The constituents of *T. purpurea* include alkaloids, saponins, glycosides, tannins, flavonoids, etc. Roots contain tephrosin, dengulin, quercetin, isotephrosin and rotenone. In roots and leaves, 2.5% rutin is also found. Recently, β-hydroxychalcopurpurnone, isolonchocarpin, pongamol, lanceolatin A, lanceolatin B, karanjin, kanjone and β-sitosterolis have been isolated from roots. The major flavonoids are tephrosin, pongaglabol, semiglabrin, purpuritenin, purpureamethide, pongamol, karanjin, lanceolatin, (+)-tephrorins a, (+)-tephrorins b, (+)-tephrosone, purpurenone, and (+)-purpurin. (Touqeer *et al.* 2013, Akanksha *et al.* 2014, Dalwadi *et al.* 2014)

Medicinal Properties: The plant is bitter, astringent, acrid, thermogenic, anthelmintic, digestive, laxative, diuretic, uterine tonic, anti-inflammatory, depurative, styptic, debostruant, alexiteric, alterant and antipyretic. The roots are useful in inflammations, skin diseases, scrofula, elephantiasis, dyspepsia, stomachalgia, flatulence, haemorrhoids, asthma, bronchitis, anaemia, hepatosplenomegaly, verminosis, strangury, dysmenorrhoea, chronic fever, boils, pimples, odontalgia and gingivitis. The leaves are useful in dyspepsia, pectoral diseases, haemorrhoids, syphilis, gonorrhoea and bruises. The seeds are useful in skin diseases and rat poisoning. (Warrier *et al.* 2010, Touqeer *et al.* 2013, Akanksha *et al.* 2014, Dalwadi *et al.* 2014).

Medicinal Uses (Traditional/Folklore): According to Ayurveda, plant is digestible, anthelmintic, alexiteric, antipyretic, alternative and cures diseases of liver, spleen, heart, and blood, as well as ailments such as tumours, ulcers, leprosy, asthma, poisoning, etc. According to Unani system of medicine, root is diuretic, allays thirst, enriches blood, cures diarrhoea, useful in bronchitis, asthma, liver, spleen diseases, inflammations, boils and pimples; leaves are tonic to intestines and a promising appetizer and effective in piles, syphilis and gonorrhoea.

Traditionally, *T. purpurea* has been used to cure several types of external wounds and gastro-duodenal disorders. Drug is used in cough, angina pectoris (tightness of chest). Decoction of root is useful in enlargement and obstruction of liver, spleen and kidney. Whole plant is used to cure tumours, ulcers, leprosy, allergic and inflammatory condition such as rheumatism, asthma and bronchitis. An extract of pods is effective as analgesic, anti-inflammatory, and their decoction is used in vomiting like symptoms. Leaves are useful in diseases of lungs and the chest, tonic to intestines, improves the appetite, good in piles, syphilis and gonorrhoea. In French Guiana, roots are used as fish poison and antidiarrhoeal and for treating enlargement of spleen. It is also used in tympanitis, dyspepsia, chronic diarrhoea, wounds and pimples. (Shukla *et al.* 2010, Touqeer *et al.* 2013, Akanksha *et al.* 2014, Dalwadi *et al.* 2014, Shaheen *et al.* 2014).

17.36 *Tephrosia villosa* (L.) Pers.

(Syn. *Cracca villosa* L.; *Tephrosia hirta* (Buch.-Ham.) Benth.)

Habit

Common Name: Shaggy Wild Indigo, Hoary Tephrosia.

Vernacular Name: None known.

Habit: Perennial herb.

Habitat: Along roadsides, railway lines, open and waste places.

Distribution: Possibly, native to tropical Africa and India, naturalized in tropical and subtropical Asia. Now it is a widespread species found in southern and eastern Africa, southern Asia and the Arabian Peninsula.

Flowering and Fruiting Time: August-October.

Taxonomic Description: An erect bushy herb, 40-120 cm tall. Stem branched, silky pubescent. Leaves imparipinnate, stipulate; leaflets 7-17, obovate, oblanceolate, glabrescent above, silky beneath, apex often emarginated, base cuneate. Stipules lance-shaped. Inflorescence raceme, 10-20 cm long. Flowers pale pink or pale red. Calyx densely silky villous. Style glabrous, stigma penicillate. Pods deflexed, densely silvery tomentose, up to 4 cm long. Seeds 6-8.

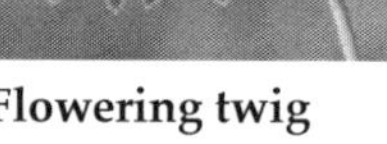

Flowering twig

Silky villous calyx

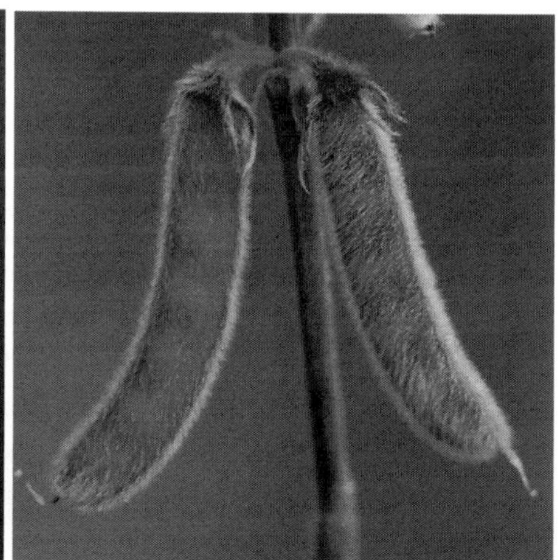

Silvery tomentose pods

Chemical Constituents: Presence of various phytoconstituents such as carbohydrates, saponins, coumarin glycosides, flavonoids and phytosterol has been reported. The roots and leaves contain retinoid. (Ahmad *et al.* 2016).

Medicinal Properties: *T. villosa* possesses anthelmintic, alexiteric, antipyretic properties. (Bobbarala *et al.* 2009).

Medicinal Uses (Traditional/Folklore): Root, leaves and bark are useful to cure diseases of liver, spleen, heart, blood, and effective against asthma, leprosy, ulcers, etc. (Bobbarala *et al.* 2009).

17.37 *Teramnus labialis* (L.f.) Spreng.

(Syn. *Glycine labialis* L.f.)

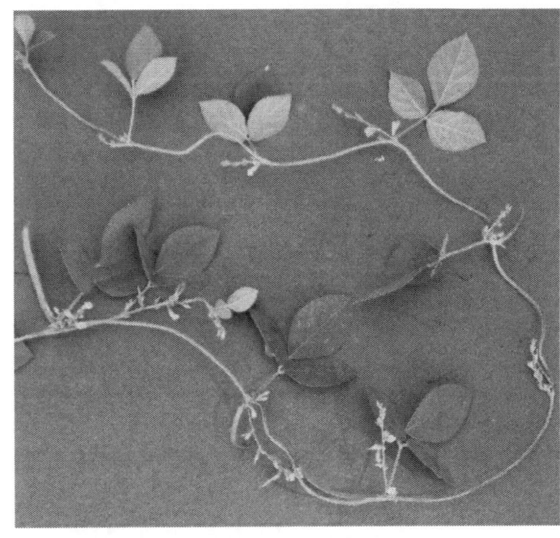

Habit

Common Name: Rabbit Vine, Horse Vine.

Vernacular Name: *Hindi*: Mashani (मशानी), Van Udad (वन उड़द); *Sanskrit*: Mashaparni (माषपर्णी), Kalyani (कल्याणी).

Habit: Climber.

Habitat: Generally found twining on herbs and bushes in wastelands, unused grounds and along roadsides.

Distribution: The species is distributed in tropical and southern Africa, India, Malaysia, Indonesia, the Philippines, New Guinea, the eastern Indian Ocean, the West Indies and Guyana.

Flowering and Fruiting Time: August-November.

Taxonomic Description: A spreading climber. Stem slender, twining, appressed hairy. Leaves alternate, 3-foliate; leaflets membranous, 3-6.5 cm long, elliptic or ovate-oblong, tip sub-acute, base rounded, margin entire, densely hairy beneath, lateral leaflets slightly smaller than the terminal one, stipules deciduous. Inflorescence loose axillary raceme. Flowers small; bracts and bracteoles minute, linear-subulate. Calyx 5-partite, pubescent. Corolla pinkish-white or purplish. Stamens monadelphous. Pods flat, 4-6 cm long, more or less pubescent, linear, slightly curved, 9-12 seeded.

Flowering and fruiting twig

Pods

Chemical Constituents: Seeds contain potassium, magnesium, calcium, and, phosphorus crude protein, crude fat, and ash. The essential amino acids lysine, leucine, +isoleucine, arginine, valine and histidine were present in relatively large quantities; antinutritional factors such as total freephenols, tannins, L-DOPA, hydrogen cyanide and phytic acid were present in minute quantities. Fatty acids such as palmitic, linoleic, oleic and stearic acids were present in high concentrations whereas linolenic acid levels were low. (Viswanathan *et al.* 1999; http://medplants.blogspot.in).

Medicinal Properties: The fruit is bitter, cooling and sweet. It is used as aphrodisiac, stomachic, nervine tonic, astringent to the bowels, antipyretic and galactagogue. (http://medplants.blogspot.in, https://easyayurveda.com).

Medicinal Uses (Traditional/Folklore): *T. labialis* is a well-known medicinal plant in the Ayurvedic system of medicine. It has been reported to be useful in treating rheumatism, tuberculosis, nerve disorders, paralysis and catarrhs. It is also used in inflammation, biliousness, blood diseases, gout, fevers, bronchitis, thirst, burning sensation, paralysis, affections of the nervous system, haemoptysis. As an Ayurvedic medicine the plant is used for treating fatigue, muscle wasting, *Vata* and *Pitta* imbalance disorders, bleeding disorders, etc. It is also used in the treatment of sexual debility. Root is used in the treatment of fever. (https://easyayurveda.com, http://medplants.blogspot.in).

17.38 *Trifolium tomentosum* L.

(Syn. *Trifolium curvisepalum* Tackh.)

Habit

Common Name: Woolly Clover, Cottonball Clover.

Vernacular Name: Tipatiya Ghaas (तिपतिया घास).

Habit: Annual herb.

Habitat: Moist and waste places, rice fields.

Distribution: Native to northern Africa, the Azores, the Madeira Islands, the Canary Islands, southern Europe, the middle-east and western Asia.

Flowering and Fruiting Time: February-April.

Taxonomic Description: A prostrate to decumbent-ascending, glabrous, annual herb. Stem striate, branched from the base. Leaves trifoliate, stipulate, upper short petioled, lower long petioled; leaflets obovate, rounded, truncate or emarginated at apex, dentate-serrate at margin, sub-sessile, up to 2 cm long. Heads globose, 0.5-1.0 cm across, peduncles shorter than the subtending leaves. Flowers pinkish, shortly pedicellate. Calyx tubular, 5-lobed. Corolla longer than the calyx, lobes-5. Style glabrous. Pods in cluster, resemble cotton balls, ovoid, ellipsoid. Seeds 2, small, yellowish-brown.

Flowering twig with ovoid clustered pods

Flowers

Chemical Constituents: None known.

Medicinal Properties: None known.

Medicinal Uses (Traditional/Folklore): None known.

17.39 *Vicia sativa* L.

(Syn. *Vicia alba* Moench)

Habit

Common name: Common Vetch, Garden Vetch, Black-Pod Vetch.

Vernacular name: *Hindi:* Chatri (छतरी), Atri (अत्रि), Matri (मात्री).

Habit: Annual sub-erect herb.

Habitat: A common winter weed, mixed with grasses in cultivated fields, moist waste places.

Distribution: *V. sativa* is widespread in the Mediterranean basin, in West and Central Asia, China, eastern Asia and in North America. In India, it is distributed in Assam, Bihar, Delhi, Haryana, Himachal Pradesh, Punjab, Madhya Pradesh, Meghalaya, Rajasthan, Uttar Pradesh, and West Bengal.

Flowering and Fruiting Time: January-March.

Taxonomic Description: An annual, sub-erect herb. Stem angular, branched, pubescent. Stipules small, obliquely lanceolate, deeply toothed. Leaflets 4-6 pairs, entire, oblong-lanceolate, mucronate to emarginated at apex, up to 3 cm long. Flowers axillary, solitary or in paired. Calyx teeth pubescent, lanceolate, subulate. Corolla reddish-blue to purple, twice the length of calyx. Stamens diadelphous. Pods 3-5 cm long, linear, glabrous with beaked apex. Seeds 8-10.

Flowering twig

Flower

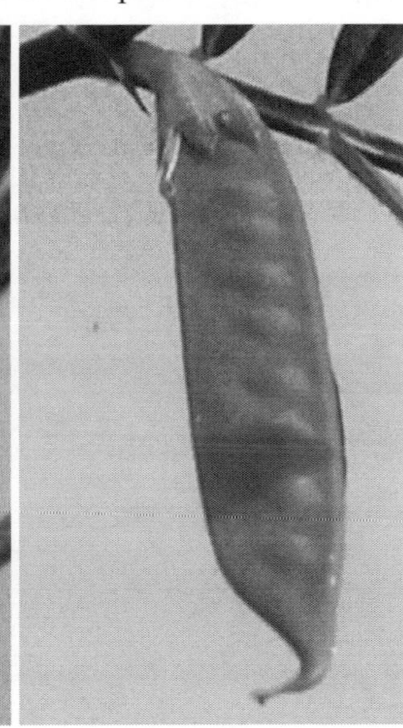

Pod

Chemical Constituents: *V. sativa* contain important constituents likes chromatotropic acid, gallic acid, quercitin, vanillic acid, syringic acid, vitamin C, trans-4-hydroxy-3-methoxy cinamic acid and kaempferol. (Saleem *et al.* 2014)

Medicinal Properties: The plant is used as an antiasthmatic, aphrodisiac, antipyretic, appetizer antirheumatic, digestion tonic, diuretic, demulcent, expectorant, laxative, cholagogue and refrigerant. It is mostly used for the treatment of asthma, coughs, bronchitis, skin diseases, piles and urinary troubles. (Gulshan *et al.* 2012, Saleem *et al.* 2014).

Medicinal Uses (Traditional/Folklore): The Common Vetch is traditionally used as a medicinal plant in skin infections, asthma, bronchitis, urinary diseases and also used as antiseptic, antipoison, aphrodisiac, antirheumatic and antipyretic. It is considered to be antidote of poison. In Pakistan (semi-tribal areas of Makerwal and Gulla Khel in Khyber), powder of the plant is dissolved in bathing water for cleaning the dandruff. (Shah *et al.* 2013, Saleem *et al.* 2014).

18. MYRTACEAE

Myrtaceae – Myrtle family

The family contains about 145 genera and 5,970 species that are widely distributed in the tropics. All species are woody, with essential oils, and flower parts in multiples of four or five. They have rather leathery evergreen leaves with oil glands. Some plants of economic importance belong to this family are *Eucalyptus*, Guava, Rose Apple, Clove, Jamun, etc. One species of this family was recorded in the flora.

18.1 Syzygium cumini (L.) Skeels

(Syn. *Eugenia jambolana* Lam.)

Fruiting branch of tree

Common Name: Java Plum, Jamun, Black Plum.

Vernacular Name: *Hindi:* Jamun (जामुन); *Sanskrit*: Jambula (जंबूल).

Habit: Evergreen tropical tree.

Habitat: Found in forests up to 1,800 m usually along river banks and moist localities; also cultivated in orchards, house compounds and as shade trees along roadsides.

Distribution: Naturally growing in India, Sri Lanka, Malesia and northern Australia. Trees are found growing throughout the Asian subcontinent, Eastern Africa, South America, Madagascar and have also naturalized in Florida and Hawaii in the United States of America.

Flowering and Fruiting Time: March-June.

Flowers

Young fruits

Ripe and unripe fruits

Taxonomic Description: An evergreen, medium sized tree. Stem branched, bark greyish brown. Leaves up to 15 cm long, petiolate, opposite, elliptic-oblong, coriaceous, smooth, glossy, gland-dotted, with many close parallel veins, entire, acute to acuminate or obtuse at apex, cuneate or obtuse at base. Inflorescence panicled cymes. Flowers greenish-white, fragrant. Calyx teeth-4, tube campanulate. Petals-4, united, calyptrate, deciduous. Stamens numerous, filaments filiform. Ovary inferior. Fruit a berry, edible, ovoid-oblong, dark-purple when ripe. Seed one, oblong.

Chemical Constituents: The plant is rich in compounds containing anthocyanins, glucoside, ellagic acid, isoquercetin, kaemferol and myrecetin. The seeds are claimed to contain alkaloid, jambosine, and glycoside jambolin or antimellin, which halts the diastatic conversion of starch into sugar besides vitamin C, vitamin A, riboflavin, nicotinic acid, Choline, Folic acid, Malaic acid, Sugar, Amino acids, Na, K, Ca, P, Fe, Zn, Mn.

Different parts of the plant have the following phytochemicals: *Flower:* Terpenoids; *Stem bark:* Betulinic acid, friedelin, friedelinol, daucosterol, kaempferol, kaempferol-3-O-glucoside, quercetin, myricetin, astragalin, β-sitosterol, β-sitosterol glucoside, sucrose, gallic acid, ellagic acid, cuminiresinol, 5′-hydroxy- methyl piperitol, syzygiresinol A, syzygiresinol B, demethyl-5-hydroxypinoresinol, dimethylpinoresinol, didemethoxypinoresinol, pinoresinol, 4′-methyl-5′-hydroxypinoresinol, bergenin; *Fruit:* Delphinidin-3-gentiobioside, malvidin-3-laminaribioside, petunidin-3-gentiobioside, pconidin, pelargonidin, petunidin, mallic acid, oxalic acid, tannins, cyanidin diglycoside, waxy component, triterpenhydroxy acid, oleanolic acid; *Seeds:* Glycoside, jamboline Gallic acid, ellagic acid, corilagin, ellagitannins, 3,6- hexahydroxydiphenoyl glucose, 4,6-hexahydroxydiphenoyl glucose, 1-galloyl glucose, 3-galloyl glucose, quercetin, 3,3′,4′ tri-O-methyl ellagic acid, 3,4′-di-O-methyl ellagic acid, caffeic acid, ferulic acid, guaiacol, resorcinol dimethyl ether, veratrole, lignanglucoside, medioresinol 4″-O-β-glucoside, (+)-pinoresinol-o-β-glucoside, (+)-syringaresinol-o-β- glucoside, dihydrodehydrodiconiferyl alcohol-4′-O-β-glucoside, 5- hydroxy methyl furfural, betulinic acid, 3,5,7,4′-tetrahydroxy flavanone. (Jadhav *et al.* 2009, Salim and Paarakh 2009, Ayyanar and Babu 2012, Chaudhary and Mukhopadhyay 2012, Swami *et al.* 2012).

Medicinal Properties: It is effective in the treatment of diabetes mellitus, inflammation, ulcers and diarrhoea and preclinical studies have also shown to possess chemopreventive, radioprotective and antineoplastic properties. Different parts of the plant are also reported for its antioxidant, anti-inflammatory, neuropsycho-pharmacological, antimicrobial, antibacterial, anti-HIV, antileishmanial and antifungal, nitric oxide scavenging, free radical scavenging, antidiarrhoeal, antifertility, anorexigenic, gastroprotective and antiulcerogenic and radioprotective activities. The bark is astringent, sweet, sour, acrid, refrigerant, carminative, diuretic, digestive, anthelmintic, febrifuge, constipating, stomachic and antibacterial. It is useful in diabetes, leucorrhoea, stomachalgia, fever, gastropathy, strangury and dermatopathy. The leaves are antibacterial and are used for strengthening the teeth and gums; tender leaves are used for preventing vomiting. The fruits and seeds are sweet, acrid, sour, tonic and cooling, and are used in diabetes, diarrhoea, pharyngitis, splenopathy, urethrorrhea and ringworm. (Jadhav *et al.* 2009, Salim and Paarakh 2009, Warrier *et al.* 2010, Chaudhary and Mukhopadhyay 2012, Swami *et al.* 2012).

Medicinal Uses (Traditional/Folklore): During the last four decades, numerous folk medicine and scientific reports on the antidiabetic effects of this plant have been cited in the literature. Clinical and experimental studies suggest that, different parts of the plant especially fruits, seeds and stem bark possess promising activity against diabetes mellitus. It is one of the significant antidiabetic plant and it has long been reported for its use in many pharmacological activities mainly diabetes. As a traditional and folklore medicine seeds are ground to make powder which is taken three times daily for controlling diabetes. Juice extracted from the leaf is mixed with honey or cow's milk and fresh fruits are taken orally which have been found to be effective against diabetes in southern India. The dosage includes taking two teaspoon of juice twice a day after food for three months. The eating of its fruit in season is also very useful for diabetes. Hot water extract of dried fruits is used for stomach ulcers, reduce acidity and for diabetes. Traditional medical healers in Madagascar have been using the seeds of jambolan for generations as the centerpiece of an effective therapy for counteracting the slow debilitating impacts of diabetes.

Young leaves (250 g) along with tender stem are washed with water, crushed and mixed with 125 g of pure butter to make a paste. This paste is applied to the wounds caused due to burning. Young leaves are crushed and filtered through a fine cloth and one teaspoon of the filtrate is given to children and three teaspoons to elders three times daily to cure kidney and urine troubles. Juice of tender leaves of this plant, leaves of mango and myrobalan are mixed and administered along with goat's milk and honey to treat dysentery with bloody discharge, whereas juice of tender leaves alone or in combination with carminatives such as cardamom or cinnamon is given in goat's milk to treat diarrhoea in children. The seed extract is used to treat cold, cough, fever and skin problems such as rashes and the mouth, throat, intestines and genitourinary tract ulcers (infected by *Candida albicans*) by the villagers of Tamil Nadu, India.

A syrup prepared from the juice of the ripe fruit is a very pleasant drink. Syrup or vinegar prepared from the ripe fruit is useful in spleen enlargement and efficient astringent in chronic diarrhoea. (Ayyanar *et al.* 2008, Jadhav *et al.* 2009, Salim and Paarakh 2009,Shukla *et al.* 2010, Marwat *et al.* 2011b, Ayyanar and Babu 2012, Chaudhary and Mukhopadhyay 2012, Swami *et al.* 2012, Shaheen *et al.* 2014).

19. LYTHRACEAE

Lythraceae – Loosestrife family

The family includes 31 plant genera and 604 accepted species. Lythraceae has a worldwide distribution, with most species in the tropics, but ranging into temperate climate regions as well; many species occur in aquatic or semi-aquatic habitats. Species are most often herbs, and less often shrubs or trees; the shrubs and trees often have flaky bark. The leaves are simple usually in pairs (opposite or whorled) and stipules typically reduced, appearing as a row of minute hairs, or absent. Flower petals often appear crumpled and emerge from the rim of the calyx tube. The stamens commonly are twice the number of petals, and are usually in two whorls, one with the filaments longer than the other. Single species from this family was recorded in the flora.

19.1 *Ammannia baccifera* L.

(Syn. *Ammannia vescicatoria* Roxb.)

Common Name: Blistering Ammannia, Acrid Weed, Monarch Redstem, Tooth Cup.

Vernacular Name: *Hindi:* Agnibuti (अग्निबूटी), Dadmari (दादमारी), Jungli Mehendi (जंगली मेंहदी); *Sanskrit:* Agnigarbha (अग्निगार्भ), Brahmasoma (ब्राह्मसोम).

Habit: Annual herb.

Habitat: Commonly found in marshes, swamps, rice fields and water courses at low elevations.

Distribution: It is widespread in the tropical regions of Asia, America and Africa: Tropical Africa, through Asia to India, Thailand, Cambodia, Laos, Vietnam, Malaysia, Philippines to Australia. A common weed distributed throughout India.

Flowering and Fruiting Time: August-November.

Taxonomic Description: An erect, 10-40 cm tall, branched, annual herb with 4-angled stem. Leaves opposite-decussate, sessile, glabrous, 2-5 cm long, oblong to linear-lanceolate or narrowly elliptic-lanceolate, margin entire, recurved, base generally attenuate, apex acute-obtuse. Flowers minute, pinkish or greenish-pink, bracteate, in axillary clusters. Calyx usually 4-lobed, campanulate, sepals triangular. Petals absent. Capsules exceeding the calyx, globose, 1.5-2 mm in diameter with ovoid, black seeds.

Chemical Constituents: Phytochemical screening yielded tannins, saponins, flavonoids, steroids, and alkaloids. Leaves contain an acrid principle, resin, glucose and lawsone and carbohydrates, glycosides, saponins, proteins, steroids, flavanoids, tannins and phenolic constituents. (Jani *et al.* 2012, Manoharan 2014, Vasudevan *et al.* 2014, Al-Snafi 2015c).

Habit

Medicinal Properties: Plants are stomachic, laxative, antirheumatic, febrifuge, and rubefacient. The leaves are irritant, rubefacient, vesicant, appetizer, laxative and stomachic. Plant extract also possesses antibacterial properties. (Manoharan 2014, Vasudevan *et al.* 2014, Al-Snafi 2015c; http://medplants.blogspot.in).

Medicinal Uses (Traditional/Folklore): In India, the leaves are used to stimulate appetite, relieve the bowels of costiveness, promote digestion, and stimulate venereal desire (Ayurveda). The plant is used in the treatment of fevers, malaria, and jaundice and to eliminate blood clots. The plants are used in treatment of scabies, ringworm, parasitic skin infections, common cold, typhoid, strangury, spinal disease, gastroenteropathy and aphrodisiac. It is useful in vitiated condition of *Kapha* and *Pitta*. The leaves are beneficial for removing phlegm from the lungs and trachea. The leaves or the ashes of the plant, mixed with oil, are applied to cure herpetic eruptions. The fresh, bruised leaves have been used in skin diseases as a rubefacient and as an external remedy for ringworm and parasitic skin affections. The freshly bruised leaves are applied as a plaster to the skin in order to increase local blood supply in the treatment of skin diseases, wounds and swellings. Leaves are used externally for ringworm, herpes eruptions and other skin diseases and in strangury, cause biliousness. The juice causes ulcers and blebs; they are used universally by the natives to raise blisters in rheumatic pains. The leaves or ashes of the plant, mixed with oil, are applied to herpetic eruptions.

In Orissa, (Jhara, Keuta, Dhivara people) five-six shoots are ground along with seven fruits of *Piper nigrum*, and the mixture is applied to cure ring worm. Leaf juice with honey is given in empty stomach to treat typhoid whereas leaf paste with common salt is useful to cure skin diseases like scabies, ringworms, skin itching etc in south Orissa. In Andhra Pradesh (Sugali), oral administration of 20 g whole plant powder for every four hours with hot cow milk works against *Bungarus fasciatus* (banded krait) bite. In North-West Himalayas, Gujjars and Bhotiya tribes use the decoction prepared from 10 g of fresh leaves with 10 g of *Cyperus rotundus* roots and 5g of fresh ginger twice a day against intermittent fever. Paste of leaves and inflorescence is applied over skin to check itching.

A. baccifera is also extensively used in traditional Chinese herbal formulations for treatment of spinal disease, gastroenteropathy, haemorrhoids, common cold, abscess, sore, itching and other skin infection. In the Philippines, the fresh leaves are bruised and applied to the skin to raise blisters and treat biliousness. (Panda and Mishra 2011, Jani *et al.* 2012, Kumar and Mishra 2012, Manoharan 2014, Vasudevan *et al.* 2014, Al-Snafi 2015c; http://medplants.blogspot.in).

20. ONAGRACEAE

Onagraceae – Evening Primrose family

The family, named after the genus *Onagra* (now known as *Oenothera),* include about 832 species of herbs, shrubs, and trees in 45 genera. The family is widespread, occurring on every continent from boreal to tropical regions. The family is characterized by flowers with parts mostly on the plan of four (four sepals, four petals, four or eight stamens), but there are some exceptions. The ovary is inferior (i.e., below the flower proper). The family includes a number of popular garden plants, including evening primroses (*Oenothera*) and fuchsias (*Fuchsia*). The family was represented by two species in the flora.

20.1 *Ludwigia hyssopifolia* (G.Don) Exell

(Syn. *Jussiaea hyssopifolia* G.Don; *Jussiaea linifolia* Vahl)

Common Name: Linear Leaf Water Primrose, Primrose Willow, Swamp Primrose, Water Primrose.

Vernacular Name: *Hindi:* Ban Long (बन लौंग), Bishkatali (बिष्टकाटली); *Sanskrit:* Bhu Lavangah (भूलवंग).

Habit: Annual herb.

Habitat: A widespread weed of rice fields; also growing abundantly along the edges of canals, rivers and wetlands.

Distribution: Pantropical in distribution. *L. hyssopifolia* has probably been introduced to tropical Africa; East Asia - southern China, India, Bangladesh, Bhutan, Sri Lanka, Nepal, Myanmar, throughout southeast Asia to New Guinea and Australia. In India, it is distributed in Andaman & Nicobar Islands, Assam, Andhra Pradesh, Bihar, Gujarat, Goa, Kerala, Manipur, Mizoram, Orissa, Rajasthan, Tamil Nadu, and Uttar Pradesh.

Habit

Flowering and Fruiting Time: August-December.

Flowering and fruiting twig

Flower

Capsule

Taxonomic Description: An annual, erect herb, 30-70 cm high. Branches angled or slightly winged. Leaves alternate, shortly petiolate, oblong-lanceolate, 4-8 x 1-2 cm, entire, base cuneate, apex acute to acuminate, somewhat pubescent along nerves beneath. Flowers 6-7 mm across, axillary, solitary. Calyx 4-partite, teeth ovate-lanceolate, persistent. Corolla 4-lobed, lobes obovate, bright yellow. Stamens-8. Ovary 4-celled, style very short, stigma 4-lobed. Capsules cylindric, 2-3 cm long, slightly hairy. Seeds numerous, minute.

Chemical Constituents: It is known to contain piperine, vitexin, isovitexin, orientin and isoorientin. (Das *et al.* 2007, Rao *et al.* 2013).

Medicinal Properties: The plant is considered as anti-inflammatory, analgesic, astringent, antitumour, antidiarrhoeal, anthelmintic, antibacterial, carminative and diuretic. (Das *et al.* 2007, Rao *et al.* 2013, Das *et al.* 2014a).

Medicinal Uses (Traditional/Folklore): It is used in diarrhoea, dysentery, enteritis and sprue. Plant decoction is used in diarrhoea and dysentery, flatulence, leucorrhoea, spitting of blood, and as a vermifuge and purgative. Decoction of leaves and stem is used to wash out infected wounds, sores, ringworm. A poultice of the plant is used to treat pimples, boils and other infections. Pounded leaves are used for dysentery; leaf paste is applied on boils, cuts and wounds and sores. The leaves are used in a febrifuge decoction and for poulticing in orchitis and glands in the neck. In Kerala, whole plant is used for burns, gonorrhoea and catarrhal affections of children; it promotes sleep, improves strength, increases semen and helps easy delivery. A cold infusion of the roots is used to treat syphilis. (Das *et al.* 2007, Rao *et al.* 2013, Deepa *et al.* 2016; http://tropical.theferns.info).

20.2 *Ludwigia octovalvis* (Jacq.) P.H.Raven

(Syn. *Jussiaea suffruticosa* L.)

Common Name: Mexican Primrose-Willow, Swamp Primrose, Water Primrose.

Vernacular Name: *Hindi:* Ban Long (बन लौंग); *Sanskrit:* Bhu Lavangah (भूलवंग).

Habit: Perennial herb.

Habitat: Mostly found in humid places, damp grassland, rice-fields and along the banks of canals and rivers.

Flower

Capsule

Upper portion of plant

Distribution: *L. octovalvis* is native of tropical Africa; found throughout the tropics, from sea level to 1,500 m elevation. It occurs in Central America, Australia, South-East Asia, the Middle East, the Central-West African regions and spreads easily to become naturalized; in India, recorded from Uttar Pradesh, Andaman and Assam.

Flowering and Fruiting Time: July-October.

Taxonomic Description: An erect, perennial herb. Stem branched, pubescent, reddish-brown in colour. Leaves up to 7 cm long, sessile or sub-sessile, elliptical, ovate-lanceolate, villous, entire, acute or acuminate at apex. Flowers shortly stalked. Calyx lobes-4, ovate, lanceolate. Petals-4, bright yellow, obovate. Stamens-8. Ovary inferior. Capsules about 5 cm long, linear, cylindrical, 8-ribbed, appressed pubescent, terminated by persistent calyx. Seeds numerous, brown.

Chemical Constituents: Thirteen compounds have been isolated from the plant: β-sitosterol, oleanolic acid, 2α-hydroxy ursolic acid, tormentic acid, daucosterol, maltol, luteolin, quercetin, apigenin, methyl brevifolincarboxylate, gallic acid, 3, 4, 8, 9, 10-pentahydroxydibenzo[b, d]pyran-6-one, and ellagic acid. (Yan and Yang 2005).

Medicinal Properties: The plant is known for its antioxidant, anti-ageing and antibacterial properties. (Yakob *et al.* 2012, Lin *et al.* 2014).

Medicinal Uses (Traditional/Folklore): A poultice prepared from the entire plant is applied externally on the skin for the cure of dermatitis, boil, ulcer, impetigo and pimple. Decoction made from the plant is taken as a health drink and to treat gastrointestinal complaints such as diarrhoea and flatulence. The whole plant is given in fever, toxemia; boiled plant is applied on body in fever to reduce body ache. Leaf decoction is given to treat dysentery in Andhra Pradesh. For the treatment of diabetes in North-Eastern states of India boiled extract of whole plant is given. (Kumar and Narain 2010, Yakob *et al.* 2012, Mondal *et al.* 2013, Savithramma *et al.* 2015).

21. CUCURBITACEAE

Cucurbitaceae – Cucumber, Cucurbit or Gourd family

A family of herbaceous climbing or trailing plants with tendrils like cucumbers, melons and gourds consisting of about 965 species in around 134 genera. Most of the plants are annual vines, but some are woody lianas, thorny shrubs, or trees. The stems are hairy and pentangular. Many species have large, yellow or white unisexual flowers, in some species male and female flower grow on different plants. The family recorded five species belonging to five genera in Chithara Village flora.

21.1 *Citrullus colocynthis* (L.) Schrad.

(Syn. *Colocynthis vulgaris* Schrad.; *Cucumis colocynthis* L.)

Habit

Common Name: Colocynth, Desert Gourd.

Vernacular Name: *Hindi:* Indrayan (इन्द्रायन); *Sanskrit:* Gavakshi (गावकशी), Chitra (चित्रा), Indravaruni (इंद्रवारुणी).

Habit: Trailing herb.

Habitat: Commonly found in dry areas, open and wastelands.

Distribution: Plant is native to the Mediterranean basin and Asia, especially Turkey (regions such as İzmir), Nubia and Trieste.

Flowering and Fruiting Time: April-October.

Taxonomic Description: A trailing, perennial herb. Stem angular, scabrid. Tendrils slender, glabrous. Leaves alternate, petiolate, cordate at base, ovate or triangular in outline, many-lobed; lobes sinutely pinnatified, scabrous beneath. Inflorescence solitary, axillary. Flowers monoecious, long pedicellate, yellow. Calyx and corolla 5-lobed in both male and female flowers. Ovary inferior, ovoid. Fruits 4-8 cm in diameter, globose, longitudinally dark green striped, intensively bitter; pulp white, spongy. Seeds yellowish-brown, numerous, compressed, ovate-oblong.

Flower

Twig with Fruit

Seeds

Chemical Constituents: The plant contains saponin, flavonoids, terpenoids and alkaloids. Seeds contain phytosterolin, two phytosterols and hydrocarbons, polysaccharides and glycosides; it also contains anti tumeruos agents like cucurbitacin B and cucurbitacin E and are rich in fatty acids such as myristic, palmitic, stearic, oleic, linoleic and linolenic acids. Flavon c-glucosides is present in fruit and aerial parts. Fruit contains iso-vitexin, isoorientin and iso-orientin-3-methylether, while aerial parts contain C-p-hydroxybenzyll derivatives. The fruit pulp contains certain bitter chemicals like colocynthin and colocynthenin, fixed or stable oils, resins that are not soluble in ether, sticky substances that is, gums, pectin, and minerals like Ca and Mg, including traces of magnesium phosphate and lignin. Their active ingredients are phytosterol glycoside, elaterin, and albuminoids. (Rahuman *et al.* 2008, Gurudeeban *et al.* 2010, Borhade *et al.* 2013, Hussain *et al.* 2014, Al-Snafi 2016b).

Medicinal Properties: Bitter, hot in nature, purgative, cathartic, anthelmintic, antipyretic, anti- inflammatory, antibacterial, antidiabetic, strong laxative and carminative. Useful in rheumatic pains and arthritis. The plant is known to possess beneficial effects as purgative, vermifuge, blood purifier, and for treating tumour in stomach, ascites, menstrual disorders, etc. It is also used to initiate abortion. In therapeutic doses, it is used in the treatment of jaundice, splenomegaly, ascites, Asthma, COPD, wheezing, breathing difficulty, cough, cold, skin diseases, abdominal tumour, bloating, and boils and nodules. (Gurudeeban *et al.* 2010, Borhade *et al.* 2013, Hussain *et al.* 2014, Al-Snafi 2016b; https://easyayurveda.com).

Medicinal Uses (Traditional/Folklore): Colocynth is an important Ayurvedic herb. Due to its usefulness in curing so many ailments it is called as bitter apple. Colocynth has been widely used in folk medicines for centuries. Johann Weyer, in *De praestigiis daemonum* (1563), offers it as a cure for lycanthropy. Interest in its anti-inflammatory properties has been renewed in modern times. Another property of colocynth is hair growth stimulation and its high efficacy to cure baldness. It is used in the treatment of urinary tract infection and colic. Plant extract is given to cure breast cancer, skin ailments, warts, scraps and corns; it is also helpful in liver disorders such as jaundice. The plant is an antidote for scorpion and snake bite. In villages, the plant is used effectively in treating menstruation disorders such as menopause, leucorrhoea, amenorrhea, menses, etc. Pulp of fruit is used for varicose vein and piles. A paste of the root is applied to the enlarged abdomen of children which is also effective in treating ascites and piles.

Colocynth is used as an ingredient in the following Ayurvedic medicines: *Maha Manjishtadi kashayam* – used as blood purifier in skin diseases; *Abhayarishta* – used in haemorrhoids and constipation; *Maha Vishagarbha Taila* – used in sciatica and joint disorders with stiffness as a symptom. (Gurudeeban *et al.* 2010, Borhade *et al.* 2013, Hussain *et al.* 2014, Al-Snafi 2016b, Upasani *et al.* 2017; https://easyayurveda.com, https://www.gbif.org/species/113598580).

21.2 *Coccinia grandis* (L.) Voigt

(Syn. *Bryonia grandis* L.; *Cephalandra grandis* Kurz; *Coccinia indica* Wight & Arn.)

Habit

Common Name: Ivy Gourd, Scarlet Gourd.

Vernacular Name: *Hindi:* Kundru (कुंदरू).

Habit: Climber.

Habitat: Commonly found climbing on other herbs and shrubs.

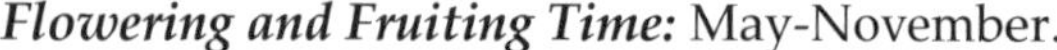

Distribution: Its native range extends from sub-Saharan Africa and South and Southeast Asia, including India, the Philippines, Cambodia, China, Indonesia, Malaysia, Myanmar, Thailand, Vietnam, eastern Papua New Guinea, and the Northern Territories, Australia. Found throughout India, cultivated as well as wild.

Flowering and Fruiting Time: May-November.

Flower **Pubescent ribbed corolla** **Fruit**

Taxonomic Description: A perennial, extensive climbing herb, dioecious. Leaves alternate, long petioled, 4-9 cm across, 3-5 lobed, entire or somewhat sinuate denticulate, base cordate, apex obtuse or mucronate margin, glandular tipped. Tendrils simple, filiform. Flowers incomplete, pedicellate, white. Male and female flowers axillary, solitary. Calyx lobes linear-oblong, reflexed, broadly campanulate. Corolla nearly 2.5 cm long, campanulate, triangular, pubescent, many ribbed. Stamens-3, filaments short. Ovary inferior, style short, stigma 3-partite. Fruits fleshy, ovoid or oblong, 4-5 cm long, bright red when ripe. Seeds many, embedded in red pulp, oblong, compressed.

Chemical Constituents: Major biochemical constituents present in the plant are alkaloids, phenols, carbohydrates, glycosides, flavonoids, tannins and saponins. (Selvam *et al.* 2014).

Medicinal Properties: Leaves of the plant are reported to have antidiabetic, antibacterial activity, hepatoprotective activity. The fruits are cooling, sweet, astringent, depurative, antipyretic, galactagogue and expectorant. Fruit extract shows antioxidant activity while stem extracts are antispasmodic. Extract of fresh leave is reported to be anti-inflammatory. It also found to be useful for treatment of bronchitis, and is effective for larvicidal action and antiulcer activity. The roots are cooling and aphrodisiac.

In Ayurveda, the plant is used in various diseases/ailments due to its wide variety of properties such as bitter, hypoglycaemic, alexiteric, amebicide, antidiabetic, antidyslipidemic, antiemetic, antiprotozoal, antipyretic, antiseptic, antispasmodic, anti-hyperuricaemia and antitussive, aphrodisiac, astringent, adenopathy, dysuria, depurative, diaphoretic, emetic, expectorant, hepatoprotective. (Warrier *et al.* 2010, Selvam *et al.* 2014; http://medplants.blogspot.in).

Medicinal Uses (Traditional/Folklore): The plant is effective in treating, anaemia, asthma, biliousness, bronchosis, catarrh, constipation, convulsion, cramp, dermatosis, diabetes, dysgeusia, earache, enterosis, fever, fungus, gas, glossosis, glycosuria, gonorrhoea, gout, gravel, halitosis, inflammation, itch, jaundice, laxative, leprosy, lowering blood cholesterol, menorrhagia, mycosis, ophthalmia, parasiticide, protisticide, parturition, psoriasis, ringworm, sinusosis, smallpox, snakebite, sore, sore throat, stomatosis, syndrome-X, syphilis, tuberculosis, and vomiting.

The leaves are bitter, sweet, astringent and cooling, and are useful in vitiated condition of *Kapha* and *Pitta.* The leaves and fruits are used in skin diseases, fever, asthma, cough and jaundice. The fruits are useful in burning sensation, leprosy, skin diseases, intermittent fever, agalactia, asthma, cough, bronchitis, consumption and jaundice. The roots are used in vomiting, burning sensation and uterine discharges. In Bangladesh, the leaves of this plant are used to treat scabies, fruits are used to treat infertility of females and roots are used to treat joint pain.

In traditional medicine, fruits have been used to treat leprosy, fever, asthma, bronchitis and jaundice. The fruit possesses mast cell stabilizing, anti-anaphylactic and antihistaminic potential. It is claimed that the plant helps to regulate blood sugar levels (antidiabetic). There is some research to support that certain compounds in the plant inhibit the enzyme glucose-6-phosphatase, which is one of the key liver enzymes involved in regulating sugar metabolism. (Warrier *et al.* 2010, Zakaria *et al.* 2011, Ashwini *et al.* 2012, Selvam *et al.* 2014; http://medplants.blogspot.in).

21.3 *Cucumis melo* L. var. *agrestis* Naudin

(Syn. *Cucumis melo* subsp. *agrestis* (Naudin) Pangalo; *Cucumis pubescens* Willd.)

Habit

Common Name: Small Gourd, Wild Musk Melon, Wild Melon.

Vernacular Name: *Hindi:* Kachari (कचरी), Kachariya (कचरिया); *Sanskrit*: Karkati (करकटी).

Habit: Climber.

Habitat: Commonly found in cultivated fields.

Distribution: Throughout the old-world tropics, adventive in neotropics; distributed almost throughout India and neighboring countries. In India, Vindhya hills and Aravalli hills extending northwards to Indo-Gangetic plains and southwards to Deccan Plateau touching rain shadow areas of Western Ghats and North-East Himalayas.

Flowering and Fruiting Time: June-October.

Taxonomic Description: An annual, scabrid, hispid, trailing herb with angular stem. Leaves alternate, long-petiolate, reniform, triangular, ovate, slightly 5-lobed, scabrid, base cordate. Tendrils slender, simple. Flowers yellow, gamopetalous, monoecious. Sepals and petals in whorls of 5. Male flowers fasciculate. Stamens-3. Female flowers solitary, axillary. Ovary inferior, densely hairy. Fruit oval or ellipsoid, up to 5 cm long, longitudinally dark-green stripped, yellow when ripe.

Flower

Fruit

Chemical constituents: The plant contains tannins, alkaloids, carbohydrates, flavonoid, saponins, proteins, glycosides, steroids and triterpinoids. (Kaur and Arora 2011, Alagar Raja *et al.* 2015, Shankar *et al.* 2015).

Medicinal Properties: The plant possesses antidyslipidemic and antihyperglycemic activity along with the antiadipogenic activity. The flowers are expectorant and emetic. The root has diuretic and emetic properties. The fruits are stomachic and seeds have antitussive, antioxidant, digestive, febrifuge and vermifuge properties; seed oil extract is reported to have antifungal activity. (Kaur and Arora 2011; http://pfaf.org).

Medicinal Uses (Traditional/Folklore): In folklore medicines, a paste of the plant is applied as a poultice around the naval when there is difficulty in urinating. The fruits are used as a cooling light cleanser or moisturiser for the skin and as a first aid treatment for burns. When used as a vermifuge, the whole seed complete with the seed coat is ground into a fine flour, then made into an emulsion with water and eaten. It is then necessary to take a purge in order to expel the tapeworms or other parasites from the body. (Kaur and Arora 2011; http://pfaf.org).

21.4 *Mukia maderaspatana* (L.) M.Roem.

(Syn. *Cucumis maderaspatanus* L.; *Melothria maderaspatana* (L.) Cogn.)

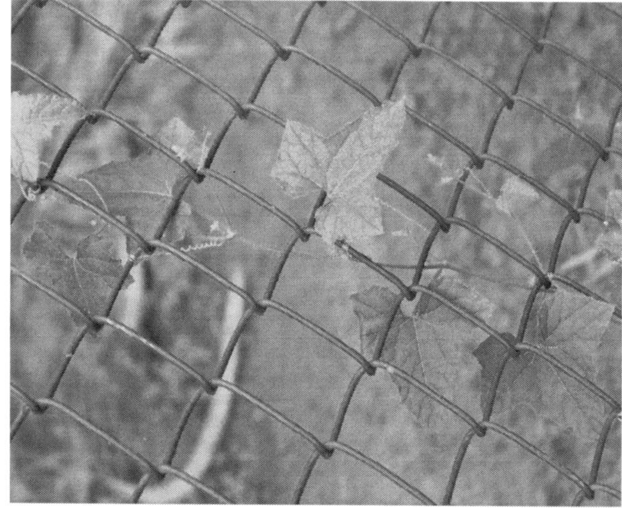

Habit

Common Name: Madras Pea Pumpkin, Rough Bryony.

Vernacular Name: Aganaki (अगनकी), Agumaki (आगुमकी), Bilari (बिलाड़ी).

Habit: Climber.

Habitat: Climbing on other plants, wires along roadsides, wastelands.

Distribution: Throughout tropics, tropical Africa, Asia, Australia; India, up to 1,800 m.

Flowering and Fruiting Time: August-November.

Taxonomic Description: A climbing annual herb. Stem angular, scabrous. Leaves 4-8 cm long, alternate, hastate, ovate-triangular, 3-5 lobed, denticulate, cordate at base, acute at apex, petiole scabrous. Tendrils simple, filiform, hairy. Flowers yellow, about 7 mm across, pedicellate. Male flowers axillary, solitary or in clusters; calyx and corolla 5-lobed, campanulate; stamens 3, filaments short. Female flowers solitary or in fascicled; ovary ovoid, style straight with an annular disk at the base, stigma 3-fid. Fruits small, shortly stalked, globose, about 1 cm across, bristly when young, bright red when ripe. Seeds numerous, margined, ovoid.

Flowering twig **Flower** **Unripe and ripe fruits**

Chemical Constituents: The plant is known to contain phytochemical constituents such as tannins, saponin, flavonoids, steroid, terpenoids and cardiac glycerides. (Petrus 2013).

Medicinal Properties: The plant is bitter, sweet, refrigerant, carminative, aperient, vulnerary, sudorific, expectorant, anodyne and tonic. It is useful in vitiated condition of pitta, burning sensation, dipsia, flatulence, colic, constipation, ulcers, cough, asthma, neuralgia, notalgia (chronic sensory neuropathy–prurites), odontalgia and vertigo. The plant is also reported to have the properties like hepatoprotective, antirheumtic, antiflatulent, anti-inflammatory, anticancer, antidiabetic, diuretic and stomachic and used for toothache and recommended in vertigo and biliousness also. (Warrier *et al.* 2010, Petrus 2013).

Medicinal Uses (Traditional/Folklore): Fresh leaves are ground to paste and applied externally to treat scabies and ringworm infection. Fresh leaf juice is consumed orally to get relief from cold and cough. Leaf extract is consumed internally to cure piles, applied to the hair as a hair dye (to blacken grey hair). Leaf and seed extract is found to be useful in toothache, piles and tuberculosis. Leaf powder is mixed with boiled rice and taken orally to treat cold and cough. Decoction of leaves of this plant is being used by siddha practitioners in Tamil Nadu for the treatment of hypertension.

In Andhra Pradesh (India), the leaves, pounded with garlic, pepper and cumin are used for treating hygroma in cattle. (Sivaperumal *et al.* 2009, Vethanarayanan *et al.* 2011, Petrus 2013, Senthamizhselvi 2013, Rajalakshmi *et al.* 2016).

21.5 *Trichosanthes cucumerina* L.

(Syn. *Trichosanthes cucumerina* var. *cucumerina*). There are two varieties of *Trichosanthes* namely var. *cucumerina* (wild) and var. *anguina* (domesticated). The details for the wild variety (*Trichosanthes cucumerina* var. *cucumerina*) which was found in Chithara Village is as follows;

Habit

Common name: Wild Snake Gourd.

Vernacular name: *Hindi:* Jangali Chachinda (जंगली चचिंडा), Jangali Chichonda (जंगली चिचौड़ा).

Habit: Climber.

Habitat: Spreading on other plants, boundaries along roadsides, railway lines.

Distribution: Found in the wild across much of South and Southeast Asia, including India, Bangladesh, Nepal, Pakistan, Sri Lanka, Indonesia, Malaysia, Myanmar (Burma), and southern China (Guangxi and Yunnan).

It is also regarded as native in northern Australia and naturalized in Florida, parts of Africa and on various islands in the Indian and Pacific Oceans. Distributed throughout India; wild and cultivated.

Flowering and Fruiting Time: July-October.

Taxonomic Description: An annual, monoecious climber with scabrous stem. Tendrils branched, 2-3 fid. Leaves 5-10 cm across, alternate, petiolate, sub-orbicular or reniform, 3-7 lobed, hairy on both the surfaces, base cordate, margin toothed. Flowers white. Male flowers axillary, clustered on long peduncle; calyx tube broadened at apex, pubescent, teeth short, triangular; corolla lobes-5, divided nearly up to the base, lanceolate-oblong, fimbriate; stamens-3, inserted on calyx tube, filaments short, anthers included. Female flowers axillary, solitary; calyx and corolla as in male; ovary inferior. Fruits ovoid, 5-6 cm long, green when young with white stripes, finally scarlet when ripe. Seeds enclosed in red pulp, oblong, compressed.

Fruiting twig **Flower** **Ripe fruit**

Chemical Constituents: The phytochemical screening of *T. cucumerina* var. *cucumerina* extract contains various secondary metabolites of therapeutic importance, namely alkaloids, flavanoids, cardiac glycosides, phenols, sterols, tannins, saponins, caretenoids, phenolic acids and lignans. β -sitosterol its glucoside and Luteolin-7-glucoside were reported from leaves. Also contains cucurbitacin B, cucurbitacin E; isocucurbitacin B; 23, 24-dihydroisocucurbitacin B; 23,24-dihydrocucurbitacin E; β -sitosterol and stigmasterol, 23, 24-dihydroisocucurbitacin D, 3β -hydroxy-olean-13(18)-en-28-oic acid; 3-oxo-olean-13(18)-en-30-oic acid and the sterol 3-O- β -glucopyranosyl-24ξ -ethylcholest-7,22-dien-3 –ol. A novel flavones glucosidal 5,6,6′-trimethoxy-3′; 4′-methylene-dioxyisoflavone-7-O-β-D-(2″-O-p-coumaroylglucopyranoside present, sterols *viz.*, cycloartenol A, parkeol H and litsomentol G also have been isolated from the plant. (Chopra *et al.* 1956, Kage *et al.* 2009, Tiwari 2016).

Medicinal Properties: The plant is known to have anti-inflammatory, antibacterial, analgesic, antidiabetic, antifertility, antihelminthic, antiovulatory, abortifacient, vermifuge, stomachic, refrigerant, purgative, antimalarial, laxative,.and hypoglysimic properties. Leaf is alexiteric, astringent, diuretic and emetic. Extracts of *T. cucumerina* L. var. *cucumerina* show varying degrees of antibacterial activity against gram-negative bacteria (*E. coli, K. pneumoniae, P. aruginosa*). (Kage *et al.* 2009, Devendra *et al.* 2009; 2010, Aryakrishna *et al.* 2016, Tiwari 2016; http://www.yourproduceman.com/news_august_22_05.html).

Medicinal Uses (Traditional/Folklore): *T. cucumerina* var. *cucumerina*, is exploited on a largescale for preparing Ayurvedic formulation- 'padavaladi kashayam' and is also one of the ingredients in several other herbal medicines prepared by 'Kottakkal' and 'Oushadhi' in Kerala, India. It is used in the treatment of headache, alopecia, fever, abdominal tumours, bilious, boils, diarrohea, and bronchitis and skin allergy. The leaves and stems are used for bilious disorders and skin diseases and as an emmenagogue. Root and seeds are cooling, anthelminthic, purgative and vermifuge and are useful in vitiated condition of Pitta, syphilis and verminosis. Leaves are cooling. The fruit is sweet, cooling digestive, carminative, purgative, anthelmintic, febrifuge and tonic. Fruits are useful in vitiated condition of Pitta, burning sensation, anorexia, dyspepsia, flatulence, constipation, helminthiasis, fever and general weakness. Decoction with ginger and honey is used as febrifuge, laxative and alterative. In Konkan region leaf juice rubbed over liver in liver congestion. Root juice taken as purgative. It is strong gastrointestinal irritant. Seeds are given in disorder of stomach.

In ethnomedicinal uses, Irulas tribe apply ripe fruit extract on forehead to cure headache while Bede, Chakma, Garo, Khasia, Marma, Murong, Santal tribes in Tripura use a combination of seed and fruit orally administered for diabetes and tuberculosis. It is also topically applied to treat dermatitis on the affected part of the skin. Other tribes such as Gond, Panika, Bhaira, Muria, Kol, Baiga, Bhil, Bhilala, Barela, Patelia use fruits kept in water for 12 hours and to give bath to patient suffering from jaundice. Fruits are used to cure stomach worms and diabetes by some tribes; fruit and seeds are given orally to control irregular menstruation (Lushai tribe) and fruits for curing skin diseases (Yerukulas, Lambadis tribes). (Chopra 1956, Warrier *et al.* 2003, Devendra *et al.* 2009, Kage *et al.*2009, Devendra *et al.* 2010, Aryakrishna et al 2016, Tiwari 2016).

22. MOLLUGINACEAE

Molluginaceae – Carpet-weed family

The plants belonging to this family were previously included in the larger family Aizoaceae. There are 103 species under 11 plant genera included in this family. The plants are herbs or dwarf shrubs with flashy or succulent leaves which are pseudoverticillate, often having scarious stipules. Flowers lack coralline whorl and seeds lack long funicle. The family recorded one species in Chithara Village flora.

22.1 *Mollugo nudicaulis* Lam.

(Syn. *Lampetia nudicaulis* (Lam.) Raf.)

Habit

Common Name: Naked-Stem Carpetweed, Daisy-leaved Chickweed.

Vernacular Name: None known.

Habit: Annual herb.

Habitat: Common in moist-waste places, cultivated fields.

Distribution: Found on sandy seashores in India, Pakistan and tropical Africa.

Flowering and fruiting time: July-September.

Taxonomic Description: A slender, acaulescent, small, annual herb. Leaves radical, glabrous, forming a rosette on the ground, up to 5 cm long, obovate-oblong to spathulate, entire, tapering to the base, obtuse at apex. Inflorescences dichasium arising from tuft of radical leaves, spreading. Flowers dull white, 3-4 mm across, pedicel and peduncle filiform. Tepals-5, elliptic, oblong, dull-white. Stamens 3 to 5. Ovary oblong, sub-globose, 3-lobed, stigma 3. Capsules ellipsoid, about 2 mm long. Seeds numerous, minute, blackish.

Chemical Constituents: The phytochemical screening revealed the presence of alkaloid, steroids, flavonoids and reducing sugar in the both aqueous and methanolic extracts. (Rameshkumar and Sivasudha 2012).

Medicinal Properties: The plant has antipyretic, antidiabetic, antioxidant, antimicrobial, antihepatic and wound healing properties. (Sindhu *et al.* 2010, Rameshkumar and Sivasudha 2012, Larbie *et al.* 2014).

Medicinal Uses (Traditional/Folklore): *M. nudicaulis* is a medicinal herb, is used by traditional practitioners to cure, jaundice, whooping cough, intertrigo, cold, fever, body pain and wounds. It is useful mainly in preventing small ring worms in stomach, to control diabetes and to increase the life energy of blood cells. Extract of the whole plant is used in cloudy vision and whooping cough. The leaf of this plant is made into a paste in the medium of butter and applied over the ulcers. (Apparanantham and Chelladurai 1986, Sindhu *et al.* 2010, Sindhu *et al.* 2012, Larbie *et al.* 2014).

23. AIZOACEAE

Aizoaceae – Fig-Marigold or Ice Plant family

The family comprise of 146 genera and about 2,271 species, commonly known as stone plants or carpet weeds. Several species are known as "ice plants" because of the glistening globular bladder cells covering their stems, fruit and leaves. Most of these species are succulents, herbaceous, rarely somewhat woody, with stems growing either erect or prostrate. One species of this family was recorded in the flora.

23.1 *Trianthema portulacastrum* L.

(Syn. *Trianthema monogyna* L.)

Common Name: Giant Pigweed, Horse-Purslane.

Vernacular Name: *Hindi:* Vishakhapara (विषखपरा), Sabuni (साबुनी); *Sanskrit:* Chiratika (चिराटिका), Dhanapatra (धानपात्रा).

Habit: Annual herb.

Habitat: In is found throughout in cultivated fields, fallow lands, river beds, waste ground, etc.; also common in waste low-lying areas, along railway embankments, roadsides, etc.

Habit

Distribution: It is widely distributed in many parts of India such as Andhra Pradesh, Haryana, Karnataka, Kerala, Madhya Pradesh, Maharashtra, Odisha, Rajasthan, Tamil Nadu, Uttar Pradesh and West Bengal. It grows throughout many tropical countries in the world.

Flowering and Fruiting Time: May-November.

Flowering twig **Flower** **Capsule**

Taxonomic Description: A prostrate, succulent annual herb. Stem much-branched, glabrous or slightly pubescent. Leaves opposite, petiolate, entire, unequal, petiole connected into a membranous sheath at the base; leaf blade obovate or obcordate, 1-4 cm long, base cuneate. Inflorescence solitary, axillary. Flowers sessile, white to pink in color enclosed in petiolar sheaths. Stamens 10-20, anthers pink or white. Ovary truncate, bilobed, covered by petiolar sheath. Fruit a capsule, 5 mm long. Seeds many, black, reniform.

Chemical Constituents: Photochemical screening has revealed the presence of steroids, flavonoid, fats, terpenes, carbohydrates, tannins, alkaloids, phenolic compounds and carbohydrates; plant contains alkaloid, trianthemine.

Roots contain a glucoside similar in properties to saponin, ecdysterone, 3, 4-dimethoxy cinnamic acid and β-cyanin. (Shivhare *et al.* 2012, Narayana and Narasimharao 2015; http://bsienvis.nic.in).

Medicinal Properties: Roots are laxative, sweet, irritant, abortifacient, emmenagogue, antiphlogistic, diuretic; roots also have cathartic and stomachic properties. Leaves are diuretic, vermifuge, mild laxative, stomachic, antioxidant, diuretic, and analgesic. They have hepatoprotective, anticarcinogenic and antifertility properties. (Shivhare *et al.* 2012, Narayana and Narasimharao 2015; http://bsienvis.nic.in).

Medicinal Uses (Traditional/Folklore): The plants are used for the treatment of blood diseases, anaemia, inflammation and night blindness, and treatment of jaundice and dropsy, edema, painful discharge of urine, obstructions of the liver. Leaf-extract is used in the treatment of asthma, strangury, amenorrhoea, dropsy, oedema, ascites and beri-beri; and also as antidote to alcohol-poisoning, in rheumatism and as vermifuge. Roots are used as cathartic, irritant, and abortifacient and for the treatment of asthma, amenorrhoea and obstruction of liver, calculi, etc. The aqueous extract of root is applied to the eye to cure corneal ulcers, itching, dimness of sight and night blindness. Aqueous extract of root and leaves is also used as eye drops for eye diseases. Leaves ground with musk, saffron and pepper are given orally to treat sores.

Trianthema species are used as adulterant of the *Boerhavia* species (Punarnavaa). (Shivhare *et al.* 2012, Gaddeyya and Ratna Kumar 2015, Narayana and Narasimharao 2015, Patel and Dabgar 2016; http://bsienvis.nic.in).

24. APIACEAE

Apiaceae – Parsley family

Apiaceae or Umbelliferae comprise of 418 plant genera and 3,257 species distributed principally in the North temperate regions of the world. Most of the species are annual, biennial or perennial herbs (frequently with the leaves aggregated toward the base), though a minority are shrubs or trees. The family includes the well-known plants: asafoetida, caraway, carrot, celery, *Centella asiatica*, coriander, cumin, dill, fennel, parsley, parsnip, etc. One species belonging to this family was recorded in the flora.

24.1 *Centella asiatica* (L.) Urb.

(Syn. *Hydrocotyle asiatica* L.)

Common Name: Indian Pennywort, Gotu kola.

Vernacular Name: *Hindi:* Brahmi (ब्राह्मी), Brahma Manduki (ब्रह्म मंडुकी); *Sanskrit*: Bhandi (भाण्डी), Bhandiri (भण्डीरी), Mandukaparni (मण्डूकपर्णी), Manduki (मण्डूकी).

Habit: Creeping herb.

Habitat: Common along water channels and moist grassy areas.

Habit

Distribution: Native to wetlands in Asia, the species is widespread throughout tropical and subtropical countries worldwide. Distributed throughout India. Commonly found in shady as well as open wastelands, grassy fields, gardens, bank of ponds, etc. in moist places, especially along bunds and canals, up to an altitude of 2,300 m.

Flowering and Fruiting Time: February-July.

A single plant

Fruits

Taxonomic Description: A slender herb; stem creeping, rooting at the nodes. Leaves 3-6 cm across, several at each node, orbicular or reniform, margin crenate-dentate, deeply cordate at base, minutely pubescent beneath; petiole long, sheathing at base,. Inflorescence umbels. Flowers small, axillary, nearly sessile, pink, in 2-5 flowered umbels; bracts-2, ovate. Stamens-5. Ovary subglobose, 2-loculed. Fruits ovoid, indehiscent, laterally compressed. Seeds-2, oblong, brown.

Chemical Constituents: The major constituents of the plant are triterpenoid saponins- brahmosides, asiaticoside, thankunisides, centelloiside; alkaloids-hydrocotyline; bitter principles-vellarin. Other chemicals present are 3-glucosylquercetin, 3-glucosyl-and 7-glucosyl-kaemfrol, polyacetylenes I-V and nine other acetylenes; amino acids, asiatic, centic, centellic, centoic, pectic, madasiatic acids, carotene, centellose, hydrocotylin (alkaloid), lipid, protein, pectin, saponins, vallerine, asiaticoside, aslaticosides; asiaticoside, indocentelloside, brahmoside, brahminoside, isothankuniside, medecassoside, thankcuriside, brahmic, and thankunic acids; β-caryophyllene, trans-β-farnesene, germacrene D, mesoinositol, polyphenols, α-terpinene, thymol methyl ether, betulic acid, β-sitosterol, campesterol, stigmasterol, tannins, amino acids, vitamins B1, B2 and C are also present. Brahmoside exhibit tranquilizing and anabolic activity. Asiaticoside gave encouraging results in the treatment of leprosy as it dissolves the waxy covering of *Bacillus leprae.* (Gohil *et al.* 2010, Bhutya 2011, Chong and Aziz 2011; http://bsienvis.nic.in).

Medicinal Properties: Plant is bitter, acrid, sweet, cooling, antidysenteric, soporific, cardiotonic, nervine/brain tonic, stomachic, carminative, antileprotic, diuretic, febrifuge, antioxidant, antimicrobial, neuroprotective and brain tonic, sedative, antiulcer, anticonvulsant and cytotoxic. The plant also has cooling, hypotensive, central nervous system relaxant, peripheral vasodilator, detoxifier, emmenagogue, improving memory, sudorific, cardio tonic, anti leprotic, tranquilizer, and beneficial to physical strength, anti-ageing/ rejuvenating properties; it is used for hypertension, insomniac, cardiac debility, epilepsy, bronchitis, asthma, hiccough, skin diseases, amentia, leprosy, strangury, hoarseness, fever and abdominal disorders, also in overcoming mental confusion, stress, fatigue and stress-induced stomach and duodenal ulcers; topically to accelerate healing, particularly in case of post-surgical and post trauma wounds, second and third degree burns and venous insufficiency. (Gohil *et al.* 2010, Bhutya 2011; http://bsienvis.nic.in, http://medplants.blogspot.in).

Medicinal Uses (Traditional/Folklore): This herb is said to have a direct action on lowering the blood pressure and is often referred to as a rejuvenating medicament. The leaves are eaten raw or finely cut and roasted with scraped coconut. They are believed to purify the blood and cure indigestion, nervousness and dysentery. In some parts of India, the leaves are dried, powdered and taken with milk, to improve the memory. The plant is useful both internally and externally for skin diseases, chronic and obstinate eczema, secondary and tertiary syphilis with ulceration, enlargement of glands, leprosy, abscesses and chronic rheumatism. It has a special influence on the urino-genital tract and urinary and ovarian irritation. It has been employed as a diuretic in several diseases. An ointment prepared from the liquid extract of the plant with ghee is used as an external application on elephantiasis, enlarged scrotum, while the liquid extract is given internally. The cold poultice of the fresh herb is used as an external application in rheumatism, elephantiasis and hydrocele. Internally, also used in leucorrhoea, kidney troubles, urethritis, dropsy. Decoction of young shoots is given in haemorrhoids. In paste form applied on boils and tumours. Plant is aphrodisiac and used in venereal diseases. Leaves are useful in abdominal disorders. Leaf juice is rubbed on the fore head to treat severe head ache, and used in eczema. It is also used as diuretic. Leaf extract is used in the preparation of medicated oil for bone fracture. Leaf extract is used to treat stress-induced stomach and duodenal ulcers; topically used to accelerate healing, particularly in cases of chronic postsurgical and post trauma wounds; also used to treat second and third degree burns. Patients suffering from venous insufficiency have been treated with a titrated extract of the drug. (Gohil *et al.* 2010, Kaul and Dwivedi 2010, Shukla *et al.* 2010; http://bsienvis.nic.in, http://medplants.blogspot.in).

25. ARALIACEAE

Araliaceae – Aralia family, Ivy family

The family includes 40 plant genera. 1,533 species of trees, shrubs, lianas, and perennial herbaceous plants. Species usually bear pinnately or palmately compound leaves, and usually have small flowers produced in large panicles. Several members of the family are economically important. Ivies (*Hedera* species) are grown as ornamental plants and houseplants. One species belonging to this family was recorded in Chithara Village.

25.1 *Hydrocotyle sibthorpioides* Lam.

Syn. *Hydrocotyle rotundifolia* Roxb. ex DC., *Hydrocotyle tenella* Buch.-Ham. ex D. Don

Habit

Common Name: Lawn Pennywort, Lawn Marsh pennywort.

Vernacular Name: None known.

Habit: Herb.

Habitat: Commonly found in shady, marshy localities, along banks of water channels.

Distribution: Tropical Africa, East Asia-China, Japan, India, Malaysia, Indo-China. It is occasionally found as a weed in damp lawns, marshy wet areas.

Flowering and Fruiting Time: March-May.

Taxonomic Description: A slender, filiform, diffuse herb. Stem creeping, rooting at nodes. Leaves 1.5-2 cm across, petioled, lobed, orbicular-reniform, base cordate, margin crenate-serrate, glabrous or somewhat pubescent on both surfaces. Inflorescence umbels; umbels 8-10 flowered. Flowers greenish white in short peduncles; bracts minute, ovate. Fruits compressed, greenish-yellow, smooth, broader than long; pericarp thin, hard.

Fruiting twig

Fruits

Chemical Constituents: The chemical compounds reported from the plant are: stigmasterol, daucosterol, hibalactone, genistein, daidzein etc. (Lan and De-Zhi 2007, Huang *et al.* 2013).

Medicinal Properties: Whole plant is abortifacient, astringent, anthelmintic, analgesic, diuretic, depurative, febrifuge and vermifuge. *H. sibthorpioides* also has property to inhibit dengue virus-2 replication. (Husin *et al.* 2015; http://medplants.blogspot.in).

Medicinal Uses (Traditional/Folklore): The juice of the plant is used in the treatment of fevers. A paste made from the plant is applied externally to wounds and boils. The plant is decocted and used in the treatment of abscesses, boils, bruises, cirrhosis, colds, coughs, hepatitis, hepatoma, influenza, itch, jaundice, sinusitis and sore throat. The plant is useful in rheumatism, sore throat, fungal infection in mouth, thrush, pulmonary and digestive troubles, syphilis, skin diseases, a cough remedy for children; decoction is given in diarrhoea and body pain; juice dropped into ear in earache and also given to kill intestinal worms; pounded mixture of the plant with flowers of *Musa balbisiana* is applied externally on ulcers. Leaves are used for treating skin diseases, hepatitis, bladder stones, flu and cough.

The plant is a Chinese herbal drug for hepatoma and it has been used as a folk remedy for the treatment of fever, edema, detoxification, throat pain, and psoriasis and hepatitis B virus infections in China. (Huang *et al.* 2013, Husin *et al.* 2015; http://medplants.blogspot.in).

26. RUBIACEAE

Rubiaceae – Coffee family

The family contains about 13,673 species under 609 genera, which makes it the fourth-largest Angiosperm family. It consists of trees, shrubs, lianas, or herbs that are recognizable by simple, opposite leaves with interpetiolar stipules. Some of the well-known plants are ornamentals such as *Gardenia* (food plant of the Bee Hawkmoth) and *Ixora,* and Coffea, Cinchona, etc. The family was represented by three species under three genera in Chithara Village.

26.1 *Mitragyna parvifolia* (Roxb.) Korth.

(Syn. *Stephegyne parvifolia* (Roxb.) Korth.)

A branch of tree

Common Name: Kaim, True Kadamb.

Vernacular Name: *Hindi:* Kaim (कैम), Kadamb (कदम्ब); *Sanskrit:* Kadamba (कदम्बा), Vitanah (वितान:).

Habit: Deciduous tree.

Habitat: Grows gregariously in deciduous and evergreen forests up to 1,200 m; cultivated in the parks and gardens.

Distribution: Throughout the drier parts of India, Pakistan and Sri Lanka.

Taxonomic Description: A small to large, deciduous tree with greyish bark. Leaves opposite-decussate, petiolate, variable in shape and size, 4-15 cm long, orbicular, obovate, elliptic-oblong, coriaceous, entire, sub-acute a apex, rounded or sub-cordate at base; stipules large, caducous, interpetioler, pinkish in colour. Flowers creamy white or light yellow, in globose, axillary heads of 2-3 cm across. Calyx tube short, 5-toothed. Corolla about 4-5 mm long, campanulate, creamy white, hairy inside. Stamens-5, inserted. Style filiform, white, exserted. Stigma mitriform. Capsules globular with 2 dehiscent cocci. Seeds many, winged.

An aggregate fruit

Chemical Constituents: The aerial parts, stem-bark and roots of the tree contain various indolic (tetrahydroalstonine, akkuamigine, hirsuteine, etc.) and oxindolic (mitraphylline, isomitraphylline, pteropodine, isopteropodine, etc.) alkaloids. Also contain pyroligneous acid, methyl acetate, ketones and aldehydes. (Sengupta *et al.* 2012, Choudhary and Jain 2016).

Medicinal Properties: The plant has diverse medicinal and therapeutic properties such as analgesic, antipyretic, anti-inflammatory, antiarthritic, anthelmentic and antioxidant properties. The roots and bark are acrid, bitter, stomachic and febrifuge, and are useful in gastropathy, colic and fever. The leaves are acrid, bitter, sweet, styptic, vulnerary, stomachic, anti-inflammatory, anodyne, depurative and febrifuge. It is useful in internal and external haemorrhages, wounds, colic, flatulence, dyspepsia, inflammations, myalgia, skin diseases, leprosy, erysipelas and fever. (Warrier *et al.* 2010, Gong *et al.* 2012, Sengupta *et al.* 2012, Choudhary and Jain 2016).

Medicinal Uses (Traditional/Folklore): The bark and root are used in folk medicine for treatment of fever, colic, muscular pain, burning, gynaecological disorders, poisoning, edema and cough. Extract as external application used for snake bites. Infusion of stem bark is given orally to the diabetic patients. The juice of stem bark is applied externally on abdomen in dropsy for three days.

As a veterinary medicine a handful of fresh and healthy leaves and same quantity of Gondhan (*Cordia gharaf*) tender leaves along with 4-5 small Kanda (*Allium cepa*) bulbs are fed to the cows and buffaloes twice a day to increase fertility and chances of conception prior to heat period. (Gong *et al.* 2012, Dadhiach and Rathore 2015, Choudhary and Jain 2016, Punjaji 2016).

26.2 *Oldenlandia corymbosa* L.

(Syn. *Hedyotis corymbosa* (L.) Lam.)

Habit

Common Name: Diamond Flower, Corymbose Hedyotis.

Vernacular Name: Hindi: Daman Pappar (दामन पप्पड), Pitpapra (पित्तपापड़ा); *Sanskrit:* Parpatah (पर्पटः).

Habit: Annual herb.

Habitat: Cultivated fields, gardens and wastelands.

Distribution: Native to tropical Africa, including Madagascar and India. It is frequently found throughout India, Sri Lanka and tropical East Asia to Java and the Philippines.

Flowering and Fruiting Time: August-October.

Taxonomic Description: A sub-erect, annual herb, 8-15 cm high. Leaves 1.5-4 cm long, opposite, sessile, linear or linear-lanceolate, margin revolute; stipules short, bristly. Flowers axillary, solitary or usually 2-3 flowered, peduncle 1-1.5 cm long. Calyx teeth-4, subulate. Petals-4, white. Stamens-4, inserted at throat of corolla. Ovary 2-celled, ovules many. Capsules 1.5-2 mm across, globose or pyriform, didymous. Seeds minute, numerous.

Flowering twig

Capsules

Chemical Constituents: The plant contains proteins, carbohydrates, phenols, tannins, flavanoids, saponins, steroids, terpenoids and glycosides. Some of the isolated compounds from whole plant are geniposide, iridoid glycosides, 6 α– hydroxygeniposide, scandoside methyl ester (6 β - hydroxygeniposide), 10-o-benzoylscandoside methyl ester, asperulosidicacid, asperuloside, deacteylasperuloside, 10-o-p-hydroxy benzoylscandoside methyl ester, rutinand (+) - lyoniresinol-3- α-o-beta glucopyranoside; urosilic acid, oleanolic acid and gamma sitosterol have also been isolated. (Noiarsa *et al.* 2008, Sivapraksam *et al.* 2014).

Medicinal Properties: The decoction of the entire plant is used as an anthelmintic, antirheumatic, depurative, diaphoretic, digestive, diuretic, febrifuge, pectoral and stomachic. It exhibits antibacterial, antioxidant, analgesic, hepatoprotective, antioxidant, radical-scavenging, anti-inflammatory, cytotoxic and anticancer properties. (Patel *et al.* 2014, Sivapraksam *et al.* 2014; http://tropical.theferns.info).

Medicinal Uses (Traditional/Folklore): The juice is given internally with a little milk and sugar to cool the burning sensation in the stomach. The decoction is given to cure remittent fever, heat eruptions and also applied to the surface of the body. In India, it is a common ingredient in mixtures used internally to treat remittent fevers, gastric irritation, and nervous depression and as a tonic. The plant is used to clear toxins and heat, thereby activate blood circulation and promote diuresis. In Chinese medicine, it is used to treat viral infections, cancer, acne, boils, appendicitis, hepatitis, eye problems and bleeding. In Africa, the plant is used to facilitate childbirth. (Patel *et al.* 2014, Sivapraksam *et al.* 2014; http://tropical.theferns.info).

26.3 *Spermacoce articularis* L.f.

(Syn. *Borreria articularis* (L.f.) F.N.Williams)

Common Name: Shaggy Buttonweed, Jointed Buttonweed, False Buttonweed.

Vernacular Name: *Hindi:* Guthari (गुथरी); *Sanskrit:* Madanaghanti (मदनघण्टी).

Habit: Annual herb.

Habitat: Commonly found in fallow fields and gardens.

Distribution: Tropical America, tropical Africa and South-East Asia; growing in the plains as a weed throughout India.

Flowering and Fruiting Time: August-October.

Habit

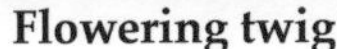

Flowering twig

Flowers

Taxonomic Description: A procumbent, annual herb. Stem quadrangular, thickened at nodes, pubescent. Leaves up to 4 cm long and 2 cm wide, opposite-decussate, obovate or ovate-oblong, sessile or sub-sessile, acute to obtuse at apex; stipules membranous, bristly, pubescent. Flowers axillary, in few flowered sessile cluster. Calyx pubescent; lobes linear-lanceolate. Corolla lobes-4, pinkish-white, ovate, oblong, acute, hairy, tube funnel-shaped. Stamens-4, inserted on throat of corolla tube. Stigma 2-lobed. Capsules ovoid or ellipsoid, hispid, about 4-6 mm long. Seeds minute, oblong, brown.

Chemical Constituents: The plants contain alkaloids, glycosides, steroids, flavonoids and tannins. (Soosairaj *et al.* 2013, Gnanadeebam and Viswanathan 2014).

Medicinal Properties: Antimicrobial, anti-inflammatory, antioxidative, antihepatic, antihypertensive, stomachic, etc. (Vadivelan *et al.* 2007, Sultana *et al.* 2008, Gnanadeebam and Viswanathan 2014).

Medicinal Uses (Traditional/Folklore): The whole plant is used for headaches, constipation, hypertension, mouth ulcers, and abscesses. The plant is also prescribed to cure haemorrhoids. The leaves are particularly useful in ulcers, wounds, scabies, skin diseases and skin eruptions. Leaf extract of the plant is used for treating haemorrhoids, gall stones, jaundice, and earache, inflammation of eyes, intermittent fever and conjunctivitis by many tribes in India. A poultice of this plant is used to heal leg ulcers, wounds, headaches, and toothaches. The juice/extract of leaves is used in ophthalmic problems, inflammation of eye and gums, blindness, fever, spleen complaints, sore, conjunctivitis, haemorrhage, gall stones, dysentery and diarrhoea; the decoction of the leaves, roots, and seeds is used in India for dropsy. Decoction of the herb is used to relieve headache, while seeds are demulcent in diarrrhea, dysentery and antimicrobial activity. The root infusion of plant is given in loss of appetite as an appetizer or for stimulating the appetite. Roots are also used as a mouthwash to relieve toothache. (Vadivelan *et al.* 2007, Sultana *et al.* 2008, Gnanadeebam and Viswanathan 2014).

27. COMPOSITAE

Compositae – Sunflower family

It is the largest family of dicot plants comprising 1,911 genera and about 32,913 species including about 1,000 species from India. Members of the family have flower heads composed of many small flowers, called florets that are surrounded by bracts (leaf like structures). Bell-shaped disk florets form the centre of each head. Strap-shaped ray florets extend out like petals from the centre and are sometimes reflexed (bent back). Some species have flowers with only disk or only ray florets. The sepals have been reduced to a ring of hairs, scales, or bristles that is called the pappus on the mature fruit. The one-seeded fruit (an achene) has a hard outer covering. A total of 24 species belonging to 20 genera of this family were recorded in Chithara flora, including the campus of Shiv Nadar University.

27.1 *Acmella paniculata* (Wall. ex DC.) R.K. Jansen

(Syn. *Spilanthes paniculata* Wall. ex DC.)

Habit

Common Name: Toothache Plant, Panicled Spot Flower, Buzz Buttons.

Vernacular Name: *Hindi* Pipulka (पीपलका).

Habit: Annual herb.

Habitat: Common in moist-waste places, roadsides and canal sides.

Distribution: It is reported from South and Southeast Asia, and southern China, Colombia, Brazil, Ecuador, Papua New Guinea, Peru and Solomon Islands. In India, it is distributed throughout.

Flowering and Fruiting Time: November-April.

Taxonomic Description: A decumbent-ascending, annual herb, more or less hairy, sometimes glabrous. Leave opposite, petioled, ovate-lanceolate, acute at apex, cuneate at base, irregularly crenate-serrate or sometimes entire at margin. Heads yellow, conical shaped, axillary, mostly with long peduncled. Involucral bracts green, pubescent, bi-seriate, oblong-lanceolate, sub-acute. Ligules when present are small, whitish or yellow. Disk floret bisexual. Achenes compressed.

Flowering twig

Flower head

Chemical Constituents: The qualitative phytochemical screening showed that the whole plant is a rich source of alkaloids, saponins, flavonoids and tannins. The chloroform extracts contains alkaloids and saponins. The ethyl acetate extract contains flavonoids, alkaloids, and saponins. Methanol extract contains alkaloids,flavonoids, tannins and saponins. The other constituents isolated from *A. paniculata* are spilanthol, isobutylamides and triterpenoids. (Mamidala and Gujjeti 2013).

Medicinal Properties: The plants have multiple pharmacological actions such as antimicrobial, antibacterial, antihepatic analgesic, antipyretic, toothache relieve and anti-inflammatory effects. (Mamidala and Gujjeti 2013, Krishna *et al.* 2015).

Medicinal Uses (Traditional/Folklore): The plants are rich source of therapeutic constituents. It is called the toothache plant because by chewing the leaves or flowers, it produces a numbing effect to the tongue and gums. Young stem and leaves are used for the cure of constipation, liver trouble, toothache, worm infection, and as tonic during jaundice in Assam, India. Apatani tribe of Arunachal Pradesh, India use leaves for treating constipation, intestinal worms, and toothache. Leaf, water decoction is used for curing toothache, cough, and fever in the Eastern Himalayan region of Arunachal Pradesh, India. Crushed flower is applied to treat toothache by Chakma Community of Tripura, India. The plant is known to be used in controlling skin diseases. *A. paniculata* is also used as a fish poison, in dysentery and against scurvy. (Berghe and Vlietinck 1991, Kala 2005, Srivastava *et al.* 2010, Tangjang *et al.* 2011, Guha and Chakma 2015).

27.2 *Ageratum conyzoides* (L.) L.

(Syn. *Eupatorium conyzoides* (L.) E. H. Krause)

Common Name: Billygoat-Weed, Chick Weed, Goatweed.

Vernacular Name: *Hindi:* Jangli Pudina (जंगली पुदीना), Gha Buti (घा बूटी), Bhakumbar (भंकुबर).

Habit: Annual herb.

Habitat: Common along roadsides, canal banks, agriculture fields and waste places.

Distribution: Native to tropical America, especially Brazil but the center of origin is in Central America and the Caribbean; ranges from Southeastern North America to Central America. It is considered as an invasive weed in many other regions. Distributed throughout India, in plains and hills and in forests as undergrowth, canal/stream banks, moist areas.

Flowering and Fruiting Time: January-May.

Habit

Upper portion of plant bearing flowers

Flower heads

Taxonomic Description: An erect, hairy, annual herb, 20-60 cm high. Stem branched, cylindrical, hispid. Leaves opposite, petiolate, aromatic, up to 8 cm long, broadly ovate or rhomboid-ovate, base rounded or sub-cordate, margin crenate-serrate, apex acute. Inflorescence terminal corymbs. Flower heads numerous, white or faintly blue. Involucral bracts linear, 2-3 seriate, sparsely pubescent. Corolla tubular, 5-lobed, 1-2 mm long. Stamens-5. Style exserted. Pappus scales-5. Achenes angled, ribbed, black.

Chemical Constituents: The plant contains flavonoids, pyrrolizidine alkaloids, saponins, Lycopsamine, cumarins, essential oils (0.7 - 2.0%), and tannins and echinatine. The essential oil contains hexadecanoic, linoleic, linolenic, palmitic, oleic and stearic acids have antibiotic properties. Leaves contain dotriacontane, conyzorigun (=eupalestin), 5, 6, 7, 8, 3′, 4′, 5′-heptamethoxy flavone (5′-methoxynobiletin), kaemferol. Kaemferol-3-rhamno-glucoside, kaemferol-3, 7-diglucoside, quercetin, caffeic and fumaric acids, a-spinasterol, stigmasterol, stigmasa-7-en-3-ol. Seeds contain ß-caryophyllene and d-cadinene, eugenol, methyleugenol, ocimene, and d-a-pinene. (Singh *et al.* 2013c; http://www.pfaf.org, http://www.rain-tree.com).

Medicinal Properties: Plant is antiseptic, anti-inflammatory, antiphlogistic, antiallergic, diuretic, stimulant, febrifuge antinematicidal, and anticoagulant, smooth muscle relaxant, haemostatic, analgesic, antifungal, antibacterial, antiallergic and hypothermic. The leaves are styptic and antidysenteric, and are commonly used for haemorrhoids, wounds and sores. The roots are bitter, acrid, thermogenic, digestive, lithontriptic, appetizer and ophthalmic; useful in dyspepsia, anorexia, renal and vesical calculi and pharyngopathy. Root juice has antidysenteric and antilithic properties. *A. conyzoides* can cause liver lesions and tumours. There was a mass poisoning incident in Ethiopia as a result of contamination of grain with *A. conyzoides*. (Durodola 1977, Borthakur and Baruah 1987, Warrier *et al.* 2010, Wiedenfeld 2011, Singh *et al.* 2013c; http://www.pfaf.org).

Medicinal Uses (Traditional/Folklore): *A. conyzoides* is widely utilized in traditional medicine by various cultures worldwide, although applications vary by region. The plant is used to treat constipation, infective hepatitis, eczema, epilepsy, sore eyes, fever, headaches, intestinal worms, filariasis, vomiting and nausea, wounds and cuts. Whole plant decoction or fresh juice of plant is employed in rhinitis, sinusitis, diarrhoea, dysentery, rheumatism, fever, dyspepsia, anorexia and calculi. Juice of the fresh plant is useful in postpartum uterine haemorrhage. Leaf paste is applied on boils, wounds, piles, sores, for fomentation in leprosy and other skin diseases. Leaf paste with rhizome of *Zingiber officinale* is used to cure snake bites in Ambala district, India. A poultice of the leaves is effective for treating boils; leaf juice is used as a lotion for eyes. Traditional communities in India use this species as an antidysenteric, and antilithic.

The leaves are dried and applied as a powder to cuts, sores and the ruptures caused by leprosy. The powder absorbs the moisture of the disease and forms a layer that is removed after 1 - 2 days. An effective cure for most cuts and sores, though it does not affect a complete cure for leprosy. The leaves are also used externally in the treatment of ague. The juice of the plant is used to treat cuts, and wounds. A paste of the leaves is used as a poultice to remove thorns from the skin. A paste made of the leaves mixed with equal amounts of *Bidens pilosa, Drymaria cordata, Galinsoga parviflora* and the rhizome of *Zingiber officinale* is used to treat snakebites. In Rajasthan, a paste made from leaves, ginger and mustard seeds is placed on skull, over the position of the brain, to bring down the temperature of the patient during typhoid fever and high body temperature. Whole plant is pounded, boiled and the solution is used in massaging body swelling, tumours, etc. Root extract is used in skin diseases. The juice of the root is antilithic. The juice of the flower heads is used externally to treat scabies, whilst a paste of flower heads is used to treat rheumatism. A tea made from the flower heads mixed with *Ocimum tenuifolium* is used to treat coughs and colds.

In Central Africa, the plant is used to treat pneumonia, wounds and burns. In Cameroon and Congo, traditional use is to treat fever, rheumatism, headache and colic. In Reunion, the whole plant is used as an antidysenteric. The use of this species in traditional medicine is extensive in Brazil where aqueous extracts of leaves or whole plants have been used to treat colic, colds and fevers, diarrhoea, rheumatism, spasms, or as a tonic; it also prevents tetanus. *A. conyzoides* has quick and effective action in burn wounds and is recommended by Brazilian Drugs Central as an antirheumatic. The leaves are useful in the treatment of boils, leprosy, skin diseases as well as eye inflammation. A hair wash consisting of a decoction of the fresh plant makes the hair fragrant soft and dandruff-free. (Durodola 1977, Borthakur and Baruah 1987, Kumar 2010, Warrier *et al.* 2010, Singh *et al.* 2013c, Vashistha and Kaur 2013, Ahirwar and Kujur 2015; http://www.pfaf.org, http://medplants.blogspot.in, http://www.rain-tree.com).

27.3 *Ageratum houstonianum* Mill.

(Syn. *Ageratum mexicanum* Sims)

Habit

Common Name: Floss Flower, Blue Billygoat Weed.

Vernacular Name: *Hindi:* Raktarodhi (रक्तरोधी); *Sanskrit:* Nilima (नीलिमा).

Habit: Annual herb.

Habitat: Often gregarious along roadsides, open waste places.

Distribution: The plant is native to tropical America and adjacent parts of Mexico, but has become an invasive weed throughout the world, including India.

Flowering and Fruiting Time: January-May.

Taxonomic Description: An erect, 30-80 cm high, annual herb. Stem and branches densely hirsute, aromatic. Leaves opposite, petiolate, up to 10 cm long, deltoid-ovate, crenate-serrate, sub-cordate at base, acute at apex, pubescent on both the surfaces. Flower heads many, violet, discoid in terminal and axillary dense corymbs. Involucral bracts 2-seriate, densely hairy. Corolla 2-3 mm long, funnel-form. Style exceeding the corolla. Stamens-5. Pappus of 5 scales. Achenes angled, 5-ribbed, black, hairy.

Leaves

Flower heads

Chemical constituents: The plant contains essential oil which is a mixture of about 50 chemical compounds, of which precocene I (22.45 %), precocene II (52.64 %) and β-caryophyllene (9.66 %) form as a major component of the total volatile constituents (94.51 %). (Menut *et al.* 1993, Kurade *et al.* 2010).

Medicinal Properties: Plant extracts have insecticidal activity against immature stages and adult mosquitoes. The essential oil of *A. houstonianum* has antibacterial activity against *Micrococcus luteus* and *Rhodococcus rhodochrous.* (Kurade *et al.* 2010, Tennyson *et al.* 2015).

Medicinal Uses (Traditional/Folklore): The juice of the plant is used externally to treat cuts and wounds as an antiseptic. (Dangol 2008).

27.4 *Artemisia scoparia* Waldst. & Kitam.

(Syn. *Artemisia capillaris* Miq., *Artemisia elegans* Roxb.)

Common Name: Redstem Wormwood.

Vernacular Name: *Hindi:* Seeta-Bani (सीता-बानी).

Habit: Annual or Perennial herb.

Habitat: Often gregarious on roadsides, vacant fields, canal banks, railway lines and forest edges.

Distribution: *A. scoparia* is widespread across much of Eurasia from France to Japan, including China, India, Russia, Germany, Poland, central and southwest Asia, etc. In India, commonly found growing up to 2,000 m.

Flowering and Fruiting Time: September-October.

Taxonomic Description: A annual or perennial, strongly aromatic, branched herb with woody root stock. Stem purplish, 40-150 cm tall, grooved. Radical leaves petioled, 2-3 pinnatisect; upper leaves sessile, 1-2 pinnatisect; segments filiform, linear. Heads many, yellow, heterogamous, oblong-ovoid, 2 mm in diameter, on short axillary peduncle. Involucral bracts oblong, obtuse. Outer florets female, fertile, corolla filiform. Disk florets bisexual, corolla tubular, limb 5-fid. Achenes minute, brown.

Habit

Leaves on twig

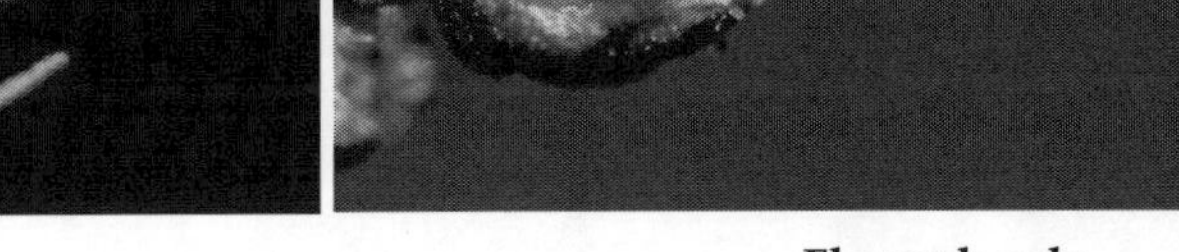

Flower heads

Chemical Constituents: The plants contain flavonoids, coumarins, and essential oils. The main compounds isolated are apillarisin, chlorogenic acid butyl ester, 6,7-dimethylesculetin, isosabandin, magnolioside (isoscopoletin-β-D-glucopyranoside), 7-methoxycoumarin, 7-methylesculetin, A, sabandin B, scoparone (6,7-dimethoxycoumarin), scopoletin, and β-sitosterol. *A. scoparia* also contains an antimalarial drug artemisinin. (Singh and Sarin 2010, Jang *et al.* 2015).

Medicinal Properties: The plant has been recognized with therapeutic efficacy in liver diseases and widely used as an alternative therapy in Asia. The plant is bitter, acrid, anticholesterolemic, antipyretic, antiseptic, antihypertensive, anti-inflammatory, antiparasitic, antiviral, immunity booster, liver tonic, cholagogue, diuretic and vasodilator. It has an antibacterial action, inhibiting the growth of *Staphylococcus aureus, Streptococci* sp., *Bacillus dysenteriae, B. typhi, B. subtilis, Pneumococci, Corynebacterium diphtheriae, Mycobacterium,* etc. (Jang *et al.* 2015; http://temperate.theferns.info, http://www.naturalmedicinalherbs.net).

Medicinal Uses (Traditional/Folklore): The aerial part of this plant is used in traditional medicine as an antiphlogistic, as a diuretic, for the treatment of hepatitis (Jaundice) and urticaria, and as an antimold agent, treatment of jaundice, hepatitis and inflammation of the gall bladder. The plant is also used in a mixture with other herbs as a cholagogue. In Rajasthan, the whole plant is used in burns. It is very useful in febrile eruptions and as a vermifuge. Infusion of the plant is purgative and in Pakistan it is used to cure earache as well as snake and scorpion bite. Aqueous extract of flowering shoots is given for malaria and as an antihelminthic. (Ahmad *et al.* 2006, Singh and Sarin 2010, Khan and Singh 2012, Jang *et al.* 2015; http://temperate.theferns.info, http://www.naturalmedicinalherbs.net).

27.5 *Bidens pilosa* L.

(Syn. *Bidens orientalis* Velen. ex Bornm.)

Common Name: Black-Jack; Beggar Tick, Cobbler's Pegs, Spanish Needle.

Vernacular Name: Kumra (कुमरा), Kumur (कुमुर).

Habit: Annual herb.

Habitat: Roadsides, wastelands, banks of rivers and canals.

Distribution: Native to the tropical America but it is known widely as an introduced species of other regions, including Eurasia, Africa, Australia, and the Pacific Islands. In India, found along the roadsides, wastelands, crop fields, forest paths and disturbed lands from 400-1,500 m.

Flowering and Fruiting Time: Most part of the year.

Habit

Taxonomic Description: An erect, annual herb, up to 1 m high. Stem 4-angular. Leaves opposite, petiolate, variable, 3-foliate or pinnatified; leaflets ovate-lanceolate, coarsely serrate, acute at apex. Inflorescence axillary or terminal corymbose panicles. Flower heads 1-1.5 cm across with two rows of floral bracts. Disk florets yellow, numerous. Ray florets white, 1-1.5 cm long. Achenes black, 1-2 cm long with bristly awns.

Flower head

Fruiting head (achenes)

Chemical Constituents: Almost two hundred chemical compounds (70 aliphatics, 60 flavonoids, 25 terpenoids, 19 phenylpropanoids, 13 aromatics, 8 porphyrins, and 6 other compounds) have been isolated from *B. pilosa*. Besides chalcone okanin, ethyl caffeate and ahydroxycinnamic acid, the plant contains polyacetylenes, polyacetylenic glycosides, aurons, auron glycosides, p-coumeric acid derivatives, caffeoylquinic acid derivatives, pheophytins, diterpenes, tannins, phytosterols, ascorbic acid, carotene, essential oils, saponins, steroids and flavonoids, etc. (Silva *et al.* 2011, Bartolome *et al.* 2013).

Medicinal Properties: The plants have antibacterial, anticancer, anti-inflammatory, antiallergic activity, antihyperglycemic, antihypertensive, antimalarial activity, antiulcerogenic, antileukaemic, antipyretic, antivirus, antiangiogenic, antirheumatic, immunosuppressive, hepatoprotective, and T helper cell modulator activity. (Silva *et al.* 2011, Cortés-Rojas *et al.* 2012, Bartolome *et al.* 2013).

Medicinal Uses (Traditional/Folklore): All parts of *B. pilosa* plant, the whole plant, the aerial parts (leaves, flowers, seeds, and stems), and/or the roots, fresh or dried, are used as ingredients in folk medicines for treating a wide variety of ailments. Traditionally, different preparations of whole plant and/or parts have been purported to treat over 40 categories of illnesses. It is frequently prepared as a dry powder, decoction, maceration or tincture or taken directly. Generally, this plant is applied as dry powder or tincture when used externally, and as a powder, maceration, or decoction when used as an internal remedy. It has been reported to be useful in the treatment of inflammation, immunological disorders, digestive disorders, infectious diseases, cancers, metabolic syndrome, wounds, and many others. *B. pilosa* is used to treat snake bites and wounds. In Trinidad and Tobago, the aqueous solution of the leaves of *B. pilosa* is used to bathe babies and children. An extract of whole herb with honey is used in cough and bronchitis and vegetable is considered useful in leucoderma. *B. pilosa* has been used to treat a variety of illnesses including inflammation and tumours in China, Cuba, the Bahamas, Amazonia and southern Brazil. *B. pilosa* was included in the official list of medicinal plants with potential for herbal use by the public health system and represents, from an economic standpoint, a valuable option as remedy for the community of Criciúma/Santa Catarina. In Chinese Medicinal Herbs of Hong Kong, *B. pilosa* (the only species listed) is used for treating influenza, colds, fever, sore throat; acute appendicitis; acute infectious hepatitis; gastroenteritis, dyspepsia; rheumatic arthralgia; malaria; and haemorrhoids, and pruritis. (Dharmananda 2000, Dhiman 2004, Silva *et al.* 2011, Borges *et al.* 2012, Bartolome *et al.* 2013).

27.6 *Blainvillea acmella* (L.) Philipson

(Syn. *Blainvillea alba* Edgew)

Common Name: Para Cress Flower, Paniculated Spot Flowers.

Vernacular Name: *Hindi:* Kanghi (कंघी).

Habit: Annual herb.

Habitat: Common in moist waste places.

Distribution: Pantropical. Native of tropical America. Introduced in to India during the 18th century and now distributed throughout the country; also occurs in Nepal, Sri Lanka, Java (Indonesia), Malaysia, Myanmar, Philippines, Thailand, Vietnam, Australia and Africa.

Flowering and Fruiting Time: August-October.

Habit

Flowering twig

Flower head

Taxonomic Description: An erect, up to 1 m tall herb. Stem dichotomously branched, covered with stiff hairs. Leaves 4-10 x 2-4 cm, petioled, ovate-lanceolate, shallowly crenate-serrate, base cuneate, apex acute or acuminate, scabrous on both the surfaces. Flower heads white, heterogamous, solitary, 8-12 mm across, borne in terminal or axillary cymes. Involucral bracts 3-4 seriate, ovate-lanceolate, pubescent. Ray florets ligulate. Disk florets tubular, campanulate, corolla 5-toothed. Pappus of 2-5 scales. Achenes triquetrous, compressed, dark brown.

Chemical Constituents: The aerial parts of *B. acmella* contain some widely distributed compounds, desacetylovatifolin, desacylgraxielic acid, four new lactones related to ovatifolin, β- [2-methylbutyryloxy]-9 β- hydroxy-14-oxo-acanthospermolide]. The extract of the roots contain stigmasterol, sitosterol, the germacranolide ovatifohn, the thiopheneacetylenes 13 and 14, rudbeckianone, l-hydroxy-acurcumene and two carotol derivatives, the cinnamates 15 and 16. (Singh *et al.* 1985).

Medicinal Properties: The plant have been found effective as ovicidal, larvicidal and pupicidal, diuretic, antifungal, antipyretic, anti-inflammatory, and for few other uses. (Sharma *et al.* 2014b).

Medicinal Uses (Traditional/Folklore): *B. acmella* has been used traditionally for various purposes, including medicinal uses. Leaves are used in rheumatism, and flowers are used for headache. For alcohol deaddiction one cup of fresh leaf juice is given during morning for 10 days by Pawra Tribe of Satpura Hills, Maharashtra, India. (Jagtap *et al.* 2009; http://www.efloraofgandhinagar.in).

27.7 *Blumea laciniata* (Wall. ex Roxb.) DC.

(Syn. *Blumea acutata* DC.)

Apical portion of plant

Common Name: Cutleaf Blumea, Cutleaf False Oxtongue.

Vernacular Name: None known.

Habit: Annual herb.

Habitat: Common along roadsides and in waste places.

Distribution: Native to India and South-East Asia.

Flowering and Fruiting Time: February-May.

Taxonomic Description: An erect, up to 1.2 m tall herb. Stem and branches grey-tomentose with glandular trichomes. Leaves alternate, hairy; basal leaves lyrate-pinnatified; cauline leaves smaller, coarsely toothed. Flower heads numerous, 6-8 mm in diameter, borne in lax corymbs or terminal panicles. Involucral bracts many-seriate, linear-lanceolate, pubescent. Florets yellow, pappus white, receptacle hairy. Achenes striated, pubescent.

Inflorescence

Chemical Constituents: Besides, cineole, carotene, vitamin C, citral, fenchone, camphor, five other chemical compounds have been isolated and identified as: protocatechuic acid, chrysoeriol, apigenin, 4-hydroxy-3, 5-dimethoxbenzoic acid and scopolet. (Yuan *et al.* 2007).

Medicinal Properties: *B. laciniata* is hot, pungent, bitter and has antivenom, antiseptic and antipyretic properties. (Rahman 2013a).

Medicinal Uses (Traditional/Folklore): The plant is useful to cure bronchitis, blood diseases, fevers, burning sensation and thirst. The root kept in the mouth cures diseases of the mouth in Rajshahi, Bangladesh. Traditionally, leaves and seeds are used in cuts, wounds, piles, leucorrhoea and dog bite in Rae-Bareli District, Uttar Pradesh, India. The plant is a cardiac tonic. Root paste is used against snake bite; simultaneously, leaves are crushed and the juice is applied externally on the bitten part. (Rai and Lalramnghinglova 2010, Rahman 2013a, Singh and Shukla 2015).

27.8 *Blumea obliqua* (L.) Druce

(Syn. *Erigeron obliquus* L.; *Blumea pubiflora* DC.)

Common Name: Common Floss Flower.

Vernacular Name: *Hindi:* Sakarkandi (सकरकंदी), Sankooli (संकूली).

Habit: Herb.

Habitat: Dry waste places and in the fallow fields.

Distribution: Native of tropical America; occurs in wastelands in India, Pakistan, Myanmar and Sri Lanka.

Flowering and Fruiting Time: August-March.

Taxonomic Description: An erect, up to 60 cm high, annual herb. Stem branched, scabrous. Leaves alternate, 3-7 x 1.5-2 cm, sessile or half-amplexicaul, oblong, obovate, apiculate, usually with dentate margin, pubescent on both the surfaces. Heads about 7 mm across, disciform, solitary, axillary on long peduncles. Involucral bracts slender, linear-lanceolate, densely hairy. Florets pinkish or yellowish. Achenes minute, oblong, brownish, pubescent, pappus pinkish-white.

Habit

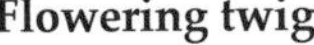

Flowering twig

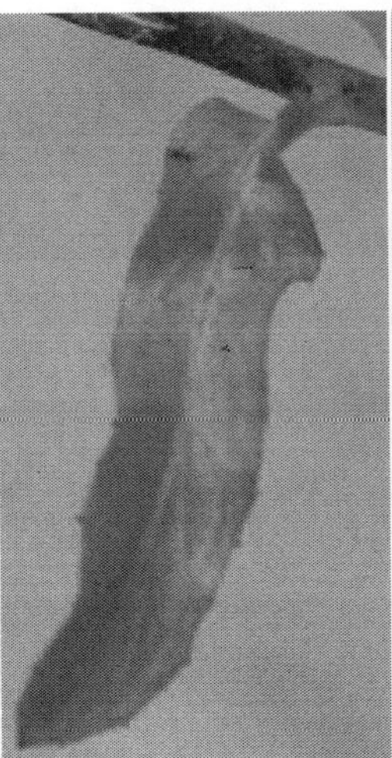

Leaf

Flower head

Chemical Constituents: The plant contains ß-sitosterol, taracasterol and its acetate, ß-amyrin and apigenin. Plants also contain thiophene acetylene derivatives, such as terthiophenes and amplectol. (Ahmad and Alam 1995, Pandya 2009).

Medicinal Properties: In action the plant is antipyretic and the leaves are vermifuge, astringent and antifungal. (Ahmad and Alam 1995, Pandya 2009).

Medicinal Uses (Traditional/Folklore): The plant has antifungal properties and used for treating venereal diseases. In folk medicines, it is used for treating malaria, bronchitis, influenza and asthma. The juice of the leaves (two teaspoonful for a period of 8-10 days) is given to expel intestinal worms, to stop bleeding, to reduce fever and to promote the flow of urine. (Ahmad and Alam 1995, Pandya 2009).

27.9 *Blumea viscosa* (Mill.) V.M.Badillo

(Syn. *Blumea aurita* (L.f.) DC.; *Laggera aurita* (L.f.) Benth. ex C.B.Clarke)

Common Name: Sticky Blumea, Clammy False Oxtongue.

Vernacular Name: None known.

Habit: Strong smelling herb.

Habitat: Along roadsides and open waste places.

Distribution: A pantropical weed distributed throughout India, Bangladesh, Myanmar and tropical Africa.

Flowering and Fruiting Time: March-May.

Taxonomic Description: A densely hairy, much-branched, annual herb, 30-80 cm high. Leaves alternate, up to 6 cm long, strong smelling when crushed, sessile, oblong, oblanceolate, shortly sub-pinnatified or toothed. Heads about 8 mm in diameter, discoid, pinkish, heterogamous in terminal, panicled corymbs. Involucral bracts multi-seriate, linear, silky hairy. Ray florets female. Disk florets bisexual; corolla 5-lobed, tubular. Pappus hairy, white, persistent. Achenes oblong, red, pubescent.

Habit

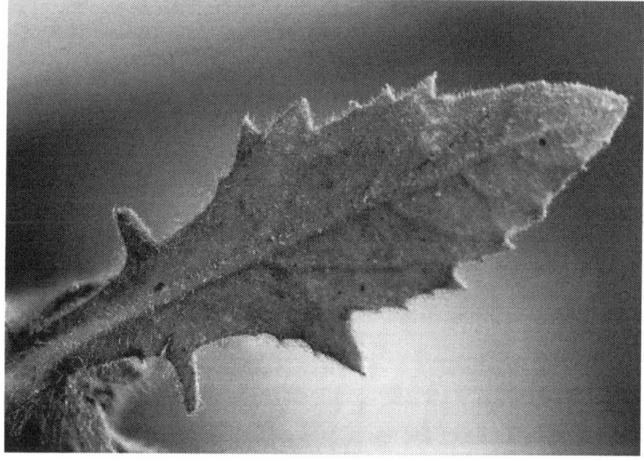

Leaf

Flower heads

A single head

Chemical Constituents: Phytochemical analysis showed the presence of saponins, tannins, carbohydrates and a number of sesquiterpenoids, monoterpenoids and flavonoids. (Dibala *et al.* 2014).

Medicinal Properties: Strongly scented plant have odour of turpentine. The leaves are diuretic having cicatrisant properties to heal cuts and wounds and bruises. The plants are rich in antinociceptive principle and antihyperalgesia due to the presence of saponins, tannins and carbohydrates. Plants have antioxidant, antimicrobial/antibacterial, anti-inflammatory, antiviral, and hepatoprotective properties. (Quattrocchi 2012, Dibala *et al.* 2014; http://tropical.theferns.info).

Medicinal Uses (Traditional/Folklore): Used in *Sidha* for arthritis, rheumatism, diarrhoea, dysentery, and stomach troubles. Powdered plant or its infusion is given for dyspepsia and indigestion. The whole plant is prepared as an enema to cure constipation and dysentery. (Quattrocchi 2012).

27.10 *Caesulia axillaris* Roxb.

(Syn. None known)

Common Name: Pink Node Flower.

Vernacular Name: *Hindi:* Gathila (गठीला).

Habit: Annual marshy herb.

Habitat: *C. axillaris* is common in rice fields and marshy areas.

Distribution: The plant occurs from east of Pakistan through India and Nepal to Bangladesh and Myanmar. Native to India, Nepal, Bangladesh, Sri Lanka and Myanmar.

Flowering and Fruiting Time: October-April.

Taxonomic Description: A sub-erect, glabrous, annual herb; stem up to 30 cm long. Leaves alternate, sessile, 5-15 x 1-1.7 cm, narrowly lanceolate, acuminate, base auricled, margin remotely serrulate. Flower heads discoid, sessile, axillary, homogamous. Involucral bracts crenate-serrate, broadly ovate or orbicular. Florets purplish-white, flowers bisexual. Pappus of 2 ovate scales. Achenes compressed, one-ribbed.

Habit

Flowering twig

Flowers

Chemical Constituents: Leaves contain glycoside, flavonoids, tannins, steroids and phenols. (Farooqui *et al.* 2016).

Medicinal Properties: It has antimalarial, antifungal and antibacterial properties. (Pradhan *et al.* 2017).

Medicinal Uses (Traditional/Folklore): Whole plant paste with camphor and mustard oil is applied on chest and throat to cure cold, cough and nasal congestion. Paste of inflorescence with black pepper and cow milk is given to cure dysentery. Certain quantity of fresh latex is mixed with a pinch of soil from the termite mound and applied externally to heal cuts and wounds fast. Plant extract is used to cure baldness, and goitre in traditional Indian System of Medicine. The whole plant is crushed and juice is extracted, which is given orally thrice a day along with curd to cure amoebic dysentery, indigestion and loss of appetite. Whole plant extract is also used internally to cure malaria. (Panda and Mishra 2011, Punjaji 2012, Ahirwar and Kujur 2015, Pani *et al.* 2015, Pradhan *et al.* 2017).

27.11 *Carthamus oxyacantha* M.Bieb.

(Syn. *Carthamus polyacantha* M.Bieb.)

Habit

Common Name: Wild Safflower, Jewelled Distaff Thistle.

Vernacular Name: *Hindi:* Peeli Kateri (पीली कटेरी), Kandiari (कंडिअरी).

Habit: Spiny annual herb.

Habitat: Occurs in open waste places, harvested wheat fields, fallow grounds and along roadsides.

Distribution: Found in India, Pakistan, Afghanistan, Iran, Azerbaijan, Iraq and Kyrgyzstan. It is a common weed along the boundaries of wheat fields.

Flowering and Fruiting Time: April-June.

Taxonomic Description: An erect, much-branched, spiny, annual herb, grows 30-80 cm in height. Stem terete, whitish. Leaves alternate, sessile, rigid, oblong-lanceolate with spinose margin, base semi-amplexicaul. Flower heads yellow, homogamous, nearly 2.5 cm in diameter. Involucral bracts multi-seriate with spiny tip. Florets yellow, pappus absent. Achenes nearly 5 mm long, smooth, polished.

Spiny leaf **Flower head** **Achenes**

Chemical Constituents: Seeds yields two types of oils: oleic oil and linoleic oil. Fatty acid oil composition of oleic oil is, palmitic acid, stearic acid, oleic acid, linoleic acid and traces of linoleic acid and longer chain fatty acids. The fatty acid composition of linoleic oil, palmitic acid, stearic acid, oleic acid, linoleic acid and longer chain fatty acids. Flowers also contain carthamin. Flavonoids, glycosides, sterols and serotonin derivatives have also been identified from flowers and seeds. Seeds of *C. oxyacantha* contain crude fats and crude fiber as the major components along with Ca, K, Mg, and P. (Ahmad *et al.* 2007, Bukhsh *et al.* 2007).

Medicinal Properties: The plants have diuretic, antimicrobial, antioxidant, antiworm/anthelmintic and laxative properties. (Ahmad *et al.* 2007, http://medplants.blogspot.in).

Medicinal Uses (Traditional/Folklore): Plant are diuretic, mainly used in dressing ulcers and against itch. Seed oil used as a dressing for bad ulcers, itch, joint pains, and liver diseases. Flowers of this species are also used to treat jaundice, while the seeds are considered as laxative. The sap is believed to reduce salivation. The oil is applied to treat scabies. Flowers are used to treat cerebral thrombosis, male infertility, rheumatism and bronchitis. Plant or flowers decoction is used as anthelmintic in children. It also induces labour and is used as a tonic tea to invigorate heart and blood circulation. *C. oxycantha* based medicines also show beneficial effect on pain and swelling associated with trauma. (Ahmad *et al.* 2007, http://medplants.blogspot.in).

27.12 *Cirsium arvense* (L.) Scop.

(Syn. *Cnicus arvensis* (L.) Hoffm.; *Serratula arvensis* L.)

Habit

Common Name: Creeping Thistle, Californian Thistle.

Vernacular Name: Kandai (कंदई).

Habit: Annual herb.

Habitat: Moist waste places, farmlands, roadsides, often becoming gregarious at few places.

Distribution: *C. arvense* is one of the world's most troublesome and persistent weeds, native to Europe and the East northern hemisphere but introduced to North America and the southern hemisphere. It is believed to have originated in the temperate Middle East and its spread has closely followed human migration and agricultural activity. Latitudinal distribution in North or South is limited by low winter and summer maximum temperatures and by a long day requirement for flowering. It is now widespread and scheduled agricultural weed in both arable crops and pastures and also a constituent in Europe's plant communities, occurring mainly on waste neglected land, roadsides, hedgerows and disturbed areas.

Flowering and Fruiting Time: February-April.

Taxonomic Description: An erect herb, 30-100 cm tall. Leaves alternate, sessile, oblong-linear, pinnatified, margin spine-tipped; basal leaves up to 20 cm long, upper smaller. Heads dioecious, 1.2-2 cm in diameter, in solitary or terminal corymbose. Female heads longer than the males. Outer involucral bracts ovate, spine-tipped. Florets pink to pink-purple. Pappus dirty white, feathery. Achenes about 3 mm long, compressed.

Chemical Constituents: Plant is known to contain flavonoid compounds, phenolic acids, tannins, sterols and triterpenes. Recent studies have shown the presence of ciryneol C, scopoletin, pectolinergenin 7-O-β-glucopyarnoside, acacetin, and 6, 7-dimethaxycoumarin. (Khan *et al.* 2011c, Banaras *et al.* 2017).

Medicinal Properties: Plants possess antibacterial, antifungal, antidiabetic, anti-inflammatory and anticancer properties. The roots have demulcent, stimulating and astringent properties. The roots are reported to contain arsenic-resistant bacteria. (Khan *et al.* 2011c, Khan *et al.* 2014c, Banaras *et al.* 2017; http://www.henriettes-herb.com).

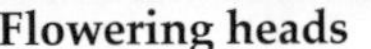
Flowering heads

A single head

Medicinal Uses (Traditional/Folklore): The plant has medicinal uses in the treatment of peptic ulcer, lung cancer and leukaemia in folk medicines. A small quantity of plant material (app. 30 g) is simmered in app. 500 ml of milk is used as an effective remedy for mild diarrhoea and dysentery, after the acute symptoms have subsided. Whole plant has been used to cure metrorrhagia, syphilis, eye infections, skin sores, gonorrhoea, bleeding piles and has also been found to be effective against diabetes. An infusion of plant is known to expedite labour, especially when the nervous system has become fatigued. Syrup made from the root is very effective against long-standing/chronic coughs. The leaf juice makes a soothing wash (or ointment) for irritable sores, tender eyes, and piles. Roots are used for the cure of mild leucorrhoea and prolapses. American Indians use an infusion of *C. arvense* roots for mouth diseases, worm infection and in the treatment for tuberculosis. (Khan *et al.* 2011c, Khan *et al.* 2014c; http://www.henriettes-herb.com).

27.13 *Cyanthillium cinereum* (L.) H.Rob

(Syn. *Conyza cinerea* L.; *Vernonia cinerea* (L.) Less.)

Habit

Common Name: Vernonia, Little Ironweed, Purple Feabane.

Vernacular Name: *Hindi:* Sahadevi (सहदेवी), Ankari (अंकारी), Ankta (आंकता).

Habit: A variable herb in shape and size.

Habitat: Roadsides, open and wastelands.

Distribution: *C. cinereum* is native to tropical Africa, tropical and temperate Asia, and Australia. It is widely naturalized in tropical regions in America and islands in the Pacific.

Flowering and Fruiting Time: Most part of the year.

Taxonomic Description: An erect, variable herb, 15-70 cm high with terete and striated stem. Leaves alternate, up to 7 cm long, petioled or sub-sessile, elliptic or ovate-lanceolate, entire or slightly crenate-serrate, obtuse or acute at apex, cuneate at base, sparsely pubescent. Heads discoid, pink, homogamous, 5-6 mm in diameter in terminal panicles. Involucral bracts many, linear-lanceolate, densely hairy. Pappus hairs white. Achenes terete, hairy.

Chemical Constituents: Plant contains triterpenes and sterols and other medicinally active biocompounds like alkaloids, phenols, tannins, saponins, steroids, glycosides, flavonoids, carbohydrates, phlobtannins and terpenoids. The seeds have a fatty oil which contains vernolic acid. (Varsha *et al.* 2015; http://medplants.blogspot.in, http://tropical.theferns.info).

Inflorescence

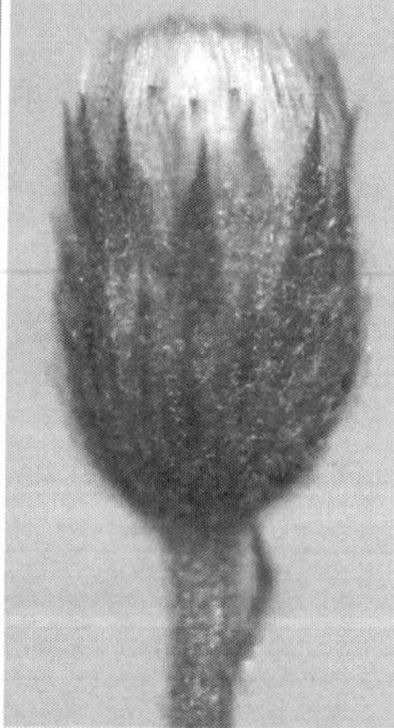

A single head

Medicinal Properties: Plants have alterative, alexipharmic, analgesic, antipyretic, anti-inflammatory, antiviral, anticancer, diuretic, diaphoretic, abortifacient and febrifuge properties. Seeds are alterative, anthelmintic and antidote. (Varsha *et al.* 2015; http://medplants.blogspot.in, http://tropical.theferns.info).

Medicinal Uses (Traditional/Folklore): The plant possesses numerous medicinal values in diverse traditional usage. The plant is used in Ayurveda herbal medicine as an alternative, diaphoretic, diuretic and febrifuge. The whole plant, including the seed and the roots, is used in decoction or infusion to treat fever, eye infections, kidney disorders, swellings, inflammation, lower abdominal pains diarrhoea, eczema and menstrual pains. It has been used as remedy for spasms of the urinary bladder and strangury, also often combined with quinine to treat malaria. Seeds are used as a source for alexipharmic and anthelmintic drugs. Leaves of *C. cinereum* have analgesic, antipyretic and anti-inflammatory effects. Juice of this plant is given to children to treat bed-wetting. Recently, some research is being undertaken on this plant as an aid to give up cigarette smoking.

A decoction of the whole plant is employed to promote perspiration in the treatment of fevers. The juice of the plant is used to treat piles. A paste made from the plant is used as a poultice on cuts, wounds and skin diseases. Decoction of stem and leaf is used to expel the placenta and as an abortifacient. It is considered a more potent abortifacient and better treatment for menstrual pains when *C. cinereum* is decocted with *Justicia secunda*. Leaves boiled with *Stachytarpheta jamaicensis* are used for treating albuminuria and sprains. In Guyana, leaves are used in a tea for cleansing the blood. The flowers are used in treating conjunctivitis.

The seeds are used as an anthelmintic and alexipharmic; they are said to be quite effective against roundworms and threadworms. They are also given for coughs, flatulence, intestinal colic and dysuria and for leucoderma, psoriasis and other chronic skin-diseases. The seeds are made into a paste with lime juice and used for destroying pediculi (lice). It is known to be used traditionally to treat many aliments like fever, respiratory disorders, skin and gastrointestinal diseases, etc. They are used as an antidote to poison and in the treatment of leprosy and chronic skin diseases.

The root is used in treating dropsy and scorpion stings. The juice of the root is used in the treatment of haemorrhoids, diarrhoea and stomach disorders. The root is valued as an anthelmintic. A paste of the root is applied externally to treat boils. (Shukla *et al.* 2010, Varsha *et al.* 2015; http://medplants.blogspot.in, http://tropical.theferns.info, http://www.flowersofindia.net).

27.14 *Eclipta prostrata* (L.) L.

(Syn. *Eclipta alba* (L.) Hassk.; *Verbesina prostrata* L.)

Habit

Common Name: False Daisy, Trailing Eclipta.

Vernacular Name: *Hindi:* Bhringaraj (भ्रिंगराज), Kesharaj (केशराज); *Sanskrit:* Bhringarajah (भ्रिंगराजः), Tekarajah (तेकराजः).

Habit: Annual herb.

Habitat: Moist places, roadsides, fallow fields and along the water channels.

Distribution: The plant is widely distributed in the warm humid tropics with plenty of rainfall. It grows commonly in moist places as a weed all over plains of India, at all elevations in waste places and on roadsides.

Flowering and Fruiting Time: Most part of the year.

Flowering twig

Flower head

Taxonomic Description: An erect or sub-erect, annual herb, up to 40 cm high. Stem terete, purple colour, hirsute. Leaves simple, opposite, sessile, 3-6.5 x 0.5-1.5 cm, linear or oblong-lanceolate, sub-entire, acute at apex, appressed hispid on both the surfaces. Heads white, heterogamous, 5-8 mm in diameter, solitary or 2-3 together on axillary and terminal peduncles. Involucral bracts ovate-lanceolate, pubescent. Ray florets 2-3 seriate. Disk florets numerous; corolla campanulate, 4-toothed, tube short. Achenes oblong, dark brown or black.

Chemical Constituents: The aerial part is reported to contain a phytosterol, b-amyrin in the n-hexane extract and luteolin-7-glucoside, b-glucoside of phytosterol, a glucoside of a triterpenic acid and wedelolactone in polar solvent extract. The leaves contain stigmasterol, a-terthienylmethanol, wedelolactone, dismethylwedelolactone and dismethylwedelolactone-7-glucoside. The roots contain hentriacontanol and heptacosanol, and polyacetylene substituted thiophenes. The polypeptides isolated from the plant yield cystine, glutamic acid, phenyl alanine, tyrosine and methionine on hydrolysis. Nicotine and nicotinic acid are also reported to occur in this plant. (Mithun *et al.* 2011, Neeraja and Margaret 2012, Sharma *et al.* 2012, Jahan *et al.* 2014, Paliwal 2017; http://www.bsienvis.nic.in).

Medicinal Properties: The plant is bitter, acrid, thermogenic, anti-inflammatory, anthelmintic, nematicidal, ovicidal, spasmolytic, diuretic, aphrodisiac, depurative, febrifuge, anticatarrhal, antidontalgic, absorbent and antihepatic; it is useful in hepatosplenomegaly, elephantiasis, gastropathy, anorexia, helminthiasis, skin diseases, wounds, ulcers, debility, hypertension, leprosy, pruritus, fever, jaundice, odontalgia, otalgia and cephalalgia. The alcoholic extract of entire plant has been reported to show antiviral activity against Ranikhet virus disease. (Warrier *et al.* 2010, Mithun *et al.* 2011, Neeraja and Margaret 2012, Sharma *et al.* 2012, Jahan *et al.* 2014, Paliwal 2017; http://www.bsienvis.nic.in).

Medicinal Uses (Traditional/Folklore): *Eclipta* is one of the ten auspicious herbs that constitute the group *dasapuspam* which is considered to destroy the causative factors of all unhealthy and unpleasant features and bestow good health and prosperity. The members of this group cure wounds and ulcers as well as fever caused by the derangement of the *tridosas - Vata, Pitta* and *Kapha*. It is used in hepatitis, spleen enlargements, chronic skin diseases, tetanus and elephantiasis. The plant has been mentioned in ancient scriptures describing its uses for various ailments: In *Atharvaveda*: affects intelligence and memory, cures bile (pitta) disorders, prevents graying and falling of hairs; In *Bhavaprakasha*: cures problems caused by phlegm and wind, beneficial for hair, skin, teeth and eyes, removes worms, and also effective in jaundice and oedema; In *Rajanighantu*: beneficial for hairs, eyes, oedema and phlegm; In *Kaiyadevanighantu*: it removes the problems caused by phlegm and wind and worms, beneficial for hair, teeth, skin, cures cough, jaundice and oedema; In *Nighanturatnakaram*: plant invigorates sex; In *Vaidyamanorama:* drinking juice of the plant strengthens the body and secures the foetus in womb.

The juice of the plant with honey is given to infants for expulsion of worms. Plant (powder) has curative effect on infective hepatitis, jaundice and viral hepatitis. Leaf (aqueous extract/juice) is found to be effective as myocardial depressant, hypotensive and is known to cure shoulder pain caused by heavy load. Bhringraj is mainly used in hair oils, but it has been considered an effective drug in hepatotoxicity. The plant is found to be effective in blackening and strengthening of the hair, for stopping haemorrhages and fluxes, and for strengthening the gums. The leaf promotes hair growth. The seeds are useful in increasing sexual vigour. Aqueous extract of the plant provides subjective improvement of vision in the case of refractive errors.

The herbal drug Trefoli, containing extracts of the plant in combination with others, when administered to the patients of viral hepatitis, produced excellent results. It is also used to prevent habitual abortion and miscarriage and also in cases of post-delivery uterine pain. A decoction of leaves is used in uterine haemorrhage. The juice of the plant with honey is given to infants with castor oil for expulsion of worms. For the relief in piles, fumigation with *E. prostrata* is considered beneficial.

In Ayurveda, it is mainly used in hair oil, while in Unani system, the juice is used in "Hab Miskeen Nawaz" along with aconite, triphala, *Croton tiglium, Piper nigrum, Piper longum, Zingiber officinale* and minerals like mercury, sulphur, arsenic, borax, etc. for various types of pains in the body. It is also a constituent of "Roghan Amla Khas" for applying on the hair and of "Majun Murrawah-ul-arwah". Other uses in Ayurveda are: cures headache, migraine; leaves remove lice, blacken skin, cure pyorrhoea, chronic dysentery, oedema, nervous weakness, jaundice, anorexia, gum troubles and remove intestinal worms.

Other traditional uses of the plant by different ethnic communities include: Leaf: in gastric troubles, hepatic disorders (by *Garo*); Stem-decoction: in liver enlargement; Leaf-extract: in fever and cough (*Manipuri tribes*); Leaf: as antidote to scorpion sting (*Toto tribes*); Plant: in itching, Leaf: in conjunctivitis and other eye problems, in promoting hair growth (*Ethnic Communities of Orissa*); Leaf: in jaundice *(Irulas tribal community)*; Leaf: in malaria and other fevers (*Ethnic Communities of Bihar*); Leaf: in conjunctivitis, eye troubles *(Ethnic Communities of Araku Valley,*Madhya Pradesh); Leaf: in sores, ulcers, wounds, spleen disorders *(Tribal Societies of Eastern Rajasthan)*; Leaf: as antiseptic *(Tribes of Kurukshetra,* Haryana); Leaf: in malaria (*Kol*); Leaf: for promoting hair growth *(Tribes of Chhindwara,* Madhya Pradesh); Leaf: as cooling*(Ethnic Communities of Salsette Island)*; Plant-decoction with paste of black pepper in fever, Leaf-juice: on wounds (*Santal*); Plant: against antifertility, Root: as antidote to snake bite, Flower bud: in fever, headache *(Tribal Societies of Anaikatty Hills,* Tamil Nadu); Whole Plant: in asthma, bronchitis, leucoderma *(Tribal Societies of Saurashtra,* Gujarat); Whole plant: in jaundice, spleen disorders, Leaf: in leucoderma, skin diseases *(Garhwali*); Whole plant: in hepatic problems, spleen troubles, Whole plant: in liver complaints *(Tribes of Bastar,* Madhya Pradesh); Plant: in toothache, headache, gland swelling, elephantiasis *(Tribes of Sagar,* Madhya Pradesh). The paste prepared by mincing fresh plants has got an anti-inflammatory effect and may be applied to insect bites, stings, swellings and other skin diseases. Leaves are also used as an antidote in scorpion sting. (Warrier *et al.* 2010, Sivarajan and Balachandran 1994, Singh *et al.* 2010d, Mithun *et al.* 2011, Manvar *et al.* 2012, Neeraja and Margaret 2012, Sharma *et al.* 2012, Jahan *et al.* 2014, Ahirwar and Kujur 2015, Paliwal 2017; http://www.bsienvis.nic.in).

27.15 *Erigeron bonariensis* L.

(Syn. *Conyza bonariensis* (L.) Cronquist)

Habit

Common Name: Asthmaweed, Flax-Leaf Fleabane, Wavy-Leaf Fleabane.

Vernacular Name: Makshikaa-Visha (मक्षिका-विशा).

Habit: Annual herb.

Habitat: Common along roadsides, in gardens and open waste places.

Distribution: *E. bonariensis* is native to South America; found throughout the tropics and subtropics as a pioneer plant. It has become naturalized in many other regions, including India.

Flowering and Fruiting Time: April-October.

Taxonomic Description: An erect, deep-rooted, up to 1 m high, hirsute, annual herb. Stem simple or branched, ribbed, appressed hairy. Leaves alternate, 5-11 x 0.5-1.5 cm, linear or linear-lanceolate, base cuneate, apex acute, margin coarsely serrate or entire. Flower heads numerous, discoid, 5-12 mm across, heterogamous, borne in corymbose panicles. Involucral bracts oblong, linear, hairy, 3-4-seriate. Ray florets whitish, filiform. Disk florets pale yellow. Pappus hairy, dirty-white. Achenes 1-1.5 mm long, brown, pubescent.

Upper portion with flower heads

Few disk floret

Involucral bracts

Chemical Constituents: The plant contains monoterpenoids, sesquiterpenoids, diterpenoids, triterpenoids; phenolic derivatives contain essential oil, tannic and gallic acids, bitter extractive, etc. Specifically, the chemical compounds are matricaria ester and caryophyllene oxide, besides stigmasterol, freideline, quercitrin, caffeic acid, etc. The aromatic glycoside, erigoside G is reported as a new natural compound from this plant. (Zahoor *et al.* 2012, Benzarti *et al.* 2013, Sharma *et al.* 2014c).

Medicinal Properties: Plants have astringent, diuretic, pungent, tonic and insecticidal properties. (Gulshan *et al.* 2012, Sharma *et al.* 2014c).

Medicinal Uses (Traditional/Folklore): Since Roman times *Erigeron* has been used as a natural insecticide to repel fleas. By growing the plants in the garden insects can be controlled; by rubbing the leaves on clothing or paths also repels mosquitoes, flies, gnats and ticks. It is a traditional medicine for the treatment of indigestion, enteritis, epidemic hepatitis, and hematuria. In Asian traditional medicine, flowers of *E. bonariensis* are used topically to treat gastrointestinal problems. The leaves have been reported as laxative. Roots have been used in diarrhoea and dysentery treatment. In Pakistan, the plant, locally called "gulava" or "mrich booti", is traditionally used in urine problems, diarrhoea, haemorrhage, excessive menstruation, piles, kidney disorders, bronchial complaints, and skin diseases like eczema and ringworm. (Gulshan *et al.* 2012, Zahoor *et al.* 2012, Benzarti *et al.* 2013, Sharma *et al.* 2014c).

27.16 *Gnaphalium purpureum* L.

(Syn. *Gamochaeta purpurea* (L.) Cabrera)

Habit

Common Name: American Cudweed, Early Cudweed, Purple Cudweed.

Vernacular Name: None known.

Habit: Winter season annual herb.

Habitat: Common winter weed in moist, waste places and cultivated fields.

Distribution: A native of tropical America; the plant is now a widespread weed throughout the tropics.

Flowering and Fruiting Time: December-March.

Taxonomic Description: An erect or ascending herb, up to 50 cm high. Stem branched from the base and appressed with white tomentum. Leaves alternate, sessile, 2.5-8 x 1-2 cm, spathulate, obovate, entire, rounded or mucronate at apex, attenuate at base, white-woolly beneath. Heads in spicate globose clusters, heterogamous, 2-3 mm across, yellow, crowded in upper leaves. Involucral bracts 3-4 seriate, oblong, brown. Pappus hairs white, united at the base. Achenes 0.5 mm long.

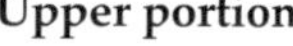

Upper portion

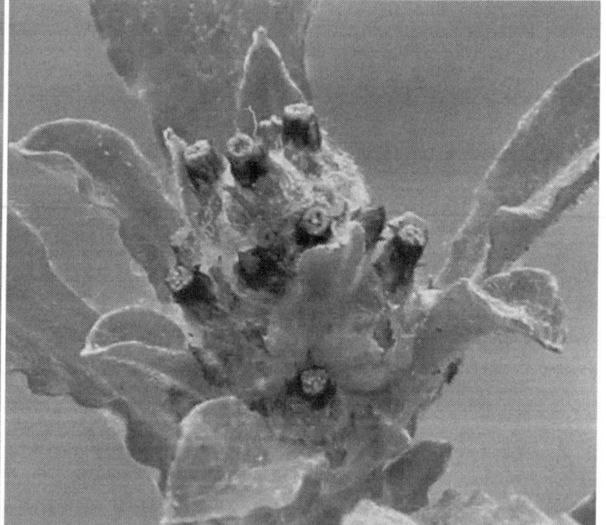

Flower heads

Chemical Constituents: Flavonoids. (Thomas *et al.* 2012, Zheng *et al.* 2013).

Medicinal Properties: Plant is known to have anticancer property. (Thomas *et al.* 2012, Zheng *et al.* 2013).

Medicinal Uses (Traditional/Folklore): In the region of tropical Atlantic Forest, Sao Paulo, Brazil, *G. purpureum* is used in diarrhoea, intestinal complaints and bellyache. In southern Uganda, leaf decoction is used in anorexia (lack of appetite). In the Andean regions of South America, the hot beverage obtained by decoction of *G. purpureum* and *G. elegans* is recommended for the cure of cancer. (Stasi *et al.* 2002, Ssegawa *et al.* 2007, Thomas *et al.* 2012, Zheng *et al.* 2013).

27.17 *Grangea maderaspatana* (L.) Poir

(Syn. *Artemisia maderaspatana* L.)

Habit

Common Name: Madras Carpet.

Vernacular Name: Mustaru (मुस्तरु), Bhediachim (भेदियाछिम).

Habit: Prostrate herb.

Habitat: Roadsides, wastelands and damp places.

Distribution: It is native of America, found in Indo-Malesia and Africa.

Flowering and Fruiting Time: Most part of the year.

Taxonomic Description: A prostrate, much-branched, densely villous, annual herb, spreading in circular patches. Leaves alternate, sessile or sub-sessile, up to 5 cm long, pilose, sinuately pinnatified, base auriculate. Flower heads 7-10 mm across, yellow, globose, heterogamous, usually solitary, axillary, shortly peduncled. Outer flowers female and inner flowers bisexual. Involucral bracts 2-3 seriate, oblong, densely pubescent. Achenes 2-3 mm long, compressed.

Flowering twig

Flower head

Chemical Constituents: Phytochemical screening revealed the presence of steroids, polyphenols, tannins, carbohydrate and flavonoids in the plant. The major constituents are γ- gurjunene, terpinyl acetate, hinesol, 10-epi-γ-eudesmol, α-guaiene and epicubebol; steroidal constituents are hardwickiic acid, the corresponding 1, 2- dehydro-derivative, acetylenic compound. A clerodane derivative, 15-hydroxy-16- oxo-15,16H-hardwickiic acid (15-hydroxycleroda – 3, 13-dien-15, 16-olide-18-oic acid) has been isolated from the aerial parts of *G. maderaspatana*. Three other components *viz.*, eudesmanolide, (-) frullanolide, (-) -7-alpha-hydroxyfrullanolide and a new eudesmanolide (+) -4 alpha, 13-dihydroxyfrullanolide have also been isolated from the whole plant of *G. maderaspatana*. (Chandrul and Singh 2015, Galani and Rachchh 2015, Rachchh and Galani 2015).

Medicinal Properties: The plant possess analgesic, anti-inflammatory, antinociceptive and antirheumatic activity. The leaves of *G. maderaspatana* are stomachic, sedative, carminative, emmenagogue and antiflatulent. The root is an appetizer; astringent to the bowels, diuretic, anthelmintic, galactagogue, stimulant; useful in griping, in troubles of the chest and lungs, headache, paralysis, rheumatism in the knee joint, piles, pain in the muscles, diseases of the spleen and the liver, troubles of the ear, the mouth and the nose. (Chandrul and Singh 2015, Rachchh and Galani 2015).

Medicinal Uses (Traditional/Folklore): *G. maderaspatana*, a popular Indian medicinal plant is traditionally used for rheumatism in the knee joint and pain in the muscles. Leaves are regarded in India as a valuable stomachic possessing deobstruent and antispasmodic properties, and are prescribed as an infusion and an electuary in cases of obstructed menses and hysteria. The plant has a good reputation in traditional medicine as analgesic and is used in pains of the eyes and ears. Plant paste with cow milk and sugar candy is given to cure menstrual disorders. Leaf paste with honey is given to cure stomach disorder. Roots are useful in lessening excess of perspiration (Unani). (Panda and Mishra 2011, Chandrul and Singh 2015, Rachchh and Galani 2015).

27.18 *Launaea procumbens* (Roxb.) Ramayya & Rajagopal

(Syn. *Launaea fallax* (Jaub. & Spach) Kuntze; *Prenanthes procumbens* Roxb.)

Habit

Common Name: Creeping Launaea.

Vernacular Name: *Hindi:* Peeli Duddhi (पीली दुद्धि), Jangali Gobhi (जंगली गोभी).

Habit: Annual or perennial herb.

Habitat: Common along roadsides, grasslands and fallow fields.

Distribution: Egypt, Arabia, Iraq, Iran, Turkmenistan, Uzbekistan, Tadzhikistan, Afghanistan, Pakistan, India, Nepal, Burma and China.

Flowering and Fruiting Time: Major part of the year.

Taxonomic Description: A procumbent or sub-erect, annual or perennial herb with yellow latex. Leaves mostly radical, forming rosette, sessile, up to 20 cm long, pinnatified or irregularly sinuate lobed; lobes sharply toothed, margin white-cartilaginous. Flowering shoot up to 40 cm long, simple or branched. Flower heads 1.2-1.8 cm in diameter, cylindric, yellow, borne on flowering shoot. Outer involucral bracts ovate, inner linear-oblong; margin white, membranous. Pappus 6-9 mm long, white. Stigma bi-lobed, longer than the corolla tube. Achenes 3-4 mm long, striated, much smaller than the pappus.

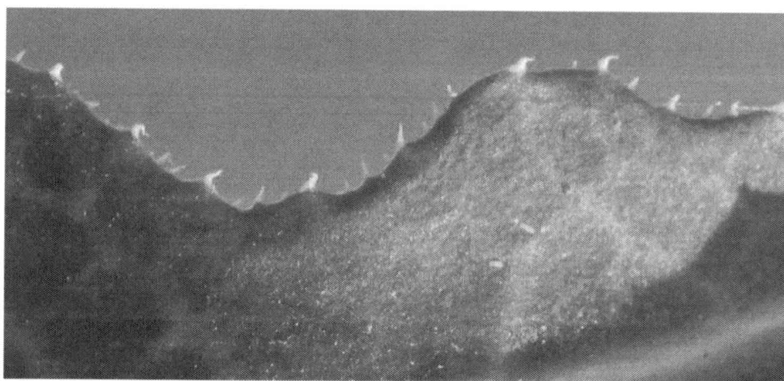

Cartilaginous leaf margin

Flower head

Chemical Constituents: *L. procumbens* contains salicylic acid, vanillic acid, 2-methyl-resercinol and gallic acid. Leaves contain xyloside, β-sitosterol and its acetate, taraxasterol, campesterol, stigmasterol, cholesterol, 28-isofucosterol, stigmast-7-enol, 24-methylcholest-7-enol; also present are hydrocarbons, higher fatty acids esters of primary alcohols, triterpene alcohols, Apigenin-7-glucoside and its 7-gentiobioside; luteolin-7-glucoside and its 7-gentiobioside, 7-rutinoside, 7,3′-diglucoside, 7,4′-diglucoside and 7-gentiobioside-4′-glucoside, aesculetin and cichoriin. (Khan 2010a, Khan 2012, Rathod and Dhale 2014).

Medicinal Properties: Plants possess diuretic, neuroprotective, cardioprotective, anticarcinogenic, anti-inflammatory, and antioxidant properties. (Khan 2012, Rathod and Dhale 2014).

Medicinal Uses (Traditional/Folklore): Ayurvedic and herbal medicines prepared from *L. procumbens* promote self-healing, good health and longevity, as well as used as a food ingredient. Plant is used in the treatment of rheumatism, inflammation and oxidative dysfunction in the kidney, reproductive disorders, hormonal imbalances and liver dysfunction. It is also useful in the treatment of nephritis, pulmonary fibrosis, and sexual diseases by local healer in Pakistan. Extracted juice from leaves is given in jaundice. Plant is used for the treatment of kidney disorders like painful urination, gonorrhoea, and sexual diseases. The whole plant is ground in water along with candy (*Misri*) and is used orally for painful micturition. Leaves are used for the treatment of fever, cancer, and swellings. (Khan 2010a, Qureshi *et al.* 2010, Khan 2012, Rathod and Dhale 2014).

27.19 *Parthenium hysterophorus* L.

(Syn. *Argyrochaeta bipinnatifida* Cav.)

Habit

Common Name: Carrot Grass, Congress Grass, Wild Carrot Weed.

Vernacular Name: *Hindi:* Gajar Ghas (गाजर घास).

Habit: Annual herb.

Habitat: Found as a common weed along roadsides, railway lines, waste places, cultivated and fallow fields.

Distribution: It is native to the American tropics, now a pantropical weed.

Flowering and Fruiting Time: Throughout year.

Taxonomic Description: An erect, strong smelling, branched, annual herb, 30-100 cm high. Stem ribbed, appressed-hairy. Leaves alternate, up to 18 cm long, deeply pinnatified or lobed; lobes oblong-lanceolate, acute, entire, tomentose on both the surfaces. Flower heads white, many, 3-5 mm across, heterogamous, borne in axillary or terminal leafy panicles. Involucral bracts 2-seriate, ovate. Ray florets-5, white, female. Disk florets numerous, corolla 5-lobed, tubular, light yellow. Achenes minute, about 2 mm long, black, obovoid.

Leaf

Flowering twig

Flower heads

Chemical Constituents: It contains several important chemical constituents mainly histamine, saponin, glucosides and triterpene (sesquiterpene). Two types of sesquiterpene lactones, hysterin and dihydroisoparthenin have been isolated from plant. Three ambrosanolides, α-epoxymethylacrylyloxy parthenin, its 11α13-dihydro derivatives and 8α- eposeymethylacrylylos-ambrosin have been characterised from chloroform extract of the aerial parts of the *P. hysterophorus*. (Chhabra *et al.* 1999, Patel 2011, Kushwaha and Maurya 2012).

Medicinal Properties: Antipyretic, antirheumatic, besides, there are some potential of this weed as insecticidal, herbicidal, anti-amoebic and also as therapeutic purpose. The plant has been found to be pharmacologically active as analgesic in muscular rheumatism, therapeutic for neuralgia and as vermifuge. (Patel 2011, Kushwaha and Maurya 2012, Marwat *et al.* 2015).

Medicinal Uses (Traditional/Folklore): Different parts of *P. hysterophorus* are used in traditional medicines. A decoction prepared from the whole plant is taken internally as a remedy for a wide variety of ailments such as fever, diarrhoea, neurologic disorders, urinary tract infections, dysentery, malaria and as emmenagogue. Ethnobotanically, it is also used as remedy for eczema, skin rashes, herpes, and rheumatic pain. The decoction of root is used in curing amoebic dysentery. Leaf juice is also carminative and gives strength to the stomach; some people use it in fever also. In the Caribbean and Central America, it is applied externally on skin disorders. In Jamaica, the decoction is used as a flea (insect) repellent both for dogs and other animals. (Oudhia 2001, Patel 2011, Marwat *et al.* 2015).

27.20 *Pluchea lanceolata* (DC.) C.B.Clarke

(Syn. *Berthelotia lanceolata* DC.)

Common Name: Rasna.

Vernacular Name: *Hindi*: Phaar (फार); *Sanskrit:* Rasna (रसना), Sugandha (सुगंधा), Surabhi (सुरभी), Yukta (युक्ता), Surasa (सुरसा), Suvaha (सुवाहा).

Habit: Undershrub.

Habitat: Common in dry areas, wastelands, along roadsides and railway lines.

Habit

Distribution: *P. lanceolata* is distributed throughout the North-Western part of India, neighbouring Asian countries, Pakistan, Afghanistan and also in North Africa. In India, it occurs in Punjab, Upper Gangetic plains, Gujarat and Rajasthan. It is a common weed of cultivated fields in sandy or saline soils.

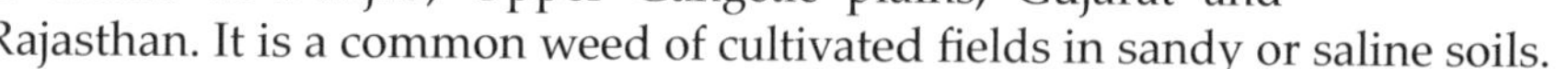

Flowering and Fruiting Time: March-July.

Taxonomic Description: An erect, undershrub, up to 1 m high with terete stem. Leaves alternate, sessile, coriaceous, up to 7 cm long, oblong, oblanceolate, apex apiculate, margin entire or obscurely toothed towards apex, appressed pubescent on both the surfaces. Flower heads pink or purple, about 3-4 mm across, in compound terminal corymbs. Involucral bracts multi-seriate, obtuse, silky pubescent. Pappus hairs united at base.

Upper portion

Flower heads

Chemical Constituents: *P. lanceolata* contains taraxosterol, b-sitosterol, pitichine, betaine hydrochloride, quercetin, isorhamentin, triterpenoids, sterols, flavonoids and lactones. The leaves contain quecetin, quercitrin, isorhamnetin, pluchioside, pleuchiol, 23-methyldotriacontan-3-one, 22-methylhentriacontane-3 and 19-dione. (Khan *et al.* 2010b, Bhutya 2011).

Medicinal Properties: The leaves are thermogenic, digestive, analgesic, febrifuge, alexiteric, laxative, nervine tonic, anti-inflammatory and antipyretic; used in asthma, gout, neuralgia, gastropathy, bronchitis, pityriasis versicolor, pains and rheumatism. Anti-inflammatory activity has been reported in the neolupinol isolated from the flowers of this plant. Flavonoids are reported as scavengers of free radicals and potent inhibitors of lipid peroxidation. Oral pre-treatment of the ethanolic extract significantly attenuated cadmium chloride, induced oxidative stress, and genotoxicity. (Bhagwat *et al.* 2010, Khan *et al.* 2010b, Bhutya 2011, Mohanty *et al.* 2013).

Medicinal Uses (Traditional/Folklore): It is widely used in the treatment of rheumatoid arthritis in Ayurveda, an Indian System of Medicine. This plant is indicated in Ayurvedic texts for its therapeutic usefulness in diseases similar to rheumatoid arthritis and other afflictions of joints. The decoction of plant is used to prevent the swellings of joint in arthritis, rheumatism and neurological diseases. Plant is useful in constipation and respiratory diseases. The decoction of the whole plant has been used traditionally in arthritis; a poultice of leaves is applied to the inflamed areas of the body.

In Ayurveda, the management of malaria considered as 'Visham jwar' and *P. lanceolata* is a one of the ingredients of poly-herbal formulation used to treat *jwar* (fever), including '*Visham jwar*'. (Bhagwat *et al.* 2010, Khan *et al.* 2010b, Mohanty *et al.* 2013).

27.21 *Sonchus asper* (L.) Hill

(Syn. *Sonchus spinosus* Lam.)

Habit

Common Name: Prickly Sow-Thistle, Rough Sow-Thistle.

Vernacular Name: *Hindi:* Dudhi (दुधी).

Habit: Annual herb.

Habitat: Common on damp, waste places, roadsides, near water channels and canal banks.

Distribution: *S. asper* is native to Mediterranean region; also become naturalized on other continents and is regarded as a noxious, invasive weed in many places. In India, found in waste-places and along roadsides and margin of the fields up to 3,000 m.

Flowering and Fruiting Time: December-March.

Taxonomic Description: An erect, annual herb, 20-80 cm tall. Stem mostly unbranched, glabrous or sparsely glandular hairy at top. Leaves variable, ovate-lanceolate or oblanceolate; radical leaves up to 22 cm long, forming a rosette, runcinated-pinnatified; cauline leaves half-amplexicaul, sharply dentate or spinose toothed, auricled at base. Flower heads cylindric, yellow, borne in umbellate cymes. Involucral bracts multi-seriate, oblong-lanceolate, glabrous or sparsely glandular hairy. Pappus hairs white. Achenes about 3 mm long, flattened, 3-ribbed.

Apical portion with inflorescence

Flower heads

Chemical Constituents: It contains large quantities of phenolic compounds, flavonoids, ascorbic acid, carotenoids and a variety of other antioxidants. (Khan *et al.* 2010a, Upadhyay *et al.* 2013).

Medicinal Properties: The plant is a general tonic. It possesses antioxidant, diuretic, laxative, emollient, antiasthamtic, anticancer and anti-inflammatory properties. The roots and leaves are used as a tonic and febrifuge. (Dhiman 2004, Khan *et al.* 2010a, Khan *et al.* 2012, Upadhyay *et al.* 2013).

Medicinal Uses (Traditional/Folklore): Roots, stem, leaves and its juice/latex have been used to treat a vast variety of conditions, ailments and diseases. These include treatment of wounds, boils, asthma, bronchitis, gastrointestinal infections, malaria, venereal diseases, inflammation, diabetes and cardiac dysfunction, kidney and liver disorders and reproductive disorder like impotence. The whole plant is pounded and its paste/extract is applied on fresh injuries, old wounds, boils and swellings; extract taken internally acts as blood purifier and tonic. An infusion is made of the entire plant and taken as a drink. Paste of plant is applied on breasts to increase lactation. Herb is used as an emollient. All parts of the plants are used as coolant. (Dhiman 2004, Katewa and Galav 2005, Khan *et al.* 2010a, Khan *et al.* 2012).

27.22 *Sonchus brachyotus* DC.

(Syn. *Sonchus arvensis* var. *glaber* Haines)

Habit

Common Name: Sadhi.

Vernacular Name: *Hindi*: Dudhali (दुधली).

Habit: Perennial herb.

Habitat: A common weed growing in waste places, fields and along canal banks.

Distribution: It is widespread across much of northern Asia, found in Japan, Korea, China, Russia, Mongolia, Thailand, Kazakhstan, Caucasus, Kyrgyzstan, etc. In India, found in waste-places, along road-sides and agricultural fields up to 2,500 m.

Flowering and Fruiting Time: January-June.

Taxonomic Description: An erect, perennial herb, 40-100 cm high. Stem angular, hollow. Leaves variable in shape and size; radical leaves up to 15 cm long, sub-sessile, forming a rosette, oblanceolate, runcinate-pinnatified or spinous-toothed; cauline leaves smaller, lanceolate, sessile, half-amplexicaul with short auricles. Heads 1.5-2 cm across, yellow, crowded in umbellately corymbose. Involucral bracts multi-seriate, densely glandular hairy. Pappus nearly 8 mm long, white-silky. Achenes 3-4 mm long, brownish, ribbed, compressed.

Flower head

Chemical Constituents: *S. brachyotus* are known to possess a number of phytochemical constituents like β-sitosterol, luteolin, apigenin and quercetin. (Zheng 2006, Mazumdar and Choudhury 2015).

Medicinal Properties: The plant has antioxidant, antihepatitits, antimicrobial, anti-inflammatory, antipyretic, antirheumatic, and detoxication and antitumour properties. (Zheng 2006, Mazumdar and Choudhury 2015; http://www.pfaf.org).

Medicinal Uses (Traditional/Folklore): It is traditionally used to treat gastric disorders, stomach pain, waist pain, cancer, rheumatism and diarrhoea. Roots are used in folk medicines against cough and bronchitis. The plant has been used in traditional Chinese folk medicines for hepatitits, inflammation, detoxication, pain-relief, fevers, and dysentery and curing snake-bite. (Dhiman 2004, Zheng 2006, Mazumdar and Choudhury 2015).

27.23 *Tridax procumbens* (L.) L.

(Syn. *Balbisia divaricata* Cass.)

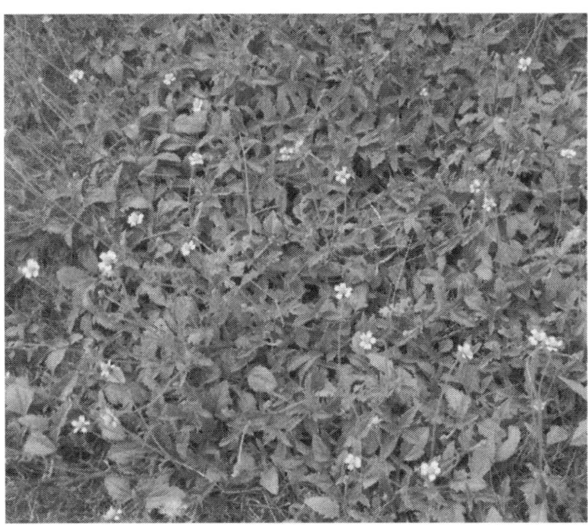
Habit

Common name: Tridax Daisy, Coat Buttons, Mexican Daisy.

Vernacular name: *Hindi:* Khal-muriya (खाल-मुरिया), Tal-muriya (ताल-मुरिया), Ghamra (घमरा); *Sanskrit:* Jayanti Veda (जयंती वेदा).

Habit: Procumbent herb.

Habitat: Common weed found along roadsides, in wastelands, and gardens.

Distribution: The plant is native of tropical America and naturalized in tropical Asia, Africa, Australia, and India. It is found commonly along paths, roadsides, wastelands, and crop fields and in the crevices of walls and rocks up to 1,300 m. It is a great favorite with low flying butterflies. In the areas where there is a great concentration of these flowers one will find plenty of butterflies too.

Flowering and Fruiting Time: Almost throughout the year.

Taxonomic Description: A procumbent, hispid-pubescent herb. Leaves opposite, petioled, 2-5 x 1.5-2.5 cm, ovate-lanceolate, irregularly toothed, cuneate at base, acute at apex, densely hairy on both the surfaces. Heads 1-1.5 cm across, heterogamous, solitary on 10-30 cm long peduncle. Involucral bracts multi-seriate with long spreading hairs. Ligulate flowers usually 5, 3-lobed, light yellow or white. Disk flowers numerous; corolla tubular, 5-lobed, yellow. Pappus dirty-white, bristly. Achenes about 2 mm long, brown, hairy.

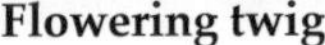
Flowering twig

Flower head

Chemical Constituents: A number of chemical constituents are reported from the plant *viz.*, alkaloids, flavonoids, carotenoids, β-sitisterol, n-hexane, fumaric acid, luteolin, quercitin, oxoester, lauric acid, myristic acid, palmitic acid, arachidic acid, linoleic acid and tannin, etc. Leaves of *T. procumbens* contains minerals such as calcium, magnesium, potassium, sodium and selenium. The plant is a good source of plant protein and potassium supplement. (Ankita and Jain 2012, Saini *et al.* 2016).

Medicinal Properties: The plant shows various pharmacological activities like immunomodulatory, antidiabetic, antihepatotoxic, antioxidant, anti-inflammatory, analgesic and marked respiratory depressant. The leaves of the herb are known for their antidiarrhoeal, antiparasitic and antidysenteric properties and are used in bronchial catarrh and their juice to control the bleedings. Leaf-juice is also used as insecticidal and pesticidal. (Dhiman 2004, Ankita and Jain 2012, Sawant and Godghate 2013, Saini *et al.* 2016).

Medicinal Uses (Traditional/Folklore): The leaf juice/extract of the herb is commonly used in Indian traditional medicine as anticoagulant, hair tonic, antifungal and insect repellent, in bronchial catarrh, diarrhoea, dysentery, wound healing, malaria, high blood pressure and to check haemorrhage/bleeding from cuts, bruises and wounds and to prevent falling of hair. Leaf extracts is used to treat infectious skin diseases in folk medicines. It is well known Ayurveda medicine for liver disorders besides gastritis and heart burn. The leaves are considered to be effective in treating piles, kidney troubles, muscular pain; their juice is used for earache and ophthalmia. The flower is bitter; astringent, carminative, stomachic; good for the teeth and the gums; lessens inflammation; useful in scabies, belching, scorpion and snake poisoning, liver complaints, bleeding piles. The flowers are employed in diseases of the eyes and for unhealthy ulcers, internally they are said to purify the blood. (Rahman 2013, Sawant and Godghate 2013, Saini *et al.* 2016).

27.24 *Xanthium strumarium* L.

(Syn. *Xanthium indicum* Roxb.)

Habit

Common name: Common Cocklebur, Clotbur.

Vernacular name: *Hindi:* Chota Dhatura (छोटा धतूरा), Ghaghra (घाघरा), Sankhahuli (संख़ाहुली); *Sanskrit:* Arishta (अरिष्ट), Medhya (मेध्य), Sarpakshi (सर्पक्षी).

Habit: Herb.

Habitat: Common in moist waste places, fallow fields, along roadsides and railway lines.

Distribution: *X. strumarium* is native of America; most often found in the temperate zone, but also occurs in subtropical and Mediterranean climates. Throughout India, up to an altitude of 2,500 m.

Flowering and Fruiting Time: Most part of the year.

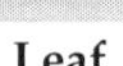

Leaf

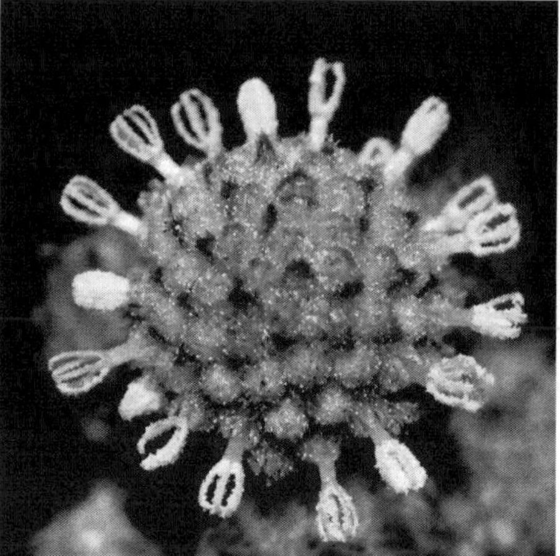

Flower head

Fruits

Taxonomic Description: An annual herb, up to 1 m tall. Stem scabrous, ribbed. Leaves alternate, 5-11 x 3-10 cm, broadly triangular, generally 3-5 lobed, hispid-pubescent, cordate at base, acute at apex, irregularly toothed; petiole long, purple in colour. Heads 5-10 mm across, unisexual, greenish-yellow in axillary and terminal clusters. Male heads above, ovoid. Female heads globose, fruiting involucre spinous all over in fruit and terminating in two hooked prickles at the apex. Achenes brownish, elliptic.

Chemical constituents: The leaves contain sesquiterpene lactones- xanthinin, xanthumin, xanthanol, iso-xanthanol, isohexcosane, chlorobutanol, stearyl alcohol, beta-sitosterol and palmitic acid; β-sitosterol glucoside is anti-inflammatory and xanthumin is a central nervous system depressant. The roots contain *n*-heptacosanol, stigmasterol, 3, 4,-dihydro oxycinnamic acid, β-sitosterol and campesterol. Alcoholic solution of *Xanthium* shows strong antibacterial activity against Gram negative bacteria and fungi. (Kamboj and Saluja 2010, Bhutya 2011).

Medicinal properties: The plant has antitumour, antitussive, anti-inflammatory, antinociceptive, hypoglycaemic, antimitotic, antioxidant, antibacterial, antifungal, insecticidal, antitrypanosomal, depressant and diuretic properties. Leaves are sialagogue, diuretic, given internally in scrofula and herpes in powder form, also for leucoderma. Root is bitter, tonic, useful in cancer and scrofula. Its extract is applied locally over ulcers, boils and abscesses. The fruits are diuretic, diaphoretic, sedative, cooling and demulcent and tonic, used for smallpox, and general debility. ((Kamboj and Saluja 2010, Bhutya 2011, Rahman 2013a).

Medicinal Uses (Traditional/Folklore): The whole plant, especially roots and fruits, is used as traditional or folk medicine. According to Ayurveda, *X. strumarium* is cooling, laxative, fattening, anthelmintic, alexiteric, tonic, digestive, antipyretic, and improves appetite, voice, complexion and memory. It cures leucoderma, biliousness and poisonous bites of insects, epilepsy, salivation and fever. Extracts of the whole plant, especially leaves, roots, fruits and seeds are applied in traditional medicine for the treatment of long-standing cases of malaria, rheumatism, tuberculosis, allergic rhinitis, sinusitis, urticaria, rheumatoid arthritis, constipation, diarrhoea, leprosy, lumbago, pruritis, bacterial and fungal infections.

The whole plants as well as leaves are used by *Hakims* as blood purifier and in scabies. The leaf powder mixed with lime is usually applied on the skin against *Tinea versicolor.* Ash is applied to sores on lips and mouth. The young stem is used in diabetes. The root is a bitter tonic, useful in cancer and strumous diseases. The prickly fruit is considered cooling and demulcent and is given in small-pox. The herb is prescribed in snake-bite and scorpion-sting. In China, the burs are used as a tonic, diuretic, diaphoretic and sedative. The herb is used as an emollient and astringent in Indo-China. An extract of the root is applied to ulcers, boils and abscesses. In Europe, the leaves are administered internally in treating scrofula and herpes. (Kamboj and Saluja 2010, Nibret *et al.* 2011, Rahman 2013a).

28. SPHENOCLEACEAE

Sphenocleaceae

Sphenocleaceae consists of 1 plant genus, *Sphenoclea* and 2 accepted species. *Sphenoclea* is a genus of succulent, helophytic, non-laticiferous, erect annual herbs. The plants occur in damp habitats throughout the tropics. Leaves are alternate, spiral, non-sheathing, simple, pinnately veined and exstipulate. Flowers are bracteolate (two bracteoles), and tetracyclic. There was single species recorded in the flora.

28.1 *Sphenoclea zeylanica* Gaertn.

(Syn. *Gaertnera pangati* Retz.; *Pongatium indicum* Lam.)

Habit

Common Name: Goose Weed, Chickenspike, Wedgewort.

Vernacular Name: *Hindi;* Phulanghas (फूलनघास).

Habit: An annual herb.

Habitat: *S. zeylanica* thrives in damp conditions including paddy fields, swamps or depressions which are periodically inundated.

Distribution: *S. zeylanica,* originally native to tropical Africa, is now distributed as a weed across the world. In India, it is found in swampy areas, along the banks of water courses and in rice fields in Kerala, Goa, Karnataka, Andhra Pradesh, Assam, Delhi, Punjab, Rajasthan, Uttar Pradesh and West Bengal.

Flowering and Fruiting Time: August-October.

Taxonomic Description: An erect, 20-80 cm tall, annual herb with hollow stem. Leaves alternate, 4-10 x 0.5-2 cm, glaucous, elliptic or oblong-lanceolate, shortly petioled, base cuneate, margin entire, apex obtuse-acute. Flowers numerous, in 2-5 cm long, dense cylindric spikes. Calyx lobes-5, lobes triangular. Corolla white, nearly 2 mm long, 5-lobed, lobes ovate-triangular. Capsules subglobose, 3-4 mm across, surrounded by the persistent calyx. Seeds minute, yellowish-brown.

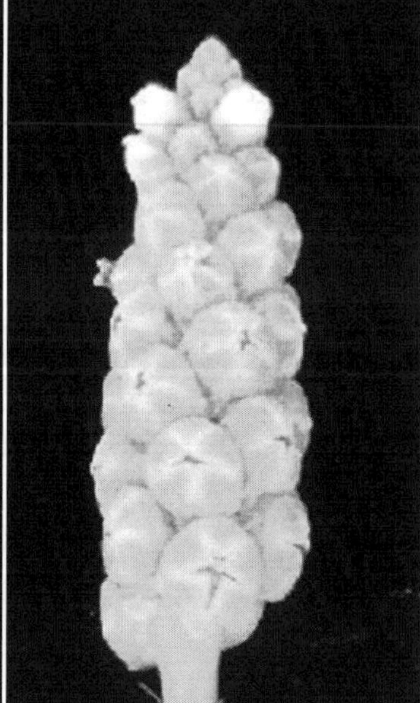

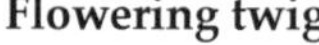

Flowering twig

Inflorescence

Chemical Constituents: Terpenoids, alkaloids, flavonoids, steroids, and phenols have been reported from the plant. (Koetpanya 1992, Gowrie *et al.* 2016).

Medicinal Properties: Antimicrobial, antiulcer, and antivenom properties. (Quattrocchi 2012, Gowrie *et al.* 2016).

Medicinal Uses (Traditional/Folklore): The leaves are used in a poultice against the stings of venomous animals and to cure the ulcers. In Sri Lanka, the plant is used to enhance the growth of hair; oil is prepared from entire plant of *S. zeylanica* and sesame oil according to *Thaila Parbhasha* and applied on hair. (Ediriweera 2007, Quattrocchi 2012).

29. PRIMULACEAE

Primulaceae - Primrose family

Family of herbaceous and woody flowering plants with about 68 genera with 2,788 species, including some common garden plants and wildflowers. Most of the species are perennial though some species, such as scarlet pimpernel, are annuals. The Primroses are herbs or slightly woody plants with usually basal or opposite leaves arranged in rosettes at the base of the plant, but sometimes whorled, like the four-leaved loosestrife. Some species are aquatic, while many others will be found near the water or at least in very moist soils. The flowers are regular and bisexual, often formed in clusters above a bract (modified leaf) on the main stem. The family was represented by a single species in Chithara Village.

29.1 *Anagallis arvensis* L.

(Syn. *Anagallis arabica* Duby)

Habit

Common Name: Blue Pimpernel.

Vernacular Name: *Hindi:* Neel (नील), Dharti Dhak (धरती धक्), Krishnaneel (कृष्णनील).

Habit: Annual herb.

Habitat: Found in cultivated fields, waste places, gardens and along small water channels during winter season.

Distribution: The Blue Pimpernel (*A. arvensis*) is native to Europe. In India, it is found as a weed up to 1,800 m.

Flowering and Fruiting Time: December-April.

Taxonomic Description: A sub-erect, annual herb, 10-30 cm high. Stem 4-angled, much-branched from the base. Leaves opposite-decussate, sessile, yellowish-green, 1-1.8 x 0.5-1.2 cm, ovate-lanceolate, gland-dotted, margin entire, apex acute. Inflorescence solitary, axillary. Flowers blue on long pedicels; pedicels slender, usually as long as or little longer than the leaves. Calyx 5-lobed; lobes linear-lanceolate, 3.5-4 mm long. Corolla lobes-5, lobes obovate-elliptic with denticulate margin. Stamens-5, filaments hairy, anthers yellow. Ovary globose, style persistent. Capsules about 3.5 mm in diameter. Seeds numerous, trigonal.

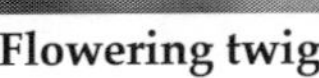

Flowering twig

Flower

Capsule

Chemical Constituents: *A. arvensis* contains tanning agents and the proteolytic enzyme primverase. The plant and its seeds contain saponins, which could explain why fresh material is strongly haemolytic; the other constituents are tannins, steroids, glycosides, alkaloids and anthrequinones, bitter cucurbitacin glycosides, arvenin, I and II and cucurnitacin, B, D, E, I and L, sterols, ß-amyrin, rutin, n-hexacosane, lacceric acid, triterpene sanagalligenone, anagalligenin, carbohydrates and enzymes. Among other possible glycosides, the root yields the triterpenoid glycosidecyclamin, which is highly toxic. (López *et al.* 2011, Al-Snafi 2015b).

Medicinal Properties: The whole herb is antitussive, antimicrobial, anti-inflammatory, antioxidant, analgesic, behavioral, cholagogue, cytotoxic, diaphoretic, diuretic, molluscicidal, nervine, purgative, stimulant, taenicidal, oestrogenic and vulnerary; acts as an expectorant for the treatment of rheumatism, dropsy, hepatic and renal complaints. The ethanol extract of the herb is known to possess hypotensive, antiviral and diuretic properties and its methyl alcohol extract inhibits herpes and poliomyelitis virus development. (Dhiman 2004; https://en.wikipedia.org/wiki/Anagallis_arvensis, http://medplants.blogspot.in)

Medicinal Uses (Traditional/Folklore): *A. arvensis* was used as an antidepressant in ancient Greece, and to treat various mental disorders in European folk medicine, leading to the German name *Gauchheil* (*Gauch* meaning «fool» or «cuckoo», and *heil* meaning «heal»). However, documented evidence for clinical efficacy is lacking. The plant is used for the treatment of several ailments such as gout, epileptic attacks, in snake-bite, dropsical affections, leprosy, hydrophobia, mania, cerebral affections and other complains of the nervous system. It can be taken internally or applied externally as a poultice. An infusion is used in the treatment of dropsy, skin infections and disorders of the liver and gall bladder. It is used internally to treat itchy skins and externally to remove warts.

In various countries plant preparations of this herb are applied externally to treat slow-healing ulcers and wounds. It also has been used as an expectorant and as a remedy for pruritis, rheumatism, haemorrhoids, rabies, leprosy and snake-bite. *Anagallis* is used in the treatment of phthisis of non-specified types, and of kidney-related conditions such as dropsy and chronic nephritis. *A. arvensis* and *A. foemina* are traditionally used in Navarra (Spain) for dermatological purposes regarding wound healing properties. In some cases they are also used to treat internal infections although they are known to be toxic at high doses. *A. arvensis* also is known in homeopathy as *Anagallis Arvensis* Herba, and used against various rashes and nervous complaints.

The plant is best harvested in June and can be dried for later use. The plant was at one time highly regarded as a medicinal herb, especially in the treatment of epilepsy and mental problems, but there is little evidence to support its efficacy and it is no longer recommended for internal use because it contains toxic saponins and cytotoxic cucurbitacins. The plant should be used with a caution as large doses can cause polyuria and tremor. (López *et al.* 2011; https://en.wikipedia.org/wiki/Anagallis_arvensis, http://medplants.blogspot.in)

30. EBENACEAE

Ebenaceae – Persimmon or Ebony family

It includes trees and shrubs placed in three genera, with about 751 species found throughout the tropics and also in some temperate regions. One species of this family was recorded in Chithara Village.

30.1 *Diospyros montana* Roxb.

(Syn. *Diospyros cordifolia* Roxb.)

Fruiting twig

Common Name: Mountain Persimmon, Bombay Ebony.

Vernacular Name: *Hindi:* Bistendu (बिसतेंदु); *Sanskrit:* Tumala (तुमला).

Habit: A deciduous tree.

Habitat: Very common in a wide range of habitats and resistant to sun and drought. Found in plains and hills.

Distribution: The plant is widely distributed in East Asia and Indian subcontinent, Sri Lanka Nepal, Myanmar, Thailand, Cambodia, Laos, Vietnam, Malaysia, Indonesia, and Philippines to tropical Australia. In India, found distributed all along the Western Ghats and northern parts in plains and hills up to 1,200 m.

Flowering and Fruiting Time: March-July.

Taxonomic Description: A small deciduous tree, sometimes large shrub with long trunk and dark grey bark. Leaves alternate, petioled, exstipulate, up to 10 cm long, oblong-elliptic or ovate-lanceolate, bluntly acuminate, entire, sub-cordate or rounded at base, acute to acuminate at apex, softly pubescent when young. Male and female flowers borne on separate trees. Male flowers greenish yellow in small, few flowered axillary panicles; calyx lobes-4, sparsely pubescent, ovate-lanceolate, margin ciliate; corolla spreading, bell-shaped, lobes-4; stamens 60, paired, united at the base. Female flowers: solitary, axillary, calyx accrescent, corolla as in male flowers. Fruit a berry, ovoid, 2-2.5 cm across, yellow when ripe.

Fruits **Seeds**

Chemical Constituents: Plant contains bio-active compounds such as saponins, tannin, terpenoids, flavonoids, alkaloids, and essential oils. The major phytoconstituents isolated are diospyrin, kaempferol diglucoside, quercetine , gallic acid , leupeol, betulin , betulinic acid , ursolic acid, α-amyrin, β-amyrin, oleanolic acid, β-sitosterol, taraxerol, sigmasterol diospyrin, β- dihydrodiospyrin , 8-hydrodiospyrin, chromenone acid, betulin, oleanolic acid, and betulinic acid. Fruit pulp yields viscous, yellow liquid, containing fatty acid, of crude proteins, pentosan and water soluble mucilage, esters of alpha-amyrin, ursolic acid, oleanolic acid and betulinic acid. (Maridass *et al.* 2008, Kumar *et al.* 2011c).

Medicinal Properties: The plants of *D. montana* show CNS depressant, spasmolytic and anticancer agent activity. ß-sitosterol is reported to help in the management of ageing, hyperlipidaemia, cholesterol absorption, and immunomodulation activity. The fruits are poisonous, bitter, astringent, febrifuge and abortifacient. (Kumar *et al.* 2011c, Quattrocchi 2012).

Medicinal Uses (Traditional/Folklore): In Ayurveda and various traditional system of medicines, *D. montana* has been used in the treatment of various diseases like delirium, vomiting, high fever, pneumonia, jaundice, diarrhoea, tuberculosis, cancer and leishmaniasis. It has been used in treatment for fever, dysuria and used as a fish poison. The plant has antibacterial properties and it is effective against organisms like *Bacillus subtilis.* The bark is an important ingredient of many medicines. Bark paste of this tree mixed with *Tylophora asthamatica* and *Alangium salviifolium* is used in snake bite by tribes in Karnataka; one teaspoonful of this paste is given orally in Chikamangalur district, Karnataka. The bark paste is given with boiled rice for swelling of the body. The extract of fresh stem bark is taken in stomachache. The fruits are applied externally to boils and spider stings. Tender shoots are boiled with salt and tied over aching joints. Ethnomedicinally, it is used in treatment of boils, fever, dysuria, neuralgia, pleurisy, menorrhagia, diarrhoea, spider bite, jaundice, deep wounds, etc. It is also beneficial in the treatment of breast cancer and cancer of the prostate gland and in certain gynaecological disorders.

The crushed leaves and fruits are used for stupefying fish; crushed leaves are used as fish poison in Wyanad, Kerala. Tender shoots are cooked and eaten. The wood is used for agricultural implements and handles. Wood is not used for fuel as it is believed that burning this wood causes feuds in the family. (Kumar *et al.* 2011c, Quattrocchi 2012).

31. APOCYNACEAE

Apocynaceae – Oleander family

This family contains 5,556 species in 410 plant genera of trees, shrubs, vines, woody climbers and some herbs, which have poisonous milky sap. Flowers are often showy and fragrant. Members of the family have milky, often poisonous juice; smooth-margined leaves; and flowers in clusters (rarely solitary). There are several plants of medicinal value. The study area, Chithara village including SNU Campus was represented by three species belonging to two genera of this family.

31.1 *Calotropis gigantea* (L.) Dryand.

(Syn. *Asclepias gigantea* L.)

Habit

Common Name: Giant Indian Milkweed, Crown Flower.

Vernacular Name: *Hindi:* Safed Aak (सफेद आक); *Sanskrit:* Sadapushpa (सदापुष्प), Arki (अर्की), Hrasvagnih (हस्वअग्निः), Arka (अर्क), Alarka (अलारका).

Habit: Shrub.

Habitat: Grows wild in open and wastelands and also found cultivated in temples as well as home gardens due to religious beliefs.

Distribution: World-wide distribution but most abundant in the sub-tropics and tropics; throughout India commonly found along roadside, railway track in open waste-lands near village surroundings and in dry wastelands.

Flowering and Fruiting Time: Most part of the year.

Flowering twig **Flowers** **Follicle**

Taxonomic Description: An erect shrub, 1-3 m tall, contains milky latex. Young branches and under surface of leaves covered with white tomentum. Leaves opposite-decussate, sub-sessile, 8-25 x 3-11 cm, obovate or elliptic-oblong, margin entire, apex acute or acuminate, base cordate-auriculate. Inflorescence umbellate cymes with white flowers. Calyx lobes-5, lobes ovate-lanceolate. Corolla 5-lobed, white, lobes ovate-lanceolate. Corona lobes elongated. Stamens 5. Mature follicles up to 10 cm long, recurved, turgid, broadly oblong. Seeds many, flat, ovate-oblong, with silky white coma.

Chemical Constituents: Root contains cardiac glycoside *viz.*, calotropin, uscharidin calactin, calotroposides, uzarigenin, etc. Root-bark contains isovalerates of a kundrol, mudarol, giganticine (nonprotein amino acid), β-amyrin, β-amyrinacetate, tetracyclic resinols, sterols, acetic, isovaleric acids, isogiganteol, taraxasterol and its ψ-isomer, triterpenes, etc. Flowers contain flavonol glycosides, hyperoside, rutin, amyrin, stigmasterol, anthocyanin,

α and β-calotropeols, etc. Leaves contain holarrhetine, cyanidin-3-rhamnoglucoside, tarasasterol isovalerate, taraxasteryl acetate, β-sitosterol, and amino acids. Latex contains α and β-calotropeol, β-amyrin, gigantin (toxic principle), proteases: calotropins DI & DII; calotropins FI & FII, ester of acetic and isovaleric acids, glutathione, 3′-methylbutanoates of α-amyrin and ψ-taraxasterol, uscharin, etc. (Kumar *et al.* 2013b, Chandrawat and Sharma 2016).

Medicinal Properties: Whole plant is considered to be toxic and poisonous. According to Ayurveda, dried whole plant is a good tonic, expectorant, depurative, anthelmintic, antiseptic, emetic and antiphlogistic. Leaves are antiphlogistic and acrid. Latex is antiseptic, vesicant, prophylactic, and purgative. Root bark is febrifuge, anthelmintic, depurative, expectorant, laxative, substitute for ipecacuanha, antidysenteric, antispasmodic and diaphoretic, and is useful in cutaneous diseases, intestinal worms, cough, ascites and anasarca. The flowers are bitter, digestive, astringent, stomachic, anthelmintic and tonic. (Warrier *et al.* 2010, Chandrawat and Sharma 2016).

Medicinal Uses (Traditional/Folklore): *C. gigantea* is used as a traditional medicinal plant with unique properties. Traditionally, *Calotropis* is used alone or with other medicinal plants to treat common diseases such as fevers, rheumatism, indigestion, cough, cold, eczema, asthma, elephantiasis, nausea, vomiting and diarrhoea. The dried root bark is a substitute for ipecacuanha. The powdered root used in asthma, bronchitis, and dyspepsia. The leaves are useful in the treatment of paralysis, arthralgia, swellings, and intermittent fevers. *Calotropis* is also a reputed Homoeopathic drug. Root-bark is useful in cutaneous diseases, intestinal worms, cough, ascites, anasarca, dysentery, constipation, piles, syphilis, elephantiasis, hydrocele; root smoke inhaled in case of migraine. Roasted leaves are applied to painful joints on swellings; powder boiled in oil is useful in eczema, skin eruptions, toothache, ulcers, wounds; tincture in intermittent fevers. Latex in low doses employed in asthma, bronchitis, allergy, prophylactic against malaria, produces abortion when a stick smeared with the latex is applied locally to induce uterine contractions. Flowers are useful in asthma, catarrh, anorexia, inflammations, and tumours. In large doses it is purgative and emetic.

C. gigantea is considered to be a good remedy for poisonous snake bites. If bitten by poisonous snakes, few leaves of the plant are plucked and chewed. Alternatively, the roots of the plant are crushed and the juice is applied over the bitten area. Juice is extracted from leaves and mixed with equal quantity of honey when consumed, it cures fever and the body temperature comes down immediately. For intestinal worms juice extracted from the leaves is mixed with honey and consumed. The leaves are dried and powdered and the powder is applied externally over the stomach to heal ulcers. It cleans the unwanted waste contained in the stomach and makes the stomach clean. The plant is also very effective in curing various stomach disorders and very effective in curing dysentery. The herb boosts the functions of the various organs in the body. The latex from the plant is taken and applied over the body to cure nerve disorders. The herb impacts a calm nature to the mind. The flowers of *Calotropis* have high medicinal value in treating asthma. The herb increases the blood circulation in the body and enhances the energy levels. It keeps the mind calm, energizes the nerves and provides youthful feeling. The herb has been used for inducing abortions in India. The leaves of the plant are inserted into the cervix so that it causes labour pains and induces abortion. The oil extracted from the herb is used to treat skin problems like eczema, ringworm, etc. It also cures the bites caused due to insects or scorpions. (Pratap *et al.* 2010, Chandrawat and Sharma 2016).

31.2 *Calotropis procera* (Aiton) Dryand.

(Syn. *Asclepias procera* Aiton)

Common Name: Rubber Bush, Sodom Apple.

Vernacular Name: *Hindi:* Aak (आक), Madar (मदार); *Sanskrit*: Arka (अर्का), Mandara (मंदरा), Vasuk (वासुका), Sadapushpa (सदापुष्पा), Alarka (अलरका).

Habit: Undershrub.

Habitat: Common in dry areas, roadsides, railway lines and waste places.

Distribution: Native to tropical Africa. Occur throughout India in hot dry wastelands, predominantly in sub-Himalayan tracts, mostly in west and central India, Punjab and Rajasthan.

Habit

Flowering and Fruiting Time: March-July.

Taxonomic Description: An erect much branched undershrub with milky latex. Young parts clothed with white tomentum. Leaves 7-18 x 5-11.5 cm, entire, opposite-decussate, ovate-oblong, leaf base sub-amplexicaul, apex mucronate. Flowers in terminal or axillary corymbose or umbellate cymes. Calyx 5-lobed, lobes ovate. Corolla 5-lobed, campanulate, purple tipped inside, silvery-white outside, coronal appendages-5. Follicles 5-11 cm long, greyish-green, recurved, turgid. Seeds numerous, flat, brown, with long silky-white coma at one end.

Flowering twig **Flowers** **Follicle**

Chemical Constituents: Root bark contains benzoylline-oline and benzolisolineolone, β- amyrine and giganteol, isogiganteol. Root, stem, leaves and latex contain β-amyrine, calotroponin and calotropangenin. Flowers contain evanidin, 3-rhmnoglucosides. Whole plant contains acardenolide, proceragenin, an antibacterial compound. The milky sap contains a complex mix of chemicals, some of which are steroidal heart poisons known as "cardiac aglycones". The glycosides present are calotropin, calotoxin, calactin, uscharidin and voruscharin. (Bhutya 2011, Chandrawat and Sharma 2016).

Medicinal Properties: Different parts of the plant have been reported to be toxic and poisonous and possess a number of biological activities such as proteolytic, antimicrobial, larvicidal, nematicidal, anticancer and anti-inflammatory. The plant has analgesic, antitumour, anthelmintic, antinociceptive, antioxidant, antidiarrhoeal, anticonvulsant, antimalarial, antimicrobial, hepatoprotective and oestrogenic properties. The bark is febrifuge, anthelmintic, depurative, expectorant and laxative. Flowers are bitter, digestive, astringent, stomachic, anthelmintic and tonic. Latex is emetic, produce burning sensation and useful for expectoration, cough and bronchitis and possesses spasmogenic, carminative, antidysenteric, antisyphilitic, antirheumatic, antifungal, mullusccide and diaphoretic properties. (Bhutya 2011, Chandrawat and Sharma 2016).

Medicinal Uses (Traditional/Folklore): The root bark and leaves of *C. procera* are used by various tribes of central India as a curative agent for jaundice and for treating bronchitis, asthma, leprosy, eczema and elephantiasis while the latex is used to treat vertigo, baldness, hair loss, toothache, intermittent fevers, rheumatoid/joint swellings and paralysis. The leaves are used to treat joint pain and reduce swelling, cure paralysis, arthralgia and intermittent fevers. The latex is used as an abortifacient and used for the treatment of leprosy, bronchial asthma and skin affections. Its flowers possess digestive and tonic properties and are useful to cure asthma, catarrh, anorexia, inflammations and tumours. The powdered root promotes gastric secretions and useful in asthma, bronchitis, dyspepsia, neuralgia, splenomegaly, piles, abdomen diseases, inflammation. The powdered root bark has been reported to give relief in diarrhoea and dysentery. The root of the plant is used as a carminative in the treatment of dyspepsia. The bark is useful to cure skin diseases, intestinal worms, cough, ascites and anasarca. (Kaur *et al.* 2015, Chandrawat and Sharma 2016).

31.3 *Oxystelma esculentum* (L. f.) Sm.

(Syn. *Oxystelma wallichii* Wight; *Periploca esculenta* L. f.)

Habit

Common Name: Rosy Milkweed Vine.

Vernacular Name: *Hindi:* Dudhialata (दुधिया लता), Dudhi Bel (दुध बेल).

Habit: Climber (twining herb).

Habitat: Often found twining on bushes, shrubs, wires in open and waste places. Also grows along the margin of ponds, lakes and canals.

Distribution: Native to China, South Asia, South-East Asia, North-Eastern Africa, and South-West Asia.

Flowering and Fruiting Time: July-September.

Taxonomic Description: A perennial glabrous climber or twinning herb with milky latex. Branches twinning, numerous, slender. Leaves 4-11 cm long, opposite, petioled, linear-lanceolate, tip acuminate, base rounded or cordate, margin entire. Flowers attractive, white-purple, drooping, in sub-umbellate cymes. Calyx lobes-5, acute, oblong-lanceolate. Corolla rotate, lobes-5, triangular, edges ciliate, gamopetalous with short tube. Staminal corona 5-lobed. Follicles 4-6 cm long, ovoid-oblong, tapering towards the apex. Seeds numerous, ovate, compressed, with long silky white coma.

Flowering twig

Flower

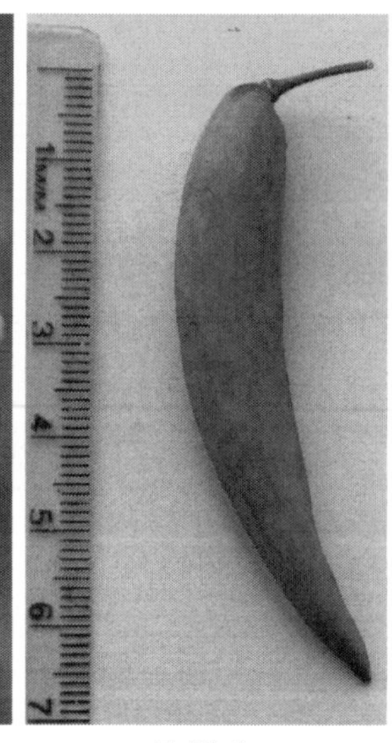

Follicle

Chemical Constituents: The plant contains phytoconstituents like cardenolides, flavonoids, phenolics, sterols and triterpenoids. Parenchymatous tissues contain large amount of calcium oxalate rosette crystals. (Pandya and Anand 2011).

Medicinal Properties: The plant is hot, bitter, tonic, expectorant, pungent, dry and indigestible, and diuretic, laxative, antiulcer, anthelmintic, aphrodisiac, useful in leucoderma and bronchitis; causes flatulence. The herb is reported to possess antiseptic, depurative and galactagogue properties. (Poornima 2009, Elumalai and Krishnappa 2015).

Medicinal Uses (Traditional/Folklore): A decoction of the plant is useful as a gargle in infections of throat and mouth. Fresh roots are prescribed for jaundice. In combination with turpentine it is prescribed for itch. It is used traditionally as a diuretic, laxative, antiulcer, galactagogue and anthelmintic. It is also used in throat infections, skin diseases and jaundice. Decoctions of all parts are used against cancer, menoxenia, and traumatic injury. *O. esculentum* leaf extract provides an excellent, potential phytopesticide for controlling *Aedes aegypti* mosquito. (Poornima 2009, Pandya and Anand 2011, Marwat *et al.* 2012, Elumalai and Krishnappa 2015).

32. GENTIANACEAE

Gentianaceae

Family of flowering plants of 96 genera and about 1,682 species distributed worldwide, but most species occur in temperate zones. The family consists of trees, shrubs and herbs showing a wide range of colours and floral patterns. Flowers are actinomorphic and bisexual with fused sepals and petals. Plants are usually rhizomatous. The family was represented by a single species in the flora of Chithara Village.

32.1 *Centaurium pulchellum* (Sw.) Druce

Syn. *Erythraea ramosissima* (Vill.) Pers.; *Gentiana pulchella* Sw.)

Habit

Common Name: Pink Centaury, Field Centaury.

Vernacular Name: *Hindi:* Barik Chirayata (बारीक चिरायता), Khet Chirayata (खेत चिरायता).

Habit: Annual herb.

Habitat: Common along the boundary of wetlands, rivers, canals and small water channels. Also found in agricultural fields.

Distribution: Native to Eurasia. It is found in marshy places, lake margin, bog meadows, and shallow waters.

Flowering and Fruiting Time: February-May.

Taxonomic Description: An erect, up to 20 cm high, annual herb with quadrangular stem. Leaves opposite, sessile, entire, 1-3 cm long, ovate-oblong to elliptic-lanceolate, entire, rounded or clasping at base, acute or mucronate at apex. Flowers bracteate, axillary and terminal in leafy dichotomous cymes, pedicels 4-5 mm long. Calyx 5-lobed, lobes linear, nearly equal to the corolla tube. Corolla 5-lobed, spreading, pink to purple; lobes oblong, lanceolate. Capsules oblong, ellipsoid, longer than the calyx. Seeds minute, numerous, globose, blackish.

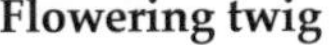

Flowering twig

Flowers

Chemical Constituents: Plants contain alkaloid, gentianine; a bitter principle kantaurin, a triterpene; oleanolic acid, erythrosterol, and xanthones; 1-hydroxy-3, 7, 8-tri-methoxyxanthone, 8-dihydroxy-3, 7-dimethoxyxanthone and 1, 8-dihydroxy-3, 5-dimethoxyxanthone. (http://www.uicnmed.org/nabp/web/documents/book/chapter3.pdf).

Medicinal Properties: The plant possesses antidiabetic, antipyretic, anti-inflammatory, antihypertensive properties and it is useful in curing appetite loss, worm infestation, kidney stones, indigestion, snake-bite and poisoning. (http://www.uicnmed.org/nabp/web/documents/book/chapter3.pdf).

Medicinal Uses (Traditional/Folklore): A decoction is used for gastric and abdominal pain, hypertension, renal colic, rheumatic pains and for the elimination of stones from the kidney and urethra; acts as a healing agent for wounds in ointments for sciatica. An infusion of herb is used for the treatment of diabetes. (http://www.uicnmed.org/nabp/web/documents/book/chapter3.pdf).

33. MENYANTHACEAE

Menyanthaceae

Family has 7 plant genera and 55 species of aquatic and wetland plants distributed worldwide. Plants have simple or compound leaves arising alternately from a creeping rhizome. The sympetalous, insect-pollinated flowers are five-parted and either yellow or white. Many species are of economic importance as ornamental water garden plants. A single species of this family was recorded in the flora.

33.1 *Nymphoides cristata* (Roxb.) Kuntze

(Syn. *Limnanthemum cristatum* (Roxb.) Griseb.)

Habit

Common Name: Crested Floating Heart.

Vernacular Name: *Hindi:* Kumudini (कुमुदिनी).

Habit: Aquatic floating herb.

Habitat: Abundant in shallow ponds, wetlands, and other aquatic bodies. Often found littoral areas of water reservoir.

Distribution: Subtropical/tropical regions; native of India, Vietnam, Taiwan and southern China.

Flowering and Fruiting Time: September-December.

Flowering and fruiting twig

Abaxial surface of leaf

Flower

Taxonomic Description: An aquatic, floating herb. Stem cylindric, with a terminal leaf. Leaves petiolate, about 4-10 cm across, ovate-orbicular to sub-orbicular, deeply cordate at base, margin entire, glossy green above, purplish and gland-dotted beneath; petiole cylindric. Flowers many, clustered at base of petiole (nodes), bisexual, pentamerous, white, 1.5-2 cm across. Calyx 3-5 mm long, deeply divided, apex obtuse. Corolla lobes white with yellow throat and long white hairs within, winged. Stamens-5, short, filaments yellow. Capsules sub-globose, 3-5 mm in diameter, with light brown seeds.

Chemical Constituents: None known.

Medicinal Properties: Hepatoprotective. (Marwat *et al.* 2011a).

Medicinal Uses (Traditional/Folklore): It has protective effect against liver damage. (Marwat *et al.* 2011a).

34. BORAGINACEAE

Boraginaceae – Borage or Forget-me-not family

Family includes a variety of shrubs, trees, and herbs, totaling about 2,686 species in 155 genera found worldwide. Most plants have hairy aromatic leaves. Flowers of the family are favorite with butterflies, bees and other insects. Milkweed butterflies are seen around these plants, which yield an essential alkaloid that is necessary for the butterflies' reproduction. The family was represented by two species belonging to two genera in the flora.

34.1 *Cordia dichotoma* G.Forst.

(Syn. *Cordia obliqua* Willd.; *Cordia wallichii* G.Don)

Common Name: Indian Cherry, Glue Berry.

Vernacular Name: *Hindi:* Lasoda (लसोड़ा), Tenti (टेंटी), Dela (डेला); *Sanskrit:* Bahuvarah (बहुवराह), Bahuvaraka (बाहुवरका).

Habit: Deciduous tree.

Habitat: Grows wild in deciduous forests and planted in the gardens and fruit orchards.

Fruiting twig

Drupes and sticky mesocarp

Habit

Distribution: Native to tropical and subtropical regions, Indomalaya ecozone, northern Australia, and western Melanesia. Throughout India, in the warmer regions; grows in the sub-Himalayan tract and outer ranges, ascending up to about 1,500 m elevation. It is found in a variety of forests ranging from the dry deciduous forest of Rajasthan to the moist deciduous forests of Western Ghats and tidal forests in Myanmar.

Flowering and Fruiting Time: March-June.

Taxonomic Description: A small to medium-sized, deciduous tree with drooping branches. Branches lenticellate, bark brownish with longitudinal wrinkles. Leaves 5-13 x 4-10 cm, alternate, broadly ovate to elliptic, coriaceous, glabrous, entire or subundulate to sinuate toothed margin, apex obtuse to mucronate, base rounded to broadly cuneate, petiole 2-5 cm. Inflorescence dichotomously branched in corymbose cymes. Flowers white, shortly stalked, fragrant. Calyx 4-6 mm; lobes-5, unequal, glabrous outside, silky within, accrescent. Corolla campanulate, white, lobes oblong-obtuse, recurved. Stamens-5, filaments hairy at the base. Drupes sub-globose, about 1.5 cm in diameter, pale yellow when ripe, apiculate, surrounded by saucer-shaped persistent calyx; mesocarp sticky, cream colored. Seed-1.

Chemical Constituents: The plant contains alkaloids, glycosides, proteins, amino acids, carbohydrates, triterpenoids, tannins, flavonoids, saponins, phenols, steroids, arabinoglucan, pyrrolizidine alkaloids, coumarins, fats, gum and fatty acids (palmitic acid, stearic acid, arachidic acid, behenic acid, oleic acid and linoleic acid), flavonoid glycosides (robinin, rutin (rutoside), datiscoside and hesperidin), flavonoid aglycone (dihydrorobinetin), phenolic derivatives (chlorogenic acid and caffeic acid), sugar (D-glucose, L-Arabinose). Bark contains a large amount of tannic acid. Fruit yields saponins, amino acids, flavonoids, sugar, gum, proteins, palmitic, stearic, linoleic acids, oleic, arachidic, behenic acids. Bark contains a principle similar to "cathartin". Two compounds isolated from the seeds, α-amyrin and 5-dirhamnoside, have been shown to have anti-inflammatory activity. The bark is medicinal and several chemicals such as tannic acid, allantoin, β-sitosterol and 3′, 5-dihydroxy-4′-methoxy flavanone-7-O-α-L-rhamnopyranoside have been identified. (Dighe and Dighe 2014).

Medicinal Properties: The medicinal attributes of *C. dichotoma* have been known since a long time. The leaves are aphrodisiac, and are useful in gonorrhoea and ophthalmodynia. The fruits are sweet, cooling, emollient, anthelmintic, purgative, vulnerary, diuretic, expectorant, aphrodisiac, depurative, febrifuge, antiulcer, wound healing, anti-inflammatory, analgesic, antidiabetic, antimicrobial, antiageing, hepatoprotective, anthelmintic (chloroform extract), laxative (fresh fruit), gonorrhoea, expectorant. Seeds are anti-inflammatory and contain α-amyrin and 5-dirhamnoside. The bark is bitter, astringent, acrid after digestion, constipating, anthelmintic, cooling and depurative, and is useful in dyspepsia, fever, diarrhoea, burning sensation, vitiated conditions of *Kapha* and *Pitta,* helminthiasis, leprosy and skin diseases. Bark is used as antidyspeptic and as febrifuge. (Warrier *et al.* 2010, Dighe and Dighe 2014).

Medicinal Uses (Traditional/Folklore): All parts of the plants have medicinal value and being used in traditional medicines. The leaves are useful in vitiated conditions of *Vata* and *Pitta,* ulcers, leprosy, skin diseases, hyperdipsia, burning sensation, bronchitis, dry cough, pectoral diseases, strangury, urethralgia, urethritis, chronic fever, arthralgia, pharyngopathy, splenopathy and ring worm. In Ayurveda, leaves and stem bark are used for dyspepsia, fever, diarrhoea and leprosy. The bark is moistened an applied to boils and tumours to hasten ripening; also used for headaches and stomachaches. Powdered bark is used for mouth ulcers while infusion of bark used as gargle. The bark juice, mixed with coconut milk, is used to relieve colicky pains. In Java (Indonesia) and Bengal (India), the bark is used as a tonic. In Java, the bark is also used for dysentery; and with pomegranate rind, used for fevers. The bark is rubbed on the teeth to strengthen them. Leaves are used for ulcers and headaches. The highly mucilaginous fruit is used for coughs and ailments of the chest, uterus and urethra. In large quantities, the plant is used as a laxative. In India, traditionally used for ulcerative colitis, ulcers, and colic pain. In Bengal, fresh fruit is used as laxative and pectoral. In Java, fruits are used for gonorrhoea. In Punjab and Kashmir, dried fruits are used as expectorant; kernels, powdered and mixed with oil, are used for ringworm. In Pakistan, fruits are used as tonic and refrigerant; for irritation of urinary passages, alleviation of thirst and dry cough. (Husain *et al.* 2008, Shukla *et al.* 2010, Singh *et al.* 2010d, Warrier *et al.* 2010, Chaudhary and Gupta 2011, Dighe and Dighe 2014)

34.2 *Heliotropium ellipticum* Ledeb.

(Syn. *Heliotropium strictum* Ledeb.)

Common Name: Heliotrope.

Vernacular Name: *Hindi:* Hathisunda (हाथीसूंडा).

Habit: Annual herb.

Habitat: Common in semi-arid areas; often a weed in cultivated and fallow fields.

Distribution: East Asia-Himalayas; wild in dry arid zones.

Flowering and Fruiting Time: March-September.

Taxonomic Description: An erect, 30-60 cm high, much-branched, annual herb. Stem terete, branchlets woolly-tomentose. Leaves alternate, petiolate, 3-6 x 2-3.5 cm, elliptic-oblong or ovate-lanceolate, base attenuate, apex obtuse, margin entire, densely pubescent on both the

Habit

surfaces. Flowers white, in dense-villous, short spikes. Calyx 5-lobed, ovate-lanceolate, densely hairy. Corolla 5-partite, campanulate, apex rounded, tube short. Style very short, stigma bi-lobed. Nutlets-4, about 2 mm long, ellipsoid, enclosed by the persistent calyx.

Flowering twig

Flowers

Chemical Constituents: Seven sterols and triterpenoids have been isolated from *H. ellipticum* and tested for their antimicrobial activity: β sitosterol, stigmasterol, β amyrin, friedelan β ol, cycloartenone, β amyrin acetate and friedelin. Pyrrolizidine alkaloids isolated from the plant are europinc, heliotridine, lasiocarpine and lasioearpine-N-oxide; Pyrrolizidine alkaloids (PAs) are responsible for various biological activities, including antitumour, antimicrobial, and antiviral effects. Phenolic compounds, terpenoids, and quinones are the other phytoconstituents reported from this plant. Cycloartenone and friedelin present in the plants are known to have antibiotic properties. (Jain and Singh 1998, Goyal and Sharma 2014).

Medicinal Properties: The plants have anti-inflammatory, antimicrobial, antifungal, and antibacterial properties. The plants are emetic, and utilized in snake bites. The seeds are toxic. (Jain and Singh 1998, Sharma *et al.* 2009, Goyal and Sharma 2014).

Medicinal Uses (Traditional/Folklore): *H. ellipticum* is used in the treatment of boils, ulcers, earache and the bites of snakes and mad dogs, and scorpion sting. In folk medicine, the plant is used for the treatment of rheumatism, gout and as febrifuge, antiseptic, cholagogue, anti-inflammatory and healing agent. (Sharma *et al.* 2009, Goyal and Sharma 2014).

35. CONVOLVULACEAE

Convolvulaceae – Morning glory family

Family of about 67 genera and 1,296 species of mostly herbaceous vines, but also trees, shrubs and herbs. Plants can be recognized by their funnel-shaped, radially symmetrical corolla. The stems of these plants are usually winding, hence their Latin name (from *convolvere*, "to wind"). Fourteen species belonging to six genera of this family were recorded in Chithara Village including SNU Campus.

35.1 *Convolvulus arvensis* L.

(Syn. *Convolvulus chinensis* Ker Gawl.)

Habit

Common Name: Field Bindweed.

Vernacular Name: *Hindi:* Hiranpug (हिरनपग).

Habit: Climbing or trailing herb.

Habitat: Moist waste places, cultivated fields.

Distribution: Native to Mediterranean Europe and Asia. Introduced throughout most temperate and sub-tropical climates, including, Africa, Australia, India, New Zealand, Hawaii, Chile and North America. Occurs in cultivated fields, fallow lands, roadsides and disturbed open sites.

Flowering and Fruiting Time: April-October.

Flowering twig

Flower

Taxonomic Description: A climbing or trailing herb; stem branched from the base. Leaves 2-5 x 0.5-1.5 cm, petiolate, hastate or oblong-lanceolate, entire. Inflorescence cymose, axillary. Flowers gamopetalous, 1-3 on long peduncle; bracts-1 pair, linear. Calyx-5 lobed, somewhat pubescent, sepals unequal. Corolla widely funnel shaped, white to pink, limb 5-lobed. Stamens-5, included. Ovary ovoid, style-1; stigma-bilobed, silky white. Capsules 2-celled, ovoid to sub-globose, glabrous. Seeds usually 4, dark brown or black, tuberculate.

Chemical Constituents: Phytochemical studies on this plant showed the presence of saponins, flavonoids, caffeic acid, alkaloids, lipids, and δ-amino levulinic acid. Field bindweed was found to contain the tropane alkaloids tropine, pseudotropine, and tropinine and the pyrollidine alkaloids cuscohygrine and hygrine. Seeds contained 6.7-16.5% oil. The oil consists of palmitic acid, stearic acid, oleic acid, linoleic acid, arachidic and, behenic acid. (Kaur and Kalia 2011).

Medicinal Properties: The root and its resin are cholagogue, diuretic, laxative and strongly purgative. Plant is also known to possess immunostimulant, antihemorrhagic, antihepatic and emetic properties. (Kaur and Kalia 2011; http://www.pfaf.org/user/plantsearch.aspx).

Medicinal Uses (Traditional/Folklore): A plant that is considered a bane to agriculturists also has multiple health benefits and medicinal uses. The plant is reported to have been used in traditional medicine system from as early as 1730s. Traditionally, it is used to treat skin ulcers, reducing wounds, inflammation and swelling. The whole plant (without roots) is used for abdominal pain and abdominal worms in children. The plant is also reported to have diuretic effect. In folk medicines leaves are used in asthma, jaundice and as antihemorrhagic.

The roots of the plant act as a good purgative, and if used in right doses, it can be effective even in children. Native Americans use the plant as an antidote to spider bites, and the leaves of the plant were believed to enhance the secretion of bile. The plant extract is known to arrest the growth of tumours, and its anticancer properties are presently being researched. Plant also exhibits actions similar to that of antidiabetic medications as it is considered to inhibit the action of β-glucosidase and α-galctosidase. This, in turn, aids in lesser absorption of carbohydrates into the intestine, thus checking the blood sugar levels. Similar to sweet potato, the insulin-like compound in the plant aids in effective diabetes management. Plant, especially its flowers, has antibacterial and antifungal properties against a broad spectrum of microbes, including *E. coli*, *Salmonella* sp., and *Candida albicans*. The plant also finds its therapeutic use for treating the effects of stress in individuals and soothe and calm the mind and nerves. It helps bring about a feeling of being at peace. However, similar to other tranquilizers or antipsychotic medications, the plant should be used with caution for treatment of depression, anxiety and stress.

In folklore usage, a cold tea made from the leaves is used as laxative and as a wash for spider bites or taken internally to reduce excessive menstrual flow. The leaf extract is immunostimulant. A tea made from flowers is laxative and also used to treat muscular weakness. It is also used to reduce fever and heal wounds. The root and its gum are known to increase the flow of bile and its discharge from the body also act as urine-inducing, laxative and strongly purgative. Roots have strong emetic effects and root tea is a laxative. (Shukla *et al.* 2010, Singh *et al.* 2010d, Kaur and Kalia 2011, Shaheen *et al.* 2014).

35.2 *Convolvulus prostratus* Forssk.

(Syn. *Convolvulus pluricaulis* Choisy)

Common Name: Prostrate Bindweed.

Vernacular Name: *Hindi:* Shankhpushpi (शंखपुष्पी).

Habit: Perennial herb.

Habitat: Open and wastelands, dry areas, roadsides.

Distribution: The plant is found in Egypt to Pakistan and India.

Flowering and Fruiting Time: March-November.

Habit

Flowering twig

Flower

Taxonomic Description: A sub-erect, diffuse perennial herb with woody root stock. Stem hairy, branches spreading from the base. Leaves alternate, shortly petioled, linear, oblong, oblanceolate, up to 2.5 cm long, densely appressed with spreading hairs on both sides. Flowers solitary or in pairs, sessile or nearly so in the leaf axils, bracts linear. Calyx clothed with hairs; sepals subequal, lanceolate, long pointed. Corolla white, rarely pink, funnel shaped. Stamens-5, included. Ovary 2-celled, stigma-2. Capsules globose. Seeds generally 4, dark brown.

Chemical Constituents: The chemical constituents in Shankhpushpi consist of carbohydrate-D-glucose, rhamnose, maltose, sucrose and starch. It also contains proteins, amino acids and the alkaloids convolvine, convolamine, confoline, phyllabine, convolidine, convoline, convosine, subhirsine and convolvidine along with fatty acid and wax constituents, hydrocarbons, aliphatic and sterol; certain other bio chemicals present include scopoletin, glacial acetic acid, three coumarins, β-sitosterol, kaempferol, tropane alkaloids, linoleic acid, palmitic acid and straight chain hydrocarbon hextriacontane, 20-oxodotriacontanol, tetratriacontanoic acid and 29-oxodotriacontanol. Alcoholic extract of plant contains di-oh-cinnamic acid, kaempferol and β-stosterolglucose steroid of microphyllic acid. Hydroxy cinnamic acid, octacosanol tetracosane along with glucose, sucrose also have been isolated from the plant. (Bhowmik *et al.* 2012, Bhalerao *et al.* 2014).

Medicinal Properties: It is used as a tonic, alterative and febrifuge, psychostimulant and tranquilizer. According to Ayurveda, Shankhpushpi is pungent, alternative tonic, bitter, brightens intellect, improves complexion, increases appetite, useful in bronchitis, biliousness, epilepsy and teething troubles of infants, etc. It is also a natural tonic for mental development of children. (Bhowmik *et al.* 2012, Bhalerao *et al.* 2014).

Medicinal Uses (Traditional/Folklore): *C. prostratus* is a very famous Ayurvedic medicine, as a *rasayana* which is mainly advocated for use in mental stimulation and rejuvenation therapy, improve memory, intelligence, etc. The leaves are the major constituent of an herbal drug *'Shankhapushpi'*. The drug is used as antiepileptic. It is used alone or is administered along with modern antiepileptic drugs. It is also used in other conditions like vomiting and toxic conditions. The herb induces a feeling of calm and peace, good sleep and a relief in anxiety, stresses, mental fatigue, producing a significant reduction in the level of anxiety, neuroticism arising due to various levels of stresses.

The whole herb is used medicinally in the form of decoction with cumin and milk in fever, nervous debility, and loss of memory. The extract of herb helps to decrease the cholesterol level in the blood including triglycerides, phospholipids and helps in removing certain types of fatty acids that are harmful for the body. Two grams of powder of the plant and black sesame seeds 1 gram along with milk are used to check bed urinating problem (bed-wetting). The leaves of *Shankhpushpi* were traditionally used to treat chronic bronchitis and asthma; roots are used to cure fever in children and oil extracted from plant stimulates the growth of hair. (Kaul and Dwivedi 2010, Bhowmik *et al.* 2012, Bhalerao *et al.* 2014).

35.3 *Cuscuta chinensis* Lam.

(Syn. *Cuscuta carinata* R. Br.)

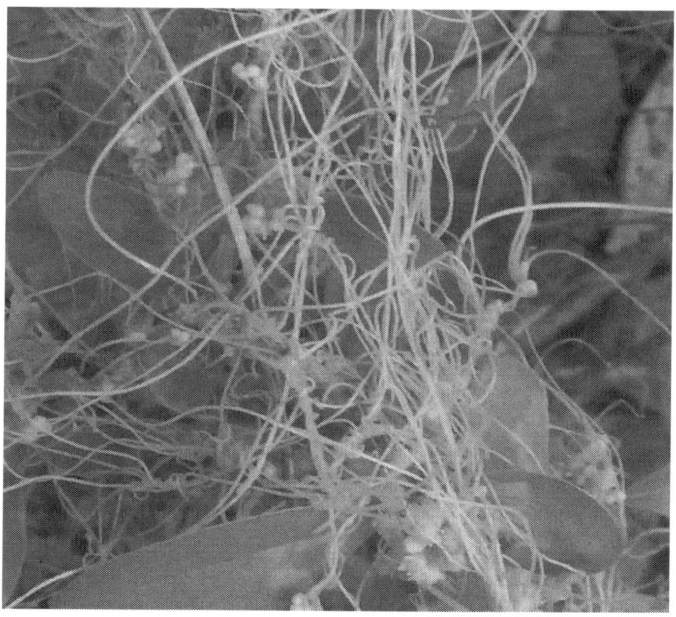

Habit

Common Name: Common Dodder.

Vernacular Name: *Hindi:* Amar Bel (अमरबेल), Makhanio (मखनियो).

Habit: Parasitic climbing/twining herb.

Habitat: Parasitic on the herbs, bushes, trees and grasses.

Distribution: Native to mediterranean region; found in Indo-Malesia and China.

Flowering and Fruiting Time: November-March.

Taxonomic Description: A leafless, stem parasite, often forming dense yellow masses on host plant. Flowers bracteate, whitish or cream-coloured, arranged in few to many flowered cymes, pedicels very short. Calyx 5-lobed, short; sepals ovate, acute, keeled on the back. Corolla about 5 mm across; lobes-5, triangular, apex acute or obtuse, scales fimbriate. Stamens-5, inserted at corolla tube. Ovary globose; styles-2, filiform. Capsules depressed-globose, nearly 3 mm in diameter.

C. chinensis **on host twig**

Fruits

Chemical Constituents: Eight compounds have been isolated from the plant and identified as quercetin 3-O-β-D-galactoside-7-O-β-D-glucoside, quercetin 3-O-β-D-apiofuranosyl-(1-->2)-β-D-galactoside, hyperoside, isorhamnetin, kaempferol, quercetin, d-sesamin and 9 (R)-hydroxy-d-sesamin. (Ye *et al.* 2002).

Medicinal Properties: It is used for the treatment of leucoderma along with other herbs and also used for the treatment of abnormal growth of bones. *C. chinensis* is an important herbal medicine that is used as tonic and aphrodisiac to nourish the liver and kidneys, and to treat impotence and seminal emission. It is considered to have antitumour effects in prostate cancer, skin carcinomas and papillomas. *C. chinensis* having antitumour properties stimulated the proliferation of epidermoid carcinoma (MCF-7) cell line, human breast cancer (T47D) cell line, and lymphocytes and rat bone cells. (Zeraati *et al.* 2010).

Medicinal Uses (Traditional/Folklore): Seeds of *C. chinensis* are used in Chinese medicine as a tonic and considered to have antitumour activity in the Unani system of medicine in India. The paste of plant is applied on chronic ulcer and wounds. Poultice is applied on painful inflammations. The stem extract or decoction of this plant, especially which is grown on *Ziziphus* given daily to increase lactation. Stem paste is rubbed onto the scalp to remove dandruff. Stem Powder mixed with borax powder (hydrated sodium tetraborate), Santalum powder (about 1gm), powdered wood of *Santalum album* L. and Sal ammonia (Ammonium chloride) (about 1g) are thoroughly missed and made into pellets which are given to patients to cure vitiligo. (Vardhana 2008, Zeraati *et al.* 2010).

35.4 *Cuscuta reflexa* Roxb.

(Syn. *Cuscuta verrucosa* Sweet)

Common Name: Giant Dodder.

Vernacular Name: *Hindi:* Amar bel (अमर बेल), Akashbel (आकाश बेल); *Sanskrit:* Amaravela (अमारवेला), Akashballi (आकाशबाली).

Habit: Parasitic climbing/twining herb.

Habitat: Often found gregariously on small trees and bushes.

Distribution: Native to mediterranean region. *C. reflexa* is mainly confined to tropical Asia, Indo-Malesia and China.

Flowering and Fruiting Time: October-March.

Taxonomic Description: A climbing or twining, glabrous, leafless, parasitic herb, forming dense yellow masses on small trees and bushes. Flowers pale white, clustered or shortly racemose, sub-sessile; bracts small, fleshy. Calyx deeply divided; sepals-5,

Habit

segments ovate-obtuse. Corolla white or cream coloured, about 7 mm long, tube almost cylindrical; lobes-5, short, deltoid, reflexed. Stamens-5. Style simple, very short; stigmas bi-lobed, diverging. Capsules depressed globose with black seeds.

Flowering twig **Flowers**

Chemical Constituents: Many phytochemicals like cuscutine, stigmasterol, kaempferol, dulcitol, myricetin and coumarin have been isolated from *C. reflexa*. Plant contains flavanoids, glycosides, alkaloids, cuscutin, cuscutalin and amarvel. cuscutin, a marbelin, β-sitosterol, stigmasterol, kaempferol, dulcitol, myricetin, quercetin. The other phytochemicals identified are kaempferol-3-0-glucoside. astragallin myrecetin, benzopyrones gluco pyranosides propanamide, flavanols, quercetin-3-0-glucoside, β– sitostorol, bergenin, Coumarine, α-amyrine, α-amyrine β-amyrine acetate, oleanolic acetate, oleanolic acid, stigma sterol and lepeol; lupeol is a pharmacologically active tri-terpenoid. (Vijikumar *et al.* 2011, Raza *et al.* 2015, Ulhas 2015).

Medicinal Properties: The plant has anticonvulsant, muscle relaxant, antioxidant, anthelmintic, carminative, alternative, purgative and diuretic, antihypertensive, cardio tonic, antiviral, and antibacterial properties. The seeds are alterative, anthelmintic, carminative, demulcent, diaphoretic, diuretic, expectorant, febrifuge, purgative, stomachic and tonic. (Vijikumar *et al.* 2011, Rahul 2013, Jagtap *et al.* 2014, Ulhas 2015; http://medplants.blogspot.in).

Medicinal Uses (Traditional/Folklore): Chinese herbalists use *Cuscuta* in formulas to treat a range of conditions, including: impotence, premature ejaculation, sperm leakage, frequent urination, ringing in the ears, lower back pain, leucorrhoea, dry eyes, blurred vision, and tired eyes. The plant (stem decoction) is useful in constipation, flatulence, liver and spleen complaints, bilious affections, diarrhoea, joint pains, paralysis and vomiting and flatulence. In Tripura, India plant juice mixed with coconut water is taken early morning for two weeks to cure jaundice. It is also used in cough and diabetes. Plant juice is also used to remove the worms in intestine. Plant juice is applied on hair to increase the length. The whole plant is ground and the juice is used for purification of blood and for controlling the heartbeat. The herb has been traditionally used for the treatment of infections of blood in Pakistan. (Mahmood *et al.* 1997, Singh *et al.* 2010d, Chaudhary and Gupta 2011,Vijikumar *et al.* 2011, Rahul 2013, Jagtap *et al.* 2014, Shaheen *et al.* 2014; http://medplants.blogspot.in).

35.5 *Evolvulus nummularius* (L.) L.

(Syn. *Convolvulus nummularius* L.; *Volvulopsis nummularium* (L.) Roberty)

Habit

Common Name: Roundleaf Bindweed, Agracejo Rastrero.

Vernacular Name: *Hindi:* Musakarni (मुसाकर्णी); *Sanskrit:* Vishnukrantha (विष्णुक्रांत:).

Habit: Prostrate herb.

Habitat: Common in grass lands, roadsides, gardens

Distribution: *E. nummularius* is native of tropical America; widely distributed in India, Sri Lanka, Nepal, Bhutan, and Africa.

Flowering and Fruiting Time: April-October.

Taxonomic Description: A small, prostrate herb. Stem densely villous, rooting at nodes. Leaves up to 2 cm long, alternate, petiolate, sub-orbicular, entire, cordate at base, obtuse-retuse at apex. Flowers tiny, white, pedicellate, axillary, and solitary or 1-2 flowered. Calyx lobes-5, ovate, oblong, ciliate at margin. Corolla gamopetalous, campanulate; lobes-5, ovate, hairy on margin. Stamens-5, filaments filiform. Ovary 2-celled, styles-2. Capsules globose, 3-4 mm long. Seeds black or brown, 1-4, glabrous, minutely tuberculate.

Villous stem with sub-orbicular leaves

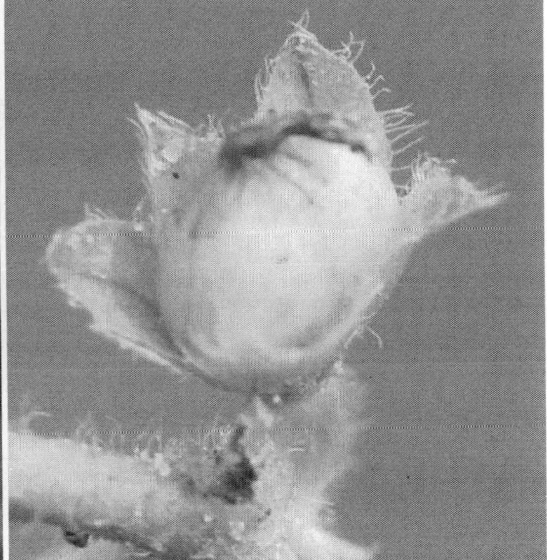

Capsule

Chemical Constituents: The plant contains phytoconstituents like alkaloid, saponin, phytosterol, flavonoids, tannins, triterpenoids, phenolic compounds, carbohydrates, etc. From the aerial parts, β-sitosterol, glucoside, stigmasterol, d-mannitol, ursolic acid and oleanolic acid have been isolated. (Namrata *et al.* 2014, Sahu and Gupta 2014).

Medicinal Properties: Antibacterial, febrifuge, anthelmintic, sedative, antirheumatic, antiepileptic, blood purifier, memory enhancer and vermifuge. (Dinda *et al.* 2007, Kaladhar *et al.* 2011, Namrata *et al.* 2014).

Medicinal Uses (Traditional/Folklore): In Indian traditional folk medicine, it is used as a medicine for hysteria, leprosy, epilepsy, cramps, to cure burns, cuts, wounds and scorpion stings. In Nepal, the paste of the plant is used to treat scabies. The whole plant is used to treat amoebic dysentery. Fresh juice obtained from young twigs is consumed with water (half teaspoonful) twice daily to treat dysentery. Whole plant paste is used externally to reduce tonsillitis pain. Plant is also used for the treatment of wounds (healing), hysteria, convulsion, bronchial asthma and as a weak sedative. The latex of the plants may cause blindness and skin irritation. (Das *et al.* 2006, Pavithra *et al.* 2009, Kumar and Satapathy 2011, Sahu and Gupta 2014).

35.6 *Ipomoea aquatica* Forssk.

(Syn. *Ipomoea reptans* Poir.)

Habit

Common Name: Water Morning Glory, Swamp Cabbage, Chinese Water Spinach.

Vernacular Name: *Hindi:* Kalmi Sag (कलमी साग), Nali (नाली).

Habit: Floating or trailing herb.

Habitat: Abundant in and along the bank of ponds, canals and wetlands.

Distribution: It is most commonly found in East and South-East Asia.

Flowering and Fruiting Time: August-December.

Taxonomic Description: A floating aquatic herb, some times trailing on wet ground. Stem branched, hollow, rooting at nodes. Leaves alternate, petiolate, 5-15 x 2-7 cm, glabrous, triangular, ovate-oblong, entire, base cordate or hastate, acute or acuminate at apex. Flowers showy, solitary or few-flowered in axillary cymes, pedicels up to 4 cm long. Calyx lobes-5, ovate, oblong-lanceolate. Corolla funnel-shaped, about 5 cm long, white with dark purple at centre. Stamens-5, included; filaments unequal, white. Stigma-bilobed. Capsules ovoid, glabrous. Seeds pubescent.

Flowering twig

Flower

Chemical Constituents: It contains alkaloids, carbohydrates, proteins and amino acids, sterols, phenols, flavonoids, gums and mucilage, glycosides, terpenes and tannins. The leaves contain amino acids like aspartic acids, glycine, alanine, etc. The leaves are a good source of minerals and vitamins (especially, carotene and tocopherol). (Prasad *et al.* 2008, Malakar and Choudhury 2015, Srivastava 2017).

Medicinal Properties: The plant has antimicrobial, antidiabetic, antioxidant and anti-inflammatory properties. Plant juice is emetic, purgative; given in piles, nervous disorders, general debility. Leaf extract exhibits potent hypoglycaemic activity and used in the treatment of diabetes, as a scorpion venom antidote, as emetic, diuretic, purgative, to treating debility, liver complaints, against nosebleed and high blood pressure. It is also used as a laxative and tonic. (Prasad *et al.* 2008, Malakar and Choudhury 2015, Srivastava 2017: http://medplants.blogspot.in).

Medicinal Uses (Traditional/Folklore): The plant is used in menorrhagia, flatulence, as a tonic and inflammation for intoxication; decoction is used as a remedy for haemorrhoids. Crushed leaves are applied to treat hardened pimples and as a poultice on sores and boils. A decoction of the leaves is used to treat coughs. A paste made from the buds is used to treat ringworm. A decoction of the roots is used as a wash against haemorrhoids.

Whole plant is used as an antidote to arsenical or opium poisoning. Plant juice is used for liver complaints. In Cambodia, *I. aquatica* is used to treat ringworm infection, combat fever, and to treat asthma. In China, the plant is used to relieve the bowels of costiveness, to counteract the putrefaction of the skin and uterus, to invigorate health, to treat food poisoning and cough. In Indonesia, the plant is used to treat asthma, to calm and to heal boils and haemorrhoids. In Africa, the Philippines and in Sri Lanka the herb is used to treat diabetes. It is also used to treat abscesses, mental illness in Tanzania and intestinal problems in Somalia. In Sudan, the herb is used to treat stomach and intestinal troubles. (Wiart 2006b, Prasad *et al.* 2008, Malakar and Choudhury 2015; http://medplants.blogspot.in, http://tropical.theferns.info).

35.7 *Ipomoea carnea* Jacq.

(Syn. *Ipomoea carnea* subsp. *carnea*)

Common Name: Bush Morning Glory.

Vernacular Name: *Hindi:* Behaya (बेहया), Besharam (बेशरम).

Habit: Shrub.

Habitat: Common along the border of ponds, lakes, roadsides, railway lines.

Distribution: Tropical and subtropical New World plant from Argentina to Mexico. It is well distributed in India, especially in Madhya Pradesh and Chhattisgarh.

Flowering and Fruiting Time: Almost throughout the year.

Habit

Taxonomic Description: A large straggling shrub with milky juice, 1-3 m tall. Stem fistular. Leaves up to 20 cm long, alternate, long-petiolate, broadly ovate-lanceolate, entire, cordate at base, acute to acuminate at apex, mid-rib with 2-small glands at base. Flowers large, bluish-purple or rose-purple in lax dichotomous, axillary and terminal cymes. Calyx lobes-5, cup-shaped. Corolla 7-9 cm long, campanulate, dark-purple inside. Stamens-5, included, filaments unequal, anthers 2-celled. Ovary 2-celled, style filiform, stigma 2-lobed. Capsules globose, 1.5-2 cm long. Seeds-4, compressed, black, silky hairy.

Flowering branch

Flower buds

Flowers

Chemical Constituents: Caffeoyl derivatives were isolated from the seeds. The leaves contain 1-3 flavonol glycosides and Ergine (D-Lysergic acid amide). Polyhydroxylated alkaloids were isolated from the leaves, flowers and seeds. Chromatographic separation of the leaf extract resulted in the isolation of swainsonine, 2-epilentiginosine and calystegines B (1), B (2), B (3) and C (1) and N-methyl-trans-4-hydroxy-l-proline and beta sitosterol. Leaves contain amino acids, steroids, flavonoids, alkaloids and tannins in methanolic and petroleum ether extracts. Main chemical constituent present in the plant are dichloromethane, ethyl acetate and methanol. It also contains nitric acid and lipopolysaccharide 3, 4 dimethoxy cinnamic acid. (Khalid *et al.* 2011, Bhalerao and Teli 2016, Srivastava 2017).

Medicinal Properties: Plants are aphrodisiac, purgative and cathartic having anti-inflammatory, antioxidant, antidiabetic, antimicrobial, antifungal, wound healing, immunomodulatory, hepatoprotective and anxiolytic properties. Aqueous extract of plant has neuromuscular blocking activity. (Khalid *et al.* 2011, Bhalerao and Teli 2016, Srivastava 2017).

Medicinal Uses (Traditional/Folklore): Plants are used in diabetes mellitus, arthritis, cancer, arsenic and opium poisoning, underdeveloped female structures, sexual debility, piles, cardiac debility, dyspepsia, hepatopathy, venereal debility, chronic bronchial asthma, epilepsy, brain tonic, nervous weakness, cerebral weakness, cardiovascular diseases, uropathy and tumour. The milky juice of *Besharam* is used for the treatment of leucoderma by applying it externally on the affected parts. It is used to decrease the teratogenic effect resulting from cyclophosphamide. Powder of dried leaves and black pepper is given in dog bite. Leaf paste is applied to treat sprains; leaves are warmed and applied to cure swellings on the body. (Vardhana 2008, Shukla *et al.* 2010, Bhalerao and Teli 2016).

35.8 *Ipomoea coptica* (L.) Roth ex Roem. & Schult.

(Syn. *Convolvulus copticus* L.; *Ipomoea dissecta* Willd.)

Habit

Common Name: Egyptian Morning Glory.

Vernacular Name: None known.

Habit: Prostrate or climbing herb.

Habitat: Common in moist, shady places, gardens, unused grounds.

Distribution: Plant has wide distribution in continents of Africa, Australia and Asia. In Asia, it is mainly distributed in India and Sri Lanka.

Flowering and Fruiting Time: July-October.

Taxonomic Description: A glabrous, prostrate or climbing annual herb. Stem branched, slender, angular. Leaves alternate, petiolate, digitate, 5-7 lobed, sub-sesssile, narrow-lanceolate, irregularly serrate, middle largest. Stipule leaf-like. Flowers small, white or pinkish-white, axillary, solitary or few-flowered on long peduncules; bracts minute, linear-lanceolate. Sepals-5, oblong-lanceolate, muricate at outer surface. Corolla narrowly funnel shaped, gamopetalous. Ovary 3-celled. Capsules sub-globose with persistent calyx, glabrous, beaked. Seeds usually 3, silky-tomentose, black.

Flowering twig

Flower

Capsule

Chemical Constituents: None known.

Medicinal Properties: None known.

Medicinal Uses (Traditional/Folklore): Cold infusion, prepared after soaking the plant in water, is given for seven days for treating dizziness and in intoxication. Herb is also used in chest complaints in children. Leaves are used as mouth wash. (Sorathia 2014, Suthari *et al.* 2014)

35.9 *Ipomoea eriocarpa* R. Br.

(Syn. *Convolvulus eriocarpus* (R. Br.) Spreng.; *Convolvulus hispidus* Vahl)

Common Name: Tiny Morning Glory, Woolly Fruited Morning Glory.

Vernacular Name: *Hindi:* Buta (बूटा).

Habit: Twining or prostrate herb.

Habitat: Twining on grasses and other herbs along roadsides, open fields.

Distribution: Paleotropics; native of Africa.

Flowering and Fruiting Time: September-November.

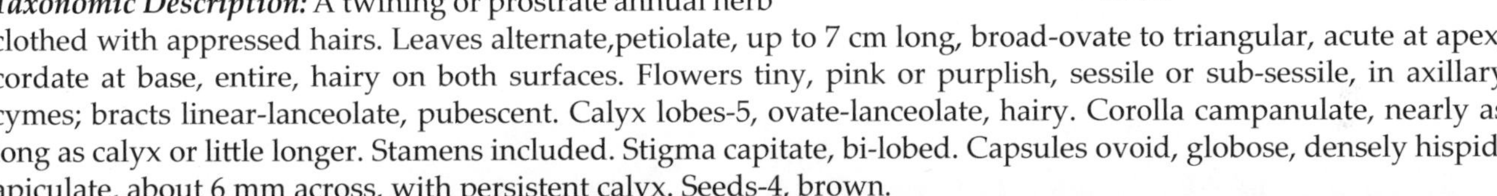

Habit

Taxonomic Description: A twining or prostrate annual herb clothed with appressed hairs. Leaves alternate,petiolate, up to 7 cm long, broad-ovate to triangular, acute at apex, cordate at base, entire, hairy on both surfaces. Flowers tiny, pink or purplish, sessile or sub-sessile, in axillary cymes; bracts linear-lanceolate, pubescent. Calyx lobes-5, ovate-lanceolate, hairy. Corolla campanulate, nearly as long as calyx or little longer. Stamens included. Stigma capitate, bi-lobed. Capsules ovoid, globose, densely hispid, apiculate, about 6 mm across, with persistent calyx. Seeds-4, brown.

Flowering twig

Flower

Capsule

Chemical Constituents: *I. eriocarpa* contains alkaloids, phenols, saponins, phytosterols and terpenoids. The GC-MS analysis of phytoconstituents of leaves revealed the presence of eleven major phytoconstituents: hentriacontane, Z,Z- 6,28-heptatriactontadien-2-one, N-hexadecanoic acid, 3,7,11,15- tetramethyl-2-hexadecen-1-ol, Methyl-8,11,14-heptadecatrienoate, hexacosanol acetate, (All-E)-2,6,10,15,19,23-hexamethyl- 2,6,10,14,18,22-tetracosahexaene, (3-Beta)-ergost-5-en-3-ol, Methyl-2-hydroxy-eicosanoate, γ-sitosterol and 2R-acetoxymethyl- 1,3,3-trimethyl-4t-(3-methyl-2-buten-1-yl)-1t-cyclohexanol. (Das and Himaja 2014, Srivastava 2017).

Medicinal Properties: The plants have insecticidal, anthelmintic, anti-inflammatory, antiarthritic and antidiabetic, antioxidant, antisecretory, antipyretic and antinociceptive properties. (Dhiman 2004, Das and Himaja 2014).

Medicinal Uses (Traditional/Folklore): A root decoction is taken by women to relieve menstrual pain. An oil extract of the plant is used in India for external application in the treatment of headache, rheumatism, leprosy, epilepsy, ulcers and fevers.

As a veterinary medicine for the treatment of endoparasites, roots are chopped and mixed with water, then the filtrate is drenched and rest of the sediments are poured on the wounds. (Shukla *et al.* 2010, Srivastava 2017).

35.10 *Ipomoea obscura* (L.) Ker Gawl.

(Syn. *Convolvulus obscurus* L.)

Habit

Common Name: Obscure Morning Glory.

Vernacular Name: *Hindi:* Pan Bel (पान बेल); *Sanskrit:* Vachagandha (वचगंधा), Lakshmana (लक्षमना).

Habit: A climbing or twining herb.

Habitat: Common along roadsides, hedges, railway lines, canal banks.

Distribution: It is native to tropical Africa, Mascarene Islands, tropical Asia, throughout Malaysia to northern Australia and Fiji. In India, it occurs in the plains up to 1,400 m.

Flowering and Fruiting Time: August-November.

Taxonomic Description: A slender, climbing or twining herb. Stem long, spreading, purple in colour, glabrous or slightly villous. Leaves up to 8 cm long, alternate, petiolate, broadly ovate to orbicular, cordate, entire, acute or acuminate, glabrous or sparsely pubescent on both surfaces. Flowers axillary, solitary or 2-3 flowered peduncles; bracts minute, linear-lanceolate. Calyx lobes ovate-lanceolate. Corolla funnel-shaped, 2-2.5 cm long, white with yellow bands, base purple. Stamens-5, included. Capsules ovoid, 6-8 mm across, apiculate. Seeds-4, dark brown or black, densely pubescent.

Flowering twig

Flower

Chemical Constituents: Phytochemicals present are various classes of secondary metabolites such as alkaloids, flavonoids, phenols, Glycosides, tannins and carbohydrates. The leaf is rich in metals like sodium, potassium, phosphorus, manganese, iron, calcium, zinc, lead, mercury, copper, nickel and magnesium. (Mungole *et al.* 2010, Hussain and Kumaresan 2014, Srivastava 2017).

Medicinal Properties: Ayurveda has identified many medicinal properties of *I. obscura* and it is effectively used against dysentery, applied to open sores and pustules. Seeds and fruits are used as cleansing agents to improve difficult breathing, relive pain and to improve vision. The plant also has antihepatic properties. (Mungole *et al.* 2010, Hussain and Kumaresan 2014).

Medicinal Uses (Traditional/Folklore): A paste of leaves is applied on ulcers, sores, haemorrhoids and swellings. The dried and powdered leaves are used to treat aphthae. A root decoction is drunk against dysentery. The leaf sap is used to treat fits of insanity. In Indonesia, a paste of leaves, mixed with *Argyreia mollis* leaves and alcohol, is applied to open sores and pustules. Leaves used in aphthous affections after roasting and boiling in ghee. Half teaspoon of leaf extract is administered thrice in two hours for treating acute stomachache. In case of horn cancer, 100 g of leaf paste is boiled in 300 g of gingelly oil (*Sesamum indicum*), and this decoction is administered daily once for a fortnight. (Christophe *et al.* 2002, Reddy *et al.* 2010a; http://uses.plantnet-project.org/en/Ipomoea_obscura_(PROTA)).

35.11 *Ipomoea pes-tigridis* L.

(Syn. *Convolvulus pes-tigridis* (L.) Spreng.)

Habit

Common Name: Tiger Foot Morning Glory.

Vernacular Name: *Hindi:* Panchpatia (पंचपतिया).

Habit: A spreading or twining annual herb.

Habitat: Commonly found spreading in open fields as well as twining on herbs and bushes.

Distribution: It is distributed in Western Ghats region of India and other places. It is native of tropical Africa.

Flowering and Fruiting Time: August-October.

Taxonomic Description: A spreading or twining annual herb. Stem covered with long spreading hairs. Leaves alternate, long petiolate, deeply palmately 5-7 lobed, cordate at base; lobes elliptic, apex acute or acuminate, margin entire, appressed hairy on both sides. Flowers whitish or pinkish, in axillary, long peduncled, few-flowered cymose heads, surrounded by large conspicuous hairy bracts; outer bracts 2-2.5 cm long, larger than the inner bracts. Calyx lobes unequal; sepals lanceolate, acuminate, pubescent. Corolla 3-4 cm long, funnel-form, tube cylindric. Stamens-5, attached to the corolla tube. Stigma 2-lobed. Capsules glabrous, globose, buried in the calyx. Seeds-4, dark brown, pubescent.

Flowering twig

Flower

Flower bracts

Chemical Constituents: The plant contains phytochemicals such as alkaloids, saponins, flavonoids and tannins in root and leaf extracts. (Selvam and Acharya 2015, Srivastava 2017).

Medicinal Properties: Leaves have antipyretic and anticancer properties. (Selvam and Acharya 2015).

Medicinal Uses (Traditional/Folklore): The tribal community in Kerala, India use the herb for various painful conditions like headaches, swellings, poisonous stings, snake bites, etc. In Java, it is used for poulticing sores, pimples, boils, carbuncles, etc. In Sri Lanka, entire creeper is crushed and the juice is given orally for treatment of prevention of rabies if bitten by a rabid dog. In India, the plant is used for wound healing. Twining stems dipped in turmeric water and worn around neck of children, in case of restlessness, to ward off evil spirits. The leaf paste is applied on the whole body and taken bath with warm water for 3-days to get relief from fever. The seed is taken as a remedy for dropsy. It is helpful to treat various kinds of ailments like headache, swellings, poisonous stings, healing wounds and snake bites. Leaf powder is smoked to get relief from bronchial spasm. *I. pes-tigridis* exhibited significant anticancer activity, particularly for liver cancer. As a part of hypnotic practices by tribes, leaf paste is warmed and cooled and applied on head. Root is used as purgative and antidote to dog bites. (Shukla *et al.* 2010, Selvam and Acharya 2015, Srivastava 2017).

35.12 *Ipomoea triloba* L.

(Syn. *Batatas triloba* (L.) Choisy; *Convolvulus trilobus* (L.) Desr.)

Habit

Common Name: Little Bell Morning Glory, Three-Lobe Morning Glory.

Vernacular Name: None known.

Habit: Twining or prostrate herb.

Habitat: Twining on grasses and bushes.

Distribution: *I. triloba* is widespread in warm tropical areas of the world.

Flowering and Fruiting Time: September-November.

Taxonomic Description: An annual twining or prostrate herb. Leaves alternate, long petiolate, broadly ovate to deeply 3-lobed, 5-10 x 2-8 cm, glabrous or sparsely pubescent, base cordate. Inflorescence umbellate cymes, several flowered, sometimes 1-flowered; peduncle about 10 cm long, angular; bracts minute, oblong-lanceolate. Calyx lobes-5, polysepalous, sub-equal, lanceolate. Corolla gamopetalous, about 2 cm long, pinkish-purple, darker at centre, funnel-form, lobes-5, mucronulate. Stamens-5, included. Stigma capitate, 2-lobed. Capsules subglobose, hairy, apiculate with persistent calyx. Seeds dark brown, hairless.

Flower

Capsules

Chemical Constituents: The *I. triloba* contains saponins, tannins, terpenes, protein, fats, fiber and carbohydrates. The powdered leaves contain tannic acid, phylate acid and oxalic acid. (Essiett and Obioboho 2014, Srivastava 2017).

Medicinal Properties: Analgesic, antibacterial and antifungal properties. (Srivastava 2017).

Medicinal Uses (Traditional/Folklore): The plant is used as a poultice in the treatment against headaches. A decoction of the leaves is used to get relief against stomachache and cure facial distortion; juice obtained from macerated leaves is applied to distorted areas. *I. triloba* is used for wound healing as it has antibacterial and antifungal properties. In Ayurveda, the whole plant is used as a pain killer. (Esha *et al.* 2012; http://tropical.theferns.info).

35.13 *Merremia hederacea* (Burm. f.) Hallier f.

(Syn. *Ipomoea chryseides* Ker Gawl.)

Habit

Common Name: Ivy Woodrose.

Vernacular Name: None known.

Habit: Climber.

Habitat: Twining on bushes along the roadsides and canal banks.

Distribution: Paleotropics: India, Bangladesh, China, Indonesia, Ryukyu Isl., Cambodia, Laos, Myanmar, Nepal, New Guinea, Pakistan, Philippines, Sri Lanka, Thailand, Vietnam, Cameroon, Chad, Gabon, Tanzania, Somalia, Sudan, South Sudan, Burkina Faso, Mali, Mauritania, Senegal, Ethiopia, D.R. Congo, Zambia, Zimbabwe, Mozambique, Angola, Guinea-Bissau, Madagascar, Sierra Leone, Ivory Coast, Ghana, Nigeria, Mauritius, Réunion, Australia, Java, South America (introduced).

Flowering and Fruiting Time: September-December.

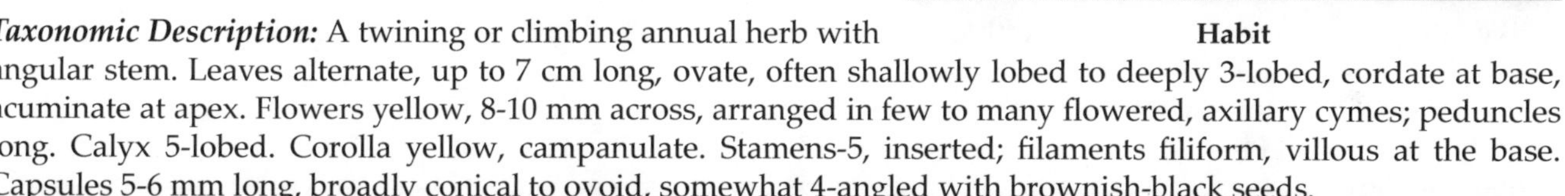

Taxonomic Description: A twining or climbing annual herb with angular stem. Leaves alternate, up to 7 cm long, ovate, often shallowly lobed to deeply 3-lobed, cordate at base, acuminate at apex. Flowers yellow, 8-10 mm across, arranged in few to many flowered, axillary cymes; peduncles long. Calyx 5-lobed. Corolla yellow, campanulate. Stamens-5, inserted; filaments filiform, villous at the base. Capsules 5-6 mm long, broadly conical to ovoid, somewhat 4-angled with brownish-black seeds.

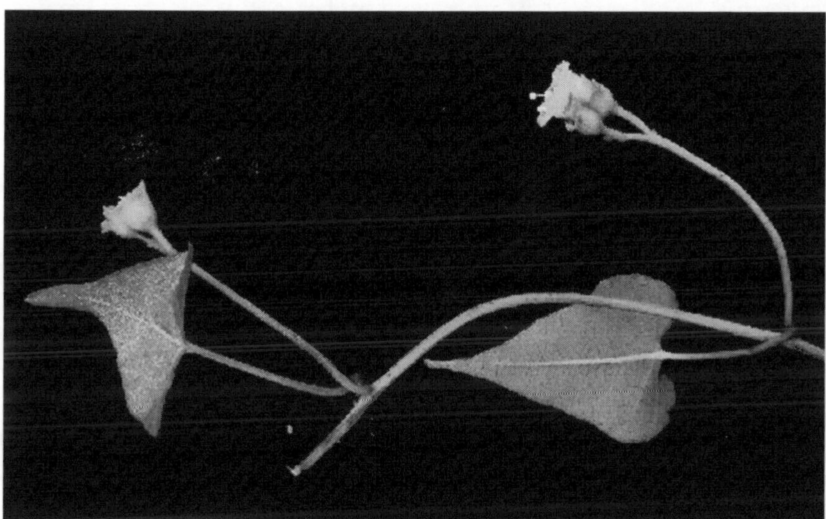

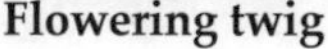

Flowering twig

Flower

Chemical Constituents: Plant contains phenols, flavonoids, terpenoids, tannins, alkaloid, amines, carbohydrates and steroids. Five new pentasaccharide resin glycosides, namely merremins A-E (1-5), two new pentasaccharide resin glycoside methyl esters, namely merremins F and G (6, 7), and four known resin glycosides, murucoidin IV, murucoidin V, stoloniferin IV, and murucoidin XVII, have been isolated from the aerial parts of *M. hederacea*. (Charles *et al.* 2012, Wang *et al.* 2014b).

Medicinal Properties: The plant is aperient, deobstruent, diuretic, and alterative. It is used to treat colds, febrile disease, sunstroke, oliguria, tonsil inflammation, laryngitis as well as leucorrhoea. The seeds are used to treat fevers, colds, sore throats, haematuria, conjunctivitis and boils. (Charles *et al.* 2012).

Medicinal Uses (Traditional/Folklore): The plant is useful to cure acute tonsillitis. A poultice of the leaves, combined with turmeric and broken rice, is used to heal cracks in the hands and feet. Whole plant is used as hair tonic/ shampoo. (Sreeramulu *et al.* 2013; http://tropical.theferns.info).

35.14 *Operculina turpethum* (L.) Silva Manso

(Syn. *Convolvulus turpethum* L.; *Ipomoea turpethum* (L.) R. Br.)

Habit

Common Name: Transparent Wood Rose, White day glory, Indian Jalap.

Vernacular Name: *Hindi:* Nisoth (निसोथ), Panila (पनिला), Pithori (पिठोरी); *Sanskrit:* Triputa (त्रिपुटा), Nishotra (निशोत्रा).

Habit: Perennial climber.

Habitat: Climbing on bushes and trees along the roadsides and railway lines.

Distribution: Found in tropical Africa, in southern Asia, Sri Lanka, Myanmar, southern China, Indo-China, Thailand, Malesia and northern Australia and Polynesia; introduced in West Indies.

Flowering and Fruiting Time: September-December.

Taxonomic Description: A large climber. Stem very long with milky latex, much twisted together, softly pubescent, angled. Leaves alternate, 5-15 cm long, broadly ovate, margin entire, apex retuse-obtuse to mucronate, base cordate or truncate, slightly pubescent on both the surfaces; petiole up to 7 cm long, hairy. Flowers about 5 cm across, white, axillary, solitary to few-flowered; peduncles stout, pubescent; bracts 2-3 cm long, oblong-lanceolate, silky-pubescent, caducous. Sepals-5, accrescent, unequal; outer two ovate-orbicular, mucronate, pubescent; inner three smaller, apiculate. Corolla funnel-form, tube short. Stamens-5, filaments slender. Stigma 2-lobed. Capsules sub-globose, enclosed in the enlarged calyx segments. Seeds globose, 5-6 mm across, black.

Abaxial side of leaf

Flower

Seeds

Chemical Constituents: Ether and chloroform extract of roots of *O. turpethum* contain carbohydrates, steroids, gums, flavonoids and saponins; ethanol, ether and chloroform extracts of leaves of contain carbohydrates, steroids, tannins, gums, flavonoids, alkaloids and saponins. (Prabhavathi *et al.* 2012, Ahmed *et al.* 2013a, Ahmad *et al.* 2017).

Medicinal Properties: The roots are bitter, acrid, sweet, thermogenic, analgesic, purgative, carminative, anthelmintic, expectorant, antipyretic, hepatic, stimulant and hydragogue. It possesses anti-inflammatory, anticancer, cytotoxic, antisecretory, ulcer protective, hepatoprotective, and antibacterial properties. The herb is useful in various ailments like flatulence, constipation, haemorrhoids and improper bowel movements. Being purgative in nature, the herb is quite useful in the treatment of arthritis and osteoarthritis. It helps to reduce the inflammation and swelling in joints and its anti-inflammatory properties help in pain reduction. (Prabhavathi *et al.* 2012, Ahmed *et al.* 2013a, Islam *et al.* 2015, Ahmad *et al.* 2017; http://tropical.theferns.info).

Medicinal Uses (Traditional/Folklore): In Ayurveda, roots are used internally to treat fevers, anorexia, edema, anaemia, ascites constipation, hepato-splenomegaly, hepatitis, intoxication, abdominal tumours, ulcers, wounds, worm infestation, pruritis and other skin disorders. Use of this herb is quite beneficial in the treatment of respiratory complications like cough, bronchitis, dyspnoea, pneumonia and tuberculosis. It helps in the clearance of mucous from air channels. It also helps to provide relief in shortening of breath, lungs congestion, nasal congestion and cough. During constipation, root decoction is administered orally twice a day for a week during indigestion. (Prabhavathi *et al.* 2012, Ahmed *et al.* 2013a, Islam *et al.* 2015; http://tropical.theferns.info).

36. SOLANACEAE

Solanaceae – Potato family, Nightshade family

Family with 115 genera and 2,678 species, contains plants with a great diversity of habitats, morphology and ecology. Members of the Solanaceae family are found throughout the world but are most abundant and widely distributed in the tropical regions of Latin America, where about 40 genera are endemic. Very few members are found in temperate regions. It is an economically important family of flowering plants including a number of important agricultural crops, medicinal plants, spices, weeds, and ornamentals. Among the most important of those are potato, eggplant, tomato, capsicum peppers, tobacco, belladonna, Datura, and nightshades. Eight species belonging to five genera of this family were recorded in Chithara Village.

36.1 *Datura innoxia* Mill.

(Syn. *Datura meteloides* DC. ex Dunal)

Common Name: Pricklyburr, Thorn-Apple.

Vernacular Name: *Hindi:* Safed Dhatura (सफेद धतूरा).

Habit: Large herb.

Habitat: Common along roadsides, railway lines and waste ground.

Distribution: It is native to Central and South America and introduced in Africa, Asia, Australia and Europe. Found throughout India, common as a weed growing in wastelands and roadsides, abundantly in Maharashtra and Rajasthan.

Flowering and Fruiting Time: Most of the year.

Habit

Flowering twig

Flower

Capsule and seeds

Taxonomic Description: A large herb, generally up to 1 m high. Stem branched, densely appressed with greyish-white tomentum. Leaves alternate, elliptic-ovate, up to 20 cm long, entire to sinuate or repand-dentate, base unequal, apex acute, pubescent on both the surfaces. Inflorescence solitary. Flowers white, erect, 10-14 cm long, axillary, pedicels pubescent. Calyx tubular, about 8 cm long, inflated at the base; lobes 5, lanceolate, acute, pubescent outside. Corolla funnel-shaped, 10-toothed. Stamens-5, included, filaments slender. Style slender, equal to the corolla tube. Capsules globose with a cap of persistent calyx, covered with sharp, straight spines, 3-4 cm across, 4-valved. Seeds numerous, reniform, compressed, pale-brown.

Chemical Constituents: *D. innoxia,* like other *Datura* species, contains the highly toxic alkaloids atropine, hyoscine (scopolamine) and hyoscyamine; scopolamine is a potent cholinergic-blocking hallucinogen, which has been used to calm schizoid patients. The alkaloids of these plants are very similar to those of mandrake, deadly nightshade, and henbane, which are also highly poisonous plants used cautiously for effective pain relief in antiquity. All parts of plant contain scopolamine, hyosayamine, hyoscine, meteloidine, atropine, withanolides (lactones), dinoxin B and northyoscyamine, as well as vitamin-C. Root contains tropane, tropine and pseudotropine. (Bhutya 2011).

Medicinal Properties: *D. inoxia* has many medicinal properties like anodyne, hallucinogenic, hypnotic, narcotic, antispasmodic, intoxicant, emetic, and anti-inflammatory. Roots are anticatarrhal, febrifuge, antidiarrhoeal. Leaves are narcotic, anodyne, and anti-spasmodic, Seeds are aphrodisiac, narcotic and antispasmodic. (Khare CP 2007, Bhutya 2011; http://tropical.theferns.info)

Medicinal Uses (Traditional/Folklore): The plant has a very special place in Ayurveda since all plant parts namely, flowers, leaves, root, stem, fruit and seeds have been employed as a pain killer and for a range of treatments such as insanity, rabies, leprosy, fevers with catarrh, diarrhoea and skin diseases, etc. However, the indiscriminate use of its extract may cause acute poisoning, delirium and can even lead to death. *D. inoxia* is used in the treatment of insanity, fevers with catarrh, diarrhoea, scabies, piles, ulcers, colds, asthma, cardiac disorders, impotency, antispasmodic, malaria, baldness, skin diseases, cough, ulcers, loss of appetite, intestinal worms, headache, hemiplegia, epilepsy, convulsions, cramps and rheumatism; also used for promoting complexion. Roots are used for hydrophobia, fevers and diarrhoea. Leaves are used for asthma, bronchitis and its poultice for otalgia, lumbago, ophthalmodynia, sciatica, neuralgia, mumps and painful swelling. Juice of leaves is used for otorrhoea, epilepsy, cephalalgia, dandruff, tumours and rheumatism. Seeds are useful for the treatment of premature ejaculation and skin diseases, and also for dandruff and lice.

In traditional medicine, it is a reputed drug in the treatment of rabid dog-bites and poisonous insect bites. The dried leaves and flowering tops are known for their narcotic and anti-spasmodic properties. They are used for the same purposes as leaves of belladonna and stramonium.

In Mexico, the plant is used as a remedy for many disorders and symptoms, particularly fevers. The Apache use the juice of the flowers and roots to disinfect wounds. Dew drops that have collected in the flowers are used as an eye wash. The Aztecs use leaves to treat broken bones and swollen joints. Leaves that had been warmed in a steam bath are placed directly on to the affected areas. Toloache (a psychoactive, hallucinogenic preparation made from the plant) is one of the most important aphrodisiacs and sedatives in Mexican folk medicine. It is given during childbirth to help with pain. In Israel, a decoction of the leaves is consumed to treat diarrhoea, and a paste of the leaves is applied externally to treat pain. In many parts of the world, the leaves of *D. innoxia* have been smoked, alone or in blends, as a most effective treatment for asthma.

Any use of this plant should be with extreme caution and under the supervision of a qualified practitioner since the toxic dose is very close to the medicinal dose. (Khare CP 2007, Singh *et al.* 2010d; http://entheology.com/category/plants, http://tropical.theferns.info).

36.2 *Nicotiana plumbaginifolia* Viv.

(Syn. *Nicotiana tenella* Cav.)

Habit

Common Name: Tex-Mex Tobacco.

Vernacular Name: *Hindi:* Jangli Tambakoo (जंगली तंबाकू).

Habit: Annual herb.

Habitat: Common in damp places, near the drains, waste places and roadsides.

Distribution: A weed native to tropical America (Mexico and the West Indies); now widely naturalized elsewhere.

Flowering and Fruiting Time: March-October.

Taxonomic Description: An erect, 60-90 cm high, viscid-pubescent herb. Leaves sessile, variable; radicle leaves up to 20 cm long, obovate, elliptic-oblong; cauline leaves smaller and narrower, clasping at the base, margin wavy. Flowers white or pinkish-white, nearly 4 cm long in lax racemes, pedicels about 2 cm long. Calyx tubular, 8-10 mm long, glandular hairy; lobes-5, linear-lanceolate. Corolla lobes-5, ovate, acute; tube long, narrow. Stamens-5, included. Capsules ovoid, about as long as calyx. Seeds numerous, yellowish brown, wrinkled.

Flowering twig

Flower

Capsule

Chemical Constituents: The plant contains carbohydrates, glycosides, tannins, phenolics, flavonoids, terpenoides and alkaloids. (Singh *et al.* 2010b).

Medicinal Properties: Antispasmodic, diuretic, expectorant and antimicrobial. (Singh *et al.* 2010b).

Medicinal Uses (Traditional/Folklore): The plant is used traditionally to treat several human ailments such as constipation, haemorrhoidal bleeding, strangulated hernia, and rheumatic swelling in order to relieve the pain. A mixture of the dried leaves along with those of *Azadirachta indica* is roasted together in a closed container; dried leaves are used in the treatment of nausea and travel sickness. The sap from the leaves is used against sores. Leaf juice is used to treat skin diseases. The steam or smoke coming out through a small hole is allowed to come in contact with the persons suffering from piles. Leaf paste along with cow's urine is applied on cuts and wounds for early healing. Leaf paste applied on tongue to check excessive growth locally known as, 'Doadjeeb' and skin ailment, "Herpes zoster". A decoction of the stem peel/bark is highly effective for treating tertiary syphilis, while the pounded roots are used against hydrocele. As a veterinary medicine, leaves are ground and the paste used as germicide for animal wounds. (Dangwal *et al.* 2011, Quattrocchi 2012, Singh *et al.* 2014b, Shahriar *et al.* 2015; http://medplants.blogspot.in).

36.3 *Physalis angulata* L.

(Syn. *Physalis lanceifolia* Nees)

Habit

Common Name: Angular Winter Cherry, Cutleaf Groundcherry.

Vernacular Name: *Hindi:* Rasbhari (रसभरी).

Habit: Annual herb.

Habitat: Common in moist waste places.

Distribution: Originally native to tropical America now widely distributed and naturalized throughout the tropical, subtropical and warm temperate regions. Found over debris, in fallows, fields and coastal sands, often on nitrate-rich soils.

Flowering and Fruiting Time: July-November.

Taxonomic Description: An erect, annual herb, 30-70 cm high. Stem much-branched, striate, angular. Leaves up to 7 cm long, ovate to elliptic, more or less entire or slightly toothed, apex acute or acuminate, base rounded or oblique, petiole up to 5 cm long. Flowers axillary, solitary. Calyx 5-toothed, teeth ovate-triangular. Corolla pale yellow, campanulate, spotted inside. Stamens-5, anthers oblong. Berries greenish-yellow, 1-1.5 cm across, enclosed within inflated calyx. Seeds discoid, numerous.

Flowering and fruiting twig

Flower

Berry

Chemical Constituents: Phytochemicals present in *P. angulata* are flavonoids, alkaloids and many different types of plant steroids. The aerial parts of the plant contain several biologically active steroidal lactones belonging to the physaline and withanolide types: physalins, physagulin, withangulatin, withanolide and physangulide, physagulmes, witaminimine, physagulines, and sitosterol. They further contain several vitasteroids, e.g. vamonolide. The pyrrolidine alkaloid phygrine (bis-hygrine) was isolated from the roots and aerial parts. (Rengifo and Arana 2013, Ogundajo *et al.* 2015).

Medicinal Properties: *P. angulata* is known to have a broad spectrum of biological activities, including antibacterial, molluscicidal, antiprotozoal, anticancer, cytotoxic, diuretic, anti-inflammatory, antispasmodic and immunomodulatory activities. The plant is diuretic, expectorant, febrifuge. The leaves are analgesic, diuretic, parasiticide and relaxant.

The physalins B and F present in the plant inhibit the growth of several types of human leukaemia cells *in vitro* and physalin F shows cytotoxicity *in vitro* on five other human cancer cell lines. In addition, physalin F has an antitumour effect in mice *in vivo*. *In vitro* withangulatin A is found to be a topoisomerase II inhibitor and a cytotoxin. In tests on rats and mice, vitanolides (isolated from aerial parts) and hyperoside (quercetin-3-O-galactoside; from the leaves) confirmed to have anti-inflammatory activity. (Rengifo and Arana 2013, Porika *et al.* 2014, Susanti *et al.* 2015; http://tropical.theferns.info).

Medicinal Uses (Traditional/Folklore): Recent ethnopharmacological studies show that *P. angulata* is used in many parts of the world in traditional medicines to treat several diseases, such as cancer, asthma, hepatitis, diabetes, malaria, anaemia, postpartum infections, scabies, dermatitis and reducing fever. In India, leaf paste is used as an external application for wounds. The whole plant cooked is recommended in baths for inflammatory processes, such as rheumatism. The leaves have been used to treat stomach disorders and Bright's disease. The plant is used in the Pacific Islands to facilitate childbirth; to treat infertility in women and dengue fever. The seeds are cooked with *Phyllanthus amarus* seeds in a preparation given to women after childbirth. The leaves are eaten, or applied as an enema, in order to cure stomachache, colic, lithiasis and anuria. Added to palm wine, they are used to cure fever and to calm attacks of asthma, vomiting and diarrhoea. Sleeping sickness is treated with a mixture of leaves combined with those of *Anchomanes difformis*. The leaves are used externally to treat a wide range of skin ailments such as itch, smallpox pustules, whitlow lesions, infected scarification wounds and rheumatic pain, and to relieve muscular stiffness and pain. The leaves are also applied to Guinea worm sores, killing the worms and easing extraction. A lotion prepared from the leaves is applied to treat ophthalmia in children. Extracts of *Physalis angulata* have proven effectiveness against both human African sleeping sickness (*Trypanosoma brucei rhodesiense*) and Chagas' disease (*Trypanosoma cruzi*). They also have an inhibitory effect against several strains of *Neisseria gonorrhoeae* and molluscicidal properties (*Biomphalaria tenagophila*).

In Central and South America *P. angulata* is also widely used as a medicinal plant to treat malaria, toothache, liver ailments including hepatitis, rheumatism, and is considered a diuretic and relaxant. Plant infusions are taken to treat gonorrhoea, indigestion, nephritis and fever. In South-East Asia, an infusion of the aerial parts, including the fruits, is taken to cure digestive and intestinal problems, and is externally applied to treat various skin problems such as sores, boils and cuts. In Papua New Guinea, a leaf decoction is taken to treat constipation. The sap of the leaves mixed in water is taken as an abortifacient, although the use of the leaves to treat sterility is also mentioned.

In Peru, it has been documented that native groups in the Peruvian Amazon, use the decoction of leaves and fruit for "Tertian" due to postpartum infections and maceration of aerial part for the treatment of malaria. Also, mestizo populations use it for treatment of diabetes, by taking a glass of macerated root combined with honey, twice daily for 60 days. The root infusion is taken for hepatitis; the leaf infusion is used as a diuretic, for asthma, malaria, inflammation and as a disinfectant; the unripe fruit is used to treat scabies. In Brazil, the sap is used for earache, the roots boiled with the genus *Bixa* and Euterpe used for jaundice. In the state of Paraíba, people use the leaf infusion as a sedative and against inflammations of bladder, spleen and kidney; in the case of inflammations a tea is taken until symptoms disappear and as a sedative it is consumed at night. In Marudá community, state of Pará, root is taken as a tea for hepatitis symptoms, anaemia, urine infection, stomachache, prostate and kidney stones. In Bolivia, the indigenous community of Tacana, the root decoction is used to treat fever. In Nigeria, the traditional use reported is broad, wherein all parts of the plant have been used for medicinal purposes; entire plant is used for childbirth, diuretic, fever, gonorrhoea, jaundice, liver diseases, malaria, nephritis, postpartum haemorrhage, rashes, skin sores, sleeping sickness, to prevent abortion and tumours. The fruits are recommended for infection, infertility, inflammation, postpartum infection, and skin diseases. The leaves are also used for asthma, dermatitis, diuretic, earache, fever, gonorrhoea, haemorrhage, hepatitis, infections, inflammation, liver disorders, malaria, postpartum infection, rheumatism, and skin diseases and to prevent abortion and worms (schistosomiasis). The root is used for treating diabetes, earache, fever, hepatitis, jaundice, liver disorders, malaria and rheumatism. In Kenya, the infusion of the whole plant is used for worms and stomach pain. In Samoa, the leaves are used as antibacterial. In the Kingdom of Tonga, the crushed leaves are topically applied for skin inflammation. In Indonesia, root decoction is a remedy for postpartum, muscle aches and hepatitis. (Rengifo and Arana 2013, Susanti *et al.* 2015; http://tropical.theferns.info).

36.4 *Physalis peruviana* L.

(Syn. *Physalis latifolia* Lam.; *Physalis tomentosa* Medik.)

Habit

Common Name: Cape Gooseberry, Peruvian Ground Cherry, Poha Berry.

Vernacular Name: *Hindi:* Rasbhari (रसभरी); *Sanskrit:* Kuntali (कुंतली), Tankari (टांकरी), Tankasi (तनकासी).

Habit: Large herb.

Habitat: Cultivated for its edible fruit and sometimes found in wild in moist waste places.

Distribution: *P. peruviana* is native to tropical America, ranging from Chile to Colombia and Venezuela. It is often said to have originated from Brazil but records from there are not abundant. In India, it is introduced and now found throughout.

Flowering and Fruiting Time: November-February.

Taxonomic Description: An erect herb, 60-90 cm high. Stem weak, sparingly branched, densely hairy. Leaves alternate, 5-14 × 5-10 cm, broadly ovate, densely pubescent, base cordate, margin irregularly sinuately lobed or toothed, apex acuminate, petiole 2-5 cm long. Inflorescence solitary, axillary. Flowers yellowish, spotted deep brown at throat, pedicel 1-1.5 cm long, slender. Calyx campanulate, accrescent, 5-lobed, much enlarged in fruit, pubescent. Corolla funnel-shaped, gamopetalous. Stamens-5, anthers 3-4 mm long. Ovary 2-celled, many ovules, stigma bi-lobed. Berries globose, 1.5-2 cm across, yellow when ripe, enclosed in balloon like fruiting calyx; fruiting calyx 4-5 cm long, pubescent. Seeds numerous, discoid, yellowish-brown, compressed.

Chemical Constituents: The plant contains alkaloids, ticloidine, phenolics, phytosterols, phygrine, kaempferol, quercetin di and tri-glycosides, withanolides and flavonoids; also present physalins (B, D, F) and glycoside, myrectin -3-O-neohesperidoside and anticancer compound 28-hydroxywithanolide. (Zhang *et al.* 2013).

A leaf with axillary flower

Inflated calyx

Flower

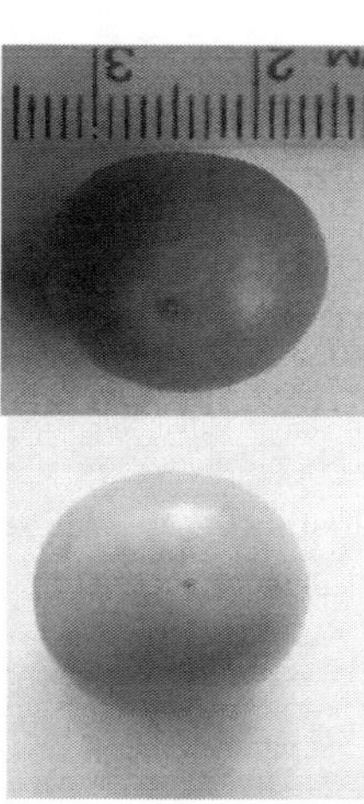
Unripe & ripe berry

Medicinal Properties: Whole plant is diuretic and vermifuge; leaves are anthelmintic. Plant is used in folk medicines as anticancer, antidiabetic, antioxidant, anti-inflammatory, hepatoprotective, antimycobacterial, antileukaemic, antipyretic, diuretic, immunomodulatory and for treating diseases such as malaria, asthma, hepatitis, dermatitis, and rheumatism. (Wu *et al.* 2009; http://medplants.blogspot.in).

Medicinal Uses (Traditional/Folklore): In traditional medicines the plant is used as a diuretic and anthelmintic and in treating abdominal disorders, cancer, malaria, asthma, hepatitis, dermatitis and rheumatism. In India, the plant is used by the Muthuvan and Tamilian tribes for the treatment of jaundice. It is used as a traditional medicine in South Africa for the cure of various diseases. It is used in diarrhoea management in Kenya. In Colombian and Peruvian traditional medicine, the fruit juice is used to treat pterygium, the growth of tissue over the cornea, may be related to its inhibiting fibroblast growth. Among 46 plant anthelminthic used in Madagascar, *P. peruviana* was the most common and found most effective. It as a popular folk medicine in Taiwan, used for treating cancer, leukaemia, hepatitis, rheumatism and leaf infusion in abdominal disorders. Zulus use an Infusion of the leaf as an enema for children with abdominal upsets and leaf juice for worms and bowel complaints. Heated leaf is used by Europeans to draw inflammations and as a poultice in Taiwan.

Some of the important medicinal uses of *P. peruviana* for curing various ailments are given below:

Anti-inflammatory: Due to the presence of polyphenols and antioxidants, the golden berry are used for many medical uses such as antiasthmatic, antispasmodic, antiseptic, antihelmenthic, anti-inflammatory, purgative, laxative and good for stomach related disorders.

Hypertension control: Rasbhari controls high blood pressure due to the presence of phyto-chemicals such as polyphenols and carotenoids and suitable percentage of potassium, which act like as high blood pressure reliever. These bio-chemicals compounds reduce cholesterol too.

Lung cancer: The fruits have appropriate percentage of polyphenols and carotenoids, which are beneficial against lung cancer cure as shown by recent research studies. The anti-hepatoma properties of the fruit are also being related to treat cancer. In folk culture, *P. peruviana* is being prescribed against cancer and leukaemia. It is still prevalent in the folk medicine of Kerala, India.

Neurological problems: Physalis peruviana has the power to control the movement of free radicals thus helps in prevention of many diseases including neurological related disorders.

Good vision: Rasbhari is rich in vitamin A and provides 14% of daily requirements of vitamin A. Vitamin A is good for eyes, prevent cataracts, and protect muscular degeneration.

Rich in nutrition: Fruits are a good source of carotene and ascorbic acid; eaten as a table fruit.

Strengthens bones: The fruit has a good proportion of pectin, which helps in calcium as well as phosphorous absorption, and thus good for making the bones stronger. It is also used in treating of rheumatism and dermatitis.

Boost immunity: The fruit contains vitamin C and fulfils the 18% of daily requirements. Having a profuse amount of vitamin C, it helps in enhancing immunity of the body.

Healthy liver: In a research, it has been confirmed that Rasbhari regulates the level of cholesterol and it is good for the health of the liver.

Atherosclerosis: The juice of the fruit contains some important phytochemicals that are good for smooth functioning of the heart.

Side effects of Rasbhari: Unripe fruit shouldn't be taken as it contains lighter form of poisonous alkaloids, which may cause allergy. (Wu *et al.* 2009; http://medplants.blogspot.in, http://tropical.theferns.info, http://www.gyanunlimited.com/health/amazing-health-benefits-and-medicinal-uses-of-rasbhari-physalis-peruviana/9353).

36.5 *Solanum americanum* Mill.

(Syn. *Solanum nigrum* L.)

Habit

Common Name: Black Nightshade.

Vernacular Name: *Hindi:* Makoy (मकोय); *Sanskrit:* Kakamachi (काकमाची), Kakamaci (काकमसि).

Habit: Annual herb.

Habitat: Roadsides, cultivated and fallow fields.

Distribution: Native to America. The distribution encompasses the tropics and subtropics of the Americas, Melanesia, New Guinea, and Australia. The plant is widely naturalized around the tropical Pacific and Indian Oceans. As a weed, found throughout in dry parts of India.

Flowering and Fruiting Time: Almost throughout year.

Taxonomic Description: An erect, nearly glabrous, annual herb, 30-80 cm high. Stem branched, angular. Leaves 3-8 x 2.8-7 cm, ovate-oblong, sub-entire or irregular toothed, acute at apex, petiole up to 3 cm long. Inflorescence extra-axillary sub-umbellate cymes, drooping. Peduncles slender nearly 1.5 cm long. Calyx lobes-5, about 2 mm long, ovate. Corolla lobes-5, 4-5 mm long; lobes ovate-lanceolate, white, sometimes bluish. Stamens-5, filaments hairy; anthers yellow, oblong. Ovary globose, stigma capitate. Berries 6-8 mm in diameter with persistent calyx, globose, glabrous, green, turning black or red when ripe. Seeds numerous, yellowish-brown.

Flowering twig

Flowers

Berries

Chemical Constituents: Phytochemicals present are alkaloids, flavonoids, hydrocyanic acid, phenols, phytic acid and tannins; other chemical constituents commonly found are glycoalkaloids, glycoproteins, polysaccharides, polyphenolic compounds such as gallic acid, catechin, protocatechuic acid, caffeic acid, epicatechin and rutin. The organic acids recorded are acetic acid, tartaric acid, malic acid and citric acid. In addition to solanine the plant also contains an unidentified alkaloid, saponin and betaine. The leaves and berries, especially when unripe, contain the alkaloid solanine. (Atanu *et al.* 2011, Chauhan *et al.* 2012).

Medicinal Properties: The plant is bitter, acrid, emollient, mildly thermogenic, antiseptic, anti-inflammatory, antispasmodic, expectorant, anodyne, vulnerary, digestive, laxative, diuretic, cardio tonic, depurative, diaphoretic and febrifuge, hydragogue, rejuvenating, sedative, alterant and tonic. Berries are laxative, aphrodisiac, diuretic, antidiarrhoeal, and antipyretic. Berries and flowers, prescribed in cough and cold, are strong sudorific, analgesic and sedative with powerful narcotic properties, used topically for the treatment of herpes zoster, dysentery, stomach complaints and fever. Plant extracts are cathartic, diuretic, and alterative. (Warrier *et al.* 2010; http://medplants.blogspot.in).

Medicinal Uses (Traditional/Folklore): *S. americanum* is used extensively in traditional and folk medicines as it possesses numerous medicinal properties. It is useful in rheumatalgia, swellings, cough, asthma, bronchitis, wounds, ulcers, flatulence, dyspepsia, hepatomegaly, otalgia, hiccough, nasal catarrh, ophthalmopathy, vomiting, cardiopathy, cardiac oedema, leprosy, skin diseases, splenomegaly, haemorrhoids, hoarseness, nephropathy, dropsy, general debility, chronic fevers, gout and rheumatoid arthritis, used in piles, liver troubles, leprosy and dysentery.

Fresh plant extract is used for sore throats, coughs, digestive problems, diarrhoea and dysentery, dermatitis, heavy female discharge, inflammatory swellings, enlargement of liver and spleen and in cirrhosis of liver. The stem, leaves and roots are used as a decoction for wounds, tumours and cancerous growths, sores and as an astringent. They are also used as a condiment, stimulant, tonic, abdominal pain, inflammation of bladder, eye ailments, itch, psoriasis, skin diseases (scabies, eczema and psoriasis), relief of cramps, rheumatism, neuralgia and expulsion of excess fluids. Whole plant is decocted for abscesses, cancer of the cervix, leucorrhoea and open sores. It is also used as a diuretic and febrifuge. A decoction of the plant depresses the central nervous system and reflexes of the spinal cord and has influence on cardiac activity and in the regulation of blood pressure. The leaves are used as poultice for rheumatic and gouty joints and skin diseases.

Unripe fruits are bitter, applied to aching teeth and squeezed on baby's gums to ease teething pain. The fruits are used as a tonic, laxative, appetite stimulant and also for treating asthma, jaundice, heart disease, liver problem, stomachache, diluted infusion of berries useful in the treatment of blindness, conjunctivitis, glaucoma, trachoma and cataract, diarrhoea, fever and eye ailments. Branches and fruits applied as a vulnerary and anti-inflammatory agent; an infusion is taken as a tonic. A decoction of berries and flowers is useful in cough, erysipelas, rat bite, bronchitis, pulmonary, tuberculosis, fever, diarrhoea, ophthalmopathy and hydrophobia. The leaves and berries are especially important as a cure for gastrohelcosis. Extract of berries, leaves and stems is used for the treatment of skin diseases. The seeds are useful in giddiness, dipsia, inflammations and skin diseases. The roots are used as an expectorant. The root bark is useful in otopathy, ophthalmopathy, rhinopathy and hepatitis.

Traditionally, the local and tribal people of the Uttarakhand use the ointment 'Gewai saag' prepared using the leaves and stem as the pain relieving agent. Garhwalis in Uttarakhand and ethnic communities of Sagar district (M.P.) use warmedup leaf to treat inflammation of testicles. They also use leaves for stomachache, female ailments and liver disorders; leaf poultice is applied to rheumatic joints, gout, rheumatic swellings and skin diseases; raw leaves are eaten to cure mouth and nose blisters. Leaf decoction is diuretic, laxative, used for swellings on the body; leaves and fruits pounded and the extract used for tonsillitis; leaves cooked/fried and eaten to cure jaundice and cough. Fresh leaves (125 g) are boiled in 3–4 cups (750–1,000 ml) of water daily for 10–15 min. When 1 cup (250 ml) of water remains, then it is filtered with a piece of cloth and the filtrate is used to treat swelling, skin diseases, inflamed and painful parts of body and to clean wounds and mouth sores. For mouth sores, half a cup (125 ml) of decoction (at one time) is applied 3–4 times per day for 6–7 days. In W. Bengal the berries are used for fever, diarrhoea, eye diseases and hydrophobia. The unripe fruits are ground into a paste and applied on ringworm.

Europeans in Africa used the plant to treat convulsions. It is used by the Africans for treating headache, ulcers and as a sedative. Young shoots are consumed as virility tonic for men and to treat dysmenorrhoea in females. In Indochina, the leaves are used as purgative and high blood pressure lowering agent while the fruits are used as laxative. Plant extract is given in liver trouble, piles and dysentery. The plant is used to treat tuberculosis, and leaves used to treat mouth ulcers. Freshly prepared leaf decoction, about 50-60 ml, mixed with rock salt and cumin seeds is given to the patient of spelenomegaly. The juice of the plant is used on ulcers and other skin diseases. In the Philippines, the fruit is a reputed cure for diabetes.

In Africa and Jamaica, the leaves have a reputation as a local anodyne for inflammation. The plant is also employed for treatment of fevers of various kinds. A paste of the green berries is used by the Zulus for ringworm. Leaves are used against asthma for which the leaves are dried in shadow and green tea prepared using dry leaves and taken 3-4 cups in a day. Extract of leaves and stems is used on piles, gonorrhoea, dropsy and enlargements of the liver and spleen. A decoction of the plant is used as fomentation for sore eyes and various skin diseases. In Zimbabwe (Rhodesia), the plant is used in the treatment of malaria, black-water fever and dysentery and in Mexico the fruit is a popular remedy for erysipelas. In Mauritius, a poultice of the plant is applied for relief of abdominal pain and inflammation of the bladder. In Europe, it is used in the treatment of headaches, ulcers, wounds and as a diuretic and emetic. Italians use it as an antispasmodic, diaphoretic, emollient and sedative. In Africa, roots are boiled in milk and given to children as tonic. Stem of *Vitis vinifera* mixed with *S. americanum* and *Cestrum parqui* and prepared into a paste which is applied to treat inflammation.

As a veterinary medicine, root mixed with pepper and ginger is fed to cows to reduce gas formation in the stomach. It has also been used as an agricultural insecticide.

Green and immature berries are toxic as they contain high concentration of glycoalkaloids. All kinds of animals can be poisoned after ingesting unripe berries of nightshade, including cattle, sheep, poultry and swine. Children have been poisoned after ingesting unripe berries in Africa. (Kaul and Dwivedi 2010, Shukla *et al.* 2010, Dangwal and Sharma 2011, Padalia 2015; http://medplants.blogspot.in/, http://medicinalplants.us/tag/febrifuge).

36.6 *Solanum surattense* Burm. f.

(Syn. *Solanum jacquini* Willd.)

Common Name: Thorny Nightshade, Yellow Berried Nightshade.

Vernacular Name: *Hindi:* Kateli (कटेली), Oonth Kateli (ऊंट कटेली); *Sanskrit:* Kantakari (कंटकारी), Vyaghri (व्याघ्री).

Habit: A prostrate or spreading herb armed with sharp prickles.

Habitat: Dry waste places, roadsides and along railway lines.

Distribution: Distributed throughout the tropical and subtropical regions of India, South East Asia, Malaysia, and Australia; in India, found growing in dry places as a weed on roadsides and in wastelands.

Habit

Flowering and Fruiting Time: December-June.

Flowering twig

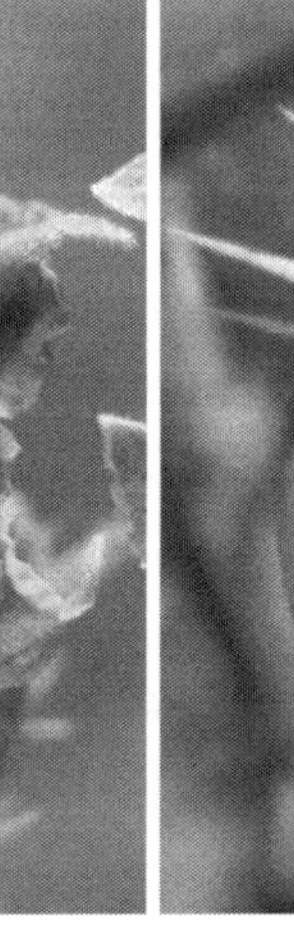

Flower

Berry

Taxonomic Description: A prickly diffuse herb. Stem prostrate, armed with 1.5 cm long sharp yellow prickles, woody at base. Leaves up to 12 cm long, ovate or elliptic, sub-pinnatified, irregularly lobed, dark green above, armed with numerous long prickles on both sides. Flowers bluish-purple, borne in few-flowered, extra axillary, peduncled cymes. Calyx lobes-5, lanceolate, prickly. Corolla 5-lobed; lobes ovate, triangular, pubescent outside, persistent. Stamens-5, anthers bright yellow, 7 mm long. Berries globose, 1.5-2 cm in diameter, variegated with green and white when young, yellow when ripe. Seeds smooth, numerous.

Chemical Constituents: *S. surattense* reported to have a number of alkaloids, sterols, saponins and flavonoids and their glycosides and especially it has high concentration of solasodine. Some of the important phytochemicals present in the plant are β-carotene, disgenin, carpesterol, solasodine, solamargine, β-solamargine, solanine, solasodino-l-rhamnosy- β-glucoside (solasurine), solanocarpine (solanine-S), tomatidienol, etc. Plant has high nutritional value and is recommended as a cheap source of plant protein, energy and mineral element such as calcium, sodium, magnesium and phosphorus. (Mali and Harsh 2015).

Medicinal Properties: It is bitter, acrid, thermogenic, anthelmintic, anti-inflammatory, anodyne, digestive, carminative, appetizer, stomachic, depurative, sudorific, febrifuge, expectorant, laxative, stimulant, diuretic, rejuvenating, emmenagogue and aphrodisiac. The stem, flowers and fruits are carminative. (Warrier *et al.* 2010).

Medicinal Uses (Traditional/Folklore): Due to its varied medicinal properties the plant is useful in helminthiasis, dental caries, inflammation, arthralgia, flatulence, colic, constipation, dyspepsia, anorexia, leprosy, skin diseases, hypertension, rheumatoid arthritis, fever, cough, asthma, bronchitis, pharyngitis, hiccough, strangury, urolithiasis, amenorrhoea, dysmenorrhoea, lumbago, haemorrhoids, cardiac disorders, rhinopathy, epilepsy and catarrh. It is also used in the treatment of dropsy, pain in chest, bronchitis, lumbago, piles, thirst, bladder stones, dropsy, heart diseases, rheumatism, gonorrhoea, pains, sore throat, toothache, watery eyes and snakebite.

In folk medicines, fresh leaf extract is applied on ringworm and other skin diseases and poured into the ear to remove the insect. The juice extract of Kantakari along with honey is used in treating dysuria (difficulty in urination). Because of its hot potency, pungent taste and *Kapha-Vatahara* property, it naturally improves digestion strength and is widely used in respiratory disorders. The fruit is tied as garland to get relieve from jaundice. The dried root is used in cough, asthma, pain in chest and fevers. The roots are used to increase fertility in women. Dried fruit smoke is effective in reducing toothache. Paste of fruits mixed with pepper and cow ghee is given for seven days as a remedy for tuberculosis. Fruits are used to cure jaundice. Powder of ripe fruits is given orally for treating piles. A paste of the root with lemon juice is applied to the affected areas of snake and scorpion bites. (Katewa and Galav 2005, Singh *et al.* 2010d, Chaudhary and Gupta 2011, Panda *et al.* 2011, Kumar 2014, Shaheen *et al.* 2014).

36.7 *Solanum torvum* Sw.

(Syn. *Solanum ficifolium* Ortega)

Common Name: Turkey Berry, Susumber.

Vernacular Name: *Hindi:* Bhurat (भूरट), Bhankatiya (भंक्तिया); *Sanskrit:* Brihati (बृहती).

Habit: A sparsely prickly shrub.

Habitat: Found growing along the canal banks, roadsides and wastelands.

Distribution: Although *S. torvum* originated in the West Indies, it has spread as a useful plant and consequently as a weed throughout the tropics and subtropics. It is distributed widely in Pakistan, India, Malaya, and China, Philippines, and tropical America.

Flowering and Fruiting Time: August-November.

Habit

Taxonomic Description: A stellatelly-tomentose shrub with sparsely prickly branchlets, 1.5-3 m tall. Leaves up to 20 cm long, ovate-elliptic, entire-sinute or pinnatilobed, oblique at base, obtuse or acute at tip, woolly tomentose on both sides, usually without prickles. Flowers white in extra-axillary, many flowered corymbose cymes, pedicels hairy. Calyx 5-toothed, without prickles. Corolla 2-2.5 cm across, rotate; lobes-5, lanceolate. Stamens-5, filaments very short; anthers oblong, yellow. Berries globose, 1-1.5 cm across, yellow when ripe. Seeds numerous, discoid.

Chemical Constituents: Phytochemical studies indicated that fruits of this species have good concentrations of various alkaloids, flavonoids, saponins, tannins, and glycosides. Antiviral isoflavonoid sulfate and steroidal glycosides have been isolated from the fruits. Aerial parts of *S. torvum* are rich sources of steroid and saponins. The root is a rich source of steroidal glycosides such as torvosides A-G. (Jaiswal 2012, Yousafa *et al.* 2013).

Abaxial side of leaf

Flower

Medicinal Properties: *S. torvum* is known to possess antiviral, antiplatelet aggregation, antiulcerogenic, antioxidant, antimicrobial, analgesic and anti-inflammatory, and cytotoxic properties. It has both a sedative and a diuretic therapeutic effect. The leaves are used as a haemostatic. (Jaiswal 2012, Yousafa *et al.* 2013; http://medplants.blogspot.in, http://tropical.theferns.info).

Medicinal Uses (Traditional/Folklore): Traditional medicinal uses of *S. torvum* have been highlighted in the Ayurveda and Chinese pharmacopeia. In India, fruits are not only used for nutritive purposes but also fruit decoctions are effectively used for cough ailments and are considered to be effective in treating liver and spleen enlargement. The ripened fruits are used in the preparation of tonic and haemopoietin agents and also for the treatment for pain. Khasi traditional healers and village folks in Meghalaya use the smoke emanated when a few drops of heated mustard oil is poured on dried seeds/fruits on red hot knife. The seeds/fruits get burnt in mustard oil due to the heat from the hot knife and fumes appear which are inhaled in the mouth to prevent dental caries or to get relieve from toothaches. For relief from colds and coughs dry leaf powder (after drying in shade) is mixed with hot water or cow milk and administrated orally in Puducherry, Karaikal, Mahe and Yanam, India. To cure cracked foot the leaf paste is applied externally on cracks. In Malaysia, a paste of roots is used to soothe cracked feet. Root extract and powdered fried fruits are used by Kurichyas in Kannur district, Kerala, India to get relief from cough. To reduce body heat, leaf juice is taken orally. Leaf extract in Kancheepuram, Tamil Nadu, India and cooked fruit are used as vermifuge in Tirunelveli, Tamil Nadu, India.

This herb is commonly used in China and considered effective in the treatment of blood stasis, menstruation, edema pain and coughs. Combination of root and leaf juice Root and leaves are used for asthma, diabetes, and hypertension by Garo tribe in Bangladesh. Root juice and roots are used to treat liver diseases, tuberculosis, and as antianaemic in Northeastern Brazil. Cubans apply the leaf juice to pimples and drink root decoction 2–3×/day for the cure of cystitis and urethritis. Dominicans use young leaf infusion for treating infantile enterosis and use root infusion for gonorrhoea and poultice of leaf and root decoction to bleeding, burns, rashes, sores, and wounds. Haitians massage crushed leaves and seed to treat fever and leaf decoction as sedative and antispasmodic for gastralgia. Jamaicans drink leaf decoction for treating colds. In Nepal, flower juice with salt water is used as eye drops and poultice of ripe fruits onto the forehead for headache as well as plant juice for asthma, chest, cough, dropsy, gonorrhoea, and rheumatism; root juice is used for vomiting due to weakness. In the Philippines, a decoction of roots is taken to counteract poisoning and to curb bleeding after childbirth. (Yousafa *et al.* 2013; http://medplants.blogspot.in, http://tropical.theferns.info).

36.8 *Withania somnifera* (L.) Dunal

(Syn. *Physalis somnifera* L.)

Habit

Common name: Winter Cherry, Indian Ginseng.

Vernacular name: *Hindi:* Aswagandha (अश्वगंधा), Asgandh (असगंध); *Sanskrit:* Aswagandha (अश्वगंधा), Ashwagandhi (अश्वगंधी).

Habit: Shrub.

Habitat: Common in dry places, along waysides.

Distribution: *W. somnifera* grows abundantly in India, Pakistan, Bangladesh, Sri Lanka, southern China and parts of northern Africa; spread from the Mediterranean region to throughout tropical region of Africa to South Africa and also from the Cape Verde Islands and Canary region to the Arabia and Middle East. In India, it grows in dry parts in sub-tropical regions of Rajasthan, Punjab, Haryana, Uttar Pradesh, Gujarat, Maharashtra and Madhya Pradesh, which are the major Ashwagandha producing states of the country. It grows wild in the Kashmir valley also.

Flowering and Fruiting Time: December-May.

Fruiting twig | **Flowers** | **Berry inside bladder like calyx**

Taxonomic Description: An erect shrub, up to 1.5 m high. Stem and branches densely stellate tomentose. Leaves petioled, up to 13 cm long, elliptic-ovate, lanceolate, entire, unequal at base, sub-acute to obtuse at tip. Flowers greenish-yellow, sub-sessile, borne in 3-5 flowered or more, axillary fascicled. Calyx (in flower) 4 mm long, accrescent, stellate tomentose, campanulate, 5-toothed. Corolla deeply 5-lobed; lobes lanceolate or triangular, tomentose outside. Stamens-5, attached at the base of corolla tube. Ovary 2-celled, stigma-bifid. Berries globose, 6-8 mm in diameter, bright red when ripe, enclosed in the much enlarged inflated bladder like calyx. Seeds numerous, sub-reniform, yellowish-brown.

Chemical Constituents: Indian ginseng's pharmacological activity has been attributed to two main withanolides, withaferin A and withanolide D. Withaferin-A is therapeutically active; withanolide reported to be present in leaves. Other phytochemicals present in the plant include alkaloids, such as choline, tropanol, pseudotopanol, cuscokygrene, 3- tigioyloxytropana, isopelletierine and several other steroidal compounds.

Aswagandha roots contain alkaloids, starch, reducing sugar, hentriacontane, glycosides, dulcital, withaniol acid and a neutral compound. The pharmacological activity of the root is attributed to the alkaloids and steroidals lactones. From the roots, 12 alkaloids, 35 withanolides and several sitoindosides have been isolated. A sitoindoside is a biologically active constituent known as withanolide containing a glucose molecule at carbon. In addition to alkaloids, the roots are reported to contain starch, reducing sugars, glycosides, dulcitol, withancil, an acid and a neutral compound. The amino acids reported from the roots include aspartic acid, glycine, tyrosine, proline, tryptophan, alanine, glutamic acid and cysteine. Leaves contain 12 withanolides, alkaloids, glycosides, glucose and free amino acids. Berries contain a milk coagulating enzyme, two esterases, free amino acids, fatty oil, essential oil and alkaloids. (Gupta and Rana 2007, Mirjalili *et al.* 2009, Supe *et al.* 2011, Umadevi *et al.* 2012, Bano *et al.* 2015, Datta *et al.* 2015).

Medicinal Properties: *W. somnifera* possesses a number of therapeutic actions which include anti-inflammatory, sedative, hypnotic, narcotic, general tonic, diuretic, aphrodisiac, alterative, deobstruent, uterine tonic and increase in production of semen. The roots are alterative, aphrodisiac, deobstruent, diuretic, narcotic, sedative and restorative in nature.

Ashwagandha is considered to be one of the best rejuvenating agents in Ayurveda. A famous Ayurvedic rejuvenative botanical used in many tonics and formulas, Ashwagandha is the best rejuvenative that helps maintain proper nourishment of the tissues, particularly muscle and bones, while supporting the proper function of the adrenals and reproductive system. Ashwagandha has anti-inflammatory, antitumour, antistress, antioxidant, mind-boosting, immune-enhancing, and rejuvenating properties. The root is a tonic, alterative, aphrodisiac and used in consumption, emaciation, debility due to old age, rheumatism, nervine, and adaptogen, etc. It has also narcotic, diuretic and deobstruent properties. Ashwagandha root has also been noted to have sex-enhancing properties and it is mentioned in the ancient Kama Sutra as an herb to be used for heightening sexual experience. Ashwagandha has the ability to restore sexual health and improve overall vitality while promoting a calm state of mind. The whole plant, but especially the leaves and the root bark, are abortifacient, adaptogen, antibiotic, aphrodisiac, deobstruent, diuretic, narcotic, strongly sedative and tonic. The plant has also been used as a liver tonic, anti-inflammatory agent, and more recently to treat asthma, ulcers, insomnia, and senile dementia. Clinical trials and animal research support the use of Ashwagandha for anxiety, cognitive and neurological disorders, inflammation, and Parkinson's disease. Incorporation of Ashwagandha in the diet may prevent or decrease the growth of tumours in human. Pharmacological studies and research so far have indicated that Ashwagandha has antitumour, antistress, antioxidant boosting, haemopoeitic and rejuvenating properties. It is also an exceptional nerve tonic and nourishes the nerves and improves nerve function to maintain calm during stressful conditions. It also nourishes crucial mind and body connection and psychological immune response (Mishra *et al.* 2000, Gupta and Rana 2007, Mirjalili *et al.* 2009, Supe *et al.* 2011, Qamaruddin *et al.* 2012, Umadevi *et al.* 2012, Rahul 2013, Bano *et al.* 2015, Datta *et al.* 2015; http://medplants.blogspot.in, http://tropical.theferns.info).

Medicinal Uses (Traditional/Folklore): Ashwagandha (*W. somnifera*), also known as Indian ginseng, and as Indian Winter Cherry is an important ancient herb in indigenous medical system for over 3,000 years; roots, seeds and leaves are used in Indian traditional systems of medicines, Ayurveda and Unani. Ashwagandha is among the most highly regarded herbal tonifiers in Ayurvedic herbalism. Since ancient times it has been considered a nervine tonic, alterative, aphrodisiac, deobstruent (having the property of removing obstruction in any system of the human body) and a sedative. It has been used in rheumatism, consumption and in debility. Ashwagandha can be used by persons of both sexes, of all ages and at all stages of their lives. In elders it provides energy, relieves inflammations, pains and aches of the back, hand and feet, and in the generative system, nervous debility and diseases due to *Vata*.

Ashwagandha is one of the most widespread tranquillizers used in India. It acts mainly on the reproductive and nervous systems, having a rejuvenative effect on the body, and is used to improve vitality and aid recovery after chronic illness. Internally, it is used to tone the uterus after a miscarriage and also in treating postpartum difficulties. It is also used to treat nervous exhaustion, debility, insomnia, wasting diseases, failure to thrive in children, impotence, infertility, multiple sclerosis, etc. It is considered a *Rasayana*, for strength, vigour and for rejuvenation. It is being prescribed as an anabolic agent, as an adaptogen and analgesic for the treatment of various arthropathies, certain forms of hypertension, insanity, etc. It imparts resistance to infection and reduces stress, has anti-inflammatory effect, boosts sexual energy in men. It stimulates sexual impulses and increases sperm counts.

The easiest way to use the root is to take one teaspoonful of the fine powder with sweetened milk, or mix the powder with ghee, lick the mixture followed by milk sweetened with sugar candy. In gynaecological practice it helps in sterility, leucorrhoea and inflammation of the vagina. It also helps in breast development. For these purposes, one teaspoonful of Ashwagandha powder is taken with half a teaspoonful of ghee, and honey before breakfast in the morning and before sleeping in the evening, followed by cold sweet milk. It must be taken for three months, particularly during winter. Externally, it has been applied as a poultice to boils, swellings and other painful parts. Leaves are applied for carbuncles, inflammation and swellings. Leaf juice is useful in conjunctivitis. Bark decoction is taken for asthma and applied locally to bed sores. The fruit is diuretic and seeds are diuretic and hypnotic.

Roots, harvested in the autumn and dried for later use, and leaves are used by many ethnic communities as a medicine in many diseases. In Ayurvedic system, powder of root is used for improvement of semen, chemopreventive efficacy against stomach and skin carcinogenesis and in many other diseases. In India, the bruised leaves and ground root are used as a local application to painful swellings, carbuncles and ulcers as the root and leaves are considered to be sedative and the root has been included in the Indian Pharmacopoeia and Codex for its narcotic and sedative properties. Ashwagandha root drug also finds an important place in treatment of rheumatic pain, inflammation of joints, nervous disorders and epilepsy. Dried roots are used as tonic for hiccup, cold, cough, female disorders, as a sedative, in case of senile debility, ulcers, etc. It helps in providing progressive, long lasting results for various health concerns like aging, anaemia and slow growth, arthritis, fatigue, waning memory, sports fitness and stress-disorders.

In Sind, Pakistan, it is used to cause abortion. The root is ground into a paste is applied on ulcers, carbuncles and painful swellings with beneficial results. The Zulus use an infusion of the root as an enema for gangrenous rectitis and the whole plant for treating syphilis. The green berries are rubbed on ringworm with beneficial effect. A decoction of the leaf is used externally and internally in the treatment of haemorrhoids. A paste of the leaf is applied on syphilitic sores. In Tanganyika, the root of the plant is used as a sexual stimulant and also as an abortifacient, while in southern and eastern Africa it is given for diarrhoea and proctitis and the leaf for nausea and rheumatism. In West African local medicine, both roots and leaves are used internally, and the freshly pounded leaves also externally, against fever, chills, rheumatism, colics, etc. The juice of the plant is said to be diuretic and emmenagogic. In local medicine in East Africa, the root is considered to have narcotic and antiepileptic actions. Some caution is advised in the use of this plant since it is toxic. (Gupta and Rana 2007, Mirjalili *et al.* 2009, Kaul and Dwivedi 2010, Supe *et al.* 2011, Qamaruddin *et al.* 2012, Umadevi *et al.* 2012, Al-shaimaa 2013, Shaheen *et al.* 2014, Datta *et al.* 2015, Gaurav *et al.* 2015; http://medplants.blogspot.in, http://tropical.theferns.info).

37. SCROPHULARIACEAE

Scrophulariaceae – Figwort family

The family Scrophulariaceae is one of the major families of Angiosperms with 76 plant genera and 1,576 accepted species. The plants are mostly annual and perennial herbs. The family is characterized by bisexual flowers with tubular corollas (fused petals) that are bilaterally symmetrical (two-lipped) and have four stamens in most, two of which are usually shorter than the other two. There were two species belonging to single genera recorded in the flora.

37.1 *Verbascum chinense* (L.) Santapau

(Syn. *Verbascum coromandelianum* (Vahl) Kuntze)

Common Name: Chinese Mullein.

Vernacular Name: *Hindi:* Kokhima (कोखिमा), Gadar-tambaku (गदर-तम्बाकू), Kalhar (कल्हार); *Sanskrit:* Bhutakeshi (भूतकेशि).

Habit: Annual or biennial herb.

Habitat: Roadsides, along the banks of canals and lakes.

Distribution: Asia - India, Pakistan, Afghanistan, Kazakhstan, Sri Lanka, Cambodia, China, Laos, Thailand. It has been introduced throughout the temperate world, and is established as a weed in Australia, New Zealand, tropical Asia, La Réunion, North America, Hawaii, Chile, Hispaniola and Argentina; also reported in Japan.

Flowering and Fruiting Time: July-October.

Taxonomic Description: An erect, glandular-hairy herb, 30-80 cm high. Leaves alternate, glandular-pubescent, irregularly dentate-serrate; radical leaves lyrate-pinnatified, petioled; cauline leaves smaller, sessile, ovate or ovate-oblong. Flowers nearly 1 cm across, arranged in simple or branched, long terminal racemes. Calyx 5-partite; lobes 4-5 mm long, ovate-lanceolate, glandular-pubescent. Corolla lobes-5, yellow. Stamens-4, filaments woolly. Capsules ovoid, glandular hairy, 6-8 mm long.

Inflorescence

Flower

Habit

Chemical Constituents: None known.

Medicinal Properties: None known.

Medicinal Uses (Traditional/Folklore): Tribal of Aurangabad district, Maharashtra use infusion of leaves internally for the treatment of gout. For reducing the swelling of joints, lukewarm leaf poultice is applied locally. Whereas Bhilla tribe in Maharashtra for treating piles use powder of leaves and seeds in a cupful of water, twice a day till the piles are cured. (Kamble *et al.* 2010, Mali and Bhadane 2011).

37.2 *Verbascum thapsus* L.

(Syn. *Verbascum simplex* Hoffmanns. & Link)

Common Name: Great Mullein, Common Mullein, Beggar's Blanket, Candlewick Plant, Adam's Flannel, Woolly Mullin.

Vernacular Name: *Hindi:* Jangli Tambaku (जंगली तम्बाकू); *Sanskrit:* Ekula Veer (एकुला वीर).

Habit: Biennial herb.

Habitat: Great Mullein is found growing along the roadsides, canal banks and near small water channels.

Distribution: Native to Europe, northern Africa, and temperate Asia as far as the Himalayas, and introduced in the Americas, Japan, Sri Lanka, Reunion, the Americas, Australia and New Zealand; it is also found in Gilgit, Afghanistan and Iran. In India, it is reported to occur in Kerala, Delhi, Uttar Pradesh and Assam.

Flowering and Fruiting Time: June-October.

Taxonomic Description: An erect, up to 1.5 m high herb, appressed with dense stellate hairs. Radical leaves rosulate, 10-40 cm long, oblanceolate, petioled; cauline leaves smaller, alternate, ovate-oblong, crenate or entire, base decurrent to stem wings. Flowers yellow, in dense, very long spicate racemes. Calyx 5-lobed, 6-7 mm long, tomentose; lobes ovate-lanceolate, persistent. Corolla 5-partite. Stamens-5; filaments of 3 short stamens hairy. Stigma capitate. Capsules 7-8 mm long, ovoid, 2-valved, numerous seeded.

Chemical Constituents: The plants contain great diversity of chemicals such polysaccharides; iridoid glycosides including harpagoside; harpagide and aucubin (especially in the leaf), flavonoids, including 3 methylguercitin, hesperedin and verbascoside; saponins and volatile oils and terpenoids, lipids, carbohydrates, minerals, etc. Some valuable constituents contained in Mullein are coumarin and hesperidin; they exhibit many healing abilities. Some valuable constituents contained in Mullein are coumarin and hesperidin, they exhibit many healing abilities. (Panchal *et al.* 2010, Riaz *et al.* 2013).

Flower

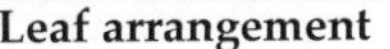

Leaf arrangement

Habit

Medicinal Properties: The whole plant possess slightly sedative, hypnotic and narcotic properties. The seeds are considered toxic. They have been historically used as a narcotic and also contain saponins. The leaves, root, and the flowers are anodyne, anti-inflammatory, antiseptic, antibacterial, antioxidant, antiviral, antispasmodic, astringent, anticancer, cardiodepressant, estrogenic, fungicidal, demulcent, diuretic, emollient, and expectorant, nervine, and vulnerary. The flowers and leaves are anodyne, analgesic, anti-inflammatory, antiseptic, antispasmodic, spasmolytic, astringent, demulcent, diuretic, emollient, expectorant and vulnerary.

Recent studies have found that great mullein contains glycyrrhizin compounds with bactericide and potential anti-tumour action and flower infusion is a strong antibacterial agent. These compounds are concentrated in the flowers. (Turker and Gurel 2005, Panchal *et al.* 2010, Riaz *et al.* 2013).

Medicinal Uses (Traditional/Folklore): Great Mullein, an herb with a long history of use in folk medicine, has been used as an alternative medicine for centuries and in many countries throughout the world. It has been employed for the treatment of asthma and other pulmonary complaints. Great mullein is a commonly used domestic herbal remedy, valued for its efficacy in the treatment of pectoral complaints. It acts by reducing the formation of mucus and stimulating the coughing up of phlegm, and is a specific treatment for tracheitis and bronchitis. Dioscorides, a Greek physician, pharmacologist, and botanist, first recommended the plant 2,000 years ago, against pulmonary diseases, and this has remained one of its primary uses, especially against cough. In olden times leaf decoctions or herbal teas were used for expectoration, consumption, dry cough, bronchitis, sore throat and haemorrhoids. Leaves were also smoked against pulmonary ailments, a tradition that in America was rapidly transmitted to Native American peoples. The Zuni people, however, use the plant in poultices of powdered root applied to sores, rashes and skin infections.

The plant has several traditional/ethnomedicinal uses. The whole plant possesses slightly sedative and narcotic properties. The combination of expectorant saponins and emollient mucilage makes the plant particularly effective for cough. All preparations are taken internally have to be finely filtered to eliminate the irritating hairs. Topical application of various *V. thapsus*-based preparations is recommended for the treatment of warts, boils, carbuncles, haemorrhoids, and chilblains, amongst others. An infusion is taken internally in the treatment of a wide range of chest complaints and also to treat diarrhoea and bleeding of the lungs and bowels. An infusion or tea of the plant is taken internally in the treatment of a wide range of chest and abdominal complaints including productive cough and diarrhoea. It is used as a tobacco substitute and rheumatic problems. Whole plant is used as a remedy in the treatment of tuberculosis in Ireland and Ear problems of cats and dogs in Canada. Whole plant is known to be active against snake bite venom in Uttar Pradesh, India.

A homeopathic remedy is made from the fresh leaves. It is used in the treatment of long-standing headaches accompanied with oppression of the ear. Topically, a poultice of the leaves is a good healer of wounds and is also applied to ulcers, tumours and piles. Ointments prepared from leaves are used for burns and earache. Leaves are used as antitussive in Albania and Italy. An ointment prepared from leaves is used for burns and earache in Eastern United States. Also, it is used for skin diseases, wound healing and gastrointestinal system ailments in sheep, mules, cattle and horses in Italy. The dried leaves are sometimes smoked to relieve the irritation of the respiratory mucus membranes, and the hacking cough of consumption. They can be employed with equal benefit when made into cigarettes, for asthma and spasmodic coughs in general. Leaves in powdered form are used effectively for the treatment of diarrhoea and dysentery in cattle and for healing wounds in Pakistan.

An infusion of flowers in olive oil is used as earache drops, or as a local application in the treatment of piles and other mucous membrane inflammations. Flower extract in hot water is used as dye to turn hair a golden color that can be changed to green dye on acidification or alkalization on to brown. Mullein oil is a very medicinal and valuable destroyer of disease germs. The oil is very effective in treating gum and mouth ulcers. Oil from the flowers was used against catarrhs, colics and, in Germany, for the treatment of piles, bruises and frostbites, earaches, eczema and other external conditions. Flowers are used as a respiratory tonic for ruminants in British Columbia, Canada and as an emollient against catarrh and coughs cataplasm applied externally to soothe gangrenous sores and topical inflammatory processes in humans and animals in Italy. Dried flower or root are used effectively as cigarettes for asthmatics by native Indians and Menominee in Michigan, USA.

The seeds, which contain saponins, are considered toxic. They have been historically used as a narcotic and aphrodisiac. A decoction of the seeds is used to soothe chilblains and chapped skin. The seeds are also used as a fish poison in different regions of Europe and Asia (Pakistan). A poultice made from the seeds and leaves is used to draw out splinters.

An infusion of the root is also used to treat athlete's foot. A decoction of the roots is used to alleviate toothache and also relieve cramps and convulsions. It is also used in alternative medicine for the treatment of migraine headaches accompanied with oppression of the ear. Infusion from root was used for dental treatment in Medieval England. In Canada, roots have been used for the treatment of symptoms of diabetes and its complications. The juice of the plant and powder made from the dried roots is said to quickly remove rough warts when rubbed on them.

An extract prepared from leaves and flowers is used for the treatment of skin diseases, cuts, wounds and swelling in Malakand, Pakistan, inflammatory ailments in respiratory tract and others in Mexico, pulmonary disease, fever and bleeding of lungs and bowel in Pakistan and as diuretic, expectorant and sedative in Turkey. (Vardhana 2008, Panchal *et al.* 2010, Riaz *et al.* 2013).

38. LINDERNIACEAE

Linderniaceae – False Pimpernel family

The family consists of about 16 genera and 164 species from worldwide, mainly in neo-tropics. The Most plants are small herbs with 4-angled stem, opposite leaves and zygomorphic flowers; lower stamens very different from upper ones being either 'Z' shaped, or long and curved with projections, or very much reduced. The family was represented by two species belonging to one genus in Chithara Village flora.

38.1 *Lindernia ciliata* (Colsm.) Pennell

(Syn. *Bonnaya brachiata* Link & Otto)

Habit

Common Name: Fringed False Pimperne.

Vernacular Name: None known.

Habit: Annual herb.

Habitat: Common along the banks of shallow water bodies, in rice fields, wet areas in gardens.

Distribution: It is restricted to the tropics and sub-tropics of Asia, Malesia, Polynesia, tropical Australia and North America. In India, it was found as an insignificant weed, mainly in rice fields

Flowering and Fruiting Time: August-October.

Taxonomic Description: A small, glabrous herb, 5-15 cm high. Stem 4-angled, branched. Leaves 1-3 x 0.3-0.8 cm, opposite, sessile, oblong, lanceolate, base amplexicaul, margin sharply spinous–serrate, apex obtuse or acute. Inflorescence terminal racemes. Flowers white or light purple, bracteate, shortly pedicelled. Calyx 4-5 mm long, deeply 5-lobed. Corolla 7 mm long, white, bi-lipped; lower lip unequally 3-lobed. Fertile stamens-2, posterior. Capsules cylindric, linear-oblong, much longer than persistent calyx. Seeds many, minute.

Flowering twig

Flower

Chemical Constituents: Three known compounds were isolated from the petroleum ether extract of the whole plant of *L. ciliata*. They are beta-sitosterol, stigmasterol and lup-20(29)-en-3 beta-ol. (Lan 1996).

Medicinal Properties: It is used as a remedy for gonorrhoea, jaundice, urinary disturbances, bronchitis, headache, liver complaints, spleen diseases, constipation, fever, loss of appetite, asthma, cough, skin diseases. (Praneetha *et al.* 2014).

Medicinal Uses (Traditional/Folklore): The sap from the crushed leaves is given after childbirth. In Taiwan, the plant is considered an effective cure for menorrhagia. It is traditionally used in the treatment of jaundice. Leaves are used as a remedy for gonorrhoea. (Kumar and Narain 2010, Praneetha *et al.* 2014; http://tropical.theferns.info).

38.2 *Lindernia crustacea* (L.) F.Muell.

(Syn. *Vandellia crustacea* (L.) Benth.)

Habit

Common Name: Brittle False Pimpernel, Malaysian False Pimpernel.

Vernacular Name: None known.

Habit: Annual herb.

Habitat: Common in marshy areas, shady places in gardens, rice fields

Distribution: The Brittle False Pimpernel distributed throughout the tropics. Found throughout India; up to 1,600 m in the Himalaya.

Flowering and Fruiting Time: August-October.

Taxonomic Description: An erect or decumbent-ascending, small, annual herb, 5-20 cm high. Stem sub-quadrangular, purplish-green. Leaves 1-2.5 x 0.5-1.2 cm, opposite, shortly petioled, ovate-lanceolate, base rounded or cuneate, margin shallowly crenate-serrate, apex sub-acute. Flowers axillary, usually in pairs, pedicel slender. Calyx 3.5-4.5 mm long, 5-lobed, lobes ovate, triangular, pubescent. Corolla purplish-white or bluish-white, 2-lipped, lower lip 3-lobed, middle lobe slightly larger. Stamens-4, didynamous. Capsules oblong-obovoid, nearly as long as persistent calyx. Seeds many, rugose, yellowish-brown

Flowering and fruiting twig

Flower

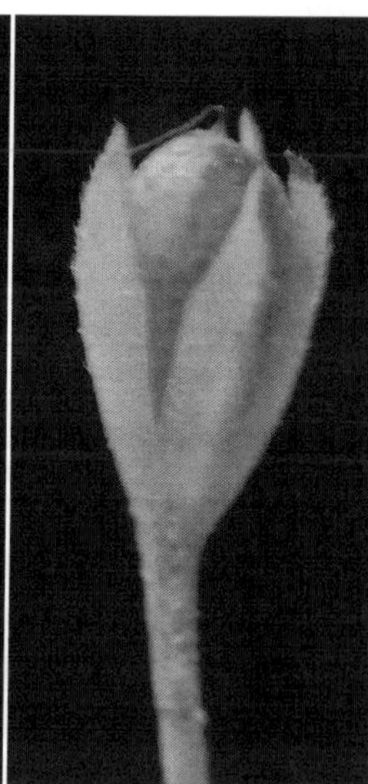

Capsule

Chemical Constituents: None known.

Medicinal Properties: The plant is considered to have antifungal, febrifuge, purgative, emetic and cathartic properties. (www.cabi.org/isc/datasheet/30864, http://tropical.theferns.info).

Medicinal Uses (Traditional/Folklore): Whole plant has given good results in treating diabetes, bilious disorders, dysentery, amenorrhoea, and hepatitis. Plant dried in shade, powdered and mixed with *Ajuga parviflora* leaf powder, and given ½ teaspoonful twice a day, for three months to treat high fever and diabetes. Decoction of herb is given ½ teaspoonful twice a day for 7 to 21 days for the treatment of abdominal ailments. Decoction of whole herb is used for the treatment of asthma, rheumatism and applied on wounds. Root is chewed to get relief from throat irritation.

The leaf decoction is applied topically to boils and itches, herpes-like sores, and to sores caused by forest ticks. The juice of the aerial parts of the plant, mixed with turmeric (*Curcuma longa*) and heated with a little water, is applied topically to treat infected fingernails. A decoction is taken as a tea for its febrifuge effect. Leaf paste with lemon juice is given orally to cure excess bile secretion; also applied externally on ringworm and boils. It is also used to treat bilious affections.

In other Asian countries, *L. crustacea* is also used as a medicinal herb, especially for umbilical infections, boils, uterine haemorrhage, leucorrhoea, jaundice and colic. It is one of the commonest plants used as a drug in Chinese pharmacies in Indonesia and Malaysia. In Malaysia, a decoction of the leaves is used as medicine after childbirth. In the Moluccas, it is applied to boils and itches and herpes-like sores. In Indo-China, it is used in treating bilious disorders, dysentery, amenorrhoea, and hepatitis. The plants is also used for bilious affections and dysentery in Indo-China. In Brunei, the powdered herb mixed with rice water is taken to relieve diarrhoea, vomiting and cholera. The leaves are used in Tahiti to treat earache. Currently, the most popular Tahitian use is in purgative remedies for infants; it is also used for leucorrhoea. In the Cook Islands, a solution of the crushed leaves is given to babies suffering from an ill-defined group of ailments known as IRA and sometimes for infection of the navel. Elsewhere, it is used in poultices for boils, ringworm and itches. (Panda and Mishra 2011, Tiwari and Rana 2015, Mathur *et al.* 2016; http://tropical.theferns.info, www.cabi.org/isc/datasheet/30864).

39. PHRYMACEAE

Phrymaceae – Lopseed family

Small family of flowering plants consisting of 10 plant genera and 199 species. Plants are mostly herbs and a few subshrubs, bearing tubular, bilaterally symmetric flowers. They can be annuals or perennials. A single species of this family was recorded in Chithara Village flora.

39.1 *Mazus pumilus* (Burm.f.) Steenis

(Syn. *Lobelia pumila* Burm.f.; *Mazus japonicus* (Thunb.) Kuntze; *Mazus rugosus* Lour.)

Habit

Common Name: Japanese Mazus, Asian Mazus.

Vernacular Name: None known.

Habit: Annual herb.

Habitat: Abundant in moist shady places, grasslands, along the banks of water bodies.

Distribution: The plant is native to the Himalayas from Kashmir to China, Japan, Korea and eastern Russia, south through Vietnam and Thailand to Indonesia and the Philippines. It grows in wet areas of lawns, in flower beds, gardens, along creeks and rivers, in greenhouses and on wet roadsides.

Flowering and Fruiting Time: November-March.

Taxonomic Description: A small, shallow-rooted, sparsely pubescent, annual herb, 5-15 cm high. Leaves mostly radical, crowded, shortly petiolate, 2-6 x 1-2.5 cm, oblong, obovate, spathulate, margin irregularly crenate, base tapering into a short petiole. Flowers few, arranged in terminal lax racemes. Calyx campanulate, pubescent, about 6 mm long, 5-lobed; lobes ovate, acute. Corolla white with violet base and yellowish centre, bi-lipped; lower lip 3-lobed, middle lobe smaller than lateral ones. Stamens-4, didynamous. Capsules sub-globose with numerous seeded.

Inflorescence

Flower

Chemical Constituents: None known.

Medicinal Properties: The plant is aperitive, antifebrile, aperient, emmenagogue, febrifuge and tonic. (Pullaiah 2006; http://www.pfaf.org/user/plantsearch.aspx).

Medicinal Uses (Traditional/Folklore): The juice of the plant is used in the treatment of typhoid. (http://www.pfaf.org/user/plantsearch.asp).

40. PLANTAGINACEAE

Plantaginaceae – Plantain family

The family consisting of 120 plant genera and 1,614 species, is a diverse, cosmopolitan family, occurring mostly in temperate zones. The group consists of herbs, shrubs and also a few aquatic plants. It contains some of the important medicinal plants such as *Digitalis*, Foxglove and garden ornamentals, *Antirrhinum* and Snapdragon. The family was represented by five species under four genera in Chithara Village.

40.1 *Bacopa monnieri* (L.) Wettst.

(Syn. *Lysimachia monnieri* L.; *Monniera cuneifolia* Michx.)

Habit

Common Name: Water Hyssop, Indian Pennywort.

Vernacular Name: *Hindi*: Brahmi (ब्राह्मी); *Sanskrit:* Brahmi (ब्राह्मी), Tiktalonika (तिक्तलोणिका), Sarasvati (सरस्वती), Kapotavanka (कपोतवान्का).

Habit: Prostrate herb.

Habitat: Wet, damp or marshy areas, bank of ponds, rivers, canals and wetlands.

Distribution: It commonly grows in India, Nepal, Sri Lanka, China, Taiwan, Pakistan, Bangladesh and is also found in Florida and other southern states of US where it can be found in damp conditions near the pond or bog garden. Occur throughout India up to 1,200 m elevation.

Flowering and Fruiting Time: June-November.

Flowering twig

Flower

Taxonomic Description: A prostrate, spreading herb. Stem creeping, purplish-red, often rooting at nodes. Leaves succulent, opposite-decussate, sessile, obovate oblong, 1-2 x 0.4-0.6 cm, glabrous, entire, cuneate at base, obtuse at apex. Inflorescence solitary, axillary. Flowers bluish-white, pedicel up to 2.5 cm long. Calyx lobes-5, sepals subequal, outer 2 large. Corolla 5-lobed; lobes sub-equal, 8-10 mm long, campanulate. Stamens-4, didynamous, filaments slender, anthers oblong. Ovary globose, stigma capitate. Capsules 3-4 mm long, ovoid, enclosed in persistent calyx. Seeds minute, oblong, yellowish-brown.

Chemical Constituents: Plants contain triterpenoid saponins, bacosides A and B, bacogenins, betulic acid, betulinic acid, d-mannitol, stigmasterol, β-sitosterol, saponin, stigmastanol, hersaponin, monnierin, nicotine, luteoline and its glucosides. Leaves contain alkaloid brahmine, which raises blood pressure when administered in therapeutic dosage, and is hypotensive when given in very strong dosage. A tranquilizing saponin resembling reserpine is also present. (Al-Snafi AE 2013; http://www.bsienvis.nic.in).

Medicinal Properties: The plant, used as nervine tonic/memory enhancer, is known to be bitter, sweet, cooling, anodyne, antispasmodic, anticancer, asthenia, astringent, anti-inflammatory, anticonvulsant, blood purifier, bronchodilator, carminative, cardiotonic, diuretic, digestive, depurative, emetic, emmenagogue, febrifuge, intellect promoting, laxative, musculature relaxant, sudorific and tonic. Brahmi has antioxidant properties. It has been reported to reduce oxidation of fats in the blood stream, which is a risk factor for cardiovascular diseases. (Warrier *et al.* 2010, Al-Snafi AE 2013; http://tropical.theferns.info).

Medicinal Uses (Traditional/Folklore): Brahmi is known to have long history of use in the Ayurveda for increasing concentration, memory capacity, anti-ageing, memory-enhancing ethnobotanical therapy and for reducing stress-induced anxiety. The plant is mentioned in different *Vedic* and ancient scripters as invigorating and life sustaining in *Charaka Samhita* and for strengthening of body and improving quality of semen in *Atharva Veda* and for increasing memory and lifespan in *Sushruta Samhita*. It has been mentioned in religious, social and medical treatises of India since the time of *Atharvan Veda* (800 BC); the first clear reference to its CNS effect is to be found in *Charak Samhita*, written in the first century AD. It is mentioned in the authentic Ayurvedic treatise, *Susrutu Samhita*, which describes Brahmi as efficacious in the loss of intellect and memory. Brahmi has been used for centuries to help benefit epilepsy, memory capacity, increase concentration, and reduce stress-induced anxiety. It is listed as a nootropic, a drug that enhances cognitive ability. It is used for the treatment of a number of disorders such as ascites, anaemia, amentia, invigorates sex; biliousness, bronchitis, constipation and indigestion, dyspepsia, dysmenorrhoea, dermatitis, diabetes, elephantiasis /filaria, flatulence, fever and general debility, erysipelas, leprosy, fever, hoarseness, insanity, epilepsy, enlargement of spleen, inflammations, leucoderma, leprosy, splenomegaly, skin diseases, syphilis, strangury, sterility, tumours, ulcers, sterility, nervous breakdown and other low adynamic conditions and general debility, etc.

Plant-juice (along with ginger, sugar and bark extract of *Moringa oleifera*) is given to children in stomach disorder and leaves (fried in ghee) as a brain tonic, in nervous weakness, hysteria, epilepsy, insanity, anxiety neurosis, and to sharpen dull memory. The plant is also used in nervous breakdown and other low adynamic conditions and as a tonic. According to Unani system of medicine, it is bitter, aphrodisiac, good in scabies, leucoderma, syphilis, etc. It is promising blood purifier and useful in diarrhoea and fevers; invigorating and effective against cold and cough. (Kaul and Dwivedi 2010, Shukla *et al.* 2010; http://medplants.blogspot.in, http://tropical.theferns.info, http://www.bsienvis.nic.in).

40.2 *Mecardonia procumbens* (Mill.) Small

(Syn. *Bacopa procumbens* (Mill.) Greenm.)

Habit

Common Name: Baby Jump Up.

Vernacular Name: None known.

Habit: Annual herb.

Habitat: Common along marshy areas, grasslands, bank of the rivers and nearby fields.

Distribution: Native to tropical and subtropical regions of the Americas. It has become widely spread in warmer regions worldwide, and is now naturalized on all continents except Antarctica.

Flowering and Fruiting Time: July-October.

Taxonomic Description: An annual, sub-erect or decumbent-ascending, 10-20 cm high, annual herb. Stem 4-angled. Leaves opposite, decussate, sub-sessile, ovate-lanceolate, margin crenate-serrate, apex acute, 1-1.6 x 0.5-0.8 cm. Flowers axillary, solitary, pedicels about 1 cm long; bracteoles-2, nearly 4 mm long. Calyx persistent, 5-lobed, lobes unequal. Corolla yellow, obscurely 2-lipped, lower lip 3-lobed. Stamens-4, included, filaments slender. Capsules 4-6 mm long, numerous seeded.

Flowering twig

Flower

Chemical Constituents: None known.

Medicinal Properties: None known.

Medicinal Uses (Traditional/Folklore): None known.

40.3 *Scoparia dulcis* L.

(Syn. *Scoparia ternata* Forssk.)

Common Name: Sweet Broom Weed.

Vernacular Name: *Hindi:* Mithi Patti (मीठी पत्ती), Ghoda Tulsi (घोड़ा तुलसी).

Habit: Undershrub.

Habitat: Common along moist waste places, near small water channels.

Distribution: It is native to the tropical America but it can be found throughout the tropical and subtropical world.

Flowering and Fruiting Time: Major part of the year.

Taxonomic Description: An erect, 30-80 cm high, undershrub. Stem much-branched, ribbed. Leaves opposite or in whorls of three, 3-4 x 1-1.5 cm, petioled, rhomboid-ovate or elliptic, base cuneate, margin serrated towards tip, apex obtuse. Flowers small, white, axillary, 3-6 in each whorl. Calyx lobes-4, ovate-oblong, persistent in fruit. Corolla 4-lobed, white, about 5 mm long, densely hairy at throat. Stamens-4, exserted, sub-equal. Capsules small, sub-globose, nearly 2 mm in diameter with many seeded.

Habit

Chemical Constituents: The phytochemical analysis of the powdered whole plant shows the presence of carbohydrates, flavonoids, saponins, tannins, alkaloids, steroids and terpenes. Other chemicals isolated from *S. dulcis* include scoparinol, scoparic acid, scopadulcic acid, scopadulciol, and scopadulin. The plant is considered to be a rich source of flavones, terpenes and steroids.

Flowering twig

Glycoside, ammelin obtained from the fresh plant shows antidiabetic activity. The methanolic and water extracts from roots of Formosan *S. dulcis* have been found to exert hypotensive activity, and 6-methoxybenzoxazolinone (MBOA) was isolated as an active principle which is reported to show anti-inflammatory property. The analgesic and anti-inflammatory activities of the ethanolic extract of this plant were recently observed, and a triterpene, glutinol, was detected as a major bioactive component by Brazilian workers. (http://medicinalplants.us/scoparia-dulcis-l-sweet-broomweed).

Medicinal Properties: The whole plant is used for ailments like diarrhoea, stomachache, kidney stones, kidney problems, antitoothache, antidiabetic, and gonorrhoea; it has hepatosis, analgesic, antipyretic, antidiabetic, antioxidant, and anti-inflammatory properties. It has cytotoxic activity against malignant brain tumour and bone cancer; Scoparic acid B has antitumour activity against human cancer cells. (Okhale *et al.* 2010, Mishra *et al.* 2013; http://medicinalplants.us/scoparia-dulcis-l-sweet-broomweed).

Medicinal Uses (Traditional/Folklore): It is traditionally used in the treatment of diabetes, dysentery, earache, fever, gonorrhoea, headaches, jaundice, snake bite, stomach problems, toothache, and warts. As a traditional medicine, *S. dulcis* has been used for diabetes in India and hypertension in Taiwan. In Brazil, it has been used for various problems such as haemorrhoids and wounds, while it is used for sickle-cell disease in Nigeria. In Nicaragua, it has been given for anaemia, burns, and headaches. A hot water infusion or decoction of the leaves or whole plant is used medicinally by indigenous tribes of Nicaragua to treat malaria, stomach disorders, menstrual disorders, insect bites, fevers and heart problems, liver disorders. In tropical and subtropical regions of Asia, Africa, and South and Central America, the fresh or dried plant of *S. dulcis* has traditionally been used as a medicament for stomach disorders, bronchitis, diabetes, hypertension, haemorrhoids and hepatosis, and as an analgesic and antipyretic. (Okhale *et al.* 2010, Mishra *et al.* 2013).

40.4 *Veronica agrestis* L.

(Syn. *Pocilla agrestis* (L.) Fourr.)

Habit

Common name: Field Speedwell, Green Field Speedwell.

Vernacular name: None known.

Habit: Annual herb.

Habitat: Common winter weed in moist waste places, gardens and along the banks of temporary water channels.

Distribution: It is found between 150 - 2,900 m in Uttarakhand, Rajasthan, Uttar Pradesh, Bihar, Orissa, Jammu & Kashmir, and Himachal Pradesh in India.

Flowering and Fruiting Time: January-March.

Taxonomic Description: A small, prostrate herb. Stem branched, densely pubescent, up to 30 cm long. Leaves shortly petioled, ovate, 1-2.5 x 1-2 cm, base cordate, margin coarsely crenate-serrate, apex obtuse. Inflorescence solitary, axillary; pedicels about as long as the leaves or little shorter, decurved in fruit. Calyx about 4 mm long, 4-lobed, lobes ciliate, ovate, enlarged in fruit. Corolla white tinged with purple, lobes-4, spreading. Stamens-2. Stigma capitate. Capsules broader than long, 2-lobed; lobes globose, densely hairy with few seeded.

Flowering twig

Flower

Capsule

Chemical Constituents: None known.

Medicinal Properties: Speedwell is an expectorant, alterative, tonic, and diuretic. It was formerly administered in coughs, catarrhs, renal, and skin diseases, jaundice, etc. (http://www.henriettes-herb.com).

Medicinal Uses (Traditional/Folklore): The decoction of *V. agrestis* is used in the treatment of haemorrhage (severe bleeding from ruptured blood vessels) and dysmenorrhoea (pain during menstruation). (Khan *et al.* 2013c).

40.5 *Veronica anagallis-aquatica* L.

(Syn. *Veronica anagallidiformis* Boreau)

Habit

Common Name: Water Speedwell, Brook-Pimpernel.

Vernacular Name: *Hindi:* Sada (सादा), Sadevi (सदेवी).

Habit: Annual herb.

Habitat: Frequent in marshy places, along the bank of water channels, ditches, wet meadows, ponds and streams.

Distribution: It is found in North Africa, Europe, Asia, the Arabian Peninsula and Kashmir to Bhutan in Himalaya.

Flowering and Fruiting Time: February-April.

Taxonomic Description: An erect, semi aquatic, shallow-rooted herb, 10-50 cm high, sparsely glandular hairy. Leaves opposite, clasping to the stem, 3-10 x 0.8-2 cm, linear-oblong or oblong-lanceolate, entire to crenate-serrate. Inflorescence racemes. Racemes axillary, many flowered, pedicel about as long as bract. Calyx 4-lobed, 3-4 mm long; lobes ovate-lanceolate, pubescent. Corolla 4-lobed; lobes broadly ovate, white or light purple. Stamens-2. Capsules slightly compressed, about as long as persistent calyx, apex notched.

Chemical Constituents: The above ground plant parts contain aromatic acids (benzoic, protocatechuic and caffeic, iridoid glucoside (aucuboside, catalpol) and a flavone (luteolin-7-glucoside), phenolic acids, sugars. In the underground parts of the plants p-hydroxybenzoic, vanillic, p-cumaric, ferulic and isoferulic acids, sugars and glucosides are present. (Komorowski and Swiatek 1982).

Medicinal Properties: Alterative, appetizer and diuretic inducing urination. (Gupta and Pandey 2014).

Flowering twig

Flower

Medicinal Uses (Traditional/Folklore): *V. anagallis-aquatica* is bruised and applied externally as a poultice on cuts, burns, sores, ulcers, whitlows, healing of wounds, etc. Decoction of plants promotes menstruation in suffering women. Crushed leaves are used as tonic and astringent. Root decoction is used in diarrhoea. Paste of leaves is applied for healing of wounds. The leaves are used in the treatment of scurvy, impurity of the blood, etc. Ethnobotanically, the leaves have been used as alterative, bladder, gargle, skin, stone and whitlow. The leaves of *Veronica* are edible (raw or cooked) and are rich in vitamin C. (Prajapati and Kumar 2003, Vardhana 2008, Gupta and Pandey 2014; http://shodhganga.inflibnet.ac.in/bitstream/10603/50574/8/08_chapter%204.pdf)

41. LENTIBULARIACEAE

Lentibulariaceae – Bladderwort family

A family of carnivorous plants containing 4 plant genera and 312 species. It is cosmopolitan in distribution, ranging from sub-arctic to tropical forests. Plants are herbaceous, terrestrial, epiphytic or aquatic carnivorous plants, characterized by small hollow sacs that actively capture and digest tiny insects, larvae, etc. One species belonging to this family was recorded in the flora.

41.1 *Utricularia stellaris* L.f.

(Syn. *Utricularia inflexa* var. *stellaris* (L.f.) P.Taylor)

Common Name: Bladderwort.

Vernacular Name: None Known.

Habit: Aquatic carnivorous plant.

Habitat: Common in less polluted ponds, water reservoirs and ditches.

Distribution: Paleo-tropics; endemic to Africa and the Indian subcontinent. Occurs in stagnant or slow running water.

Habit

Flowering and Fruiting Time: September-December.

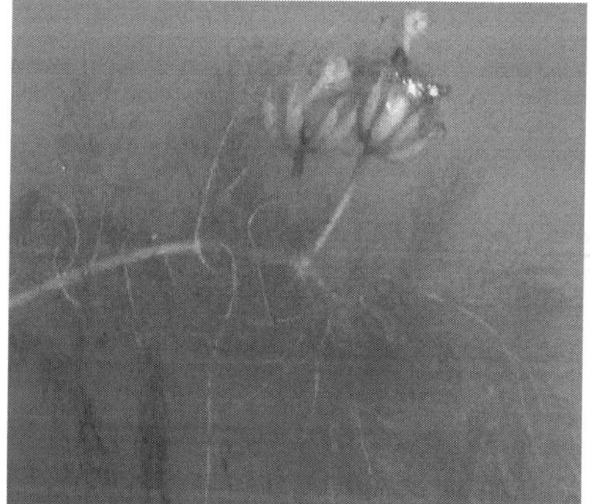

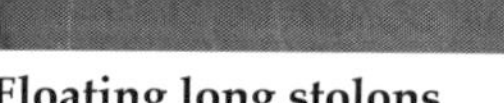

Floating long stolons

Spongy whorls

Flowers

Taxonomic Description: A slender, carnivorous, rootless, aquatic herb. Stolon long, much branched, floating under the surface of water. Leaves pinnately divided into many filiform, capillary segments, each primary segment with small roundish bladder at their base. Flowers 5-12, born in erect aerial racemes, peduncles held up above water by a whorl of spongy floats, pedicels deflexed in fruit, bracts present. Calyx 2-lobed, accrescent. Corolla spurred, yellow, bi-lipped, upper lip ovate-rounded. Capsules about 6 mm long, globose with enlarged calyx. Seeds minute, discoid, winged.

Chemical Constituents: None Known.

Medicinal Properties: None Known.

Medicinal Uses (Traditional/Folklore): None Known.

42. PEDALIACEAE

Pedaliaceae – Sesame family

The Sesame family consisting of 14 genera and 67 species has native distribution exclusively in Old World, in tropical and dry habitats. These are herbs or shrubs with spurred flowers and ovaries with axial placentation that often develop hooks or prickles as the fruit wall begins to decompose. The best-known member is *Sesamum indicum*. The family was represented by a single species in Chithara Village.

42.1 *Sesamum indicum* L.

(Syn. *Sesamum orientale* L.)

Common Name: Sesame.

Vernacular Name: *Hindi*: Safed Til (सफेद तिल).

Habit: Annual herb.

Habitat: Cultivated as well as found growing wild in open and waste places.

Distribution: Cultivated and wild throughout India in plains and up to an altitude of 1,200 m.

Flowering and Fruiting Time: August-October.

Habit

Taxonomic Description: An erect, pubescent, up to 1.5 m high, annual herb with foeted smell. Stem 4-angled. Leaves variable, pubescent beneath; lower leaves opposite, 6-15 x 2-8 cm, oblong, ovate-lanceolate, sparsely crenate-serrate, long petioled; upper leaves alternate, entire. Inflorescence axillary, solitary. Flowers whitish-pink or purplish, pedicel short with 2 yellow glands at the base. Calyx deeply 5-lobed, persistent, hairy, lobes linear-lanceolate. Corolla bilabiate, unequal, 2-2.5 cm long, pubescent outside. Stamens-4, filaments filiform, anthers oblong. Capsules oblong, nearly 2.5 cm long, hairy. Seeds numerous, oily, brownish-white.

Flowering twig

Flowers

Capsules

Chemical Constituents: Sesame has bio-active compounds like sesamin, sesaminol, gamma tocopherol, cephalin and lecithin. Sesame seeds contain many phytochemically important compounds like flavonoids, phenolic acids, alkaloids, tannins, saponins, steroids, terpenoids and minerals like calcium, iron, magnesium, manganese, copper, zinc, phosphorus. The major saturated fatty acids in *S. indicum* seed oil are palmitic, stearic acids with small arachidic acid. The main unsaturated fatty acids are linoleic and oleic acids. The oil samples contained saturated and unsaturated acids. The oil is predominantly made up of oleic and linoleic acids, respectively. (Neeta *et al.* 2015).

Medicinal Properties: The seeds are sweet, astringent, bitter, acrid, emollient, thermogenic, aphrodisiac, laxative, galactagogue, digestive, hair-restorer and tonic. They are useful in haemorrhoids, ulcers, burns, dysentery, migraine, alopecia, ophthalmopathy and obesity. Externally, it is used for dryness of the skin and leucoderma. (Warrier *et al.* 2010, Neeta *et al.* 2015; http://tropical.theferns.info).

Medicinal Uses (Traditional/Folklore): Roots and leaves are emollient and a decoction of the two forms a good hair- wash which promotes hair growth and blacken them. The leaves are useful in dysentery, cholera, vitiated condition of *kapha*, nephropathy, uropathy, ophthalmopathy and dermatopathy. A decoction is used in various traditional medicines to treat asthma and coughs. Mixed with lime water, the oil is used externally to treat burns, boils and ulcers. In ethno-medicine, oil is used for treatment of hair, skin, teeth, and bone and lung problems. Oil is used in products like hair oils, perfumes, cosmetics like skin conditioning agents and moisturizers. Sesame oil shows highest resistance to rancidity and can be used as a substitute for olive oil. (Singh *et al.* 2010d, Warrier *et al.* 2010, Chaudhary and Gupta 2011, Neeta *et al.* 2015; http://tropical.theferns.info)

43. ACANTHACEAE

Acanthaceae – Acanthus family

The Acanthus family (Greek Acanthos -a thorn, referring to the often spiny leaves and flower bracts) consists of 242 genera and about 3,947 species, mostly shrubs, annuals or perennial herbs and some climbers, mostly distributed in tropical and sub-tropical climate. Most have simple leaves arranged in opposite pairs, with cystoliths (crystals of calcium carbonate) in streaks or protuberances in the vegetative parts. The flowers are often two lipped. Members of this family are favoured food plants of the Pansy group of butterflies. Five species under four genera belonging to this family were recorded in the flora of Chithara Village.

43.1 *Barleria prionitis* L.

(Syn. *Barleria coriacea* Oberm.; *Barleria spicata* Roxb.)

Habit

Common Name: Porcupine Flower.

Vernacular Name: *Hindi:* Vajradanti (वज्रदंती); *Sanskrit:* Artagalah (आर्तगल), Dasi Kurantakah (दासी कुरण्टल).

Habit: Undershrub.

Habitat: Found in dry areas along roadsides and in waste places; also planted as a hedge plant in villages.

Distribution: Tropical Africa, tropical Asia, Sri Lanka, India through Malesia.

Flowering and Fruiting Time: October-May.

Taxonomic Description: A much-branched, armed undershrub. Spines straight and sharp, present in the axil of leaves. Leaves opposite, petiolate, up to 14 cm long, elliptic, margin entire, apex acute, base tapering to the petiole. Inflorescence axillary and terminal spikes. Bracts foliaceous, bracteoles spinescent. Calyx deeply divided, lobes spine-tipped. Corolla about 4 cm long, orange-yellow with 5 subequal parts. Stamens 2+2. Capsules ovoid, beaked. Seeds-2, black.

Young leaves with spines in the axil

Flowers

Chemical Constituents: Plants contain alkaloids, terpenoids, phenolic compounds, tannins, carbohydrates and flavonoids. (Banerjee *et al.* 2012, Sharma *et al.* 2013b, Aiswarya and Ravikumar 2014, Talukdar *et al.* 2015, Singh and Navneet 2017).

Medicinal Properties: Plants have wide range of medicinal properties like anthelmintic, antifertility, antioxidant, antidiabetic, anti-inflammatory, antiarthritic, cytoprotective, hepatoprotective, diuretic, antidiarrhoeal, antifertility, antimicrobial, antifungal, antidental decay, antipyretic, antirheumatic, anticatarrhal, expectorant, rubefacient and enzyme inhibitory effects without any toxic effects. (Banerjee *et al.* 2012, Aiswarya and Ravikumar 2014, Singh and Navneet 2017).

Medicinal Uses (Traditional/Folklore): The whole plant, root, leaves and bark of the *B. prionitis* are utilized in the indigenous system of medicine (Ayurveda) in India for treating different types of ailments such as inflammations, swellings, boils, glandular swellings, anaemia, toothache, bacterial disorders, catarrhal affections of children, boils, fever, and gastrointestinal disorders etc. The juice of has been used for curing whooping cough in Uttar Pradesh and Madhya Pradesh states of India; leaves are used for the treatment of toothache, rheumatism, and root powder to cure fever. The juice of leaves are useful in fungal infections, ulcer and fever. The decoction of aerial parts is used in whooping cough, anti-respiratory syncytial virus, antiarthritic, anti-inflammatory and antifertility activities. For the cure of tonsillitis 4 g of plant powder mixed with *Nimbuka Swarasa* is given twice a day for 10 days. Root extract is used locally on skin to expel out spine from the skin and decoction is taken orally for the cure of snakebite. *B. prionitis* is given in treating urinary infection, jaundice, hepatic obstruction and dropsy. Ash of the whole plant with honey is useful in bronchial asthma.

Leaves are chewed for the treatment of fever, rheumatism, liver diseases, and indigestion with constipation, jaundice and urinary infection. The leaves are used externally to promote the healing of wounds and to relieve joint pains and toothache. The bitter juice of the plant is given to children as a treatment for catarrh. Oil extract of the plant is used to prevent graying of hair. Its flowers are used to get rid of migraine. In medico-botanical survey of villages of Bulandshahr, Uttar Pradesh, India, rural population use this plant for the cure of asthma and whooping cough.

Bark is used in whopping cough as an expectorant; the whole plant particularly the roots are used as tonic and diuretic. The whole plant is used to cure stiffness of limbs, enlargement of scrotum and sciatica. In Maharashtra (India), crushed leaves of *B. prionitis* are applied on the wound. Pills prepared from *B. prionitis* are used as a rubefacient for massage in combination with coconut oil. The folk medicinal healers of Bangladesh use the plant for anti-inflammatory activity, and also for the treatment of cancer and tumour; also root of the plant with goat milk is given to treat rheumatic fever. Root, stem or leaves powder with cow milk is taken as remedy for dropsy and liver congestion. In Andhra Pradesh, India, *B. prionitis* is used to increase vitality by using seed extract daily once for fortnight; the plant is also being used in gout, ulcer of mouth and oedema. In Orissa (India), the plant has been used in cuts, wounds and malaria, and In Gujarat (India), leaf ash is being used for the management of leucoderma by applying with butter. The use of fresh leaf paste has also been reported to be effective against scabies in Karnataka (India).

In folklore medicines, for treating skin diseases crushed leaf is applied directly on the skin; for scabies paste of fresh leaves is applied on the affected area. For pus in ears plant extract is applied. Leaves are chewed and sap is swallowed for treating mouth ulcers. Plant decoction from freshly extracted shoots is used for treating pyorrhea. For greying of hair extract of whole plant mixed with oils is applied on hair regularly.

Ethno- medicinal survey in, Andhra Pradesh, India, reveals that local residents use this plant to increase vigour by using seed extract daily once for fortnight. (Shukla *et al.* 2010, Singh *et al.* 2010d, Banerjee *et al.* 2012, Sharma *et al.* 2013b, Talukdar *et al.* 2015, Singh and Navneet 2017; http://medplants.blogspot.in, http://tropical.theferns.info).

43.2 *Justicia adhatoda* L.

(Syn. *Adhatoda vasica* Nees; *Adhatoda zeylanica* Medik.)

Habit

Common Name: Malabar Nut, Vasaka, Adhatoda, Adulsa.

Vernacular Name: *Hindi:* Arus (अरुस), Arusa (अडूसा); *Sanskrit:* Vasaka (वसाका), Vasuka (वासुका), Atarusa (अतरुसा), Pra-Madya (प्रामाद्य).

Habit: Shrub.

Habitat: Along roadsides and railway lines, open and waste places as a weed.

Distribution: The plant's range includes Sri Lanka, Nepal, India, Pakistan, Indonesia, Malaysia, and China, as well as Panama where it is thought to have been introduced. In India, it is found throughout the plains in stony wastelands and in sub-Himalayan tracts, up to an altitude of 1,300 m. It is almost found in groups.

Flowering and Fruiting Time: January-June.

Flowers

Taxonomic Description: A branched evergreen shrub, up to 3 m tall. Young branches pubescent. Leaves simple, entire, petiolate, opposite, decussate, 7-18 x 3-7.5 cm, ovate to elliptic lanceolate, acuminate. Spikes terminal or axillary, up to 7 cm long. Flowers white in colour, bracts ovate-oblong, imbricate. Calyx 5-lobed, slightly pubescent outside. Corolla tubular, bi-lipped, white with red or purplish stripes at throat. Stamens-2. Capsules 2.5 cm long, pubescent. Seeds circular in shape.

Chemical Constituents: The leaves of *J. adhatoda* contain phytochemicals such as alkaloids, tannins, saponins, phenolics and flavonoids. The leaves contain a most important alkaloid vasicine, a quinazoline alkaloid and N-oxides of vasicine, vasicinone, deoxyvasicine and maiontone. The vasicine yield of the herbage has been measured as 0.541 to 1.1% by dry weight. The roots contain vasinolone, vasicol, peganine and 2-glucosyl-oxychalcone. The flowers contain β -sitosterol-D-glucoside, kaempferol and its glucoside and quercetin. (Bhutya 2011, Chanu and Sarangthem 2014).

Medicinal Properties: Adhatoda has been used in traditional medicine to treat respiratory disorders. This plant is a source of Vitamin C and has medicinal uses, mainly antispasmodic, fever reducer, anti-inflammatory, anti-bleeding, bronchodilator, antidiabetic, disinfectant, antijaundice and oxytocic. It is antiperiodic, astringent, diuretic, purgative and is also used as an expectorant in addition to liquefy sputum. Plant and leaves are refrigerant, expectorant, antispasmodic, febrifuge, depurative, tonic, styptic, respiratory stimulant, hypotensive, antifungal, sudorific, cardiac depressant, utero-tonic and abortifacient. The primary alkaloids, vasicine and vasicinone present in the leaves possess respiratory stimulant activity and are well established as therapeutical respiratory agents. Whereas vasicine at low concentrations, induce bronchodilation and relaxation of the tracheal muscles.

The leaves, flowers and roots of this plant used in herbal drugs against tubercular activities, cancer and possessed anti-helmintic properties. Similarly in Homeopathy, the plant has been used in the treatment of cold, cough, pneumonia, fever, jaundice, catarrh, whooping cough and asthma. (Bhutya 2011, Dhankhar *et al.* 2011, Singh and Huidrom 2013, Chanu and Sarangthem 2014).

Medicinal Uses (Traditional/Folklore): Adhatoda is a well known plant drug in Ayurvedic and Unani medicines for centuries, and is mentioned in Sanskrit scriptures. It has been used for the treatment of various diseases and disorders, particularly for the respiratory tract ailments in both children and adults. Therefore, it is a primary herb of the Ayurvedic system used in the treatment of cough, bronchitis, asthma and symptoms of common cold. The source of the drug 'Vasaka' is well known in the indigenous system of medicine for its beneficial effects, particularly in bronchitis. Its main action is expectorant and antispasmodic (bronchodilator). All the parts of the plant are used in Indian traditional medicine for the treatment of asthma, joint pain, lumbar pain, sprains, cold, cough, eczema, malaria, rheumatism, swelling and venereal diseases.

The whole plant is used as an ingredient of numerous popular formulations including cough syrup used in combination with Ginger (*Zingiber officinale*) and Tulsi (*Ocimum sanctum*) where it exerts its action as an expectorant and antispasmodic. Root is more efficient for expectoration. The plant is used for treatment of excessive phlegm and menorrhagia in Sri Lanka. It is also used for the treatment of bleeding piles, impotence and sexual disorders. Similarly, Bisolven, a branded drug containing Vasaka as an ingredient is used to clear the airways by decreasing the mucus secretions and opening the passages. There are many herbal formulations of Adhatoda used for the treatment of various kinds of respiratory disorders.

A yogic practice is to chew the leaf buds alone or with a little ginger root, to clear the respiratory passages in preparation for the vigorous breathing exercises. Leaves along with roots are useful to cure irritable cough, bleeding in diarrhoea, hemoptysis, bronchial and allergic asthma, fevers, vomiting, urinary disorders, tuberculosis, intestinal worms, skin diseases, cardiac tonic, and bronchitis. Leaves are found to activate digestive enzyme trypsin and are antifungal and sudorific. The plant leaves are used for checking postpartum haemorrhage and urinary trouble. It is found that 70 percent of the pregnant women in the Gora village of Lucknow (Uttar Pradesh) use the leaves to induce abortion. Moreover, it is observed that the Neterhat people in Bihar used a decoction of leaves to stimulate and heal before and after delivery. Flowers are used for ophthalmia, fevers and intestinal worms.

Extracts of Adhatoda's leaves and roots are useful in treating bronchitis, and other lung and bronchiole disorders, as well as common coughs and colds. A decoction of the leaves of Adhatoda has a soothing effect on irritation in the throat, and acts as an expectorant to loosen phlegm in the respiratory passages. The leaf powder boiled in sesame oil is used to stop bleeding, earaches as well as pus from ears and jaundice. Decoction and ash of leaves are used for bronchial complaints such as asthma, tuberculosis, and to relieve fever (antipyretic) and acidity. It was also used for stomach catarrh with constipation, gout, urinary stone and warmed leaves used externally for rheumatic pains and dislocation of joint. The various preparation of leaves are used for curing bleeding, haemorrahge, skin diseases, wounds, headache and leprosy in Southeast Asia. The bruised fresh leaves are used for snake-bites in India and Sri Lanka. The leaves are toxic and have insecticidal effect.

In Manipur, the plant is used as an herbal remedy for treating cold, cough, whooping cough and chronic bronchitis and asthma, as sedative expectorant, antispasmodic and anthelmintic. Usually, yellow leaves are exploited for cough and smoke from leaves is used for asthma. The leaves are used for treating respiratory disorders and the leaf juice is used as a remedy in treating diarrhoea and dysentery.

The fresh flowers are used for opthalmia and various preparations of flowers are used for treatment of cold, phthisis, asthma, bronchitis, cough, antispasmodic, fever and gonorrhoea in South-East Asia. The flowers are also used as antiseptic to improve blood circulation of blood. The fruit of *J. adhatoda* are used for curing cold, antispasmodic, chronic bronchitis, jaundice, diarrhoea, dysentery, fever and as laxative.

The extract of roots is commonly used by rural population against diabetes, cough and certain liver disorders. The paste, powder and decoction of root is used for curing tuberculosis, diphtheria, malarial fever, leucorrhoea and eye diseases in Southeast Asia. The paste of roots mixed with sugar and used for treatment of acute nightfall in Sitapur District, Uttar Pradesh, India. Moreover, the macerated roots are applied on the pubic region and vagina to help parturition as it facilitates the expulsion of foetus. The root decoction is also used for treating gonorrhoea.

The plant has also been used by the European medical practitioners. The fluid extract and tincture were used in England as an antispasmodic, expectorant and febrifuge. It was said to be beneficial in intermittent, typhus fever and diphtheria. In Germany, the leaves are used as an expectorant and spasmolytic agent. In Sweden, Adhatoda is classified as a natural remedy and some preparations against cough containing an extract of Vasaka are accessible. (Kaul and Dwivedi 2010, Dhankhar *et al.* 2011, Singh and Huidrom 2013; http://medplants.blogspot.in, http://tropical.theferns.info).

43.3 *Justicia japonica* Thunb.

(Syn. *Justicia simplex* D. Don)

Habit

Common Name: Common Small Justicia.

Vernacular Name: None known.

Habit: Annual herb.

Habitat: Roadsides, open and moist waste places.

Distribution: Nepal, North-East and western India, Pakistan, Sri Lanka, Burma, Malaysia and Thailand (plains to 2,500 m.).

Flowering and Fruiting Time: August-December.

Taxonomic Description: An annual herb, usually erect, up to 30 cm high. Stem and branchlets angled, softly hairy. Leaves opposite, up to 5 cm long, ovate-oblong or elliptic, base cuneate, apex obtuse, margin entire, sparsely pubescent, petioled. Inflorescence axillary and terminal, 3-8 cm long spikes. Bracts and bracteoles softly hairy. Calyx 4-lobed, linear, pubescent, margin scarious. Corolla bi-lipped, light purple or pink, pubescent outside, lower lip slightly 3-lobed; Capsules oblong, pubescent, 3-4 mm long with wrinkled seeds.

Chemical Constituents: None known.

Flowering twig

Flower

Medicinal Properties: Expectorant. (Quattrocchi 2012).

Medicinal Uses (Traditional/Folklore): A decoction of plant is expectorant and used to cure cough. Leaves are used in ophthalmia. (Quattrocchi 2012, Dutta and Sarma 2013).

43.4 *Peristrophe bicalyculata* (Retz.) Nees

(Syn. *Dianthera bicalyculata* Retz.)

A portion of plant

Common Name: Panicled Foldwing, Kali Anghedi.

Vernacular Name: *Hindi:* Atrilal (अत्रीलाल); *Sanskrit:* Nadikanta (नादिकांता), Kakajangha (कांकजंघा), Sulomasha (सुलोमशा).

Habit: Annual herb.

Habitat: Common along roadsides, railway lines and open and waste places.

Distribution: South East Asia, Malaysia; distributed throughout India.

Flowering and Fruiting Time: August-November.

Flowering twig

Flower

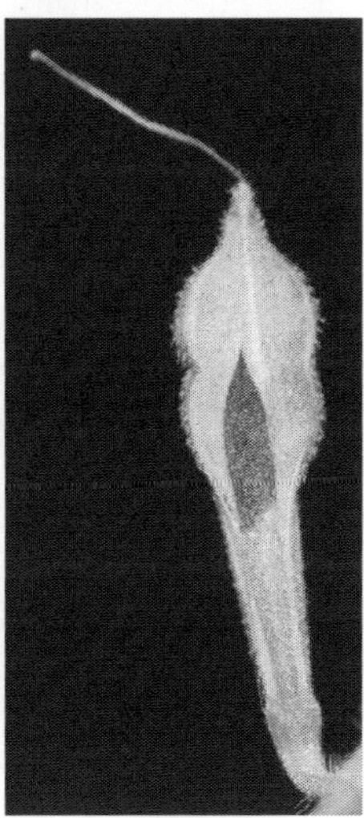
Capsule

Taxonomic Description: An erect, much-branched annual herb, 60 to 100 cm high. Stem and branches sharply 6-angled, swollen at the nodes, hairy. Leaves pubescent, opposite, shortly petioled, ovate-lanceolate, entire, acute to acuminate, 4-10 x 2-4 cm. Flowers pedicellate, in trichotomous cymes; bracts-2, unequal, opposite, linear. Calyx segments lanceolate-subulate, pubescent. Corolla bi-lipped, pink to purple, lips unequal. Stamens-2, filaments hairy. Capsules oblong, about 1 cm long, beaked, hairy.

***Chemical Constituents*:** Plants contain various secondary metabolites like steroids, alkaloids, phenols, coumarins, flavonoids, saponins, triterpenoid and tannins. Flowers contain petunidin-3-rhamnoglucoside; aerial parts contain 14-methyl-tritriacont-14-en-15-ol, 35-hydroxynonatricontanol. Besides, plants also contain volatile oil (herb), pentunidin – 3 – rhamnoglucoside (flower), sterols, fatty acids (stem and root), free amino acids and free sugars (stem). (Satynarayana *et al.* 1993, Bhutya 2011, Athokpam *et al.* 2014).

***Medicinal Properties*:** The herb is said to possess a number of therapeutic properties such as expectorant, antipruritic, febrifuge, analgesic, anti-inflammatory, antimicrobial, and antitoxic properties. The extract of the plant is antihypersensitive, hepatoprotective, hypolipidemic and antiatherogenic. The plant is known to be very effective for total normalization of diabetes induced hyperlipidemia and protection of the liver against diabetes induced injury. (Bhutya 2011, Khan *et al.* 2011b, Athokpam *et al.* 2014).

Medicinal Uses (Traditional/Folklore): A number of formulations containing root, seed or entire herb, have been mentioned in Ayurvedic formulary. *P. bicalyculata* is used for the treatment of tuberculosis, leucorrhoea, otorrhoea, fever, hemoptysis, ulcers, pruritis, and intestinal worms. Root is used for insomnia, fear psychosis in children and antidote for snake poison/snake bite. The plant is used by the traditional healers in the treatment of many skin related problems; it is also used as an antidote for snake poison when macerated in an infusion of rice, and as an insect repellant. In Uttar Pradesh, the paste of the plant is used for sprain and bone fracture. It is one of the traditional herbs recommended in case of tuberculosis. Further, it is said to be a good substitute for *Fumaria parviflora* (Fumariaceae) which is one of the reputed herbs employed in case of fever and skin diseases. The plant is used in South West Nigeria in the treatment of hypertension and other cardiovascular diseases. (Satynarayana *et al.* 1993, Abdulazeez 2010, Kaul and Dwivedi 2010; http://www.flowersofindia.net).

43.5 *Rungia pectinata* (L.) Nees

(Syn. *Rungia parviflora* Nees; *Justicia pectinata* L.)

Upper portion of stem

Common Name: Comb Rungia.

Vernacular Name: None known.

Habit: Annual herb.

Habitat: Common weed of moist waste places in gardens, roadsides and along small water channels.

Distribution: India, Sri Lanka and Myanmar. Grows as a weed throughout the warmer parts of India, in waste places and hedges.

Flowering and Fruiting Time: October-January.

Taxonomic Description: A decumbent-ascending, 10-30 cm long, annual herb. Stem terete, much-branched, sparsely pubescent. Leaves petiolate, opposite, 2-5.5 x 0.5-2 cm, narrowly oblong-lanceolate to elliptic, entire, narrowed at both the ends. Flowers minute, blue, borne in 1-2 cm long, axillary or terminal clustered spikes. Bracts dimorphic; the sterile bract elliptic-lanceolate, hairy; fertile bracts circular, pubescent. Calyx lobes linear-lanceolate. Corolla 3-4 mm long, blue, bilipped, lower lip 3-lobed. Stamens-2. Capsules ellipsoid with orbicular seeds.

Chemical Constituents: Phytosterols, terpenes, tannins, glycosides, flavonoids, phenolic compounds, amino acids, fixed oils, etc. are some of the important phytochemicals isolated from this plant. The flowers are reported to contain isosalipurposide, luteolin, glucoside, lutein, delphinidin-3, 5-diglucoside, and pigments. (Shekhawat *et al.* 2016)

Inflorescence

Abaxial side of leaves

Medicinal Properties: *R. pectinata* has cooling, anti-inflammatory, diuretic, antimicrobial, febrifuge and vermifuge properties. Leaf juice is aperient, febrifuge, refrigerant; bruised leaves are applied externally to disperse swellings. Root has febrifuge property. (Khare 2007, Swain *et al.* 2008, Quattrocchi 2012, Shekhawat *et al.* 2016; http://medplants.blogspot.in).

Medicinal Uses (Traditional/Folklore): Leaf paste is used on small-pox and fresh-leaf juice is diuretic. The paste prepared from fresh leaves mixed with castor oil is reported to cure *Tinea capitis*, a scaly fungoid infection on scalp. Roots are used as febrifuge and vermifuge by the tribal population in India. The juice of the leaves is considered cooling and aperient, and is given to children suffering from smallpox. Bruised leaves are applied to contusions to relieve pain and reduce swelling. In Bihar, the roots are used as a febrifuge by the tribal population. (Swain *et al.* 2008, Singh *et al.* 2013b, Shekhawat *et al.* 2016).

44. VERBENACEAE

Verbenaceae – Vervain family

A mixed family consisting of 34 plant genera and 1,035 species of herbs, shrubs, lianas and trees that range from creeping low Bank Mat to the mighty Teak tree; mainly tropical flowering plants notable for heads, spikes, or clusters of small flowers, many of which have an aromatic smell. The most well-known species is the common *Lantana*, an alien invasive species. The family was represented by two species under two genera in Chithara Village.

44.1 *Lantana camara* L.

(Syn. *Camara vulgaris* Benth.)

Common Name: Lantana, Common Lantana.

Vernacular Name: Raimuniya (राईमुनिया).

Habit: An aromatic shrub with short prickles.

Habitat: Common on roadsides, railway lines, and open fields.

Distribution: The native range of *L. camara,* an invasive alien species in India, is tropical America; however it has become naturalized in around 60 tropical and sub-tropical countries worldwide, including India.

Flowering and Fruiting Time: Throughout the year.

Habit

Txonomic Description: An erect or straggling, aromatic, dark green, much-branched shrub, up to 3 m tall. Stem and branches quadrangular with small recurved prickles. Leaves opposite-decussate, 2.5-7 x 1.5-2.8 cm, short-petioled, elliptic-lanceolate to ovate-oblong, crenate-serrate, acute to obtuse at apex, rugose above, sub-cordate at base, scabrous on both surfaces. Inflorescence umbel like. Flowers complete, bisexual, many coloured, orange to red, yellow or pink, sessile; bracts lanceolate, ciliate, exceeding the calyx. Calyx cup shaped, scabrous, base truncate, margin ciliate. Corolla tube slender, pubescent, obovate. Stamens-4, didynamous, epipetalous. Fruits drupaceous, globose, about 4.5 mm across, greenish when young, black shining when ripe.

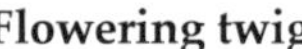

Flowering twig

Flowers

Drupes

Chemical Constituents: Phytochemical studies showed that the plant is rich in essential oils, monoterpenes, triterpenes, flavones coumarin, steroids, and iridoid glycosides. The major components of the oil are caryophyllene oxide, (-) - spathulenol, D-nerolidol, and (-) -β- caryophyllene. Others chemicals isolated are α-pinene epoxide , davana ether, 1-naphthalenol, α-caryophyllene, copaene, linalyl alcohol, ϱ-meth-1-ene-8-ol, germacrene, 4-terpineol, lilac alcohol, cis-β-terpineol, eucalyptol, lilac alcohol formate A and cis-verbenol. The bark contains lantanine. (Naz Bano 2013, Jawonisi and Adoga 2013, Murugesan *et al.* 2016).

Medicinal Properties: Leaf extracts of *Lantana* exhibit antimicrobial, fungicidal, antiprotozoal, antioxidant, antiviral and allelopathic, insecticidal and nematicidal properties. The root is sweet and bitter in taste, refrigerant and antifebrile. The leaves are aromatic with a minty taste having a cooling nature; they are antiphlogistic, diaphoretic, febrifuge, pectoral, stimulant, tonic and vulnerary. (Naz Bano 2013, Murugesan *et al.* 2016; http://medplants.blogspot.in, http://tropical.theferns.info).

Medicinal Uses (Traditional/Folklore): *Lantana* has been used for treating a variety of ailments, including cancer, skin itches, leprosy, rabies, chickenpox, measles, asthma and ulcers. *Lantana* oil is sometimes used for the treatment of skin itches, as an antiseptic for wounds and externally for leprosy and scabies. Different parts of the plant, mainly the leaves, have been used in treatment of stomachache, rheumatism, wound healing, biliary fever, toothache, bronchitis, antiseptic and other affections. A decoction of the dried flowers is used in the treatment of haemoptysis and pulmonary tuberculosis. A decoction of shoots is used to treat influenza, cough, mumps, incessant high fever, malaria, cervical lymph node tuberculosis, asthma, toothache, headache, inflammation, gonorrhoea and leucorrhoea. An infusion of the leaves and flowering tops is used in the treatment of fevers, constipation, tuberculosis, catarrh and bronchitis. (Murugesan *et al.* 2016; http://medplants.blogspot.in, http://tropical.theferns.info).

44.2 *Phyla nodiflora* (L.) Greene

(Syn. *Lippia nodiflora* (L.) Michx.; *Verbena nodiflora* L.)

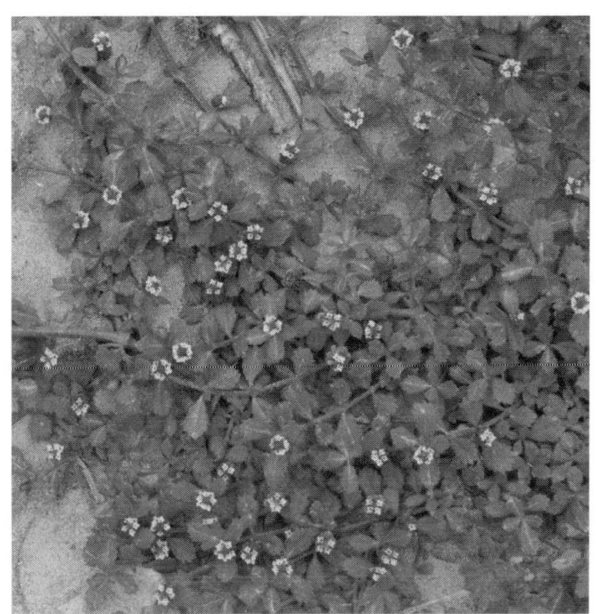

Habit

Common Name: Frog Fruit, Creeping Lip Plant.

Vernacular Name: *Hindi:* Jal Buti (जल बूटी).

Habit: Prostrate herb.

Habitat: Marshy areas, near water channels.

Distribution: Throughout India, up to 900 m in wet places along bunds of irrigation channels, canal edges and river banks.

Flowering and Fruiting Time: Most part of the year.

Taxonomic Description: A creeping or prostrate herb. Stem sub-quadrangular, branched from the base, often rooting at lower nodes. Leaves opposite, petiolate, 1.5 to 1.8 cm long, obovate-oblong, spathulate, entire towards the base, sharply serrated towards rounded or obtuse apex. Spikes axillary; flowers sessile, pink to white in dense globose heads. Bract obovate, ciliate towards the apex. Calyx small, hairy, deeply bi-lobed, linear-lanceolate. Corolla subequal, bi-lipped; upper lip bi-fid, lower 3-fid. Stamens-4, didynamous, filament short. Drupes globose, splitting into 2. Seeds-1, pyrene glabrous.

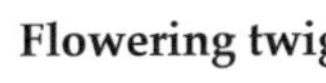

Flowering twig

Flowers

Chemical Constituents: The plant contains a variety of constituents such as triterpenoids, flavonoids, phenols, steroids, and many others. Nodifloridin-A and nodifloridin-B compounds have also been isolated from *P. nodiflora.* (Sharma and Singh 2013, Evelyne Ravindhran 2015).

Medicinal Properties: The plant is acrid, astringent, cooling, appetizing, stomachic, constipating, anthelmintic, digestive, vulnerary, aphrodisiac, ophthalmic, diuretic, alexiteric and febrifuge. (Warrier *et al.* 2010).

Medicinal Uses (Traditional/Folklore): In India, the plant has been used traditionally for the treatment of various health problems including respiratory disorders, gastrointestinal infections, diabetes, fungal infections, jaundice, and cardiovascular diseases. The plant is useful in vitiated condition of pitta, burning sensation, anorexia, flatulence, colic, dyspepsia, helminthiasis, diarrhoea, ulcers, ophthalmopathy, strangury, asthma, bronchitis, knee joint pain, gonorrhoea, irritation of internal haemorrhoids, cardiopathy, lymphadenitis, erysipelas and fever.

Phyla has been the source of traditional medicines against diseases like liver disorders, dandruff control, and indigestion in children. The juice of the plant is cooling and is used to relieve minor gastric troubles, fevers, coughs and colds. A poultice of the fresh plant is applied to ripen boils. A paste or poultice prepared from the plant is also applied to swollen cervical glands, to erysipelas, burns, and to chronic indolent ulcers. It is used for pain in knee joints, for lack of bowel movement, ulcers and boils. A decoction of leaves is given in stomachache. The roots are pounded with betel and given as a tonic after child birth. A decoction of leaves and roots is used in treating malaria and other fevers in Children in Malaysia. In Brazil and Central America, the plant is used as a menstrual disorder remedy. (Sharma and Singh 2013, Evelyne Ravindhran 2015; http://tropical.theferns.info).

45. LAMIACEAE

Lamiaceae – Mint family

The family consisting of 245 plant genera and 7,886 species of strong smelling herbs and shrubs has a cosmopolitan distribution. Many of the plants are aromatic and include widely used culinary herbs, such as basil, mint, rosemary, sage, savory, marjoram, oregano, thyme, lavender, etc. *Tulsi* is common household sacred and medicinally important plant of this family. Five species belonging to five genera of this family were recorded in the flora of Chithara Village.

45.1 *Anisomeles indica* (L.) Kuntze

(Syn. *Nepeta indica* L.)

Common Name: Indian Catmint, Malabar Catmint.

Vernacular Name: Kala Bhangra (काला भंग्रा).

Habit: Undershrub.

Habitat: Wastelands roadsides, along the bank of canals and rivers.

Distribution: *A. indica* is naturalized in some Pacific islands, South-East Asia and China. In India, it is distributed in wastelands, roadsides and scrub-jungles up to 1,500 m.

Flowering and Fruiting Time: July- October.

Flowers

Apical portion of plant bearing inflorescence

Taxonomic Description: An erect undershrub, up to 1.5 m tall, densely pubescent or woolly. Stem and branchlets 4-angled. Leaves opposite, 5-10 cm long, long stalked, ovate, acute, densely pubescent on both sides, base truncate-rounded, margin crenate-serrate. Flowers numerous, bluish-purple in long spike like inflorescence; bracts linear-lanceolate, stiff with hairs. Calyx teeth nearly as long as the tube, acute, lanceolate, ciliate. Corolla bi-lipped; upper lip pale-purple; lower lip deep-blue, bearded with red hairs. Stamens-4. Nutlets shining, black when ripe.

Chemical Constituents: The plant contains cembrane-type diterpenoid, two benzenoids, five flavonoids, and six phenyl propanoids; also present are anisomelic acid, anisomelolide, 2-acetoxymalabaric acid, anisomelyl acetate anisomelin, betulinic acid, ß sitosterol, citral, gerainic acid, malabaric acid, ovatodiolide and triterpenebetulinic acid. (Rao *et al.* 2009, Baranwal *et al.* 2012, Murthy *et al.* 2015).

Medicinal Properties: The plant has anticancer, antioxidant, anti-inflammatory, astringent, carminative, febrifuge and tonic properties. The leaves shows antimicrobial activity against a number of fungal and bacterial species such as *Bacillus anthracis, Proteus vulgaris, Salmonella stanley, Salmonella newport, Streptococcus agalactiae, Staphylococcus aureus, Escherichia coli, Aspergillus fumigatus* and *Aspergillus niger*. Several diterpenes have been isolated from the whole dried plants that have a cytotoxic action. (Dhiman 2004, Baranwal *et al.* 2012, Murthy *et al.* 2015; http://tropical.theferns.info).

Medicinal Uses (Traditional/Folklore): The plant is widely used as a medicinal plant in India, Sri Lanka and China, and on a smaller scale in Malaysia. Generally, the whole plant is used, but especially the leaves and the roots have more medicinal values. The extract of whole plant is a powerful astringent, carminative, febrifuge and tonic and used in the treatment of rheumatism, colds, fevers, abdominal pain, skin sores, and snake bites. Besides, the root-powder or extract is given to avoid nocturnal emission. An infusion of roots is useful in affections of the stomach and bowels, and in catarrh and intermittent fevers. The juice of the leaves is administrated to children for colic, dyspepsia, and fever caused by teething. Inhaling the vapor of the hot infusion induces copious perspiration. A decoction of the plant or essential oil distilled from the leaves is an excellent fomentation used externally as an embrocation in rheumatism and arthritis. A decoction of leaves is used as an antirheumatic and stomachic, and also to treat uterine infections, kidney gravel, swellings and hypertension. The dried or fresh leaves are used as a wash for external afflictions, eczema, pruritis, skin problems and snake bites. The leaves are chewed for toothaches and also used in rheumatism, cold, fever, abdominal pain, intermittent fever and dyspepsia. The dried plant is burned as a mosquito repellent. A decoction of aerial parts is used as an analgesic in Sri Lanka. In China and India, the plant is used to treat gastric dysfunction, inflammatory disorders and hypertension. *A. indica* was used in Ayurvedic traditional system of medicine for the treatment of urinary affections in Nepal. In Taiwan, it has been used for the treatment of diverse conditions such as inflammatory skin diseases, liver diseases, gastrointestinal disease, hypertension and immune system deficiencies. (Rao *et al.* 2009, Kunwar *et al.* 2010, Hsieh *et al.* 2016; http://tropical.theferns.info).

45.2 *Clerodendrum phlomidis* L.f.

(Syn. *Volkameria multiflora* Burm.f.)

Habit

Common Name: Arni (अर्नी).

Vernacular Name: *Hindi:* Arni (अर्नी); *Sanskrit*: Agnimantha (अग्निमंथा), Sriparni (श्रीपर्णी).

Habit: Large shrub.

Habitat: Roadsides, canal banks, and in fallow fields.

Distribution: India, Pakistan, Sri Lanka and Burma. Agnimantha is widely distributed throughout India. It is found mostly in dried parts. It commonly grows in upper Gangetic plains, Uttar Pradesh, Bihar, West Bengal, Orrisa, coastal areas of southern India and dry regions of western coast.

Flowering and Fruiting Time: August-November.

Taxonomic Description: A large, much-branched shrub. Branchlets 4-angled, young branches more or less hairy. Bark smooth, dark-brown. Leaves opposite, petiolate, 4-7 x 2-4 cm, broad ovate to rhomboid, cuneate or truncate at base, acute or obtuse at apex, coarsely crenate-dentate or undulate at margin, pubescent. Flowers white, borne in axillary and terminal dichotomous cymes, forming a large leafy panicle; bracts ovate-lanceolate. Calyx campanulate, nearly 1 cm long, green, 5-lobed, lobes ovate. Corolla white, tube 2-2.5 cm long, pubescent outside, lobes-5. Stamens-4, exserted, filaments slender. Drupes black, obovoid, 4-lobed.

Chemical Constituents: *Root:* β-sitosterol, γ-sitosterol, ceryl alcohol, clerodin ($C_{24}H_{34}O_7$), clerosterol ($C_{29}H_{48}O$) and clerodendrin A ($C_{27}H_{26}O_{17}$). α-L-rhamnopyranosyl-(1→2)-α-D-glucopyranosyl-7-O-naringin-4′-O-α-D-glucopyranoside-5- methyl ether ($C_{34}H_{44}O_{19}$). *Stem:* D-mannitol, β-D-glucoside of β-sitosterol, β-sitosterol and ceryl alcohol. *Aerial parts:* Lup-20(29)-en-3-triacontanoate ($C_{60}H_{108}O_2$), tetratriacontanol and 24β-ethylcholesta-5, 22E, 25-triene- 3β-ol have been isolated from aerial parts, besides pectolinarigenin, hispidium, cleodendrin, raffinosa, etc. (Raja and Mishra 2010, Rambhau *et al.* 2014).

Flowering twig

Inflorescence

Medicinal Properties: *C. phlomidis* is known for its analgesic and anti-pyretic properties besides having demulcent, alterative, stomachic properties. The alcoholic and aqueous extracts analgesic, antidiarrhoeal, antiplasmodial, hypoglycaemic, minor tranquilizers, antiasthmatic, antifungal, nematicidal, and antiarthritic. Root is bitter tonic. (Raja and Mishra 2010, Rambhau *et al.* 2014; http://medplants.blogspot.in).

Medicinal Uses (Traditional/Folklore): *Agnimantha* is very essential medicinal plant mentioned in texts since Vedic period. It is described as one of the ten herbs of *Dashmoola. Agni* means fire and *mantha* means to shake and its wood is rubbed together to produce fire in ancient times. Plant parts are used in treating dyspepsia, stomachache, colic, cholera, dysentery, postnatal fever, during convalescence from measles, haemorrhoids, cough and asthma, to boost immune system and purify blood, and urinary tract infection. Root and bark-bitter tonic, used in debility, nervous disorders, dysuria and retention of urine. The plant is reported to be useful in inflammation, glycosuria, pox, coryza, scrotal enlargement and postnatal complaints. In southern India, the juice of leaves is given in neglected syphilitic complaints in doses of half an ounce or more twice daily. Its decoction is useful to improve strength and immunity after fever. It is also useful in premature ejaculation. For freckles, the bark paste in milk is applied topically. The roots of the plant are boiled in water to make decoction and given in dose of 100 ml to treat swelling in body and for in painful micturition. Root is bitter tonic and given in convalescence of measles. Juice of leaves is alterative and given in neglected syphilitic complaints. The root is given as a demulcent in gonorrhoea, and decoction of the plant is considered as an alternative. Pulp of roots is given orally to treat urticaria and other skin disorders. As a veterinary medicine it helps cure stomach troubles and swellings in cattle. (Vardhana 2008, Raja and Mishra 2010, Rambhau *et al.* 2014; http://www.planetayurveda.com).

45.3 *Hyptis suaveolens* (L.) Poit.

(Syn. *Ballota suaveolens* L.)

Common Name: American Mint, Bush Mint.

Vernacular Name: *Hindi*: Vilaiti Tulsi (विलायती तुलसी); *Sanskrit*: Bhustrna (भुस्त्रैण).

Habit: Undershrub.

Habitat: Roadsides, dry areas, along railway lines, wastelands, watercourses, pastures and open forests where the soil is well drained.

Distribution: Native to tropical America, the plant has spread to tropical and subtropical areas worldwide from plains to an elevation of 1,600 meters. It can form dense thickets in all areas of growth. In India, *Hyptis* is found in the tropical dry

Habit

deciduous forests of the Vindhyan highlands, which lies in North Indian River Plain, between the Gangetic plains and Narmada valley in North India spreading to northern plains and hills; also occurs in Andaman and Nicobar.

Flowering and Fruiting Time: September-December.

Taxonomic Description: An erect strongly sweet smelling, branched undershrub, 1-2 m tall with 4-angled stem. Stem and branches hispid-tomentose. Leaves opposite, petiolate, generally 2-8 cm long, broadly ovate, rounded or cordate at base, acute at apex, serrate-denticulate, slightly pubescent above, densely villous or tomentose underside. Flowers blue; 2-4 flowered cluster together in axillary and terminal cymes. Calyx 5-lobed, linear; tube campanulate, hispid. Corolla bi-lipped, blue; upper lip-2, lower lip-3. Stamens-4, diadynamous, filaments pilose. Nutlets generally 2, oblong-obovate, flat, glabrous, blackish brown.

Flowering twig

Flowers

Chemical Constituents: *H. suaveolens* is an important source of essential oils, alkaloids, flavonoids, phenols, saponins, terpenes, tannins and sterols, for example diterpenes: suaveolic acid, suaveolol, methyl suaveolate, two steroids: β-sitosterol, β- sitosterol glycoside, two phenolic constituents: rosamarinic acid and methyl rosmarinate along with some other important constituents; also present are oleanolic acid or oleanic acid, ursolic acid, 3β-hydroxy-lupoic acid, 3β-urșenoloic acid, 1,19 dihydroxy-ursdienoic acid and 3β- hydroxyl lupenoic acid. Eucaliptol was the most abundant component in the oil, followed for gama-ellemene, beta-pynene, (+) 3-carene, trans-beta-cariophyllene and germacrene. (Prince *et al.* 2013).

Medicinal Properties: The plant has numerous therapeutic properties such as stimulant, carminative, sudorific, galactagogue, anticatarrhal, antimalarial, anthelmintic, antitumour, anticancer, antifertility, antifungal, anti-inflammatory, antiplasmodial, antispasmodic, antinociceptive, emmenagogic, antisoporific, antirheumatic, anticephalalgic, antipyretic (fever, flu), febrifuge, stomachic, dyspepsia, carminative and insect repellant. The principle constituent of the essential oil is menthol. (Wiart 2006b, Prince *et al.* 2013; http://tropical.theferns.info, http://medplants.blogspot.in).

Medicinal Uses (Traditional/Folklore): Traditionally, the entire plant is used for the cure of various ailments such as headache, convulsions, migraine, uterine affections, parasitical cutaneous diseases, catarrh and skin complaints; essential oil is a remedy for toothache, also applied on forehead for headache.

Roots are chewed with betel nuts as a stomachic and its decoction is used as an appetizer while some parts of the plant are used for the treatment of headache. A decoction of the root is said to be emmenagogic, and a stimulant if utilized in rheumatism. Roots are used in the treatment of a wide range of conditions including flatulence and other stomach problems and fevers associated with colds. Root extract is given for hematuria and bark for diarrhoea and dysentery.

Leaf extract or juice of *H. suaveolens* is applied on measles for early relief and infusions for colds, flu, fever, malaria, constipation. Decoction of leaves and or stems with leaves is used for coughs, asthma and respiratory infections, skin diseases, colds, stomachache; leaves crushed and applied for sprain and swellings; in convulsions, child bathed with mixture of mother's urine and crushed leaves. Squeezed leaves are used for boils, eczema, cuts and wounds, as antiseptic healer. Seeds soaked in water are applied on wounds allowing pus to flow out; blackish seeds soaked in a glass of milk and taken for spermatorrhea. Leaves of the plant and rhizome of *Zingiber officinale* are crushed together to paste which is taken (half a teaspoon) twice daily to cure dysentery. Furthermore, the plant/leaf extracts are used to cure swellings, abscesses haemorrhoids and also as memory aid and many other ailments. Plant infusion is used in infections of the uterus; leaf juice is taken in cases of colic and stomachache; the juice of leaves, mixed with lime juice, is taken for stomach aches. Applied externally, it is used as a wash or poultice on skin disorders such as dermatitis and eczema, boils, headaches, etc. A poultice of the pounded fresh material is applied as a poultice on snake bites. The juice of leaves is applied daily between the toes as a treatment for athlete's foot. The leaf paste is applied on sores and fungal skin infections. A high concentration of omega-6 lipids in the seed suggests *Hyptis* oil to be an ideal product for dry and flaky skin. In trials, the essential oil has shown many promising properties including antibacterial and antifungal activity; plus a better anti-inflammatory activity than diclofenac sodium, which is a standard marketed formulation.

American Indians make tea from the roots of *H. suaveolens* which purify the blood, and is also used as a remedy for the "diseases" of women. A leaf poultice is applied to cancers and tumours in South America. It is used in Africa for stomachache and colic, headaches, fever and as a general beverage. In West Africa, the leaves of *H. suaveolens* are employed as antifertility agent. In case of a burning sensation when passing urine (dysuria) and other urinary complains, dry seeds of *H. suaveolens* are soaked overnight in a glass of water and taken in the morning on an empty stomach along with small amounts of sugar for about a week. *H. suaveolens* serves in West Africa as an acceptable substitute in infusion for tea which is carminative, sudorific (causing or increasing sweat), lactogenic, anti-catarrhal and antiparasitic. Also, *H. suaveolens* is popularly used in the treatment of respiratory and gastrointestinal infections, indigestion, colds, pain, fever, cramps and skin diseases. The leaves are used as an anticancer and antifertility (in females) agent, while their aqueous extract has showed an antinociceptive effect and acute toxicity. Leaf sap of *H. suaveolens* with lemon juice is taken in Sierra Leone for stomachache and the leaf is applied around the head for headache. Leaves are also used in treating infection of uterus and parasitic skin diseases.

In China, *H. suaveolens* is used to combat fever and to invigorate health. In Malaysia, a decoction of 15 g of dried herb is used to combat fever, assuage headache and to promote digestion; the plant is used externally to soothe skin rashes, eczema and to resolve swellings. In Indonesia, the plant infusion is used to treat catarrhal (inflammation of mucous membranes, especially of the nose and throat) conditions, affections of the uterus, parasitic cutaneous diseases while the leaves are used as stomachic. In Philippines, the leaves are used for the antispasmodic, anti-rheumatic, antisoporific, promote menses as well as to invigorate health. A decoction of the plant is used to assuage stomachache and to treat dysentery. In Taiwan, the aerial parts are used to assuage pain and skin discomfort, whereas in Vietnam, the plant is used to increase milk secretion (lactation).

The very strong aromatic mint/thyme-like smell leads to the use of the plant (leaf juice) as an insectifuge, especially against mosquitoes. Leaves are spread on the cot to ward off bed bugs. The plant is used as a veterinary medicine. Plants are kept on eggs being incubated by hen and in the cages to protect poultry from insect and fungal infestation. Leaves are very effective against sucking insects infesting the livestock. Dried aromatic leaves are used in smoking in cattle sheds to repel flies and mosquitoes. (Wiart 2006b, Singh *et al.* 2010d, Prince *et al.* 2013; http://tropical.theferns.info, http://medplants.blogspot.in).

45.4 *Leucas cephalotes* (Roth) Spreng.

(Syn. *Leucas capitata* Desf.)

Common Name: Head Leucas.

Vernacular Name: *Hindi:* Guma (गूमा), Goma Madhupati (गोमा मधुपति); *Sanskrit:* Katumba (कुटुम्बा).

Habit: Annual herb.

Habitat: Moist and waste places and in cultivated fields.

Distribution: Throughout India, as a weed, ascending up to an altitude of 1,800 m in the Himalayas.

Flowering and Fruiting Time: July-December.

Taxonomic Description: An erect, annual, pubescent herb, 20-60 cm high. Stem and branches obtusely angular. Leaves hairy, opposite, petiolate, 4-8 cm long, ovate-lanceolate, crenate-serrate, apex acute to sub-acute. Flowers in terminal large, dense whorls. Bracts

Habit

numerous, leafy, hairy, ciliate, covering the floral heads. Calyx tubular, slightly curved, mouth oblique, throat hairy, teeth-10. Corolla 2-lipped; lower lip longer, 3-lobed; upper smaller, white woolly. Nutlets, smooth, obovoid-oblong, dark brown.

Apical shoot with floral head

Flowers

Chemical Constituents: Plants contain β-sitosterol and its glycoside, labdane, norlabdame and abietane type diterpenes, leucasdins A, B and C, respectively, and two prostostane-type triterpenes such as leucastrins A and B; other specific phytoconstituents isolated are oleanoic acid, 7-oxositosterol, 7-oxostigmasterol, 7α-hydroxystigmasterol stigmasterol, 5-hydroxy-7, 4′-dimethoxyflavone, pillion, gonzalitosin I, tricin, cosmosin, apigenin 7-O-beta-D-(6-O-p-coumaroyl) glucopyranoside, anisofolin A and luteolin 4′-O-beta-D-glucuronopyranoside. (Bhutya 2011, Chavan and Nishteswar 2013).

Medicinal Properties: *L. cephalotes* possess stimulant, anthelmintic, thermogenic, emmenogoue, carminative, digestive, sudorific, antipyretic, expectorant, antiseptic, antihepatic (jaundice), and depurative, diaphoretic and insecticidal properties. (Bhutya 2011, Ansari *et al.* 2013, Chavan and Nishteswar 2013).

Medicinal Uses (Traditional/Folklore): One of the plant's most common historical uses is treatment for snakebite. The plant is used for treating cough, colds, bleeding piles, asthma, cephalalgia, helminthiasis, loss of appetite, psoriasis, inflammations, bronchitis, dysmenorrhoea, amenorrhea, colic, intermittent fevers, skin diseases (psoriasis, scabies, chronic skin eruptions), as a blood purifier and eye diseases.

Leaf paste is applied on old boils on head by the Kol tribe of Uttar Pradesh, India. In Rewa district, Madhya Pradesh, tribal chew or massage the paste of leaves over gums and teeth to cure mouth ulcers and headache. Decoction of flower and leaves with black pepper is given as expectorant and anthelmintic. Flowers are efficacious against burns. The root is particularly useful in regularization of menstrual cycle, tuberculosis and dysentery. It is used as homeopathic drug in the diagnosis of chronic malaria and asthma in many parts of India, mainly in North India. The plant is steeped in water which is then used for bathing and for washing livestock. (Ansari *et al.* 2013, Bhoria and Kainsa 2013, Chavan and Nishteswar 2013; http://tropical.theferns.info, http://www.flowersofindia.net).

45.5 *Nepeta hindostana* (B.Heyne ex Roth) Haines

(Syn. *Glechoma hindostana* B.Heyne ex Roth; *Nepeta ruderalis* Buch.-Ham. ex Benth.)

Common Name: North Indian Catmint, Catmint.

Vernacular Name: *Hindi*: Billilotan (बिल्लीलालटेन); *Sanskrit*: Badranj Boya (बदरंज बोया).

Habit: Annual herb.

Habitat: Generally found near small water channels.

Distribution: Native to Afghanistan and tropical Asia. In India, found in Punjab, Upper Gangetic plains, Bihar, West Bengal, Madhya Pradesh, Rajasthan and Deccan Plateau up to an altitude of 2,400 m in the Himalayas.

Flowering and Fruiting Time: May-September.

Taxonomic Description: An erect, aromatic, annual herb, grows up to 60 cm in height. Stem and branches 4-angled, grooved, densely hairy. Leaves opposite, petiolate, 2-5 x 2-4.5 cm, broad-ovate, crenate-serrate, cordate or truncate at base. Flowers bluish-white, arranged in dense, many flowered axillary cymes; bracts linear-lanceolate. Calyx nearly 4 mm long, densely hairy. Corolla 2-lipped. Stamens-4, didynamous, ascending under the upper lip. Nutlets minute, ellipsoid.

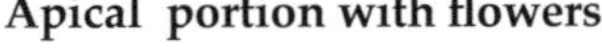

Apical portion with flowers

Habit

Chemical Constituents: The most important active ingredient in the herb is aldehyde nepehinal which is a triterpenoid; the other terpenoids found in Catmint include nepetidone, nepedinol and triterpenic acid, which render the herb its therapeutic properties. GC and GC-MS analysis of the oil revealed recognition of thirty three major and minor compounds, representing 91.0% (area percent) of the total oil sample. Of which, (E)-β- farnasene, spiro decan-1-one, 6-hydroxy, spiro decan-6-ol, 6-methyl, β-caryophyllene and ageratochromene were the major constituents while α- pinene, camphene, β-ylangene, linalool and trans-α-bisabolene epoxide were present in low amounts. The oil was characterized by higher content of sesquiterpene hydrocarbons followed by oxygenated monoterpenoids and oxygenated sesquiterpenoids and only 1.2% of monoterpene hydrocarbons. (Kainsa and Singh 2015, Pandey *et al.* 2015, Siddique *et al.* 2017).

Medicinal Properties: N. *hindostana* is used as carminative, stimulant, tonic, diaphoretic, emmenagogue, antispasmodic, and aphrodisiac. It has a number of pharmacological properties like cardioprotective, antiasthmatic, hypocholesteremic, hypotensive, antiplasmodial, antipyretic, and anti-inflammatory. (Pandey *et al.* 2015, Siddique *et al.* 2017).

Medicinal Uses (Traditional/Folklore): In Indian medicine, the cardio-protective properties of the herb have been well documented in treating various cardiovascular ailments. It is very effective in treating a variety of heart related conditions like angina pectoris, cardiac thrombosis, tachycardia and cardiac weakness. Traditionally, it is used as a cardio-tonic to support healthy heart functions. It is also used as brain tonic and in hypochondriac conditions. Also effective as a cardiac tonic and relieves rheumatic pain and breast inflammation. The herb is also used for treating and preventing asthma attacks and in hysteria, chlorosis, colic, amenorrhea and toothache; it strengthens gums and improves mouth odor. The extract of the herb is effective in reducing chronic anxiety. As a veterinary medicine the plant extract is reported to lower the blood cholesterol level in animals. (Vardhana 2008, Mall and Tripathi 2015, Siddique *et al.* 2017).

46. NYCTAGINACEAE

Nyctaginaceae – Four o'clock family

Family of around 34 plant genera and 450 accepted species names consists of herbs, shrubs and trees. It is widely distributed in tropical and subtropical regions, with a few representatives in temperate regions. Members of the family have smooth-edged leaves and petalless flowers. The flowers have a small, tubular, petal-like surrounding structure called the perianth—probably equivalent to the sepals in other plants—and coloured bracts (leaflike structures) that fuse and may resemble petals. In some species these bracts are like sepals and enclose a single, brightly coloured perianth whorl. The small dry fruit bears grooves or is winged; it consists of a single-seeded fruit surrounded by the persistent perianth tube. Herbs like *Punarnava* (*Boerhavia diffusa*) and shrubs and climbers like *Bougainvillea* belong to this family. One species belonging to this family was recorded in Chithara Village.

46.1 *Boerhavia diffusa* L.

(Syn. *Boerhavia adscendens* Willd.)

Habit

Common Name: Red Spiderling, Spreading Hog Weed.

Vernacular Name: *Hindi:* Punarnava (पुनर्नवा), Gadahpurna (गदहपूर्णा); *Sanskrit*: Punarnava (पुनर्नवा), Raktakanda (रक्तकाड़ा), Shothaghni (शोथघ्नी), Varshabhu (वर्षाभू).

Habit: Prostrate perennial herb.

Habitat: Grows as a weed on wastelands, roadsides, railway lines, crevices of old walls and foot paths after the rainy season.

Distribution: The genus *Boerhavia* has several species, and is distributed in the tropical, subtropical, and temperate regions of the world. It is found in Australia, China, Egypt, Pakistan, Sudan, Sri Lanka, South Africa, USA and in several countries of the Middle East. Out of the 40 species of this genus, six species are found in India – *B. diffusa, B. chinensis, B. erecta, B. repens, B. rependa*, and *B. rubicunda*. It is found throughout the warmer parts of the country up to an altitude of 2,000 m in the Himalayan region.

Flowering and Fruiting Time: Major part of the year.

Taxonomic Description: A prostrate, tap rooted, perennial herb. Stem trailing, purplish, much-branched, thickened at the nodes. Leaves opposite, ovate, sub-orbicular, 2-5 x 2-4 cm, rounded or truncate at base, obtuse or acute at apex, entire, glaucous beneath, petioled. Inflorescence axillary, terminal panicles, sparsely pubescent. Flowers pink to purple, about 3 mm across; 2-10 flowere cluster in long peduncle. Perianth lobes-5, lobes emarginated. Stamens 2-3, slightly exserted. Stigma capitate. Fruits oblong, 3-4 mm long, ribbed, viscid-glands.

Flowering twig

Flowers

Fruits

Chemical Constituents: Recent studies have revealed existence of a wide range of chemical compounds having good efficacy in treating different disorders. Chemically the plant contains important bioactive compounds such as flavonoids, alkaloids, steroids, triterpenoids, lipids, lignin, carbohydrates, proteins, and glycoproteins. Punarnavine ($C_{17}H_{22}N_2O$; M.P. 236–237°C), hypoxanthine 9-larabinofuranoside, ursolic acid, punarnavoside, lirodendrin, and a glycoprotein have been isolated and studied in detail for their biological activity. Punarnavine itself and the rotenoid class of molecules (Boeravinones) are the main active ingredients of the plant. (Mishra *et al.* 2014, Rao 2016).

Medicinal Properties: In Ayurveda, the plant known as *Punarnava* (in Sanskrit) means that rejuvenates or renews the body. The plant is bitter, cooling, aphrodisiac, antioxidant, astringent, anthelmintic, and anti-inflammatory, cardiac stimulant, diuretic, diaphoretic, emetic, expectorant, febrifuge, laxative and tonic. It is useful in cardiac disorders, jaundice, anaemia, bronchitis, constipation, cough, dyspepsia, leucorrhoea, lumbago, myalgia, ophthalmia, scabies, strangury and general debility. (Warrier *et al.* 2010, Mishra *et al.* 2014, Rao 2016; http://tropical.theferns.info, http://medplants.blogspot.in).

Medicinal Uses (Traditional/Folklore): Historically, *B. diffusa* is used as a single remedy and also as a chief constituent in many traditional and folk medicinal preparations. Various parts of *B. diffusa* i.e., root, leaves, aerial parts or the whole plant have been employed for the treatment of numerous disorders in Ayurveda and other system of medicine with its mystic effects. Medicinally, *Punarnava* root is useful as antidiabetic, anti-inflammatory, diuretic, hepatoprotective, etc. In different parts of the world, *B. diffusa* has also been used for pain relief, anti-inflammation, and treating indigestion. Roots are chewed as tonic. *Punarnava* can even be used to treat cataract; leaf extract is used in eye diseases, especially in the treatment of cataract. Leaf juice with honey, dropped into the eyes for chronic ophthalmia. Fresh root juice of *Punarnava* is put into eyes so as to get relief from various eye ailments like night blindness and conjunctivitis.

Punarnava acts as a carminative, increases appetite, digests *Ama* (residue of undigested or partially digested food in Ayurveda) and reduces abdominal pain. Plant infusion is given in asthma and bronchitis; it clears the lungs and reduces cough, asthma, etc. In Arthritis, it helps to reduce inflammation and pain in joints. It also relieves constipation. It rejuvenates male reproductive system and acts as an aphrodisiac. This increases libido, erection and quality and quantity of semen. It detoxifies skin and clears it of infections. It is very useful in treating skin diseases like scabies. It is also considered as an excellent natural remedy for guinea worms. The plant rejuvenates liver and detoxifies it. It helps to get rid of renal calculi (kidney stones). In Ayurvedic preparations for treating renal diseases it acts as a diuretic and cleanses the kidneys and is used for renal calculi (kidney stone), cystitis and nephritis. In menorrhagia, *B. diffusa* is used as an active ingredient in Ayurvedic preparations for menorrhagia. In Brazilian herbal medicines, it is employed to stimulate the emptying of the gallbladder and for all types of liver disorders.

Punarnava rejuvenates the whole body and gives a new life and health. It strengthens the body and normalizes all the *doshas*. This helps to boost the immunity to diseases. Thus, the plant *Punarnava* acts on all important organ systems and rejuvenate them, giving a new life. Hence, the name *Punarnava*. (Shukla *et al.* 2010, Rahul 2013, Mishra *et al.* 2014, Rao 2016; http://medplants.blogspot.in, http://tropical.theferns.info, http://uses.plantnet-project.org/en/Boerhavia_diffusa_(PROTA).

47. AMARANTHACEAE

Amaranthaceae – Amaranth/Cockscomb family

This is a large family known as the Amaranth or Cockscomb family consisting of 178 plant genera and 2,052 accepted species of herbs and shrubs, distributed nearly worldwide. The scientific name of this family is derived from Greek *Amarantos* (unfading), referring to the long-lasting flowers. A number of species, including beet root (*Beta vulgaris*) and *Chenopodium*, are important food crops, and several are cultivated as garden ornamentals. A total of 12 species under eight genera belonging to this family were recorded in Chithara Village.

47.1 *Achyranthes aspera* L.

(Syn. *Achyranthes acuminata* E.Mey. ex Cooke & Wright; *Achyranthes canescens* R.Br.)

Habit

Common Names: Prickly Chaff Flower, Devil's Horsewhip.

Vernacular Name: *Hindi:* Chirchita (चिरचिटा), Latjira (लटजीरा); *Sanskrit:* Apamarga (अपामार्ग).

Habit: Annual or perennial herb.

Habitat: Abundant along roadsides, railway lines, and in wastelands, and open and fallow fields.

Distribution: It is distributed throughout the tropical world, in many places growing as an introduced species and a common weed. It is an invasive species in some areas; found throughout India, along roadsides, field boundaries and wastelands as a weed as well as on hills up to 2,100 m. Plants of rainy season and moist shady areas are much luxuriant and less hairy compared to plants of open dry areas and summer plants.

Flowering and Fruiting Time: Throughout the year.

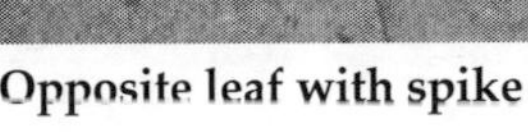

Opposite leaf with spike

Flowers

Taxonomic Description: An erect, up to 1.2 m high, annual or perennial herb with woody base. Stem branched, angular, thickened at the nodes, pubescent. Leaves opposite-decussate, variable, obovate or elliptic to sub-orbicular, 2.5-8 x 1.5-5 cm, tomentose, acute to acuminate or rounded at apex, shortly petioled. Flowers about 4 mm long, bisexual, numerous in axillary or terminal, 10-35 cm long tomentose spikes. Bracts and bracteoles ovate, persistent, awned, shorter than perianth. Perianth 5-partite, greenish-white, about 4 mm long, ovate-lanceolate. Stamens-5; filaments purplish, filiform. Utricle oblong, enclosed by the persistent perianth.

Chemical Constituents: *A. aspera* contains triterpenoid saponins which possess oleanolic acid as the aglycone. Ecdysterone, an insect moulting hormone, and long chain alcohols are also found. Medicinally important chemicals like achyranthine, betaine, pentatriaontane, 6-pentatriacontanone, hexatriacontane and tritriacontane are also present.

Root contains triterpenoid, saponins, betaine, achyranthine, hentriacontane, ecdysterone (polypodine A) and ecdysone. Fruit contains two oleanolic acid base saponins; two new saponins C & D also have been isolated from fruits. Seed oil contains linoleic, oleic, palmitic, stearic, behenic, arachidic, myristic and lauric acids. (Srivastav *et al.* 2011, Hasan 2014).

Medicinal Properties: The plant is acrid, bitter, thermogenic, expectorant, revulsive, carminative, digestive, stomachic, laxative, anodyne, depurative, anthelmintic, diuretic, lithontriptic, sudorific, demulcent, and anti-inflammatory. It is useful in cough, asthma, dyspepsia, dropsy, vomiting, leprosy and skin diseases. Extracts from roots have been reported to possess spermicidal activity. (Warrier *et al.* 2010, Srivastav *et al.* 2011, Hasan 2014; http://tropical.theferns.info, http://medplants.blogspot.in).

Medicinal Uses (Traditional/Folklore): *A. aspera* has occupied a pivotal position in Indian culture and folk medicine. Since ancient times the tribal and rural people of India commonly use this herb in various disorders. The plant is known for its mood enhancing effects and is used to treat depression and forms of anxiety. Plant is useful in cough, asthma, bronchitis, dyspepsia, flatulence, colic, painful inflammation, dropsy, opthalmopathy, vomiting, leprosy, skin diseases, pruritus, helminthiasis, strangury, renal and vesical calculi, cardiac disorders, anaemia and general debility. Seeds are used as emetic and in treatment of hydrophobia. Unripe fruits are used in the treatment of respiratory diseases.

In Western India, the juice prepared from the plant is applied to relieve toothache. The ashes of dried and slow-burnt plant with honey are given to relieve cough; the root in dose of *ca.* 10-12 g is given at bedtime for night blindness. Root paste prepared with water is used as an *anjan* (eye salve) in opacities of the cornea. In Uttar Pradesh, the plant is used for a great many medicinal purposes, especially in obstetrics and gynecology, including abortion, induction of labour, and cessation of postpartum bleeding. In Rajasthan, the plant is used in tooth ache and snake bites. The Maasai people of Kenya use the plant medicinally to ease the symptoms of malaria. (Srivastav *et al.* 2011, Hasan 2014; http://tropical.theferns.info, http://medplants.blogspot.in).

47.2 *Alternanthera paronychioides* A.St.-Hil.

(Syn. *Alternanthera leucantha* Moq.)

Habit

Common Name: Smooth Chaff Flower.

Vernacular Name: None Known.

Habit: Prostrate herb.

Habitat: Common in damp places, along the banks of the wetlands, canals and ponds.

Distribution: Native of tropical America, getting established in tropical Asia as a weed.

Flowering and Fruiting Time: March-October.

Taxonomic Description: A prostrate, appressed pubescent, tufted, perennial herb. Stem much-branched, terete, rooting at the nodes. Leaves opposite, 1-3 x 0.3-0.7 cm, elliptic, spathulate, base attenuate, margin entire, apex acute or obtuse. Flowers white, arranged in dense, 5-10 mm long, globose, axillary spikes. Bracts and bracteoles present. Tepals-5, sub-equal. Stamens-5, anthers yellow. Utricle orbicular. Seed-1, black, sub-orbicular.

Chemical Constituents: *A. paronychioides* are known to possess phenol, flavonoids, polyphenolics, ferulic acid and quercetin. (Wu *et al.* 2013, Sufian *et al.* 2017).

Tuft of branches with spathulate leaves | Twig with spike

Medicinal Properties: The leaves of *A. paronychioides* have thrombolytic and membrane stabilizing property. Antioxidant activity of this plant can protect pancreatic β-cells from glucotoxicity. (Sufian *et al.* 2017).

Medicinal Uses (Traditional/Folklore): Traditionally, the plant is used as carminative, poultice and dermatitis. It improves the ability of insulin to lower the blood sugar levels in both type 1 and type 2 diabetes. (Wu *et al.* 2013, Sufian *et al.* 2017).

47.3 *Alternanthera philoxeroides* (Mart.) Griseb.

(Syn. *Achyranthes philoxeroides* (Mart.) Standl.); *Bucholzia philoxeroides* Mart.)

Habit

Common Name: Alligator Weed, Pig Weed, Smooth Chaff Flower.

Vernacular Name: None Known.

Habit: Aquatic to semi aquatic herb.

Habitat: Commonly growing in ponds, lakes, along the banks of rivers and canals.

Distribution: *A. philoxeroides* is native to tropical South America; it originated in the Parma river region of South America. In India, it is widely distributed in most of the States, except in Rajasthan, Gujarat, Punjab, Himachal Pradesh, and Jammu & Kashmir. It is one of the greatest threats to rivers, wetlands and irrigation systems in the world. Naturalized populations of the plant exist in California and the south-eastern United States, Puerto Rico, France, Italy, India, Sri Lanka, China, Myanmar, Thailand, Singapore, Indonesia, Papua New Guinea, New Zealand and Australia.

Flowering and Fruiting Time: November-April.

Taxonomic Description: A prostrate or ascending, aquatic to semi aquatic herb. Stem fistular, creeping and ascending, much-branched, spreading, farming dense masses, often rooting at the nodes. Leaves opposite, 5-12 x 1-2 cm, elliptic, oblong-obovate, attenuate at base, margin entire, apex acute or obtuse, glabrous, sub-sessile with prominent mid-rib on lower surface. Flowers bisexual, arranged in globose heads; heads 0.8-1.5 cm in diameter, axillary, long peduncled; bracts and bracteoles sub-equal, ovate-lanceolate. Perianth white; tepals-5, sub-equal, oblong-lanceolate, longer than the bracts. Stamens-5, united at the cup shaped base; anthers 1-celled, staminodes distinct. Ovary ovoid, stigma capitate.

Flowering twig

Open flower in globose heads

Chemical Constituents: The plant contains bioconstituents such as flavonoids, glycoside, saponin, tannin, free reducing sugar and amino acids; however, alkaloids were found to be absent. Nine specific chemical compounds characterized from the plant are: phaeophytin a, pheophytin a′, oleanoic acid, β-sitosterol, 3β-hydroxystigmast-5-en-7-one, α-spinasterol, 24-methylenecycloartanol, cycloeucalenol and phytol. (Dutta 2015, Pulipati *et al.* 2015, Sivakumar and Sunmathi 2016).

Medicinal Properties: Whole plant has allelopathic, laxative, anticonstipation, and antiasthmatic properties. (Dutta 2015, Pulipati *et al.* 2015).

Medicinal Uses (Traditional/Folklore): The young stem and leaf of plant are used to treat dysentery, asthma, etc. in various regions of Assam. Juice prepared from young shoots and leaves (50-100 ml) with equal amount of water is given before bedtime for stomach pain regularly for a week by Bodo, Santhal, and Goreswar tribes of North East. In Manipur, the plant is used to treat anaemia, constipation, fever and diseases of eyes and also used as laxative. (Dutta 2015, Pulipati *et al.* 2015).

47.4 *Alternanthera pungens* Kunth

(Syn. *Alternanthera echinata* Sm.)

Habit

Common Name: Chaff-Flower.

Vernacular Name: Khaki-Weed (खाकी-वीड), Kante Wali Santhi (काटे वाली सांठी), Isharmul (ईशरमूल).

Habit: Prostrate herb.

Habitat: Waste places, path verges, fallow fields, along railway lines, usually on sandy, rocky soils.

Distribution: *A. pungens* is native to tropical America and now is pantropical. It is thought to have come from Central and South America, and to have become widely established in Australia, New Guinea and Southern Africa. In India, it is a weed of pebbly soils and dry locations.

Flowering and Fruiting Time: July-December.

Taxonomic Description: A prostrate, spreading, annual herb. Stem much-branched, terete, hispid, up to 30 cm long, often rooting at nodes. Leaves opposite, simple, 1.5-3.5 x 1-2.5 cm, unequal, sub-orbicular, entire, apex obtuse, tapering at base, sparsely pubescent, petioled. Heads sessile, axillary, 8-15 mm long, greenish-white. Bracts and bracteoles lanceolate, acute, about 4 mm long. Tepals-5, unequal, lanceolate. Stamens-5. Utricle ellipsoid. Seed-1, brown.

Flowering twig

Flower head

Chemical Constituents: From the plant α and β-spinasterols have been isolated. Flower oil contains many bioconstituents such as azulene, α-borneol, bornylacetate, camphene, camphore, 1,8-cineol, α-cucumene, p-cymene, elemol acetate, eudesmol, genaniol, limonene, linalool, myrcene, β-ocimene, α-pinene, β-pinene, α-terpineol, α-terpinolene and α-thujone. (Jakhar and Dahiya 2017; http://bsienvis.nic.in/Database/bsi_3946.aspx.).

Medicinal Properties: *A. pungens* is diuretic, antibacterial and antioxidant. (Jakhar and Dahiya 2017).

Medicinal Uses (Traditional/Folklore): A decoction of whole plant is used internally to treat gonorrhoea; decoction of whole plant with neem leaves is used in gonorrhoea in Ambala district, India. The plant is used in digestive disorders, diarrhoea, "hot diarrhoea" (caused by sun), and "empacho". The powder of plant is also used in jaundice. (Vashistha and Kaur 2013, Amjad *et al.* 2017; http://bsienvis.nic.in/Database/bsi_3946.aspx.).

47.5 *Alternanthera sessilis* (L.) R.Br. ex DC.

(Syn. *Gomphrena sessilis* L.)

Habit

Common Name: Sessile Joyweed, Dwarf Copperleaf, Joyweed

Vernacular Name: Garundi (गरुंडी), Guroo (गुरू).

Habit: Annual or perennial herb.

Habitat: Common in rice fields, rivers and canal banks, near small water bodies and along other moist and wet localities.

Distribution: Throughout India, in moist places up to 2,100 m, often cultivated. Very common along the margin of water bodies, drainage canal, roadside waste places; highly variable according to habitat.

Flowering and Fruiting Time: Most part of the year.

Taxonomic Description: A prostate or ascending herb with stem rooting at lower nodes, branchlets 20-50 cm long. Leaves opposite, shortly petioled, variable in shape and size, elliptic, linear-oblong to oblanceolate, base cuneate, margin entire, apex acute or obtuse. Spikes axillary, sessile, globose, 4-6 mm in diameter, solitary or 1-4 together. Flowers white in condensed spikes, bracts and bracteoles present. Tepals-5, ovate, about 2.5 mm long. Stamens-3. Utricle nearly 2-2.5 mm long, obovoid.

Flowering twig

Spikes

Chemical Constituents: The plant contains α and β-spinasterols; lupeol isolated from roots. Isolation of 24-methylenecycloartanol, cycloeucalenol, campesterol, sitosterol, 5 α-stigmast-7-enol and their respective palmitates, nonacosane, 16-hentriacontane, β-sitosterol, stigmasterol and handianol isolated. A saponin having oleanolic acid as aglycone and glucose and rhamnose as sugars has been isolated from leaves. (Sinha *et al.* 1984, Sivakumar and Sunmathi 2016; http://www.bsienvis.nic.in).

Medicinal Properties: The plant is sweet, astringent, slightly bitter, acrid, cooling, digestive, constipating, depurative, cholagogue, galactagogue and febrifuge. (Warrier *et al.* 2010, Anitha and Kanimozhi 2012; http://tropical.theferns.info).

Medicinal Uses (Traditional/Folklore): Plant is useful in vitiated condition of cough, burning sensation, diarrhoea, leprosy, skin diseases, dyspepsia, haemorrhoids, agalactia, splenomegaly and fever. Stems and leaves are useful in eye trouble. Decoction is taken with little salt to check vomiting of blood. Shoots with other ingredients are used to restore virility. Poultice made from the leaves is used for boils. Decoction of the fruits relieves itching and leaves are cooked as vegetable in Ambala district, India. (Warrier *et al.* 2010, Anitha and Kanimozhi 2012; http://tropical.theferns.info).

47.6 *Amaranthus viridis* L.

(Syn. *Amaranthus gracilis* Desf.)

Habit

Common Name: Slender Amaranth, Green Amaranth.

Vernacular Name: *Hindi:* Jangli Chaulai (जंगली चौलाई).

Habit: Annual herb.

Habitat: Abundant weed along the roadsides, wastelands, gardens and cultivated fields.

Distribution: Found throughout India in wastelands and roadsides as a weed.

Flowering and Fruiting Time: Most part of the year.

Taxonomic Description: An erect, glabrous, 20-80 cm long, annual herb. Stem-branched. Leaves alternate, long petioled, 2-9 cm long, 1.5-5.5 cm wide, ovate, rhomboid or oblong, base truncate or cuneate, usually emarginate at apex. Inflorescence spike. Flowers unisexual, pale green, small, arranged in small clusters at the end of branches. Bract and bracteole whitish, shorter than the perianth, ovate, acute, membranous with green keel. Stamens shorter than perianth. Perianth-3, mucronate. Stigma 2-3. Utricles 1-2 mm long, ovoid, beaked. Seeds minute, black, lenticular.

Fruiting twig

Ovoid and beaked utricles

Chemical Constituents: Principal constituents of this plant are saponins, flavonoids, rutin and quercetin. It is rich in minerals and contains sterols and fatty acids in the seeds. The stem and leaves contain oxalic acid. The powdered leaves yield tannin, with some reducing sugar and resin but no alkaloids were isolated. Spinosterol (24-ethyl-22-dehydrolathosterol) was identified as major component along with 24-methyllathosterol 24- ethyllathosterol, 24-methyl-22- 52 dehydrolathosterol, 24-ethyl cholesterol and 24-ethyl-22-dehydrocholesterol as minor components in sterol fraction. The roots of the plant contain amasterol (24-methylene-20-hydroxycholesta-5, 7-dien-3β-ol), a steroidal constituent. (Khan *et al.* 2011a, Ahmed *et al.* 2013c, Ragasa *et al.* 2015; http://shodhganga.inflibnet.ac.in/bitstream/10603/11731/5/chapter%203.pdf).

Medicinal Properties: Plants are vermifuge, diuretic, astringent, emollient, antirheumatic, antidiabetic, antiulcer, analgesic, antihyperlipidaemic and antileprotic, antiemetic and laxative. Plants are known to improve appetite and have antiviral, anticancer properties and show antinociceptive activity. (Ahmed *et al.* 2013c, Sarwar *et al.* 2016; http://tropical.theferns.info).

Medicinal Uses (Traditional/Folklore): The herb is used as astringent, emollient, in dysentery, and as an anthelmintic. It is used in inflammation, constipation, eczema, bronchitis, antidiabetic, anaemia and leprosy; it reduces labour pain and acts as an antipyretic. Other traditional uses range from an anti-inflammatory agent of the urinary tract, in venereal diseases, treatment of respiratory and eye problems and treatment of asthma. Roots/ shoots are used to control excessive menstruation, and as a blood purifier, digesting agent, and against piles.

The Negrios of the Philippines apply the bruised leaves directly to eczema, psoriasis, and rashes with good results. The leaves make a good emollient preparation available in some of the Filipino villages for insect bites, sunburn, and regular burns. The reddish-brown fibre from the leaves are soaked and used for eye treatments. The decoction of young roots is used for the treatment of respiratory complaints and asthma. (Ahmed *et al.* 2013c, Sarwar *et al.* 2016; http://tropical.theferns.info).

47.7 *Celosia argentea* L.
(Syn. *Celosia cristata* L.)

Habit

Common Name: Silver Cock's Comb, Quail Grass, Silver Spiked Cockscomb.

Vernacular Name: *Hindi*: Safed Murga (सफेद मुर्गा).

Habit: Annual herb.

Habitat: Common along roadsides, and in cultivated and fallow fields.

Distribution: *C. argentea* grows as a weed during rainy season throughout India and other tropical regions of the world such as Sri Lanka, South Asia, Africa and America.

Flowering and Fruiting Time: September-January.

Taxonomic Description: An annual herb, up to 1.5 m tall. Stem erect, green, ribbed. Leaves simple, alternate, sub-sessile, often tinged with red, 5-10 x 1-3 cm, oblong-lanceolate to linear-lanceolate, base attenuate, margin sub-entire, apex acute to acuminate; lower leaves larger than upper leaves. Flowers silvery-pink, borne on terminal, many flowered dense spikes; spikes 4-15 cm long with a conic apex. Perianth segments 5; segments oblong-lanceolate, 8-10 mm. Bracts and bracteoles lanceolate, shorter than perianth. Stamens-5; filaments delicate, cream coloured, anthers pink to purple. Capsules about 3.5 mm, ellipsoid. Seeds very minute, black, shining.

Flowering twig

Spike

A single flower

Chemical Constituents: Roots, Leaves and stem of *C. argentea* are known to contain alkaloids, amino acids, carbohydrate, proteins, flavonoids, glycosides, phenolic compounds, saponins, steroids, triterpenoids, gums, mucilage, and tannins. (Rani and Raju 2014).

Medicinal Properties: Seeds are antipyretic, improve vision, relieve fever associated with liver ailments; seeds are antidiarrhoeal and aphrodisiac, and antidote for snake-poison. (Rani and Raju 2014; http://medplants.blogspot.in, http://tropical.theferns.info).

Medicinal Uses (Traditional/Folklore): *C. argentea* is a medicinal plant having extremely small seeds which are used traditionally for the treatment of jaundice, gonorrhoea, and wounds and treating fever. In Indian folk medicine, whole plant is used for diabetes. Whole plant is also used as an antidote for snake-poison and in the traditional medicine for sores, ulcers, skin eruptions, mouth sores and skin eruptions. The leaves are used for the treatment of inflammations, fever and itching. They are efficacious remedy in diarrhoea. Poultice of leaves, smeared with honey, used as cooling application to inflamed areas and painful affections such as buboes and abscesses. Flowers and seeds are used for bloody stools, haemorrhoidal bleeding, leucorrhoea and diarrhoea. Traditionally, seeds are used for treatment of jaundice, gonorrhoea, wounds and fever. Seeds when in decoction or finely powdered, are considered antidiarrhoeal and aphrodisiac. The juice of the seeds applied into the nostrils cures epistaxis. Decoction of the seeds with sugar is prescribed against dysentery. Root is used for colic, gonorrhoea and eczema. The root, seed, and flower of *C. argentea* are used for redness, swelling of the eyes, photophobia, lacrimation, headache, colic, gonorrhoea and eczema.

In Sri Lanka, leaves are used for inflammations, fever and itching. Seeds used for fever and mouth sores. In China, flowers and seeds are used in the treatment of gastroenteritis and leucorrhoea. (http://medplants.blogspot.in, http://tropical.theferns.info).

47.8 *Chenopodium album* L.

(Syn. *Atriplex alba* (L.) Crantz; *Atriplex viridis* (L.) Crantz)

Common Name: Goosefoot, Wild Spinach, Lamb's-Quarters.

Vernacular Name: *Hindi:* Bathua (बथुआ).

Habit: Annual herb.

Habitat: The plant is common along roadsides, cultivated and non-cultivated fields.

Distribution: Native to Europe; found throughout India, in areas up to 4,200 m elevation, wild as well as cultivated.

Flowering and Fruiting Time: November-March.

Taxonomic Description: An annual herb, 20-150 cm high. Stem erect, branched, green to purplish-red, ribbed, mealy pubescent. Leaves variable, alternate, up to 8 cm long, rhomboic to oblong-lanceolate, margin entire to irregularly toothed, base cuneate; lower leaves long petioled, upper ones short petioled. Inflorescence terminal spikes. Flowers minute, bisexual, greenish-white, in axillary and terminal paniculate spikes. Perianth lobes-5, oblong-lanceolate. Stamens generally 5, slightly exserted, anthers dithecous, yellow. Ovary superior, stigma bilobed. Utricle enclosed in perianth lobes. Seeds minute, black, shining.

Habit

Flowering twig

Close up of inflorescence with few open flowers

Chemical Constituents: The plant contains essential oils, phenols, lignan, ecdysteroids, alkaloids, trigonelline, chenopodine and flavonoid as phenolic amide, saponin, cinnamic acid amide, alkaloid chinoalbicin, apocortinoid, xyloside, phenols and lignans as active phytoconstituents. Leaves are rich in potassium and vitamin C. Phytochemical analyses revealed the presence of alkaloids, apocarotenoids, flavonoids, phytoecdysteroids and an unusual xyloside in the plant. The essential oil from leaves of *C. album* contains abundance of limonene, α-terpinyl acetate, α- terpinene and cis-ascaridole. (Gohar and Elmazar 1997, DellaGreca et al 2004, DellaGreca *et al.* 2005a, Agrawal *et al.* 2014, Kumar and Kumar 2015).

Medicinal Properties: It is sweet, acrid, oleaginous, digestive, carminative, anthelmintic, diuretic, antiphlogistic, antirheumatic, aphrodisiac, contraceptive, odontalgic and tonic; useful in vitiated conditions of *Pitta*, peptic ulcer, helminthiasis, dyspepsia, flatulence, strangury, seminal weakness, pharyngopathy, splenopathy, haemorrhoids, ophthalmopathy, cardiac disorders and general debility. Pharmacological studies of plant reveal its hypoglycaemic, antibacterial, spasmolytic, antipruritic, anti-inflammatory, hepatoprotective, antioxidant, and anticancer properties. As therapeutic agent, it is used as laxative, anthelmintic against round and hook worms, as blood purifier in hepatic disorders, spleen enlargement, intestinal ulcers and burns. Various bioactivities such as antifungal, antipruritic, antinociceptive and hypotensive properties of crude and isolated compounds from plant justified its uses in traditional medicine. (Warrier *et al.* 2010, Agrawal *et al.* 2014, Kumar and Kumar 2015).

Medicinal Uses (Traditional/Folklore): *C. album* is an important medicinal plant in Ayurveda and folklore medicines used in diseases of blood, heart, spleen, and eye and in biliousness conditions, cough, abdominal pain, pulmonary obstruction and in nervous affections; the herb is laxative, anthelmintic and cardio tonic. The plant is traditionally used for the cure of dysentery, diarrhoea, wound healing, hepatic disorder, skin diseases, snake bite, animal bites, stimulant, tonic (for debility), sedative, laxative, anthelminthic (round and hook worm). It is useful in *Vata-Kaph* and eye disease. It is used in the form of pot herb in piles. The leaves are used in the treatment of bug bites, sunstroke, urinary problems, skin problems, jaundice, irregular menses, anaemia, and as a blood purifier. The finely powdered leaves are used as a dusting powder for the external genitalia in children. An infusion is taken in the treatment of rheumatism. The leaves are applied as a wash or poultice to bug bites, sunstroke, rheumatic joints and swollen feet, whilst a decoction is used for carious teeth. The seeds are chewed in the treatment of urinary problems and are considered useful for relieving the discharge of semen through the urine. The juice of the stems is applied to freckles and sunburn. The juice of the root is used in the treatment of bloody dysentery. The powdered plant (25-50%), when mixed with normal food, was reported to suppress the oestrous cycle. (Agrawal *et al.* 2014, Kumar and Kumar 2015; http://medplants.blogspot.in, http://tropical.theferns.info).

47.9 *Chenopodium murale* L.

(Syn. *Atriplex muralis* (L.) Crantz)

Habit

Common Name: Nettle-Leaved Goosefoot, Australian Spinach.

Vernacular Name: Jangli Bathua (जंगली बथुआ), Khartua (खरतुआ).

Habit: Annual herb.

Habitat: Fairly common weed in moist waste places, roadsides, shady areas and agriculture fields.

Distribution: It is widespread worldwide, particularly in tropical and subtropical areas due to its being easily introduced; found in U.S.A., Mexico, Brazil, Argentina Republic, Barbados, Porto Rico and India.

Flowering and Fruiting Time: December-February.

Taxonomic Description: Annual herb, up to 80 cm high with young parts mealy. Stem green, rarely tinged with red, ribbed. Leaves dark green, alternate, up to 10 cm long, rhomboic-ovate, lobed or irregularly toothed, cuneate at base, obtuse or acute at apex, petioled. Inflorescence axillary and terminal, composed of divaricately branched cymes. Perianth segments-5, greenish, keeled, segments oblong. Stamens-5, anthers yellow. Fruit enclosed in persistent perianth. Seeds black, 1-1.5 mm in diameter, keeled.

Flowering twig

An open flower

Chemical Constituents: Chemical studies have shown that the plant contains essential oils, a wide variety of sterols and steroidal oestrogens like substances. Recently, four more compounds have been identified in the plant as 3,7-dihydroxy-3`-(4-hydroxy-3-methylbutyl)-5,6,4`-trimethoxyflavone, 5,7-dihydroxy-3`-(2-hydroxy-3-methyl-3-butenyl)-3,6,4`-trimethoxyflavone, β-Sitosterol 3-O-β-D-glucoside, stigmasterol 3-O-β-D-glucoside. (Ahmad *et al.* 2003, Ahmed *et al.* 2017).

Medicinal Properties: Plant is sweet and acrid, digestive, carminative, aphrodisiac, nutritive, diuretic, mild purgative, emollient, anthelmintic, tranquilizer, stomachic, antispasmodic, diaphoretic, emmenagogue, tonic for liver; leaves are edible. (Ahmad *et al.* 2003, Ahmed *et al.* 2017).

Medicinal Uses (Traditional/Folklore): The plant is used for the pain of amenorrhea, as an abortifacient and for the relief of asthma, catarrh and migraine and antihypertensive.

In India, it is used as tonic, laxative and in vitiated conditions of *Pitta*, peptic ulcer, dyspepsia, flatulence, seminal weakness, haemorrhoids, cardiac disorder, ophthalmopathy, spleenopathy, pharyngopathy and general debility. Leaves are used for gastrointestinal tract pains, fever, and kidney diseases in Tunisia. (Ahmad *et al.* 2003, Ahmed *et al.* 2017; http://medplants.blogspot.in).

47.10 *Digera muricata* (L.) Mart.

(Syn. *Achyranthes muricata* L.; *Digera alternifolia* (L.) Asch.; *Digera arvensis* Forssk.)

Habit

Common Name: False Amaranth.

Vernacular Name: *Hindi:* Lahsuva (लहसुआ), Latmahuria (लटमहूरिया); *Sanskrit*: Aranyavastuka (अरण्यवस्तुका), Kuranjara (कुरंजना).

Habit: Annual herb.

Habitat: Fairly common in gardens, waste places, cultivated and fallow fields.

Distribution: Native to Asia; widespread in eastern tropical Africa (from Sudan and Ethiopia south to Tanzania), Madagascar and subtropical Asia (from Yemen to Afghanistan, Pakistan, India, Malaysia and Indonesia). In India, it is widely distributed in Uttar Pradesh, Rajasthan, Maharashtra, Andhra Pradesh and Kerala.

Flowering and Fruiting Time: June-October.

Taxonomic Description: An annual, erect herb, 20-60 cm high. Stem angular, profusely branched near the base. Leaves alternate, dark green,

3-8 x 2-5 cm, ovate-lanceolate, apex acute, margin undulate or entire, rounded or cordate at base, petioled. Flowers small, bisexual, pink, born in 4-14 cm long axillary pedunculate spikes. Bracts ovate-lanceolate. Perianth 5-partite; outer 2 large, acute. Stamens-5; filaments free, filiform. Ovary 1 celled, sigma 2-lobed. Fruits globose, about 3 mm across, indehiscent, one seeded.

Leaf

Flowers

Chemical Constituents: The plant contains flavonoids, alkaloids, terpenoids, saponins, coumarins, tannins, cardiac glycosides, anthraquinones and α- and β- spinasterol. Rutin and hyperoside flavonoids have also been isolated and identified in the plant extracts of this plant. (Mathad and Mety 2009, Sharma and Vijayvergia 2013).

Medicinal Properties: The plants have cooling, refrigerant, astringent, antioxidant, laxative, diuretic, antidiabetic and hepatoprotective properties. (Mathad and Mety 2009, Sharma and Vijayvergia 2013).

Medicinal Uses (Traditional/Folklore): *D. muricata* is popularly known for herbal remedy for various ailments. In Ayurveda, this herb is considered as cooling, astringent of bowels and also used as a laxative. The whole plant is used in digestive system disorders. The leaves and young shoots of this plant are locally used as a vegetable and given to relieve constipation. The leaves are used for treatment of diabetes. Boiled root infusion given to mothers after child birth to increase lactation purpose need to be evaluated. The whole plant is used in urinary tract disorders. The extract of this plant is used in biliousness and in urinary discharges; flowers and seeds are also used to treat urinary discharges. As a folk medicine the plant has been used to treat renal disorders. The decoction of leaves is given once a day for kidney stone treatment. This plant is also used as an alternative for secondary infertility; leaf paste is applied locally to prevent pus formation. Ethyl alcohol extract of plant is diuretic. (Mathad and Mety 2009, Sharma and Vijayvergia 2013; http://tropical.theferns.info).

47.11 *Dysphania ambrosioides* (L.) Mosyakin & Clemants

(Syn. *Chenopodium ambrosioides* L.)

Common Name: Mexican Tea.

Vernacular Name: *Hindi*: Sugandha Vastooka (सुगंध वास्तूका); *Sanskrit:* Sugandha Vastooka (सुगंध वस्तूका).

Habit: An aromatic undershrub.

Habitat: Commonly growing gregariously along roadsides, moist waste places near water.

Distribution: Native species of tropical America. It is distributed throughout the tropical, subtropical and temperate regions.

Flowering and Fruiting Time: June-October.

Taxonomic Description: An erect to sub-erect, strongly aromatic, 30-100 cm high, undershrub. Stem profusely branched, leafy, angular-ribbed. Leaves alternate, variable, 3-11 x 2-4 cm, ovate, oblong to lanceolate, irregularly serrate or deeply incised, base attenuate, apex acute or obtuse; cauline leaves small, entire or sub-entire. Flowers greenish, sessile, bracteate, axillary, forming a leafy branched panicle at the top of the stem. Perianth 5-lobed, green; lobes ovate, rounded, gland dotted. Stamens usually 5. Utricle membranous, enclosed by persistent perianth.

A branch with irregularly serrate leaf

Flowering twig **Fruiting twig**

Chemical Constituents: Leaves contain kaempferol--rhamnoside and ambroside. The plant contains an essential oil having ascaridole, p-cymene, neral and geraniol as the main constituents; other constituents present are kaempferol-7-O-alpha-L-rhamnopyranoside, kaempferol-3,7-di-O-alpha-L-rhamnopyranoside, patuletin, quercetin-7-O-alpha-L-rhamnopyranoside, grasshopper ketone, 4-hydroxy-4-methyl-2-cyclohexen-1-one, syringaresinol, benzyl beta-D-glucopyranoside, dendranthemoside B, N-trans-feruloyl tyramine, N-trans-feruloyl 4′-O-methyldopamine, and 4-hydroxy-N-[2-(4-hydroxyphenyl) ethyl] benzamide. (Boutkhil *et al.* 2009, Song *et al.* 2014; http://medplants.blogspot.in).

Medicinal Properties: The plant is attributed to be bitter, amoebicidal, analgesic, diuretic, antispasmodic, pectoral, haemostatic, anthelmintic, emmenagogue, antiasthmatic, antifungal, carminative, stomachic and vermifuge. (Blanco *et al.* 2014; http://medplants.blogspot.in, http://tropical.theferns.info).

Medicinal Uses (Traditional/Folklore): In folk medicines it is used to treat asthma, malaria, and various nervous diseases, but its main use has been as a vermifuge. In the early 1900s, *D. ambrosioides* was used as an anthelmintic to treat intestinal worms such as ascarids and hookworms in humans, cats, dogs, horses, and pigs. Dried herb is a good anthelmintic effective against round and hookworms. Ascaridole, an active constituent of the oil, is highly active against roundworms, hookworms and small, but not large, tapeworms. The oil has been found useful also in amoebic dysentery and intestinal infections (should be used with caution). In traditional/folk uses fresh juice of the leaves is combined with coconut milk and taken internally for intestinal parasites and other gastrointestinal disorders. The seed oil paralyses, but does not kill intestinal worms, which must be expelled by a purgative. The oil is particularly effective in getting rid of hookworms, but is a dangerous remedy as overdose may cause poisoning. Mild reactions to the oil are headache, dizziness and nausea, while more severe symptoms include vomiting (sometimes bloody), deafness, tingling of the extremities, and depression. Acute poisoning may be fatal. Nevertheless it is widely used in the world as a vermifuge. In western medicines, the oil is now seldom used in pharmaceutical preparations because it has been replaced by synthetic anthelmintic. The plant continues to be employed medicinally in Asia for a variety of ailments.

In the Dominican Republic the leaves are used as an antiseptic for treating wounds or skin ulcers, and a tea of the leaves is used for treating asthma, colic, conjunctivitis and stomach ache. Herbal teas and infusions are also used as postpartum depurants. The plant has the potential for causing abortion. It is also employed in treating nervous affections, particularly chorea. This plant also has culinary uses as a condiment and is considered a medicinal food because of its anti-flatulent effects. The plant is used as a leaf vegetable and herb for its pungent flavor and its claimed ability to prevent flatulence caused by eating beans but also in the treatment of amenorrhea, dysmenorrhoea, malaria, chorea, hysteria, catarrh, and asthma.

The plant is highly toxic and can cause serious side effects. Touching the plant may cause dermatitis and allergic reactions, and a case of vertigo from contact with essential oil released during harvest has been described. Chickens have died from eating the seeds, and while grazing animals usually avoid Mexican tea; it is known to have caused fatalities in cows. (Blanco *et al.* 2014; http://medplants.blogspot.in, http://tropical.theferns.info).

47.12 *Gomphrena celosioides* Mart.

(Syn. *Gomphrena alba* Peter)

Habit

Common Name: Prostrate Globe-Amaranth.

Vernacular Name: None known.

Habit: Annual herb.

Habitat: Abundant along roadsides, in gardens, open waste places and unused grounds.

Distribution: The plant *G. celosioides* is native to tropical America. It has naturalized in tropical Asia, Africa and Australia.

Flowering and Fruiting Time: Most part of the year.

Taxonomic Description: A prostrate to ascending, deep-rooted annual herb. Stem much-branched from the base, white-villous, up to 30 cm long. Leaves opposite, 1.5-5 x 0.5-2 cm, oblong-elliptic, appressed with densely long silvery hairs beneath, base cuneate, apex acute-mucronate. Flowers silvery-white, borne in dense cylindrical, terminal spikes, subtended by pair of leaves. Bracts and bracteoles deltoid-ovate. Tepals-5, lanceolate. Staminal tube 5-fid, anthers yellow. Style very short, stigma bilobed. Utricles enclosed by perianth. Seeds minute, brown, shining.

Chemical Constituents: The plant contains various secondary metabolites including alkaloids, tannins, saponins, steroids, glycosides and reducing sugars. (Tarnam *et al.* 2014).

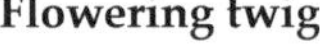

Flowering twig

Terminal spike

Medicinal properties: The plants show anti-inflammatory, analgesic and antimicrobial activity. Root extract has weak antibacterial activity, while extracts of other parts are active on nervous autonomic system of rats with sympathetic and parasympathetic symptoms. (Olutola *et al.* 2014, Tarnam *et al.* 2014).

Medicinal Uses (Traditional/Folklore): *G. celosioides* is most effective and commonly used in the treatment of urinary tract and kidney stones. Whole plant juice along with *Piper nigrum* and lemon juice is taken twice a day for 10 days to cure urolithiasis. This plant is also used in folk medicine for treatment of gastrointestinal, respiratory or infectious diseases. Tender shoots of the plant eaten regularly as vegetable are good blood purifier. To treat leucorrhoea one cup of decoction of whole plant is taken twice daily for one week. (Prachi *et al.* 2009, Buragohain 2011, Olutola *et al.* 2014, Shalini *et al.* 2014, Tarnam *et al.* 2014).

48. POLYGONACEAE

Polygonaceae – Buckwheat, Knotweed or Smartweed family

This is a mixed group of herbs and shrubs consisting of 59 plant genera and 1,384 accepted species. Family name is derived from Greek *Poly* (many) and *gony* (a knee) referring to the jointed stems. Plants are characterized by a membranous, sheathing stipule called an ocrea. Family consists of popular vegetables and ornamentals such as Buckwheat (*Fagopyrum esculentum*) and Rhubarb (*Rheum rhaponticum*). Five species of herbs belonging to four genera of this family were recorded in Chithara Village flora.

48.1 *Emex australis* Steinh.

(Syn. *Emex centropodium* Meisn.)

Habit

Common Name: Three Corner Jack, Devil's Thorn, Double Gee.

Vernacular Name: None Known.

Habit: Prostrate herb.

Habitat: Waste places, fallow fields and pastures.

Distribution: It occurs indigenously in South Africa and is an introduced weed in many countries. It has become an invasive species in Australia and Texas in the USA.

Flowering and Fruiting Time: January-April.

Taxonomic Description: A prostrate, monoecious, annual herb. Stem ribbed, much-branched from the base. Leaves ovate-lanceolate, 5-9 x 2-5 cm, entire, base cordate or truncate, apex acute or obtuse; ochrea loose, lacerate; petiole up to 6 cm long. Flowers unisexual. Male flowers sessile, axillary and in terminal clusters; tepals greenish-yellow, ovate-lanceolate. Female flowers in axillary clusters, sessile and in short leafless spikes; tepals ovate-oblong. Fruiting perianth 8-12 mm long with 3 sharp, spreading spines at the apex. Achenes 4-6 mm long.

Flowering and fruiting twig

Flowers

Nuts

Chemical Constituents: Nonc Known.

Medicinal Properties: Laxative. (http://tropical.theferns.info).

Medicinal Uses (Traditional/Folklore): In South Africa, an aqueous decoction of the root is administered to treat constipation and restlessness in infants and adults. Roots are also used for stomach cramps. A decoction of roots of *Galenia secundica* and *E. australis* is applied for the treatment of kidney pain. Unspecified parts are used to treat stomach and intestinal problems. When eaten in quantity, the leaves have a laxative effect. (Dubey 2015; http://tropical.theferns.info).

48.2 *Persicaria barbata* (L.) H.Hara

(Syn. *Polygonum barbatum* L.)

Apical portion of plant

Common Name: Bearded Knotweed.

Vernacular Name: None Known.

Habit: Annual herb.

Habitat: Common in marshy places and along water channels.

Distribution: Distribution of *P. barbata* ranges from Saudi Arabia and Yemen to eastern Asia, Himalayas to western and South China, Bhutan, India, Indonesia, Malaysia, Myanmar, Nepal, New Guinea, Philippines, Sri Lanka, Thailand and Viet Nam. The plants are recorded in Katerniaghat Wildlife Sanctuary, a tropical moist deciduous forest along the Indo-Nepal border; depressions in shaded situations at elevations from 1,700 - 1,900 m in the western Himalayas. Also found in Palampet, Warangal District, Andhra Pradesh, India. The plant is a well-known reed swamp that typically grows along the margin and banks of streams up to 1,800 m.

Flowering and Fruiting Time: August-November.

Taxonomic Description: An erect, annual herb, up to 80 cm high. Stem simple or branched, terete, pubescent. Leaves 5-15 x 1.5-3.5 cm long, linear-lanceolate, pubescent on both surfaces, apex acuminate, tapering towards the base, margin ciliate. Stipules ochreate; ochreae tubular, hispid; tube with 1.5-2 cm long bristles. Inflorescence terminal, 4-10 cm long, many flowered spicate raceme. Perianth-5, white or greenish, tepals 1.5-2.5 mm long. Stamens 5-8; filaments long, equal. Style 3-fid, stigma capitate. Achenes enclosed in persistent perianth; nuts trigonous, black, shining.

Flowering twig

Ochreate stipule

Flowers

Chemical Constituents: Plant is known to contain polyphenols such as flavanone.

Medicinal Properties: The plant has stimulant, diuretic, carminative, anthelmintic, tonic, purgative and emetic properties. The root is astringent and cooling. Seeds possess tonic, purgative and emetic properties. (Kamaluddin *et al.* 2014, Tonny *et al.* 2017).

Medicinal Uses (Traditional/Folklore): The infusion of leaves is a stimulant which is used primarily for the obstructions of menses in young women. The seeds are used to relieve the gripping pains of colic and ulcers. They are used in the treatment of dysentery and cholera. A paste of the root is used externally in the treatment of scabies. In Bangladesh, seeds having emetic properties are used as tonic and purgative. Decoction of the shoots is used to wash ulcers. In China, the leaf juice is used for itch and also as a diuretic, carminative and anthelmintic. Powdered leaves are used in ulcers of goats caused by infection due to fly. The whole plant is used as a fish poison. (Kamaluddin *et al.* 2014; http://tropical.theferns.info).

48.3 *Persicaria lanigera* (R.Br.) Soják

(Syn. *Polygonum lanigerum* R.Br.)

Apical portion of plant

Common Name: African Persicaria, Smartweed, Knotweed, Lady's Thumb.

Vernacular Name: None known.

Habit: Herb (densely clothed with white tomentum).

Habitat: Abundant along the bank of rivers, canals, ponds and marshy places.

Distribution: Open wet sites, near streams, etc. at low to medium elevation in Africa to India, China, Taiwan, Philippines, and southwards to Australia.

Flowering and Fruiting Time: September-November.

Taxonomic Description: An erect, robust, 80-150 cm high herb. Stem and branches clothed with white tomentum, swollen above the nodes. Leaves shortly petioled, up to 20 cm long, alternate, narrowly lanceolate, acuminate, densely white woolly beneath. Stipules ochreate, long, membranous. Inflorescence racemose. Flowers small, bracteate, white to pink, arranged in dense panicled racemes; peduncles white tomentose. Perianth eglandular, 4-partite. Stamens usually 6. Nutlets minute, orbicular, black, shining.

Flowering twig **Ochreate stipule** **Flowers**

Chemical Constituents: It contains condensed tannin and steroids. (Ekunseitan *et al.* 2013).

Medicinal Properties: None known.

Medicinal Uses (Traditional/Folklore): None known.

48.4 *Polygonum plebeium* R.Br.

(Syn. *Polygonum roxburghii* Meisn.)

Habit

Common Name: Small Knotweed.

Vernacular Name: Machechi (मचेची).

Habit: Prostrate herb.

Habitat: Usually found in gardens, open areas, near ponds, and ditches.

Distribution: It is found throughout much of South Asia and in Madagascar. In India, it grows from sea level to 2,200 m in the Himalayas.

Flowering and Fruiting Time: January-May.

Taxonomic Description: A prostrate, annual herb with woody root stock. Stem terete, much-branched from base, up to 40 cm long. Leaves alternate, sessile or sub-sessile, 1.0-2.5 x 0.2-0.4 cm, linear-oblong, obovate, entire, acute or apiculate at apex, recurved. Stipules short, 3-4 mm long, ciliate. Flowers pinkish, axillary, solitary or 2-3 flowered, sessile or shortly pedicellate. Perianth-5; tepals 1.5- 2 mm long, divided to the base. Stamens 5-8, filaments united at the perianth base. Style 3-fid. Nutlets trigonous, enclosed in persistent perianth, shining, black-brown.

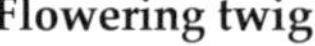

Flowering twig

Flower

Chemical Constituents: Phytochemical screening of *P. plebeium* confirmed the presence of flavonoids, alkaloids, glycosides, steroids, saponins and carbohydrates. The fresh root contains 11% tannin, and oxymethylanthraquinone. (Nazmul-Hasan *et al.* 2015; http://tropical.theferns.info).

Medicinal Properties: The herb is antioxidant, carminative and antidiarrhoeal; aqueous extract of this plant has antibacterial properties. (Nazmul-Hasan *et al.* 2015).

Medicinal Uses (Traditional/Folklore): The crushed seeds are cooked and eaten as a remedy for bowel complaints and diarrhoea. Plant decoction is given in colic complaints. Plant ash with oil is applied on eczema. Root paste is applied twice a day for inflammations. Roots and whole plant is helpful in respiratory tract infection, pneumonia, indigestion, vomiting and diarrhoea. Decoction of leaves with rhizome of *Zingiber officinale* and seeds of *Piper nigrum* are taken once a day for five days to treat swelling on body. (Jagtap *et al.* 2006, Kamaluddin *et al.* 2014).

48.5 *Rumex dentatus* L.

(Syn. *Rumex nipponicus* Franch. & Sav.)

Habit

Common Name: Toothed Dock, Aegean Dock.

Vernacular Name: *Hindi:* Jangli Palak (जंगली पालक).

Habit: Annual herb.

Habitat: Common in roadsides, disturbed areas and cultivated fields.

Distribution: *R. dentatus* is native to parts of Eurasia and North Africa. It is widely known elsewhere as an introduced species.

Flowering and Fruiting Time: January-April.

Taxonomic Description: Annual, deep-rooted herb, up to 70 cm high. Root purple in colour. Stem erect, grooved, branched from the base, lower stem purplish. Leaves variable, oblong-elliptic to linear-lanceolate, 8-25 x 2-5 cm, base sub-cordate, margin entire-undulate, apex acute or obtuse; upper leaves smaller, short-petioled; lower long-petioled. Ochreae membranous, partially deciduous when mature. Inflorescence racemose, paniculate. Flowers hermaphrodite, green, in whorls arranged on long racemes. Perianth 4-5 x 2-3 mm; inner fruiting tepals enlarged in the fruit with much-toothed wings. Nutlets brown, trigonous.

Fruiting twig

Capsules

Chemical Constituents: *R. dentatus* contains alkaloids, flavonoids, saponins, anthraquinones and tannins. Six C-glucosyl anthrones identified from the roots of *R. dentatus* by column chromatography are rumejaposide E (10R-C-β-d-glucosyl-10- hydroxyemodin-9-anthrone) and rumejaposide F (10S-C-β-d-glucosyl-10-hydroxyemodin-9- anthrone), rumejaposide G (10R-C-β-d-glucosylemodin-9-anthrone) and rumejaposide H, (10SC-β-d-glucosylemodin-9-anthrone), cassialoin, (10S-C-β-d-glucosyl-10 hydroxychrysophanolanthrone) and rumejaposideI (10R-C-β-d-glucosyl-10 hydroxychrysophanol-9-anthrone). Also identified from the roots are another ten compounds, helonioside A, gallicacid, isovanillic acid, p-hydroxycinnamic acid, succinic acid, n-butyl-beta-D-fructopyranoside, quercetin, hexadecanoic acid 2, 3-dihydroxy propyl ester, β-a-sitosterol and daucosterol. (Fatima *et al.* 2009, Rao *et al.* 2011, Najafabadi *et al.* 2016).

Medicinal Properties: *Rumex* plants are used as bactericidal, antidermatitis, anti-inflammatory, antitumour, astringent, diuretic, cholagogue, tonic and laxative agent. Plants have cytotoxicity and used for antiscorbutic, cutaneous disorders and as a dye. (Fatima *et al.* 2009, Rao *et al.* 2011).

Medicinal Uses (Traditional/Folklore): *R. dentatus* is traditionally used medicinally for its antidermatitis, bactericidal, anti-inflammatory, and antitumour properties. The root is used as an astringent in the treatment of cutaneous disorders. Whole plant decoction is given against sunstroke for cooling effect to the body in Rewa district, Madhya Pradesh, India. In Haryana, leaf juice and paste is used to cure skin allergy; also fine powder made from the roots is taken to relieve constipation. Seeds are used as aphrodisiac by tribes in Pakistan. People with a tendency to rheumatism, arthritis, gout, kidney stones or hyperacidity should take especial caution if including this plant in their diet since it can aggravate their condition. (Fatima *et al.* 2009, Najafabadi *et al.* 2016).

49. EUPHORBIACEAE

Euphorbiaceae – Castor/Spurge family

The family which includes more than 228 genera and about 6,547 species, mostly monoecious herbs, shrubs, and trees is distributed worldwide, particularly in the tropics; with the majority of the species in the Indo-Malayan region and tropical America. In India, it is represented by more than 70 genera and about 450 species. In several species of *Euphorbia*, the stem is succulent cladode resembling a cactus and leaves as thorns/spines. Almost all the plants have latex which is either milky or watery. A number of plants of the family are of considerable economic importance. Prominent plants include cassava *(Manihot esculenta)*, castor oil plant (*Ricinus communis*), Barbados nut (*Jatropha curcas*), and the Para rubber tree (*Hevea brasiliensis*). Many are grown as ornamental plants, such as poinsettia (*Euphorbia pulcherrima*). Eight species belonging to four genera from this family were recorded in Chithara Village including SNU Campus.

49.1 *Croton bonplandianus* Baill.

(Syn. *Croton sparsiflorus* Morong)

Common Name: Croton, Rushfoil.

Vernacular Name: *Hindi:* Ban Tulsi (बन तुलसी), Kala Bhangra (काला भांगरा).

Habit: An erect herb.

Habitat: Roadsides, railway lines, open and wastelands.

Distribution: Native to temperate South America; found throughout India as a weed.

Flowering and Fruiting Time: September-January.

Habit

Taxonomic Description: An erect, much branched, up to 80 cm high herb, possess milky sap. Leaves simple, petiolate, alternate, 2 to 5 cm long, ovate-lanceolate with serrated margin. Stem branched, woody, cylindrical, clothed with multi armed trichomes. Flowers small, white, borne in long raceme at the top of branches. Male flowers in upper part of the raceme and female flowers at the base of the raceme. Sepals and petals-5. Stamens mostly 15. Capsules triangular, hairy.

Flowering twig **Flowers** **Capsules**

Chemical Constituents: Phytochemically, the plant is rich in secondary metabolites including alkaloids and terpenoids, the latter including irritant co-carcinogenic phorbol esters, flavonoids, saponins, steroids, resins and phenols; it contains rutin as main constituent besides crotsearinine, crotasparine and its methyl derivative, aphorbol. (Divya *et al.* 2011, Jeeshna *et al.* 2011, Singh *et al.* 2014a).

Medicinal Properties: *C. bonplandianus* is known to possess antioxidant, antimicrobial, antimalarial, antitumour, antiseptic and antidote and wound healing properties. Bark and roots are alterative and cholagogue. (Divya *et al.* 2011, Jeeshna *et al.* 2011, Singh *et al.* 2014a).

Medicinal Uses (Traditional/Folklore): Traditionally, the plant is used to treat liver and skin diseases, including ring worm infection and also to cure the swelling of body. Bark and roots of *C. bonplandianus* are cholagogue promoting the discharge of bile from the system. Leaves of this plant being antiseptic and antidote are highly medicinal and used to treat high blood pressure, cuts and wounds, venereal sores and cholera. The leaf extract has been proved to have wound healing effect and external application has shown to cure the ringworm infection. The seeds are used for the treatment of jaundice, acute constipation, abdominal dropsy and internal abscesses. The leaf extract of the plant is found to be effective against larvae of mosquito. (Das *et al.*2008, Bapuji *et al.* 2009, Jeeshna *et al.* 2011, Kuppuswamy *et al.* 2013, Singh *et al.* 2014a, Das *et al.*2017).

49.2 *Euphorbia heterophylla* L.

(Syn. *Poinsettia heterophylla* (L.) Klotzsch & Garcke)

Habit

Common Name: Wild Poinsettia, Painted Spurge, Milkweed, Lesser Green Poinsettia.

Vernacular Name: None known.

Habit: Annual herb.

Habitat: Shady places in gardens, roadsides agricultural fields.

Distribution: *E. heterophylla* originated in the tropical and subtropical regions of America; now distributed throughout tropical Africa, Asia and the Pacific. It has been introduced to South and South-East Asia as an ornamental plant, having become a weed in India and Thailand, where it has invaded cotton fields and other agricultural fields.

Flowering and Fruiting Time: September-April.

Taxonomic Description: An erect, monoecious, annual herb, 30-90 cm high. Stem hollow with long inter nodes. Leaves linear-lanceolate to oblong-ovate, up to 14 cm long, margin entire or coarsely toothed, apex acute, base rounded or cuneate; petiole 1-4 cm long with scattered hairs. Male and female flowers crowded at top of the stem. Cyathia in dense terminal cymes; involucre about 2.5 mm with 1 cup-shaped gland. Capsules sub-globose, 3-lobed, about 4 mm long and 6 mm wide. Seeds dark brown.

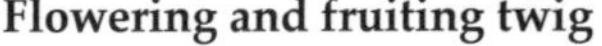

Flowering and fruiting twig

Capsules

Chemical constituents: The plant contains alkaloids, terpenoids, phenolics, tannins, flavonoids and cardiac glycosides. The leaves of *E. heterophylla* have been reported to contain quercetin; diterpenoids is present in the roots. All parts of the plant contain latex. The plant also contains lectins and carbohydrates. A dimeric N-acetylgalactosamine-specific lectin was isolated from the seeds. The red coloured pigment of the coloured leaves and bracts is identified as porcetin. (Falodun *et al.* 2006, Okeniyi *et al.* 2012).

Medicinal Properties: *E. heterophylla* leaf has a rubefacient, anti-inflammation, skin irritant, haemostatic, purgative, tumour-promoting, antimicrobial, antitumour/anticancer and anti-HIV properties. Leaf is used in traditional medical practices as laxative, migraine and wart cure. The roots are cathartic, emetic and galactagogue. The flowers are pectoral. (Falodun *et al.* 2006, Okeniyi *et al.* 2012; http://medplants.blogspot.in, http://tropical.theferns.info).

Medicinal Uses (Traditional/Folklore): *E. heterophylla* is reported to be poisonous; leaf is used in traditional medical practices as laxative, anti-gonorrhoeal, migraine and wart cure. The leaf extract is taken to treat body pain. They are used in small doses in the treatment of gonorrhoea, syphilis and to increase lactation (milk production in breast-feeding women) in East Africa. Decoction of stem and leaf is used for stomachache and regulating menses, stopping bleeding. In peninsular Malaysia, a leaf extract is taken to treat body pain; bark and root decoction is taken for ague. The latex is irritant to the skin and eyes and may be employed as a rubefacient and to remove warts and corns. Milky juice is applied to treat wounds, pimples, vitiligo, and skin diseases. Although, *E. heterophylla* contains latex with a caustic effect upon the skin and mucous membranes, this plant is widely used in traditional African medicines. Extracts of the decoction of the leaves is used in the treatment of respiratory tract infections and asthma by traditional medicine practitioners in Nigeria. In Africa, a decoction or infusion of the stems and fresh or dried leaves is taken as a purgative and laxative to treat stomachache and constipation, and to expel intestinal worms. A leaf infusion is used as a wash to treat skin problems, including fungal diseases, and abscesses. In Nigeria, the latex and preparations of the leaves and root are applied to treat skin tumours.

As a veterinary medicine, feeding of this plant increases the production of milk in cattle. Dry plants are toxic to rabbits. The latex is used in the preparation of arrow poison and fish poison. The plant latex has been used as an insecticide and ordeal poison. The latex is also used as an antidote against the irritation caused by the latex of other *Euphorbia* species. (http://medplants.blogspot.in, http://tropical.theferns.info, https://www.prota4u.org/database/protav8.asp?g=pe&p=Euphorbia+heterophylla+L.).

49.3 *Euphorbia hirta* L.

Synonym: *Chamaesyce hirta* (L.) Millsp.; *Euphorbia capitata* Lam.

Common Name: Asthma Weed, Asthma Plant.

Vernacular Name: *Hindi:* Dudhi (दुधी), Bara Dudhi (बड़ा दुधी); *Sanskrit:* Dugadhika (दुग्धिका), Chara (चारा), Amampatchairasi (अममपतचयाइरसी).

Habit: Annual herb.

Habitat: Common weed found everywhere along roadsides, railway lines, open and waste fields.

Distribution: *E. hirta*, a pantropical weed native to tropical America, is distributed throughout the hotter parts of India and Australia, often found in waste places along the roadsides and in open grasslands and pathways.

Habit

Flowering and Fruiting Time: Most part of the year.

Taxonomic Description: An erect or decumbent, annual herb, up to 30 cm high. Stem greenish-purple, clothed with long yellowish hairs, milky latex present. Leaves opposite, hispid, shortly petioled, 1.5-3.5 x 0.7-2 cm, oblong, elliptic-lanceolate, stipulate, base oblique, margin serrate, apex acute or sub-acute. Cyathia axillary or terminal, aggregated in dense peduncled cymes. Involucres campanulate, stalked, gland minute. Capsules globose, about 1 mm long, hairy, 3-lobed. Seeds trigonous, reddish.

Abaxial side of leaves

Capsules

Chemical Constituents: The herb contains several phytoconstituents: flavonoids - euphorbianin, leucocyanidol, camphol, quercitrin and quercitol; polyphenols- gallic acid, myricitrin, 3,4-di-O-galloylquinic acid,2,4,6-tri-O-galloyl-D-glucose, 1,2,3,4,6-penta-O-galloyl-β- D-glucose; tannins- euphorbins A, B, C, D, E; triterpenes and phytosterols-β-Amyrin, 24-methylenecycloartenol, and β-Sitosterol; and alkanes: heptacosane, n-nonacosane. Other phytochemicals present in the plant are terpenes, anthocyanins, alcohols and steroids, shikimic acid, choline, L-inositol and free sugars and quercetin, which is responsible for antidiarrhoeal activity. Dimeric-hydrolysable-tannins, euphorbians, have also been isolated from the plant. (Bhutya 2011; http://medplants.blogspot.in).

Medicinal Properties: *E. hirta* has a wide spectrum of medicinal properties which include aldose-reductase-inhibitor, amebicide, analgesic, aphrodisiac, anthelmintic, antiaggregant, antiallergic, antiasthmatic, antibacterial, antifungal, anticancer, anticonvulsant, antidiabetic, antidiarrhoeal, antiedemic, antiemetic, antifertility, antihistaminic, anti-inflammatory, antileukaemic, antimalarial, antiplasmodial, antiplatelet, antipyretic, antiprostaglandin, antiseptic, antispasmodic, antiviral, anxiolytic, aphrodisiac, astringent, bactericide, bronchodilator, bronchorelaxant, carcinogenic, cardiodepressant, cicatrizant, curare, cytotoxic, diuretic, emmenagogue, expectorant, febrifuge, gram-icide, hemostat, hydragogue, hypoglycaemic, hypotensive, immunosuppressive, insecticide, lactagogue, laxative, litholytic, mastogenic, molluscicide, myorelaxant, narcotic, oxytocic, pectoral, parasiticide, plasmodicide, purgative, respirostimulant, sedative, stimulant, vermifuge, and vulnerary. Latex is vermifuge. (Kumar *et al.* 2010, Bhutya 2011; http://medplants.blogspot.in, http://tropical.theferns.info).

Medicinal Uses (Traditional/Folklore): The plant has traditionally been used to treat bronchial and respiratory diseases (bronchitic asthma, laryngeal spasm asthma, bronchitis, hay fever, etc.), and in conjunctivitis, though in modern herbalism it is more used in the treatment of gastrointestinal disorders (diarrhoea, dysentery, intestinal parasitosis, amoebic dysentery, etc.). In Ayurveda an extract of the *E. hirta* is given to bronchitis patients; other symptoms like colds, asthma and coughs are also relieved. The extract of this plant is effective in controlling venereal diseases like gonorrhoea and syphilis. This herb is useful in relieving sexual disorders which are seen in men. Impotency and premature ejaculation are controlled by using this herb. *Euphorbia* also relieves several female disorders. A decoction made from the root of the plant increases lactation in women nursing their babies. The herb is very effective in controlling diarrhoea and dysentery. A decoction made from the root of *Euphorbia* plant is used for gargling to relieve thrush and mouth ulcers. The plant is also used for relieving toothache, severe headache, rheumatism and colic, for treating urogenital diseases like kidney stones, menstrual problems and venereal diseases. The extract of the plant is used as an antidote for snakebites and scorpion stings.

Latex is used to cure diseases of urino-genitory tract, dysmenorrhoea, amenorrhea, helminthiasis, ring worm, wounds, piles, treating warts, female disorders, respiratory ailments (cough, coryza, bronchitis, and asthma), and worm infestations in children, dysentery, jaundice, pimples, gonorrhoea, digestive problems, and tumours alopecia, chronic cough, cardiac debility, acne vulgaris, laryngitis, chronic nasal and bronchial catarrh, diarrhoea, intestinal parasitosis, postnatal complaints, failure of lactation, excreted an inhibitory effect on platelet aggregation and as a tonic. It kills various types of pathogenic bacteria and malarial parasite, *Plasmodium*. The latex of the plant is used to remove thorns from the skin and when applied on lower eyelids, like *Surma* to cure eye sores. Leaf tea is given to people with dengue fever. The whole plant is decocted for athlete's foot, dysentery, enteritis, fever, gas, itch, dyspnoea eruptions, excrescences, influenza, fractures and skin conditions.

In folklore use, in central Province of Papua New Guinea, the plant is boiled and the solution is taken by patient who pass blood in the urine. The Chinese use the plant to treat fever, dysentery and skin conditions. In the Philippines and Indonesia, the plant is used to treat bowel problems. The latex is used on warts and abscesses. In Indonesia, the latex is used to treat ringworm infection and heal wounds. A decoction of this herb is used to treat asthma and assuage bronchial discomfort. The leaves are chewed to facilitate abortion, and the flowers are used to treat infertility (barrenness). In Malaysia, the latex is used to treat diseased eyes, heal wounds and soothe bruises. A decoction consisting of about 70 g of the fresh herb is drunk to promote urination, stop dysentery, remove blood from urine, assuage urethral pain and treat asthma. The plant is applied externally to treat dermatitis, eczema and irritated skin. A paste is used to soothe sores, and heal boils. In the Philippines, the leaves are mixed with those of *Datura metel* to make cigarettes which are smoked to treat asthma. The plant is used to check bleeding, to calm a person, and stimulate sweating. In Vietnam, *E. hirta* is used to stop dysentery. The dried entire plant was used in Britain in the form of a liquid extract for the treatment of cough and asthma. It has been warned that the plant should not be used without expert guidance since large doses cause gastrointestinal irritation, nausea and vomiting. (Kumar *et al.* 2010; http://medplants.blogspot.in, http://tropical.theferns.info).

49.4 *Euphorbia hypericifolia* L.

(Syn. *Chamaesyce hypericifolia* (L.) Millsp.)

Common Name: Graceful Spurge, Large Spotted Spurge, Milk Purslane.

Vernacular Name: *Hindi:* Dudh Mogra (दूध मोगरा).

Habit: Annual herb.

Habitat: Moist shady waste places, roadsides, cultivated and fallow fields.

Distribution: *E. hypericifolia* is native to tropical America, where it has been recorded from southern USA, the Caribbean and South America. The plant has been widely introduced to the Pacific Islands, and to some extent in Africa and South and Southeast Asia, including Madagascar, India, New Zealand, Thailand, Sri Lanka and Japan. Found throughout the warm regions of India and tropics of both the hemisphere.

Flowering and fruiting twig

Cyathia

Habit

Flowering and Fruiting Time: May-October.

Taxonomic Description: An erect, monoecious, shallow rooted, annual herb. Stem purplish, sparingly pubescent, up to 40 cm long. Leaves opposite, 1.5-3.5 x 0.5-1.5 cm, linear to elliptic-oblong, entire or minutely serrulate, base rounded, apex obtuse, shortly petioled, stipules triangular. Cymes axillary, clustered in dense peduncled cymes. Involucre funnel shaped, nearly 1 x 1 mm, glands-4, white to light pink. Capsules triangular, 1.5 x 2 mm, with quadrangular seeds.

Chemical Constituents: The plant contains caoutchouc, resin, tannin, gallic acid, etc.; also contains the sterols, taraxerol, β-sitosterol, stigmasterol, campestol and the flavonoids kaemferol, quercetin, quercetrin (quercetin-3-rhamnoside), rhamnetin-3-galactoside, rhamnetin-3-rhamnoside and ellagic acid. Aliphatic alcohols have been isolated from the aerial parts. (http://database.prota.org/PROTAhtml/Euphorbia%20hypericifolia_En.htm, http://tropical.theferns.info, https://www.cabi.org/isc/datasheet/119929).

Medicinal Properties: Plants have antipyretic, astringent, diaphoretic, diuretic, emmenagogue, hemostat, laxative, narcotic, and antitumour properties; seeds are purgative. (http://medplants.blogspot.in, http://tropical.theferns.info).

Medicinal Uses (Traditional/Folklore): The herb acts principally upon the digestive tract and the sympathetic nervous system. It is widely used for gastro-intestinal ailments: abrasion, bleeding, buccal ulcer, callus, cancer, colic, coma, constipation, diarrhoea, dysentery, dysuria, fever, gynecopathy, irritation; cholera infantum; mucoid-enteritis; dysentery; vertigo, with constipation, wart, water retention. It is used in the treatment of marasmus, menorrhagia, ophthalmia, ringworm, stomatosis, toothache; whole plant is used to treat leucorrhoea, syphilitic blennorrhoea and cephalalgia.

Other medicinal uses include for treating toothache, asthma, bronchitis, conjunctivitis, dysentery, dysuria, fever and vaginitis. It is also used for the treatment of measles and skin rashes. *E. hypericifolia* products are widely traded in the USA in tablets and powders, mainly to treat bowel disorders. (http://database.prota.org/PROTAhtml/Euphorbia%20hypericifolia_En.htm, http://medplants.blogspot.in, http://tropical.theferns.info, https://www.cabi.org/isc/datasheet/119929).

49.5 *Euphorbia prostrata* Aiton

(Syn. *Chamaesyce prostrata* (Aiton) Small)

Habit

Common Name: Prostrate Sandmat, Prostrate Spurge.

Vernacular Name: *Hindi:* Choti Dudhi (छोटी दुधी).

Habit: Prostrate herb.

Habitat: Open waste places, roadsides, cultivated and fallow fields.

Distribution: It is widely distributed throughout the tropics and subtropics; occur widely in India, Pakistan and Java.

Flowering and Fruiting Time: Most part of the year.

Taxonomic Description: A prostrate annual herb, monoecious. Stem purplish, branched from the base, branchlets hairy. Leaves opposite, stipulate, shortly petioled, 4-8 x 3-4 mm, oblong, obovate, obtuse at apex, oblique at base, somewhat denticulate at tip. Cyathia axillary, involucre pubescent outside; glands-4, each with a petaloid limb. Capsules about 1.5 mm long, beaked, sub-globose with spreading hairs. Seeds very minute, brown, quadrangular.

Chemical Constituents: The active principles in *E. prostrata* are chiefly flavonoids, phenolic acid and tannins. Phenolic compounds like gallic acid activates Hageman factor which causes hypercoagulability and ellagic acid suppresses histamine release. It also contains flavonoids like apigenin, which inhibits I kappa B and suppresses inflammatory mediators and luteolin, which inhibits protein tyrosin phosphorylation and IKB mediated inflammation. From different fractions of extracts of the dried leaves a range of hydrolyzable ellagitannins were isolated, including prostratins A, B and C, euphorbins G and H, tellimagradin I and II, and rugosins A, D, E and G. Flavonoids isolated from the aerial parts include: kaempferol, cosmosiin (apigenin-7-glucoside), rhamnetin-3-galactoside, quercetin and quercetin-3-rhamnoside. Other constituents of the aerial parts include the sterols β-amyrine acetate, β-sitosterol, campesterol, stigmasterol and cholesterol. The aerial parts also contain the terpene alcohol β-terpineol, gallic acid, corilagin, 1, 2, 3-tri-O-galloyl-D-glucose, geraniin, and various amino acids, including n-valeramide and N, N-dimethyl-4-benzoxybutylamine. From the roots a myricylic alcohol and two triterpenes, taraxerol and tirucallol, have been isolated. (Gupta 2011, Porwal *et al.* 2015; http://tropical.theferns.info).

Medicinal Properties: Flavonoids and phenolic acid have been reported to have anti-inflammatory, antiviral, antiallergic, antiplatelet, antitumour, analgesic, antioxidant, antidiabetic, astringent, haemostatic, antithrombotic and vasoprotective actions. Tannins are known to possess astringent and haemostatic properties. The latex is irritant and blistering to the skin and mucous membranes and is reported to cause blindness. (Dhiman 2004, Gupta 2011, Porwal *et al.* 2015; http://tropical.theferns.info).

Flowering and fruiting twig **Capsules**

Medicinal Uses (Traditional/Folklore): *E. prostrata* extract is very effective in treating grade I and II haemorrhoids, bleeding haemorrhoids and diabetes due to its contents of flavonoids, phenolics and phenolic acids. In India, the latex is used to treat diabetes, as it is considered to have hypoglycaemic and anti-inflammatory activities. The plant is also used to treat asthma and an infusion is taken as a blood purifier. *E. prostrata* extract tablets have been marketed in India by Panacea Biotec Ltd.

All parts of *E. prostrata* are widely used in African traditional medicine. In Burkina Faso the leaves are rubbed onto wasp stings and scorpion stings. In Togo, a leaf decoction is taken orally to treat threatened abortion. Small balls of meshed plants are inserted into the vagina to treat female sterility and painful menstruation. In Benin, the pounded aerial parts with pounded shells are taken to treat irregular menstruation. A paste of leaves grounded in water is used against difficult childbirth. In Nigeria, a plant decoction is taken for its astringent, vulnerary and anthelmintic properties, and crushed plants are used by the Igbo people as a poultice for broken arms. In Cameroon, crushed leaves are eaten to treat amoebic dysentery. In Gabon, a leaf extract is applied as an enema to treat inflammation. Leaf powder mixed with palm oil is rubbed on the head to treat headache. In Congo, the whole plant or only the leaves are warmed over a fire, crushed and squeezed on the body to cure insect bites and fungal infection; an infusion or decoction of the leaves is also taken orally to treat fungal infections. The crushed whole plant is eaten with bread against kidney stones. In Kenya, Maasai people chew the plant to treat gonorrhoea. In Uganda, crushed fresh leaves in water are used as a mouthwash and gargle to treat oral sores. Pregnant women eat the boiled shoots, mixed with sesame, to reduce the risk of miscarriage. Plant juice is taken to induce labour during childbirth. A bath with the plant infusion is recommended to treat insanity. In Angola, a vapour bath of the whole plant is taken to treat scabies; the plant is also crushed and applied on the affected areas of the body. Throughout the Indian Ocean islands an infusion of the leaves or aerial parts is taken either alone or combined with other plants to treat diarrhoea, dysentery and stomachache. In the Comoros, the plant is used in association with other plants to treat deformations of the spinal column. In Réunion, a bath with the whole plant is taken to facilitate healing of measles and other skin eruptions. In Mauritius, a decoction of the whole plant is taken orally to treat painful menstruation and used as an eye wash against conjunctivitis. The latex is applied to warts and abscesses. It is also used as an arrow poison.

Similar uses as above have been reported from other parts of the world. In the United States the latex is applied to snakebites, in Mexico and Venezuela to treat tumours. In French Guyana the aerial parts in decoction are taken as a bitter diuretic. Crushed fresh plants are applied as an embrocation to heal sprains and strains. (http://tropical.theferns.info, https://www.prota4u.org/database/protav8.asp?g=pe&p=Euphorbia+prostrata+Aiton).

49.6 *Euphorbia serpens* Kunth

(Syn. *Chamaesyce serpens* (Kunth) Small; *Euphorbia orbiculata* var. *jawaharii* Rajagopal & Panigrahi)

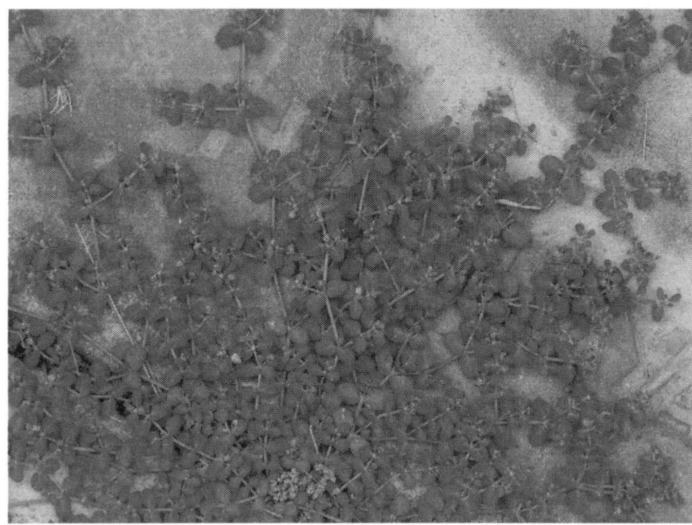

Habit

Common Name: Matted Sandmat, Roundleaf Spurge.

Vernacular Name: *Hindi:* Dudhi (दुधी).

Habit: Prostrate herb.

Habitat: Common along roadsides, in moist waste places, gardens and cultivated fields.

Distribution: *Euphorbia serpens* is native to South America and introduced to the most continents as a weed species.

Flowering and Fruiting Time: Most part of the year.

Taxonomic Description: An annual, monoecious, prostrate herb with fibrous root. Stem glabrous, much-branched, spreading from the base; branchlets somewhat purple. Leaves opposite, 4-6 x 3-4 mm, oblong, ovate-suborbicular, base cordate, margin entire, apex obtuse-retuse, petiole nearly 1 mm long; stipules triangular, caducous. Cythia solitary, axillary, involucre campanulate; glands-4, appendages white. Capsules sub-globose, about 1.5 mm long, smooth, glabrous. Seeds very small, brown.

Flowering and fruiting twig

Capsule

Chemical Constituents: None known.

Medicinal Properties: Antioxidant, antifungal. (Payne *et al.* 2015).

Medicinal Uses (Traditional/Folklore): The plant is useful in diarrhoea, menstrual problems, and slow milk production in nursing mothers, tumours and skin and eye diseases. Contact with the plant's latex can cause dermatitis in some individuals. (Moran 2009; https://gobotany.newenglandwild.org/species/euphorbia/serpens).

49.7 *Jatropha gossypiifolia* L.

(Syn. *Manihot gossypiifolia* (L.) Crantz)

Habit

Common Name: Bellyache Bush, Cotton-Leaf Physic Nut.

Vernacular Name: *Hindi:* Ratanjoti (रतनजोती); *Sanskrit:* Rakta-Vyaaghrairanda (रक्त-व्याघ्रेरांदा), Dravanti (द्रवन्ति).

Habit: Shrub.

Habitat: Roadsides, open waste places and also cultivated in the gardens.

Distribution: It is widely distributed in countries of tropical, subtropical, and dry tropical weather and tropical semiarid regions of Africa and America.

Flowering and Fruiting Time: Most part of the year.

Taxonomic Description: An erect, much-branched, glandular hairy shrub, 1-2 m high. Young branches purplish-red. Leaves alternate, stipulate, 3-5 deeply lobed; lobes elliptic, apex acute, margin gland-tipped; petiole 5-12 cm long, covered with glandular bristles. Inflorescence terminal corymbose cymes. Flowers monoecious, about 5 mm across, reddish-purple. Sepals-5, margin with gland-tipped hairs. Petals-5, obovate. Capsules oblong, 1-1.5 cm long, 3-lobed, truncate at both the ends. Seeds greyish-red, carunculate.

Flowering and fruiting twig

Flowers

Fruit

Chemical Constituents: Phytochemical analysis reveals the presence of various, bioactive phytochemical compounds. In methanol extract carbohydrate, alkaloids, glycosides, phenol, starch, organic acid, steroids were present. Alkaloids, phenol, flavonoid, organic acid, saponin, diterpene were present in acetone extract. The petroleum ether extract contained carbohydrate, alkaloids, phenol, flavonoid, protein, diterpene. In chloroform extract, phenol, protein, diterpene, flavonoid, were present and in aqueous extract phenol, starch, saponin were present. The seeds contain crude oil of iodine, protein, fibre, carbohydrate and saponins. (Ogbobe and Akano 1993, Saini *et al.* 2015b).

Medicinal Properties: *J. gossypiifolia* possess wide pharmacological activity such as antihypertensive, anti-inflammatory, anticancer, analgesic, antipyretic, antimicrobial, antibacterial activity, haemostatic, molluscicidal, wound healing, antianaemic, antidiabetic, insecticidal, and antihemorrhagic properties. (Saini *et al.* 2015b; http://medplants.blogspot.in, http://tropical.theferns.info).

Medicinal Uses (Traditional/Folklore): The leaf decoction of *J. gossypiifolia* is used for bathing wounds. The leaf bath is used for sores, sprains, rash and bewitchment in Latin America and the Caribbean; the poultices are used for sores and pain in Trinidad. In Southern Nigeria, the extract from fresh leaf applied with crushed leaf is routinely used by herbalists and local people to stop bleeding from the skin and nose. Latex from leaf and cut stem is applied externally over infected parts to treat eczema and scabies. Seed oil is applied over joints to treat rheumatic pain. Other uses are also related to this plant, such as biodiesel production, pesticide, insecticide, vermifuge, ornamentation, and even its use in religious ritual. (Saini *et al.* 2015b; http://medplants.blogspot.in, http://tropical.theferns.info).

49.8 *Ricinus communis* L.

(Syn. *Cataputia major* Ludw.)

Habit

Common Name: Castor Bean, Castor Oil Plant.

Vernacular Name: *Hindi*: Arandi (अरंडी), Chiyan (चियां); *Sanskrit:* Gandharva Hasta (गंधर्व हस्ता).

Habit: Shrub.

Habitat: Open, waste and disturbed places; abundant along the roadsides and railway tracks.

Distribution: *R. communis* is native to Africa. Currently, the plant is naturalized across the African continent, from the Atlantic coast to the Red Sea, from Tunisia to South Africa and on islands in the Indian Ocean. It is widely cultivated and naturalized in tropical and subtropical regions of America and Asia and in many temperate areas of Europe. The plant occurs throughout India.

Flowering and Fruiting Time: December- May.

Inflorescence Capsules Seeds

Taxonomic Description: An evergreen, monoecious shrub, 3-5 m high. Stem fistular, purplish, young branchlets glaucous. Leaves alternate, large, up to 25 cm long, palmately 6-9 lobed; lobes oblong-lanceolate, margin coarsely serrate, apex acuminate; petiole up to 40 cm long, hollow with 2 glands at apex. Flowers greenish-white or yellowish, in terminal or leaf opposed paniculate racemes. Female flowers occupying the upper portion and male flowers below. Stamens many, filaments branched. Capsules ovoid, 1.5-2 cm across, softly spinous. Seeds oblong, about 8 mm long, smooth, carunculate, brownish-black, shining.

Chemical Constituents: Phytochemicals found in the plant are steroids, saponins, alkaloids, flavonoids, and glycosides. The seeds contain glycosides of ricinoleic, isoricinoleic, stearic and dihydroxystearic acids and also lipases and a ricinine. The major phytoconstituents reported in this plant are rutin, gentistic acid, quercetin, gallic acid, and kaempferol-3-O-β-d-rutinoside, kaempferol-3-O- β-d-xylopyranoid, tannins, ricin A, B & C, ricinus agglutinin, indole-3-acetic acid and an alkaloid ricinine. (Jena and Gupta 2012, Rana *et al.* 2012).

Medicinal Properties: Roots are acrid, sweet, astringent, thermogenic, carminative, purgative, anthelmintic, emollient, diuretic, and aphrodisiac, galactagogue, sudorific, expectorant and depurative. They are useful in gastropathy, constipation, inflammations, fever, ascites, strangury, bronchitis, cough, leprosy, skin diseases. The leaves are diuretic, anthelmintic and galactagogue, and are useful in burns, nyctalopia, strangury, urodynia and arthralgia. Flowers are useful in urodynia and glandular tumours. Seeds are acrid, thermogenic, digestive, cathartic and aphrodisiac. They are useful in dyspepsia and for preparing a poultice to treat arthralgia. The oil obtained from seeds is slightly bitter, acrid, sweet, antipyretic, thermogenic and viscous. Stems of *R. communis* have anticancer, antidiabetic and antiprotozoal activity, and analgesic, anodyne, antirheumatic, antiscrofulous, cathartic, emollient, febrifuge, galactagogue, parturient, purgative, vermifuge, and vulnerary properties. Seed oil is used as a purgative. (Warrier *et al.* 2010, Jena and Gupta 2012, Rana *et al.* 2012; http://medplants.blogspot.in, http://tropical.theferns.info).

Medicinal Uses (Traditional/Folklore): In Ayurveda, the whole plant and its parts are used for formulations for treating pain in the urinary bladder, lumbago, diseases of the abdomen and inflammations, helminthiasis, dysuria, arthritis, pain in the urinary bladder, and abscesses; also used for treating constipation, rheumatism, diseases of the liver and spleen, piles, lumbago, sciatica, and induction of labour. Oil is useful in constipation, fibromyalgia , nerve/ back pain, pain relief and parasites/worms, etc.

In Manipur, leaves are warmed, crushed and used as a remedy for bleeding piles. Leaf-paste is used as poultice on sores, gout or rheumatic swellings. Decoction of root is given in lumbago. For lactation, leaves of the plant are heated and applied to a woman's breasts to improve secretion of milk. The leaves are used as a poultice to relieve headaches and treat boils. It is so effective that it is regularly used to clear the digestive tract in cases of poisoning. The oil has a remarkable antidandruff properties. The oil is well-tolerated by the skin and so is sometimes used as a base for medicinal and cosmetic preparations. The leaf preparations are used in burns, nyctalopia, and strangury and for bathing and fermentation and vitiated conditions of *Vata*, especially in rheumatoid arthritis. Leaves and seeds are useful in swellings. As an abortifacient, 20 g fresh seeds paste is prepared and two tea spoonful of this seed paste/ decoction is taken twice daily for three days.

The seeds of castor bean or castor oil plant, are very poisonous to people, animals and insects; just one milligram of ricinine (one of the main toxic proteins in the plant) can kill an adult; commercially prepared castor oil contains none of the toxins. The castor oil is extracted from the beans, is used for medicinal purposes. (Jena and Gupta 2012, Rana *et al.* 2012, Singh *et al.* 2013b, Das *et al.* 2014b, Ghoshal 2015; http://medplants.blogspot.in, http://tropical.theferns.info).

50. PHYLLANTHACEAE

Phyllanthaceae – Phyllanthus family

The family, comprising of 2,099 species under 58 plant genera, is mostly distributed in the tropics, with many in the South Temperate Zone, and a few ranging as far as north in the middle of the North Temperate Zone. Family is characterized by monoecious or dioecious trees, shrubs or herbs with finely cracking bark, latex absent. Nearly all species are trees, shrubs, or herbs and a few are climbers, or succulents, and one species, *Phyllanthus fluitans*, is aquatic. The family was represented by four species under one genus in the flora of Chithara Village.

50.1 *Phyllanthus amarus* Schumach. & Thonn.

Syn. *Phyllanthus nanus* Hook.f.; *Phyllanthus niruri* var. *amarus* (Schumach. & Thonn.) Leandri)

Habit

Common Name: Black Catnip, Stone Breaker.

Vernacular Name: *Hindi:* Bhuiaonla (भुईआंवला), Hajarmani (हजारमणी); *Sanskrit:* Bahupatra (बहुपात्रा), Bhumyaamalaki (भूम्यामलाकि).

Habit: Annual herb.

Habitat: Common in moist/wet places, cultivated and wastelands; also found in crevices of stones.

Distribution: It is native to China, India, and South/Central America. The plant also has been found in the Philippines, Cuba, and Nigeria. It grows like a weed in most tropical areas of the World. In India, *P. amarus* is widely distributed as a weed in cultivated and wastelands.

Flowering and Fruiting Time: June-October.

Taxonomic Description: An erect, annual herb, up to 80 cm high. Stem often angular, woody at base; branchlets slender, spreading. Leaves alternate, distichous, shortly petiolate, oblong-obovate, often overlapping, 8-15 x 3-5 mm, oblong, entire, rounded at base, obtuse at apex; stipules linear-lanceolate. Flowers minute, greenish white or yellowish, axillary, numerous, hanging down below the branchlets. Male flowers small; sepals-5, ovate; disk glands-5; pedicels 0.5-1 mm; stamens-3. Female flowers solitary; sepals-5, obovate or ovate-lanceolate. Capsules depressed globose, smooth, obscurely lobed. Seeds yellowish brown, 3-angled.

Staminate flowers and fruits on the lower side

Chemical Constituents: The active phytochemicals, flavonoids, alkaloids, terpenoids, lignans, ellagitannins, polyphenols, tannins, coumarins and saponins, volatile oil, and tritepene have been identified from various parts of the plant. (Verma *et al.* 2014; http://tropical.theferns.info).

Medicinal Properties: *P. amarus* is an important plant of Indian Ayurvedic system of medicine which is used in the problems of stomach, genitourinary system, liver, kidney, gall bladder and spleen. It is known to be very useful for relieving mainly hepatitis A. It is bitter, analgesic, astringent, stomachic, diuretic, febrifuge, spasmodic, acrid, cooling, alexipharmic, antihepatotoxic, antihepatitis-B, antihyperglycemic and antiseptic. (Verma *et al.* 2014; http://medplants.blogspot.in, http://tropical.theferns.info).

Medicinal Uses (Traditional/Folklore): *P. amarus* has been used in traditional medicines for more than 3,000 years. The plant is used worldwide as an additive in various medicinal formulations and found to be very effective in relieving liver and kidney problems. Ayurveda recommends its use for bronchitis, leprosy, anaemia, urinary discharge, asthma, etc. The herb has found its traditional usefulness in several health problems such as anaemia symptoms, hypertension, diarrhoea, dysentery, dropsy, jaundice, flu, cold, tuberculosis, intermittent fevers, urinogenital disorders, ophthalmopathy, scabies and wounds. Further, the plant is found to be very effective in the treatment of kidney problems, urinary bladder disturbances, pain, gonorrhoea, diabetes and chronic dysentery; it is used to remove bladder stones, and is a remedy for jaundice. Topically, it is used for several skin problems ranging from skin ulcers, sores, swelling and itchiness, wounds, bruises, scabies, ulcers and sores, edematous swellings, tubercular ulcers, ringworm, scabby and crusty lesions. Use of roots relieves skin diseases like leprosy, ring worm as well as hair disorders. The root extract is added to massage oils, body lotions and hair oils. Its effect in excretory system is due to its antiurolithic property and is used in the treatment of kidney/gallstones, other kidney related problems, appendix inflammation and prostate problems. Because of its efficacy in the field of gastro-intestinal disorders it is used in the treatment of disorders like dyspepsia, colic, diarrhoea, constipation and dysentery. The herb has found use in several female problems such as in vaginitis, leucorrhoea, menorrhagia and mammary abscess and can act as galactagogue. The plant is also used to relieve bacterial infections such as prostatitis, venereal diseases, urinary tract problems as well as cystitis. The plant is used to treat syphilitic blennorrhoea. The whole plant is used in gonorrhoea, menorrhagia and other genital affections.

Fresh leaves are ground and mixed with a cup of cow or goat's milk and taken internally to cure jaundice in Kanchipuram district of Tamil Nadu, as well as in Kerala, India. A poultice of the leaves, combined with salt, is used to treat scabby affection of skin, used without salt it is applied on bruises and wounds. In the Unani System of medicine this herb is used for sores and chronic dysentery. Its seeds are used in the treatment of ulcers, wounds, scabies and ringworms. The root of this plant is considered to be an excellent remedy for liver diseases; the root extract is used to cure stomach pain. The young shoots of plant are administered in the form of an infusion for the treatment of chronic dysentery. Fresh leaf paste has wound healing capacity and used to cure white spots on skin and jaundice. The stem juice is used as wound healers. The whole plant extract is used in urinary problems and swelling of liver. The paste of flowers of this plant is applied externally as antidote against snake bite. (Verma *et al.* 2014; http://medplants.blogspot.in, http://tropical.theferns.info).

50.2 *Phyllanthus maderaspatensis* L.

(Syn. *Diasperus maderaspatensis* (L.) Kuntze)

Upper portion of plant

Common Name: Madras Leaf-Flower.

Vernacular Name: *Hindi:* Hajarmani (हजारमणी); *Sanskrit:* Bhumyaamalaki (भूम्यामलकि).

Habit: Annual herb.

Habitat: Common in gardens, waste places and cultivated fields.

Distribution: It is native in a wide range of African countries, India and Sri Lanka, and naturalized elsewhere.

Flowering and Fruiting Time: May-September.

Taxonomic Description: An erect or ascending, glabrous, annual herb, 30-50 cm high, monoecious. Leaves alternate, shortly petioled, up to 2.5 cm long, obovate, oblanceolate, entire, cuneate at base, apiculate at apex; stipules membranous, narrowly lanceolate. Male flowers axillary, minute, clustered, greenish-white, sub-sessile; tepals-6; stamens-3. Female flowers solitary, longer, shortly stalked; tepals-6. Capsules depressed globose, greenish, about 2.5 mm across. Seeds 3-sided, brownish.

Chemical Constituents: Extracts of the above-ground parts were found to contain resins, steroids, triterpenoids, alkaloids, phenolic compounds, tannins and saponins, but no glycosides. The plants contain essential oil, maderin, mucilage and β- sitosterol. Seeds contain long chain fatty acids and β- sitosterol. Defatted seed cake contains mucilage, which yields galactose, arabinose, rhamnose and aldobionic acid, niruriside, phyllanthin, hypophyllanthin and cinnamoyl sucrose acetate. (Ravichandran *et al.* 2012, Rani and Raju 2014).

Fruiting twig

Capsules

Medicinal Properties: The plants possess several medicinal properties such as astringent, deobstruent, stomachic, diuretic, febrifuge and antiseptic. The leaves are expectorant, diaphoretic useful in strangury and excessive sweats. The seeds having a bad taste used as carminative, laxative, astringent, bowels tonic, liver tonic, diuretic, and useful in bronchitis. Extract of aerial parts is used as a hepatoprotective agent. (Ravichandran *et al.* 2012, Rani and Raju 2014; http://tropical.theferns.info).

Medicinal Uses (Traditional/Folklore): The plant is found to be useful in headache, bronchitis, earache, ophthalmia, griping, cough, ascites, incipient, blindness, sores, ulcers, stomachache, inflammations, intestinal spasms, gonorrhoea, antimicrobial and viral infections. Traditional healers use this plant in treating fever and burns. A root decoction is taken to cure constipation, diarrhoea, lack of appetite, intestinal pain, menstrual problems, gastrointestinal disorders, testicular swelling, chest complaints and snakebites. Powder from dried plant material mixed with milk is taken to treat jaundice. The whole plant is pounded and boiled and the decoction used for washing areas of the body affected by scabies. Pounded and meshed leaves are rubbed on the skin with lemon juice as a treatment for rheumatism. (Ravichandran *et al.* 2012; http://tropical.theferns.info).

50.3 *Phyllanthus reticulatus* Poir.

(Syn. *Kirganelia reticulata* (Poir.) Baill.)

Common Name: Black Honey Shrub, Black Berried Featherfoil.

Vernacular Name: *Hindi:* Kale Madhu Ka Per (काले मधु का पेड़); *Sanskrit:* Krishna-Kamboji (कृष्ण-कामबोजी).

Habit: Straggling shrub.

Habitat: Roadsides, waste places and in thickets along streams and canals.

Distribution: Widespread in old world tropics from Africa to India, China, South–East Asia and northern Australia. Found throughout India.

Upper portion of plant

Flowering and Fruiting Time: Major part of the year.

Taxonomic Description: A large, much-branched, straggling shrub, 1-2 m high with drooping branchlets. Leaves alternate, shortly petioled, 1.5-3.5 x 1-2 cm, membranous, oblong to elliptic, glabrous, entire, sub-acute or obtuse at apex, rounded at base. Stipules ovate, deciduous. Male flowers greenish-purple in small fascicles. Stamens-5. Female flowers solitary. Berries dark-purple or black, globose, shining, 4-5 mm across. Seeds many, trigonous.

Flowering twig

Mature berries

Chemical Constituents: Phytochemical screening of *P. reticulatus* revealed the presence of terpenes, phytosterols, phenolic compounds, carbohydrates, flavonoids and minute amount of alkaloids. The stem bark contains pentacosane, 21-A-hydroxyfriedelan-3-one, taraxerol and lupene-24-diol. (Ram *et al.* 2009; http://tropical.theferns.info).

Medicinal Properties: The plant is astringent, sweet, cooling, diuretic, alterant, stomachic, constipating and attenuant. It is useful in vitiated conditions of *Pitta*, burning sensation, strangury, gastropathy, ulemorrhagia, ophthalmodynia, sores, burns, suppuration, diarrhoea, skin eruptions and obesity. (Jamal *et al.* 2008, Ram *et al.* 2009, Warrier *et al.* 2010).

Medicinal Uses (Traditional/Folklore): The plant is used for a variety of ailments, including smallpox, syphilis, asthma, and diarrhoea and bleeding from gums. Decoction of bark is used as astringent, diuretic and alterative. The juice of leaves is used medicinally in the Konkan region in India. It is made into a pill with camphor and cubebs, which is allowed to dissolve in the mouth as a remedy for bleeding gums. The roots are used in Tamil Nadu, India as a red eye remedy. For this 10 ml root decoction is given orally in cold water. The leaves are also used in diabetes. In Indo-China, the whole plant is used in the treatment of small pox and syphilis. A decoction of leaves is used in the treatment of hookworms; also, leaves are applied locally against pinworms. A root decoction is used to treat gonorrhoea and other venereal diseases and also against diarrhoea accompanied by bleeding. In Zanzibar, the plant is considered as a remedy for anaemia and intestinal haemorrhage. A decoction of the roots is taken against dysmenorhoea and to increase fertility; leaves and roots also used against paralysis. The powdered leaf is used externally for topical application to sores, including venereal sores, burns, suppurations. The mashed leaves are rubbed over the body of a malaria patient. Decoction of the root is taken as a male aphrodisiac, to increase fertility and also to treat headache, for dysmenorrhoea, and for hard abscesses. In Tanzania, the fruits and roots are used for poisoning purpose. (Ram *et al.* 2009, Ramakrishna and Saidulu 2014; http://tropical.theferns.info).

50.4 *Phyllanthus tenellus* Roxb.

(Syn. *Diasperus tenellus* (Roxb.) Kuntze; *Phyllanthus minor* Fawc. & Rendle)

Habit

Common Name: Mascarene Island leaf-flower, Long Stalked Phyllanthus.

Vernacular Name: None known.

Habit: Annual herb.

Habitat: Commonly found near wetlands, ditches and disturbed places.

Distribution: Native of Mascarene Islands; now established in many tropical countries.

Flowering and Fruiting Time: May-January.

Taxonomic Description: An erect, glabrous, annual herb, 30-60 cm high. Leaves alternate, shortly petioled, stipulate, 1-2.5 x 0.5-1 cm, elliptic or obovate, glaucous beneath, margin entire, base rounded, apex sub-acute. Flowers white, nearly 1.5 mm across, axillary, borne in 1-few flowered small cymes; pedicels 2-6 mm long, slender. Male flowers: sepals-5, white with greenish midrib, disk segments-5, stamens-5. Female flowers: sepals as in male, disk present; styles-3, persistent in the fruit. Capsules depressed-globose, green, 1.5-2 mm across, smooth, numerous seeded with persistent sepals.

Flowering and fruiting twig

Flowers and fruits

Chemical Constituents: The plants contain many known metabolites, such as nirantin, nirtetralin, hinoquinine and geranin (pertaining to the alkaloid group), carotenoids and phenolic compounds. (Silva *et al.* 2012).

Medicinal Properties: *P. tenellus* possesses anti-inflammatory, anti-lithic, analgesic, hypotensive, antispasmodic, antiviral, antibacterial, antifungal, diuretic, anti-mutagenic, immunomodulatory, hypoglycaemic, and antihepatotoxic, antihepatitis properties. (Morebise 2015; http://shodhganga.inflibnet.ac.in/bitstream/10603/21353/10/13.%20%20chapter%203.pdf).

Medicinal Uses (Traditional/Folklore): In Dominica, the nature isle of the Caribbean, plant (stem and leaves) is boiled in water to make tea which is consumed for the treatment of flu, cold, cough, high blood pressure and high blood glucose. (Morebise 2015).

51. ULMACEAE

Ulmaceae – Elm family

It comprises 8 plant genera and 64 species of trees and shrubs, distributed primarily throughout temperate regions. Species of the family have watery sap, and its leaves alternate along the stem. The leaves usually have toothed edges and often are lopsided at the base. The flowers lack petals. Male and female flowers are borne together or apart on the same plant. Elms (*Ulmus*) and hackberries (*Celtis*) are important shade and ornamental trees. One species belonging to this family was recorded in Chithara Village Panchayat.

51.1 *Holoptelea integrifolia* Planch.

(Syn. *Ulmus integrifolia* Roxb.)

Habit

Common Name: Indian Elm, Indian Beech Tree, Jungle Cork Tree.

Vernacular Name: *Hindi:* Chilbil (चिलबिल), Papri (पापड़ी); *Sanskrit:* Chirivilva (चिरिविलवा), Pootikaranja (पूतिकरंजा).

Habit: Medium or large deciduous tree.

Habitat: Planted along roadsides and in gardens; also found wild.

Distribution: It is distributed in temperate and tropical areas of northern hemispheres. The plant is naturalized in the tropical regions of Asia including India, Nepal, Sri Lanka, Indo-China, Cambodia, Laos, Myanmar, Vietnam, Myanmar, and China. In India, found throughout, especially in deciduous forests; it is found in outer Himalayan region from Jammu eastward up to 700 m extending to Assam and Myanmar, and southwards from Bengal to central, western and southern India to dry regions of Sri Lanka.

Flowering and Fruiting Time: February-April.

Leaves

Fruits (samara)

Taxonomic Description: A medium-sized to large deciduous tree. Stem and branches grey. Leaves 5-13 x 5-7 cm, simple, alternate, petiolate, slightly oblique at base, entire, ovate or elliptic, acuminate, stipules lanceolate. Flowers small, greenish-yellow; male and female flowers mixed in numerous fascicles on leafless branches; male flower with 8 stamen and bisexual flower contains 5 stamens. Sepals-4, pubescent. Stamens 4-8, anthers hairy. Fruits samara, about 2.5 cm in diameter, light brown, orbicular or sub-orbicular, notched at apex, reticulately veined wings. Seed 1 per samara, whitish.

Chemical Constituents: The plant contains wide range of phytochemicals such as terpenoids, sterols, saponins, tannins, proteins, carbohydrates, and alkaloids. In addition, *H. integrifolia* also contains flavonoids, phenols, cardiac glycosides, coumarins, and quinines, holoptelin-A, holoptelin-B, friedlin, epifriedlin, 2-aminonaphthoquinone, β-sitosterol, β-D-glucose, β-amyrin, stigmasterol, and hederagenin which have been isolated from heart wood and bark; while hexacosanol, octacosanol, β-sitosterol, and α-amyrin were isolated from the leaves. Seeds are the major sources of fatty acids like palmitic acid, myristic, stearic, linoleic, linolenic acid and steroids like stigmasterol, α and β-sitosterol, β-amyrin, friedel-1-en-3-one, lupeol, β-sitosterol-D-glucoside. Histamine, 5-hydroxy tryptamine are derived from pollen grains and 24-ethyl-cholest- 22-en-3α-ol is isolated from the roots. (Sharma and Singh 2012, Ganie and Yadav 2014, Sutar *et al.*2014, Mondal and Bandyopadhyay 2016).

Medicinal Properties: The bark and leaves are bitter, astringent, acrid, thermogenic, digestive, carminative, laxative, anthelmintic, depurative, repulsive and astringent, antidiabetes, antihelminthiasis, anti-inflammatory, and antirheumatic. (Warrier *et al.* 2010, Sharma and Singh 2012, Ganie and Yadav 2014, Sutar *et al.*2014, Mondal and Bandyopadhyay 2016).

Medicinal Uses (Traditional/Folklore): Ethnomedicinally, the leaves and stem bark of *H. integrifolia* are used by tribal people for the treatment of various ailments. The plant is used traditionally for the treatment of inflammation, gastritis, dyspepsia, colic, intestinal worms, vomiting, wound healing, leprosy, diabetes, haemorrhoids, dysmenorrhoea, and rheumatism. The mucilaginous bark is boiled and the juice squeezed out and applied to rheumatic swellings. Paste of the stem bark is externally applied to treat the inflammation of lymph glands, ringworm, and scabies. Stem bark acts as an anti-inflammatory agent specifically for eyes. Stem bark paste is externally applied on forehead of the patient suffering from common fever. Bark and leaf paste of the plant are applied externally on the white patches or leucoderma. Bark boiled in coconut oil and mixed with garlic is applied externally to eczema. For treatment of herpes simplex infection, bark paste is applied over the affected part until it disappears. Bark cut in the shape of a coin is tied on left arm below the shoulder for treatment of malaria. It is also used for the treatment of intestinal cancer. Powdered bark is mixed with lemon juice and made into a paste which is given for weakness.

Leaves are used for treating leprosy, leucoderma, facial paralysis, boils, inflammation of lymph glands, skin diseases such as ringworm, eczema, scabies, and other cutaneous diseases, alopecia, termination of pregnancy and scorpion sting. For rickets young leaves are applied on back bone and tied for 1 hr. Leaf bud mixed with lime juice is applied externally to affected area for treating herpes infection. Decoction of the leaves is used to regulate fat metabolism, treat ringworm, eczema, and cutaneous diseases. Leaf bud mixed with lime juice is applied externally to affected areas for treatment of hair loss due to infection.

Seeds are applied externally in the form of poultice on the injured part for uncontrolled bleeding, and fresh wounds and extract as a nasal drop. Seeds are used to cure ringworm and dried fruits for the treatment of polyuria and urinary disorders. (Rajakumar and Shivanna 2009, Shukla *et al.* 2010, Singh *et al.* 2010d, Sharma and Singh 2012, Sushama and Nishteswar 2013, Ganie and Yadav 2014, Sutar *et al.*2014; http://tropical.theferns.info).

52. URTICACEAE

Urticaceae – Nettle family

The family has 54 plant genera and 1,465 accepted species. Urticaceae include a number of well-known and useful plants, including nettles in the genus *Urtica*, mostly as herbs, shrubs, lianas and rarely as trees. Leaves are entire or serrated and bear stipules with urticating (stinging) hairs. Flowers are unisexual and can be both monoecious or dioecious and borne in cymose clusters. The family was represented by one species in the flora.

52.1 *Gonostegia pentandra* (Roxb.) Miq.

(Syn. *Pouzolzia pentandra* (Roxb.) Benn.; *Urtica pentandra* Roxb.)

Habit

Common Name: Melastome Pouzolz's Bush.

Vernacular Name: None known.

Habit: Perennial herb.

Habitat: Common along the banks of canals, ditches and other aquatic bodies.

Distribution: It is distributed from Afghanistan, India to the Philippines and throughout South-East Asia, especially Subtropical Himalaya, India, Sri Lanka, China, and Malaysia.

Flowering and Fruiting Time: July-November.

Taxonomic Description: An erect or ascending, monoecious herb. Stem 30-80 cm high, terete below, angular above. Leaves stipulate, sessile or shortly petioled, 2-10 x 0.5-2.0 cm, ovate-lanceolate to oblong-lanceolate, base rounded or sub-cordate, margin entire, apex acute or acuminate, 3-nerved from the base, lower leaves opposite; upper leaves alternate, much-smaller than the lower ones. Flowers greenish-white, axillary, clustered in upper leaves. Male flowers stalked; perianth 5-partite, about 2 mm long; stamens-5. Female flowers sessile, perianth with 2 or 3 lateral wings. Achenes shining black, ovoid.

Basal opposite leaves

Flowering and fruiting twig

Chemical Constituents: *G. pentandra* contains tannins, flavonoids, anthraquinones, alkaloids and terpenoids. (Joycharat *et al.* 2012).

Medicinal Properties: Plants have diuretic, analgesic, anticancer, anti-inflammatory, emmenagogue properties. The plant exhibits the growth inhibiting effects against bacterium named *Helicobactor pylori* known to cause various gastric ailments. (Joycharat *et al.* 2012, Lovleen and Saggoo 2014; http://tropical.theferns.info).

Medicinal Uses (Traditional/Folklore): A decoction of the whole plant is taken as an emmenagogue and diuretic. The ground fresh leaves are used topically on abscesses and as an anti-inflammatory agent. Decoction of whole plant is administered orally for relieving from fever, while that decoction is also topical administered by keeping it in mouth to cure toothache. Plants are also effective against menstrual disorders, skin inflammation, and periodontitis. (Joycharat *et al.* 2012; http://tropical.theferns.info).

53. MORACEAE

Moraceae – Mulberry family

The family consists of about 40 plant genera and 1,217 species of deciduous or evergreen trees and shrubs, distributed mostly in tropical and subtropical regions. Plants of the family contain a milky latex and have alternate or opposite leaves and small, petal-less male or female flowers. The fruits of many species are multiple because fruits from different flowers become joined together. Four spècies belonging to three genera from this family were recorded in Chithara Village flora.

53.1 *Ficus palmata* Forssk.

(Syn. *Ficus pseudocarica* Miq.)

Habit

Common Name: Indian Fig.

Vernacular Name: *Hindi:* Jangli Anjir (जंगली अंजीर).

Habit: A large shrub, occasionally a small tree.

Habitat: Roadsides, and edges of canals and rivers.

Distribution: Asia: Afghanistan, India, Iran, Iraq, Nepal, Pakistan; Africa: Egypt, Ethiopia, Somalia, Sudan.

Flowering and Fruiting Time: March-June.

Taxonomic Description: A large shrub or small tree. Young branches tomentose, bark smooth. Leaves alternate, stipulate, petioled, 10-15 x 5-12 cm, membranous, orbicular or broad-ovate, usually undivided, occasionally lobed, acute, dentate-serrate, cordate at base, appressed hairy on both the surfaces. Hypanthodia stalked, axillary, usually solitary, sub-globose or pyriform, pubescent, about 2 cm long, yellow at maturity, peduncles tomentose. Male flowers numerous, pedicellate, pedicels hairy. Female flowers many, styles long, perianth 5-partite.

Abaxial side of leaf

Hypanthodia

Chemical Constituents: Phytochemical analysis of *F. palmata* plant extracts showed the presence of alkaloids, tannins, flavonoids, terpenoids and cardiac glycosides. (Chauhan *et al.* 2014, Joshi *et al.* 2014, Bhandari 2016).

Medicinal Properties: Plants have antitumour, antiulcer, antidiabetic, lipid lowering and antifungal properties. The fruit is demulcent, emollient, laxative and used as a poultice. (Chauhan *et al.* 2014, Joshi *et al.* 2014, Bhandari 2016).

Medicinal Uses (Traditional/Folklore): Plant is very useful in the treatment of gastrointestinal disorders and hypoglycaemia. Traditionally, stem latex is applied to extract spines deeply lodged in the flesh. Fruit is laxative and also used in diseases of lungs and urinary bladder. Latex used to cure toothache and moles. Fruit is given to breast feeding women for increasing milk (lactation), used as part of diet to treat constipation. (Ajaib *et al.* 2014, Joshi *et al.* 2014, Bhandari 2016; http://tropical.theferns.info).

53.2 *Ficus religiosa* L.

(Syn. *Ficus caudata* Stokes; *Ficus peepul* Griff.)

Branch of tree

Common Name: Peepal, Bo Tree, Bodhi Tree, Holy Fig Tree, Sacred Fig Tree.

Vernacular Name: *Hindi:* Peepal (पीपल); *Sanskrit:* Plaksha (प्लक्ष), Bodhivriksha (बोधिवृक्ष), Ashwattha (अश्वथामा).

Habit: Large tree.

Habitat: Commonly found planted as a holy tree in temples; also grows wild.

Distribution: Native to the Indian subcontinent, South-West China and Indochina, northern Thailand and Vietnam, but has also been introduced and cultivated elsewhere; throughout India, wild as well as cultivated.

Flowering and Fruiting Time: December-May.

Taxonomic Description: A large tree with whitish or dark grey bark. Leaves up to 18 cm long, heart-shaped, alternate, long petioled, shining, glabrous, margin entire, base cordate or truncate, apex caudate-acuminate, stipules deltoid. Male flowers few, sessile, tepals 2-3, ovate-lanceolate, stamen-1. Gall flowers many. Female flowers like gall flowers, sessile, tepals 3-4. Syconium globose, 0.8-1.5 cm across, axillary, sessile, dark-purple at maturity.

Flowering twig

Fruits (Syconium)

Chemical Constituents: The stem bark of *F. religiosa* is reported of have phytoconstituents such as phenols, tannins, steroids, alkaloids and flavonoids, β-sitosteryl-D-glucoside, vitamin K, n-octacosanol, methyl oleanolate, lanosterol, stigmasterol, and lupen-3-one. The active constituent of the root bark is β-sitosteryl-D-glucoside; fruits contain phenols and flavonols namely kaempeferol, quercetin, and myricetin and 4.9% protein having the essential amino acids, isoleucine, and phenylalanine. Seeds contain phytosterolin, β-sitosterol, and its glycoside, albuminoids, carbohydrate, fatty matter, coloring matter, caoutchoue (0.7–5.1%). Leaves and fruits contain carbohydrate, protein, lipid, calcium, sodium, potassium, and phosphorus. The aqueous extract of dried bark has been reported to contain phytosterols, flavonoids, tannins, furanocoumarin derivatives namely bergapten and begaptol. (Chandrasekar *et al.*2010, Makhija *et al.* 2010, Bhalerao and Sharma 2014).

Medicinal Properties: The bark is sweet, cooling, antibacterial, antiprotozoal, antiviral, antiseptic, antidiarrhoeal, astringent, alterative, laxative, haemostatic, vaginal disinfectant, anti-inflammatory and aphrodisiac. Leaves and tender shoots have purgative properties and also recommended for wounds and skin diseases. Fruits are laxative and digestive; the dried fruit pulverized and taken in water cures asthma; seeds are refrigerant and laxative. (Chandrasekar *et al.*2010, Makhija *et al.*2010, Warrier *et al.* 2010, Bhalerao and Sharma 2014).

Medicinal Uses (Traditional/Folklore): Traditionally and in folk medicines, plant and its parts are used in the treatment of gonorrhoea, gastric problems, diarrhoea, dysentery, haemorrhoids, and gastrohelcosis, asthma, diabetes, epilepsy, inflammatory disorders, bacterial and fungal infections, sexual disorders, controlling palpitation of heart and thereby curing the cardiac weakness and as a body tonic, especially useful for patients suffering from jaundice. The latex is good for neuralgia, inflammations and haemorrhages. The bark is used for gonorrhoea, diarrhoea, dysentery, diabetes, leucorrhoea, menorrhagia, nervous disorders, skin diseases, haemorrhoids, gastrohelcosis and burns. Bark decoction is cooling and used for gonorrhoea, skin diseases, scabies, hiccups and vomiting. A paste of powdered bark is a good absorbent for inflammatory swellings and effective in curing burns. The bark paste is applied externally on unhealthy ulcers and wounds; leaves and twigs are laxative. Traditionally, the bark is used in the treatment of gonorrhoea and ulcers, and the leaves for skin diseases. The leaves reported to have antivenom activity and regulate the menstrual cycle. In Bangladesh, leaves are used in the treatment of various diseases such as cancer, inflammation, or infectious diseases. Fruits are used as laxatives and fruit powder to treat asthma and latex as a tonic. Leaves and tender shoots are purgative and their paste is used for curing wounds and skin diseases. Leaf juice is given in asthma, cough, sexual disorders, diarrhoea, haematuria, toothache, migraine, eye troubles, gastric problems and scabies; leaf decoction is analgesic and used for toothache. Dried fruits are given for treating tuberculosis, fever, paralysis and haemorrhoids and fresh fruits, which are laxative and digestive given in asthma.

In Ayurveda, *F. religiosa* belongs to a class of drugs called *Rasayana. Rasayana* are rejuvenators, antioxidants and relieve stress in the body, *Prameha, Raktapitta, Vrana, Vatarakta,* and *Yonidosha.* The bark of *F. religiosa* is an important ingredient in many Ayurvedic formulations, such as *Nalpamaradi tailam, Chandanasavam, Nyagrodhadi churna* and *Saribadyasavam.* (Chandrasekar *et al.* 2010, Gulecha *et al.* 2010, Kaul and Dwivedi 2010, Makhija *et al.*2010, Singh *et al.* 2010d, Rahul 2013, Bhalerao and Sharma 2014, Shaheen *et al.* 2014; http://tropical.theferns.info).

53.3 *Morus alba* L.

(Syn. *Morus intermedia* Perr.)

Common Name: White Mulberry, Silkworm Mulberry.

Vernacular Name: Shahtoot (शहतूत), Kattha (कट्ठा).

Habit: Small tree.

Habitat: Cultivated for its edible fruits and also grows wild along roadsides, open and waste places.

Distribution: The plant, *M. alba* is cultivated throughout the world wherever silkworms are raised, and is occasionally cultivated elsewhere in Europe, North America, and Africa. Having escaped, trees often appear on roadsides, along fencerows, and as ornamentals. Occur throughout India, wild and cultivated.

Flowering and Fruiting Time: February-May.

Habit (Blooming tree)

Taxonomic Description: A small, monoecious, deciduous tree. Leaves alternate, petioled, stipulate, variable, up to 8 cm long, broad ovate or irregularly lobed, base truncate or cordate, margin serrated, apex obtuse-acute or acuminate, sparsely pubescent beneath. Inflorescence catkin. Male catkin elongate, yellowish-green, calyx 4-partite, stamens-4, peduncles hairy. Female catkin usually smaller than male, sepals-4, ovary 1-celled. Fruits sorosis type, edible, greenish white or light to dark purple.

Fruiting twig Fruit

Chemical Constituents: *M. alba* is reported to contain the phytoconstituents such as tannins, phytosterols, sitosterols, saponins, triterpenes, flavonoids, benzofuran derivatives, morusimic acid, anthocyanins, anthroquinones, glycosides and oleanolic acid as the main active principles. Chemical constituents of leaves include prenylflavanes, glycoside, isoquercitrin, astragalin, scopolin, skimmin, roseoside II and benzyl D-glucopyranoside. The plant is a very good source of ascorbic acid, of which over 90% is present in a reduced form, and also contains carotene, Vitamin B1, folic acid, folinic acid, isoquercetin, quercetin, tannins, flavonoids and saponins, which act as a good source of natural antioxidant. Leaves contain triterpenes (lupeol), sterols (β- sitosterol), bioflavonoids (rutin, moracetin, quercetin-3-triglucoside and isoquercitrin), coumarins, volatile oil, alkaloids, amino acids and organic acids. *M. alba* leaves contain rutin, quercetin and apigenin as bioactive constituents. One of the major constituents of *M. alba* is 1-deoxynojirimycin; leaf extract contains nitric acid, prostaglandin E2 and cytokines in macrophages and root bark, a polysaccharide. (Devi *et al.* 2013, Salihoglu *et al.* 2013).

Medicinal Properties: The roots are astringent, anthelmintic and purgative. The leaves are diaphoretic, emollient and antibacterial, and are good for pharyngodynia, scabies and hoarseness. The fruits are sour, sweet, cooling, aphrodisiac, laxative, diuretic, anthelmintic and brain tonic. (Warrier *et al.* 2010, Devi *et al.* 2013).

Medicinal Uses (Traditional/Folklore): Root, stem barks, twigs and leaves of mulberry have long been used in traditional medicines to treat fever, inflammation, hepatitis, cancer, diabetes, dislipidemia, diarrhoea, and dyspepsia, hypertension, burning sensation, small pox, pharyngopathy, lumbago, dyspepsia, ulcerated intestine, strangury and lower blood pressure. It is useful in protection of liver, hepatopathy, splenopathy and melancholia, and improving eyesight, strengthening joints and facilitating discharge of urine. The leaves are also useful in treating pharyngodynia, scabies and hoarseness.

Leaves are boiled in water for a few minutes and decoction is taken for cough and throat irritation; leaves are used to promote sweating. The branches and bark help lower blood pressure and also very effective in treating chronic diseases of the digestive tract as it help to improve digestion, stimulate the appetite, promote gastric juice secretion and eliminate constipation. The leaves of the plant stimulate the nervous system and have a positive effect on the respiratory and cardiovascular system, as well as the uterine system. Tea made from root is used twice a day for 5-7 days to treat diarrhoea. The bark is used as a remedy for cough, wheezing, edema, fever, headache, and red, dry and sore eyes. It also promotes urination. Mulberry fruit is used to treat weakness, dizziness, tinnitus, fatigue, anaemia, and incontinence. (Sultana *et al.* 2006, Tomar 2009, Kaul and Dwivedi 2010, Devi *et al.* 2013, Rahul 2013, Emniyet *et al.* 2014, Shaheen *et al.* 2014; http://tropical.theferns.info).

53.4 *Streblus asper* Lour.

(Syn. None known)

Habit

Common Name: Sand Paper Tree, Toothbrush Tree.

Vernacular Name: *Hindi:* Dahia (दहिया), Sihora (सीहोरा); *Sanskrit:* Shakotakah (शाखोटक:).

Habit: Small tree.

Habitat: Found along roadsides, railway lines and along canal banks. Not common in the study area.

Distribution: Native to dry regions in Cambodia, Thailand, India, Sri Lanka, Malaysia, and Vietnam.

Flowering and Fruiting Time: February-June.

Taxonomic Description: A small, much-branched, usually dioecious tree with greyish-white bark. Leaves alternate, shortly petioled, 4-9 x 2-5 cm, elliptic-obovate or rhomboid, rough, scabrid on both the surfaces, somewhat crenate-serrate, cuneate at base, acute or obtuse at apex, stipules caducous. Flowers unisexual, greenish-white. Male flowers bracteate, in short peduncled spherical heads, tepals-4, stamens-4. Female flowers pedunculate, bracteate, generally solitary, axillary, tepals-4. Drupes sub-globose, yellow when ripe, enclosed in persistent tepals.

Leafy twig

Abaxial side of leaf

Chemical Constituents: It is a rich source of cardiac glycosides. From the stem bark of this plant, α-amyrin acetate, lupeol acetate, β-sitosterol, α-amyrin, lupeol and diol, strebloside and mansonin have been isolated; a pregnane glycoside named sioraside has also been isolated. From the aerial parts, n-Triacontane, tetraiacontan-3-one, β-sitosterol, stigmasterol, betulin and oleanolic acid were identified. An unidentified cardenolide, β-sitosterol, α-amyrin and lupeol were isolated from root bark and leaves. The volatile oil from fresh leaves of *S. asper* contained phytol, α-farnesene, trans-farnesyl acetate, caryophyllene and trans-trans-a-farnesene. (Rastogi *et al.* 2006, Sanjay 2015).

Medicinal Properties: Possesses antioxidant, anti-allergic, antimalarial, anti-cancer, cardio-tonic and insecticidal properties. The leaves are depurative and laxative. The seed is considered tonic and carminative, as well as an appetizer. The leaf extract has been shown to possess insecticidal activity towards mosquito larvae. (Rastogi *et al.* 2006, Sanjay 2015; http://tropical.theferns.info, https://www.bimbima.com/ayurveda).

Medicinal Uses (Traditional/Folklore): Plant are used in Ayurveda and other folk medicines for the treatment of different ailments such as filariasis, leprosy, toothache, diarrhoea, dysentery and cancer. It is used in the treatment of skin diseases, piles, urinary disorders, excessive urination, scrofula, and cervical lymphadenitis. In Ayurvedic medicine, leaves are used to help promote the milk-flow of breast-feeding mothers, as well as being used as poultices for boils and skin ulcers. For the treatment of jaundice, *Streblus asper* leaf + *Psidium guajava* tender shoot tip + *Artocarpus heterophyllus* tender shoot + *cumin* seeds ground in milk is given at early morning in empty stomach for nine days. The seeds are useful in epistaxis, piles and diarrhoea. Externally, the paste of seeds is applied in treating leucoderma.

A decoction of the stem bark is used in the treatment of dysentery, diarrhoea and fevers. The bark is chewed as an antidote in snake poisoning. The bark extract has been used in fever, dysentery, relief of toothache and antigingivitis. The powdered root is used for diarrhoea and dysentery and can be made into a poultice for ulcers on the skin. The powdered root bark is used to treat toothache and to cure peritonitis. It is used traditionally in leprosy, piles, diarrhoea, dysentery, elephantiasis and cancer. The bark of the tree can be boiled in water and the resultant liquid is used as a disinfectant for wounds and skin problems. (Jain 1991, Rastogi *et al.* 2006, Sanjay 2015; http://herbs-treatandtaste.blogspot.in, http://tropical.theferns.info, https://www.bimbima.com/ayurveda).

54. CANNABACEAE

Cannabaceae – Hemp family

A small family consisting of 8 plant genera and 102 accepted species of erect herbs, climbing vines, shrubs, and trees. Plants, often dioecious (distinct male and female plants), have often more or less palmately lobed or compound leaves and always bear stipules with petalless flowers and dry, one-seeded fruits. Cystoliths are always present and some members of this family possess laticifers. Hemp (*Cannabis*) and hop (*Humulus*) are the only economically important species. The family was represented by a single species in the flora.

54.1 *Cannabis sativa* L.

(Syn. *Cannabis generalis* E.H.L.Krause; *Cannabis indica* Lam.)

Common Name: Marijuana, Hemp.

Vernacular Name: *Hindi*: Bhang (भांग), Ganja (गांजा), Charas (चरस); *Sanskrit:* Bahuvadini (बहुवादिनी), Bhanga (भंगा).

Habit: Annual herb.

Habitat: Commonly growing gregariously along roadsides, railway lines, waste places, cultivated and fallow fields.

Distribution: Throughout India, wild in North West Himalayas; grows wild in Uttar Pradesh, Orissa, Bihar, Madhya Pradesh and Rajasthan.

Flowering and Fruiting Time: February-September.

Habit

Male Flowers **Achenes**

Taxonomic Description: An erect, dioecious, 80-150 cm tall, branched annual herb. Stem angular; branchlets strong-smelling, resinous, densely white-pubescent. Leaves alternate, stipulate, palmately 3-7 foliate; leaflets sessile, linear-lanceolate, coarsely serrate, acuminate, pubescent, petiole 2-8 cm long. Male plant: male flowers 4-6 mm across, whitish-green or yellowish, arranged in axillary, short peduncled, cymose clusters. Perianth 5-lobed, tepals 3-5 mm long, oblong-lanceolate, pubescent. Stamens-5, anther greenish-yellow. Female plant: female flowers greenish, crowded in leaf like bracts, sparsely pubescent. Fruits achene, ovoid.

Chemical Constituents: *Cannabis* contains 421 chemicals of various classes, such as cannabinoids, cannabispirans and alkaloids. More than 60 cannabinoids have been isolated; the most important one is δ -9-tetrahydrocanabinol. Most varieties contain cannabinol and cannabinin; resin contains crystalline compound cannin. Trigonelline occurs in the seed. *Cannabis* also contains choline, eugenol, guaiacol, nicotine, and piperidine, all listed as toxins by the National Institute of Occupational Safety and Health. A β -resercyclic acid derivative has antibiotic and sedative properties. (Bhutya 2011; https://www.hort.purdue.edu/newcrop/duke_energy/Cannabis_sativa.html).

Medicinal Properties: Arab physicians knew and used *Cannabis* for its diuretic, antiemetic, antiepileptic, anti-inflammatory, painkilling and antipyretic properties, among others. Medicinally, plants are tonic, intoxicant, stomachic, antispasmodic, analgesic, narcotic, sedative and anodyne. Leaves are astringent, hallucinogenic, hypnotic, sedative, analgesic, anti-inflammatory, tranquilizer, aphrodisiac, tonic, stomachic and abortifacient. (Bhutya 2011; http://tropical.theferns.info, https://www.hort.purdue.edu/newcrop/duke_energy/Cannabis_sativa.html).

Medicinal Uses (Traditional/Folklore): In traditional medicines, *C. sativa* has been used as hallucinogenic, hypnotic, sedative, analgesic, and anti-inflammatory agent. Leaves are used as an intoxicating agent. Plants are known to have a greater array of folk medicine uses: alcohol withdrawal, anthrax, asthma, blood poisoning, bronchitis, burns, catarrh, childbirth, convulsions, coughs, cystitis, delirium, depression, diarrhoea, dysentery, dysmenorrhoea, epilepsy, fever, gonorrhoea, gout, inflammation, insomnia, jaundice, lockjaw, malaria, mania, menorrhagia, migraine, morphine withdrawal, neuralgia, palsy, rheumatism, scalds, snakebite, swellings, tetany, toothache, uterine prolapse, and whooping cough. Plant and its parts are also useful in the treatment of abdominal disorders, otalgia, and stomatalgia, loss of appetite, penile dysfunction, premature ejaculation, hypertension, indigestion, piles, epilepsy, amenorrhea, dysmenorrhoea, neuritis, headache, insanity, hiccough and cholera.

Leaves boiled with butter are given to prevent vomiting. Drops of leaf juice are poured in the ear to cure earache. The leaf, prepared in various manners, is said to alleviate cancerous sores, scirrhous tumours, cold tumours, and white tumours. Seeds and leaves are used to treat old cancer and scirrhous tumours. The seed, either as a paste or as an unguent, is said to be a folk remedy for tumours and cancerous ulcers. The decoction of the root is considered to be a good remedy for hard tumours and knots in the joints; the plant is also used for mammary tumours and corns. Seeds ground and mixed with porridge are given to weaning children.

Contra indication: It's excessive use causes dyspepsia, cough, sleeplessness, delirium, headache, impotence, melancholy, dropsy, hyperpyrexia and insanity. (Singh *et al.* 2010d, Rahul 2013; http://tropical.theferns.info, https://www.hort.purdue.edu/newcrop/duke_energy/Cannabis_sativa.html).

55. SALICACEAE

Salicaceae – Willow family

Traditional family Salicaceae, which includes the willows, poplar, aspen, and cottonwoods, has 54 plant genera and 1,269 accepted species. Widespread, from tropical to north temperate and boreal regions many species are planted for ornamental or architectural value, especially the Weeping Willows. The family consists of bushes and trees sometimes dwarf with simple, alternate leaves often with salicoid teeth. The flowers are unisexual with male and female flowers appearing in catkins on separate plants (dioecious). The sepals are greatly reduced or absent, and there are no petals. Male flowers have 2 or more stamens. In the pistillate (female) flower, the ovary is positioned superior and consists of 2 to 4 united carpels (syncarpous) forming a single chamber. It matures as a capsule, usually with silky "cotton" to help transport the seeds by air. The family was represented by a single plant species in the flora.

55.1 *Salix tetrasperma* Roxb.

(Syn. *Pleiarina tetrasperma* (Roxb.) N. Chao & G.T. Gong)

Common Name: Indian Willow.

Vernacular Name: *Hindi:* Laila (लैला), Bains (बैंस); *Sanskrit:* Jalavetasa (जलवेतस), Naadeya (नादिया), Baishi (बेसी).

Habit: Small deciduous tree.

Habitat: Common in swampy areas, along the banks of canals at low and medium altitudes and also planted in the gardens.

Distribution: It grows in many parts of South and southeastern Asia, including India, Pakistan, Nepal, Laos, Myanmar, China, Thailand, Indonesia, Malaysia, Philippines and Vietnam. In India, it is recorded from Maharashtra, Karnataka, Kerala, Tamil Nadu and Uttar Pradesh.

Flowering and Fruiting Time: November-March.

Upper branch of plant

Leafy twig showing silvery abaxial side of leaves

Catkin inflorescence

Taxonomic Description: A much-branched, deciduous, small tree. Stem greyish brown, vertically fissured, young branchlets pubescent. Leaves 6-15 x 2-4 cm, alternate, petioled, oblong-lanceolate, glabrascent, glaucous beneath, base rounded, tip acuminate, margin finely serrulate, occasionally entire; stipules deciduous, ovate-orbicular.

Flowers bracteate, dioecious, small, in dense pedunculate catkins. Perianth absent. Male catkins yellow, 5-10 cm long, rachis densely silky hairy; stamens unequal, 5-12; anthers yellow, disc with 2 glands. Female catkin yellow, nearly as long as male catkin, ovary ovoid, stigma bi-lobed. Fruit a capsule.

Chemical Constituents: Phytochemical screening of plants yield various types of sapogenins: quinovic acid, salicortin, saligenin, phenolic glycosides and pyrocatechol from the bark and leaves and tannins, triterpenes, *viz.*, ß-amyrin, lupeol and chalcinasterol, steroids *viz.*, ß-sitosterol and stigmasterol, with very low concentrations of free salicylaldehyde. Stem bark contains salicin, friedelin, 3 β-friedelinol acetate, β- sitosterol, β-sitosteryl-3-O-D-glucopyranoside, β-sitostenone, cis-1,2- cyclohexanediol, and long chain ester. (Chhetree *et al.* 2012, El-Wakil *et al.* 2015, Prashith *et al.* 2017; http://www.stuartxchange.org/Malatiki.html).

Medicinal Properties: The bark of Indian Willow has febrifuge and analgesic, sedative, antirheumatic, anti-inflammatory, antipyretic properties. Dried leaves are cardiotonic and neurotonic. (Chhetree *et al.* 2012, Prashith *et al.* 2017; http://medplants.blogspot.in, http://www.stuartxchange.org/Malatiki.html).

Medicinal Uses (Traditional/Folklore): Dried leaves mixed with sugar are given in rheumatism, arthritis, fever, hepatitis, spermatorrhea, epilepsy, venereal diseases, and for treating stone in the bladder, piles and for swellings. Dried leaves are also used as anti-inflammatory agent. As a folkloric medicine bark is used to treat fever, decoction of leaf and root for whooping cough in children, paste of both leaf and root applied externally for scorpion stings, bug bites, sores and warts. Decoction of dried root is taken internally for treatment of hepatitis, sap of stem taken orally for dysmenorrhoea and hot water extract of entire plant instilled in vaginal cavity as abortifacient; and also used for treating local rectal sores. In Egyptian folk medicine, the plant is used as antirheumatic, sedative and analgesic. (Chhetree *et al.* 2012, Prashith *et al.* 2017; http://www.stuartxchange.org/Malatiki.html).

56. CERATOPHYLLACEAE

Ceratophyllaceae – Hornwort or Coontail family

Ceratophyllaceae is a primitive small family of Dicotyledons* comprising of 1 plant genus and 4 accepted species of rootless aquatic plants that grow underwater. They have long, branching stems and leaf-like thin branches, whorled on the stem, that are forked into two or more linear sections. Flowers lack petals and they have either pollen-bearing or ovule-bearing parts, but not both, with both types borne on a single plant. The family was represented by a single species in the Chithara Village.

*APG IV system of flowering plant classification, mostly molecular-based system of plant taxonomy includes Ceratophyllaceae under Eudicots.

56.1 *Ceratophyllum demersum* L.

(Syn. *Ceratophyllum verticillatum* Roxb.)

Common Name: Hornwort, Coon's Tail.

Vernacular Name: *Hindi:* Brihatpushpi (बृहतपुष्पी).

Habit: Aquatic herb.

Habitat: Common in ponds, wetlands, ditches and canals.

Distribution: Cosmopolitan distribution; native to all continents except Antarctica. *C. demersum* grows in lakes, ponds and quiet streams having still or very slow-moving water with water temperatures of 15-30 °C.

Flowering and Fruiting Time: September-December.

Habit

Whorled leaf

Apical portion

Taxonomic Description: An aquatic, much-branched, submerged, root-less herb. Leaves whorled, dark-green; segments linear-filiform, 2-2.5 cm long, apex terminated with 1 or 2 sharp bristles. Flowers minute, greenish, axillary, solitary, unisexual, male and female flowers at different nodes, apex ended with 2 bristles. Male flowers sessile or sub-sessile, stamens 8-30. Female flowers sessile, ovary 1-celled. Fruit a nutlet, about 3 mm long, ovoid, subtended by 2 lateral and 1 apical spine.

Chemical Constituents: The secondary metabolites isolated from the plant are flavonoid glycosides and apigenin-7-O-glucoside and seven sterols, the main one sitosterol. Seven compounds *viz.,* tricin-7-O-β-D-glucoside, naringenin-7-O-β-D-glucoside, esculetin, β-sitosterol, 7-α-hydroxyl-β-sitosterol, 7-α-methoxyl-β-sitosterol and palmitic acid were isolated recently. The oil extracted from the plant contains 2-methylpropanoic acid 3-hydroxy-2, 4, 4-trimethylpentyl ester, 2-methylpropanoic acid 2, 2-dimethyl-1-(2-hydroxy-1-methylethyl) propyl ester, *β*-ionone-5, 6-epoxide, toluene, hexanal and 1, 2-benzenedicarboxylic acid di (2-methylpropyl) ester were the major components. Volatile compounds identified are mainly n-paraffin, together with benzyl acetate and a sesquiterpene. (Bankova 1999, Abu 2017).

Medicinal Properties: The plant is bitter, sweet, cooling, antiperiodic, styptic, astringent, carminative, antiulcer, antidiarrhoeal, wound healing activities and antipyretic. (Quattrocchi 2012, Chakraborty 2014, Abu 2017).

Medicinal Uses (Traditional/Folklore): In traditional medicine, whole plant is used to treat ulcer, diarrhoeal and wound conditions. Leaf extract is used to cure diarrhoea, fever, ulcers, dipsia, dysentery, piles, and hemoptysis. It is also useful in the treatment of biliousness and scorpion stings. Leaf juice is cooling and used to stop vomiting in Mau district of UP. (Gond *et al.* 2012, Quattrocchi 2012, Chakraborty 2014, Abu 2017).

II. MONOCOTYLEDONS

Fig. 30: A creeping monocotyledon, *Commelina* sp.

Monocotyledons, commonly referred to as monocots, are one of the two major groups of flowering plants, or angiosperms comprising of approximately 60,000 species of plants. Most of them are distinguished by the presence of only one embryonic leaf or cotyledon. The group includes not only major grain producing plants (rice, wheat, maize, etc.), but also many species of Commelina (Fig. 30), sugarcane, palm (Fig. 31 A), grasses (Fig. 31 B), sedges, and the bamboos. Other economically important monocot crops include banana, gingers and their relatives, turmeric and cardamom, asparagus and the onions and garlic family. Additionally, most of the horticultural bulbs such as lilies, daffodils, irises, amaryllis, cannas, bluebells and tulips, are monocots.

The APG III system of 2009 identifies a clade called "monocots" but does not assign it to a taxonomic rank.

Fig. 31: A palm (A - *Phoenix sylvestris*) and a grass species of monocot (B - *Saccharum spontaneum*)

57. HYDROCHARITACEAE

Hydrocharitaceae

Hydrocharitaceae is a Monocotyledonous family including 16 known plant genera with a total of 133 accepted species, including a number of aquatic species. The family includes both fresh water and marine aquatics found throughout the world in a wide variety of habitats, but are primarily tropical. The leaves are simple and usually found submerged, though they may be found floating or partially immersed. The flowers are arranged in a forked, spathe-like bract or between two opposite bracts. Two species under two genera of this family were represented in Chithara flora.

57.1 *Hydrilla verticillata* (L.f.) Royle

(Syn. *Serpicula verticillata* L.f.; *Vallisneria verticillata* (L.f.) Roxb.)

Habit

Common Name: Waterthyme, Hydrilla.

Vernacular Name: *Hindi:* Sevar (सेवार), Jhangi (झंगी).

Habit: Aquatic herb.

Habitat: Common in ponds, wetlands, canals and other aquatic bodies.

Distribution: *H. verticillata* has a wide and rather disjointed geographical range. It is native to the cool and warm waters in Asia, Europe, Africa and Australia, with a sparse, scattered distribution.

Flowering and Fruiting Time: July-October.

Dense mass of plant

Leafy twig

Taxonomic Description: A submerged, monoecious or dioecious, aquatic herb, forming a dense masses. Stems slender, profusely branched, leafy. Leaves whorled, often 4-8 at each node, sessile, 1-2 x 0.2-0.3 cm, linear-oblong, margin serrated, apex acute-apiculate. Male spathe shortly stalked, subglobose, flower solitary, sub-sessile, sepals-3, petals-3, stamens-3. Female spathe sessile, 1-flowered, flower similar to male. Fruit cylindrical.

Chemical Constituents: Five phytochemicals such as 1-(5′-Hydroxy-4′-hydroxymethyl-1′-methyl-1*H*-pyrrol-2′-yl)-henicosa-2, 12, 15-trien-1-one and thymidine, crystal loliolide, sulfur-gamma, and octadecanedioic acid have been isolated from the plant. (Xiao *et al.* 2007).

Medicinal Properties: *H. verticillata* has detoxifying, cicatrizing, pyogenes, antioxidant, free radical scavenging, anti-ageing, and antipollution properties. (Pal and Nimse 2006).

Medicinal Uses (Traditional/Folklore): The whole plant is used in the treatment of abscesses, boils and wounds, especially if there is debris in the wound. A dried powder of the plant is applied to cuts and wounds to help accelerate healing and drying. Whole plant is slightly warmed and tied over the painful affected area for relief from joint pain in Odisha. The plant is also used in sugar refineries for purification. (Pal and Nimse 2006, Panda *et al.*2016; http://www.pfaf.org/user/plantsearch.aspx).

57.2 *Vallisneria spiralis* L.

(Syn. *Vallisneria aethiopica* Fenzl)

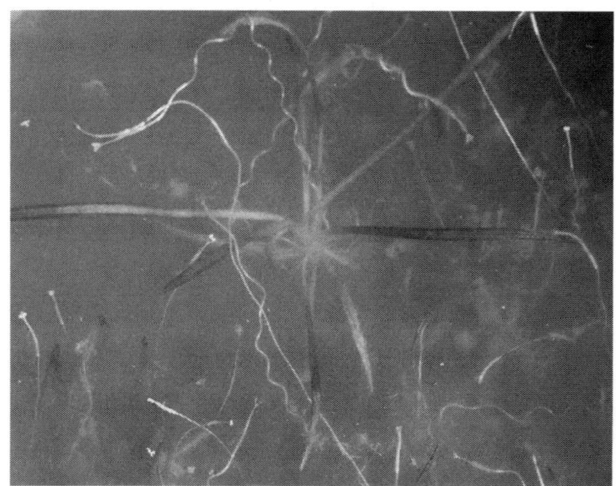

Habit

Common Name: Tape Grass, Straight Vallisneria.

Vernacular Name: *Hindi:* Sival (सिवाल), Shaival (शैवाल).

Habit: Perennial submerged aquatic herb.

Habitat: Common in muddy bottom of wetlands, rivers, ditches, ponds, canals and waterlogged swamps.

Distribution: Widespread in tropical and subtropical areas of both hemispheres. Native to southern Europe, northern Africa, the Middle East and southwest Asia, but now spread to the north and northwest in France, Belgium, Netherlands, southern Britain, Poland and Russia. Introduced to North America, West Indies, New Zealand, New Caledonia and the Hawaiian Islands. In India, it is reported from Uttar Pradesh and West Bengal.

Flowering and Fruiting Time: July-November.

Lower portion with spirally coiled peduncle **Female flower with spathe** **Perianth with 3 stigmas**

Taxonomic Description: An aquatic, submerged, stoloniferous, dioecious herb. Leaves basal, sessile, up to 40 cm long, narrowly linear, ribbon-shaped, margin entire or narrowly serrulate towards the tip. Flowers unisexual. Male flowers very minute, shortly pedicellate, several in short stalked spathe, which breaks off and each flower liberated and floating at the water surface, sepals-3; petals-3, minute or rudiments. Female flowers solitary in 1.5-2 cm long spathe, spathe borne at the apex of a long, slender, leafless scape, which gets spirally coiled after anthesis; sepals and petals as in male, stigmas fringed. Fruits numerous seeded.

Chemical Constituents: Plants contain essential oils consisting of 2-Methylpropanoic acid 3-hydroxy-2,4,4-trimethylpentyl ester (>15%), 2-methylpropanoic acid 2,2-dimethyl-1-(2-hydroxy-1-methylethyl)propyl ester (>3%), β-ionone-5,6-epoxide (>7%), toluene (>6%), hexanal (>5%) and 1,2-benzenedicarboxylic acid di(2-methylpropyl) ester (>5%) as the major components. (Xian *et al.* 2006).

Medicinal Properties: The plant is aperitif, demulcent, refrigerant, stomachic and is also used in the treatment of women's ailments. (http://www.pfaf.org/user/plantsearch.aspx).

Medicinal Uses (Traditional/Folklore): *V. spiralis* is used in folk medicine, Siddha, and traditional Chinese medicine. The plant is used in the treatment of leucorrhoea, anorexia, bleeding, hematospermia and is made into a tea with Sesame seeds (*Sesamum indicum*) to improve the appetite. (http://www.pfaf.org/user/plantsearch.aspx).

58. PONTEDERIACEAE

Pontederiaceae- Pickerel-weed family

Family Pontederiaceae comprising of 6 plant genera, 33 accepted species is a small family of heterostylous aquatic plants, free-floating or rooted in mud, occurring in tropical and subtropical waters. It is best known for the water hyacinth (*Eichhornia crassipes*), which is an invasive species in many waterways. The partially immersed leaves usually consist of a basal sheath, a distinct stalk (i.e., petiole), and an expanded blade, but one species has flaccid, frequently submersed, grass-like leaves. The flowers vary from solitary to arrayed in inflorescences and may be zygomorphic (i.e., bilaterally symmetrical) or actinomorphic (i.e., radially symmetrical). They have both pollen-bearing and ovule-bearing parts. The tepals are blue, purple or white, are fused at the base, and have one or two spots on the upper tepal in some species. The fruit is a capsule or achene. The family was represented by a single plant species.

58.1 *Eichhornia crassipes* (Mart.) Solms

(Syn. *Pontederia crassipes* Mart.)

Habit

Common Name: Water Hyacinth.

Vernacular Name: *Hindi:* Jal Kumbhi (जल कुम्भी); *Sanskrit:* Jalakumbhi (जल कुम्भी), Variparni (वरिपर्णी).

Habit: An invasive aquatic herb.

Habitat: A serious weed of aquatic fields; growing gregariously in ponds, wetlands, canals, rivers and other water reservoirs.

Distribution: *E. crassipes*, a native of tropical America, is a major freshwater weed naturalized in most of the frost-free regions of the world and is generally regarded as the most troublesome aquatic plant in Africa, Australia, India and many other countries. Some of the phenolic compounds and cyanide contribute to the negative impact associated with the proliferation of *E. crassipes,* in aquatic systems in addition to blocking water ways. It is a serious threat to aquatic biodiversity in India.

Flowering and Fruiting Time: May-October.

Close up of population

Inflorescence

Taxonomic Description: A perennial, free floating aquatic herb. Leaves radical, rosette forming, glabrous, thick, dark green, shining, 8-20 cm across, suborbicular, spathulate or paddle-shaped, entire; petioles long, spongy, swollen. Flowers sessile, lilac or violet blue, showy, arranged in many-flowered terminal spikes. Perianth segments-6, obovate, 3-4 cm long, one segment tinged with yellow spot. Stamens-6; filaments of 3 longer stamens are curved, glandular-hairy; anthers bluish-black, oblong. Ovary 3-celled. Capsules ovoid-oblong.

Chemical Constituents: *E. crassipes* contains flavonoids (luteolin, apigenin, tricin, chrysoeriol, kaempferol, azaeleatin, gossypetin and orientin), amino acids (methionine, valine, glutamic acid theonine, tryptophan, tyrosine, leucine and lysine), tannins, alkaloids and saponins, sterols, glycosides, resins, phosphorous, protein, organic matter and cyanide. (Nyananyo *et al.* 2007, Tulika and Mala 2015).

Medicinal Properties: Plants are known to have anticancer, anti-inflammatory, antimicrobial, antiviral, antioxidant, wound healing, antitumour and larvicidal properties. (Tulika and Mala 2015).

Medicinal Uses (Traditional/Folklore): In the Philippines, after cleaning the plant thoroughly, the decoction of the whole plant is used to treat blood disorders, emaciation and weakness. The decoction is also used to treat goitre. Along with this decoction, a poultice is prepared from the crushed plant and also applied onto the enlarged part of the neck. The leaf petioles are eaten as a treatment for diarrhoea and an infusion of the inflated petioles is used in a bath to treat fevers in the Guianas (Guyana, Surinam, and French Guiana). In Kedah (Malaysia), the flowers are used for medicating the skin of horses and used as a tonic. (Vardhana 2008, DeFilipps 2004; http://tropical.theferns.info).

59. COMMELINACEAE

Commelinaceae – Spiderwort family

The members of Commelinaceae are annual herbs comprising 41 plant genera and 728 accepted species that are often somewhat succulent and frequently seen around human dwellings like the common garden. Flowers, in cymose inflorescences, are ephemeral and lack nectar, usually surrounded by a boat-shaped spathe (bract). Well known genera include *Commelina* (dayflowers) and *Tradescantia* (spiderworts). Four herb species belonging to 3 genera from this family were recorded in Chithara Village.

59.1 *Commelina benghalensis* L.

(Syn. *Commelina canescens* Vahl; *Commelina hirsuta* R.Br.)

Common Name: Bengal Dayflower, Whiskered Commelina, Tropical Spiderwort.

Vernacular Name: *Hindi:* Kankawwa (कनकव्वा), Kana (काना), Buchna (बुचाना).

Habit: Annual herb.

Habitat: Occurs in moist-waste places, shady areas and in gardens.

Distribution: Native to tropical Asia and Africa. It has been widely introduced to areas outside its native range, including to the neotropics, Hawaii, the West Indies and to both coasts of North America. It flowers from spring into the fall and is often associated with disturbed soils. In India, Nepal, Sri Lanka very widely distributed in scrub jungles, wayside puddles. Weed of arable lands, marshy plains from the coast up to 1,100 m.

Flowering and Fruiting Time: July-October.

Habit

Flowering twig

Chasmogamous flower

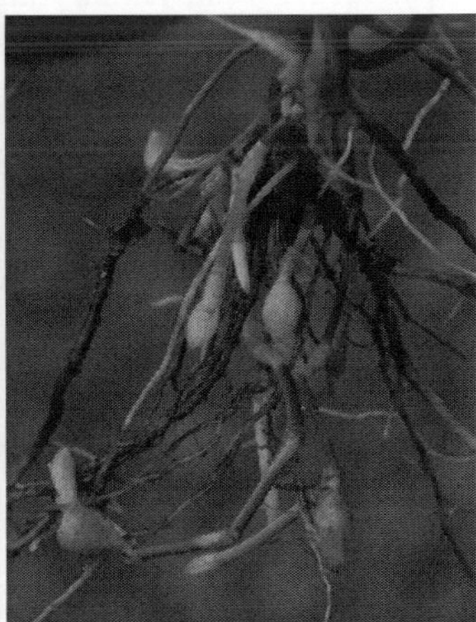

Cleistogamous flowers

Taxonomic Description: A prostrate or decumbent, annual herb. Stem slender, rooting at lower nodes. Leaves alternate, shortly petioled with sheathing leaf base, up to 6 cm long and 3.5 cm wide, ovate-lanceolate, entire, base cuneate-truncate or subcordate, apex acute or obtuse, sheaths hairy, about 1.5 cm long. Flowers two types; aerial chasmogamous flowers and underground cleistogamous flowers. Aerial chasmogamous flowers blue, subtended by spathe. Sepals-3, ovate-oblong. Petals-3, unequal, obovate, orbicular. Stamens-6, fertile-3, staminodes-3. Capsules pyriform, 5-seeded. Seeds wrinkled. Underground flowers smaller than the aerial ones, generally solitary in a pear shaped spathe.

Chemical Constituents: The phytochemical screening revealed the presence of phlobatannins, carbohydrates, tannins, glycosides, volatile oils, resins, balsams, flavonoids and saponins, while terpenes, sterols, anthorquinones and phenols were absent. (Ibrahim *et al.* 2010, Thombre *et al.* 2016).

Medicinal Properties: Febrifugal, bitter, demulcent, laxative, emollient, astringent, anti-inflammatory, cooling, diuretic, antileprotic, antibacterial, antimicrobial, antipyretic, antiedematic and antitoxic. (Thombre *et al.* 2016; http://medplants.blogspot.in, http://tropical.theferns.info).

Medicinal Uses (Traditional/Folklore): The plant is used in Ayurveda and *Sidha* to cure skin inflammations, leprosy, burn wounds; whole plant paste is applied against pain, piles and chest pain. Roots are effective in fever and liver complaints. Sap is used for treatment of eye ailments, sore throat and burns and topical application for thrush in infants. Leaves have antimicrobial properties; fresh leaf extract is applied on fungal infection caused by *Tinea pedis* and for treating pneumonia; leaf decoction is used to stop profuse menstrual bleeding; leaves pounded and soaked in warm water and the extract is used to treat diarrhoea. Whole plant is also used for epidemic influenza and upper respiratory tract infection. Knee-joint problems are treated after prolonged use of the plant extract.

In China, *C. benghalensis* is used medicinally as a diuretic, febrifuge and anti-inflammatory. In Pakistan it is used as animal fodder and also eaten by humans as a vegetable. It is also used there medicinally, but with different purported effects, including as a laxative and to cure inflammations of the skin as well as leprosy. The people of Nepal eat the young leaves as a vegetable, use a paste derived from the plant to treat burns, and treat indigestion with a juice prepared from the roots. In South-East Asia and Africa the plant is used as fodder and also medicinally as a poultice.

As a veterinary medicine, leaf extract is applied for cataract treatment; leaves made into a paste and applied for yoke bite and used as anthelmintic. (Thombre *et al.* 2016; http://medplants.blogspot.in, http://tropical.theferns.info).

59.2 *Commelina forsskalii* Vahl

(Syn. *Commelina falcata* Hassk.)

Habit

Common Name: Kanpet, Day flower, Rat's ear.

Vernacular Name: *Hindi:* Kankawwa (कनकव्वा).

Habit: Creeping herb.

Habitat: It grows in cultivated fields, gardens and moist localities along waysides.

Distribution: *C. forsskalii* is distributed in tropical Africa, Vietnam and India.

Flowering and Fruiting Time: July-October.

Taxonomic Description: A prostrate or creeping, much-branched, annual herb with two types of flowers; aerial chasmogamous and underground cleistogamous. Stem often rooted at lower nodes. Leaves 4-7 x 1-1.5 cm, shortly petioled with sheathing leaf base, oblong, linear-lanceolate, entire or undulate, acute at apex, sparsely pubescent. Aerial spathes ciliate, about 1.5 cm long, axillary. Chasmogamous flowers deep blue. Sepals-3, boat-shaped. Petals-3, 2-larger, spatulate or orbicular. Fertile stamens-3, staminodes-3. Capsules 3 seeded. Seeds subglobose, smooth. Cleistogamous flowers reduced, generally 1-flowered in each spathe.

Chemical Constituents: None known.

Flowering twig

Chasmogamous flower

Cleistogamous flowers

Medicinal Properties: None known.

Medicinal Uses (Traditional/Folklore): None known.

59.3 *Cyanotis axillaris* (L.) D.Don ex Sweet

(Syn. *Commelina axillaris* L.; *Tradescantia axillaris* (L.) L.)

Habit

Common Name: Creeping Cradle Plant.

Vernacular Name: *Hindi:* Kana (काना).

Habit: Annual herb.

Habitat: Occurs in moist places, gardens, shady areas.

Distribution: Indo-Malaysia. In India, distributed in degraded deciduous forests and in moist waste places.

Flowering and Fruiting Time: August-November.

Taxonomic Description: A decumbent, branched, annual herb. Stem terete, ribbed, sparsely pubescent. Leaves alternate, sessile, up to 12 cm long, linear-lanceolate, entire, acute or sub-acuminate at apex, sheaths minutely hairy. Flowers about 7 mm across, axillary, borne in leaf sheath. Calyx connate, pubescent. Corolla pinkish-blue, broadly ovate, sub-acute. Stamens-6, filaments bearded, anthers oblong. Capsules ellipsoid-oblong, 5-6 mm long. Seeds brownish, flattened.

Flowering twig

Flower in axillary sheath

Tepals

Chemical Constituents: None known.

Medicinal Properties: None known.

Medicinal Uses (Traditional/Folklore): *C. axillaris* is used to treat boils, ascites and abortions as an external/local application. Decoction of the whole plant is used in the treatment of swellings above the abdomen. Warm leaf juice is dropped into the ear to get relief from inflammation in the eardrum. The plant is a remedy for tympanites in Nadia district of WB. Raw decoction of leaf SMSD (for swelling) and DID-1 (for itching) is used orally by Irula tribe, Walayar, Tamil Nadu. (Swapna *et al.* 2011, Chakravorty and Ghosh 2013, Venkatachalapathi *et al.* 2016; http://www.stuartxchange.org/Sabilau.html).

59.4 *Murdannia nudiflora* (L.) Brenan

(Syn. *Commelina nudiflora* L.)

Habit

Common Name: Naked-Stem Dewflower, Doveweed.

Vernacular Name: *Hindi:* Kansura (कंसुरा); *Sanskrit:* Koshapushpi (कोशपुष्पी).

Habit: Annual herb.

Habitat: Wet places of gardens, cultivated and fallow fields.

Distribution: Native to the Old World tropics, possibly India, and is a common weed occurring from India to China. It has now spread to other tropical and sub-tropical parts of Africa, Asia, Oceania, Central, North and South Americas, invading both the agricultural and non-agricultural areas.

Flowering and Fruiting Time: August-October.

Taxonomic Description: A decumbent-ascending, simple or branched, annual herb. Leaves alternate, sessile, 5-11 x 0.5-1.3 cm, linear-lanceolate, entire, acute to acuminate, sheath ciliate. Flowers bluish-purple, arranged in axillary and terminal, few-flowered, sub-corymbose panicles. Sepals-3, about 3.5 mm long, obtuse. Petals-3, nearly 4 mm long, obovate, rounded. Fertile stamens-2, filaments bristly hairy. Capsules subglobose, apiculate, 3-celled. Seeds brownish.

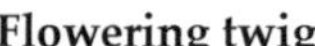

Flowering twig

Flower

Capsules

Chemical Constituents: Phytochemical compounds present are phenolic compounds, phytosterols, alkaloids and flavonoids. (Patwari *et al.* 2014).

Medicinal Properties: *M. nudiflora* is astringent and antiasthmatic. (Patwari *et al.* 2014).

Medicinal Uses (Traditional/Folklore): Used locally in the treatment of asthma, leprosy, piles, stomach complaints, giddiness, burns, boils and sores. Root paste with goat milk is prescribed orally to cure asthma. Whole plant paste with common salt is applied on the affected area of skin to cure leprosy. Leaf paste is also used on the forehead to cure headache. (Patwari *et al.* 2014, Chowdhury and Das 2015).

60. ARECACEAE

Arecaceae – Palm family

There are 185 plant genera and 2,522 accepted species in the family most of which are restricted to tropical, subtropical, and warm-temperate climates. Most of the species are perennial climbers, shrubs, acaules and trees commonly known as palm trees, restricted to tropical, subtropical, and warm temperate climates. Palms are widely used in landscaping for their exotic appearance, making them one of the most economically important plants. The family was represented by a single palm species in the study area.

60.1 *Phoenix sylvestris* (L.) Roxb.

(Syn. *Elate sylvestris* L.)

Common Name: Wild Date Palm, Silver Date Palm.

Vernacular Name: Khajoor (खजूर).

Habit: Palm tree.

Habitat: Open, moist and dry areas. It is found in groves and individual palm tree at several places in Chithara Village including SNU Campus.

Distribution: All over India, mostly in Andhra Pradesh, Karnataka, Madhya Pradesh, Uttar Pradesh and Rajasthan.

Flowering and Fruiting Time: May-July.

Fruiting spadix

Fruits

Habit

Taxonomic Description: A moderate-sized, dioecious palm. Stem unbranched, erect, marked with the persistent leaf base. Leaves 3-4 m long, form a dense crown, pinnate, bright green. Leaflets numerous, rigid, sessile, linear, ensiform, entire, spinous at apex, lower leaflets modified into thorns. Flowers small, arranged in spadix. Spadix up to 90 cm long. Male flowers white, fragrant, calyx and corolla 3-lobed, stamens usually 6. Female flowers white or greenish, sepals-3, petals-3. Fruits a drupe, about 2 cm long, orange-yellow, oblong-ellipsoid, edible. Seed-1, longitudinally grooved.

Chemical Constituents: Fresh unfermented sap (*Niraa*) is a good source of ascorbic acids, nicotinic and isonicotinic acids, riboflavin, thiamine, sugars, cysteine, leucine, isoleucine, lysine, phenylalanine, threonine, tyrosine. Fruits contain protein, fat, calcium and phosphorus. Half ripe fruits (*Chuhara*) contain protein, fat, carbohydrates and calcium, phosphorus and iron. Presence of sterols of ergosterol group and ester-1 has been reported from dried date palms. (Bhutya 2011).

Medicinal Properties: The fruits of wild Date Palm are restorative, antitussive, expectorant, demulcent, laxative, diuretic, and used for general debility, sexual debility, urinary disorders, fatigue, bronchitis, cough, cold, loss of appetite, cardiac disorders, hemoptysis, bleeding diathesis, pyrexia, diarrhoea, excessive thirst, asthma; also beneficial for physical strength and gaining body weight. Sap is cooling laxative, gastric stimulant, digestive, appetizer, used for burning sensation, loss of appetite and weakness. (Bhutya 2011).

Medicinal Uses (Traditional/Folklore): Sap of the plant is nutritious, cooling and laxative whereas central tender part is useful to treat gonorrhoea. Root is used in toothache and in nervous debility and helminthiasis. Gum is useful in diarrhoea and genito-urinary diseases. Fruit is tonic and restorative. Seed paste is good for the opacity of the cornea, inflammation and wounds. Roots are used for nervous debility and gum for diarrhoea, and genital urinary disorders.

Fruits are eaten fresh or in dried form. After the rains sapling sprout up and young tender shoots are cooked as vegetable. Sap is obtained by giving a cut near the apex of the tree. It is boiled down to sugar which is used as a health food. During scarcity periods, the farinaceous deposit from the apex of the tree is eaten by the tribal. (Katewa *et al.* 2003, Sinhababu and Banerjee 2013, Islam *et al.* 2014; http://tropical.theferns.info).

61. TYPHACEAE

Typhaceae – Cattail Family

The family having large marsh/aquatic herbs includes 2 plant genera and 65 accepted species. Typically, the plants are rhizomatous with long, slender two-ranked leaves. Plants are monoecious, producing a distinctive brownish compact spike of unisexual flowers with the terminal portion staminate and the basal portion pistillate. There was single species belonging to this family present in Chithara Village.

61.1 *Typha domingensis* Pers.

(Syn. *Typha angustata* Bory & Chaub.)

Common Name: Elephant Grass, Reed, Lesser Indian Reed Mace, Small Bulrush, Narrow leaf cattail.

Vernacular Name: *Hindi:* Patera (पटेरा), Pater (पटेर).

Habit: Perennial marshy herb.

Habitat: Abundant in marshy areas, along the bank of wetlands, canals, and other aquatic bodies.

Distribution: Found throughout temperate and tropical regions worldwide. In India, the species is distributed in Andhra Pradesh, Jharkhand, Himachal Pradesh, Karnataka, Kerala, Madhya Pradesh, Maharashtra, Tamil Nadu, Uttar Pradesh and West Bengal.

Flowering and Fruiting Time: April-October.

Habit

Long leaves

Basal portion of plant

Inflorescence

Taxonomic Description: A robust, perennial herb, 2-3 m high with creeping rhizome. Leaves sheathing at the base, very long, linear, spongy, nearly reaching or exceeding the flowering stem. Flowers numerous, unisexual, arranged in terminal, reddish-brown, long, terete spikes. Male spikes above and female spikes below; both are usually separated by 5-15 cm, long intervals. Male flowers closely packed, covered with linear silky hairs in the spike, stamens-3, filaments short. Female flowers bracteate, style filiform, stigma linear. Nutlets minute.

***Chemical Constituents*:** Phytochemical screening revealed the presence of alkaloids, polyphenols, flavonoids, tannins, saponins, hydroxycinnamic acid and proanthocyanidins. (Al-Saeed 2012, Rao *et al.* 2016b).

Medicinal Properties: Leaves are diuretic. The pollen is astringent, desiccant, diuretic, haemostatic and vulnerary. It is contraindicated for pregnant women. The seed down is haemostatic. The rootstock is astringent and diuretic. (Al-Saeed 2012, Rao *et al.* 2016b; http://tropical.theferns.info).

Medicinal Uses (Traditional/Folklore): The plant and its parts are used in the treatment of nose bleeds, haematemesis, haematuria, uterine bleeding, dysmenorrhoea, postpartum abdominal pain and gastralgia, scrofula and abscesses. Root extract, about two tea spoonful is mixed with half a cup of curd. This mixture is given orally three times in a day for three days to cure dysuria. A mixture is prepared from root extract, about two spoonful mixing with 5 g of sugar candy and 2 g of *Cuminum cyminum* Linn. This mixture is taken orally three times a day for one week for reducing intestinal heat. (Ghorband *et al.* 2012, Rao *et al.* 2016b; http://shodhganga.inflibnet.ac.in/bitstream/10603/152758/13/13_chapter5.pdf, http://tropical.theferns.info).

62. ARACEAE

Araceae – Arum family (often colloquially known as aroids).

This family of 117 plant genera and 3,368 accepted species has all perennial rhizomatous or tuberous herbs like taro, *Suran* or elephant foot and yam. Many of the species are aquatic or wetland plants, but a few grow in uplands. Plants are usually with calcium oxalate crystals or raphides and commonly with milky sap. The leaves are usually with a sheathing base. The inflorescence is a fleshy spike or spadix subtended or commonly partially enveloped by a bract or spathe which is sometimes petaloid or brightly colored. Three aquatic species belonging to three genera from this family were recorded in the flora.

62.1 *Lemna perpusilla* Torr.

(Syn. *Hydrophace perpusilla* (Torr.) Lunell)

Plants in aquatic habitat

Common Name: Common Duckweed, Common Waterlens, Duck Meat.

Vernacular Name: None known.

Habit: Free-floating aquatic herb.

Habitat: Common in stagnant water of ponds, ditches and wetlands.

Distribution: *L. perpusilla* is pantropical in distribution, occurring in Africa, Asia and America, and is locally naturalized in some more temperate areas, including France and Italy in southern Europe. In India, it has been recorded from Uttar Pradesh and Jammu & Kashmir states.

Flowering and Fruiting Time: July-October.

Taxonomic Description: An aquatic, small, floating herb. Fronds solitary or 2-5 together, 2-3 x 1.5-2.5 mm, obovate or ovate-oblong, pale green, asymmetrical, entire, usually flat on the both surfaces, Root solitary, 1 per frond, sheathed near base. Spathe with one pistillate and two staminate flowers. Stamen-1, filament filiform. Ovary 1-celled. Utricles ovoid, 1-seeded.

Floating population of plants | **Individual plant with root**

Chemical Constituents: None known.

Medicinal Properties: None known.

Medicinal Uses (Traditional/Folklore): Plants ground into a paste with rice water applied externally to treat boils and herpes. (http://tropical.theferns.info, shodhganga.inflibnet.ac.in/bitstream/10603/131737/15/15_appendix%202.pdf).

62.2 *Pistia stratiotes* L.

(Syn. *Pistia aegyptiaca* Schleid.)

Habit (gregariously covering entire surface of water)

Common Name: Water Cabbage, Water Lettuce.

Vernacular Name: Jalkumbhi (जलकुम्भी).

Habit: Aquatic herb, floating on water surface.

Habitat: Found gregariously floating in lakes, ponds, ditches and other aquatic bodies.

Distribution: It is distributed in the tropical and subtropical regions of Asia, Africa, and America. Found throughout India.

Flowering and Fruiting Time: August-November.

Taxonomic Description: A stoliniferous, aquatic herb, floating on surface of water. Roots fibrous, white, filiform, pendulous under water. Leaves rosulate, yellowish-green, 4-13 x 2-6 cm, sessile, closely spiral, obovate, base tapering, margin undulate, apex truncate, softly hairy. Inflorescence spadix, hidden by the leaves. Spathe small, up to 2 cm long, white, tomentose outside, constricted; male flowers above, female flower below.

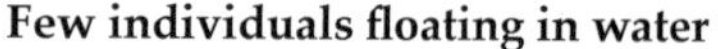

Few individuals floating in water

Root system

Chemical Constituents: The biologically active chemical constituents of *P. stratiotes* are alkaloids, glycosides, flavonoids, and steroids. Other phytochemicals reported are stigmasta-4, 22-dien-3-one, stigmasterol, stigmasteryl stearate, palmitic acids and 2-di-C-glcosylflavones of vicenin and lucenin type, anthocyanin-cynidin-3-glucoside, luteolin-7-glycoside and mono-C-glcosyl flavones – vitexin and orientin. Leaves are rich in vitamin A and C, and also contain vitamin B. The ash is rich in potassium chloride and sulfate. (Tripathi *et al.* 2010, Tulika and Mala 2015).

Medicinal Properties: A large number of medicinal properties are attributed to *P. stratiotes*, particularly the leaves. The plant is bitter and pungent flavored and has antiseptic, antitubercular, and antidysenteric, antidiabetic, antidermatophytic, antifungal, antimicrobial, cooling, diuretic and laxative properties. The root is laxative, emollient, and diuretic. It has been reported that ethanol and hot water fractions of the plant exerts antimicrobial action on a few pathogenic bacteria while chloroform fraction of the same plant possess both antifungal and antibacterial activities on some pathogens. (Tripathi *et al.* 2010, Khan *et al.* 2014b, Tulika and Mala 2015; http://medplants.blogspot.in).

Medicinal Uses (Traditional/Folklore): *P. stratiotes* is commonly used in Ayurvedic medicines. It is useful in "Tridosha," fever, and diseases of blood. Leaves infusions have been mentioned in the folklore to be used for dropsy, bladder complaints, kidney afflictions, hematuria, dysentery, and anaemia. In Gambia, the plant is used as an anodyne for eyewash. Juice of plant is used by Mundas in ear complaints. The ash of plant is applied to the ringworm of the scalp. Leaf extract is used in eczema, leprosy, ulcers, piles, and syphilis. Juice of leaves boiled in coconut oil is applied externally in chronic skin diseases. The leaves are used in the treatment of dysentery, asthma, coughs, and skin diseases in Chitwan, Nepal. Its extract is used as an anodyne for eyewash and for relieving ear complaints. Its ash is applied to scalp for curing ringworm. Leaf extract boiled in coconut oil is applied to the skin in chronic dermatitis. Its concoction is useful for relieving nervous disorders, fever and intestinal bacterial infections. *P. stratiotes* is also reported to be useful in the treatment of stomach disorder, throat and mouth inflammation. (Tripathi *et al.* 2010, Tulika and Mala 2015; http://tropical.theferns.info).

62.3 *Spirodela polyrrhiza* (L.) Schleid.

(Syn. *Lemna polyrrhiza* L.)

Plants in aquatic habitat

Common Name: Common Duck Meat, Duckweed.

Vernacular Name: None Known.

Habit: Aquatic herb.

Habitat: Generally growing in dense colonies in ponds, canals, wetlands, ditches and water-logged rice fields.

Distribution: Found nearly worldwide in stagnant water bodies.

Flowering and Fruiting Time: July-September.

Taxonomic Description: A small, free floating, aquatic herb, forming a mat on the surface of water. Fronds in groups, orbicular or broad obovate, up to 1 cm long, upper surface flat, lower surface purplish, margin entire, apex obtuse-rounded; each frond with several roots. Daughter fronds connected to the margin of the mother fronds. Flowers minute, enclosed in membranous spathe, staminate flowers-2, pistillate flower-1. Fruits ovoid, winged.

Habit

Close up of plants

Chemical Constituents: Low-molecular weight metabolites of *S. polyrhiza* include manool (25.11%), biformen (12.83%) and phytol (8.05%). (Kurashov *et al.* 2016).

Medicinal Properties: *S. polyrhiza* has diuretic, diaphoretic, antipyretic, antipruritic, cardiotonic and carminative properties. It animates the existence of measles rash. (http://www.pfaf.org/user/plantsearch.aspx).

Medicinal Uses (Traditional/Folklore): *S. polyrhiza* is utilized to cure colds, measles, oedema and trouble in urination. (Heim 2015).

63. ALISMATACEAE

Alismataceae – Water-Plantains family

The family comprising of 17 plant genera and genera 120 accepted species of aquatic or wetland herbs, annual or perennial, are commonly found in shallow water and swamps or on muddy banks or wet sand. Juvenile and submerged leaves are often linear, whilst more mature and immersed leaves can be linear to ovate or even sagittate - arrowhead-shaped. The family was represented by a single species in Chithara Village.

63.1 *Sagittaria guayanensis* Kunth

(Syn. *Lophiocarpus guayanensis* (Kunth) Micheli)

Habit

Common Name: Guyanese Arrowhead.

Vernacular Name: None known.

Habit: Aquatic herb.

Habitat: Rice fields in water-logged condition, shallow ponds, lakes and ditches.

Distribution: It is native to Mexico, Central America, the West Indies, and much of South America, as well as West Africa (from Senegal to Cameroon), south and South-East Asia, Sudan and Madagascar.

Flowering and Fruiting Time: August-October.

Taxonomic Description: An aquatic, fibrous rooted herb. Leaves floating, membranous, long petioled, sagittate or broadly ovate, base deeply cordate, margin entire, apex obtuse or rounded. Flowering scapes 10-30 cm long. Flowers white, bracteate, pedicels slender. Male flowers towards the apex and bisexual flowers towards the base of the inflorescence. Sepals-3, triangular. Petals-3, obovate, white, with yellowish spot inside. Stamens 6-12, filaments very short. Achenes numerous, winged.

Inflorescence Male flower

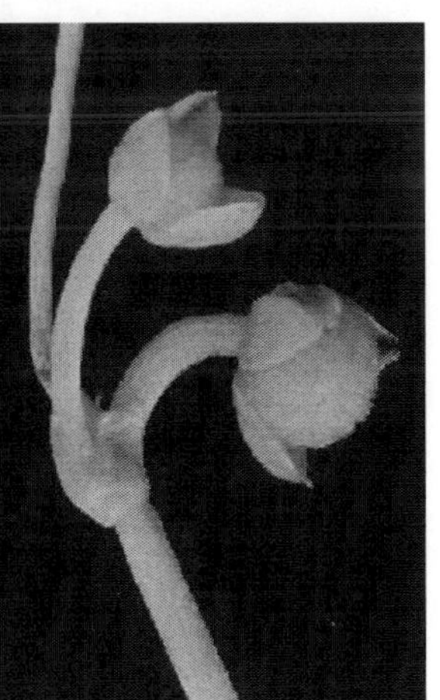

Fruiting heads

Chemical Constituents: None known.

Medicinal Properties: None known.

Medicinal Uses (Traditional/Folklore): Rhizome is used in skin diseases. (Ahmad *et al.* 2010).

64. POTAMOGETONACEAE

Potamogetonaceae – Pondweed family

The aquatic family of monocotyledonous flowering plants having 6 plant genera and 186 accepted species, is a worldwide family with the greatest distribution in the northern hemisphere. The leaves in this family may be flaccid and submerged or thicker and floating; a given species may have one or both types of leaves, often with narrow leaves underwater and broad leaves on the surface. There were two species belonging to two genera recorded in Chithara Village.

64.1 *Potamogeton crispus* L.

(Syn. *Potamogeton crenulatus* D.Don)

Common Name: Curly-Leaf Pondweed.

Vernacular Name: None known.

Habit: Aquatic herb.

Habitat: Common in aquatic water bodies such as in ponds, canals, lakes and ditches.

Distribution: Native to Eurasia, Africa, and Australia. This aquatic invasive species is now reported in all of the continents, except for Maine and South Carolina in US.

Flowering and Fruiting Time: November-February.

Leafy twig

Plant in aquatic habitat

Taxonomic Description: A submerged, aquatic herb. Stem long, terete to somewhat angular. Leaves alternate, sessile, stipulate, 3-8 x 0.5-1 cm, linear, oblong, crispate, undulate or finely serrate at margin, obtuse-rounded at apex, translucent. Spikes aerial with long peduncle. Flowers bisexual, perianth 4-partite, stamens-4, carpels-4. Fruits about 4 mm long, ovoid, beaked.

Chemical Constituents: It contains carotenoids, fatty acids, lignan, labdane di-terpenoids, flavonoids, and phytosterins. Two major flavonoids, luteolin-3'-O-β-d-glucopyranoside and flavone-6-C-β-d-glucopyranoside have been isolated from *P. crispus.* (Du *et al.* 2015).

Medicinal Properties: None known.

Medicinal Uses (Traditional/Folklore): None known.

64.2 *Stuckenia pectinata* (L.) Börner

(Syn. *Potamogeton pectinatus* L.)

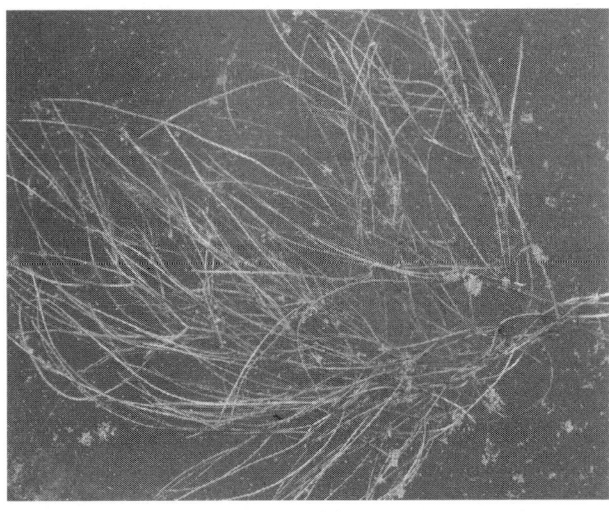

Habit

Common Name: Sago Pondweed, Fennel Pondweed.

Vernacular Name: None known.

Habit: Aquatic herb.

Habitat: Common in aquatic habitats such as ponds, canals, lakes and other water bodies.

Distribution: Cosmopolitan water plant species that grows in fresh and brackish water on all continents except Antarctica; found throughout India.

Flowering and Fruiting Time: December-April.

Taxonomic Description: A submerged, grass-like, aquatic herb. Stem terete, filiform, dichotomously branched. Leaves alternate, sessile, up to 15 cm long and 2-3 mm wide, narrowly linear-filiform, apex acute or mucronate, stipules fused to the leaf sheath. Inflorescence spikes. Flowers bisexual, few, whorled on aerial spikes. Perianth 4-lobed. Pistil-4. Fruits oblique, shortly beaked.

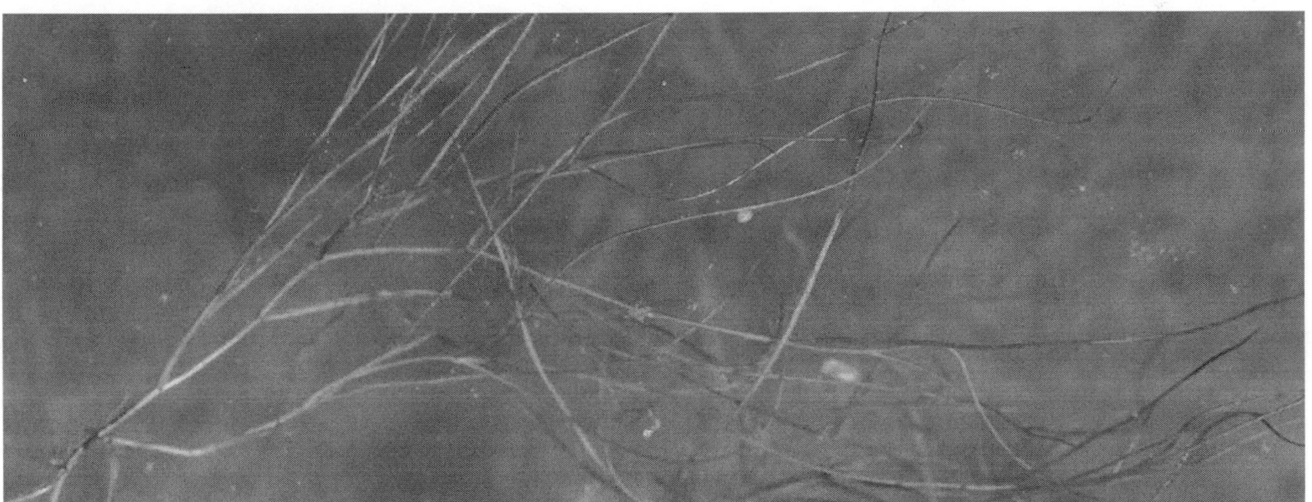

A portion of shoot showing long needle-like leaves

Chemical Constituents: None known

Medicinal Properties: Astringent and tonic. (Quattrocchi 2012).

Medicinal Uses (Traditional/Folklore): A decoction of the plant is used in the treatment of feverish liver. Root stock is used as a tonic, astringent and nutritious drink. (Quattrocchi 2012; http://tropical.theferns.info).

65. CYPERACEAE

Cyperaceae – Sedge family

Cyperaceae having 110 plant genera and 5,784 accepted species names, which superficially resemble grasses and rushes and found especially in wet regions throughout the world. The leaves, parallel-veined, strap-shaped blade, are arranged such that they spiral around the three-sided stem in 3-ranks and have a basal portion that sheaths the stem. Seven species belonging to 2 genera from this family were recorded in the flora of Chithara Village.

65.1 *Cyperus alopecuroides* Rottb.

(Syn. *Juncellus alopecuroides* (Rottb.) C.B.Clarke)

Common Name: Nut Grass, Yellow Nut Sedge. Chufa Sedge.

Vernacular Name: None known.

Habit: Perennial sedge.

Habitat: Common in marshy areas along the bank of wetlands, canals, ponds and rivers.

Distribution: Native to most of the Western Hemisphere as well as southern Europe, Africa, Madagascar, the Middle East and the Indian subcontinent. It has become naturalized in many other regions, including Ukraine, China, Hawaii, Indochina, New Guinea, Java, New South Wales and various oceanic islands.

Habit

Flowering and Fruiting Time: August-November.

Inflorescence

Spike

Taxonomic Description: An erect, perennial, marshy sedge, 80-120 cm high. Stem green, solid, glabrous, triangular. Leaves 1-1.5 cm wide, nearly as long as stem, linear, apex tapering-pointed, sheath brown. Spikelets pale brown, 4-6 mm long, numerous, densely crowded in the spikes, borne in a compound inflorescence. Bracts leaf-like, 5-7, few bracts longer than the inflorescence. Glumes ovate, keeled, mucronate. Stamens-2. Nuts yellowish-brown, obovoid.

Chemical Constituents: Benzoquinone and flavonoids - benzoquinone, named alopecuquinone, were isolated from the inflorescences of *C. alopecuroides*. The known flavonoids, vicenin 2, orientin, diosmetin, quercetin 3, 3′-dimethyl ether and its 3, 4′-dimethyl ether, have also been isolated and characterized from the plant. (Nassar *et al.* 2002).

Medicinal Properties: None Known.

Medicinal Uses (Traditional/Folklore): None Known.

65.2 *Cyperus compressus* L.

(Syn. *Cyperus brachiatus* Poir.)

Habit

Common Name: Poorland Flat Sedge.

Vernacular Name: Mothi (मोठी).

Habit: Annual sedge.

Habitat: Common in wet sandy soil, near small water channels and in paddy fields.

Distribution: *C. compressus* has very wide distribution at global level across the tropics and sub-tropics in warm countries, except Australia. It is a very prominent weed of paddy in India (throughout the northern plains), Pakistan, and Bangladesh and throughout South-East Asian countries.

Flowering and Fruiting Time: July-October.

Spikes

Taxonomic Description: An erect, tufted, annual herb with 3-angled, 10-30 cm long stem. Leaf blades up to 25 cm long, narrowly linear, tapering towards the apex, sheaths purplish or pale brown. Bracts 2-6, foliose, long, spreading, unequal, often longer than the primary rays. Spikelets up to 2 cm long, digitately clustered and compressed, in spikes. Glumes 3-4 mm long, ovate-mucronate, keeled. Stamens-3. Nuts obovoid, dark-brown.

Chemical Constituents: None known.

Medicinal Properties: None known.

Medicinal Uses (Traditional/Folklore): None known.

65.3 *Cyperus difformis* L.

(Syn. *Cyperus oryzetorum* Steud.)

Habit (image from herbarium specimen)

Common name: Small-Flowered Nut Sedge, Variable Flat Sedge, One-Arm Sedge, Rice Sedge, Small-Flower Umbrella-Sedge.

Vernacular Name: *Hindi:* Dila (दिला), Motha (मोथा).

Habit: Annual sedge.

Habitat: Common in paddy fields, and marshy areas of small water channels, ponds, lakes and canals.

Distribution: *C. difformis* is distributed in the Old World (subtropical and tropical areas), southern Europe, most of Africa and Asia, and Australia. It is native of tropical America.

Flowering and Fruiting Time: July-November.

Taxonomic Description: An annual, 15-40 cm high, erect sedge. Stem tufted, triquetrous, smooth. Leaves few, usually shorter than the stem, nearly 3 mm wide, linear, smooth. Inflorescence of 3-6, somewhat globose heads of spikelets. Rays few, up to 3 cm long. Bracts leafy, 2-3, unequal, erect or spreading, longer than inflorescence. Glumes ovate-orbicular, keeled, apex rounded. Stamen 1-2. Stigmas-3. Nutlets yellowish, elliptic-obovoid.

Inflorescence

Chemical Constituents: There is predominance of sesquiterpenes compounds in addition to monoterpenes in *C. difformis* plants. Several chemical components have been identified in the oil of *C. difformis*; cyperene (34.8 %), cyperotundone (7.8 %), isorotundene (7.8 %) were the main constituents. Oils is very rich in sesquiterpene hydrocarbons and oxygenated sesquiterpenes. (Feizbakhsh *et al.* 2013).

Medicinal Properties: None known.

Medicinal Uses (Traditional/Folklore): In Pir Nasoora National Park, Jammu and Kashmir, the plant is crushed and made into paste which is externally applied on skin infections. (Amjad *et al.* 2015).

65.4 *Cyperus iria* L.

(Syn. *Chlorocyperus iria* (L.) Rikli)

Habit

Common Name: Rice Flat Sedge.

Vernacular Name: *Hindi:* Morphula (मोरफुला).

Habit: Annual sedge.

Habitat: Often grows in rice fields and marshy places as a weed.

Distribution: *C. iria* is most often found as a weed in Japan, the Pacific Islands and Australia to the south, and through India to the West. Outside Asia, it has been reported in southern and western Africa and in the USA. It has also been recorded in Kenya, Uganda and Brazil. Found throughout India; in Karnataka it is widespread. From Iran, Afghanistan, China and Japan to Australia, also in eastern Africa; common throughout Indo-China, Thailand and Malesia. Introduced and naturalized in the south-eastern United States and the West Indies.

Flowering and Fruiting Time: July-October.

Inflorescence

Close up of inflorescence

Taxonomic Description: An erect, 20-60 cm high, annual sedge. Leaves linear, finely acuminate, shorter than the culm, sheaths reddish-brown. Inflorescence a compound umbel, with 4-12 cm long spikes. Spikelets yellowish, 6-20 flowered, compressed, oblong or narrowly ovoid. Primary bracts 3-6, very long, leaf-like, unequal. Glumes obovate, nearly 2 mm long, 3-5 nerved, keeled, apex retuse-mucronate. Stamens 2-3. Stigmas-3. Nuts dark brown, obovate.

Chemical Constituents: None known.

Medicinal Properties: Astringent, stimulant, stomachic, tonic. (http://tropical.theferns.info, http://uses.plantnet-project.org/en/Cyperus_iria_(PROSEA)).

Medicinal Uses (Traditional/Folklore): In India the plant is used as a tonic, stimulant, stomachic, febrifuge, astringent, and for amenorrhoea. In Vietnam, the whole plant is used to treat rheumatism, and to regulate menstruation; the rhizomes are used as a diuretic. (http://tropical.theferns.info, http://uses.plantnet-project.org/en/Cyperus_iria_(PROSEA)).

65.5 *Cyperus rotundus* L.

(Syn. *Cyperus tuberosus* Rottb.; *Pycreus rotundus* (L.) Hayek)

Common Name: Nut Sedge, Common Nut Sedge, Nut Grass, Jawa Grass, Coco Grass.

Vernacular Name: *Hindi:* Bara Nagar Motha (बड़ा नागर मोथा), Motha (मोथा); *Sanskrit*: Mustaka (मुस्तका), Varida (वरिदा), Chakranksha (चक्रांक्षा), Charukesara (चारुकेसरा).

Habit: Perennial sedge.

Habitat: Common weed of agricultural fields, roadsides, gardens and fallow fields.

Distribution: Nagarmotha (*C. rotundus*) is one of the most widespread, cosmopolitan, problematic, and economically damaging agronomic weeds, growing wildly in various tropical, subtropical and temperate regions of the world. In India, it is common in open, as a weed in disturbed habitats and wastelands to an elevation of about 1,800 m.

Flowering and Fruiting Time: July-December.

Taxonomic Description: An erect, perennial, stoloniferous sedge. Stem triangular, smooth, 20-50 cm high, tuberous at base. Leaves basal, usually shorter than the stem, linear-lanceolate, acuminate, 4-6 mm wide, sheath up to 6 cm long. Inflorescence compound spikes. Bracts-3, leaf-like, unequal, exceeding the inflorescence. Spikelets pinkish-purple, 2-3 cm long, linear, 4-10 on a spike. Glumes 3-4 mm long, boat-shaped, apex obtuse. Stamens-3, anthers 1.5 mm long. Stigmas-3. Nuts ovoid-ellipsoid, pale-brown.

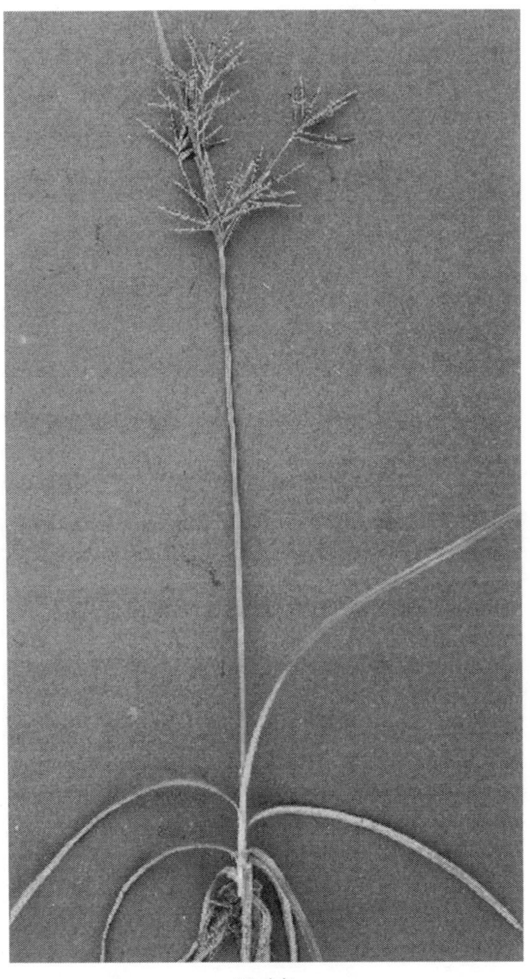

Habit

Chemical Constituents: The major chemical components of this herb are essential oils, flavonoids, terpenoids, and mono sesquiterpenes. Cyperine is the major constituent in the plant, which gives Nut Grass its pharmacological properties. The plant contains the following chemical constituents: cyprotene, acopaene, cyperene, aselinene, rotundene, valencene, cyperol, gurjunene, trans-calamenene, dcadinene, gcalacorene, cadalene, amuurolene, gmuurolene, cyperotundone, mustakone, isocyperol, acyperone, 4, 11-selinnadien-3-one and 1, 8-cineole. The oil of *C. rotundus* was mainly composed of cyperol, α-cyperene, rotundine, α-cyperone, α-copaene, valerenal, myrtenol, β-pinene, α-pinene and α-selinene, sesquiterpene hydrocarbons (caryophyllene). (Imam *et al.* 2013).

Spikes

Medicinal Properties: The plant possesses several pharmacological properties such as carminative, emmenagogue, anthelminthic, analgesic, antipyretic, anti-inflammatory, antidysenteric, antirheumatic, antimicrobial, antimutagenic, radical scavenging, antimalarial, ovicidal and larvicidal, anticonvulsant and antioxidant, wound healing, insect repellent hypotensive, antihyperglycemic, and α-amylase inhibitory activity. The tubers are acrid, bitter, astringent, cooling, anti-inflammatory, galactagogue, depurative, intellect promoting, digestive, anthelmintic, stomachic, and constipating; they also have lithontriptic, emmenagogue, febrifuge, hypotensive, vulnerary, and nervine tonic properties. (Warrier *et al.* 2010; http://medplants.blogspot.in, http://tropical.theferns.info).

Medicinal Uses (Traditional/Folklore): The plant is mentioned in the ancient Ayurvedic treatise, *Charaka Samhita*. Ayurvedic physicians use the plant, known as *Musta* or *Musta moola churna,* for treating fevers, digestive system disorders, dysmenorrhoea and other maladies. The plant is useful in vitiated condition of *Kapha* and *Pitta,* hyperdipsia, inflammations, leprosy, scabies, erysipelas, neurasthenia, epilepsy, flatulence, colic, diarrhoea, dysentery, renal and vesical calculi, bronchitis, amenorrhoea, dysmenorrhoea, wounds, ulcers, fever, intermittent and vomiting and general debility. In Ayurvedic medicine, Nut Grass is taken daily as a tonic for the liver and heart, a digestion stimulant, and an aid against hypertension. It is extremely useful in cases of blocked or infrequent menstruation and premenstrual syndrome (PMS).

C. rotundus has been used as an herbal remedy for pain, cancer, and for high blood pressure. There are numerous ethno-pharmacological reports on the ehtnobotanical use of *C. rotundus* in atherosclerosis, aging, apoptosis, cancer, cystitis, epilepsy, hirsutism, and nociception, prostatitis, and genotoxicity disorders. The herb has a positive effect on the liver, facilitating its detoxification and treatment. *Dried entire plant, infusion*: used for all types of fever, anti-inflammatory agent; anti-emetic, astringent, analgesic, and antitussive. *Rhizome*: aphrodisiac, astringent, nutritive tonic, for the treatment of dysmenorrhoea and menstrual irregularities. Rhizome juice is given in the dose of 25 ml thrice daily for 3 days to treat constipation. The rhizomes are scraped and pounded with green ginger mixed with honey prescribed in dysentery, gastric and intestinal troubles. *Tuber essential oil*: antibacterial, hypochondriac pain and distension, dysmenorrhoea, liver protection, emmenagogue. It is traditionally used to treat various clinical conditions at home such as diarrhoea, diabetes, pyresis, and inflammation, malaria and stomach and bowel disorders. *Fresh tubers*: are applied to the breast as a galactagogue. Paste is applied to scorpion stings and when dried, to spreading ulcers. The tubers are also useful in the treatment of irregular menstruation, gastralgia, and diarrhoea and vomiting. Modern alternative medicine recommends using the plant to treat nausea, fever and inflammation and digestive problems; for pain reduction, analgesic for menstruation pains, for muscle relaxation and many other disorders. The plant is a potential source of natural mosquitocidal agents; tuber extracts are very effective for repelling of the mosquito vector even at a low dose. (Imam *et al.* 2013; http://medplants.blogspot.in, http://tropical.theferns.info).

65.6 *Fimbristylis dichotoma* (L.) Vahl

(Syn. *Fimbristylis diphylla* (Retz.) Vahl)

Habit

Common Name: Common Fringe Sedge, Tall Fringe Rush, Forked Fimbry.

Vernacular Name: None known.

Habit: Annual or perennial sedge.

Habitat: Common in rice fields and bank of ponds, canals and lakes.

Distribution: Widely distributed in Asia and Africa, as well as in other parts of the tropics.

Flowering and Fruiting Time: July-October.

Taxonomic Description: An annual or perennial sedge, up to 50 cm high, with triquetrous, tufted stem. Leaves flat, shorter than stem, linear, glabrous or sparsely pubescent, sheath hairy. Leafy bracts 3-5, unequal, linear-lanceolate, up to 10 cm long. Spikelets ovate-oblong in compound umbel. Rays subequal. Glumes 2-3 mm long, ovate, apex obtuse, keeled. Stamen 1-2, anthers linear-oblong. Stigma 2-fid. Nutlets obovoid.

Inflorescence

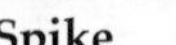

Spike

Chemical Constituents: Cardiac glycosides, saponins and alkaloids are found in aerial parts of plants. (Padalia and Chanda 2015).

Medicinal Properties: Antipyretic, diuretic and antihepatic. (Barik 2017).

Medicinal Uses (Traditional/Folklore): The rhizomes are diuretic and its paste is taken for treating kidney problems. The plant is effective in controlling bleeding, fever and bowel complaints. (Mishra *et al.* 2015e, Barik 2017).

65.7 *Fimbristylis ferruginea* (L.) Vahl

(Syn. *Scirpus ferrugineus* L.)

Common Name: Rusty Sedge, West Indian Fimbry.

Vernacular Name: None known.

Habit: Annual herb.

Habitat: Abundant in swampy and wet areas, paddy fields and in open grasslands.

Distribution: Pantropical in distribution. The plant is common along the coast line and estuaries of Australia. It is also native to parts of Africa, southern Asia, and South America.

Flowering and Fruiting Time: June-October.

Taxonomic Description: A tufted, 40-80 cm high, erect sedge with striated stem. Leaves linear, narrow, much shorter than the stem, margin scabrid, sheath brown. Spikelets 5 to many, pale brown, 7-10 x 3-4 mm, ellipsoid, borne in simple or compound umbel. Involucral bracts 3 to 4, leafy, usually overtopping to umbel. Glumes 3-4 mm long, broadly ovate, pubescent, mucronate. Stigmas-2. Nuts biconvex, sub-orbicular, apiculate.

Lower portion of plant (Herbarium photo) **Inflorescence**

Chemical Constituents: None known.

Medicinal Properties: None known.

Medicinal Uses (Traditional/Folklore): Whole plant is used for stomach ailments in coastal mangrove areas of West Bengal. In Kutch it is edible. (Chaudhury and Das 2009; http://shodhganga.inflibnet.ac.in/bitstream/10603/75475/12/12_chapter%206.pdf).

66. POACEAE

Poaceae – Grass family

Poaceae or Gramineae, comprising of 759 plant genera and 11,554 accepted species, is a large and nearly ubiquitous family of monocotyledonous flowering plants known as grasses. They rank among the top five families of flowering plants in terms of the number of species. The leaves have parallel veins and consist of a cylindrical sheath that wraps around the stem with a blade above it. The stems have conspicuous nodes where the base of the leaf attaches. The flowers are generally small, without obvious sepals or petals, and are arranged in structures called spike-lets that consist of the flowers with specialized subtending scales (called glumes, lemmas, and paleas). A total of 31 species belonging to 24 genera from this family were recorded in Chithara Village flora including the campus of Shiv Nadar University.

66.1 *Avena sterilis* L.

(Syn. *Avena fatua* var. *sterilis* (L.) Fiori & Paol.)

Common Name: Sterile Oat, Winter Wild Oat.

Vernacular Name: *Hindi:* Jangali Jai (जंगली जइ).

Habit: Annual grass.

Habitat: Common weed along roadsides and in agricultural fields.

Distribution: It is native to northern Africa, temperate Asia, Pakistan and India in the Indian sub-continent and throughout Europe. It is widely naturalized elsewhere. *A. sterilis* is highly invasive in cultivated and disturbed ground and has probably already invaded many suitable regions of the world.

Habit

Flowering and Fruiting Time: January-April.

Fibrous root system

Inflorescence

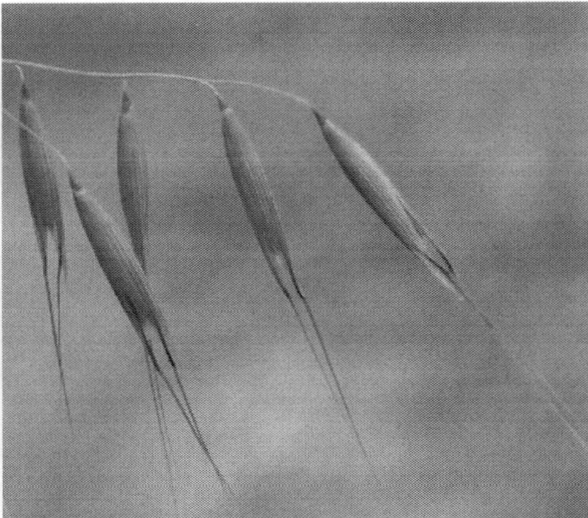

Spikes

Taxonomic Description: An annual, erect grass, up to 1 m tall. Culms fistular, scabrous. Leaves up to 40 cm long, flat, linear-lanceolate, acuminate, scabrid with long sheath. Ligules prominent. spikelets generally 2-4 flowered, borne in 10-30 cm long, loose, terminal panicle. Floret callus bearded. Glumes several nerved, oblong-lanceolate. Lemma usually 7-nerved, lanceolate, scabrid; paleas 2-keeled, shorter than the lemmas; stamens-3; stigmas-2. Caryopsis silky, oblong.

Chemical Constituents: None Known.

Medicinal Properties: Seeds are antidiabetic. (Katiri *et al.* 2017).

Medicinal Uses (Traditional/Folklore): Seeds of *A. sterilis* are given in powder form for the treatment of diabetes in Morocco. A porridge made from seeds is used by Bedouin women in Sinai desert to strengthen the bodies of nursing mothers after the childbirth. In Lebanon, decoction of flowers is used for the treatment of renal insufficiency-diuresis, arthritis, rheumatism and skin diseases. (Abu-Rabia 2015, Safaa *et al.* 2015, Katiri *et al.* 2017).

66.2 *Brachiaria ramosa* (L.) Stapf

(Syn. *Panicum ramosum* L.)

Upper portion (herbarium photo)

Common Name: Browntop Millet.

Vernacular Name: *Hindi:* Pedda-Sama (पेड्डा-समां), Popti (पोपटी).

Habit: Annual grass.

Habitat: Often abundant in moist fallow fields; wet places such as rice fields, along ditches and canals and in marshes.

Distribution: Distributed in Africa and tropical Asia. *B. ramosa* is native to Bangladesh, Bhutan, Nepal, India (Andhra Pradesh, Kerala, Madhya Pradesh, Maharashtra, Rajasthan, Uttarakhand), Pakistan, Cambodia, China, Myanmar, Malaysia, Thailand, Malawi, Senegal, South Africa, Yemen, Zimbabwe up to 2,000 m elevation.

Flowering and Fruiting Time: June-October.

Taxonomic Description: A decumbent-ascending, annual grass. Culms tufted, branched from the base, 20-60 cm long. Leaves narrowly lanceolate, pubescent, up to 12 cm long, flat, base rounded or clasping, apex finely acuminate, ligule very short, sheath hairy. Racemes 5-14, erect or spreading in 7-15 cm long panicle; rachis angular, hispid. Spikelets pubescent, 3 x 2 mm, ovoid or elliptic, cuspidate, often in pairs, one sessile other shortly pedicelled. Glumes ovate, unequal, 5-7 nerved. Lower floret empty, upper bisexual. Stamens-3. Fruit a caryopsis, elliptic.

Inflorescence

Close up of spikelets

Chemical Constituents: None known.

Medicinal Properties: None known.

Medicinal Uses (Traditional/Folklore): Nimar region of Madhya Pradesh, India has a sizable tribal population of Bhil and Bhilalas residing in the forest of Satpura and Vindhya ranges who use the ash of the plant as an ointment in edible oil on burns. The plant is used as fodder for animals in Pakistan. (Zereen *et al.* 2013; http://shodhganga.inflibnet.ac.in/bitstream/10603/120517/3/ethnobotany).

66.3 *Brachiaria reptans* (L.) C.A.Gardner & C.E.Hubb.

(Syn. *Brachiaria prostrata* (Lam.) Griseb.)

Habit

Common Name: Running Grass, Creeping Panic Grass, Sprawling Signal Grass.

Vernacular Name: *Hindi:* Para Ghas (पारा घास).

Habit: Annual grass.

Habitat: Abundant in roadsides, moist waste places, gardens and cultivated fields.

Distribution: Kenya, United Republic of Tanzania, tropical Asia and introduced throughout the tropics.

Flowering and Fruiting Time: July-October.

Taxonomic Description: An annual grass with tufted and decumbent culms. Culms up to 50 cm long, often rooting at lower nodes. Leaves 2-7 x 0.5-1 cm, ovate-lanceolate, amplexicaul at base, acuminate, margin ciliate towards base, sheath shorter than internode. Inflorescence a panicle of 5-12 alternate racemes, pedicels with long soft hairs. Spikelets pale, glabrous, about 2 mm long, ovate-elliptic. Glumes ovate, up to 7-nerved. Lower lemma 5-nerved, ovate-oblong, upper mucronate. Grains (caryopsis) elliptic.

Ciliate leaf margin at base | Inflorescence | Close up of raceme

Chemical Constituents: None known.

Medicinal Properties: None known.

Medicinal Uses (Traditional/Folklore): Leaf juice is used to cure anaemia; leaf juice is obtained by crushing and squeezing the leaves and taking it once daily for a fortnight. (Patil and Patil 2007a).

66.4 *Cenchrus ciliaris* L.

(Syn. *Pennisetum ciliare* (L.) Link)

Habit

Common Name: Buffel Grass, African Foxtail Grass.

Vernacular Name: *Hindi:* Anjan (अंजन), Dhaman (धमान), Kusa (कुसा), Dhamanio (धमनियो).

Habit: Perennial grass.

Habitat: Common along roadsides, railway lines and in open wastelands.

Distribution: Native to most of Africa, southern Asia (east to India), southern Iran, and the extreme south of Europe.

Flowering and Fruiting Time: Most part of the year.

Taxonomic Description: A perennial grass. Culms erect or decumbent-ascending, stout, tufted, much-branched from the base. Leaves up to 25 cm long, pubescent, linear; ligules membranous, hairy. Inflorescence cylindric, 5-10 x 0.5-1 cm, straw-coloured or purplish, densely bristly; rachis angular, hairy. Spikelets oblong-lanceolate, subtended by involucre of bristles. Lower and upper glume unequal, 1-3 mm long, ovate. Lemmas lanceolate. Grains oblong, pale-brown.

Inflorescence **Close up of spike**

Chemical Constituents: Plants contain crude fiber, lipids, sodium, potassium, nitrogen and proteins. (Ashraf *et al.* 2013).

Medicinal Properties: It can be used as anodyne (pain reliever), lactogogue (increase milk flow), and diuretic and as an emollient. (Ashraf *et al.* 2013).

Medicinal Uses (Traditional/Folklore): *C. ciliaris* has also been used as folk medicine for lactation, kidney pain, tumours, sores and wounds. (Ashraf *et al.* 2013).

66.5 *Chloris barbata* Sw.

(Syn. *Chloris inflata* Link)

Habit

Common Name: Finger Grass, Feather Finger Grass, Purple Top Chloris, Airport Grass.

Vernacular Name: None known.

Habit: Annual or perennial grass.

Habitat: Common in lowland cultivated fields, lawns, by roadsides and railway tracks and in disturbed situations.

Distribution: The Finger Grass, *C. barbata* is native to tropical America. It is distributed throughout the tropics of South-East Asia.

Flowering and Fruiting Time: June-October.

Taxonomic Description: An erect, tufted grass, 15-60 cm high. Leaves linear-lanceolate, flat, finely acuminate, sheaths glabrous, ligule very short. Spikes 5-15, purplish, 3-6 cm long, digitately arranged in a terminal fascicle. Spikelets 2-3 mm long, 2-seriate, usually 3-awned, awns 4-7 mm long. Involucral glumes unequal, lanceolate, acute. Floral glumes of hermaphrodite florets broad elliptic, awned, with densely bearded margin, paleate. Stamens-3. Caryopsis trigonous, brownish.

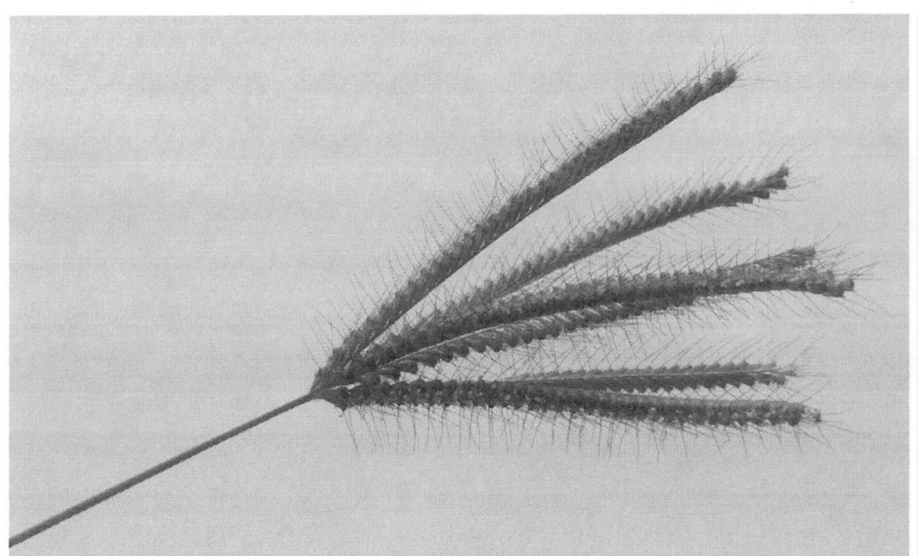

Spikes

Close up of spikes

Chemical Constituents: Phytoconstituents such as phytosterols, flavonoids, tannins, phenols, carbohydrates, proteins and amino acids have been detected in the preliminary phytochemical analysis. (Natrajan *et al.* 2012).

Medicinal Properties: Analgesic, anti-inflammatory, antipyretic, antidiabetic, antimicrobial agent. (Swathy *et al.* 2010, Kumar *et al.* 2012b, Jena and Satapathy 2015).

Medicinal Uses (Traditional/Folklore): Traditionally, *C. barbata* has been used in the treatment of many types of pain and inflammatory conditions. The entire plant is used to treat rheumatism in traditional medicine. The leaf paste/juice is used externally for skin disorders and leaf juice in fever, diarrhoea and diabetes. The juice from the plant possesses antidiabetic, and antimicrobial properties. (Swathy *et al.* 2010, Kumar *et al.* 2012b).

66.6 *Chrysopogon zizanioides* (L.) Roberty

(Syn. *Andropogon zizanioides* (L.) Urb.; *Phalaris zizanioides* L.; *Vetiveria zizanioides* (L.) Nash)

Habit

Common Name: Vetiver, Khas-Khas Grass.

Vernacular Name: *Hindi*: Khas-Khas Ghas (खस-खस घास).

Habit: Perennial grass.

Habitat: Commonly growing along the bank of wetlands, canal and rivers.

Distribution: Native to India and grows throughout the country; but also distributed throughout Africa, Myanmar, Sri Lanka, South-East Asia.

Flowering and Fruiting Time: July-October.

Taxonomic Description: A perennial grass with rhizomatous and aromatic root stock. Culms densely tufted, erect, 1-2 m high, nodes glabrous. Leaves up to 1 m long and 0.5-1 cm wide, erect, linear, convolute, acuminate with sheathing base. Ligules membranous, reduced to a scarious rim. Inflorescence panicle, with sessile and pedicelled spikelets, 20-40 cm long, contracted. Spikelets arranged in many pairs in whorled racemes, whorls 6-12. Sessile spikelets about 4 mm long, linear-lanceolate, yellowish, turning purplish to black. Glumes 3-5 mm long, lower tuberculate. Lower floret empty, upper floret bisexual, stamens-3. Pedicelled spikelets 3-4 mm long, lanceolate. Grains oblong.

Inflorescence

Close up of spikelets

Chemical Constituents: The chemical constituents present in the plant are vetiverol, vetivone, khusimone, khusimol, vetivene, khositone, terpenes, benzoic acid, tripene-4-ol, ß-humulene, epizizianal, vetivenyl vetivenate, iso khusimol, vetiver oils, vetivazulene, zizaene, prezizaene and bvetispirene. Among these the major active constituents identified are khusimol, vetivone, eudesmol, khusimone, zizaene, and prezizaene which are considered to be the fingerprint of the oil. (Snigdha et al 2013).

Medicinal Properties: The essential oil obtained from the roots is used medicinally as a carminative, diaphoretic, diuretic, emmenagogue, refrigerant, stomachic, and tonic, antispasmodic and sudorific. The roots are aromatic, antifungal action, cooling, antiemetic, diaphoretic, haemostatic, expectorant, stimulant, and antioxidant. (Vardhana 2008, Snigdha et al 2013, Sathyamurthi 2015; http://tropical.theferns.info, http://www.pfaf.org/user/plantsearch.aspx).

Medicinal Uses (Traditional/Folklore): The plants are useful in hysteria, insomnia, skin diseases, and asthma, amentia, amenorrhoea, kidney and gall stones problems. Root decoction is used as cooling agent in high fever, inflammation, and sexual diseases. Root ash, leaf paste, and stem juice used by Oraon, South Indian tribes for acidity, boils, burn, epilepsy, scorpion bites, mouth ulcer, rheumatism and sprain. It is used as an antiseptic to treat acne and sores. The essential oil, and the roots, which have insecticidal and insect-repellent properties are used as a mosquito repellent.

The oil has a wide range of applications, being used in perfumery, cosmetics, deodorants, soaps and other toilet articles. An essential oil obtained from the roots is used as a flavouring in sherbets, syrup sweets, fruit drinks and canned *Asparagus*. (Snigdha et al 2013, Sathyamurthi 2015; http://tropical.theferns.info, http://www.pfaf.org/user/plantsearch.aspx).

66.7 *Cynodon dactylon* (L.) Pers.

(Syn. *Panicum dactylon* L.)

Habit

Common Name: Bermuda Grass, Bahama Grass.

Vernacular Name: *Hindi:* Doob (दूब), *Sanskrit:* Durva (दूर्वा).

Habit: Perennial grass.

Habitat: Common along roadsides, moist areas, bank of small water bodies, gardens, agricultural and fallow fields.

Distribution: Although it is not native to Bermuda, it is found in abundance as an invasive species. *C. dactylon* is very widely distributed throughout the world; found throughout India.

Flowering and Fruiting Time: Most part of the year.

Taxonomic Description: A perennial, prostrate or decumbent, stoloniferous grass. Culms 10-30 cm long, rooting from lower nodes and often forming matted tufts. Leaves 3-12 x 0.2-0.3 cm, convolute or flat, linear-lanceolate. Ligules fimbriate. Spikes 4-5, digitate. Spikelets 1.5-2 mm long, sessile, laterally compressed, imbricate. Glumes linear-lanceolate. Lemmas oblong, apiculate, 3-nerved. Paleas oblong-lanceolate, 2-keeled. Stamens-3, anthers oblong, nearly 1 mm long. Grains oblong, about 1 mm long.

Inflorescence

Spikelet

Chemical Constituents: The chemical constituents present in *C. dactylon* are – β-sitosterol, β- carotene, vitamin C, palmitic acid, triterpenoids, arundoin, friedelin, selenium, alkaloids- ergonovine and ergonovinine, and ferulic, syringic, p-coumaric, vanilic, p-hydroxybenzoic and o-hyroxyphenyl acetic acids; cyanogenic hyperoside, cyanogenic glucoside- triglochinin, furfural, furfural alcohol, phenyl acetaldehyde, acetic acid, phytol, β- ionone; mono and oligosaccharides, lignin (whole plant); hydrocarbons (tritriacontane) esters, eicosanoic and docosanoic acids,free alcohol, free aldehydes (hexadecanal) and free acids (hexadecanoic acid) (surface cuticular wax); flavone – apigenin, luteolin, flavone glycosides – orientin (8-C-β-D-glycosyl luteolin), vitexin (8-C-β-D-glycosyl apigenin), iso –orientin (6-C-β-D-glycosyl luteolin) and iso- vitexin (6-C- β-D-glycosyl apigenin) (aerial parts). Bermuda grass is also reported to contain cynodin, hydrocyanic acid, and triticin. (Asthana *et al.* 2012).

Medicinal Properties: Bermuda grass is reported to be sweet, cooling, haemostatic, alexipharmic, anthelmintic, antidiarrhoeal, anticatarrhal, antiseptic, antipyretic, alexiteric, alterative, anabolic, aperient, astringent, carminative, constipating, cyanogenetic, depurative, demulcent, depurative, emetic, emmenagogue, expectorant, tonic and vulnerary; it is reported to be photosensitizing in animals, to cause contact dermatitis, and hay fever. (Warrier *et al.* 2010, Asthana *et al.* 2012, Pappan and Thomas 2017).

Medicinal Uses (Traditional/Folklore): In Ayurveda, it has numerous uses for the cure of *Raktapitta, Trushnaroga, Daharoga, Visarpa, Tvakaroga, Arocaka, Duhsvapna, Bhutaroga, Raktapitta, Chardi, Murccha, Raktapradara, and Mutra Daha.* The grass is reputed as a remedy in epistaxis, haematuria, inflamed tumours, whitlow, fleshy excrescences, cuts, wounds, cystitis, nephritis and in skin diseases. The dried fibrous root is effective in menorrhagia, metrorrhagia, eye diseases, renal diseases, diarrhoea, diseases due to menstrual disorders in female and urinary disorders in male and skin infections.

In traditional medicines, it is used for indigestion and the treatment of wounds. Plant is useful in vitiated conditions of *Pitta* and *Kapha,* burning sensation, haemoptysis, conjunctivitis, cephalalgia, erysipelas, leprosy, skin diseases, vomiting, diarrhoea, dysentery, strangury, colporrhagia, abortion and general debility. It is also used to treat eye disorders and weak vision; it destroys foulness of breath, useful in leucoderma, bronchitis, piles, asthma, tumours, and enlargement of the spleen. It is folk remedy for anasarca, calculus, cancer, carbuncles, cough, cystitis, epilepsy, haemorrhage, hypertension, hysteria, laxative, measles, snakebite, sore stones, tumours and urogenital disorders. The plant is used as a laxative, brain and heart tonic, aphrodisiac, useful against grippe in children, and for inflammations, and toothache. (Khare 2007, Warrier *et al.* 2010, Asthana *et al.* 2012; http://www.pfaf.org/user/plantsearch.aspx, http://tropical.theferns.info).

66.8 *Dactyloctenium aegyptium* (L.) Willd.

(Syn. *Cynosurus aegyptius* L.)

Habit

Common Name: Crowfoot Grass, Finger Comb Grass, Four-Finger Grass.

Vernacular Name: *Hindi:* Makra Ghas (मकड़ा घास).

Habit: Annual grass.

Habitat: Frequently common in cultivated and fallow fields and along roadsides.

Distribution: The Crowfoot Grass distributed throughout Indonesia, Central America, northern part of South America, Africa, the Arabian Peninsula, South-East Asia and Australia. It is found throughout India.

Flowering and Fruiting Time: July-October.

Taxonomic Description: An annual, erect or decumbent grass. Stem often rooting from decumbent culms, nodes 2-3. Leaf blades flat, up to 20 cm long, narrowly linear, apex acute or acuminate, margin ciliate, sheath nearly glabrous, shorter than internodes. Ligules ciliate. Inflorescence 2-6, digitate, up to 3 cm long terminal spikes. Spikelets 3-4 mm long, sessile, closely packed. Lower glumes 2-2.5 mm long, upper glumes as long or little shorter than lower, awned. Lemmas mucronate or shortly awned. Palaes 2-keeled, keels winged, ciliate. Anthers oblong, yellow. Grains globose, about 1.5 mm long.

Chemical Constituents: *D. aegyptium* contains carbohydrates, proteins, amino acids, saponins, flavonoids and tannins. Whole plant is a rich source of secondary metabolites. (Babu and Savithramma 2013, Al-Snafi 2017b).

Inflorescence

Medicinal Properties: Plants are astringent, bitter tonic, and anthelmintic and have anti-oxidant, anti-inflammatory, anticancer and antipyretic properties. (Jebastella and Reginald 2015, Al-Snafi 2017b, Pappan and Thomas 2017; http://medplants.blogspot.in).

Medicinal Uses (Traditional/Folklore): In Indian medicine, the grass is used for imparting medicinal properties of Takra (buttermilk) in intestinal, biliary and urinary diseases. Plants are used for treating polyuria and externally for wounds and ulcers. Juice of fresh plants is prescribed in fevers. Decoction of the plant is given as a cure for small pox. An infusion of the leaves, mixed with the seeds of *Cajanus cajan*, is used to accelerate childbirth. A decoction of the leaves, combined with *Scoparia dulcis*, is used as a remedy for dysentery. Decoction of seeds is used as alleviator of pains in the region of kidney. (Al-Snafi 2017b; http://tropical.theferns.info, http://medplants.blogspot.in).

66.9 *Desmostachya bipinnata* (L.) Stapf

(Syn. *Briza bipinnata* L.; *Eragrostis cynosuroides* (Retz.) P.Beauv.)

Habit

Common Name: Daabh, Salt Reed-Grass, Halfa Grass.

Vernacular Name: *Hindi:* Dabh (दाभ), Davoli (दावोली).

Habit: Perennial grass.

Habitat: Common in dry areas, fallow fields, along roadsides and railway lines.

Distribution: Native to North-East and West tropical, and northern Africa and countries in the Middle East, and temperate and tropical Asia.

Flowering and Fruiting Time: June-October.

Taxonomic Description: A perennial grass, 40-120 cm high, with creeping root stock and tufted culms. Leaves 20-50 x 0.5-1 cm, linear, flat or rolled, tapering to a filiform tip, scabrid, sheaths leathery. Panicles reddish-brown, 20-50 cm long, erect; racemes clustered or spaced, up to 3 cm long. Spikelets 3-13 mm long, biseriate, linear-oblong, laterally compressed, sessile. Glumes unequal, ovate-lanceolate; lower 1-nerved, keeled. lemmas 3-nerved, somewhat scabrid on keels. Paleas slightly shorter than Lemmas. Stamens-3. Grains ovoid.

Inflorescence

Spikelets

Chemical Constituents: It contains various types of chemical compounds which provide the base for the occurrence medicinally active constituents like alkaloids, carbohydrates, phytosterols, saponins, tannins and phenolic compounds, flavonoids, lignin, protein and free amino acids. (Packialakshmi and Alwin 2014, Subramaniam and Sivasubramanian 2015).

Medicinal Properties: Diuretic, emollient and galactagogue. The culms are considered diuretic and are used in the treatment of dysentery and menorrhagia. (http://tropical.theferns.info, https://herbpathy.com/Uses-and-Benefits-of-Desmostachya-Bipinnata-Cid1167).

Medicinal Uses (Traditional/Folklore): *D. bipinnata* is useful in asthma, indigestion, strangury, dipsia, jaundice, dysentery and leucorrhoea. In folk medicines, the plant has been used to treat dysentery and menorrhagia, and as a diuretic. Roots are used in the Indian traditional system of medicine as cooling, sweet, astringent, diuretic and galactagogue and also useful in dysentery, diarrhoea, urinary calculi, dysuria and other diseases of bladder and skin diseases. Flowers and root are used against persistent fever; flowers are also used against loose motion. Root are also used against retention of urine; 10-15 g of root is boiled until the colour of water turned golden or yellowish and after cooling 1 glass of this water is taken, urine will be passed within 10-20 minutes. (Atta-ur-Rahman 2006, Subramaniam and Sivasubramanian 2015; http://tropical.theferns.info, https://herbpathy.com/Uses-and-Benefits-of-Desmostachya-Bipinnata-Cid1167).

66.10 *Dichanthium annulatum* (Forssk.) Stapf

(Syn. *Andropogon annulatus* Forssk.)

Common Name: Marvel Grass, Ringed Dichanthium.

Vernacular Name: *Hindi:* Karad (कराड़), Delhi Grass (डेल्ही ग्रास), Sheda Grass (षेड ग्रास), Jargu Grass (जरगू ग्रास).

Habit: Perennial grass.

Habitat: Frequently growing in gardens, unused grounds, and cultivated fields and along roadsides.

Distribution: It is native to tropical Asia, the Middle East, and parts of Africa.; also in tropical and North Africa extending east through Southeast Asia to China, Australia, Fiji and New Guinea. Common throughout the plains and hills of India up to 1,660 m.

Flowering and Fruiting Time: September-March.

Habit

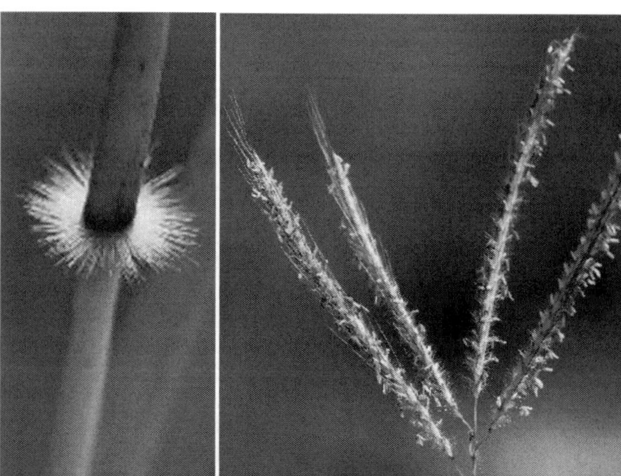

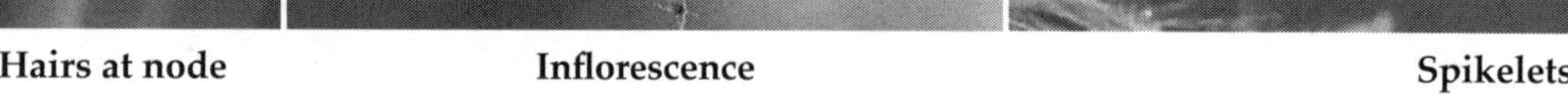

Hairs at node **Inflorescence** **Spikelets**

Taxonomic Description: An erect or decumbent-ascending, perennial grass, 40-100 cm high. Culms tufted with purplish, bearded nodes. Leaves up to 20 cm long, linear or narrowly lanceolate, scabrid, acuminate; sheaths hairy, bearded at mouth, ligules fimbriate. Inflorescence of 3-9, sub-digitate, purplish, 3-5 cm long racemes. Joints of racemes silky hairy. Spikelets sessile and pedicelled, 3-5 mm long. Sessile spikelets oblong, awned. Glumes 3-5 mm long, pubescent. Lower lemma lanceolate, epaleate; upper lemma reduced to long awn. Stamens-3; anthers linear, 1-1.5 mm long. Stigmas feathery, longer than the anthers. Pedicelled spikelets ciliate. Grains ellipsoid, obovate-oblong.

Chemical Constituents: Phytochemicals isolated and identified are two pairs of lignoflavone derivatives stereoisomers; an epimers of tricin 4`-O-(threo-β- guaiacylglyceryl) ether (salcolin A) and tricin 4`-O-(erythro-β-guaiacylglyceryl) ether (salcolin B) and an epimer of tricin 4`-O-[threo-β-guaiacyl-(7``-O-methyl-9``-O-acetyl)-glyceryl] ether and tricin 4`-O-[erythro-β- guaiacyl-(7``-O-methyl-9``-O-acetyl)-glyceryl] ether, one flavone; tricin , two flavone glycosides; tricin 7-O-β-D-glucopyranoside and tricin 7-O-neohesperidoside, one flavone C-glucoside; isoorientin, one phenolic acid; p-coumaric acid, one lignan; 4-ketopinoresinol and two sterols; stigmasterol and β-sitosterol-3-O-β-D-glucoside. (Awad *et al.* 2015).

Medicinal Properties: None known.

Medicinal Uses (Traditional/Folklore): Dysentery and menorrhagia. (Nisar *et al.* 2014).

66.11 *Digitaria ciliaris* (Retz.) Koeler

(Syn. *Digitaria adscendens* (Kunth) Henrard; *Panicum ciliare* Retz.)

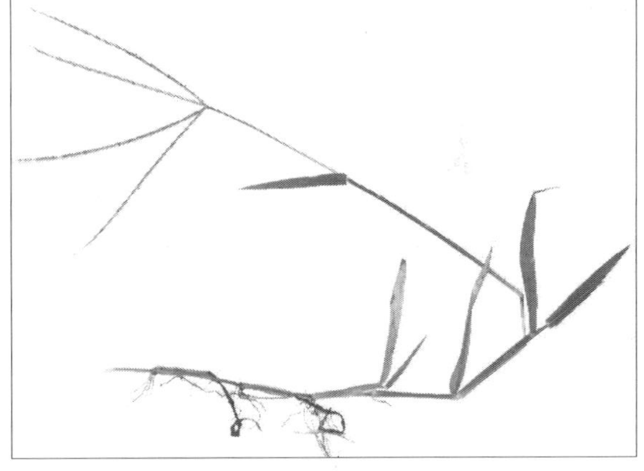

Habit

Common Name: Wild Crabgrass, Bamboo Grass, Summer Grass, Tropical Finger Grass.

Vernacular Name: None known.

Habit: Annual grass.

Habitat: Common in wastelands, gardens, cultivated fields and along roadsides.

Distribution: Paleotropics.

Flowering and Fruiting Time: July-November.

Taxonomic Description: An erect or sub-erect, up to 60 cm high, annual grass. Leaves linear-lanceolate, 3-10 x 0.4-1 cm, finely acuminate, somewhat pilose, base rounded or subcordate, sheaths sparsely long hairy to glabrous, ligule membranous. Inflorescence of 2-6 or more, digitate or subdigitate racemes. Spikelets awnless, elliptic, about 3.5 mm long. Glumes unequal; lower glume absent or reduced to a minute scale; upper ovate-lanceolate, ciliate. Lower floret sterile, upper floret bisexual. Stamens-3. Grains oblong.

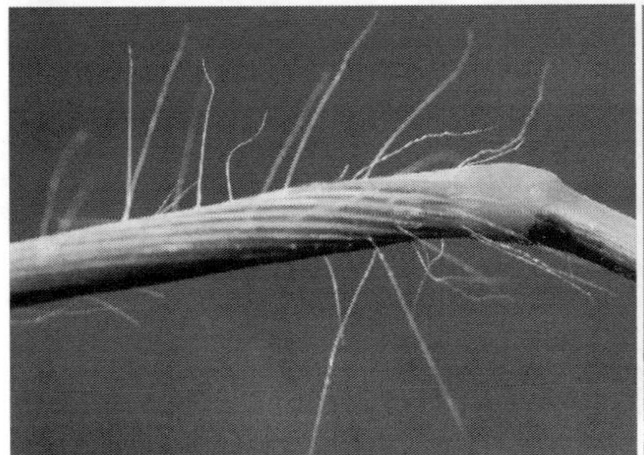

Hairy sheath

Portion of raceme

Chemical Constituents: None known.

Medicinal Properties: None known.

Medicinal Uses (Traditional/Folklore): None known.

66.12 *Echinochloa colona* (L.) Link

(Syn. *Panicum colonum* L.)

Habit

Common Name: Jungle Rice, Shama Millet, Small Barnyard Grass.

Vernacular Name: *Hindi:* Jangli Jhangora (जंगली झंगोरा), Jharwa (झरवा), Shama (शामा).

Habit: Annual grass.

Habitat: Commonly growing at the bank of canals, ponds and lakes; also abundant along roadsides and in paddy fields.

Distribution: Native of the America; introduced into the Mediterranean, South-West Asia and Africa. The grass occurs throughout tropical Asia and Africa.

Flowering and Fruiting Time: July-November.

Taxonomic Description: A densely tufted, annual grass. Culms erect, 30-70 cm long, nodes glabrous. Leaves flat, up to 20 cm long, linear-lanceolate, tapering to a fine points. Spikes up to 3 cm long, 6-20, sessile or sub-sessile, often distant, borne in 10-20 cm long panicle. Rachis angular. Spikelets clustered, hairy, 3-5 mm long, ovoid-elliptic, apiculate. Glumes not equal, ovate, pubescent. Lower floret sterile; upper floret bisexual, cuspidate. Stamens-3. Grains ellipsoid or plano-convex, brown.

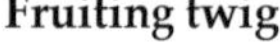

Fruiting twig

Close up of spike

Chemical Constituents: None known.

Medicinal Properties: None known.

Medicinal Uses (Traditional/Folklore): Decoction of whole plant is digestive, useful in constipation; increases bile secretion. (Youssef 2013).

66.13 *Echinochloa crus-galli* (L.) P.Beauv.

(Syn. *Panicum crus-galli* L.)

Common Name: Barnyard Grass, Japanese Millet, Jungle Rice, Cockspur Grass, Prickly Grass.

Vernacular Name: *Hindi:* Sanwak (संवक), Kayada (कायदा); *Sanskrit:* Varuka (वारुका).

Habit: Annual grass.

Habitat: Commonly found in rice fields and along roadsides, railway lines, and in disturbed areas. It also invades riverbanks and the shores of wetlands and ponds.

Distribution: *E. crus-galli* native to America. Barnyard grass commonly occurs throughout tropical Asia and Africa. It is widespread in Europe, Asia and Australia. Found throughout India.

Flowering and Fruiting Time: July-October.

Taxonomic Description: An erect or decumbent-ascending, 30-100 cm in length, tufted, annual grass. Leaves up to 30 cm long and 1 cm wide, linear-lanceolate, margin scabrous, nodes glabrous, sheaths ribbed, ligules absent. Spikelets awned, 2-4nate, 3-4 mm long, ovate-elliptic, greenish or purple-tinged, crowded in 5-15 cm long inflorescence. Lower glumes 5-nerved, broadly ovate; upper ovate-lanceolate, cuspidate. Lower lemma ovate-lanceolate, awned; upper cuspidate. Caryopsis ovoid.

Inflorescence

Spikelets

Chemical Constituents: The plants contain alkaloids, glycosides, flavonoids, phytosterols, terpenes, proteins, saponins; rich in amino acids. (Al-Snafi 2017a).

Medicinal Properties: Plants have antidiabetic, antioxidant, antimicrobial, hypolipidemic and anti-obesity effects. (Al-Snafi 2017a).

Medicinal Uses (Traditional/Folklore): Barnyard grass is a folk remedy in India; used as preventive and tonic for carbuncles, haemorrhages, sores, spleen troubles, cancer and wounds. The roots are boiled to cure indigestion in the Philippines. The plant extract is used in diseases of the spleen. Whole plant is reported to be preventative and tonic. (Kumar *et al.* 2013, Al-Snafi 2017a; https://en.wikipedia.org/wiki/Echinochloa_crus-galli, http://tropical.theferns.info).

66.14 *Eleusine indica* (L.) Gaertn.

(Syn. *Cynosurus indicus* L.)

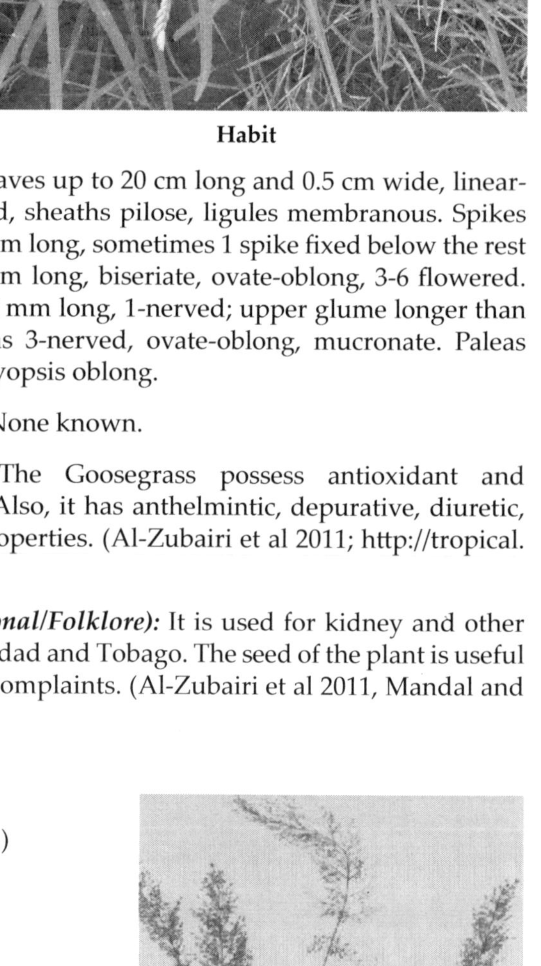

Habit

Common Name: Goosegrass, Indian Goosegrass, Crowfoot Grass.

Vernacular Name: *Hindi:* Jangali Marua (जंगली मरुआ); *Sanskrit:* Nandimukhi (नन्दी मुखी).

Habit: Annual grass.

Habitat: Commonly found in disturbed sites, lawns, gardens and along roadsides.

Distribution: It is pan tropical found throughout the tropics up to 2000 m altitude and spreads significantly into the sub-tropics, mainly in North America, Europe and Africa.

Flowering and Fruiting Time: July-October.

Taxonomic Description: An annual grass, 20-60 cm high. Culms growing in tufts, erect or sometimes geniculate, nodes glabrous. Leaves up to 20 cm long and 0.5 cm wide, linear-lanceolate, usually folded, sheaths pilose, ligules membranous. Spikes 2-7, usually digitate, 3-8 cm long, sometimes 1 spike fixed below the rest one. Spikelets about 5 mm long, biseriate, ovate-oblong, 3-6 flowered. Lower glume about 1.5-2 mm long, 1-nerved; upper glume longer than lower, 5-nerved. Lemmas 3-nerved, ovate-oblong, mucronate. Paleas 2-keeled. Stamens-3. Caryopsis oblong.

Inflorescence

Chemical Constituents: None known.

Medicinal Properties: The Goosegrass possess antioxidant and antibacterial properties. Also, it has anthelmintic, depurative, diuretic, febrifuge and laxative properties. (Al-Zubairi et al 2011; http://tropical.theferns.info).

Medicinal Uses (Traditional/Folklore): It is used for kidney and other urinary problems in Trinidad and Tobago. The seed of the plant is useful in the treatment of liver complaints. (Al-Zubairi et al 2011, Mandal and Mandal 2016).

66.15 *Eragrostis amabilis* (L.) Wight & Arn.

(Syn. *Eragrostis tenella* (L.) P.Beauv. ex Roem. & Schult.; *Poa tenella* L.)

Habit (herbarium photo)

Common Name: Japanese Lovegrass.

Vernacular Name: *Hindi:* Bharbhusi (भरभूसी).

Habit: Annual grass.

Habitat: It occurs in cultivated fields, waste places, lawns, roadsides, and gardens.

Distribution: It is found in tropical Asia, introduced and now widely distributed in tropical regions of America and Africa.

Flowering and Fruiting Time: June-November.

Taxonomic Description: An annual grass. Culms loosely tufted, slender, 10-40 cm in height, erect or geniculate at base, nodes glabrous. Leaves up to 12 cm long, narrowly linear-lanceolate, finely acuminate, sheaths bearded at mouth.

Inflorescence panicle, 10-15 cm long, green or purplish, loose, branches spreading. Spikelets 2-3 mm long, 3-9 flowered, ovate-oblong. Glumes membranous, unequal, 1-nerved, ovate-lanceolate. Lemmas ovate-oblong, obtuse at apex. Paleas ciliate along the keels. Stamens-3. Caryopsis ovoid.

Bearded sheath at mouth

Spikelets

Chemical Constituents: None known.

Medicinal Properties: None known.

Medicinal Uses (Traditional/Folklore): Decoction of leaves and roots is given orally for menorrhagia in Garhwal, Uttarakhand, India. (Ghildiyal *et al.* 2014).

66.16 *Eragrostis japonica* (Thunb.) Trin.

(Syn. *Poa interrupta* Lam.; *Poa japonica* Thunb.)

Common Name: Pond Lovegrass.

Vernacular Name: Panghas (पानघास).

Habit: Annual grass.

Habitat: *E. japonica* found to occur in dry paddy fields, moist and wet places, and along the banks of streams.

Distribution: *E. japonica* is widely distributed in tropical and subtropical Asia, Africa and Australia. In India, it is reported from Andhra Pradesh, Assam, Bihar, Gujarat, Himachal Pradesh, Jammu and Kashmir, Kerala, Maharashtra, Nagaland, Orissa, Punjab, Sikkim, Rajasthan, Tamil Nadu, Uttar Pradesh, and West Bengal.

Flowering and Fruiting Time: October-March.

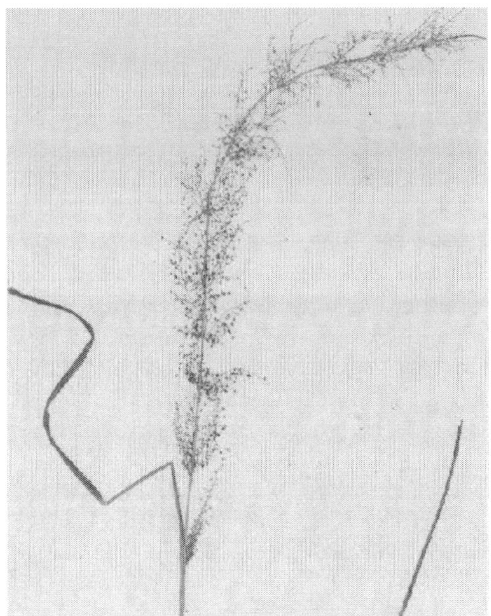

Upper potion with inflorescence (photo from herbarium sheet)

Taxonomic Description: An erect or sub-erect, slender, annual grass with glabrous noded culms. Leaves linear-lanceolate, 10-25 x 0.3-0.5 cm, acuminate, sheaths long and loose, ligules membranous. Panicles loose, elongated up to 30 cm long or more, usually purplish, branches clustered or veriticillate. Spikelets ovate, 1.5-2 mm long, 4-10 flowered. Glumes subequal, 1-nerved, ovate-oblong, obtuse at apex. Lemmas 3-nerved, ovate, apex obtuse. Paleas ciliate on the keels. Caryopsis ovoid.

Chemical Constituents: None known.

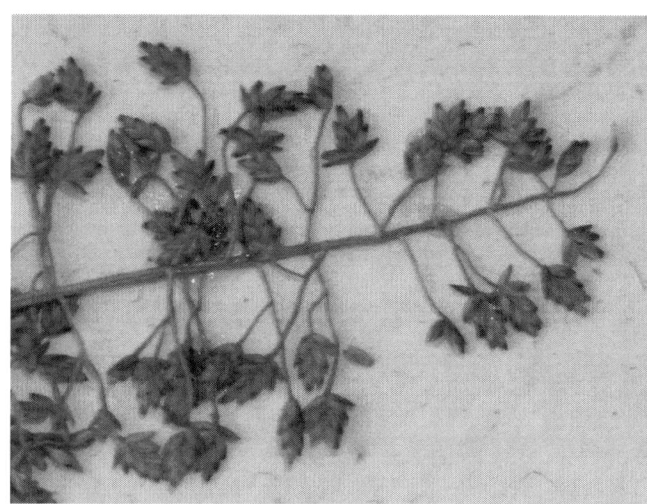

Inflorescence

Close up of spikelets

Medicinal Properties: None known.

Medicinal Uses (Traditional/Folklore): It is used as a lactation stimulant (lactogen) in animals/cows in Nigeria. (https://www.cabi.org/isc/datasheet/21634).

66.17 *Eriochloa procera* (Retz.) C.E.Hubb.

(Syn. *Agrostis procera* Retz.)

Common Name: Tropical Cupgrass, Cup Grass, Spring Grass.

Vernacular Name: None known.

Habit: Semi-aquatic grass.

Habitat: Common in paddy fields, plains near damp areas, along wetlands, ditches and roadsides.

Upper portion with inflorescence

Distribution: *E. procera* is distributed in South and South-East Asia, including India, Bangladesh, Bhutan, Sri Lanka, Nepal, Cambodia, China, Myanmar, Indonesia, Lao People's Democratic Republic, Malaysia, Fiji, and Australia and introduced into tropical America and Africa. In India, it is reported from Andhra Pradesh, Bihar, Gujarat, Karnataka, Maharashtra, Orissa, Madhya Pradesh, Punjab, Tamil Nadu, Uttar Pradesh and Kerala from 200 to 1,700 m elevation.

Spikes

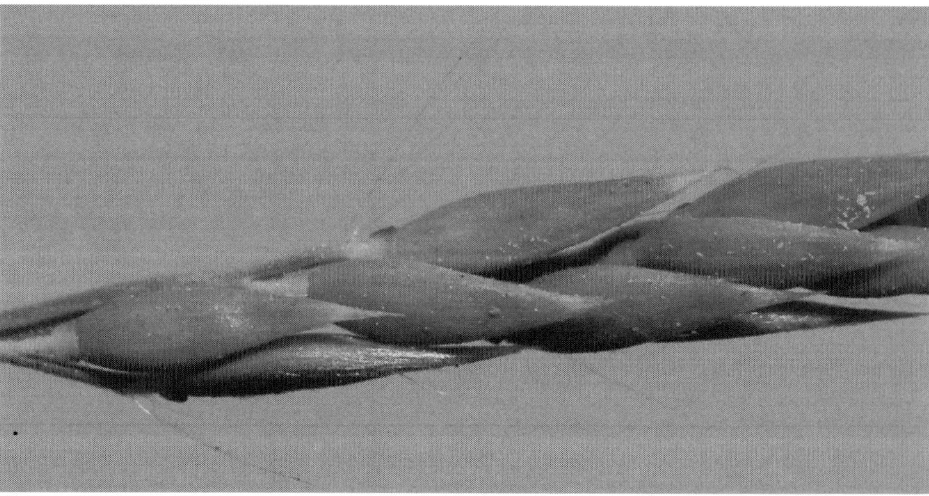

Spikelets

Flowering and Fruiting Time: September-December.

Taxonomic Description: A tufted, erect or geniculately ascending, annual grass, culms 40-120 cm long. Leaf blades linear-lanceolate, up to 30 cm long, finely acuminate, sheath keeled, ligules fimbriate. Racemes many, alternate in 10-20 cm long panicles. Spikelets 3-4 mm long, ovate-lanceolate, shortly pedicellate. Lower glume absent, upper ovate-lanceolate, appressed silky. Lower floret empty, upper bisexual. Stamens-3. Caryopsis oblong.

Chemical Constituents: None known.

Medicinal Properties: None known.

Medicinal Uses (Traditional/Folklore): None known.

66.18 *Hemarthria compressa* (L.f.) R.Br.

(Syn. *Rottboellia compressa* L.f.)

Common Name: Jove Grass, Whip Grass.

Vernacular Name: *Hindi:* Biksa (बिक्सा), Baika (बैका), Buksha (बुकशा), Pansheru (पनशेरु).

Habit: Perennial grass.

Habitat: Abundant in marshy places; often along the bank of ponds, lakes, canals and in paddy fields.

Distribution: *H. compressa* is distributed in warm and tropical Asia, Australia and North Africa, Italy and Turkey, southwards to Nigeria and Ethiopia; Zambia southwards to the Cape; also in Madagascar, with isolated records in Burma, Thailand and Borneo; introduced to America. In India, it is known to occur in Delhi, Himachal Pradesh, Karnataka, Madhya Pradesh, Maharashtra, Gujarat, Rajasthan, Tamil Nadu, West Bengal, Uttar Pradesh, and Andhra Pradesh.

Habit

Flowering and Fruiting Time: July-October.

Taxonomic Description: An erect or ascending, perennial, marshy grass. Culms 30-100 cm, glabrous, rooting at lower nodes. Leaves flat, up to 12 cm long and 5 mm wide, linear, scabrid along margin, sheaths shorter than internodes; ligules membranous, very short. Racemes 5-10 cm long, slightly compressed, usually curved. Spikelets 2-nate, sessile and pedicelled. Sessile spikelets 3-5 mm long, linear-lanceolate, lower glume oblong-lanceolate, upper glume nearly as long as the lower. Pedicelled spikelets similar to the sessile. Caryopsis oblong.

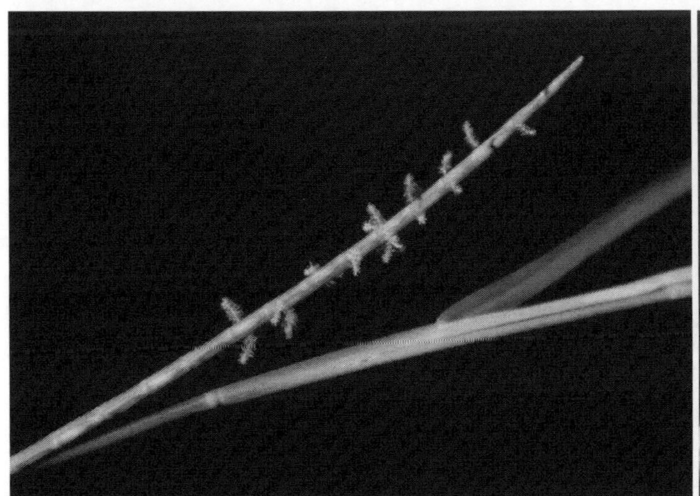

Flowering twig

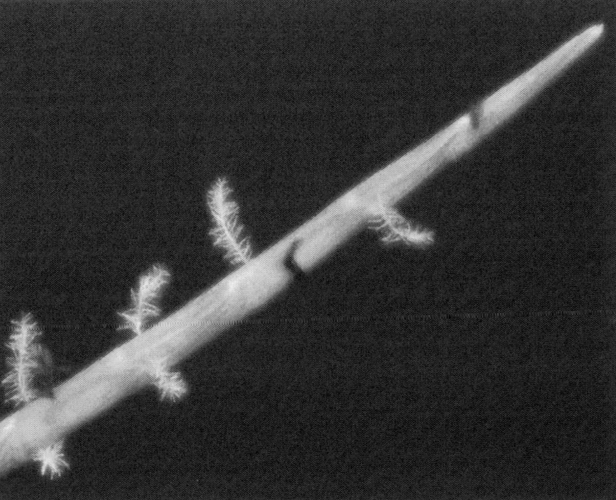

Close up of raceme

Chemical Constituents: None known.

Medicinal Properties: None known.

Medicinal Uses (Traditional/Folklore): Root paste with a pinch of common salt is taken orally during indigestion by tribes in Madhya Pradesh. Plant is used in dyspepsia and treating cuts. (Adhikari *et al.* 2010, Mishra *et al.* 2015e; http://shodhganga.inflibnet.ac.in/bitstream/10603/120517/3/ethnobotany).

66.19 *Imperata cylindrica* (L.) Raeusch.

(Syn. *Imperata arundinacea* Cirillo; *Lagurus cylindricus* L.)

Habit

Common Name: Blady Grass, Cogon Grass, Kunai Grass, Japanese Blood Grass.

Vernacular Name: *Hindi*: Dabh (दाभ), Uloo (उलू); *Sanskrit:* Darbha (दरभा).

Habit: Perennial grass.

Habitat: Common in moist waste places, roadsides, railway lines and along the bank of small water channels.

Distribution: It is ubiquitous in the humid tropics; distributed in America, East and South- East Asia, India, Micronesia, Melanesia, Australia, and eastern and southern Africa. A widespread weed, found in warm temperate to tropical areas of the world.

Flowering and Fruiting Time: May-October.

Taxonomic Description: An erect, 15-60 cm high, perennial grass with creeping rhizomatous rootstock. Leaves flat, mostly basal, 5-15 x 0.3-0.5 cm, linear-lanceolate, acuminate, nodes bearded, margin scabrid, ligules membranous. Spikelets 2-4 mm long, numerous, silky hairy, clustered, linear-lanceolate, borne in 4-10 cm long, silvery-hairy, cylindrical panicle. Callus covered by long silvery hairs. Lower glume oblong-lanceolate, membranous. Lower floret sterile, upper floret bisexual. Stamens-2; anthers oblong, orange-yellow. Stigmas purplish-black. Grains oblong-ovoid.

Lower portion with leaves

Bearded node

Close up of spikelets

Chemical Constituents: The plant contains the triterpenoids arundoin, cylindrin and fernenol. Rhizomes contain flavonoids, lignans, graminone A and B, sesquiterpenoid, cylindrene, biphenylether compounds, cylindol A and B, glucose, fructose and organic acids. Metabolites that have been found in roots consist of arundoin, fernenol, isoarborinol, silindrin, simiarenol, kampesterol, stigmasterol, ß-sitosterol, scopoletin, skopolin, p-hidroksibenzaladehida, katekol, chlorogenic acid, isoklorogenat acid, p-acid kumarat, neoklorogenat acid, acetic acid, oxalic acid, d-malic acid, citric acid, potassium (0.75% of dry weight) and large amounts of calcium and 5-hidroksitriptamin. (Bhutya 2011; http://baliherbs.blogspot.in).

Medicinal Properties: The rhizomes are diuretic, anti-inflammatory, antibacterial, anthelmintic, galactagogue and febrifuge. The flowers and the roots are antibacterial, diuretic, febrifuge, sialagogue, styptic and tonic. The root is astringent, antifebrile, antiviral, diuretic, emollient, haemostatic, restorative and tonic. The root bark is febrifuge, restorative and tonic. Extracts of the plant have shown viricidal and anticancer activity. (Bhutya 2011, Parkavi *et al.* 2012; http://tropical.theferns.info).

Medicinal Uses (Traditional/Folklore): *I. cylindrica* is used in the treatment of urodynia, pollakiuria, hematuria, inflammations, dyspepsia, dysuria, renal and vesical calculi and fever.

The flowers are used in the treatment of haemorrhages, wounds, etc. They are decocted and used to treat urinary tract infections, fevers, thirst, etc. It is used in the treatment of nose bleeds, haematuria, haematemesis, oedema and jaundice. Root is used in the treatment of nose bleeds, haematuria, haematemesis, oedema and jaundice. The root has antibacterial action against *Staphylococcus aureus, Bacillus dysenteriae,* etc. A decoction of the root is used as an anthelmintic and also to treat digestive disorders such as indigestion, diarrhoea and dysentery. Rhizome is used as a skin softener, facilitating urine, blood cleansers, appetite enhancer, stopping the bleeding. In addition, it is also used in the treatment of venereal diseases (gonorrhoea, blood urine), kidney disease, injury, fever, high blood pressure and nerve diseases. All parts of plants used as animal feed, paper materials, and for the treatment of ringworm. In Malay, the plant is used in folk remedies for cancer, colds, diarrhoea, dysentery, gonorrhoea, myalgia, night sweats, piles, rheumatism, and tumours. (Parkavi *et al.* 2012; http://tropical.theferns.info, shodhganga. inflibnet.ac.in/bitstream/10603/7149/10/10_result%20part%20ii.pdf).

66.20 *Leptochloa panicea* (Retz.) Ohwi

(Syn. *Cynodon filiformis* (Roxb.) Voigt; *Poa panicea* Retz.)

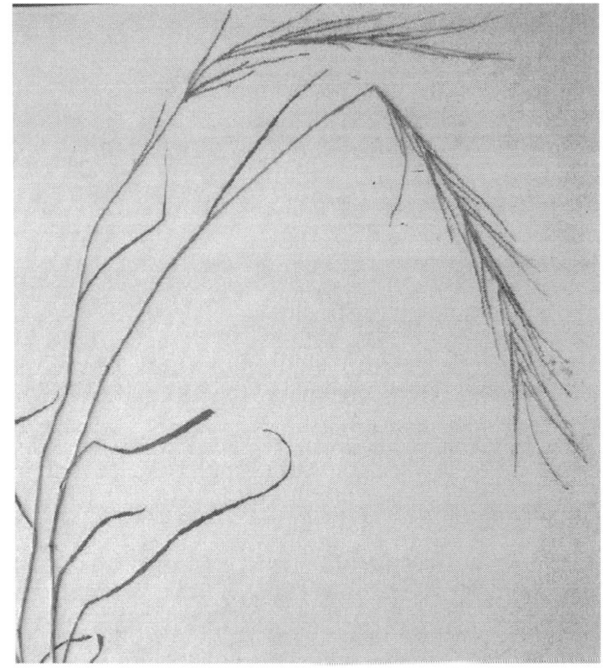

Habit (herbarium photo)

Common Name: Mucronate Sprangletop, Sprangle Top, Thread Sprangletop.

Vernacular Name: None known.

Habit: Annual grass.

Habitat: Found in damp or periodically flooded places and also grows along riverbanks, lake margin, sloughs, and in ditches and paddy fields.

Distribution: It is distributed in Africa, Asia-temperate, Asia-tropical, Australasia and America. In India, it is distributed in Madhya Pradesh, Andhra Pradesh, Tamil Nadu and Maharashtra.

Flowering and Fruiting Time: July-October.

Taxonomic Description: An erect or geniculately ascending, slender, annual grass. Culms tufted, 30-100 cm long. Leaves up to 30 cm long, linear-lanceolate, finely acuminate, scabrid, sheaths ciliate, ligules membranous. Inflorescence panicle, 10-25 cm long, diffuse. Spikes 5-12 cm long, filiform, ascending or spreading. Spikelets purplish green, 1.5-2 mm long, usually 3 flowered. Lower involucral glume 1-nerved, nearly 1 mm long; upper linear-lanceolate, about 1.5 mm long. Lemmas elliptic-oblong, 3-nerved, ciliate. Stamens-3. Caryopsis elliptic.

Ciliate sheath | Part of inflorescence (herbarium photo) | Close up of spikelets

Chemical Constituents: None known.

Medicinal Properties: None known.

Medicinal Uses (Traditional/Folklore): None known.

66.21 *Panicum antidotale* Retz.

(Syn. *Panicum miliare* Lam.)

Habit

Common Name: Blue Panic Grass, Giant Panic.

Vernacular Name: *Hindi:* Kutki (कुटकी), Hadjodi (हड़जोड़ी), Ghirano (घिरानो).

Habit: Perennial grass.

Habitat: Common along roadsides, open wastelands and at the bank of aquatic bodies.

Distribution: Native to India now introduced to many countries as a pasture grass. Distributed in East Asia-Himalayas from Afghanistan to Nepal, and the Upper Gangetic Plains in India.

Flowering and Fruiting Time: July-November.

Taxonomic Description: An erect, up to 1.5 m tall, perennial grass. Culms tufted, terete, solid, usually branched with creeping rootstock, nodes thickened. Leaves flat, 20-50 x 0.5-2 cm, linear, tapering to a fine tips. Panicle pyramidal, 10-30 cm long, rachis angular, branches spreading and drooping. Spikelets 2-flowered, about 3 mm long, elliptic-ovoid, laxly crowded on the branchlets. Glumes ovate, lower 3-5 nerved, upper 7-9 nerved. Lower floret sterile or male, upper bisexual. Styles-2, obvious. Stamens-3. Grains ellipsoid.

Chemical Constituents: None known.

Twig with inflorescence

Spikelets

Medicinal Properties: Plants have antipyretic, antibacterial, antiviral, and antihepatic properties. (Gothandam *et al.* 2010; http://www.instituteofayurveda.org/plants/plants_detail.php?i=1043&s=Local_name).

Medicinal Uses (Traditional/Folklore): Entire Plant is used for leucorrhoea, urinary tract infections, liver diseases, splenomegaly and jaundice. The smoke of the burning plant is used to fumigate wounds and as a disinfectant in the treatment of smallpox and throat infections. Whole plant used as a decoction against cough, throat infection, fever and general debility. Plants may contain sufficient oxalate to cause oxalate poisoning in sheep or osteodystrophia in horses. Also reported as a cause of atypical interstitial pneumonia. (https://medical-dictionary.thefreedictionary.com/Panicum+capillare, http://tropical.theferns.info, http://www.instituteofayurveda.org/plants/plants_detail.php?i=1043&s=Local_name).

66.22 *Paspalum distichum* L.

(Syn. *Digitaria paspalodes* Michx.; *Paspalum paspalodes* (Michx.) Scribn.)

Common Name: Knot Grass, Ginger Grass, Seashore Paspalum, Salt Jointgrass.

Vernacular Name: *Hindi:* Besak (बेसक).

Habit: A perennial grass with long stolons.

Habitat: Common along the bank of canals, ponds, wetlands, rice fields, and other marshy areas.

Distribution: *P. distichum* is widely distributed throughout the tropics.

Flowering and Fruiting Time: June-September.

Habit (upper portion of plant)

Taxonomic Description: A perennial, stoloniferous grass. Culms decumbent-ascending, 20-50 cm long, rooting at lower nodes. Leaf blades 4-12 cm long, linear-lanceolate, acuminate, glabrous, sheaths hairy at throat, ligules membranous. Spikelets 2-seriate, about 3 mm long, acute, elliptic-lanceolate, arranged in 2-4 subdigitate racemes, rachis about 2 mm wide. Lower glume absent; upper elliptic-lanceolate, slightly hairy. Lower floret barren, upper hermaphrodite. Upper lemma greenish-white. Stamens-3. Caryopsis elliptic.

Chemical Constituents: The plant contains only alkaloids that are hallucinogenic. (Ladewig and Pletscher 1993).

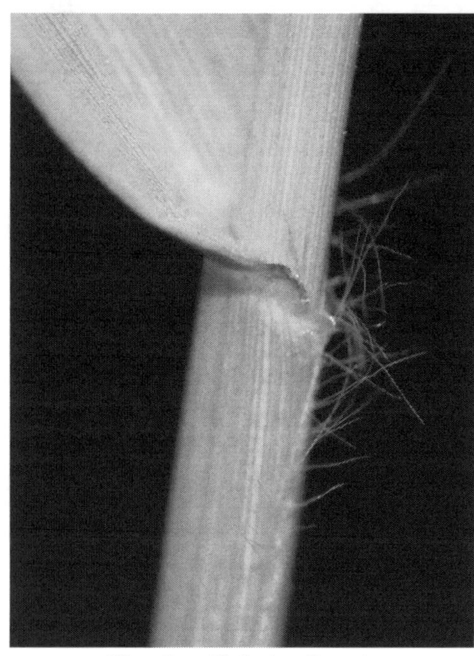
Sheath with hairy throat

Racemes

Spikelets

Medicinal Properties: Hallucinogenic. (Ladewig and Pletscher 1993).

Medicinal Uses (Traditional/Folklore): The juice from young leaf buds is used to treat gonorrhoea. (Cambie and Ash 1994).

66.23 *Paspalum scrobiculatum* L.

(Syn. *Paspalum orbiculare* G.Forst.)

Habit

Common Name: Kodo Millet, Scrobic Paspalum.

Vernacular Name: *Hindi:* Kodo (कोदो).

Habit: Annual grass.

Habitat: Common along the bank of ponds, rivers, lakes, canals and in paddy fields.

Distribution: This species is widespread in Africa and the Indian sub-continent. It has been introduced to New Zealand, North America, Australia, Mexico and Puerto Rico.

Flowering and Fruiting Time: June-September.

Taxonomic Description: An annual grass. Culms erect or decumbent-ascending, 30-80 cm high, often rooting at lower nodes, nodes glabrous. Leaves 10-30 x 0.5-1 cm, flat, linear-lanceolate, apex acuminate; sheaths compressed, glabrous; ligule minute, membranous, ciliate. Inflorescence of 2-5, distant, 3-7 cm long racemes; rachis nearly 2 mm wide, flat. Spikelets usually 2-ranked, 2-3.5 mm long, ovate-elliptic or orbicular. Lower glume absent; upper up to 3.5 mm long, membranous, orbicular. Stamens-3. Grains biconvex.

Inflorescence

Spikelets

Chemical Constituents: None known.

Medicinal Properties: Plant has antidiabetic, and antiseptic properties. (Jain *et al.* 2010, Barik *et al.* 2017).

Medicinal Uses (Traditional/Folklore): The leaf pastes are applied externally in cutaneous affections as an antiseptic. It is also used for carbuncle, intoxication, narcotics, opthalmia, parturition and sores. (Barik *et al.* 2017).

66.24 *Phalaris minor* Retz.

(Syn. None known)

Apical portion of plant

Common Name: Dwarf Canary Grass, Little-Seed Canary Grass, Small Canary Grass.

Vernacular Name: *Hindi:* Mandusi (मंडूसी), Kanaki (कनकी), Gulli Danda (गुल्ली डंडा).

Habit: Annual grass.

Habitat: Common winter weed of moist waste places and in the fields of wheat and barley.

Distribution: The Dwarf Canary Grass is native to North Africa, Europe-Mediterranean to northwest France. It is widely naturalized in many regions.

Flowering and Fruiting Time: December-March.

Taxonomic Description: A tufted, annual grass. Culms erect, 15-80 cm long with thickened and glabrous nodes. Leaves 5-30 x 0.5-1.2 cm, glabrous, linear-lanceolate, long-tapered and finally acuminate at apex, ligules membranous. Spikelets 4-6 mm long, laterally compressed, densely packed in the 2-7 x 1-2 cm, terete, erect panicle. Glumes 3-nerved, keels intensely winged. Lemmas unequal, hairy, ovate-lanceolate. Stamens-3; anthers oblong, about 1.5 mm long. Grains oblong, 2-2.5 mm long.

Flowering twig

Inflorescence

Close up of spikelets

Chemical Constituents: None known.

Medicinal Properties: The *P. minor* acts as carminative, flushes out the worms and above all it improves the immunity of the animals. (Kaur *et al.* 2006).

Medicinal Uses (Traditional/Folklore): Root paste is used in treating wounds. (Bhat *et al.* 2013).

66.25 *Phragmites karka* (Retz.) Trin. ex Steud.

(Syn. *Phragmites roxburghii* (Kunth) Steud.)

Habit

Common Name: Tall Reed.

Vernacular Name: *Hindi:* Narkul (नरकुल), Nal (नाल), Doka-Ghas (डोका-घास).

Habit: A tall perennial grass.

Habitat: Common along swampy areas; bank of rivers, canals and wetlands.

Distribution: It is distributed from Africa through India and South-East Asia to Australia and Oceania. In tropical Africa it occurs from Senegal East to Eritrea and South to Kenya and Uganda; it is also reported from Madagascar. Throughout India, found in marshy places, wetlands and along banks of lakes and streams.

Flowering and Fruiting Time: April-October.

Taxonomic Description: A perennial grass with elongated creeping rhizome. Culms stout, very long, erect or decumbent-ascending, cylindric, fistular. Leaves 20-70 x 1-2.5 cm, linear-lanceolate, more or less glabrous, base rounded, apex long acuminate, ligules ciliate; sheaths ribbed, hairy at throat. Inflorescence 20-60 cm long, terminal panicle with filiform branches, rachilla hairy. Spikelets 0.8-1.2 cm long, linear-lanceolate, many-flowered. Glumes oblong-lanceolate, membranous. Lemmas glabrous, upper shorter than the lower one, epaleate. Grains free, oblong.

Chemical Constituents: *P. karka* is capable of producing lignocellulosic biomass for bio-ethanol production. It contains phenolic compounds. Oven-dry reed (stem with inflorescences) of this plant contains about 55% cellulose, 23% pentosans, 26% lignin and 3% ash. (Faraji and Jamei 2016; http://uses.plantnet-project.org/en/Phragmites_karka_(PROTA)).

Inflorescence

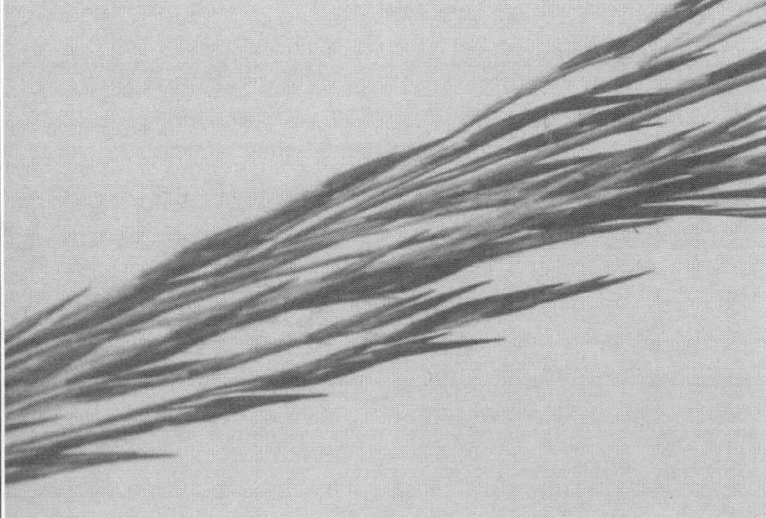
Spikelets

Medicinal Properties: The rhizomes and roots are diuretic, antiemetic and diaphoretic and are used for the treatment of diabetes. (Faraji and Jamei 2016; http://uses.plantnet-project.org/en/Phragmites_karka_(PROTA)).

Medicinal Uses (Traditional/Folklore): In traditional medicine in India, the roots are used to heal broken/fractured bones, and the whole plant is used against rheumatic pain. This plant species is traditionally used as a remedy for diabetes and as a diuretic. The plant is useful in biliousness, urinary troubles, vaginal and uterine complaints and heart diseases. (Sarwar 2015, Faraji and Jamei 2016; http://uses.plantnet-project.org/en/Phragmites_karka_(PROTA)).

66.26 *Poa annua* L.

(Syn. *Ochlopoa annua* (L.) H.Scholz)

Common Name: Annual Bluegrass, Annual Meadow Grass.

Vernacular Name: None known.

Habit: Annual grass.

Habitat: Common in the moist places of the parks and gardens, shady areas in the fields, and near small water channels.

Distribution: Cosmopolitan distribution throughout temperate regions of both the hemispheres, and in tropical highland regions.

Flowering and Fruiting Time: December-March.

Habit

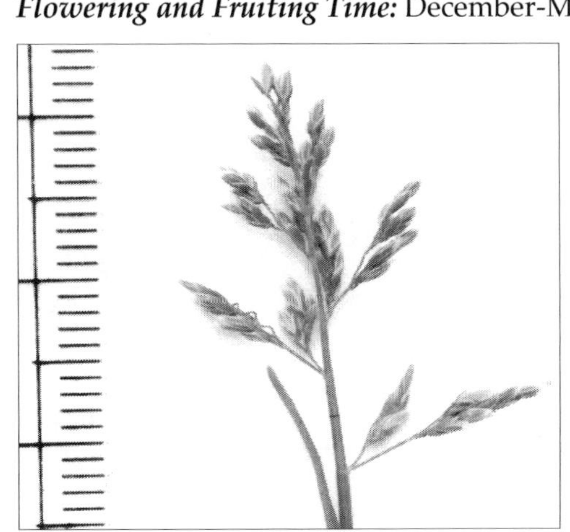
Inflorescence

Taxonomic Description: A tufted, annual grass. Culms sub-erect, often decumbent, 5-20 cm tall, nodes glabrous. Leaves flat, up to 10 cm long and 2-3.5 mm wide, linear, glabrous, apex acute or sub-obtuse, sheaths long, compressed, ligules membranous, 2 mm long. Spikelets usually 2-5 flowered, greenish, 4-6 mm long, oblong-ovate, borne on 2-8 cm long panicle. Glumes unequal, lower about 2 mm long, glabrous, lanceolate; upper ovate-lanceolate, longer than the lower one. Lemmas 3 mm long, ovate-oblong, 5-nerved, slightly hairy along the nerves. Paleas shorter than lemmas, keels ciliate. Stamens-3, anthers oblong. Grains elliptic-oblong.

Chemical Constituents: None known.

Medicinal Properties: None known.

Medicinal Uses (Traditional/Folklore): None known.

66.27 *Saccharum bengalense* Retz.

(Syn. *Erianthus munja* (Roxb.) Jeswiet; *Saccharum munja* Roxb.)

Habit

Common Name: Munj, Sarkanda, Sweet Cane.

Vernacular Name: *Hindi:* Moonj/Munj (मूंज), Sarkanda (सरकन्डा), Sarpat (सरपत).

Habit: Perennial, very tall grass.

Habitat: Common along the bank of canals, wetlands, rivers.

Distribution: The plant is native to northern India, Afghanistan, Pakistan, and Iran. A primary native distribution area is northeastern India, particularly in Assam in the foothills of the Himalayas. In northern India, one of its main ranges is, specifically in the grasslands of the Terai-Duar. The plant is distributed throughout India, in the plains and lower hills, along river-beds.

Flowering and Fruiting Time: September-December.

Taxonomic Description: A densely tufted, robust, perennial grass, up to 4 m high. Leaves 1-1.5 m long, linear, flat, somewhat glaucous beneath, mid-rib white, margin scabrid, sheaths silky hairy at throat; ligules ciliate, membranous. Panicle 30-70 cm long, erect, silvery-white, densely clothed with soft silky hairs, branches whorled with ciliate joints. Spikelets paired, one sessile and other pedicelled, enveloped by long whitish hairs. Sessile spikelets 4-6 mm long, lower glume ciliate at the back and upper glabrous; lower lemma empty, epaleate; upper lemma ovate-lanceolate, bisexual, epaleate. Pedicelled spikelets similar to sessile one. Stamens-3. Grains sub-globose.

Inflorescence

Spikelets

Chemical Constituents: The stem contains furfural yielding reducing sugars; glucose, xylose, galactose and rhamnose have been identified in the hydrolysate which contains fermentable sugars. (Bhutya 2011).

Medicinal Properties: The roots are aphrodisiac, diuretic, demulcent, depurative, cooling, used for burning sensation, dipsia, erysipelas, skin diseases, ophthalmopathy, dysuria, anuria and general debility. (Bhutya 2011).

Medicinal Uses (Traditional/Folklore): Dry roots are boiled in water and the decoction is used in the morning and evening with a teaspoon of pure ghee for severe stomach ache. Decoction of 100 g fresh leaves is taken orally by the tribal ladies for three days to cure menstrual complaint. It is also a good remedy for abortion.

As a veterinary medicine, 500 g fodder of young leaves and green leaves are fed to animals for removal of retained placenta, especially in buffaloes. (Tareen *et al.* 2010, Katewa *et al.* 2011, Yadav *et al.* 2014).

66.28 *Saccharum spontaneum* L.

(Syn. *Imperata spontanea* (L.) P.Beauv.)

Common Name: Kans Grass, Thatch Grass, Tiger Grass.

Vernacular Name: *Hindi:* Kaans (कांस).

Habit: Perennial grass.

Habitat: Common along the bank of wetlands, canals and in fallow fields.

Distribution: It is a polymorphic grass species native to tropical West Asia. Now present across southern Asia, east Africa and in the Mediterranean. The plant is found throughout India, up to 1,800 m.

Habit

Flowering and Fruiting Time: August.-January.

Taxonomic Description: An erect, rhizomatous, perennial grass. Culms tufted, 1-3 m high, solid, smooth. Leaves 50-150 x 0.4-0.6 cm, linear, usually erect, somewhat glaucous beneath; margin scabrid, involute; sheaths auriculate, hairy at throat; ligules ovate, ciliate, membranous. Inflorescence panicle, up to 60 cm long, densely silvery-white with whorled branches, pedicels ciliate. Spikelets paired, one pedicellate and other sessile, 3-5 mm long, oblong-lanceolate, callus with long silky hairs. Lower glume 4-6 mm long with ciliate margin, upper nearly as long and similar to lower ones. Lower floret sterile, lemma epaleate. Upper floret bisexual, lemma epaleate. Stamens-3. Caryopsis oblong to sub-globose.

Inflorescence **Spikelets**

Chemical Constituents: The phytochemical screening has shown the presence of quinines, alkaloids, saponins, tannins, flavonoids, anthraquinones, carbohydrates, protein, coumarin, phenol, steroid and glycosides. It possesses strong allelochemicals and allelopathic properties. (Kumar *et al.* 2010a).

Medicinal Properties: The roots of *S. spontaneum* are sweet, astringent, emollient, refrigerant, and diuretic, lithotriptic, haemostatic, laxative, aphrodisiac and tonic. (Kumar *et al.* 2010a, Warrier *et al.* 2010).

Medicinal Uses (Traditional/Folklore): The stems (culm) are useful in vitiated conditions of *Pitta* and *Vata,* burning sensation strongly, renal and vesicle calculi, dyspepsia, haemorrhoids, menorrhagia, dysentery, agalactia, phthisis and general debility. Roots are also useful in the treatment of dyspepsia, burning sensation, piles, sexual weakness, gynaecological troubles, respiratory troubles. Indigenously, a decoction of its roots along with that of *Gokhru (Tribulus terrestris)* is given to remove urinary bladder's calculi. A paste of its fresh roots with that of *Cynodon dactylon* in 5.0 g doses is given to women with cow milk and sugar, early in the morning daily for one month to cure the leucorrhoea. (Kumar *et al.* 2010a, Warrier *et al.* 2010, Chaudhary 2011, Shah *et al.* 2013, https://easyayurveda.com/2017/10/08/kans-grass-saccharum-spontaneum-kasha).

66.29 *Setaria pumila* (Poir.) Roem. & Schult.

(Syn. *Setaria pallidifusca* (Schumach.) Stapf & C.E.Hubb.)

Habit

Common Name: Yellow Fox-Tail Grass, Yellow Bristle Grass.

Vernacular Name: *Hindi:* Bandra (बंदरा), Bandri (बंदरी), Ban Kauni (बन कौनी), Siun (सीऊं).

Habit: Annual grass.

Habitat: Common in moist waste places, marshy localities and along roadsides.

Distribution: Paleotropics.

Flowering and Fruiting Time: July-September.

Taxonomic Description: A loosely tufted, annual grass. Culms erect or decumbent, 20-80 cm tall, nodes glabrous. Leaves up to 25 cm long, narrow, linear, apex acuminate, upper surface scabrous (more near leaf base) with long silky hairs, sheaths keeled. Inflorescence erect, terete, 3-10 cm long, spiciform panicle. Spikelets 2-2.5 mm long, broadly ovate or elliptic, subtended by 4-7 mm long, yellowish bristles. Glumes unequal, shorter than spikelets. Stamens-3. Grains ovoid-ellipsoid, plano-convex, enclosed by lemma and palea.

Twig with inflorescence **Scabrid leaf** **Close up of inflorescence**

Chemical Constituents: The plant contains different phytochemicals belonging to alkaloids, lignin, quinones, saponins, steroids and tannins. The seeds contain protein, fat, and carbohydrates. (Babu and Savithramma 2013; http://tropical.theferns.info).

Medicinal Properties: Plants are used in skin diseases, worm infection and rheumatic pain. (Peter and Thomas 2015).

Medicinal Uses (Traditional/Folklore): Awns of inflorescence are collected and crushed in milk and filtered; one spoonful of the filtrate is given to children to get rid of from tapeworm infestation. The whole aerial plant is boiled along with the leaves of *Malvastrum coromandelianum* and *Sida cordata* and the hot extract is used to cure skin allergy and other skin diseases. The whole plant is boiled along with the leaves of *Eupatorium cannabinum*, *Toona ciliata*, and *Phragmites karka* and the decoction is used as bath in various skin diseases including allergy. Freshly prepared root paste is warmed gently and applied for reducing the rheumatic pain by the Lodhas community.

As a veterinary medicine, grains are boiled and pounded to paste after mixing with black pepper (*Piper nigrum*) seeds at a ratio of 3:2 and applied on the bull's neck sores by the Lodhas community. (Patil and Patil 2006, Saha *et al.* 2013, Singh *et al.* 2015b, Peter and Thomas 2015).

66.30 *Setaria verticillata* (L.) P.Beauv.

(Syn. *Panicum verticillatum* L.; *Pennisetum verticillatum* (L.) R.Br.)

Habit

Common Name: Bristly Foxtail, Bur Bristle Grass, Bur Grass, Foxtail.

Vernacular Name: *Hindi:* Laptuna (लपतुना), Latkaunya (लटकौन्या).

Habit: Annual grass.

Habitat: Common in moist places, shady areas in gardens, disturbed lands and also along roadsides.

Distribution: *S. verticillata* is a native of the Old World but has been introduced into North and South America; found in India, Pakistan, tropical and warm regions.

Flowering and Fruiting Time: August-December.

Taxonomic Description: An annual grass. Culms ascending, up to 1 m high, nodes glabrous. Leaves 5-30 x 1.5-2 cm, broadly linear-lanceolate, flat, margin scabrous, sheaths glabrous or sparsely hairy, ligules 0.5-1 mm. Inflorescence a panicle. Panicle 3-10 cm long, spike-like, linear, dense-cylindrical. Spikelets 2-2.5 mm long, ellipsoid, greenish with apparent darker veins, subtended by long bristles, bristles 4-8 mm long, retrorsely barbed. Glumes apiculate, lower shorter than the spikelets, upper nearly as long as the spikelets. Lower lemma sterile. Stamens-3. Grains oblong-ellipsoid.

Inflorescence

Portion of the inflorescence-close up

Chemical Constituents: *S. verticillata* contains alkaloids, carbohydrates, flavonoids, sterols, cardiac glycosides, saponins, tannins, phenols, terpenoids and steroids. (Shivakoti and Ramesh 2015).

Medicinal Properties: Antirheumatic, antidiarrhoeal, and antiseptic properties. (Shivakoti and Ramesh 2015).

Medicinal Uses (Traditional/Folklore): Plants are useful in rheumatism, psoriasis and chronic eruptions. For diarrhoea, leaves and inflorescence are pounded, cold water added and the juice taken three times a day. In South Africa, the seeds of *S. verticillata* are used to produce malt for beer. In Nambia, the Topnaar people harvest *S. verticillata* seeds and use them in making porridge.

As a veterinary medicine root powder is applied over the septic wounds of cattle for early cure by the Santhal tribe. The mats prepared from the inflorescence portion (without grains) are used as flooring mat in the granary. It is said that this carpet functioned as a rodent repellent. (Mitra and Mukherjee 2009, Shivakoti and Ramesh 2015, Mall 2016, Panda and Mishra 2016; https://www.cabi.org/isc/datasheet/49775).

66.31 *Sorghum halepense* (L.) Pers.

(Syn. *Andropogon halepensis* (L.) Brot.)

Habit

Common Name: Johnson Grass, Aleppo Grass, Aleppo Millet Grass.

Vernacular Name: *Hindi:* Jangli Jowar (जंगली जोवार), Baru (बारू), Chinna (चिन्ना).

Habit: Perennial grass.

Habitat: Common in open waste places and unused grounds.

Distribution: The native status of the Johnson grass is uncertain but it is naturalized throughout the world.

Flowering and Fruiting Time: July-November.

Taxonomic Description: A perennial, rhizomatous grass. Culms usually erect, 1-2 m high, polished. Leaves broadly linear-lanceolate, 10-60 x 1-3.5 cm, margin scabrid, base rounded or sub-amplexicaul, apex acuminate, sheaths glabrous, ligules membranous. Inflorescence 10-40 cm long panicle. Panicle pyramidal, yellowish brown or purplish in colour; branches 1-3 nate, distant. Spikelets 2-nate, one sessile and other pedicelled. Sessile spikelets 4-5.5 mm long, elliptic, callus densely hairy, glumes sub-equal; upper lemma mucronate, generally with long awns. Pedicelled spikelets 4-6 mm long, pubescent, narrowly lanceolate or elliptic. Stamens-3; anthers yellow, oblong, 2-3 mm long. Grains obovoid, brownish-black, shining.

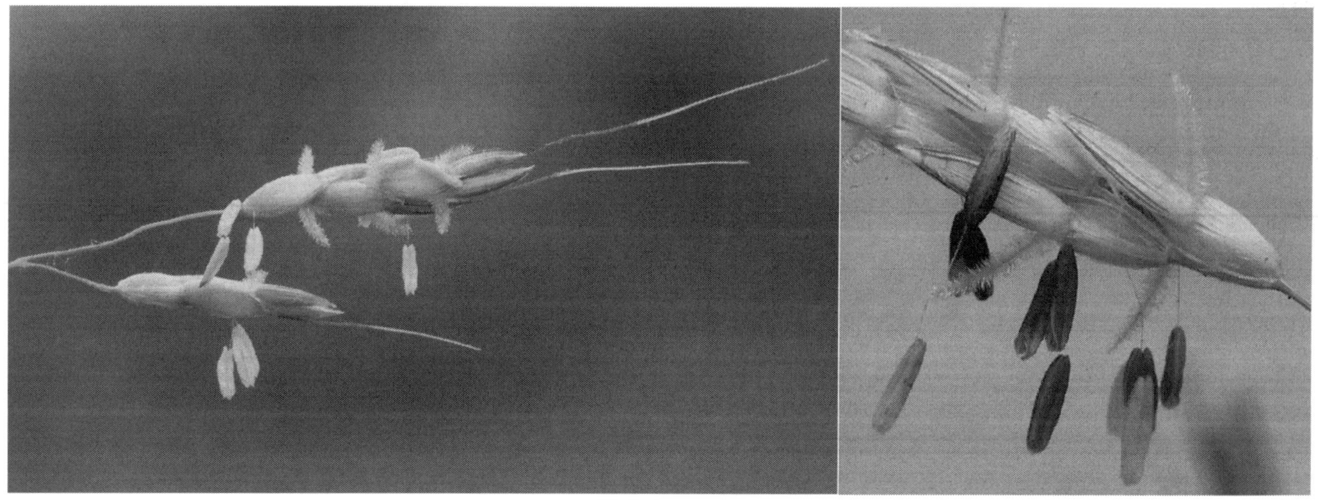

Inflorescence Spikelets

Chemical Constituents: The plant is a rich source of various phytochemicals including tannins, phenolic acids, anthocyanins, phytosterols and policosanols. The plant yields a volatile oil which contains twenty-six chemicals. The main component are sesquiterpenes, such as trans-α-berg-amotene, α-patchoulene, α-himachalene, etc. (Duke and Wain 1981, Awika and Rooney 2004).

Medicinal Properties: The seeds are demulcent and diuretic; a decoction of seeds is made into a paste and used for skin problems and irritation. (Duke and Wain 1981, Mall 2016; http://herbs-treatandtaste.blogspot.in/2011/07/).

Medicinal Uses (Traditional/Folklore): Johnson grass is a folk remedy for blood and urinary disorders.

As a veterinary medicine seed flour mixed with water is given to the animals to cure diarrhoea and weakness. Used as fodder for cattle but in dry places it become tough and dangerous to animals and most of cattle die after eating it. (Galav *et al.* 2010, Khan *et al.* 2013c, Mall 2016).

III. PTERIDOPHYTES (Ferns and Fern Allies)

Pteridophytes (Fig. 32), a group of vascular plants (with xylem and phloem) that reproduce via spores, are included under cryptogams due to the absence of flowers or seeds. The members include the ferns, horsetails, and the lycophytes (clubmosses, spike mosses, and quillworts). There is a rich fossil record to show that Pteridophytes have ancestors dating back nearly four hundred million years from Silurian and Devonian periods of the Paleozoic era. The group has a total of 48 plant families containing 10,620 accepted species under 587 plant genera.

Fig. 32: Pteridophytes (A) Terrestrial fern *Pteris*; (B) Aquatic fern *Marsilea*; (C) Strobilus of horsetail - *Equisetum*

67. EQUISETACEAE

Equisetaceae - Horsetail family

The Equisetaceae is a well-defined group of extinct and extant plants not clearly related to any other group of Pteridophytes. The group dates back to the Devonian, ± 300 million years ago. Members of the family are terrestrial, herbaceous, homosporous vascular plants represented by three plant genera and 27 accepted species of horsetails and scouring rushes. The plant consists of a rhizome with adventitious roots, and an aerial portion with ribbed, silicified stems that are hollow except at the jointed nodes which bear whorls of reduced scale like megaphyllous leaves, sometimes alternating with branches. Clusters of 5-10 eusporagia hang from the margin of umbrella like sporangiophores that are aggregated into a compact strobilus at the tip of the shoot axis. The living species of *Equisetum* is distributed worldwide except in Australia, New Zealand, and Antarctica. In Chithara flora (SNU Campus and around) two species of *Equisetum i.e., Equisetum ramosissimum* (*E. ramosissimum* subsp. *ramosissimum)* and *E. ramosissimum* subsp. *debile* were recorded. Since, *Equisetum* is highly variable, morphologically the two species have many common features. Besides others, the following characters distinguish the two species.

- *Equisetum ramosissimum* (subsp. *ramosissimum)*: **stem with double common endodermis (an inner endodermis surrounding all bundles);** stomata in one to several lines; **sheath longer, teeth persistent**; from southern and eastern Africa, southern and central Europe to most of Asia except Malesia.
- subsp. *debile:* **stem with individual endodermis (an endodermis surrounding each vascular bundle);** stomata in one to several lines; **sheath, short, truncated, teeth regularly breaking off**; India, southern China, throughout Malesia (except Peninsular Malaysia) to New Caledonia.

As per The Plant List 2013, the accepted name of these two subspecies are as follows;

1. *Equisetum ramosissimum* Desf. (Syn. *Equisetum ramosissimum* subsp. *ramosissimum; Hippochaete ramosissima* (Desf.) Börner)
2. *Equisetum ramosissimum* subsp. *debile* (Roxb. ex Vaucher) Hauke (Syn. *Hippochaete ramosissima* subsp. *debilis* (Roxb. ex Vaucher) Á. Löve & D. Löve)

The confidence level was medium for the first species and low for the second one.

67.1 *Equisetum ramosissimum* Desf.

(Syn. *Equisetum ramosissimum* Desf. subsp. *ramosissimum; Hippochaete ramosissima* (Desf.) Börner)

Habit

Common Name: Horsetail, Branched Horsetail, Droukgrass.

Vernacular Name: Putod, Sumbok.

Habit: Perennial fern.

Habitat: Found in marshy grounds near streams, banks of rivers, wetlands, and forests, forest margins, under bushes, meadows. It is sometimes also found in shallow waters. It is often found dormant and can form extensive stands.

Distribution: Cosmopolitan in warm temperate and tropical regions. The plant is widespread throughout Africa, except western parts of Africa; also in Europe, Asia and introduced in America. In India, recorded from Lachiwala, Dehradun, Mossey falls and Burning Ghat in Mussoorie in Uttarakhand, northern and western Himalayas, and Pahlgam and Ladakh in Kashmir, Delhi, Uttar Pradesh (Surajpur wetlands, Gautam Buddha Nagar district), Haryana, Rajasthan, Maharashtra, Tamil Nadu, Kerala and Assam. Found growing scattered but abundantly in open sunny sandy marshy areas or wet grounds along rivers, streams, canals and lakes up to 3,600 m of altitude.

The authors failed to relocate the species in Surajpur wetlands.

Growth phases:

February-April: Young shoots started to arise/develop from the ground from the underground rhizomes.

April-August: Growth continues – lateral branches started to appear from the bottom nodes first – fewer branches at the top.

September-December: Production of strobilus in terminal and lateral branches.

Taxonomic Description: A small to medium-sized herb with subscandent stem. Rhizome long creeping, copiously branched, underground, 2-3 mm in diameter, blackish brown, rooting at the nodes, roots few, long yellowish brown. Aerial stem perennial, monomorphic, dark green, 1-2 m tall, 2-3 mm in diameter, rough texture with ridges and furrows, weak, initially unbranched later branches arise from the lower part of the stem, lower portion of main stem often highly branched, tufted; internodes 2.8-5.2 cm long. Leaves minute scaly, at the nodes; sheaths narrow and up to 9 mm long, lower portion grayish green, upper portion grayish brown; sheath teeth deltoid, teeth of middle portion grayish white or few teeth blackish brown, margin (sometimes upper part) membranous, caducous or persistent, with conspicuous stomatal bands on sheath teeth. Lateral branches hard, cylindrical 8 to 10 ridged. Leaf sheath teeth of upper portion brown, lanceolate, leathery but membranous at margin, persistent. Strobilus terminal, shortly clavate or ellipsoid 1-1.5 cm long, 3-5 mm in diameter at middle, apex with small acute tip, sessile. Sporophylls peltate, closely packed.

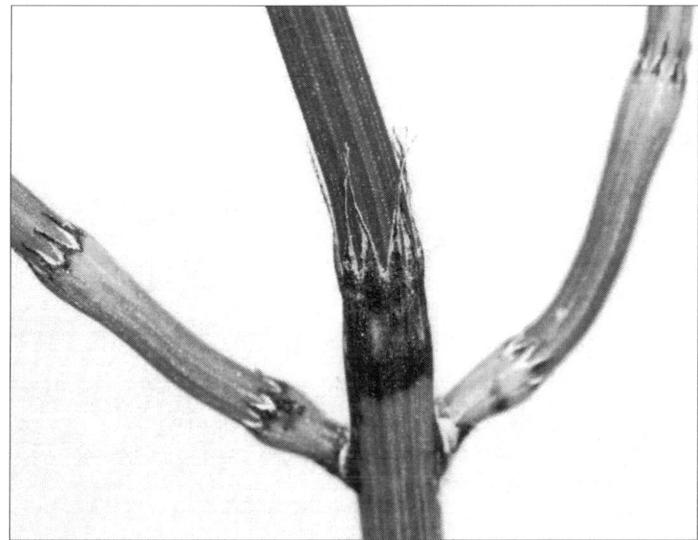

Branching

Mature strobilus

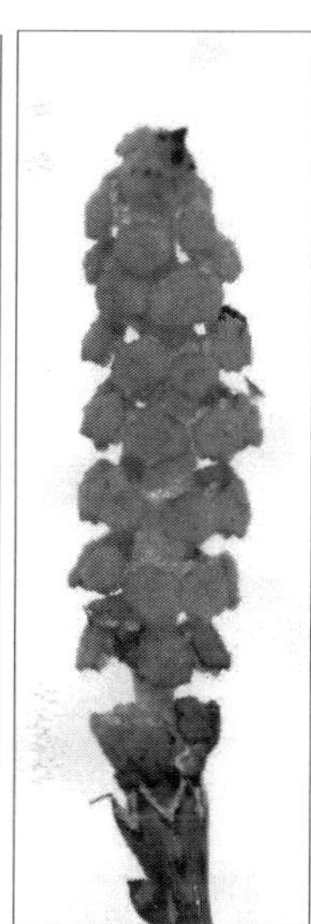

Ripe strobilus

Chemical Constituents: Seventeen chemical compounds have been isolated from the whole plant of *Equisetum ramosissimum* by column chromatography, thin-layer chromatography and recrystallization, and structurally identified by modern spectroscopic technology: 5α,6α-expoxy-β-ionone-3-O-β-D-glucopyranoside, loliolide, cycloart-24(30)-ene-3β-ol, cycloart-22(23)-ene-3β-ol, ergost-6,22-diene-3β,5α,8α-triol, friedelinol, apigene, genkwanin, genkwanin-5-O-β-D-glucopyranoside, apigene-5-O-β-D-glucopyranoside, luteolin, quercentin-3-O-β-D-glucopyranoside, kaempferol-3-β-D-glucopyranoside, kaempferol-3-O-β-D-glucose-7-O-β-D-glucopyranoside, adenine, β-sitosterol and β-daucosterol; α-bisabolol Oxide A (12.3%) and cuminaldehyde (9.8%) were identified as major constituents of the oil. (Xiao-xiong and Zhong-jian 2005).

Medicinal Properties: Plant is known to have antioxidant, central nervous system (CNS) depressant, antihypertensive, antidiabetic, diuretic, haemostatic, antirheumatic, antifungal and antiviral properties. (Benítez *et al.* 2010, Jain *et al.* 2016).

Medicinal Uses (Traditional/Folklore): Whole plant or the stem is utilized in India for several centuries as a medicine for treating skin wounds. The plant is used in the treatment of hypertension, reddening and swelling pain in the eye, pterygium of the cornea, enteritis, diarrhoea, jaunditic hepatitis and renal lithiasis; it is used as cooling medicine for gonorrhea. The stem, which is sweet and slightly bitter tasting, is diuretic and antitussive. Powdered stem dissolved in water (decoction) is given for enema during stomach disorders in children. It is said to cleanse the liver and clear the eyesight. Traditional use of the plant include: application of plant paste in bone fracture in Uttarakhand and Vindhya (MP); use of entire plant for cooling in gonorrhea by tribal community of Melaghat area

(Maharashtra) India; and use of young cones to treat kidney trouble in Vindya (MP). In Gonda district, Uttar Pradesh whole frond is powdered and mixed with mustard oil and used as treatment for bone fracture and muscular pain.

In China, decoction of whole plant is used for the cure of wounds and ulcers. In Macao, China it is used to clear heat, promote diuresis, clear liver-fire, and to improve vision. Also indicated to be useful for acute conjunctivitis, icteric hepatitis and urethral stone. Infertile women are given rhizome decoction to facilitate fertilization in South Africa. A few pieces of the branches mixed with leaves of other specific plants and made into a paste is used as local application for the treatment of fracture and the dislocation of bones. The juice of the roots is used to relieve the toothache. In Indonesia and Malaysia, *E. ramosissimum* stems are powdered and made into an ointment which is used externally to treat bruises, fractures and arthritis. A decoction is taken as a diuretic and astringent to treat dysentery and haemorrhoids. In Nepal, juice of the roots (about 4 teaspoons 4 times a day for a week), is given to relieve fever. In South Africa, juice from the plant is used to relieve toothache and applied to the wounds after tooth extraction. In Papua New Guinea, horsetail is crushed together with leaves of *Ficus* sp., *Impatiens* sp. and *Stellaria* sp. and the sap is taken to soothe a bad cough. The plant is used by the islanders of the Indian Ocean for treating diabetes, baldness, bone fracture and pneumonia. The Sesotho tribe use the plant for various ailments and rituals. The rhizomes are cooked and a decoction is taken orally by woman for getting pregnancy. The Zulu use the sap from the plant to relieve toothache and is also applied to the wound after tooth extraction. The plant is used to drive caterpillars from the maize fields. This is done by encapsulating some of the caterpillars in clay balls which are then burnt in the fields.

***Equisetum ramosissimum* growing on a soil mound near the boundary in SNU Campus, June 2015**

E. ramosissimum is known to be toxic at high dosage. One study on horses indicates that *Equisetum* species have a high content of the enzyme thiaminase. Upon being eaten, the enzyme is released in the animal's gut and causes degradation of Vitamin B_1 or thiamine. This causes livestock to appear drunk and they may harm themselves. A high dose can lead to avitaminosis and death. The stems of *Equisetum* accumulate crystals of silica and the fine abrasive action of these crystals makes it a useful cleansing agent. Throughout New Guinea *E. ramosissimum* is used to clean cooking and eating utensils. The sandpaper-like qualities lead to its use in shaping and smoothing tools, ornaments and weapons. *E. ramosissimum* is often cultivated as a pot plant. The stems are sometimes made into handicraft objects, e.g. picture frames. (Singh *et al.* 2005, Anon. 2014, Jain *et al.* 2016; http://uses.plantnet project.org/en/Equisetum_ramosissimum_(PROSEA), http://www.biodiversityofindia.org/index.php?title=Property:Modern_scientific_knowledge_description;http://www.biodiversityofindia.org/index.php?title=Property:Medicinal_use_description, http://plants.jstor.org/stable/10.5555/al.ap.flora.sffa001370184100001).

67.2 *Equisetum ramosissimum* subsp. *debile* (Roxb. ex Vaucher) Hauke

(Syn. *Hippochaete ramosissima* subsp. *debilis* (Roxb. ex Vaucher) Á. Löve & D. Löve)

Common Name: Horsetail.

Vernacular Name: None Known.

Habit: Perennial Fern.

Habitat: Damp soil along the bank of streams, wetlands and marshy areas.

Distribution: *E. ramosissimum* subsp. *debile* occurs in India and southern China through Malaysia to New Caledonia and Fiji. It is distributed in South East Asia, Thailand, Southern China, Taiwan, Hong Kong, Philippines, Papua New Guinea, Indonesia, India, Nepal, Sri Lanka and Bangladesh. In India, it is distributed in northeastern India (Arunachal Pradesh, Assam, Meghalaya), West Bengal, Dehradun, Saharanpur, and Banks of Yamuna in Delhi, Singhbhum, Manbhum and Ranchi.

Habit

Growth phases:

February-April: Young shoots started to arise/develop from the ground from the underground rhizomes.

May-July: Growth continues – lateral branches started to appear from the bottom nodes first; fewer branches at the top.

August-December: Production of strobilus in terminal and lateral branches.

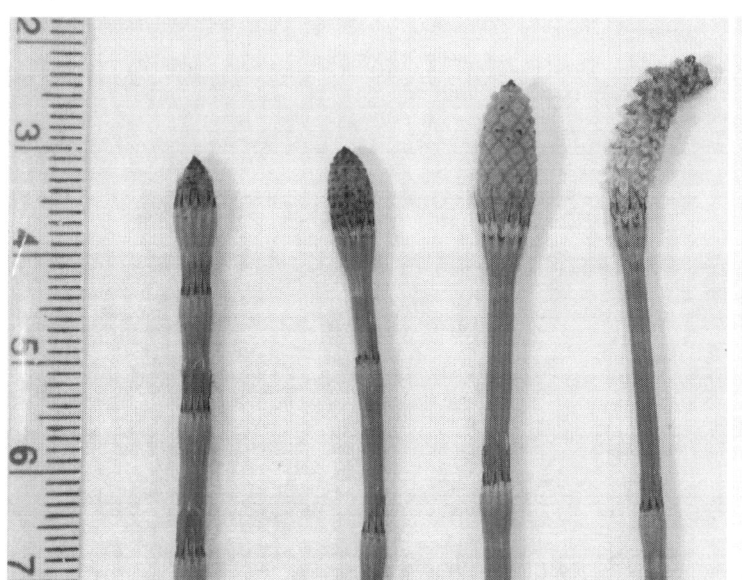

Various stages of strobilus

Mature strobilus

Ripe strobilus

Taxonomic Description: A small to medium-sized terrestrial fern. Rhizomes deep underground, dark brown to black, branched, rough 3-4 mm in diameter. Roots numerous, wiry, thin arising from nodes. Stem dark green, erect, simple, irregularly branched; internodes articulate, hollow, cylindrical 3-4 mm in diameter and 2-4 cm in length; ridges 16-17 in number on main stem. Leaves small scale like in whorls fused to form sheath at the stem nodes. Sheath cylindrical to slightly funnel shaped, 6-7 mm long, margin whitish or colourless, drying or deciduous. Strobilus apical ellipsoidal, 1.4-1.9 cm long and 3-5.5 mm in diameter in the middle, initially greenish turning yellowish on maturity. Sporangiophores hexagonal.

Chemical Constituents: Several phytochemicals have been isolated from *E. ramosissimum* subsp. *debile*. These include pinocembrin, chrysin, β-sitosterol, β-D- glycosylsitosterol, β-D-glucose, flavonoid glycosides and fatty acids12,13, flavonoid glycosides (kaempferol 3-O-sophoroside, kaempferol 3,7-O-β-D-diglucopyranoside, kaempferol 3-O-sophoroside-7-O-β-D-glucopyranoside, kaempferol 3-O-sophoroside and Caffeoyl-methylate-4-β-gluco-puranoside), megastigmane diglucoside (3S,5R,6S,7E,9S)- megastigman-7-ene-5,6-epoxy-3,9-diol 3,9-O-β-D-diglucopyranoside, (6R,9S)-3-oxo-a-ionol 9-O-β-D-glucopyranoside, (3S,5R,6R,7E,9S)-9-[(β-D-glucopyranosyl) oxy] megastigm-7- ene-3,5,6-triol), phenylethyl-O-β-D-glucopyranoside, (Z)-3-hexenyl-O-β-D-glucopyranoside, L-tryptophan14 and debilosides A-C; also isolated Blumenol A, corchoinoside C, sammangaoside A15. (http://www.stuartxchange.org/Sumbak.html).

Medicinal Properties: Cooling, astringent, anticancer, antiulcer, antihepatic, antimalarial, and diuretic. (Heim E 2015; http://www.stuartxchange.org/Sumbak.html).

Medicinal Uses (Traditional/Folklore): Medicinal use in Pachmarhi by tribal: 10-15 ml of decoction of rhizomes is used orally for 7-14 days for curing gonorrhoea. The same dose is taken orally twice a day for a month to facilitate fertilization in women. *E. debile* is medicinally important in Nepal and it is applied to treat burns, scabies, malarial fever, gonorrhoea, dislocated bones, and liver and chest complaints. In Malaysia, it is used for pains in the joints. In Africa, juice of *E. debile* is taken orally to cure jaundice and hepatitis; also reported use as diuretic and treatment of kidney infections. In Nepal, paste from pounded plants is applied to cure old ulcers while in India it is used for the treatment of urinary tract and kidney stones. In Bangladesh, the whole plant is used as cooling medicine; given for gonorrhoea and bone fractures; decoction of plant is used for nasal polyps, and for the treatment of various cancers (breast, liver, intestine, stomach, kidneys, and tongue). (Singh and Singh 2012; http://www.stuartxchange.org/Sumbak.html).

68. MARSILEACEAE

Marsileaceae - Water Clover/Pepperwort family

The Marsileaceae are a small family of heterosporous aquatic and semi-aquatic ferns with long-creeping, slender, glabrous or hairy rhizomes; leaves with two or four leaflets at the petiole tip or lacking a blade altogether, the venation of leaflets dichotomously branched; sori enclosed in stalked bean-shaped sporocarps. Plants often grow in dense clumps in mud along the shores of ponds or streams, or they may grow submerged in shallow water with some of the leaves extending to float on the water surface. The family has 5 plant genera and 32 accepted species distributed nearly worldwide. A single species of this family was recorded in Chithara Village.

68.1 *Marsilea quadrifolia* L.

(Syn. None known)

Common Name: Four Leaf Clover, Water Shamrock, Common Water Clover.

Vernacular Name: *Hindi:* Chaupatiya (चौपतिया), Susni (सुस्नि); *Sanskrit:* Chatuh Patri (चतु:पत्री), Chatuspatri (चतुष्पत्री), Svastika (स्वास्तिका).

Habit: Aquatic or sub-aquatic fern.

Habitat: Common along marshy areas, paddy fields, ditches, shallow ponds and wetlands.

Distribution: *M. quadrifolia* is found naturally in central and southern Europe, Caucasia, western Siberia, Afghanistan, South-West India, China, Japan, and Vietnam. Throughout India, in marshy places and along the banks of canals, wetlands, and ponds.

Habit

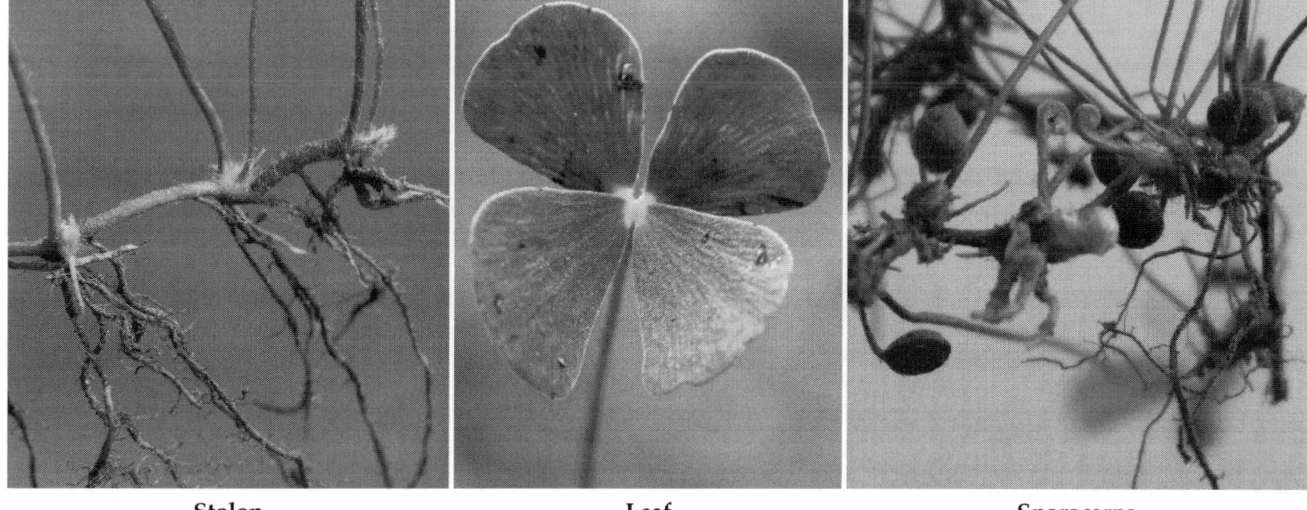

Stolon | Leaf | Sporocarps

Taxonomic Description: An aquatic or sub-aquatic, small fern with long, creeping rhizome. Roots arises at nodes and few on internodes, on the underside of the hairy rhizome. Leaves compound, arranged alternately at the upper surface of the creeping rhizome. Leaflets-4, triangular-obovate or fan-shaped, 1.5-2.5 cm long, clustered at the tip of petiole, sparsely pubescent, petiole 3-12 cm long, slender. Sporocarps dark brown, 4-5 mm long, oblong, ovoid, densely leathery to glabrate, attached at the base of the petiole.

Chemical Constituents: Fresh leaves contain high amount of antioxidant due to the presence of phytochemicals such as steroids, phenols, resins, tannins, flavonoids, terphenoids, carbohydrates, amino acids and protein. The plant is also known to contain marsilin (1- triacontanol-cerotate), 3-hydroxy-tri-acontan-11-one, hentriacotan-6-ol, methylamine, Beta-sitosterol, marsileagenin A, flavonol-O-mono and diglycoside, C- glucoflavones and C-glucosyl xanthones. (Arul and Agnes 2015, Balakrishnan and Subramanian 2016).

Medicinal Properties: The plant is sweet, astringent, refrigerant, acrid, emollient, anodyne, hypnotic, ophthalmic, diuretic, constipating, expectorant, aphrodisiac, depurative and febrifuge. (Warrier *et al.* 2010, Arul and Agnes 2015).

Medicinal Uses (Traditional/Folklore): *M. quadrifolia* is used to treat cough, bronchitis, diabetes, psychiatric diseases, eye diseases, and diarrhoea and skin diseases. As per the traditional claims, *M. quadrifolia* plant has been used for astringent, hypnotic, psychopathy, ophthalmia, strangury, diarrhoea, cough, bronchitis, skin diseases expectorant, aphrodisiac, anodyne, constipating, dyspepsia and also used in the treatment of leprosy, haemorrhoids, fever and insomnia. A juice made from the leaves is used to treat diuretic, febrifuge, snakebite and also applied to abscesses. It is highly nutritious, good for reducing body heat, thirst and act as anti-inflammatory drug. Crushed plant is taken with sugar candy or honey to cure infantile diarrhoea in Bargarh district of Orissa.

The plant is also applied externally in the treatment of snakebites and skin injuries, including abscesses. Juice or paste of whole plant is applied externally on the head to get a relief from sleep disorder and hypertension. Two drops of juice of crushed leaves is put in the nostrils twice a day for Migraine. Washed and chopped 250 g of petiole and leaves (Jaduguda village, Jharkhand) cooked in 50 ml warm mustard oil with garlic (2-5 g) are used by tribal for its nerve relaxant nature and curative properties for various other ailments of nervous system and its nutritional value. Petiole and leaves are cooked in seasoning with 4-5 g *desi masala* (mixture of red chilies, turmeric, coriander and cumin seeds) and salt and consumed only in the evening for sound tension free night sleep (minimum 12-14 hours), physical and mental relaxation, hypertension and other nervous disorders, all type of body aches and insomnia. (Soni and Singh 2012, Gopalakrishnan and Udayakumar 2014, Balakrishnan and Subramanian 2016).

69. THELYPTERIDACEAE

Thelypteridaceae

The family has 38 plant genera and 951 accepted species more at lower elevations, very few tropical species above 4500 m: widespread in all tropical and subtropical zones of the world, less common in temperate zone, particularly more common in Asia. The plants of this family are distinguished from others by having greyish white unicellular acicular hairs and pubescence throughout the plant. Members of Thelypteridaceae are distributed nearly worldwide, but species are most diverse in tropical regions. Nearly all of the species are terrestrial, and most occur in moist or wet habitats. Leaf size and morphology are extremely variable, but the lamina is typically one or more times pinnately compound. A characteristic feature of the family is the presence of small needle like hairs on the leaves; these hairs sometimes occur in clusters and appear to be branched. The sori vary from round to linear and sometimes have a kidney-shaped, membranous protective flap of tissue (indusium). The spores are bean-shaped (bilateral). There was a single species represented in the flora.

69.1 Ampelopteris prolifera (Retz.) Copel.

(Syn. *Dryopteris prolifera* (Retz.) C. Chr.; *Thelypteris prolifera* (Retz.) C.F. Reed)

Common Name: Walking Fern.

Vernacular Name: *Hindi:* Bhaisag (भैसाग).

Habit: Pteridophytic herb.

Habitat: Commonly found up to 1,250 m elevation, amongst tall grasses and sedges in freshwater swamps; also abundantly occurring along the banks of canals, rivers, wetlands and ponds.

Habit

Distribution: *A. prolifera* is widely distributed in old world tropics - from West Africa, through mainland Asia to New Guinea, north-eastern Australia and New Caledonia. It occurs throughout South-East Asia; it is absent in southern Peninsular Malaysia.

Frond

Sori (Herbarium photo)

Taxonomic Description: A fern, up to 1 m tall with creeping rhizome, scales on rhizome black in colour. Fronds narrowly spaced, sparsely coriaceous. Stipe yellowish brown, 30-40 cm long. Pinnae 4-15 cm long, oblong-lanceolate, subopposite, shortly petioled, truncate at base, margin undulate or somewhat incised-lobed, slightly pointed at apex, softly pubescent beneath. Sori usually rounded, attached along the veins on underside of the leaves.

Chemical Constituents: Plants contain phenolics, triterpenoids and flavonoids. (Bharti and Pravesh 2012).

Medicinal Properties: Fronds are apirients and alterative and have antibacterial, expectorant, antiviral, antiseptic, blood purifier, anthelmintic properties; leaf paste is applied over skin to cure skin diseases, itches, and wounds. (Benniamin 2011, Bharti and Pravesh 2012, Quattrocchi 2012; http://www.moef.gov.in/sites/default/files/Volume-II.pdf).

Medicinal Uses (Traditional/Folklore): *A. prolifera* is used in traditional medicine in India. The leaves are used as a laxative. Five gram of leaf paste is taken for 30 days with milk as a tonic by tribal in Hadoti Plateau, South-eastern Rajasthan, India. Fresh fronds are cooked with *Methi* seeds and eaten regularly to treat diabetes in Jalpaiguri district of West Bengal, India. The young fronds are used widely as laxative. In Tanzania, the leaf-sap is drunk in a mixture with *Hypoestes aristata* to treat meningitis and encephalitis. *A. prolifera* is also used for treating wounds and eczema and for the cure of jaundice. Roots of the plant have also been employed against dysmenorrhoea and treating spermetorrohea and skin diseases such as eczema. Leaf and root paste is taken to cure stomach pain, and blood dysentery. (Bharti and Pravesh 2012, Quattrocchi 2012; http://tropical.theferns.info, http://uses.plantnet-project.org/en/Ampelopteris_prolifera_(PROTA)).

4

Alien Plant Species Recorded in the Flora

An alien plant is an exotic, introduced, and non-indigenous species that has been introduced intentionally or otherwise through human agency or accidentally from one region/country to another. From an ecological perspective, when an alien plant species spreads and establishes in larger populations in an ecosystem beyond their natural distribution limits, they become invasive (Mack *et al.* 2000). Invasive weeds, considered as a primary cause of global biodiversity loss, usually have faster rates of growth and biomass production, high competitive ability, high reproductive efficiency including production of a large number of seeds, efficient dispersal, easy vegetative reproduction, rapid establishment and other traits as compared to native species that help them adapt to new introduced habitats within a short period of time (Simberloff *et al.* 2005, Sharma *et al.* 2005, Huang *et al.* 2009).

At least ten percent of the world's vascular plants (300,000) have the potential to become invasive in other ecosystems and affect native vegetation in direct or indirect ways (Singh *et al.* 2006). About 18 percent of the Indian flora constitutes aliens, of which 55 percent are American, 30 percent Asian and Malaysian and 15 percent European and Central Asian species (Nayar 1997). In India, 1,599 vascular alien plants (including cultivated) have been recorded, belonging to 842 genera in 161 families (Khuroo *et al.* 2012). It has been observed that about 25 percent of the introduced species became invasive in India within a short period of time of 50-100 years. A total of 173 species (10.8 percent of the total) belonging to 117 genera are invasive plants representing 1 percent of the Indian flora (Reddy *et al.* 2008). A comprehensive inventory of the invasive alien flora of India's most populous state, Uttar Pradesh, recorded 152 plant species belonging to 109 genera and 44 families (Singh *et al.* 2010). In Delhi, which is much smaller in geographical area than Uttar Pradesh, Mishra *et al.* (2015d) recorded a total of 102 alien plant species belonging to 69 genera and 33 families based on secondary data and field observations.

The flora of Chithara Village Panchayat recorded a total of 272 species belonging to 69 families, which included herbs, undershrubs, grasses, sedges, shrubs, climbers, creepers and trees. Exhaustive literature search was made to gather relevant information on the nativity of the plant species recorded in the flora. The species were also searched in CABI (Invasive Species Compendium) for their nativity.

Preponderance of Alien Species in the flora

Of the 272 vascular plant species recorded in Chithara, a total of 92 species (about 34 percent of the flora) belonging to 69 genera and 28 families were identified as alien species for they are native of other countries and regions in the world. They were somehow introduced and eventually got naturalized in the country surviving as alien weeds. Of the 92 alien species, 63 (approx. 68 percent) are reported to be invasive in India by Reddy *et al.* (2008) (Table 7); about 23 percent of the species in the flora are invasive alien species. This is very high proportion of occurrence of invasive species in Chithara flora as compared to Indian invasive flora where only 10 percent of the species are invasive. This probably indicates high degree of disturbance in the Chithara ecosystem due to various anthropogenic factors resulting in extensive land use and land cover changes, and subsequently large scale invasion of alien species. In a study of invasive alien flora of Uttar Pradesh, Singh *et al.* (2010c), reported 152 plant species from 109 genera and 144 families. In the present study, 92 alien plant species accounts for approximately 60 percent of the alien flora of Uttar Pradesh. Recently, Mishra *et al.* (2015d) documented 102 alien plant species belonging to 69 genera and 33 families in Delhi, which is very close to the present study. In a few other similar studies, Khuroo *et al.* (2007) reported 571 alien plant species from Kashmir Himalayas, Dogra *et al.* (2011) 159 species from temple courtyard of Himachal Pradesh and Sekar (2012) 190 alien plant species from Indian Himalayan regions. The alien species recorded in the flora are given in Table 7.

Table 7. Alien and alien invasive (#) with their nativity recorded in the flora

SN	Name of plant species	Family	Nativity	Habit
1	*Ageratum conyzoides* (L.) L. #	Compositae	America (Trop. America)	H
2	*Ageratum houstonianum* Mill. #	Compositae	America (Trop. America)	H
3	*Alternanthera paronychioides* A.St.-Hil. #	Amaranthaceae	America (Trop. America)	H
4	*Alternanthera philoxeroides* (Mart.) Griseb. #	Amaranthaceae	America (Trop. America)	H
5	*Alternanthera pungens* Kunth #	Amaranthaceae	America (Trop. America)	H
6	*Alternanthera sessilis* (L.) R.Br. ex DC.	Amaranthaceae	America (Trop. America)	H
7	*Anagallis arvensis* L.	Primulaceae	Europe	H
8	*Argemone mexicana* L. #	Papaveraceae	America (South America)	H
9	*Bidens pilosa* L. #	Compositae	America (Trop. America)	H
10	*Blainvillea acmella* (L.) Philipson #	Compositae	America (Trop. America)	H
11	*Blumea obliqua* (L.) Druce #	Compositae	America (Trop. America)	H
12	*Calotropis gigantea* (L.) Dryand. #	Apocynaceae	Africa (Trop. Africa)	SH
13	*Calotropis procera* (Aiton) Dryand. #	Apocynaceae	Africa (Trop. Africa)	US
14	*Cardamine hirsuta* L. #	Brassicaceae	America (Trop. America)	H
15	*Celosia argentea* L. #	Amaranthaceae	Africa (Trop. Africa)	H
16	*Chenopodium album* L.	Amaranthaceae	Europe	H
17	*Chenopodium murale* L.	Amaranthaceae	America (Trop. America)	H
18	*Chloris barbata* Sw.	Poaceae	America (Trop. America)	GR
19	*Cleome viscosa* L. #	Cleomaceae	America (Trop. America)	US
20	*Corchorus aestuans* L. #	Malvaceae	America (Trop. America)	H
21	*Corchorus olitorius* L.	Malvaceae	Africa (Trop. Africa)	H
22	*Corchorus tridens* L.	Malvaceae	Africa (Trop. Africa)	H
23	*Croton bonplandianus* Baill.	Euphorbiaceae	America (South America)	H
24	*Cuscuta chinensis* Lam. #	Convolvulaceae	Mediterranean	CL/CR
25	*Cuscuta reflexa* Roxb. #	Convolvulaceae	Mediterranean	CL/CR
26	*Cyperus difformis* L. #	Cyperaceae	America (Trop. America)	SE
27	*Cyperus iria* L. #	Cyperaceae	America (Trop. America)	SE
28	*Datura innoxia* Mill. #	Solanaceae	America (Trop. America)	H
29	*Digera muricata* (L.) Mart. #	Amaranthaceae	Asia (South West Asia)	H
30	*Dysphania ambrosioides* (L.) Mosyakin & Clemants	Amaranthaceae	America (Trop. America)	US
31	*Echinochloa colona* (L.) Link #	Poaceae	America (Trop. America)	GR
32	*Echinochloa crus-galli* (L.) P.Beauv. #	Poaceae	America (South America)	GR
33	*Eclipta prostrata* (L.) L. #	Compositae	America (Trop. America)	H
34	*Eichhornia crassipes* (Mart.) Solms #	Pontederiaceae	America (Trop. America)	H
35	*Emex australis* Steinh.	Polygonaceae	Africa (South Africa)	H
36	*Erigeron bonariensis* L.	Compositae	America (South America)	H
37	*Euphorbia heterophylla* L. #	Euphorbiaceae	America (Trop. America)	H
38	*Euphorbia hirta* L.	Euphorbiaceae	America (Trop. America)	H
39	*Euphorbia hypericifolia* L.	Euphorbiaceae	America (Trop. America)	H
40	*Euphorbia serpens* Kunth	Euphorbiaceae	America (South America)	H
41	*Evolvulus nummularius* (L.) L. #	Convolvulaceae	America (Trop. America)	H
42	*Gnaphalium purpureum* L.	Compositae	America (Trop. America)	H

Table 7. continued...

SN	Name of plant species	Family	Nativity	Habit
43	*Gomphrena celosioides* Mart.	Amaranthaceae	America (Trop. America)	H
44	*Grangea maderaspatana* (L.) Poir. #	Compositae	America (Trop. South America)	H
45	*Hyptis suaveolens* (L.) Poit. #	Lamiaceae	America (Trop. America)	US
46	*Imperata cylindrica* (L.) Raeusch. #	Poaceae	America (Trop. America)	GR
47	*Indigofera astragalina* DC. #	Leguminosae	America (Trop. America)	US
48	*Indigofera linifolia* (L.f.) Retz. #	Leguminosae	America (Trop. South America)	H
49	*Indigofera linnaei* Ali #	Leguminosae	Africa (Trop. Africa)	H
50	*Ipomoea carnea* Jacq. #	Convolvulaceae	America (Trop. America)	SH
51	*Ipomoea eriocarpa* R. Br. #	Convolvulaceae	Africa (Trop. Africa)	CL/CR
52	*Ipomoea obscura* (L.) Ker Gawl.	Convolvulaceae	Africa (Trop. Africa)	CL/CR
53	*Ipomoea pes-tigridis* L. #	Convolvulaceae	Africa (Trop. Africa)	CL/CR
54	*Jatropha gossypiifolia* L.	Euphorbiaceae	America (Brazil-South America)	SH
55	*Lantana camara* L. #	Verbenaceae	America (Trop. America)	SH
56	*Lathyrus aphaca* L.	Leguminosae	Mediterranean	H
57	*Lepidium didymum* L.	Brassicaceae	America (Trop. America)	H
58	*Leucaena leucocephala* (Lam.) de Wit #	Leguminosae	America (Trop. America)	T
59	*Ludwigia octovalvis* (Jacq.) P.H.Raven #	Onagraceae	Africa (Trop. Africa)	H
60	*Malvastrum coromandelianum* (L.) Garcke #	Malvaceae	America (Trop. America)	H
61	*Mecardonia procumbens* (Mill.) Small #	Plantaginaceae	America (Trop. America)	H
62	*Melilotus albus* Medik.	Leguminosae	Europe	H
63	*Melochia corchorifolia* L. #	Malvaceae	America (Trop. America)	H
64	*Nicotiana plumbaginifolia* Viv. #	Solanaceae	America (Trop. America)	H
65	*Oxalis corniculata* L. #	Oxalidaceae	Europe	H
66	*Oxalis debilis* var. *corymbosa* (DC.) Lourteig	Oxalidaceae	America (Trop. South America)	H
67	*Parthenium hysterophorus* L. #	Compositae	America (Trop. North America)	H
68	*Phyllanthus tenellus* Roxb. #	Phyllanthaceae	Mascarene Islands (Indian Ocean)	H
69	*Physalis angulata* L.	Solanaceae	America (Trop. America)	H
70	*Physalis peruviana* L.	Solanaceae	America (Trop. America)	H
71	*Pistia stratiotes* L. #	Araceae	America (Trop. America)	H
72	*Portulaca oleracea* L. #	Portulacaceae	America (South America)	H
73	*Portulaca quadrifida* L. #	Portulacaceae	America (Trop. America)	H
74	*Prosopis juliflora* (Sw.) DC. #	Leguminosae	America (Mexico-Central America)	T
75	*Ricinus communis* L.	Euphorbiaceae	Africa	SH
76	*Saccharum spontaneum* L. #	Poaceae	Asia (Trop. West Asia)	GR
77	*Scoparia dulcis* L. #	Plantaginaceae	America (Trop. America)	US
78	*Senna obtusifolia* (L.) H.S.Irwin & Barneby #	Leguminosae	America (Trop. America)	H
79	*Senna occidentalis* (L.) Link #	Leguminosae	America (Trop. South America)	US
80	*Sesbania bispinosa* (Jacq.) W.Wight #	Leguminosae	America (Trop. America)	SH
81	*Sida acuta* Burm.f. #	Malvaceae	America (Trop. America)	US
82	*Solanum americanum* Mill. #	Solanaceae	America (Trop. America)	H
83	*Solanum torvum* Sw. #	Solanaceae	West Indies	SH
84	*Sonchus asper* (L.) Hill #	Compositae	Mediterranean	H
85	*Spermacoce articularis* L.f.	Rubiaceae	America (Trop. America)	H
86	*Stellaria media* (L.) Vill.	Caryophyllaceae	Europe	H

Table 7. continued...

SN	Name of plant species	Family	Nativity	Habit
87	*Tribulus terrestris* L. #	Zygophyllaceae	America (Trop. America)	H
88	*Tridax procumbens* (L.) L. #	Compositae	America (Trop. Central America)	H
89	*Triumfetta rhomboidea* Jacq.	Malvaceae	America (Trop. America)	US
90	*Typha domingensis* Pers.	Typhaceae	America (Trop. America)	H
91	*Urena lobata* L. #	Malvaceae	Africa (Trop. Africa)	SH
92	*Xanthium strumarium* L. #	Compositae	America (Trop. America)	H

H-Herb; SH-Shrub; US-Undershrub; CL/CR-Climber/Creeper; GR-Grass; SE-Sedge; T-Tree; Trop.-Tropical
 # Species recorded as Alien invasive species by Reddy et al. (2008)

Among a total of 203 genera recorded from Chithara Village Panchayat, 69 genera are alien as they are native of other continents/regions; *Alternanthera, Euphorbia, Ipomoea* had the highest number of alien species (4 species each) in the flora followed by *Corchorus* and *Indigofera* (3 species each). Ten genera had two species each and all the rest of the genera (54) were represented by a single species (Table 7).

Table 8. Contribution of different geographical regions to the alien flora of Chithara

SN	Native distribution of alien plant species	Alien Species
1	**Africa** (Tropical Africa, South Africa)	13
2	**America** (Tropical America, North America, Central America, South America, Brazil, Mexico)	66
3	**Asia** (South West Asia, Tropical West Asia)	02
4	**Europe**	5
5	**Mascarene Islands** (Indian Ocean)	1
6	**Mediterranean** (Mediterranean Basin)	4
7	**West Indies**	1
	TOTAL	92

The largest number of species (66; 72 percent) were of American origin, followed by Africa (13 species; 14 percent), Europe (5 species; 6 percent), Mediterranean (4 species; 4 percent) and Asia (2 species; 2 percent) (Table 8). There was a single species i.e., *Phyllanthus tenellus* originated from Mascarene Islands (Indian Ocean) and one species, *Solanum torvum* from West Indies (Tables 7). Altogether, the large number of species (91%) were of Africa, America and European origin (Fig. 33). A similar trend of results was obtained in a study of alien flora of Delhi by Mishra *et al.* (2015d) and Uttar Pradesh (Singh *et al.* 2010c), where the maximum number of alien species were introduced from tropical America followed by Africa. In India flora also tropical American region contributes the greatest alien species accounting to 74 percent followed by tropical Africa (11 percent) (Reddy *et al.* 2008).

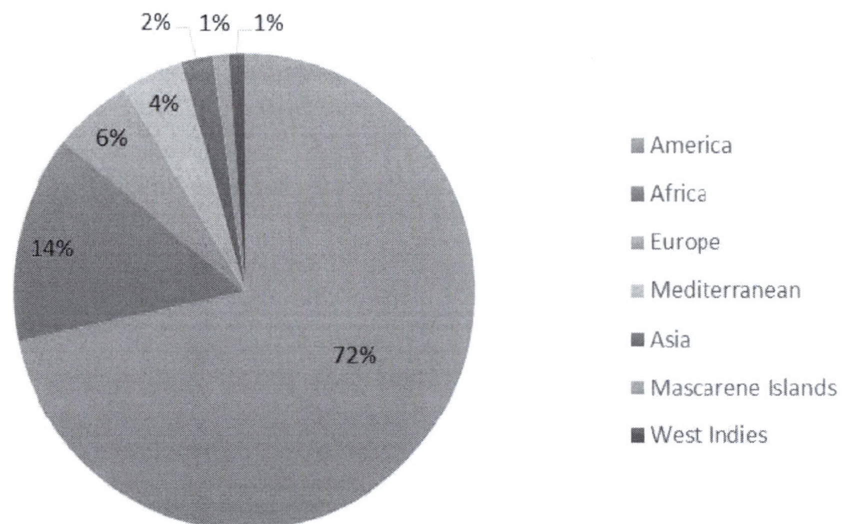

Fig. 33: Pie chart showing the contribution of different geographical regions to the alien flora

Family dominance

With respect to families, the family Compositae was the most dominant with 13 alien species in the flora.

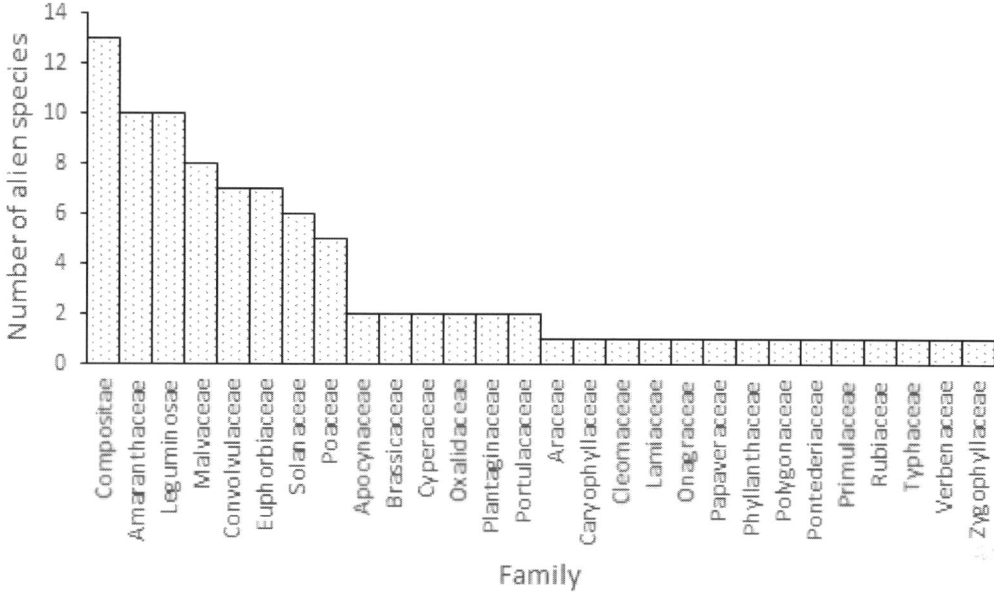

Fig. 34: Bar diagram showing distribution of number of alien species in various families

The co-dominant families were Amaranthaceae and Leguminosae (10 species each), Malvaceae (8 species), Convolvulaceae and Euphorbiaceae (7 species each), Solanaceae (6 species) and Poaceae (5 species); six families had 2 alien species and fourteen families had a single species (Fig. 34). The dominance of the family Compositae, Amaranthaceae and Leguminosae was also reported in alien flora of Delhi (Mishra *et al.* 2015d) and Uttar Pradesh (Singh *et al.* 2010c). Of the 44 families of invasive alien species in Indian flora Compositae is the most dominant, followed by Leguminosae, Convolvulaceae, Solanaceae, Amaranthaceae, Poaceae, Euphorbiaceae and Tiliaceae. These families contributed to 63.6 percent of invasive alien species in Indian flora (Reddy *et al.* 2008).

Habit of plants

In Chithara flora, the number of exotic herb species were greater in numbers than the perennials. Herbs with 61 species contributed maximum percentage (66 percent) of alien species followed by undershrubs (10 percent), shrubs (9 percent), grasses and sedges (8 percent), climbers/creepers (5 percent) (Fig. 35). Trees were least dominant with 2 alien species i.e., *Prosopis juliflora* and *Leucaena leucocephala*. Mishra *et al.* (2015d) and Singh *et al.* (2010c) also reported the dominance of herbaceous alien flora in Delhi and Uttar Pradesh, respectively. They also reported a very few number of tree species as aliens, a similar trend is reflected in the present study.

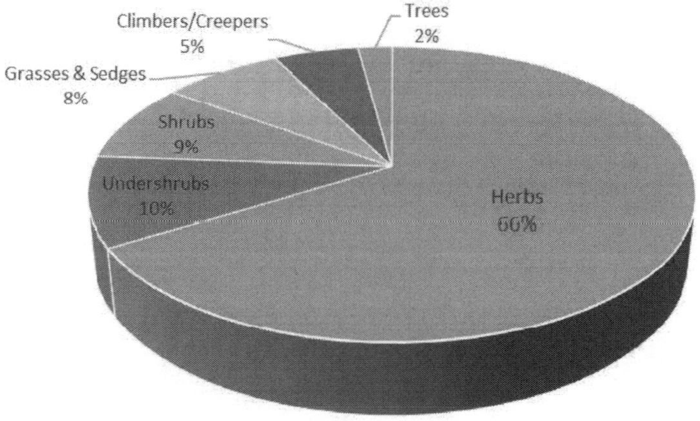

Fig. 35: Pie chart showing the habit wise distribution of alien species

Habit-wise analysis of the Indian flora of invasive alien species also shows that herbs with 87.3 percent predominate followed by shrubs, climbers and trees (Reddy et al. 2008).

Not all alien species in the flora are invasive

Of 272 species recorded in Chithara flora there is preponderance of alien invasive species contributing to about 23 percent of the alien species. The rest of the species though alien did not become invasive. The possible reason for this is ecological. Only some of the alien species possibly become invasive because they lack control by natural enemies, that is the predators, parasites, and pathogens that control the species in its native home range. Often invasive species leave behind the natural enemies that regulate their population growth and in their absence, the invader's populations grow rapidly and it helps to spread such species widely (IUCN 2001). While there is considerable research on the characteristics and traits which allow a species to become invasive, the reasons in individual cases are not very clear. Probably likelihood of an alien species becoming invasive is increased if it is introduced in a geographic area with conducive climatic conditions intentionally and cultivated over a long period of time, since it is more likely to become established and therefore invasive (CBD 2009).

Detection, management and utilization of alien flora

A coordinated planning is required at the national level for early detection and reporting of infestation and spread of new and naturalized weeds by establishment of plant detection network in prominent research institutions, especially all the regional stations of Botanical Survey of India through communication links between taxonomists, ecologists and foresters to monitor and timely eradication/control of alien species. This is important as it will probably be easier to manage invasion of alien species in the initial stages of introduction rather than when it has already spread in a larger geographic area.

Since, many of the alien species have excellent medicinal properties being used for the cure of a large number of diseases, both in traditional and folklore medicines in India and elsewhere their sustainable utilization for medicinal purposes will not only provide livelihood to people, and augment dwindling herbal resources in the country but will also help to manage the invasive population of alien species (Mandal 2011).

5

Medicinal Plants of the Flora and their Properties

Medicinal plants in traditional medicines

The use of natural products with therapeutic properties is as ancient as human civilization and, for a long time, mineral, plant and animal products were the main sources of drugs used for prolonging the life of man and for combating various ailments (De Pasquale 1984). Even today, the traditional systems of medicine continue to be widely practiced and herbal plants have remained main components of the traditional systems of medicine, namely, Ayurveda, Unani, Siddha, Chinese and Homeopathy not only in India but in other countries as well. It is estimated that more than 50,000 species are used for medicinal purposes worldwide, of which almost 13 percent are flowering plants (Schippmann *et al.* 2002). Nevertheless, there is a growing resurgence in the traditional and alternate health care all around the globe. Traditional herbal medicines are being promoted in national health care programmes as such drugs are easily available at low cost and inherently safer than the potent synthetic drugs. This recent global resurgence of interest in herbal medicines has also led to an increased demand for medicinal plants for the manufacture of drugs, their direct use and drug development.

Medicinal plants and drug discovery

It has only been during the past few decades that natural products became secondary in drug discovery and drug development as a result of phenomenal advances in molecular biology and combinatorial chemistry making possible the rational design of chemical compounds to target specific molecules. However, in the past few years there has been a renewed interest in the use of natural compounds and, more importantly, their role as a basis for drug development. There has been growing interest in alternative therapies and the therapeutic use of natural products, particularly those which are derived from plants (Goldfrank *et al.* 1982, Vulto and Smet 1988). This interest in drugs of plant origin is due to several reasons, namely, conventional medicine can be inefficient (e.g. side effects and ineffective therapy), abusive and/or incorrect use of synthetic drugs results in side effects and other problems, a large percentage of the world's population does not have access to conventional pharmacological treatment, and folk medicine and ecological awareness suggest that "natural" products are harmless. However, often the use of these herbal medicines is not always authorized by legal authorities dealing with efficacy and safety procedures, and many published papers point to the lack of quality in the production, trade and prescription of herbal products.

Even then the medicinal herbs continue to play a role both as sources of current drugs and in drug discovery as they are a good source of anti-infective agents, which are not only cost-effective but have fewer side effects as compared to allopathic medicines. In this context, attempts have been made by the World Health Organization to identify all the medicinal plants used globally and listed more than 20,000 species. Recently, the WHO (World Health Organization) estimated that 80 percent of people worldwide rely on herbal medicines for some aspects. Many developing countries all over the world have intensified their efforts in documenting the ethnomedical data and scientific research on medicinal plants. There has been a growing interest in the analysis of plant products which has stimulated intense research on their potential health benefits as they can provide potential bioactive compounds for the development of new "leads" to combat various diseases. Also, there has been preference for synthetic products for pharmacological treatment which led to synthesize the biomolecules obtained from these medicinal plants. About 25 percent of the drugs prescribed worldwide come from plants and many others are synthetic analogues built on prototype compounds isolated from plant species in modern pharmacopoeia. Of the

252 drugs considered as basic and essential by the World Health Organization (WHO), 11 percent are exclusively of plant origin and a significant number are synthetic drugs obtained from natural precursors. It is estimated that 60 percent of anti-tumour and anti-infectious drugs already on the market or under clinical trial are of natural origin (Yue-Zhong Shu 1998). The vast majority of these cannot yet be synthesized economically and are still obtained from wild or cultivated plants.

In 1991, almost half of the best-selling pharmaceuticals were either natural products or their derivatives (O'Neill and Lewis 1993). Based on a World Health Organization survey (Farnsworth *et al.* 1985) and more recent review articles based on literature searches for drugs of natural plant product origin (Farnsworth and Soejarto 1991, Cragg *et al.* 1997), there are at least 121 pharmaceutical compounds currently in use as medicines worldwide that are of plant origin.

Weeds as a source of pharmaceuticals

The role of weeds in the present pharmacopeia has been overlooked, despite significant evidence that weeds in particular, could be an important source of medicines and have a highly significant over representation in indigenous pharmacopoeias in relation to other types of plants (Stepp and Moerman 2001). There is increasing evidence to support the hypothesis that weeds are relatively high in bioactive secondary compounds and are thus likely to hold promise for drug discovery (Stepp 2004). Secondary compounds in weeds are important for a variety of ecological functions. Chief among these are allelopathy, where secondary compounds inhibit germination and growth of other plants (Harborne 1993). The significant representation of weeds in the flora from which drugs are currently derived has been reviewed by Stepp (2004) and he suggested that disturbed habitats may be important areas to search for novel compounds in drug discovery.

Use and diversity of medicinal plants in India

In India, the reference to the curative properties of some herbs in the *Rigveda* appears to be the earliest record of the use of plants as medicines. Ayurveda, the oldest medical system in Indian sub-continent, has alone reported approximately 2,000 medicinal plant species, followed by *Siddha* and *Unani*. The *Charak Samhita*, an age-old written document on herbal therapy, reports on the production of 340 herbal drugs and their indigenous uses. Of the 17,000 species of higher plants, about 7,500 plants are known for their medicinal value; however, only 800 plant species are used in the preparation of herbal drugs. This proportion of medicinal plants is probably the highest proportion of plants used for medical purposes in any country of the world considering the existing flora (Kala *et al.* 2006) (Table 9).

Table 9. Distribution of medicinal plants in the world (After Kala *et al.* 2006)

Country or region	Total number of native species in flora	No. of medicinal plant species reported	% of medicinal plants	Source
World	297000	52885	10	Schippmann *et al.* (2002)
India	17000	7500	44	Shiva 1996
Indian Himalayas	8000	1748	22	Samant *et al.* (1998)

As per Planning Commission (2000) over 8,000 plant species are used in traditional and modern medicine in India, and 90-95 percent collection of medicinal plants is from the wild, of which more than 70 percent collection involves destructive and unscientific extraction. Over exploitation of trade species, destructive way of collection, vulnerability due to anthropogenic pressure are some of the major threats to medicinal plants.

Weeds as medicinal plants in flora

Of the 272 plants recorded in the flora 260 plant species (approx. 96 percent) possessed medicinal properties against one or more number of diseases; the twelve plant species which did not record any medicinal uses were monocots *viz., Commelina forsskalii, Cyperus alopecuroides, Cyperus compressus, Digitaria ciliaris, Eriochloa procera, Leptochloa panicea, Poa annua, Potamogeton crispus, and dicots Mecardonia procumbens, Persicaria lanigera, Trifolium tomentosum, Utricularia stellaris* (Table 12). This is certainly a very high percentage of occurrence of medicinal plants in a small geographic area of Chithara Village Panchayat indicating that the weeds possess high potential of having medicinal properties (Stepp 2004). The highest number of diseases and medicinal properties (>60) were recorded for twenty five plant species *viz., Euphorbia hirta* (121 number of diseases and medicinal properties), *Solanum americanum* (107), *Coccinia grandis* (80), *Cissampelos pareira* (78), *Hyptis suaveolens* (78), *Eclipta prostrata* (76), *Azadirachta indica* (74), *Justicia adhatoda* (73), *Cannabis sativa* (72), *Ziziphus jujuba* (72), *Bacopa monnieri* (71), *Centella asiatica* (71), *Cynodon dactylon* (71), *Tribulus terrestris* (67), Withania *somnifera* (67), *Melia azedarach* (66), *Cyperus rotundus* (66), *Ageratum conyzoides* (65), *Oxalis corniculata* (65), *Phyllanthus amarus* (65), *Argemone mexicana* (63), *Portulaca oleracea* (63), *Solanum surattense* (63), *Verbascum thapsus* (63) and *Sida cordifolia* (62).

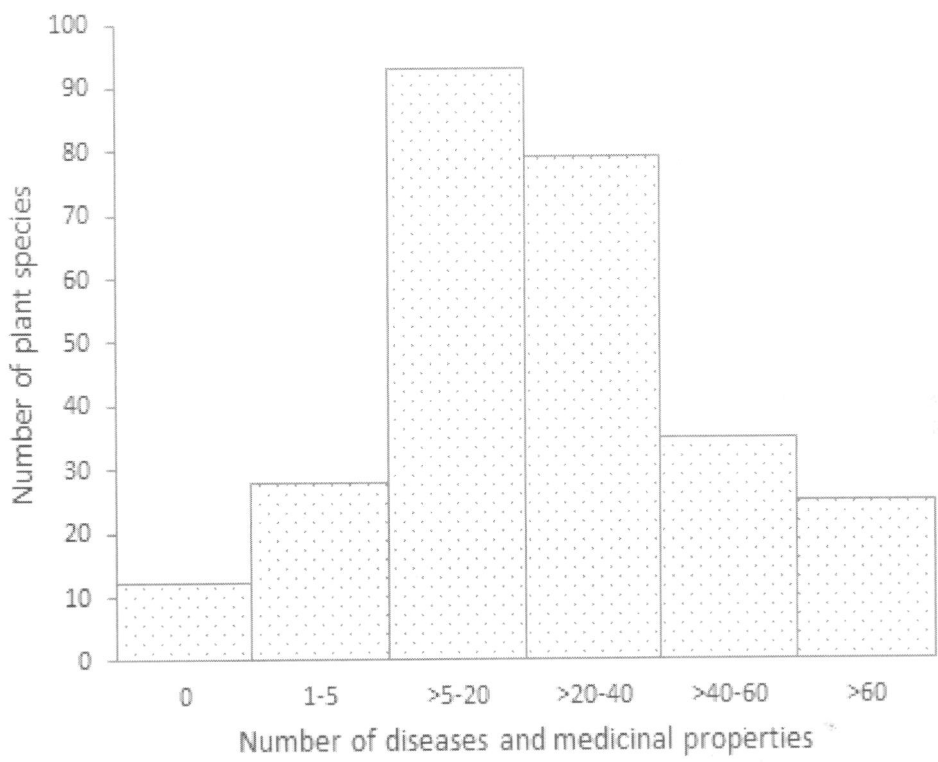

Fig. 36. Number of diseases cured/treated and medicinal properties

Table 10. Medicinal properties and uses of plants recorded in the flora

S.N.	Medicinal Properties and Uses			Number of plants	% of plants
	Legend	**Number of diseases and medicinal properties**	**Star rating**		
1	Outstanding	>60	*****	25	9.19
2	Excellent	>40 - 60	****	35	12.87
3	Very good	>20 - 40	***	80	29.41
4	Good	>5 - 20	**	92	33.82
5	Marginal	1 - 5	*	28	10.29
6	None Known	0	-	12	4.41
	Total plants			272	100

The plant species which were used for treating fewer diseases probably had limited efficacy due to their chemical constituents, especially the secondary metabolites.

Of the top fifteen species, five species *viz., Euphorbia hirta, Solanum americanum, Hyptis suaveolens, Eclipta prostrata* and *Tribulus terrestris* were dominant aliens growing as weed in the study area; which shows that in addition to indigenous, the alien species are equally or more important as medicinal plants for they are being used to treat a number of diseases in various systems of medicines and folklore (Fig. 37).

The medicinal properties and uses of these plants were rated on 1 to 5 star (*) scale (marginal to outstanding) depending upon cure/effective against a number of diseases given in Tables 10 and 12. Twenty five plant species were rated as outstanding having more than 60 number of medicinal properties and uses, 35 as excellent (>40 – 60 number of medicinal properties and uses) and 79 species as very good (>20 – 40 number of medicinal properties and uses) accounting for about 51 percent of the total species; the largest number of species (96) came under Good category having more than 5 to 20 number of medicinal properties and uses (Fig. 36). There were 28 plant species which are known to have very few number of medicinal properties and uses (1 to 5) came under marginal category. There were a total of 12 plant species (approx. 4%) which did not possess any medicinal properties or their potential is not yet explored. The top 15 species having largest number of medicinal properties and uses are depicted in Fig. 37.

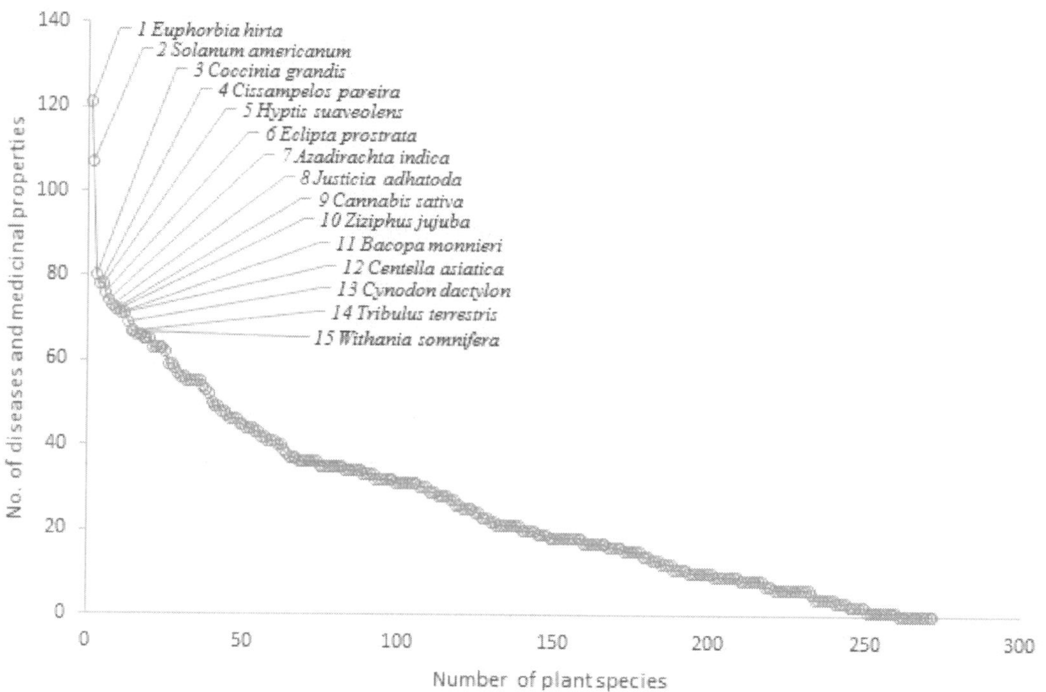

Fig. 37. The dominance curve showing use of 272 plant species against number of diseases and their medicinal properties *(The top 15 dominant species are marked in the curve)*

Medicinal properties and uses of alien plant species

Out of 272 plant recorded in the flora, a total of 92 plants were alien species. Except one plant species, *Mecardonia procumbens* rest of 91 alien species are known to have medicinal properties and have been used in traditional and folklore medicines in India and various other countries. The medicinal properties and uses of these alien plants were also rated on 1 to 5 star (*) scale (marginal to outstanding) depending upon cure/effective against a number of diseases (Tables 11, 12).

Table 11. Medicinal properties and uses of alien plants recorded in the flora

SN	Medicinal Properties and Uses			Number of alien plants	% of alien plants
	Legend	**Number of diseases and medicinal properties**	**Star rating**		
1	Outstanding	>60	*****	9	9.78
2	Excellent	>40 - 60	****	16	17.39
3	Very good	>20 - 40	***	26	28.26
4	Good	>5 - 20	**	35	38.04
5	Marginal	1 - 5	*	5	5.43
6	None Known	0	-	1	1.09
	Total plants			92	100

Nine alien species *viz., Euphorbia hirta, Solanum americanum, Hyptis suaveolens, Eclipta prostrata, Tribulus terrestris, Ageratum conyzoides, Oxalis corniculata, Argemone mexicana* and *Portulaca oleracea* were rated outstanding as they are known to have large number of medicinal properties and uses (>60) (Table 16), both in traditional and folklore medicines in India and elsewhere. Sixteen alien species (17.39 percent) had 'Excellent' medicinal properties and uses, 26 species (28.26 percent) 'Very Good' and 35 alien species (38.04 percent) had 'Good' medicinal properties and uses (Table 11; Fig. 38).

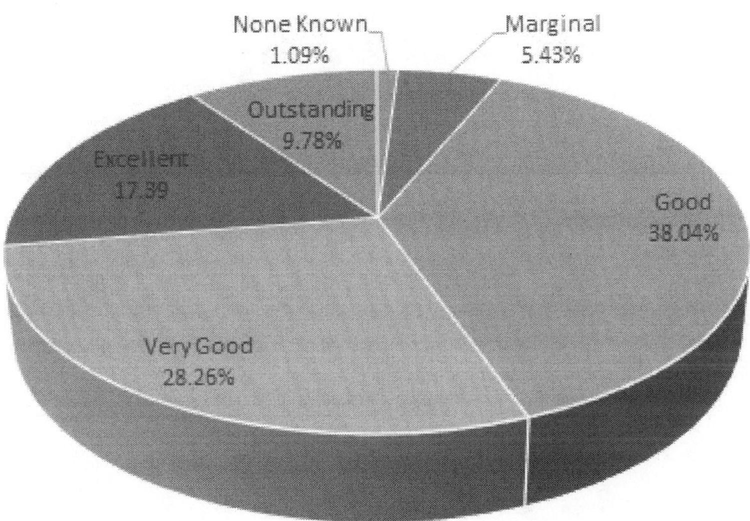

Fig. 38. Pie chart showing the medicinal properties and uses of the alien flora

Five plant species namely *Cyperus difformis, Echinochloa colona, Emex australis, Gnaphalium purpureum* and *Lathyrus aphaca* and *Oxalis debilis* var. *corymbosa* have marginal (least) medicinal properties and uses (Table 12) as reported in the literature. Singh *et al.* (2010c) also found that a large number of alien plant species in Uttar Pradesh are medicinally useful and being utilized by local people for medicinal purposes.

Table 12. Plant species used for the cure of Number of Diseases (ND) and their Medicinal Properties (MP) as recorded in the literature

SN	Name of plant species	ND and MP	Star rating	SN	Name of plant species	ND and MP	Star rating	SN	Name of plant species	ND and MP	Star rating
1	*Abrus precatorius*	55	****	32	*Blumea obliqua* #	10	**	63	*Commelina benghalensis*	32	***
2	*Abutilon indicum*	26	***	33	*Blumea viscosa*	19	**	64	*Commelina forsskalii*	0	-
3	*Acacia nilotica*	35	***	34	*Boerhavia diffusa*	48	****	65	*Convolvulus arvensis*	29	***
4	*Achyranthes aspera*	42	****	35	*Brachiaria reptans*	1	*	66	*Convolvulus prostratus*	29	***
5	*Acmella paniculata*	16	**	36	*Brachiaria ramosa*	1	*	67	*Corchorus aestuans* #	10	**
6	*Ageratum conyzoides* #	65	*****	37	*Butea monosperma*	45	****	68	*Corchorus olitorius* #	19	**
7	*Ageratum houstonianum* #	6	**	38	*Caesulia axillaris*	11	**	69	*Corchorus tridens* #	6	**
8	*Albizia lebbeck*	33	***	39	*Cajanus scarabaeoides*	21	***	70	*Cordia dichotoma*	48	****
9	*Alhagi maurorum*	18	**	40	*Calotropis gigantea* #	55	****	71	*Crateva adansonii subsp. odora*	43	****
10	*Alternanthera paronychioides* #	6	**	41	*Calotropis procera* #	59	****	72	*Crotalaria medicaginea*	4	*
11	*Alternanthera philoxeroides* #	8	**	42	*Cannabis sativa*	72	*****	73	*Crotalaria mysorensis*	1	*
12	*Alternanthera pungens* #	6	**	43	*Capparis sepiaria*	23	***	74	*Croton bonplandianus* #	22	***
13	*Alternanthera sessilis* #	25	***	44	*Capsella bursa-pastoris*	50	****	75	*Cucumis melo var. agrestis*	17	**
14	*Alysicarpus monilifer*	11	**	45	*Cardamine hirsuta* #	7	**	76	*Cuscuta chinensis* #	18	**
15	*Alysicarpus vaginalis*	16	**	46	*Carthamus oxyacantha*	20	**	77	*Cuscuta reflexa* #	39	***
16	*Amaranthus viridis*	37	***	47	*Cayratia trifolia*	36	***	78	*Cyanotis axillaris*	8	**
17	*Ammannia baccifera*	36	***	48	*Celosia argentea* #	28	***	79	*Cyanthillium cinereum*	47	****
18	*Ampelopteris prolifera*	21	***	49	*Cenchrus ciliaris*	6	**	80	*Cynodon dactylon*	69	*****
19	*Anagallis arvensis* #	41	****	50	*Centaurium pulchellum*	15	**	81	*Cyperus alopecuroides*	0	-
20	*Anisomeles indica*	36	***	51	*Centella asiatica*	71	*****	82	*Cyperus compressus*	0	-
21	*Arabidopsis thaliana*	1	*	52	*Ceratophyllum demersum*	18	**	83	*Cyperus difformis* #	1	*
22	*Arenaria serpyllifolia*	14	**	53	*Chenopodium album* #	59	****	84	*Cyperus rotundus*	66	*****
23	*Argemone mexicana* #	63	*****	54	*Chenopodium murale* #	34	***	85	*Cyperus iria* #	9	**
24	*Artemisia scoparia*	32	***	55	*Chloris barbata* #	8	**	86	*Dactyloctenium aegyptium*	15	**
25	*Avena sterilis*	4	*	56	*Chrysopogon zizanioides*	37	***	87	*Dalbergia sissoo*	52	****
26	*Azadirachta indica*	74	*****	57	*Cirsium arvense*	26	***	88	*Datura innoxia* #	56	****
27	*Bacopa monnieri*	71	*****	58	*Cissampelos pareira*	78	*****	89	*Desmodium gangeticum*	46	****
28	*Barleria prionitis*	57	****	59	*Citrullus colocynthis*	36	***	90	*Desmodium triflorum*	36	***
29	*Bidens pilosa* #	34	***	60	*Cleome viscosa* #	40	***	91	*Desmostachya bipinnata*	20	**
30	*Blainvillea acmella* #	9	**	61	*Clerodendrum phlomidis*	36	***	92	*Dichanthium annulatum*	2	*
31	*Blumea laciniata*	15	**	62	*Coccinia grandis*	80	*****	93	*Digera muricata* #	15	**

Table 12. Continued...

SN	Name of plant species	ND and MP	Star rating	SN	Name of plant species	ND and MP	Star rating	SN	Name of plant species	ND and MP	Star rating
94	*Digitaria ciliaris*	0	-	125	*Hemarthria compressa*	3	*	156	*Ludwigia octovalvis* #	17	**
95	*Diospyros montana*	33	***	126	*Holoptelea integrifolia*	44	****	157	*Malva parviflora*	28	***
96	*Dysphania ambrosioides* #	41	****	127	*Hydrilla verticillata*	9	**	158	*Malvastrum coromandelianum* #	28	***
97	*Echinochloa colona* #	2	*	128	*Hydrocotyle sibthorpioides*	35	***	159	*Marsilea quadrifolia*	32	***
98	*Echinochloa crus-galli* #	13	**	129	*Hyptis suaveolens* #	78	*****	160	*Mazus pumilus*	6	**
99	*Eclipta prostrata* #	76	*****	130	*Imperata cylindrica* #	44	****	161	*Mecardonia procumbens* #	0	-
100	*Eichhornia crassipes* #	17	**	131	*Indigofera astragalina* #	6	**	162	*Medicago monantha*	6	**
101	*Eleusine indica*	10	**	132	*Indigofera linifolia* #	9	**	163	*Medicago polymorpha*	9	**
102	*Emex australis* #	5	*	133	*Indigofera linnaei* #	18	**	164	*Medicago lupulina*	8	**
103	*Equisetum ramosissimum*	32	***	134	*Indigofera tinctoria*	37	***	165	*Melia azedarach*	66	*****
104	*Equisetum ramosissimum subsp. debile*	21	***	135	*Ipomoea aquatica*	31	***	166	*Melilotus albus* #	13	**
105	*Eragrostis amabilis*	1	*	136	*Ipomoea carnea* #	31	***	167	*Melilotus indicus*	27	***
106	*Eragrostis japonica*	1	*	137	*Ipomoea coptica*	3	*	168	*Melochia corchorifolia* #	16	**
107	*Erigeron bonariensis* #	23	***	138	*Ipomoea eriocarpa* #	15	**	169	*Merremia hederacea*	18	**
108	*Eriochloa procera*	0	-	139	*Ipomoea obscura* #	13	**	170	*Mimosa rubicaulis*	17	**
109	*Euphorbia heterophylla* #	38	***	140	*Ipomoea pes-tigridis* #	18	**	171	*Mitragyna parvifolia*	29	***
110	*Euphorbia hirta* #	121	*****	141	*Ipomoea triloba*	7	**	172	*Mollugo nudicaulis*	10	**
111	*Euphorbia hypericifolia* #	32	***	142	*Jatropha gossypiifolia* #	21	***	173	*Morus alba*	54	****
112	*Euphorbia prostrata*	11	**	143	*Justicia adhatoda*	73	*****	174	*Mukia maderaspatana*	35	***
113	*Euphorbia serpens* #	8	**	144	*Justicia japonica*	2	*	175	*Murdannia nudiflora*	10	**
114	*Evolvulus nummularius* #	20	**	145	*Lantana camara* #	46	****	176	*Nepeta hindostana*	25	***
115	*Ficus palmata*	17	**	146	*Lathyrus aphaca* #	4	*	177	*Nicotiana plumbaginifolia* #	18	**
116	*Ficus religiosa*	58	****	147	*Launaea procumbens*	18	**	178	*Nymphaea pubescens*	25	***
117	*Fimbristylis dichotoma*	6	**	148	*Lemna perpusilla*	2	*	179	*Nymphoides cristata*	1	*
118	*Fimbristylis ferruginea*	1	*	149	*Lepidium didymum* #	14	**	180	*Oldenlandia corymbosa*	31	***
119	*Fumaria indica*	32	***	150	*Leptochloa panicea*	0	-	181	*Operculina turpethum*	33	***
120	*Gnaphalium purpureum* #	4	*	151	*Leucaena leucocephala* #	21	***	182	*Oxalis corniculata* #	65	*****
121	*Gomphrena celosioides* #	9	**	152	*Leucas cephalotes*	36	***	183	*Oxalis debilis var. corymbosa* #	11	**
122	*Gonostegia pentandra*	12	**	153	*Lindernia ciliata*	15	**	184	*Oxystelma esculentum*	24	***
123	*Grangea maderaspatana* #	35	***	154	*Lindernia crustacea*	30	***	185	*Panicum antidotale*	14	**
124	*Heliotropium ellipticum*	17	**	155	*Ludwigia hyssopifolia*	34	***	186	*Parthenium hysterophorus* #	18	**

Table 12. Continued...

SN	Name of plant species	ND and MP	Star rating	SN	Name of plant species	ND and MP	Star rating	SN	Name of plant species	ND and MP	Star rating
187	*Paspalum distichum*	2	*	216	*Rumex dentatus*	18	**	245	*Stellaria media* #	44	****
188	*Paspalum scrobiculatum*	8	**	217	*Rungia pectinata*	10	**	246	*Streblus asper*	34	***
189	*Peristrophe bicalyculata*	28	***	218	*Saccharum bengalense*	17	**	247	*Stuckenia pectinata*	3	*
190	*Persicaria barbata*	18	**	219	*Saccharum spontaneum* #	25	***	248	*Syzygium cumini*	49	****
191	*Persicaria lanigera*	0	-	220	*Sagittaria guayanensis*	1	*	249	*Tamarix indica*	19	**
192	*Phalaris minor*	4	*	221	*Salix tetrasperma*	21	***	250	*Tephrosia pumila*	12	**
193	*Phoenix sylvestris*	32	***	222	*Scoparia dulcis* #	34	***	251	*Tephrosia purpurea*	55	****
194	*Phragmites karka*	9	**	223	*Senna obtusifolia* #	35	***	252	*Tephrosia villosa*	9	**
195	*Phyla nodiflora*	45	****	224	*Senna occidentalis* #	53	****	253	*Teramnus labialis*	23	***
196	*Phyllanthus amarus*	65	*****	225	*Sesamum indicum*	31	***	254	*Trianthema portulacastrum*	34	***
197	*Phyllanthus maderaspatensis*	41	****	226	*Sesbania bispinosa* #	10	**	255	*Tribulus terrestris* #	67	*****
198	*Phyllanthus reticulatus*	35	***	227	*Sesbania sesban*	27	***	256	*Trichosanthes cucumerina*	46	****
199	*Phyllanthus tenellus* #	19	**	228	*Setaria pumila*	6	**	257	*Tridax procumbens* #	42	****
200	*Physalis angulata* #	55	****	229	*Setaria verticillata*	7	**	258	*Trifolium tomentosum*	0	-
201	*Physalis peruviana* #	31	***	230	*Sida acuta* #	35	***	259	*Triumfetta rhomboidea* #	35	***
202	*Pistia stratiotes* #	30	***	231	*Sida cordata*	33	***	260	*Typha domingensis* #	16	**
203	*Pluchea lanceolata*	20	**	232	*Sida cordifolia*	62	*****	261	*Urena lobata* #	49	****
204	*Poa annua*	0	-	233	*Silene conoidea*	6	**	262	*Utricularia stellaris*	0	-
205	*Polygonum plebeium*	12	**	234	*Sisymbrium irio*	22	***	263	*Vallisneria spiralis*	8	**
206	*Portulaca oleracea* #	63	*****	235	*Solanum americanum* #	107	*****	264	*Verbascum chinense*	3	*
207	*Portulaca quadrifida* #	31	***	236	*Solanum surattense*	63	*****	265	*Verbascum thapsus*	63	*****
208	*Potamogeton crispus*	0	-	237	*Solanum torvum* #	43	****	266	*Veronica agrestis*	11	**
209	*Prosopis cineraria*	31	***	238	*Sonchus asper* #	24	***	267	*Veronica anagallis-aquatica*	16	**
210	*Prosopis juliflora* #	21	***	239	*Sonchus brachyotus*	17	**	268	*Vicia sativa*	20	**
211	*Ranunculus sceleratus*	40	***	240	*Sorghum halepense*	6	**	269	*Withania somnifera*	67	*****
212	*Rhynchosia capitata*	2	*	241	*Spergula arvensis*	12	**	270	*Xanthium strumarium* #	55	****
213	*Rhynchosia minima*	21	***	242	*Spermacoce articularis* #	31	***	271	*Ziziphus jujuba*	72	*****
214	*Ricinus communis* #	56	****	243	*Sphenoclea zeylanica*	4	*	272	*Ziziphus nummularia*	30	***
215	*Rorippa palustris*	4	*	244	*Spirodela polyrrhiza*	10	**				

#Alien species; ND- Number of Diseases; MP- Medicinal Properties

Number of diseases cured by plants

If we consider the efficacy of plant species in curing various diseases it is found that 260 medicinal plants in the flora are known to have 598 medicinal properties and uses (Appendix 8). There are many plants which were being used for treating the same disease. It is amazing to find that plants have cure for a wide variety of diseases/ailments from simple ones such as cuts and wounds to heart ailments, blood pressure, asthma, bronchitis, rheumatism, epilepsy, cancer, etc. A large number of plants possessed anti-inflammatory properties (140 plants) used for various types of inflammation and antipyretic properties (136 plants) used for treating fevers. A total of 120 plants are being used as diuretic. A large number of plant species (116) were used to cure various types of skin diseases (itching, skin allergy, lesions and rashes, skin sores, skin plagues, skin ulcers, blisters, skin tumour, pustules, papillomas, pruritus) followed by wounds (112) and rheumatic diseases (102). Ninety nine plant species had expectorant properties used in cough whereas ninety seven plants were vermifuge/anthelmintic used to prevent different types of worms such as roundworms, threadworms, tapeworms, hookworms, pinworms, intestinal worms etc. A total of 97 plants had antidiarrhoeal properties used to cure diarrhoea followed by laxative (93), astringent (87), antiulcer (86), antidysenteric (84), analgesic (80), antiasthmatic (78) and antioxidant (75). The flora plants possess properties to cure even ailments/diseases such as abdominal pain (73 plants), diabetes (73), bowel complaints (72), Jaundice (66), Cancer (62), Eye diseases (57), Urinary tract infections/urinary disorders (55), liver complaints/diseases (54) and Malaria (51). Altogether 69 plant species had antibacterial properties whereas 51 plants were known for antifungal and 28 plants for antiviral properties.

Besides, there were plants which were used for appetite loss (46), kidney and other stones (44), gonorrhoea (43 plants), toothache/teeth problems (37), vomiting and nausea (36), scabies (35), indigestion (33), abortion (29), anaemia (29), hepatitis (26), syphilis (24), epilepsy (23) and tuberculosis (23). The details of medicinal properties of plants and their uses in curing different ailments/diseases is given in Appendix 8.

Need for accelerated research on drug development and clinical trials

With the emerging worldwide interest in adopting and studying traditional systems and exploiting their potential based on different health care systems, the evaluation of the rich heritage of traditional medicine is essential. Hence, the efficacy of medicinal plants is required to be addressed through interdisciplinary research. The use of plant drugs also demands correct identification and characterization as their safety and efficacy depend on the use of proper plant part and its biological potency, which in turn depends upon the presence and nature of required active compounds/secondary metabolites. It is imperative to use phytochemical parameters to screen and analyze the bioactive compounds, not only for the quality control of the crude drugs but also for the elucidation of their therapeutic mechanisms.

As evident, India has great pool of diverse medicinal plant sources, a long and well characterized traditional medicinal system which opens enormous opportunities for new drug discoveries which can bring India on the forefront of drug development. A number of drugs have been obtained from the plant sources and several others have been discovered by using natural substances as lead. Potential of medicinal weeds as important sources of drugs due to their unique habitats in which they grow has been supported by Stepp (2004). Of the 260 medicinal plants in Chithara flora, five plant species *viz., Cannabis sativa, Cassia* spp., *Centella asiatica, Cissampelos pareira,* and *Ricinus communis* are being used for the manufacture of various drugs in the pharmaceutical industries (Table 13).

Table 13. Weeds in flora from which pharmaceuticals (drugs) have been derived and their clinical use/action (Modified from Stepp 2004).

SN	Plant species recorded in the flora	Drug	Action/Clinical Use
1.	*Cannabis sativa* L.	a-Tetrahydrocannabinol (THC)	Antiemetic, decrease occular tension
2.	*Cassia* spp. (*Senna occidentalis* and *Senna obtusifolia*)	Danthron	Stimulant
3.	*Centella asiatica* L.	Asiaticoside	Vulnerary
4.	*Cissampelos pareira* L.	Cissampeline	Skeletal muscle relaxant
5.	*Ricinus communis* L.	Castor oil	Stimulant and irritant laxative, Induces laxation and uterus contraction

In recent years, research on these areas is increasing and lot more drug/formulations are being investigated and also clinical trials of formulated drugs are underway in various public or private sector organizations in India. Some of the plants recorded in Chithara flora, which have been used in Indian traditional medicine since ancient time, have been reported to yield potent drugs (Sen and Chakraborty, 2017). Some of these important drug yielding weed plants are *Adhatoda vasica* (vasicine and vasicinone), *Bacopa monnieri* (bacosoids), *Tribulus terrestris* (protodioscin), *Withania somnifera* (withanolides), *Centella asiatica* (asiaticoside), *Boerrhavia diffusa* (boeravinones) and *Azadirachta indica* (nimbidin). Considering such immense potential, accelerated research on drug discovery from these and several other weeds having medicinal properties and being used in different folklore and traditional system of medicines followed by clinical trials is urgently needed.

Conservation and utilization of medicinal weeds

Considering the importance of weeds as medicinal plants as evident from their use in curing numerous disease problems, their conservation and sustainable use require immediate focus in the context of conserving biodiversity. But while the demand for medicinal plants is growing, some of them are increasingly facing constant threat of extinction under the duress of massive extraction and habitat degradation by drug companies and local people.

India's share of the world herbal trade is estimated to be less than 1 percent despite its rich biodiversity, traditional knowledge and heritage of herbal medicines. Hence, a scientific approach for exploiting the weeds in a sustainable manner and large scale cultivation and propagation of medicinal plants is needed to meet the market demand and to utilize them in a sustained manner for their use in traditional medicines and drug development. Medicinal plant species which are endangered or rare should be identified and conserved through the coordinated effort of *in situ* and *ex situ* strategies. The important wild medicinal plants growing as weeds having potent properties should be explored to bring them under cultivation at village level.

6

Horticultural Flora

A. Horticultural plants of the SNU Campus

The University has extensive landscaped gardens, tree avenues, lawns to give a green ambience to the Campus. From the aesthetic point of view and to merge with the architecture appropriate, horticultural plants belonging to 98 species have been introduced in the gardens of the Campus for landscaping. All these species belong to 41 families and 84 genera. The details of species are given in Appendix 5.

Trees which are mostly planted in avenue planting, parking areas and general landscaping have great diversity represented by 38 species belonging to 32 genera under 19 families. Among the trees the most dominant family was Leguminosae having 10 species followed by Moraceae and Myrtaceae (4 species each) and Bignoniaceae (3 species). Shrubs planted individually and used for hedge accounting for 28 species belonging to 26 genera under 16 families; the dominant families are Apocynaceae (5 species), Euphorbiaceae and Leguminosae (3 species each). Perennial herbs which are used as ground cover plants are represented by 10 species belonging to 10 genera and 9 families. The campus landscaping also have 8 species of climbers, 9 species of palms, 1 species of banana, 1 species of bamboo and 3 species of Gymnosperms (Table 14).

Table 14. Taxonomic diversity in horticultural plants of the Campus

SN	Habit of plant species	No. of species	No. of genera	No. of families	Dominant families
1.	Trees	38	32	19	Leguminosae (10 species), Moraceae and Myrtaceae (4 species each), Bignoniaceae (3 species)
2.	Shrubs/Shrubs Hedge	28	26	16	Apocynaceae (5 species), Euphorbiaceae and Leguminosae (3 species each)
3.	Perennial Herbs (Ground Cover Plants)	10	10	9	Asparagaceae (2 species)
4.	Climbers	8	6	6	Oleaceae and Passifloraceae (2 species each)
5.	Palm	9	9	2	Arecaceae (8 species)
6.	Banana	1	1	1	Musaceae (1 species)
7.	Bamboo	1	1	1	Poaceae (1 species)
8.	Gymnosperms	3	3	3	Cupressaceae, Cycadaceae, Pinaceae (1 species each)

In future expansion of the garden in faculty housing in the Campus, it is envisaged to introduce some other flowering tree species for landscaping such as *Acacia pectinata* Hoffmanns., *Senna pallida* (Vahl) H.S.Irwin & Barneby (Syn. *Cassia biflora* L.), *and Saraca asoca* (Roxb.) Willd (*Sita Ashok*), etc. Besides, some of the common fruit trees such as Java Plum (*Jamun*), Mulberry (*Shahtoot*), Guava, Star fruit (*Kamrakh*), Chinese orange, Pomegranate (*Anar*), Indian gooseberry (*Amla*), Plum (*Aarhu*), Custard apple (*Sharifa*), Moringa, Jack fruit (*Kathal*), etc. will also be planted at appropriate places. This will further expand the floristic diversity of the Campus.

B. Plants in Thematic Botanic Garden in the SNU Campus

Broadly speaking a Botanical Garden or Botanic Garden is a garden dedicated to the collection, cultivation and display of a wide range of plants labelled with their botanical names. It may contain specialized plant collections such as cacti and other succulent plants, medicinal plant gardens, plants from particular parts of the world, rare and endangered plants for conservation, and so on; there may be greenhouses, shade houses, again with special collections such as tropical plants, alpine plants, or other exotic plants. In a way Botanical Gardens help in *ex-situ* conservation of plant diversity.

According to the definition of Botanic Gardens Conservation International (BGCI) in the International Agenda for Botanic Gardens in Conservation (IABGC) (2000), *"Botanic Gardens are institutions holding documented collections of living plants for the purposes of scientific research, conservation, display and education."*

Luca Ghini was the first person to establish a Botanical Garden on scientific lines in 1543 at Pisa in Italy. At present there are over 800 important Botanical Gardens in the world. Currently, there are 1775 Botanic Gardens and Arboreta in 148 countries around the world with many more under construction or being planned. In India, there are 30 major Botanic Gardens of which the oldest is Calcutta Botanic Garden established in 1787 by Colonel Robert Kyd, an army officer of the British East India Company. The Calcutta Botanic Garden, previously known as Royal Botanic Garden, Calcutta, situated in Shibpur, Howrah near Kolkata, has been rechristened as Acharya Jagdish Chandra Bose Indian Botanic Garden. The garden exhibits a wide variety of rare plants and a total collection of over 12,000 specimens spread over 109 hectares. It is under Botanical Survey of India (BSI) of Ministry of Environment, Forests and Climate Change, Government of India.

Role of Botanic Gardens

1. *Taxonomic Studies*

Botanic Gardens provide valuable taxonomic information on various plants belonging to local as well as exotic flora, rare and endangered plant species. They act as "outdoor laboratories" for students and researchers.

2. *Botanical Research*

Botanic Gardens provide taxonomically identified plant specimens, seeds, flowers, fruits for further research in molecular characterization, isolation of bio-molecules of medicinal plants, etc.

3. *Conservation*

Botanical Gardens conserve and propagate rare species and genetic diversity.

4. *Education, Recreation and Awareness*

They supply facilities for courses in local flora, horticulture, hybridization, plant propagation, etc. Botanic Gardens frequently organize educational programs, including workshops, training programs for teachers, students, naturalists, etc.

Thematic Botanic Gardens are unique as besides playing a crucial role in conservation and research on various genotypes they provide an opportunity to people to see a range of plants belonging to the same category at one place which is otherwise not possible in classical Botanical Gardens.

Botanic Gardens Conservation International (BGCI), UK is a membership organization representing a network of 500 botanic gardens in more than 100 countries, including the largest and most prominent gardens in the world. BGCI is the largest plant conservation network in the world, which aims to collect, conserve, characterize and cultivate samples from all of the world's plants diversity. The following Gardens of Uttar Pradesh and Delhi NCR have the membership of BGCI.

i) Botanical Garden, Agra College, Agra, Uttar Pradesh
ii) Aligarh Muslim University Botanic Garden, Aligarh, Uttar Pradesh
iii) Botanical Garden, Delhi University, Delhi
iv) Botanical Garden, National Bureau of Plant Genetic Resources (NBPGR), Delhi
v) CSIR National Botanical Research Institute (NBRI), Lucknow, Uttar Pradesh
vi) Botanic Garden of Indian Republic, Noida, Gautam Budh Nagar, Uttar Pradesh

Thematic Botanic Garden

Considering the importance of Botanic Garden in education, research, genepool conservation and awareness it was decided to establish a Thematic Botanic Garden in the campus of Shiv Nadar University. Developmental activities in the area allotted for the garden were initiated in February 2014 with soil filling and levelling. In the meantime plants for the garden, mostly indigenous, were procured from Government Sundar nursery, Lodhi Garden Nursery, New Delhi, forest department nurseries situated in Ghaziabad Division, Surajpur Division and Meerut Division in Uttar Pradesh and NDMC nurseries located at Purana Quila and Lodhi Road Gardens, New Delhi and private nurseries in Greater Noida. Plants were also procured from Forest Research Institute (FRI), Dehradun; Arid Forest Research Institute and Central Arid Zone Forest Research Institute, Jodhpur; Kerala Forest Research Institute, Peechi, Kerala; Botanical Survey of India Botanical Garden, Andaman & Nicobar Islands, and private nurseries in Lal Bagh, Bangalore; Bikaner, Jaisalmer, Udaipur in Rajasthan; Srinagar, and Jammu & Kashmir.

Fig. 39: Solar water pump in the Botanic Garden (2015)

The planting in the garden started as per the layout plan in the following monsoon of 2014 and it is still continuing depending upon the availability of species. *Karonda* (*Carissa carandas*) hedge, 1 m in width, was planted all around the Garden as a bio-hedge. A boundary plantation consisting of a mix of large trees, small trees and shrubs was planted leaving a gap of 1 m from the hedge. To increase the diversity of plant species in the Garden efforts were made to procure unique and rare or uncommon species. A number of rare and economically important species have been introduced in the Garden (Table 15). Details of all the plants introduced in the garden till date, including scientific names, family, common names and Hindi names are given in Appendix 6.

Two solar water pumps (Fig. 39) have been installed in Botanic Garden for irrigation purpose as well as for lighting in the night.

Fig. 40: Diversity of plants in Boundary Plantation in Botanic Garden-1; (A) *Ceiba speciosa* **(B)** *Bombax ceiba* **(C)** *Putranjiva roxburghii* **(D)** *Pterospermum acerifolium*

Fig. 41: Diversity of plants in Boundary Plantation in Botanic Garden-2; (A) ***Delonix regia*** **(B)** ***Lagerstroemia speciosa*** **(C)** ***Cassia fistula*** **(D)** ***Markhamia lutea*** **(E)** ***Fernandoa adenophylla*** **(F)** ***Jacaranda mimosifolia***

Boundary Plantation

The Garden has been provided with a bio-fencing of *Carrisa carandus (Karonda)* plantation on all the four sides. There is a boundary plantation around the Botanic Garden consisting of a mix of large tree, small tree and shrubby species in between. The boundary plantation consists of a total of 88 plant species belonging to 32 families and 72 genera. Of the 88 plant species, 64 belong to small to large-sized tree and 24 shrub species (Appendix 6).

Table 15. Some of the rare and economically important species introduced in the Garden

SN	Name of Plant Species	Family	Common Name	Hindi Name
1	*Acacia catechu* (L.f.) Willd.	Leguminosae	Cutch Tree	खैर
2	*Acacia concinna* (Willd.) DC.	Leguminosae	Soap Pod	शिकाकाई
3	*Adansonia digitata* L.	Malvaceae	Baobab, Kalpavriksha	गोरख इमली
4	*Alangium salviifolium* (L.f.) Wangerin	Cornaceae	Sage-Leaved Alangium	अंकोल
5	*Anogeissus pendula* Edgew.	Combretaceae	Buttontree	करधई
6	*Artocarpus lacucha* Buch.-Ham.	Moraceae	Monkey Fruit, Barhal	बरहल
7	*Barringtonia acutangula* (L.) Gaertn.	Lecythidaceae	Barringtonia	...
8	*Cinnamomum camphora* (L.) J.Presl	Lauraceae	Camphor Tree	कपूर
9	*Commiphora wightii* (Arn.) Bhandari	Burseraceae	Indian Bdellium, Guggul	गुग्गुल
10	*Corymbia citriodora* (Hook.) K.D.Hill & L.A.S.Johnson	Myrtaceae	Lemon-Scented Eucalyptus	सफ़ेदा
11	*Crateva adansonii* subsp. *odora* (Buch.-Ham.) Jacobs	Capparaceae	Garlic Pear Tree, Caper Tree, Three Leaf Caper	बर्ना, बर्नी
12	*Dillenia indica* L.	Dilleniaceae	Elephant Apple	चलता करमबेल
13	*Elaeocarpus serratus* L.	Elaeocarpaceae	Rudraksha	रुद्राक्ष
14	*Equisetum ramosissimum* Desf.	Equisetaceae	Horsetail, Branched Horsetail	...
15	*Ginkgo biloba* L.	Ginkgoaceae	Ginkgo, Maidenhair Tree	...
16	*Hylocereus undatus* (Haw.) Britton & Rose	Cactaceae	Dragon Fruit	ड्रैगन फल
17	*Limonia acidissima* Groff	Rutaceae	Wood Apple	कैथ
18	*Madhuca longifolia* var. *latifolia* (Roxb.) A.Chev.	Sapotaceae	Indian Butter Tree	महुआ
19	*Mallotus nudiflorus* (L.) Kulju & Welzen	Euphorbiaceae	False White Teak	पिण्डार
20	*Mesua ferrea* L.	Calophyllaceae	Ironwood, Indian Rose Chest-nut	नागकेसर, नागचंम्पा
21	*Mitragyna parvifolia* (Roxb.) Korth.	Rubiaceae	Kaim, Kadamb	कैम
22	*Mussaenda erythrophylla* Schumach. & Thonn.	Rubiaceae	Red Flag Bush	बेदिना
23	*Operculicarya decaryi* H. Perrier	Anacardiaceae	Elephant Tree	...
24	*Osmanthus fragrans* Lour. (Kashmir)	Oleaceae	Sweet Osmanthus, Fragrant Olive	सिलंग
25	*Persea americana* Mill.	Lauraceae	Avocado, Alligator Pear	...
26	*Platanus orientalis* L.	Platanaceae	Oriental Plane, Chinar	चिनार
27	*Pterocarpus santalinus* L.f.	Leguminosae	Red Sandalwood, Saunder-swood	रक्तचंदन

Table 15. continued...

SN	Name of Plant Species	Family	Common Name	Hindi Name
28	*Pterygota alata* (Roxb.) R.Br.	Malvaceae	Buddha Coconut	...
29	*Quercus griffithii* Hook.f. & Thomson ex Miq.	Fagaceae	Oak	...
30	*Russelia equisetiformis* Schltdl. & Cham.	Plantaginaceae	Fountain Bush, Coral Plant	रसीली
31	*Salix alba* L.	Salicaceae	White Willow	...
32	*Santalum album* L.	Santalaceae	Sandalwood	चंदन
33	*Sapindus mukorossi* Gaertn.	Sapindaceae	Chinese Soapberry	रीठा, रिष्ट
34	*Salvadora oleoides* Decne.	Salvadoraceae	...	पीलू
35	*Salvadora persica* L.	Salvadoraceae	Toothbrush Tree, Miswak	मिस्वाक, पीलू
36	*Saraca asoca* (Roxb.) Willd.	Leguminosae	Sita Ashok	सीता अशोक
37	*Vitex trifolia* L.	Lamiaceae	Indian Wild Pepper, Simple-Leaf Chaste Tree	संभालू

Total plant diversity introduced

A total of 498 plant species (including varieties and subspecies) have been introduced in Thematic Botanic Garden. All these species belong to 111 plant families and 335 genera (Table 16).

Table 16. Taxonomic diversity in the Botanic Garden

SN	Introduced in	No. of taxa	No. of genera	No. of families	Dominant families
1	Boundary Plantation	88	72	32	Leguminosae (16 taxa) , Bignoniaceae (8), Apocynaceae (7), Malvaceae and Moraceae (5 each)
2	Theme Gardens	334	225	88	Arecaceae (29), Leguminosae (28), Moraceae (20), Poaceae (14), Rutaceae (13), Compositae, Euphorbiaceae and Lamiaceae (11 each).
3	Other plants in the Garden	97	69	39	Leguminosae (16), Malvaceae (11), Apocynaceae (6), Euphorbiaceae and Lamiaceae (5 each)
4	Indoor plants/Cover plants	26	22	15	Acanthaceae and Asparagaceae (4 each)
	Total	**498** (545)*	**335** (393)*	**111** (174)*	Leguminosae (48), Arecaceae (29), Malvaceae (25), Moraceae (20), Euphorbiaceae and Lamiaceae (18 each), Apocynaceae (17)

**Due to repetition of taxa, genera and families, the total numbers are higher as given in brackets.*

The total introduced plants (498 taxa, 335 genera and 111 families) in the Botanic Garden are tabulated in Appendix 6 along with their families, common name and Hindi name. Illustrations of some of the plants of the Garden are given in Fig. 40, 41 (Diversity of plants in boundary plantation), Fig. 42, 43 (Diversity of rare and uncommon plants), Fig. 45 (Diversity of *Terminalia*), Fig. 46 (Diversity of shrub species), Fig. 47 (Diversity of climbing plants), Fig. 48 (Diversity of indoor/seasonal plants), Fig. 51 (Diversity of fruit plants), Fig. 52 (Diversity of palms), Fig. 53 (Diversity of Gymnosperms), Fig. 55 (Diversity of medicinal plants), Fig. 56 (Diversity of xerophytes) and Fig. 57 (Exotic/bamboo/aquatic plants).

Fig. 42: Diversity of rare (uncommon) plants in Botanic Garden-1; (A) ***Vitex trifolia*** **(B)** ***Platanus orientalis*** **(C)** ***Saraca asoca*** **(D)** ***Madhuca longifolia*** **var.** ***latifolia*** **(E)** ***Crateva adansonii*** **subsp.** ***odora*** **(F)** ***Equisetum ramosissimum***

Fig. 43: Diversity of rare (uncommon) plants in Botanic Garden-2; (A) *Adansonia digitata* **(B)** *Cinnamomum camphora* **(C)** *Eriobotrya japonica* **(D)** *Elaeocarpus serratus*

Theme Gardens established in the Thematic Botanic Garden

Of the 18 themes proposed the following eight Theme Gardens have been established in the Botanic Garden.

1. Fruit Plants Garden
2. Palm Garden (Palmetum)
3. Gymnosperms Garden
4. Bamboo Garden (Bambusetum)
5. Medicinal Plants Garden
6. Xerophytes Garden (Arid & Semi-Arid Zone Plants)
7. Aquatic Plants Garden and Conservation Area
8. Ficus Garden

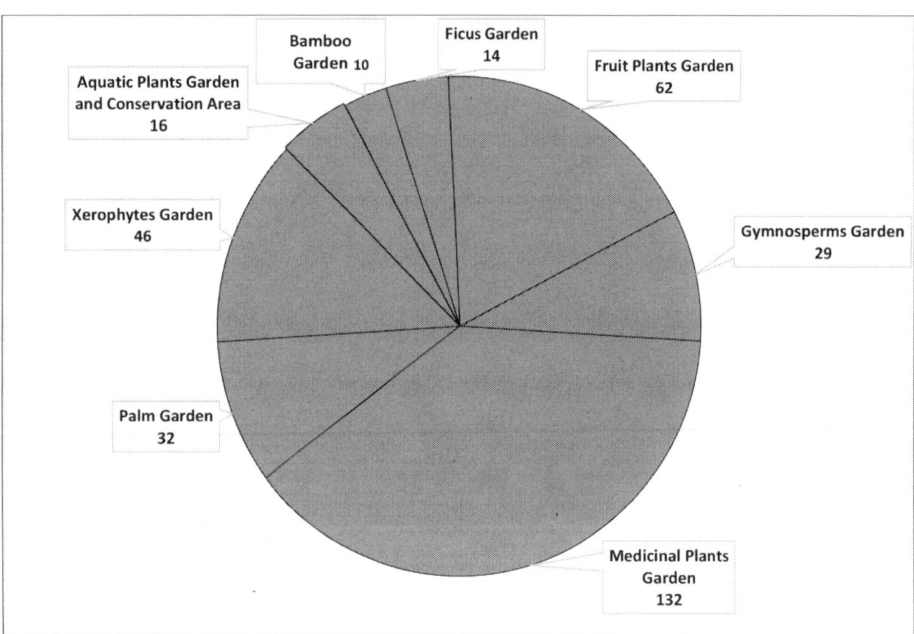

Fig. 44: Established theme in Botanic Garden along with number of taxa

Plants of various species under eight different themes such as Fruit Plants Garden (62 variety of plant species), Palm Garden (32 species), Gymnosperms Garden (29 species), Bamboo Garden (10 species), Medicinal Plants Garden (132 plant species), Xerophytes Garden (46 species), Aquatic Plants Garden (16 species), Ficus Garden (14 species) were planted (Fig. 44).

Altogether 341 taxa have been planted in eight established Theme Garden (Table 17: Fig. 49). Out of 341 taxa, seven species namely *Aloe vera, Bryophyllum pinnatum, Calotropis procera, Cordia dichotoma, Moringa oleifera, Phyllanthus emblica, Tamarindus indica* have been planted in more than 1 Theme Garden.

The total plants (334 taxa, 225 genera and 88 families) in Theme Gardens are tabulated in Appendix 7 along with their families, common name and Hindi name.

Fig. 45: Diversity of Terminalia in Botanic Garden; (A) ***Terminalia arjuna*****- Arjun (B)** ***Terminalia chebula*****- Harad (C)** ***Terminalia bellirica*****- Baheda (D)** ***Terminalia mantaly***

Fig. 46: Diversity of shrub species in Botanic Garden; (A) ***Senna alata*** **(B)** ***Plumeria alba*** **(C)** ***Plumeria pudica*** **(D)** ***Plumeria rubra*** **(E)** ***Tecoma castanifolia***

Table 17. Taxonomic diversity in Theme Gardens

SN	Theme Garden	No. of taxa	No. of genera	No. of families	Dominant families
1	Fruit Plants Garden	62	35	24	Rutaceae (11 taxa), Myrtaceae and Rosaceae (7 each), Moraceae (6)
2	Palm Garden (Palmetum)	32	27	4	Arecaceae (29)
3	Gymnosperms Garden	29	18	8	Cupressaceae (9), Pinaceae (6), Zamiaceae (5), Araucariaceae (4)
4	Bamboo Garden (Bambusetum)	10	2	1	Poaceae (10)
5	Medicinal Plants Garden	132	106	47	Legumonosae (13), Compositae and Lamiaceae (11 each), Solanaceae (9), Amaranthaceae and Malvaceae (7 each)
6	Xerophytes Garden (Arid & Semi-Arid Zone Plants)	46	31	18	Leguminosae (12), Euphorbiaceae (6), Asparagaceae (5), Cactaceae (4)
7	Aquatic Plants Garden and Conservation Area	16	16	12	Araceae (3 species), Hydrocharitaceae and Potamogetonaceae (2 each)
8	Ficus Garden	14	1	1	Moraceae (14)
	Total	**334** (341)*	**225** (236)*	**88** (115)*	Arecaceae (29), Leguminosae (28), Moraceae (20), Poaceae (14), Rutaceae (13), Compositae, Euphorbiaceae, Lamiaceae (11 each),

**Due to repetition of taxa, genera and families, the total numbers are higher as given in brackets. .*

Fig. 47: Diversity of climbing plants in Botanic Garden; (A) ***Passiflora edulis*** **(B)** ***Allamanda cathartica*** **(C)** ***Clitoria ternatea*** **(D)** ***Ipomoea cairica*** **(E)** ***Combretum indicum***

Fig. 48: Diversity of indoor and seasonal plants in Botanic Garden; (A) *Hibiscus rosa-sinensis* (red) (B) *Hibiscus rosa-sinensis* (white) (C) *Adenium obesum* (D) *Tradescantia pallida* (E) *Chrysanthemum* sp. (F) *Mesembryanthemum* sp.

(i) Fruit Plants Garden

India is the largest producer of fruits in the world and is known as fruit basket of world. The major fruits grown in India are Mango, Grapes, Apple, Apricot, Orange, Banana, Citrus, Avocado, Guava, Litchi, Papaya, Pear, Plum Pomegranate, Pineapple, Almond, Walnut, Water melons, etc. Many of the traditional fruits such as *Khirnee, Barhal, Kaith, Phalsa,* etc. are getting disappeared due to preference of younger generation to other fruits, non-availability in the market due to short supply and changing socio-economic conditions. The younger generation has not even heard the names of some of these fruits or even if they have heard about the fruit they have no idea about the plant or the fruit as such. This fruit conservatory will provide an opportunity to see most of the fruit plants at one place.

Fig. 50: A view of Fruit Plants Garden in the Botanic Garden

In Fruit Plants Garden (Fig. 50), a total of 62 variety of fruit plants such as *Aonla, Amra,* Apple, *Bel, Ber,* Banana, *Barhal,* Chinese orange, Custard apple, Dragon fruit, Fig (3 types), Grapes, Guava (4 types), Jack fruit, *Jamun,* Rose apple, *Jangal Jalebi, Karonda, Khirnee, Lasora,* Lemon, Litchi, *Kagzi neebu*/lime, Sweet lemon, Mandarin (Kinnow), Orange, Mango (*Amrapali* and *Dasheri*), Papaya, Passion fruit (purple and red), Pear , *Phalsa,* Pomegranate, Pomelo (*Chakotara*), Star fruit (*Kamrakh*), Sapota, Walnut, West Indian cherry, Wood apple (*Kaith*), etc. have been planted.

(ii) Palm Garden (Palmetum)

Palms are woody monocotyledons belonging to the family Arecaceae (Palmae). Palms are one of the multi-use monocots confined to tropical regions of the world. They are a natural group of plants with a characteristic appearance that enables to recognize them without great difficulty. Palms are widely regarded as being second only to the grasses in economic importance and form a vital component of the forest and agricultural ecosystems, providing a range of products necessary for daily life. The aesthetic value of palms is no less important than their traditional and commercial values. From India 22 genera of palms and 105 species have been reported.

Altogether, 32 species of palms have been introduced in the Botanic Garden, the largest collection in NCR region such as; *Areca catechu* (Supari palm), *Beaucarnea recurvata* (Nolina palm), *Bismarckia nobilis* (Bismarckia palm), *Butia capitata* (Jelly palm), *Caryota mitis* (Fish tail palm), *Chamaerops humilis* (Chamaerops palm), *Copernicia prunifera* (Copernicia palm), *Dypsis lutescens* (Areca palm), *Livistona sp.* (Livistona palm), *Phoenix sylvestris* (Date palm), *Ravenala madagascariensis* (Travellers palm), *Rhapis excelsa* (Rhapis palm), *Roystonea regia* (Royal palm), *Trachycarpus fortunei* (China palm), *Wodyetia bifurcata* (Fox tail palm), etc.

(iii) Gymnosperms Garden

Gymnosperms, woody perennials, either shrubs or trees, are a group of seed-producing vascular plants that includes conifers, pines cycads, Ginkgo, and Gnetales. The name Gymosperm means 'naked seeds' which is derived from the Greek composite word 'gymnos' meaning naked and 'sperma' meaning seed, based on the unenclosed condition of their seeds. Their naked condition stands in contrast to the seeds and ovules of flowering plants (angiosperms), which are enclosed within an ovary. It is widely accepted that the gymnosperms originated in the late Carboniferous period, replacing the lycopsid rainforests of the tropical region.

The Garden has a total of 29 Gymnosperms (7 cycads and 22 others including pines) belonging to 18 genera and 8 families. The important Gymnosperms are; *Abies pindrow* (West himalayan fir), *Agathis robusta* (Queensland kauri pine), *Araucaria bidwillii* (Bunya pine), *Araucaria cunninghamii* (Hoop pine), *Cedrus deodara* (Devdar), *Cryptomeria japonica* (Japanese cedar), *Cycas circinalis* (Queen sago), *Cycas revoluta* (Japanese sago palm), *Dioon edule* (Chestnut dioon), *Dioon spinulosum* (Giant dioon), *Ginkgo biloba* (Maidenhair tree), *Juniperus indica* (Black juniper), *Pinus roxburghii* (Chir pine), *Podocarpus neriifolius* (Brown pine), *Taxus baccata* (English yew), *Zamia furfuracea* (Mexican cycad), etc.

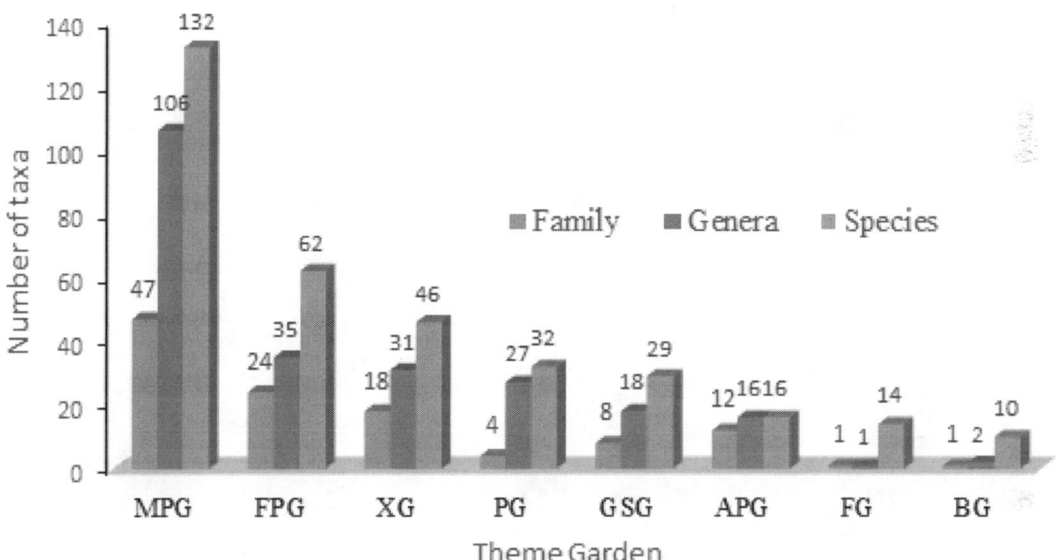

Fig. 49: Diversity of taxa in each established Theme Garden
MPG - Medicinal Plants Garden; FPG - Fruit Plants Garden; XG - Xerophytes Garden (Arid & Semi-Arid Zone Plants); PG - Palm Garden (Palmetum); GSG- Gymnosperms Garden; APG - Aquatic Plants Garden and Conservation Area; FG - Ficus Garden; BG - Bamboo Garden (Bambusetum)

Fig. 51: Some fruiting trees in Botanic Garden; (A) Karonda (B) Grapes (C) Papaya (D) Variegated Chiku (E) Pomegranate (F) Chinese Orange (G) Peach

Fig. 52: Diversity of Palm in Botanic Garden; (A) Date Palm (B) Bismarckia/Silver Blue Fan Palm (C) Travellers Palm (D) Nolina Palm (E) Fox Tail Palm (F) Red Latan Palm (G) Chamaerops Palm

Fig. 53: Diversity of Gymnosperms in Botanic Garden; (A) *Pinus roxburghii* **(B)** *Dioon edule* **(C)** *Zamia furfuracea* **(D)** *Ginkgo biloba* **(E)** *Araucaria* **sp. (F)** *Dioon spinulosum*

(iv) Bamboo Garden (Bambusetum)

Bamboos are aptly called the poor man's timber and are found in great abundance in some state of India. Their strength, straightness and lightness combined with extraordinary hardness, range in sizes, abundance, easy propagation and the short period in which they attain maturity make them suitable for a variety of purposes. The diversity of this fascinating plant has to be conserved, not just for financial reasons, but also more importantly in the revitalization of traditional sciences and technologies.

In India, a total of 136 species belonging to 36 genera of bamboos have been recorded; fifty-eight species of bamboo belonging to 10 genera are distributed in the northeastern states alone. A beginning has been made by planting a few species of bamboos in the Bambusetum to show the bamboo diversity.

A total of 10 taxa are introduced in Botanic Garden *viz., Bambusa nana* (Thai silk bamboo), *Bambusa vulgaris* (Buddha belly), *Dendrocalamus asper* (Edible bamboo), *Dendrocalamus strictus* (Desi bamboo), Golden bamboo, Variegated bamboo, etc.

(v) Medicinal Plants Garden

About 80 percent of the world's population relies solely or largely on traditional remedies for their healthcare needs. Today, about 70,000 to 80,000 plant species are used for medicinal or aromatic purposes globally. India with its ecological, geographical and climatic diversities is perhaps the richest nation with a vast herbal medicinal wealth (About 15000-20000 plants have good medicinal value). In India the therapeutic use of herbs dates back to the Vedic period. The Rigveda has documented about 67 medicinal plants, Yajurveda 81 species and Atharvaveda 290 species.

Fig. 54: A view of collection of medicinal plants in the Botanic Garden

There are more than 7000-7500 plants are used for their medicinal values by traditional communities in India. In India, drugs of herbal origin have been used in traditional systems of medicines such as *Unani* and *Ayurveda* since ancient times. The *Ayurveda* system of medicine uses about 700 species, *Unani* 700, *Siddha* 600, *Amchi* 600 and modern medicine around 30 species.

Fig. 55: Diversity of medicinal plants in Botanic Garden; (A) Brahmi (B) Bhringraj (C) Ashwagandha (D) Safed Aak (E) Chirchita/Latjira (F) Gokhru

A total of 132 species of medicinal plants from 106 genera and 47 families are planted in the Medicinal Plants Garden (Fig. 54). Some of the important medicinal plants introduced in Botanic Garden are; *Abutilon indicum* (*Kanghi*), *Achyranthes aspera* (*Latjeera*), *Artemisia scoparia* (*Seeta-bani*), *Bacopa monnieri* (*Brahmi*), *Cannabis sativa* (*Bhang*), *Chenopodium album* (*Bathua*), *Clerodendrum phlomidis* (*Arni*), *Calotropis gigantea* (*Safed Aak*), *Clitoria ternatea* (*Aparajita*), *Commiphora wightii* (*Guggul*),*Croton bonplandianus* (*Ban tulsi*), *Eclipta prostrata* (*Bhringraj*), *Euphorbia hirta* (Asthma plant), *Hyptis suaveolens* (*Vilaiti tulsi*), *Ocimum sanctum* (*Tulsi*), *Ricinus communis* (Castor), *Santalum album* (*Chandan*), *Solanum americanum* (*Makoy*), *Tribulus terrestris* (*Gokhru*), *Withania somnifera* (*Ashwagandha*), etc.

(vi) Xerophytes Garden (Arid & Semi-Arid Zone Plants)

Have you ever wondered how plants survive in areas with very little water? Xerophytes have adapted over thousands of years to live in harsh conditions. This specialized group of plants are called xerophytes that survive in very dry regions. Xerophytes have adapted to survive in an environment can survive with little liquid water, such as a desert or an ice- or snow-covered region of Earth as they possess specialized features that help them prevent water loss.

Altogether 46 xerophytic plants belonging to 31 genera from 18 families have been planted in the Botanic Garden such as; *Acacia* spp., *Cactus* spp., *Euphorbia tirucalli* (Pencil Tree), *Opuntia dillenii* (*Nagphani*), *Cordia dichotoma* (*Lasora*), *Prosopis cineraria* (*Khejri Tree*), *Salvadora oleoides, S. persica (Miswak)*, *Yucca* spp., etc.

(vii) Aquatic Plants Garden and Conservation Area

A total of 16 species of wild hydophytes such as *Ceratophyllum demersum, Hydrilla verticillata, Lemna perpusilla, Nymphaea pubescens, Nymphoides cristata, Potamogeton crispus, Pistia stratiotes, Sagittaria guayanensis, Spirodela polyrrhiza, Stuckenia pectinata, Utricularia stellaris, Vallisneria spiralis* have been introduced in the Aquatic Plants Garden. The serious aquatic weed, *Eichhornia crassipes* (Water Hyacinth) is also introduced in the Garden for education and research purpose. In addition, one species of macro algae (*Chara* sp.) and 2 species of ferns (*Ampelopteris prolifera, Marsilea quadrifolia*) are also included in the Aquatic Plants Garden.

(viii) Ficus Garden

Ficus is an important genus in the family Moraceae. Many of the Ficus or fig trees have their role as keystone species in various ecosystems and provide home to numerous species of birds, butterflies, monkeys, bats and insects. The *Ficus religiosa*, native to the Indian subcontinent, is one of the religious trees of India worshipped among Hindus a well as followers of Jainism and Buddhism.

A total of 14 species of *Ficus* have been assembled in the Ficus Garden, including *Ficus benghalensis, F. benjamina, F. elastica, F. hispida, F. lyrata, F. natalensis* subsp. *leprieurii, F. racemosa, F. religiosa, F. retusa, F. virens*, etc. In NCR this is the first Garden to have so many species of *Ficus* assembled at one place.

Fig. 56: Diversity of Xerophytes (Arid & Semi-Arid Zone Plants) in Botanic Garden; (A) *Acacia catechu* **(B)** *Dichrostachys cinerea* **(C)** *Furcraea foetida* **(D)** *Yucca aloifolia* **(E)** *Opuntia dillenii* **(F)** *Jatropha curcas*

Fig. 57: Exotic, Bamboo and Aquatic plants in Botanic Garden; (A) *Corymbia citriodora* (B) *Populus deltoides* (C) *Buddha Belly Bamboo* (D) *Golden Bamboo* (E) *Nymphoides cristata* (F) *Nymphaea pubescens*

Bibliography

Abbas MN, Rana SA, Shahid M, Rana N, Mahmood-ul-Hassan M and Hussain M 2012. Chemical evaluation of weed seeds mixed with wheat grains at harvest. *The Journal of Animal & Plant Sciences,* 22(2): 283-288.

Abbasi AM, Khan MA, Ahmad M and Zafar M 2012. Medicinal plant biodiversity of lesser Himalayas-Pakistan. Springer-Verlag, New York.

Abbasi AM, Khan MA, Ahmed M and Zafar M 2010. Herbal medicines used to cure various ailments by the inhabitants of Abbottabad district, North West Frontier Province, Pakistan. *Indian Journal of Traditional Knowledge,* 9(1): 175-183.

Abbasi AM, Khan MA, Shah MH, Shah MM, Pervez A and Ahmad M 2013. Ethnobotanical appraisal and cultural values of medicinally important wild edible vegetables of lesser Himalayas-Pakistan. *Journal of Ethnobiology and Ethnomedicine,* 9(1): 66.

Abdulazeez AM, Awasum CA, Dogo YS and Abiayi PN 2010. Effect of *Peristrophe bicalyculata* on blood pressure, kidney and liver functions of two kidney one clip (2K1C) hypertensive rats. *British Journal of Pharmacology and Toxicology,* 1: 101-107.

Abelson PH 1990. Medicines from plants. *Science,* 247: 513.

Abii TA and Onuha EN 2014. A preliminary investigation into the phytochemicals, vitamins and mineral constituents of the leaf of two traditional medicinal plants –*Urena lobata* and *Cassia alata* used in Nigeria. *International Organization of Scientific Research,* 7(2): 1-4.

Abu T 2017. A review: aquatic macrophyte *Ceratophyllum demersum* L. (Ceratophyllaceae): Plant profile, phytochemistry and medicinal properties. *International Journal of Science and Research,* 6(7): 394-399.

Abu-Rabia A 2015. Indigenous medicines among Bedouin in the Middle East. Berghahn Books. New York and Oxford, p. 232.

Adegoke AA and Adebayo-Tayo BC 2009. Phytochemical composition and antimicrobial effects of *Corchorous olitorius* leaf extracts on four bacterial isolates. *Journal of Medicinal Plants Research,* 3(3): 155-159.

Aderibigbe SA, Adetunji OA and Odeniyi MA 2011. Antimicrobial and pharmaceutical properties of the seed oil of *Leucaena leucocephala* (Lam.) de Wit (Leguminosae). *African Journal Biomedical Research,* 14: 63-68.

Adhikari BS, Babu MM, Saklani PL and Rawat GS 2010. Medicinal plants diversity and their conservation status in Wildlife Institute of India (WII) Campus, Dehradun. *Ethnobotanical Leaflets,* 14: 46-83.

Agarwal VS and Ghosh B 1985. Drug plants of India: Root drugs. Kalyani Publishers, New Delhi.

Agra MF, França PF and Barbosa-Filho JM 2007. Synopsis of the plants known as medicinal and poisonous in Northeast of Brazil. *Revista Brasileira de Farmacognosia,* 17: 114-140.

Agra MF, Silva KN, Basílio IJLD, França PF, Barbosa-Filho JM 2008. Survey of medicinal plants used in the region Northeast of Brazil. *Revista Brasileira de Farmacognosia,* 18: 472-508.

Agrawal MY, Agrawal YP and Shamkuwar PB 2014. Phytochemical and biological activities of *Chenopodium album. International Journal of Pharm. Tech. Research,* 6(1): 383-391.

Agrawal PC, Singh V, Kumar D and Bansal A 2010. Ethnomedical survey of plants used by rural community of Gonda block of district Aligarh, Uttar Pradesh. *Indian Journal of Applied and Pure Biology,* 25(1): 187-190.

Ahamed N and Gupta AK 2010. An analysis of flora of Baghpat district in Uttar Pradesh, India. *Indian Journal of Forestry,* 33(3): 405-418.

Ahamed N, Gupta AK and Neetu 2010. Ethnomedicinal plants of Deoband tehsil of Saharanpur district, Uttar Pradesh, India. *The Indian Forester,* 136(11): 1520-1528.

Ahirwar RK and Kujur M 2015. Ethnomedicinal uses of some plants species by the tribes of Amarkantak district Anuppur, Madhya Pradesh, India. *International Journal of Science and Research,* 4(8): 1648-1651.

Ahmad B, Jan Q, Bashir S, Choudhary MI and Nisar M 2003. Phytochemical evaluation of *Chenopodium murale* Linn. *Asian Journal of Plant Sciences*, 3: 1072-78.

Ahmad F, Khan MA, Ahmad M, Zafar M, Mahmood T, Jabeen A and Marwat SK 2010. Ethnomedicinal uses of grasses in salt range region of northern Pakistan. *Journal of Medicinal Plants Research*, 4(5): 362-369.

Ahmad I, Ibrar M, Barkatullah and Ali N 2011. Ethnobotanical study of tehsil Kabal, Swat district, KPK, Pakistan. *Journal of Botany*, 2011: 1-9. *DOI: 10.1155/2011/368572.*

Ahmad M, Khan MA, Zafar M and Sultana S 2006. Ethnomedicinal demography and ecological diversification of some important weeds from district Attock-Pakistan. *Pakistan Journal of Weed Science Research*, 12(1-2): 37-46.

Ahmad M, Qureshi R, Arshad M, Khan MA and Zafar M 2009. Traditional herbal remedies used for the treatment of diabetes from district Attock (Pakistan). *Pakistan Journal of Botany*, 41(6): 2777-2782.

Ahmad M, Waheed I, Khalil-ur-Rehman M, Niaz U and Saeed-ul-Hassan S 2007. A review on Carthamus oxycantha. *Pakistan Journal of Pharmaceutical Sciences*, 20-23(1&2): 37-41.

Ahmad S, Khan M, Rageeb M, Usman M, Akhtar R and Imran M 2016. Quantitative and qualitative characterization of phytoconstituents from leaves of *Tephrosia villosa* Pers. *International Journal of Pharmaceutical Research and Bio-Science*, 5(4): 247-256.

Ahmad SS, Butt A, Erum S, Malik B and Jabeen A 2015. Exploring ethnomedico secrets of the herbs of Gatwala Forest Park, Faisalabad, Pakistan. *The Pharmaceutical and Chemical Journal*, 2(2): 41-51.

Ahmad T, Husain MK, Tariq M, Siddiqui JI, Khalid M, Ahmed MW and Kazmi MH 2017. A review on *Operculina turpethum*: A potent herb of unani system of medicine. *Journal of Pharmacognosy and Phytochemistry*, 6(1): 23-26.

Ahmad V, Ali Z, Ali M and Zahid M 1998. Chemical constituents of *Silene conoidea*. *Fitoterapia*, 69: 406–408.

Ahmad VU and Alam N 1995. New Antifungal Bithienylacetylenes from *Blumea obliqua*. *Journal of Natural Products*, 58(9): 1426-1429.

Ahmed A, Howlader MSI, Dey SK, Hira A, Hossain MH and Nasiruddin MM 2013a. Phytochemical screening and antibacterial activity of different fractions of *Operculina turpethum* root and leaf. *American Journal of Scientific and Industrial Research*, 4(2): 167-172.

Ahmed E, Arshad M, Saboor A, Qureshi R, Mustafa G, Sadiq S and Chaudhari SK 2013b. Ethnobotanical appraisal and medicinal use of plants in Patriata, New Murree, evidence from Pakistan. *Journal of Ethnobiology and Ethnomedicine*, 9:13, https://doi.org/10.1186/1746-4269-9-13.

Ahmed OH, Hamad MN and Jaafar NS 2017. Phytochemical investigation of *Chenopodium murale* (family: Chenopodiaceae) cultivated in Iraq, isolation and identification of scopoletin and gallic acid. *Asian Journal of Pharmaceutical and Clinical Research*, 10(11): 70-77.

Ahmed S, Hasan MM and Mahmood ZA 2016. Antiurolithiatic plants in different countries and cultures. *Journal of Pharmacognosy and Phytochemistry*, 5(1): 102-115.

Ahmed SA, Hanif S and Iftkhar T 2013c. Phytochemical profiling with antioxidant and antimicrobial screening of *Amaranthus viridis* L. leaf and seed extracts. *Open Journal of Medical Microbiology*, 3: 164-171.

Aiswarya T and Ravikumar R 2014. A comparitive study on phytochemical analysis, antibacterial activity and antioxidant activity of *Barleria prionitis* leaves extract of petroleum ether and ethanol extract. *International Journal of Chem. Tech. Research*, 6(5): 3025-3033.

Ajaib M, Haider SK, Zikrea A and Siddiqui MF 2014. Ethnobotanical studies of shrubs and trees of Agra valley parachinar, upper kurram agency, Pakistan. *FUUAST Journal of Biology*, 4(1): 73-81.

Ajithabai MD, Rani SP, Balachandran S and Jayakumar G 2012. Review on the species of *Sida* used for the preparation of nayopayam kashayam. *International Journal of Research in Ayurveda and Pharmacy* 3(1): 99-104.

Akanksha B, Avijit M, Chakraborthy GS and Seema G 2014. Phytopharmacological uses of *Tephrosia purpurea* - A review. *Pharmacophore*, 5(4): 658-665.

Akbar S, Hanif U, Ali J and Ishtiaq S 2014. Pharmacognostic studies of stem, roots and leaves of *Malva parviflora* L. *Asian Pacific journal of tropical biomedicine*, 4(5): 410-415.

Akerele O, Heywood V and Synge H 1991. The conservation of medicinal plants. Cambridge University Press, Cambridge.

Alagar Raja M, Sahithi G, Vasanthi R, David Banji, Rao KNV, Selvakumar D 2015. Study of phytochemical and antioxidant activity of *Cucumis melo* var. *agrestis* fruit. *Journal of Pharmacognosy and Phytochemistry*, 4(2): 303-306.

Albuquerque AP, Monteiro JM, Ramos MA and Amorim ELC 2007. Medicinal and magic plants from a public market in Northeastern Brazil. *Journal of Ethnopharmacology*, 110: 76-91.

Ali A, Naveed A, Khan BA, Khan MS, Rasul A, & Zaman, Shahiq-UZ-Zaman, Khalid N, Waseem K, Mahmood T and Ali L 2012. *Acacia nilotica*: A plant of multipurpose medicinal uses. *Journal of Medicinal Plants Research* 6(9): 1492-1496.

Ali H and Qaiser M 2009. The ethnobotany of Chitral Valley, Pakistan with particular reference to medicinal plants. *Pakistan Journal of Botany*, 41(4): 2009-2041.

Ali Z, Ahmad VU, Ali MS, Iqbal F, Zahid M and Alam N 1999. Two new C-Glycosylflavones from *Silene conoidea*. *Natural Product Letters*, 13: 121–129.

Almagboul AZ, Bashir AK, Farouk A and Salih AKM 1985. Antimicrobial activity of certain Sudanese plants used in folkloric medicine- Screening for antibacterial activity. *Fito-terapia*, 56: 331-337.

Al-Qudah MA and Zarga MHA 2010. Chemical composition of essential oils from aerial parts of *Sisymbrium Irio* from Jordan. *E-Journal of Chemistry*, 7(1): 6-10.

Al-Saeed AHM 2012. The phytochemical composition and the effect of methanolic extract of *Typha domingensis* Pers. Fruite on some biochemical parameters in adult male rabbits. *Basrah Journal of Veterinary Research*, 11(1): 224-228.

Al-saffar DF, Abbas IS and Dawood AH 2016. Investigation of the main alkaloid of London Rocket (*Sisymbrium irio* L) as a wild medicinal plant grown in Iraq. *International Journal of Pharmaceutical Sciences Review and Research*, 39(1): 279-281.

Al-shaimaa HA, Ali M, Hassan S, Aboubaker D, Djama M, Asfaw Z and Kelbessa E 2013. Medicinal plants and their uses by the people in the region of Randa, Djibouti. *Journal of Ethnopharmacology*, 148:701-713.

Al-Snafi AE 2013. The pharmacology of *Bacopa monniera*. a review. *International Journal of Pharma Sciences and Research*, 4: 154-159.

Al-Snafi AE 2015a. The therapeutic importance of *Cassia occidentalis* - An overview. *Indian Journal of Pharmaceutical Science & Research*, 5(3): 158-171.

Al-Snafi AE 2015b. The chemical contents and pharmacological effects of *Anagallis arvensis* - A review. *International Journal of Pharmacy*, 5(1): 37-41.

Al-Snafi AE 2015c. The chemical constituents and pharmacological effects of *Ammannia baccifera* - A review. *International Journal of Pharmacy*, 5(1): 28-32.

Al-Snafi AE 2015d. The chemical constituents and pharmacological effects of *Capsella bursa-pastoris*- A review. *International Journal of Pharmacology & Toxicology*, 5(2): 76-81.

Al-Snafi AE 2016a. The constituents and pharmacology of *Corchorus aestuans*: A review. *The Pharmaceutical and Chemical Journal*, 3(4): 208-214.

Al-Snafi AE 2016b. Chemical constituents and pharmacological effects of *Citrullus colocynthis* - A review. *IOSR Journal of Pharmacy*, 6(3): 57-67.

Al-Snafi AE 2017a. Pharmacology of *Echinochloa crus-galli* – a review. *Indo American Journal of Pharmaceutical Sciences*, 4(1): 117-122.

Al-Snafi AE 2017b. The pharmacological potential of *Dactyloctenium aegyptium*- a review. *Indo American Journal of Pharmaceutical Sciences*, 4(1):153-159.

Al-Zubairi AS, Abdul AB, Abdelwahab SI, Peng CY, Mohan S, and Elhassan MM 2011. *Eleucine indica* possesses antioxidant, antibacterial and cytotoxic properties. *Evidence-Based Complementary and Alternative Medicine*, 2011: 1-6. doi:10.1093/ecam/nep091.

Ambasta SP 1986. The useful plants of India. Publications and Information Directorate, CSIR New Delhi.

Aminullah A, Khan MI, Yaqooburrehman, Waqas M, Khan MA, Sher A, Zaman G and Ali H 2014. Antibacterial and antifungal activities of crude methanolic and ethanolic extracts of *Medicago polymorpha*. *International Journal of Bioassays*, 4(1): 3618-3651.

Amiya T 2006. Some ethnomedicinal plants of family Fabaceae of Chattisgarh State. *Indian Journal of Traditional Knowledge*, 5(4): 551-553.

Amjad MS, Arshad M and Qureshi R 2015. Ethnobotanical inventory and folk uses of indigenous plants from Pir Nasoora National Park, Azad Jammu and Kashmir. *Asian Pacific Journal of Tropical Biomedicine*, 5(3): 234-241.

Amjad MS, Arshad M, Saboor A and Chaudhari SK 2017. Ethnobotanical profiling of the medicinal flora of Kotli, Azad Jammu and Kashmir, Pakistan: Empirical reflections on multinomial logit specifications. *Asian Pacific Journal of Tropical Medicine*, 10(5): 503-514.

Amos TN, Bashir L, Saba SE, Saba MA, Mohammed BM, Abdulsalam IH and Josiah GJ 2015. Phytochemicals and acute toxicity profile of aqueous and methanolic extracts of *Crateva adansonii* leaves in Swiss Albino Rats. *Asian Journal of Biochemistry*, 10: 173-179.

Anand RK, Khare N and Dwivedi SV 2013. Ethno-medicinal forest genetic resources of District Sonbhadra, Uttar Pradesh. *Trends in Biosciences*, 6(5): 663-668.

Anand SP, Velmurugan G and Revathi D 2016. Survey of medicinal plants from Vadachennimalai Hill, Salem district of Tamil Nadu, India. *Journal of Medicinal Plants Studies*, 4(3): 219-223.

Anbazhakan S, Dhandapani R, Anandhakumar P and Balu S 2007. Traditional medicinal knowledge on *Moringa concanensis* Nimmo of Perambalur district, Tamil Nadu. *Ancient Science of Life*, 274): 42-45.

Andrade-Cetto A and Heinrich M 2005. Mexican plants with hypoglycaemic effect used in the treatment of diabetes. *Journal of Ethnopharmacology*, 99(3): 325-348.

Angadi KK, Kandru A and Rahaman A 2013. Antihyperglycaemic, antihyperlipidaemic and antioxidant assays (*in vivo*) of *Nymphaea pubescens* leaf extract. *International Journal of Pharmacy and Biological Sciences,* 4(2): 624-630.

Anisuzzaman M, Ahsan MQ, Kuddus MR and Rashid MA 2014. Pharmacological activities of *Senna obtusifolia* Linn: A medicinal plant of Bangladesh. *Bangladesh Pharmaceutical Journal,* 17(2): 182-186.

Anitha R and Kanimozhi S 2012. Pharmacognostic evaluation of *Alternanthera sessilis* (L.) R.Br.ex.DC. *Pharmacognosy Journal,* 4(28): 31-34.

Ankita J and Jain A 2012. *Tridax procumbens* (L.): A weed with immense medicinal importance: A review. *International Journal of Pharma and Bio Sciences,* 3(1): 544-552.

Anon. 1948-1976. The wealth of India: A dictionary of Indian raw materials and industrial products. Raw Materials. CSIR, Delhi.

Anon. 1975. Herbal pharmacology in the people's republic of China, section 1. National Academy of Sciences, Washington D.C.

Anon. 1978. The promotion and development of traditional medicine. Report of a world health organization meeting. Technical Report Series 622, Geneva.

Anon. 1998. Medicinal plants in the republic of Korea. WHO Regional Office for the Western Pacific, Manila.

Anon. 2002. Indian herbal pharmacopoeia. Indian Drug Manufacturers' Association, p. 521.

Anon. 2002-2005. Traditional medicine strategy. World Health Organization, Geneva.

Anon. 2004. Medicinal plants of Myanmar (WHO). Ministry of Health Department of Traditional Medicine, pp. 122.

Anon. 2006. The Ayurvedic pharmacopoeia of India, Part- I, Vol. V. Government of India, Ministry of Health and Family Welfare, Department of Ayush, New Delhi.

Anon. 2008-2009. District brochure of Gautam Buddha Nagar, U.P. Central Ground Water. (A.A.P.: 2008-2009) by Dr. B.C. Joshi Scientist 'B'. http//:www.cgwb.gov.in/District Profile/UP/GB%20Nagar.

Anon. 2009-2010. The wealth of India, Raw materials, Vol. I-XI. CSIR, New Delhi, India.

Anon. 2014. Cumulative impact and carrying capacity study of Subansiri sub basin including downstream impacts. Final Report Annexure, Vol. II. December 2014.

Ansari MY, Wadud A, Ehteshamudddin and Bano H 2013. Pharmacognostical evaluation of root of Gumma (*Leucas cephalotes* spreng.). *Indian Journal of Natural Products and Resources,* 4(1): 88-95.

Ansari NA, Khan AA and Ram J 2016. Vascular plants of Surajpur wetland, National Capital Region, India. *Indian Journal of Plant Sciences,* 5(1): 54-69.

Apparanantham T and Chelladurai V 1986. Glimpses of folk medicines of Dharmapuri Forest division Tamilnadu. *Ancient Science of Life,* 3: 182-185.

Aqeel A, Khursheed AK, Viqaruddin A and Sabiha Q 1989. Antimicrobial activity of *Prosopis juliflora. Arzneimittel-Forschung,* 39(6): 652-655.

Arany AM, Jong TJD, Kim HK, Dam NMV, Choi YH, Verpoorte R and Meijden EVD 2005. Glucosinolates and some other chemical compounds in leaves from natural populations of *Arabidopsis thaliana* and their effects on generalist and specialist herbivore. *Chemoecology,* 18(2): 65-71.

Arora M, Sharma T, Devi A, Bainsal N and Siddiqui AA 2012. An inside review of *Cissampelos pareira* Linn: A potential medicinal plant of India. *International Research Journal of Pharmacy,* 3(12): 38-41.

Arul MK and Agnes JX 2015. Phytochemical screening and in vitro antioxidant Activity of *Marsilea quadrifolia. World Journal of Pharmacy and Pharmaceutical Sciences,* 4(5): 853-864.

Aryakrishna UK, Vinayan R and Sudha Bai R 2016. Phytochemical profiling and histochemical localization in leaf and stem of *Trichosanthes cucumerina* (L) var. *cucumerina* with reference to the influence of plant age and geographical variations. *Journal of Pharmacy Research,* 10(10): 660-665.

Ashashri S, Pankaj G and Kumar RS 2015. Conservation and sustainability of *Ashwagandha*: a medicinal plant. *Journal of Biological and Scientific Opinion,* 3(2): 94-99.

Ashraf MA, Mahmood K, Yusoff IB and Qureshi AK 2013. Chemical constituents of *Cenchrus ciliaris* L. From the Cholistan desert, Pakistan. *Archives of Biological Sciences,* 65(4):1473-1478.

Ashwini M, Lather N, Bole S, Vedamutrhy AB and Balu S 2012. In vitro antioxidant and anti-inflammatory activity of *Coccinia grandis. International Journal of Pharmacy and Pharmaceutical Sciences,* 4(3): 239-242.

Aslam MS, Choudhary BA, Uzair M and Ijaz AS 2012. The genus *Ranunculus*: A phytochemical and ethnopharmacological review. *International Journal of Pharmacy and Pharmaceutical Sciences,* 4: 15-22.

Aslam MS, Roy SD, Ghaffari MA, Choudhary BA, Uzair M, Ijaz AS and Khan TR 2014. Survey of ethno-medicinal weeds of district Rajhan pur, Punjab, Pakistan. *Indian Research Journal of Pharmacy and Science,* 1(2): 38-45.

Asolkar LV, Kakkar KK and Charke OJ 1992. Second supplement to glossary of Indian medicinal plants with active principles Part-I (A-K) (1965–1981) New Delhi, India. Publications and Information Directorate CSIR, New Delhi, India.

Asprey GF and Thornton P 1953-1954. Medicinal plants of Jamaica. Parts 1 & 11. Reprinted from the West Indian Medical Journal.

Asthana A, Kumar A, Gangwar S and Dora J 2012. Pharmacological perspectives of *Cynodon dactylon*. *Research Journal of Pharmaceutical, Biological and Chemical Sciences,* 3(2): 1135- 1147.

Atanu FO, Ebiloma UG and Ajayi EI 2011. A review of the pharmacological aspects of *Solanum nigrum* Linn. *Biotechnology and Molecular Biology Review,* 6(1):1-7.

Athokpam R, Bawari M and Choudhury MD 2014. A review on medicinal plants of Manipur with special reference to hepatoprotection. *International Journal of Advances in Pharmaceutical Research,* 5(3): 182-191.

Atiqurrahman M, Uddin SB and Wilcock CC 2007. Medicinal plants used by Chakma tribe in Hill Tracts districts of Bangladesh. *Indian Journal of Traditional Knowledge,* 6(3): 508-517.

Atta-ur-Rahman, Choudhary MI and Bullo S 2006. Medicinal plants of Sindh. Indigenous knowledge and scientific facts (Monograph). Study Sponsored by department of planning and development, government of Sindh.

Austin DF 2004. Florida ethnobotany. CRC Press, Taylor and Francis Group, p. 952.

Awad MA, Ragab EA and Hela AAE 2015. Phytochemical investigation and biological evaluation of *Dichanthium annulatum* (Forrsk). *Journal of Scientific and Innovative Research,* 4(3): 131-137.

Awika JM and Rooney LW 2004. *Sorghum* phytochemicals and their potential impact on human health. *Phytochemistry,* 65(9): 1199-1221.

Ayyanar M and Babu PS 2012. *Syzygium cumini* (L.) Skeels: A review of its phytochemical constituents and traditional uses. *Asian Pacific Journal of Tropical Biomedicine,* 2(3): 240-246.

Ayyanar M, Sankarasivaraman K and Ignacimuthu S 2008. Traditional herbal medicines used for the treatment of diabetes among two major tribal groups in south Tamil Nadu, India. *Ethnobotanical Leaflets,* 12: 276-280.

Azaizeh H, Fulder S, Khalil K and Said O 2003. Ethnobotanical knowledge of local Arab practitioners in the Middle Eastern region. *Fitoterapia,* 74: 98-108.

Azam MM, Mamun-Or-Rashid ANM, Towfique NM, Sen MK and Nasrin S 2013. Pharmacological potentials of *Melia azedarach* L. - A review. *American Journal of BioScience,* 1(2): 44-49.

Babu CR 1977. Herbaceous flora of Dehradun. Publications and Information Directorate, CSIR, New Delhi.

Babu RH and Savithramma N 2013. Phytochemical screening of underutilized species of Poaceae. *An International Journal,* 1(10): 947-951.

Babu SS, Madhuri DB and Ali SL 2016. A pharmacological review of *Urena lobata* plant. *Asian Journal of Pharmaceutical and Clinical Research,* 9(2): 20-22.

Badwaik H, Singh MK, Thakur D, Giri TK, and Tripathi DK 2011. The Botany, Chemistry, pharmacological and therapeutic application of *Oxalis corniculata* Linn–A review. *International Journal of Phytomedicine,* 3:1-8.

Bagci E, Genc H and Sahun A 2001. Fatty acid composition of four *Lathyrus aphaca* L. varieties, a chemotaxonomic approach. *Pakistan Journal of Biological Sciences,* 4: 872-874

Bailey LH 1949. Manual of cultivated plants (Rev. edn.). Macmillan, New York.

Bajpai O, Kumar A, Srivastava AK, Kushwaha AK, Pandey J and Chaudhary LB 2015. Tree species of the Himalayan Terai region of Uttar Pradesh, India: a checklist. Check List 11(4): 1718, doi: http://dx.doi.org/10.15560/11.4.1718ISSN 1809-127X.

Baker HG 1965. Characteristics and modes of origin of weeds. In: Genetics of colonizing species, Academic Press, New York, p. 147-172.

Balakrishnan S, and Subramanian M 2016. A review on *Marsilea quadrifolia* L. – A medicinally important plant. *Journal of Comprehensive Pharmacy,* 3(2): 38-44.

Balakrishnan V, Prema P, Ravindran KC and Robinson JP 2009. Ethnobotanical studies among villagers from Dharapuram Taluk, Tamil Nadu, India. *Global Journal of Pharmacology,* 3(1): 8-14.

Balasubramanian P, Rajasekaran A, and Prasad SN 1997. Folk medicine of the Irulas of Coimbatore forests. *Ancient Science of Life,* 16(3): 222-226.

Banaras S, Javaid A, Shoaib A and Ahmed E 2017. Antifungal activity of *Cirsium arvense* extracts against a phytopathogenic fungus *Macrophomina phaseolina*. *Planta Daninha,* doi: 10.1590/S0100-83582017350100014.

Bandyopadhyay S and Mukherjee SK 2005. Diversity of aquatic and wetland vascular plants of Koch Bihar district, West Bengal. In: Plant taxonomy: Advances and relevance, Eds. A.K.Pandey, Jun Wen & J.V.V.Dogra, p. 223-244.

Banerjee D, Maji AK, Mahapatra S and Banerji P 2012. *Barleria prionitis* Linn: A review of its traditional uses, phytochemistry, pharmacology and toxicity. *Research Journal of Phytochemistry,* 6(2): 31-41.

Bankova V, Ivanova P, Christov R, Popov S and Konaklieva SD 1999. Secondary metabolites of *Ceratophyllum demersum*. *Hydrobiologia,* 316(1): 59-61.

Bano A, Sharma N, Dhaliwal HS and Sharma V 2015. A systematic and comprehensive review on *Withania somnifera* (L.) Dunal- An Indian Ginseng. *British Journal of Pharmaceutical Research,* 7(2): 63-75.

Bapuji JL and Ratnam SV 2009. Traditional uses of some medicinal plants by tribals of Gangaraju Madugula Mandal of Visakhapatnam district, Andhra Pradesh. *Ethnobotanical Leaflets,* 13: 388-398.

Baranwal VK, Irchhaiya R and Singh S 2012. *Anisomeles indica*: An overview. *International Research Journal of Pharmacy,* 3(1): 84-87.

Barbosaa AF, Carvalhoa MGD, Smithb RE, Sabaa-Srurac AUO 2016. Spilanthol: Occurrence, extraction, chemistry and biological activities. *Brazilian Journal of Pharmacognosy,* 26(1): 128-133.

Barik KL 2017. Study of phytodiversity of a grassland community of Ambadiha with special reference to human welfare, Odisha, India. *Indian journal of research*: 6(12): 522- 526.

Bartolome AP, Villaseñor IM, and Yang WC 2013. *Bidens pilosa* L. (Asteraceae): Botanical properties, traditional uses, phytochemistry, and pharmacology. *Evidence-Based Complementary and Alternative Medicine,* http://dx.doi.org/10.1155/2013/340215.

Bassey ME and Effiong EO 2011. Preliminary investigation of herbs used in paediatric care among the people of Akwa Ibom State, Nigeria. *Journal of Natural Product and Plant Resources,* 1(3): 33-42.

Basu BD, Kirtikar KR and Basu K 1991. Indian medicinal plants, Vol. 1-4. Bisen Singh Mahendra Pal Singh, Dehradun.

Basumatary N, Teron R, and Saikia M 2014. Ethnomedicinal practices of the Bodo-Kachari tribe of Karbi Anglong district of Assam. *International Journal of Life Sciences Biotechnology and Pharma Research,* 3(1): 161-167.

Bekker NP, Ul'chenko NT and Glushenkova AI 2002. Lipids of the aerial part of *Capsella bursa-pastoris. Chemistry of Natural Compounds,* 38(6): 610-611.

Benítez G, González-Tejero MR, and Molero-Mesa J 2010. Pharmaceutical ethnobotany in the western part of Granada province (southern Spain): Ethnopharmacological synthesis. *Journal of Ethnopharmacology,* 129: 87-105.

Benniamin A 2011. Medicinal ferns of North Eastern India with special reference to Arunachal Pradesh. *Indian Journal of Traditional Knowledge,* 10(3): 516-522.

Bentley R and Trimen H 2008. Medicinal plants. International Book Distributors, Dehra Dun, India I-IV.

Benzarti A, Hammami S, Piras A, Falconieri D, Mokni RE, M'Henni MF, Bruno M and Mighri Z 2013. Effects of different ecological conditions and extraction techniques on the quality of volatile oils from flaxleaf fleabane (*Erigeron bonariensis* L.). *Journal of medicinal plant research,* 7(41): 3059-3065.

Berghe DA and Vlietinck 1991. Screening methods for antibacterial and antiviral agents from higher plants. In: Methods in Plant Biochemistry, Vol. 6. Acadamic Press, London, p. 47-69.

Berlin B, Berlin EA, Fernández JC and Espinoza MG 1999. The Maya ICBG: Drug discovery, medicinal ethnobiology, and alternative forms of economic development in the highland Maya region of Chiapas, Mexico. *Pharmaceutical Biology,* 37 (4):127-144.

Besra SE, Gomes A, Chaudhury L, Vedasiromoni JR, Ganguly DK 2002. Antidiarrhoeal activity of seed extract of *Albizzia lebbeck* Benth. *Phytotherapy Research,* 16(6):529-33.

Bever BO 1986. Medicinal plants in tropical west Africa. Cambridge University Press, p. 375.

Bhagwat DP, Kharya MD, Bani S, Kaul A, Kour K, Chauhan PS, Suri KA and Satti NK 2010. Immunosuppressive properties of *Pluchea lanceolata* leaves. *Indian Journal of Pharmacology,* 42(1): 21-26.

Bhalerao SA and Sharma AS 2014. Ethenomedicinal, phytochemical and pharmacological profile of *Ficus religiosa* Roxb. *International Journal of Current Microbiology and Applied Sciences,* 3(11): 528-538.

Bhalerao SA and Teli NC 2016. *Ipomoea carnea* Jacq: Ethnobotany, phytochemistry and pharmacological potential. *International Journal of Current Research in Biosciences and Plant Biology,* 3(8): 138-144.

Bhalerao SA, Verma DR, Teli NC and Trikannad AA 2014. Ethnobotany, phytochemistry and pharmacology of *Convolvulus pluricaulis*, Choisy. *Research Journal of Pharmaceutical, Biological and Chemical Sciences,* 5(3): 629-636.

Bhandari MM 1978. Flora of Indian desert. Scientific Publishers, Jodhpur.

Bhandari P 2016. A wild edible fruit- *Ficus palmata. European Journal of Biomedical and Pharmaceutical Sciences,* 3(7): 168-174.

Bharath M, Tulasi ELR, Sudhakar K and Eswaraiah MC 2013. *Dalbergia sissoo* DC -An important medicinal plant. *International Journal of Research in Pharmacy and Chemistry,* 3(2): 384-388.

Bharti M and Pravesh R 2012. Antibacterial activity of *Lygodium flexuosum* (L.) Sw.and *Ampelopteris prolifera* (Retz.)Copel of Ranchi district of Jharkhand. *The Ecoscan,* 1: 397-401.

Bhat JA, Kumar M, and Bussmann RW 2013. Ecological status and traditional knowledge of medicinal plants in Kedarnath Wildlife Sanctuary of Garhwal Himalaya, India. *Journal of Ethnobiology and Ethnomedicine,* 9(1):1-18.

Bhati A, Agarwal S, Kumar A and Singh A 2014. Diversity of medicinal flora of Moradabad district, Uttar Pradesh, India. *International Journal of Current Research,* 6(2): 5197-5204.

Bhattacharjee A, Shashidhara SC and Saha S 2013. Phytochemical and ethno-pharmacological profile of *Desmodium gangeticum* (L.) DC: A review. *International Journal of Biomedical Research,* 4(10): 508-515.

Bhattacharya M, Singh A and Ramrakhyani C 2014. *Dalbergia sissoo*-An important medical plant. *Journal of Medicinal Plants Studies,* 2(2): 76-82.

Bhoria R and Kainsa S 2013. *Leucas cephalotes* (Roth) Spreng: Review at a glance. *International Journal of Pharmaceutics,* 3(1):77-81.

Bhowmik D, Kumar KPS, Paswan S, Srivatava S, Yadav A and Dutta A 2012. Traditional Indian herb *Convolvulus pluricaulis* and its medicinal importance. *Journal of Pharmacognosy and Phytochemistry,* 1(1): 44-51.

Bhowmik S, Datta BK and Saha AK 2013. Ethno medicinal and phytochemical screening of some hydrophytes and marsh plants of Tripura, India. *World Applied Sciences Journal,* 22(10): 1453-1459.

Bhutya RK 2011. Ayurvedic medicinal plants of India, Vol. 1-2. Scientific Publishers, p. 649.

Bieski IGC, Santos FR, Oliveira RM, Espinosa MM, Macedo M, Albuquerque UP, and Martins DTO 2012. Ethnopharmacology of medicinal plants of the Pantanal region (Mato Grosso, Brazil). *Evidence-Based Complementary and Alternative Medicine,* 2012:1-36.

Bisby F 1994. Phytochemical dictionary of the Leguminosae. Chapman and Hall/CRC, p. 1180.

Biswajit P, Deenabandhu S and Misra MK 2014. Ethno-medicobotanical studies of Mohana area of Gajapati district, Odisha, India. *International Journal of Herbal Medicine,* 2(4): 40-45.

Biswas K, Chattopadhyay I, Banerjee RK and Bandyopadhyay U 2002. Biological activities and medicinal properties of neem (*Azadirachta indica*). *Current science,* 82(11): 1336-1345.

Blanco MEA, Rodríguez MGR, Duque JLM, Ortega MM, and Juárez JV 2014. Amoebicidal activity of essential oil of *Dysphania ambrosioides* (L.) Mosyakin & Clemants in an Amoebic Liver Abscess Hamster Model. *Evidence-Based Complementary and Alternative Medicine,* 2014: 1-7.

Bobbarala V, Chadaram RK, Vadlapudi V and Katikala PK 2009. Medicinal plants as alternative biocontrol agents in the control of seed borne pathogen *Macrophomina phaseolina*. *Journal of Pharmacy Research,* 2(6): 1045-1048.

Borges CC, Matos TF, Moreira J, Rossato AE, Zanette VC and Amaral PA 2012. *Bidens pilosa* L. (Asteraceae): Traditional use in a community of southern Brazil. *Revista Brasileira de Plantas Medicinais,* 15: 34-40.

Borhade P, Deshmukh T, Patil V and Khandelwal K 2013. Review on *Citrullus colocynthis*. *International Journal of Research in Pharmacy and Chemistry,* 3(1):46-53.

Borris RP 1995. Natural products research: Perspectives from a major pharmaceutical company. *Journal of Ethnopharmacology,* 51(1-3): 29-38.

Borthakur N and Baruah AKS 1987. Search for precocenes in *Ageratum conyzoides* Linn. of North-East India. *Indian Chemical Society,* 64: 580-581.

Bose D, Roy JG, Mahapatra SD, Datta T, Mahapatra SD and Biswas H 2015. Medicinal plants used by tribals in Jalpaiguri district, West Bengal, India. *Journal of Medicinal Plants Studies,* 3(3): 15-21.

Boutkhil S, Idrissi ME, Amechrouq A, Chbicheb A, Chakir S and Badaoui KE 2009. Chemical composition and antimicrobial activity of crude, aqueous, ethanol extracts and essential oils of *Dysphania ambrosioides* (L.) Mosyakin & Clemants. *Acta Botanica Gallica,* 156(2): 201-209.

Brahmachari G, Gorai D and Roy R 2013. *Argemone mexicana*: chemical and pharmacological aspects. *Revista Brasileira de Farmacognosia (Brazilian Journal of Pharmacognosy),* 23(3): 559-575.

Bremer B, Bremer K, Chase M and Fay M 2009. An update of the Angiosperm phylogeny group classification for the orders and families of flowering plants: APG III. *Botanical Journal of the Linnean Society,* 161: 105-121.

Bukhari IA 2013. The Central analgesic and anti-inflammatory activities of the methanolic extract of *Carthamus Oxycantha*. *Journal of Physiology and Pharmacology,* 64(3): 369-375.

Bukhsh E, Malik SA and Ahmad SA 2007. Estimation of nutritional value and trace elements content of *Carthamus oxyacantha, Eruca sativa* and *Plantago ovata*. *Pakistan Journal of Botany,* 39(4): 1181-1187.

Buragohain J 2011. Ethnomedicinal plants used by the ethnic communities of Tinsukia district of Assam, India. *Recent Research in Science and Technology,* 3(9): 31-42.

Caius JF 1986. The medicinal and poisonous plants of India. Scientific Publishers, Jodhpur, India. p. 397-399.

Cambie RC and Ash J 1994. Fijian medicinal plants. CSIRO Publishing, p. 365.

CBD 1992. Convention on biological diversity, Montreal Canada. https://www.cbd.int/convention.

CBD 2009. The CBD and invasive alien species. https://www.cbd.int/idb/2009/about/cbd/.

Chakraborty BK 2014. Status of diversify of medicinal plants in floodplain basin of northern Bangladesh. *Journal of Crop and Weed,* 10(2):196-204.

Chakraborty NR and Duary B 2014. Utilization of some weeds as medicine by the local people in Birbhum district of West Bengal, India. *International Journal of Bio-resource and Stress Management,* 5(1):148-152.

Chakravorty A and Ghosh PD 2013. Exploration of the otherwise-neglected weeds of Nadia district - the treasure trove of medicines. *International Journal of Pharmaceutical Science Invention,* 2(6): 19-21.

Chandra S and Rawat DS 2015. Medicinal plants of the family Caryophyllaceae: a review of ethno-medicinal uses and pharmacological properties. *Integrative Medicine Research,* 4(3): 123-131.

Chandra S, Jenkins CRF, Kumari A and Srivastava A 2008. A summary of the status of threatened Pteridophytes of India. *Taiwania,* 53(2): 170-209.

Chandrasekar SB, Bhanumathy M, Pawar AT and Somasundaram T 2010. Phytopharmacology of *Ficus religiosa. Pharmacognosy Reviews,* 4(8): 195-199.

Chandrashekar KR and Sandhyarani S 1994. Differences in chemical composition of *Alysicarpus vaginalis* (L.) DC. Growing in saline and non-saline habitats. *Biologia Plantarum,* 36(1):139-143.

Chandrawat P and Sharma RA 2016. The Genus *Calotropis*: An overview on bioactive principles and their bioefficacy. *Research Journal of Recent Sciences,* 5(1): 61-70.

Chandrul KK and Singh B 2015. Pharmacognostical, phytochemical and pharmacological profile of *Grangea maderaspatana* (L.) Poir. - An inclusive review. *World Journal of Pharmaceutical Research,* 4(11): 423-436.

Chanu WS and Sarangthem K 2014. Phytochemical constituents of *Justicia adhatoda* Linn. found in Manipur. *Indian Journal of Plant Sciences,* 3(2):19-21.

Charles A, Joseph M and Alex RV 2012. Phytochemical investigation of *Merremia hederacea. Journal of Natural Product and Plant Resources,* 2(4):486-489.

Chatterjee A and Dewanji A 2014. Effect of varying *Alternanthera philoxeroides* (alligator weed) cover on the macrophyte species diversity of pond ecosystems: A quadrat–based study. *Regional Euro-Asian Biological Invasions Centre,* 9(3): 343-355.

Chatterjee A and Pakrashi AC 1995. The treatise on Indian medicinal plants, Vol. IV, Publication & Information Directorate, CSIR, New Delhi.

Chaturvedi S 2015. *In-vitro* antioxidant activity, phenolic and flavonoid Content of aerial parts of *Rhynchosia capitata* DC. *World Journal of Pharmacy and Pharmaceutical Sciences,* 4(3): 713-724.

Chaudhary B and Mukhopadhyay K 2012. *Syzygium cumini* (l.) Skeels: A potential source of nutraceuticals. *International Journal of Pharma and Bio Sciences,* 2(1): 46-53.

Chaudhary LB, Sudhakar Jana Venkata, Kumar Anoop, Bajpai Omesh, Tiwari Rinkey and Murthy VS 2012. Synopsis of the Genus *Ficus* L. (Moraceae) in India. *Taiwania,* 57(2): 193-216.

Chaudhary RS 2010. Taxa of family Fabaceae: A potential of local medicinal values in Vindhya region Uttar Pradesh, India. *International Journal of Pharma and Bio Sciences,* 1(4): B46-B53.

Chaudhary S 2011. Medicinal plants of district Bijnor (U.P.) India with special reference to their folk medicinal uses. *Journal of Experimental Sciences,* 2(4): 19-23.

Chaudhary S and Gupta AK 2011. Folk medicinal uses of the plants of Bijnor district (UP), India. *Current Botany,* 2(1): 45-47.

Chaudhary S and Kumar R 2015. Ethnomedicinal plants of the district Bijnor (U.P.) India. *Indian Botanical Society,* 94 (1 & 2):97-103.

Chaudhary S and Kumar R 2015. Folk medicinal plants in Ghaziabad district of western Uttar Pradesh. India. *Indian Botanical Society,* 94(1&2): 73-80.

Chaudhuri AB 2007. Endangered medicinal plants. Daya Publishing House, p. 307-308.

Chaudhury S and Das I 2009. Naturally colonized mangroves in the digha-junput coast, west Bengal with their Ethno-medicinal value. http://www.indiaenvironmentportal.org.in/files/Mangroves%20Digha.pdf.

Chauhan PK, Sharma S, Chandrika, Harsh, Manisha and Mansi 2014. Evaluation of phytochemical and in-vitro antioxidant and antibacterial activities of wild plant species of *Bauhinia* and *Ficus* sp. *World Journal of Pharmacy and Pharmaceutical Sciences,* 3(4): 659-668.

Chauhan PP, Nigam A and Santvan VK 2016. Ethnobotanical study of wild fruits in Pabbar valley, district Shimla, Himachal Pradesh. *Journal of Medicinal Plants Studies,* 4(2): 216-220.

Chauhan R, Ruby, Shori A and Dwivedi A 2012. *Solanum nigrum* with dynamic therapeutic role: A review. *International Journal of Pharmaceutical Sciences Review and Research,* 15(1): 65-71.

Chavan S and Nishteswar K 2013. A review on source plants of Dronpushpi- *Leucas cephalotes* (Roth) Spreng and *Leucas aspera* Spreng. *International Journal of Ayurvedic Medicine,* 4(4): 290-303.

Chetan C, Suraj M, Maheshwari C, Rahul A and Priyanka P 2011. Screening of antioxidant activity and phenolic content of whole plant of *Barleria prionitis* Linn. *International Journal of Research in Ayurveda & Pharmacy,* 2(4): 1313-1319.

Chhabra BR, Kohli JC and Dhillon RS 1999. Three ambrosanolides from *Parthenium hysterophorus. Phytochemistry (Oxford),* 52(7): 1331-1334.

Chhatre S, Nesari T, Somani G, Kanchan D and Sathaye S 2014. Phytopharmacological overview of *Tribulus terrestris. Pharmacognosy Reviews,* 8(15): 45-51.

Chhetree RR, Dash Gk, Mondal S, and Parhi R 2012. Studies on the hypoglycaemic activity of the bark of salix *tetrasperma* Roxburgh. *International Journal of Drug Development & Research,* 2(4): 799-805.

Chinsembu KC 2015. Plants as antimalarial agents in Sub-Saharan Africa. *Acta Tropica* 152: 32-48.

Chintawar SD, Somani RS, Kasture VS and Kasture SB 2002. Nootropic activity of *Albizzia lebbeck* in mice. *Journal of Ethnopharmacology*, 81(3): 299-305.

Chithra M, Prabhu Kumar KM and Geetha SP 2016. A comparative study on ethanobotanical usage of plants for twenty selected diseases by six tribal communities in Malappuram district. *International Journal of Herbal Medicine,* 4(4): 108-113.

Chokchaisiri R, Suaisom C, Sriphota S, Chindaduang A, Chuprajob T and Suksamrarn A 2009. Bioactive flavonoids of the flowers of *Butea monosperma. Chemical & Pharmaceutical Bulletin (Tokyo)*, 57(4): 428-432.

Chong NJ and Aziz Z 2011. A systematic review on the chemical constituents of *Centella asiatica. Research Journal of Pharmaceutical, Biological and Chemical Sciences*, 2(3): 445-459.

Chopra RN, Chopra IC and Verma BS 1969. Supplement to Glossary of Indian Medicinal Plants. CSIR, New Delhi.

Chopra RN, Chopra IC, Handa KL and Kapur LD 1958. Indigenous drugs of India, Academic Publishers: Calcutta, 2nd ed., p. 508-674.

Chopra RN, Nayar SL and Chopra IC 1956. Glossary of Indian medicinal plants. Council of Scientific and Industrial Research, New Delhi, India, p. 170.

Choudhary GP and Jain AP 2016. A review on *Mitragyna parvifolia* (Roxb.) Korth. - An Indian medicinal plant. *International Journal of Pharmacy and Pharmaceutical Research,* 7(1): 175-184.

Choudhary K and Singh M 2008. Ethnobotanical Survey of Rajasthan- An update. *American-Eurasian Journal of Botany*, 1(2): 38-45.

Chowdhury A and Das AP 2015. Ethnopharmacological survey of wetland plants used by local ethnic people in Sub-Himalayan Terai and Duars of West Bengal, India. *American Journal of Ethnomedicine*, 2: 112-135.

Chowdhury AR and Banerji R 1995. Studies on Leguminous seeds. Lipid, 97: 457-458.

Chowdhury AR, Banerji R, Misra G and Nigam SK 1984. Studies on leguminous seeds. *Journal of the American Oil Chemists Society*, 61(6):1023-1024.

Chowdhury NP 1973. The Pteridophyte Flora of the Upper Gangetic Plains. Navyug Traders. New Delhi.

Christophe W, Pharm D 2002. Ethnopharmacology of medicinal plants: Asia and the Pacific: Humana Press, 1:69.

Chudamani Burlakoti and Ripu M. Kunwar 2008. Folk herbal medicines of Mahakali watershed area, Nepal. *Ecological Society*, 187-193.

Chulet R, Pradhan P, Sharma KS and Jhajharia KM 2010. Phytochemical screening and antimicrobial activity of *Albizzia lebbeck*. *Journal of Chemical and Pharmaceutical Research,* 2(5):476-484.

Cook CDK 1996. Aquatic and wetland plants of India. Oxford University Press, p. 400.

Corrêa GM and Alcântara AFDC 2012. Chemical constituents and biological activities of species of *Justicia*- A review. *Revista Brasileira de Farmacognosia,* 22(1): 220-238.

Corrêa MP 1984. Dictionary of useful plants of Brazil and exotic cultivated. Rio de Janeiro: Ministry of Agriculture / Brazilian Institute of Forest Development.

Correa WR, Tasco AJH, Marinho JVN, Pascoal ACRF, de Carvalho JE, Marchioretto MS and Salvador MJ 2016. Antioxidant and cytotoxic activities and chemical profile of five Amaranthaceae plants collected in the South of Brazil. *Natural Products Chemistry & Research*, 4(5): 1-7.

Cortés-Rojas DF, Chagas-Paula DA, Costa FBD, Souza CRF and Oliveira WP 2012. Bioactive compounds in *Bidens pilosa* L. populations: A key step in the standardization of phytopharmaceutical preparations. *Revista Brasileira de Farmacognosia,* 23(1): 28-35.

Cousins D and Huffman MA 2002. Medicinal properties in the diet of Gorillas: An ethno-pharmacological evaluation. *African Study Monographs*, 23(2): 65-89.

Cragg GM, Newman DJ, Snader KM 1997. Natural products in drug discovery and development. *Journal of Natural Products,* 60(1): 52-60.

Cutillo F, D'Abrosca B, DellaGreca M, Zarrelli A 2004. Chenoalbicin, a novel cinnamic acid amide alkaloid from *Chenopodium album. Chemistry and Biodiversity,* 1: 1579-1583.

Dadhiach LK and Rathore NK 2015. Ethno medicinal plants used in diabetes by tribals of Jhalawar district (Raj.). *Remarking,* 1(10): 26-28.

Dafni A, and Yaniv Z 1994. Solanaceae as Medicinal Plants in Israel. *Journal of Ethnopharmacology*, 44: 11-18.

Dalwadi PP, Patel JL and Patani PV 2014. *Tephrosia purpurea* Linn. (Sharpunkha, Wild Indigo): A review on phytochemistry and pharmacological studies. *Indian Journal of Pharmaceutical and Biological Research,* 2(1):108-121.

Dangol DR 2008. Traditional uses of plants of commonland habitats in western Chitwan, Nepal. *Journal of the Institute of Agriculture and Animal Science*, 29: 71-78.

Dangwal LR and Sharma A 2011. Indigenous traditional knowledge recorded on some medicinal plants in Narendra Nagar block (Tehri Garhwal), Uttarakhand. *Indian Journal of Natural Products and Resources,* 2(1): 110-115.

Dangwal LR, Rana CS and Sharma A 2011. Ethno-medicinal plants from transitional zone of Nanda Devi Biosphere Reserve, District Chamoli, Uttarakhand (India). *Indian Journal of Natural Products and Resources,* 2(1): 116-120.

Dangwal LR, Sharma A and Rana CS 2010. Ethno-medicinal plants of the Garhwal Himalaya used to cure various diseases: A case study. *New York Science Journal,* 3(12): 28-31.

Das AK, Dutta BK and Sharma GD 2008. Medicinal plants used by different tribes of Cachar district, Assam. *Indian Journal of Traditional Knowledge,* 7(3): 446-454.

Das B, Kundu J, Bachar SC, Kundu JK 2014a. Anti-inflammatory, analgesic and diuretic activity of *Ludwigia hyssopifolia* Linn. *Archives of Medical Research,* 1(4):139-146.

Das B, Kundu J, Bachar SC, Uddin MA and Kundu JK 2007. Antitumour and antibacterial activity of Ethylacetate extract of *Ludwigia hyssopifolia* Linn. and its active principle piperine. *Pakistan Journal of Pharmaceutical Sciences,* 20(2): 128-131.

Das B, Talukdar AD and Choudhury MD 2014b. A few traditional medicinal plants used as antifertility agents by ethnic people of Tripura, India. *International Journal of Pharmacy and Pharmaceutical Sciences,* 6(3): 47-53.

Das M and Himaja M 2014. Phytochemical screening, GC-MS analysis and biological activity of *Ipomoea eriocarpa* leaf extracts. *International Journal of Pharmacy and Pharmaceutical Sciences,* 6(4): 592-594.

Das NJ, Saikia SP and Devi K 2006. Medicinal plants of North-Kamrup district of Assam used in primary healthcare system. *Indian Journal of Traditional Knowledge,* 5(4): 489-493.

Das R, Das G and Pahari S 2017. Endangered dune ecology with its risky Phyto-resources – a red alarm to Mandermoni-Tajpur sector over Midnapore coast in West Bengal. *International Journal of Recent Scientific Research,* 8(6): 17371-17384.

Dashahre AK, Navaneethan B, Bhutt P and Mahato S ©2014. Medicinal plants of Sariska Tiger Reserve, (Rajasthan) India. *Journal of Medicinal Plants Studies,* 2(2): 137-146.

Datta AK, Das A, Bhattacharya A, Mukherjee S and Ghosh BK 2010. An Overview on *Withania somnifera* (L.) Dunal – The 'Indian ginseng'. *Medicinal and Aromatic Plant Science and Biotechnology,* 5(1): 1-15.

Datta SC and Banerjee AK 1978. Useful weeds of West Bengal rice fields. *Economic Botany,* 32: 297-310.

Dave PN and Bhandari J 2013. *Prosopis julifera*: A review. *International Journal of Chemical Studies,* 1(3): 181-196.

De Pascual TJ, Vincente S, Gonzalez MS, Bellido IS 1983. Nerolidol-5, 8-oxides from the essential oil of *Santolina oblongifolia. Phytochemistry,* 22: 2235-2238.

De Pasquale A 1984. Pharmacognosy: the oldest modern science. *Journal of Ethnopharmacology,* 11: 1-16.

Debta MR and Srivastava SK 2015. Flora of Sultanpur National Park, Gurgaon district, Haryana. Botanical Survey of India.

Deepa MR, Dharmapal S and Udayan PS 2016. Medicinal plants in the selected sacred groves of Kodungallur, Thrissur district, Kerala. *Journal of Medicinal Plants Studies,* 4(3): 149-155.

DeFilipps RA 2004. Medicinal plants of the Guianas (Guyana, Surinam, French Guiana). Department of Botany, National Museum of Natural History, Smithsonian Institution.

DellaGreca M, D'Abrosca B, Fiorentino A, Previtera L, Zarrelli A 2005a. Structure elucidation and phytotoxicity of ecdysteroids from *Chenopodium album. Chemistry and Biodiversity,* 2: 457-462.

DellaGreca M, Di Marino C, Zarrelli A and D'Abrosca B 2004. Isolation and phytotoxicity of apocarotenoids from *Chenopodium album. Journal of Natural Products,* 679):1492-5.

Deshpande UR and Singh NP 1986. Grasses of Maharashtra: An annotated inventory. Mittal Publications, p. 129.

Devendra NK, Deshpande Raghunandan and Seetharam Y 2010. Anti-inflammatory activity of *Trichosanthes cucumerina* L. var. *cucumerina* seeds. *Pharmacologyonline,* 2: 172-176.

Devendra NK, Vijaykumar BM, Seetharam YN, Suresh P and Saraswati BP 2009. Effect of ethanol extract of whole plant of *Trichosanthes cucumerina* L. var. *cucumerina* on gonadotropins, ovarian follicular kinetics and estrous cycle for screening of antifertility activity in albino rats. *International Journal of Morphology,* 27(1): 173-182.

Devi AD, Devi OI, Singh TC and Singh EJ 2014. A study of aromatic plant species especially in Thoubal district, Manipur, North East India. *International Journal of Scientific and Research Publications,* 4(6): 1-12.

Devi B, Sharma N, Kumar D and Jeet K 2013. *Morus alba* Linn: A phytopharmacological review. *International Journal of Pharmacy and Pharmaceutical Sciences,* 5: 14-18.

Devmurari VP, Ghodasara TJ and Jivani NP 2010. Antibacterial activity and phytochemical study of ethanolic extract of *Triumfetta rhomboidea* Jacq. *International Journal of PharmTech Research,* 2(2): 1182-1186.

Dey J, Saha PR, Debnath P and Sinha R K 2017. Comparative studies on some biochemical parameters of *Cajanus scarabaeoides* (L.) Thouars and *Cajanus cajan* (L.) Millsp. *Tropical Plant Research,* 4(2): 314-318.

Dhanalakshmi R and Manavalan R 2014. Determination of bioactive constituents of leaves of *Corchorus* aestuans (L.) by GC - ms analysis. *International Journal of Pharmacy and Pharmaceutical Sciences,* 6(9): 248-251.

Dhankhar S, Kaur R, Ruhil S, Balhara M, Dhankhar S and Chhillar AK 2011. A review on *Justicia adhatoda*: A potential source of natural medicine. *African Journal of Plant Science,* 5(11): 620-627.

Dharmananda S 2000. Bidens: A popular remedy escapes notice of western practitioners. *ITM*, 9 pages. http://www.itmonline.org/arts/bidens.htm.

Dhiman AK 2004. Medicinal Plants of Uttaranchal State. Chowkhamba Sanskrit Series Office, p. 565.

Dibala CI, Konaté K, Diao M, Ouédraogo M and Dicko MH 2014. Phytoconstituents analysis, antioxidant capacity and antimicrobial properties of extracts from *Laggera aurita* L. (Asteraceae). *International Journal of Pharmacy and Pharmaceutical Sciences*, 6(7): 172-178.

Dighe PA, and Dighe S 2014. A review on medicinal fruit Bhokar of species *Cordia dichotoma* Forst. *International Journal of Pharmaceutical & Biological Archives,* 5(3): 41-47.

Dileep P and Nair GG 2015. Taxonomic and ethanobotancial studies of grasses used by tribals of Wayanad district, Kerala, South Wetern Ghats of India. *Journal of Global Scicence*, 4(5): 2212-2235.

Dinan L 1992. The analysis of phytoecdysteroids in single (preflowering stage) specimens of fat hen, *Chenipodium album. Phytochem Analysis,* 3:132-138.

Dinda B, Ghosh B, Arima S, Sato N and Harigaya Y 2007. Chemical constituents of *Evolvulus nummularius. Indian Journal of Chemistry,* 46B: 492-498.

Divakar MC, John J, Vyshnavidevi, Poornima, Anisha, Subash A and Govindan V 2013. Herbal remedies of Madayipara hillock tribals in Kannur district, Kerala, India. *Journal of Medicinal Plants Studies,* 1(6): 34-42.

Divya S, Krishna KN, Ramachandran S and Dhanaraju MD 2011. Wound healing and *in vitro* antioxidant activities of *Croton bonplandianum* leaf extract in rats. *Global Journal of Pharmacology,* 5(3): 159-163.

Dixit Ram Das and Vohra Jitinder Nath 1984. A dictionary of the Pteridophytes of India. Botanical Survey of India, Calcutta. Series: Flora of India, Series 4.

Dogra KS, Sood SK, Dobhal PK and Sharma S 2010. Alien plant invasion and their impact on indigenous species diversity at global scale. *A Review Journal of Ecology and the Natural Environment,* 2: 175-186.

Dokosi OB 1998. Herbs of Ghana, Ghana Universities Press, Accra, p. 313-314.

Dolui AK, Sharma HK, Marein TB and Lalhriatpuii TC 2004. Folk herbal remedies from Meghalaya. *Indian Journal of Traditional Knowledge,* 3(4): 358-364.

Drabu S, Chaturvedi S and Sharma M 2012. *Tamarix gallica* – An overview. *Asian Journal of Pharmaceutical and Clinical Research,* 5(3): 17-19.

Du Y, Feng J, Wang R, Zhang H and Liu J 2015. Effects of flavonoids from *Potamogeton crispus* L. on proliferation, migration, and invasion of human ovarian cancer cells. *Public Library of Science One,* 10(6): 1-17. https://doi.org/10.1371/journal.pone.0130685.

Dubey NK 2015. Plants as a source of natural antioxidants, CABI.

Duke JA 2008. Duke's Handbook of medicinal plants of Latin America. CRC Press, p. 832.

Duke JA and Wain KK 1981. The medicinal plants of the World, Computer index with more than 85,000 Enteries. p. 3.

Duke JA with Mary Jo Bogenschutz-Godwin Judi duCellier Peggy- Ann K. Duke 2002. Handbook of medicinal herbs, Second Edition. CRC Press, p. 896.

Dular AK 2015. Study of biodiversity of Sariska Tiger Reserve in Aravallis with particular emphasis on medicinal and endemic plants. *International Journal of Science and Research,* 4(2): 420-425.

Durodola JJ 1977. Antibacterial property of crude extracts from herbal wound healing remedy – *Ageratum Conyzoides. Planta Medica,* 32:388-390.

Duthie JF 1903-29. Flora of the Upper Gangetic Plain and of the Adjacent Siwalik and Sub-Himalayan tracts. Calcutta.

Duthie JF 1960. Flora of the Upper Gangetic Plain and of the Adjacent Siwalik and Sub-Himalayan tracts. Reprint Edition, Botanical Survey of India, Calcutta.

Dutt UC, Lall Sen KBL 1989, 1995. The Materia Medica of the Hindus. Mittal Publications.

Dutta P 2015. Pharmacognostical evaluation and preliminary phytochemical analysis of *Alternanthera philoxeroides. International Journal of Medi. Pharm Research,* 1(1): 7-13.

Dutta U and Sarma GC 2013. Medicinal plants used by the local Fringe communities of Chirang Reserve Forest, BTAD, Assam. *Indian Journal of Research,* 2(2): 262-265.

Dutta, DAK 2017. A review on potency of *Butea monosperma* in the field of traditional medicinal system. *World Journal of Pharmacy and Pharmaceutical Sciences,* 6(5): 309-320.

Ebadi M 2006. Pharmacodynamic basis of herbal medicine, Second Edition. CRC Press, p. 699.

Ediriweera ERHSS 2007. A review on medicinal uses of weeds in Sri Lanka. *Tropical Agricultural Research & Extension,* 10: 11-16.

Ekundayo O 1988. Volatile constituents of *Pinus* needle oils. *Flavour and Fragrance,* 3(1): 1-11.

Ekunseitan DA, Yusuf AO and Odesanmi AF 2013. Assessment of nutritive values of some waterweeds. *Food Science and Quality Management,* 22: 23-27.

El-Gengaihi SE, Hassan EE, Hamed MA, Zahran HG and Mohammed MA 2013. Chemical Composition and biological evaluation of *Physalis peruviana* Root as Hepato-renal Protective Agent. *Journal of Dietary Supplements,* 10(1):39-53.

El-Ghazali GE, AlKhalifa KS, Saleem GA and Abdallah EM 2010. Traditional medicinal plants indigenous to Al-Rass province, Saudi Arabia. *Journal of Medicinal Plants Research,* 4(24): 2680-2683.

El-Mokasabi FM 2014. The state of the art of traditional herbal medicine in the eastern Mediterranean coastal region of Libya. *Middle-East Journal of Scientific Research,* 21(4): 575-582.

El-Sayed HM, El-Nor, SAHA, Kholif AM and Tawila MA 1999. *In vivo* evaluation of coffee bean shell as a component in ruminant rations. *Egyptian Journal of Nutrition and Feeds,* 2(1): 1-7.

Elumalai K and Krishnappa K 2015. Mosquitocidal properties of *Oxystelma esculentum* (Asclepiadaceae)-Indian medicinal plant tested against *Ades aegypti* (Linn.) (Diptera: Culicidae). *Journal of Coastal Life Medicine,* 3(4): 326-332.

El-Wakil EA, El-Sayed S. Abdel-Hameed E, Mortada M. El-Sayed MM and Abdel-Lateef EE 2015. Identification of the chemical composition of the methanolic extract of *Salix tetrasperma* Roxb. Using LC-ESI-MS and evaluation its potential as antioxidant agent. *Der Pharma Chemica*: 7(2):168-177.

Emniyet AA, Avcı E, Ozcelik B, Avcı GAA and Kose DA 2014. Antioxidant and antimicrobial activities with GC/MS analysis of *Morus alba* L. leaves. *Hittite Journal of Science & Engineering,* 1(1): 37-41.

Ereifej KI, Feng H, Rababah T, Almajwal A, Aludatt M, Gammoh SI and Oweis LI 2015. Chemical composition, phenolics, anthocyanins concentration and antioxidant activity of ten wild edible plants. *Food and Nutrition Sciences,* 6: 581-590.

Esha RT, Chowdhury MR, Adhikary S, Khan MDH, AK Acharjee M, Nurunnabi MD, Khatun Z, Lee YK and Rahmatullah MD 2012. Medicinal plants used by tribal medicinal practitioners of three clans of the Chakma tribe residing in Rangamati district, Bangladesh. *American-Eurasian Journal of Sustainable Agriculture,* 6(2): 74-84.

Esmaeili A., Rashidi B and Rezazadah S 2012. Biological activities of various extracts and chemical composition of *Trigonella monantha* C.A. Mey. sub species *monantha* grown in Iran. *Iranian Journal of Pharmaceutical Research,* 11(4): 1127-1136.

Essiett UA and Obioboho GE 2014. Phytochemical, nutrients and antinutrients of *Ipomoea triloba, Ipomoea batatas, Ipomoea involucrata* leaves. *International Journal of Research,* 1(11): 1412-1418.

Evelyne PS, Ravindhran R 2015. Pharmacognostical characterization of *Phyla nodiflora* L. (Verbenaceae): a valuable medicinal plant. *Asian Journal of Pharmaceutical and Clinical Research,* 8(3): 256-261.

Facknath S, Kawol D 1996. Antifeedant and insecticidal effects of some plant extracts on *Crocidolomia binotalis. Insect Science and its Application,* 14(5/6): 571-74.

Fagbohun ED, Asare RR and Egbebi AO 2012. Chemical composition and antimicrobial activities of *Urena lobata* L. (Malvaceae). *Journal of Medicinal Plants Research,* 6(12): 2256-2260.

Faisal M, Singh PP and Irchhaiya R 2012. Review on *Albizia lebbeck* - A potent herbal drug. *International research journal of pharmacy,* 3(5): 63-68.

Falodun A, Okunrobo LO, and Uzoamaka N 2006. Phytochemical screening and anti-inflammatory evaluation of methanolic and aqueous extracts of *Euphorbia heterophylla* Linn (Euphorbiaceae). *African Journal of Biotechnology,* 5(6): 529-531.

Faraji P and Jamei R 2016. The study of phenolic compounds antioxidant activity in methanolic and aqueous extracts of several plant species of Urmia Lake margin. *International Journal of Pharmacognosy and Phytochemical Research,* 8(2): 289-296.

Farhan H, Rammal H, Hijazi A and Badran B 2012. Preliminary phytochemical screening and extraction of polyphenol from stems and leaves of a Lebanese plant *Malva parviflora* L. *International Journal of Current Pharmaceutical Research,* 4(1): 55-59.

Farnsworth NR 1993. Ethnopharmacology and future drug development: The North American experience. *Journal of Ethnopharmacology,* 38 (2-3):145-52.

Farnsworth NR and Soejarto DD 1991. Global importance of medicinal plants. In: Akereb O, Heywood V, Synge H (eds.) Conservation of medicinal plants. Cambridge University Press, Cambridge.

Farnsworth NR, Akerele O, Bingel AS, Soejarto DD and Guo Z 1985. Medicinal plants in therapy. *Bulletin of the World Health Organization,* 63: 965-981.

Farooqui M, Pradhan V, Khan JV and Khan TA 2016. Primary phytochemical study of *Psoralea corylifolia* L. and *Caesulia axillaris* Roxb. *Journal of Chemical and Pharmaceutical Research,* 8(3): 198-201.

Fatima N, Zia M, Rehman R, Rizvi ZF, Ahmad S, Mirza B and Chaudhary MF 2009. Biological activities of *Rumex dentatus* L: Evaluation of methanol and hexane extracts. *African Journal of Biotechnology,* 8(24): 6945-6951.

Fawar S and Patil DA 2001. Herbal folk medicines of Jalgaon district (Maharashtra). *Ancient Science of Life,* 21(2): 87-91.

Feizbakhsh A, Aghassi A and Naeemy A 2013. Chemical Constituents of the Essential Oils of *Cyperus difformis* L. and *Cyperus arenarius* Retz from Iran. *Journal of Essential Oil Bearing Plants,* 15(1): 48-52.

Ferrington EA. 1990. Clinical Materia Medica (Reprint Ed.) B. Jain Publ. Pvt. Ltd., New Delhi, Ganapathm. Kalyani Publishers, Ludhiana, India. p. 347-353.

Franco MF and Narasimhan D 2009. Plant names and uses as indicators of knowledge patterns. *Indian Journal of Traditional Knowledge,* 8(4): 645-648.

Frei B, Sticher O and Heinrich M 2000. Zapotec and Mixe use of tropical habitats for securing medicinal plants in Mexico. *Economic Botany*, 54: 73-81.

Gaddeyya G and Ratna Kumar DPK 2015. Acomprehensive review on ethnobotany and photochemistry of an herbal weed *Trianthema portulacastrum* L. *Journal of Pharmacognosy and Phytochemistry*, 4(4): 25-31.

Gadgil M and Rao RPS 1993. Nurturing biodiversity: An Indian Agenda. Center for Environment Education, Ahemdabad

Gafar MK and Itodo AU 2011. Proximate and mineral composition of hairy Indigo leaves. *Research Journal of Pharmaceutical, Biological and Chemical Sciences*, 2(1): 669-682.

Gafar MK, Hassan LG, Dangoggo SM and Itodo AU 2010. Investigating the foliage of *Indigofera astragolina* for their nutritive and antinutritive composition. *Journal of Chemical and Pharmaceutical Research*, 2(5): 259-269.

Gairola S, Sharma J and Bedi YS 2014. A cross-cultural analysis of Jammu, Kashmir and Ladakh (India) medicinal plant use. *Journal of Ethnopharmacology*, 155: 925-986.

Galani VJ and Rachchh RP 2015. *Grangea maderaspatana* (L.) Poir. – A comprehensive review. *Innoriginal International Journal of Sciences*, 2(3): 1-8.

Galav P, Jain A, Katewa SS & Nag A 2010. Animal healthcare practices by livestock owners at Pushkar animal fair, Rajasthan. *Indian Journal of Traditional Knowledge*, 9(4): 660-663.

Gangrade NK, Sheorey RV, Rawal H and Chouhan S 2012. Investigation of wound healing activity of *Malvastrum tricuspidatum* syn. *Malvastrum coromandelianum* on experimental animals. *International Journal of Pharmacy and Life Sciences*, 3(12): 2259-2264.

Gangwar AK and Ghosh AK 2014. Medicinal uses and pharmacological activity of *Adhatoda vasica*. *International Journal of Herbal Medicine*. 2(1): 88-91.

Ganie SA and Yadav SS 2014. *Holoptelea integrifolia* (Roxb.) Planch: A review of its ethnobotany, pharmacology, and phytochemistry. *BioMed Research International*, (1):401213.

Ganjhu RK, Mudgal PP and Arunkumar G 2014. Pharmacological and phytoconstituent profile of *Desmodium gangeticum*- an update. *International Journal of Pharmacognosy and Phytochemical Research*, 6(3): 643-657.

Garaniya N and Bapodra A 2014. Ethno botanical and phytophrmacological potential of *Abrus precatorius* L.: A review. *Asian Pacific Journal of Tropical Biomedicine*, 4: S27-S34.

Gaur RD 1999. Flora of the district Garhwal North West Himalaya: With ethnobotanical notes. Transmedia, Srinagar Garhwal, India. p. 1-811.

Gaur RD, Sharma J and Painuli RM 2010. Plants used in traditional healthcare of livestock by Gujjar community of Sub Himalayan tracts, Uttrakhand, India. *Indian Journal of Natural Products and Resources*, 1(2): 243-248.

Gaurav N, Kumar A, Tyagi M, Kumar D, Chauhan UK and Singh AP 2015. Morphology of *Withania somnifera* (distribution, morphology, phytosociology of *Withania somnifera* L. Dunal). *International Journal of Current Science Research*, 1(7): 164-173.

Gautam AK, Bhatia MK and Bhadauria R 2011. Diversity and usage custom of plants of South-western Himachal Pradesh, India-Part I. *Journal of Phytology*, 3(2): 24-36.

Gautam P and Richhariya GP 2015. Ethnoveterinary medicinal plants used by tribal and rural communities of Chitrakoot, District Satna (M.P.). *International Journal of Pharmacy and Life Sciences*, 6(4): 4427-4430.

Gavalapu VR, Kolli P, Korra SK, Kavuri MK, Avagadda C, Singam V, Vanumu Y and Kudirella H 2013. Preliminary phytochemical screening and anthelmintic activity of *Desmodium triflorum* (L.) DC leaf and root extracts. *International Journal of Pharma Sciences*, 3(1): 156-158.

Gbolade AA, Tira-Picos V and Nogueria JMF 2010. Chemical constituents of *Chenopodium ambrosioides* var. *anthelminticum* herb essential oil from Nigeria. *Chemistry of Natural Compounds*, 46(4): 654-655.

Genest S, Kerr C, Shah A, Rahman MM, Saif-E-Naser GMM, Nigam P, Nahar L and Sarker SD 2008. Comparative bioactivity studies on two Mimosa species. *Boletín Latinoamericano y del Caribe de Plantas Medicinales y Aromáticas*, 7(1): 38-43.

Gentry AH 1993. Tropical forest biodiversity and the potential for new medicinal plants. *American Chemical Society*, 534: 13-24.

Ghaisas M, Navghare V, Takawale A, Zope V, Tanwar M and Deshpande A 2009. Effect of *Tectona grandis* Linn. on dexamethasone induced insulin resistance in mice. *Journal of Ethnopharmacology*, 122: 304-307.

Ghatapanadi SR, Johnson N and Rajasab AH 2010. Medicinal plants of North Karnataka used in treatment of kidney stones and urinary tract infections. *The Bioscan*, 2(3&4): 23-24.

Ghatapanadi SR, Johnson N and Rajasab AH 2011. Documentation of folk knowledge on medicinal plants of Gulbarga district, Karnataka. *Indian Journal of Traditional Knowledge*, 10(2): 349-353.

Ghazanfar SA 1994. Handbook of Arabian Medicinal Plants. CRC Press, p. 272.

Ghildiyal JC, Juyal P and Sadana G 2014. Indigenous uses of plants in different women ailments in Garhwal region. *Indian Journal of Pharmaceutical and Biological Research*, 2(1): 39-44.

Gholami A 2015. Biology of *Alysicarpus monilifer* (L.) DC.: An important medicinal plant. *Medicinal Plants - International Journal of Phytomedicines and Related Industries*, 7(3): 248-251.

Ghorband, Biradar DP and Sharad D 2012. Folk medicine used by the tribes of Kinwat forest of Nanded district, Maharashtra, India. *Indian Journal of Natural Products and Resources,* 3(1): 118-122.

Ghori MK, Ghaffari MA, Hussain SN, Manzoor M, Aziz M, Sarwer W 2016. Ethnopharmacological, phytochemical and pharmacognostic potential of genus *Heliotropium* L. *Turkish Journal of Pharmaceutical Sciences,* 13(2): 259-280.

Ghos NC 1988. Comparative Materia Medica. Hannemann Publ. Co. Pvt. Ltd. Calcutta, India.

Ghoshal KP 2015. Diversity, indigenous use and conservation of ethno-medicinal wealth of Gondia District of Maharashtra, India. *Journal of Natural Product and Plant Resources,* 5(1):1-5.

Girish HV and Satish S 2008 Antibacterial activity of important medicinal plants on human pathogenic bacteria- A comparative analysis. *World Applied Sciences Journal,* 5(3): 267-271.

Gnanadeebam S and Viswanathan P 2014. GC-MS Analysis of methanolic extract of *in vitro* and *in vivo* explants of *Spermacoce articularis* L. f. *Medical Science,* 13: 39-42.

Gogoi B and Zaman K 2013. Phytochemical constituents of some medicinal plant species used in recipe during 'Bohag Bihu' in Assam. *Journal of Pharmacognosy and Phytochemistry,* 2(2): 30-40.

Gohar AA, Elmazar MMA 1997. Isolation of hypotensive flavonoids from *Chenopodium* species growing in Egypt. *Phytotherapy Research,* 11:564-567.

Gohil KJ, Patel JA and Gajjar AK 2010. Pharmacological review on *Centella asiatica*: A potential herbal cure-all. *Indian Journal of Pharmaceutical Sciences,* 72(5): 546-556.

Goldfrank L, Lewin N, Flomenbaum N and Howland M A 1982. The Pernicious panacea: herbal medicine. *Hospital physician,* 10: 64-86.

Gond DK, Kumar S, Samuel CO, Saini DC, Kulshreshtha K and Abbasi P 2012. Ethno-medicinal studies on indigenous wetland plants of Mau district of Uttar Pradesh India. Proceeding of the National conference on "Climate change, Biodiversity & Conservation".

Gong F, Gu Hai Peng, Xu Qi-tai and Kang Wen-yi 2012. Genus *Mitragyna*: Ethnomedicinal uses and pharmacological studies. *Phytopharmacology,* 3(2): 263-272.

Gopalakrishnan K and Udayakumar R 2014. Antimicrobial activity of *Marsilea quadrifolia* (L.) against some selected pathogenic microorganisms. *British Microbiology Research Journal,* 4(9): 1046-1056.

Gorski PM, Jurzysta M, Burda S, Oleszek WA and Ploszynski M 1984. Studies on *Medicago lupulina* saponins 2. Isolation, chemical characterization and biological activity of saponins from *Medicago lupulina* tops. *Acta Societatis Botanicorum Poloneae,* 53(4): 527-533.

Gothandam KM, Aishwarya R and Karthikeyan S 2010. Preliminary screening of antimicrobial properties of few medicinal plants. *Journal of Phytology,* 2(4): 01–06.

Goun EA, Petrichenko VM, Solodnikov SU, Suhinina TV, Kline MA, Cunningham G, Nguyen C and Miles H 2002. Anticancer and antithrombin activity of Russian plants. *Journal of Ethnopharmacology,* 81:337-342.

Gowri J, Pragathiswaran C and Sahayaraj AP 2017. Physicochemical and heavy metal analysis of the leaf, stem, and flower extracts of *Sphenoclea zeylanica. Innovare Journal of Science,* 5(1): 28-31.

Gowri J, Sahayaraj PA, Dawood AS and Prema AA 2016. Evaluation of phytochemical qualitative and quantitative analysis of *Sphenoclea zeylanica. World Journal of Pharmacy and Pharmaceutical Sciences,* 5(4):1637-1646.

Goyal N and Sharma SK 2014. Bioactive phytoconstituents and plant extracts from genus *Heliotropium. International Journal of Green Pharmacy,* 8(4): 217-225.

Gritto MJ, Nanadagopalan V and Doss A 2015. Ethnobotanical survey of medicinal plants used by traditional healers in Shobanapuram village of Pachamalai Hill, Tamil Nadu. *Advances in Applied Science Research,* 6(3):157-164.

Guha A and Chakma D 2015. Traditional usage of ethno-medicinal plants among the Chakma community of Tripura, India. *Global Journal of Pharmacology,* 9(4): 377-384.

Guhabakshi DN, Sensarma P and Pal DC 1999. A lexicon of medicinal plant in India. New Delhi, p. 180-181.

Gulecha VS, Sivakumar T, Mahajan MS, Khandare RA and Upasani CD 2010. *Ficus religiosa* and *Ficus glomerata* Extracts exhibits antiinflammatory activity in the Rat. *Pharmacologyonline,* 1:324-331.

Gulnaz AR and Savitha G 2013. Evaluation of antimicrobial activity of leaf and stem extracts of Sidda medicinal plant *Sida cordata. International Journal of Medicine and Pharmaceutical Sciences,* 3(3): 39-50.

Gulshan AB, Dasti AA, Hussain S, Atta MI and Aminudin M 2012. Indigenous uses of medicinal plants in rural areas of Dera Ghazi Khan, Punjab, Pakistan. *Journal of Agricultural & Biological Science,* 7(9): 750-762.

Gunasekaran M and Balasubramanian P 2012. Ethnomedicinal uses of Sthalavrikshas (temple trees) in Tamil Nadu, Southern India. *Ethnobotany Research and Applications,* 10: 253-268.

Gundidza M, Gweru N, Magwa ML, Ramalivhana NJ, Humphrey G, Samie A and Mmbengwa V 2009. Phytochemical composition and biological activities of essential oil of *Rhynchosia minima* (L) (DC) (Fabaceae). *African Journal of Biotechnology,* 8(5):721-724.

Gupta A and Pandey VN 2014. Herbal remedies of aquatic macrophytes of Gorakhpur district, Uttar Pradesh (India). *International Journal of Pharma and Bio Sciences*, 5(1): 300-308.

Gupta AK and Sharma M 2007. Review on Indian Medical Plants (ICMR New Delhi, India) 5: 879-882.

Gupta GL and Rana AC 2007. *Withania somnifera* (Ashwagandha): A review. *Pharmacognosy Reviews*, 1:129-136.

Gupta PJ 2011. The efficacy of *Euphorbia prostrata* in early grades of symptomatic hemorrhoids –a pilot study. *European Review for Medical and Pharmacological Sciences*, 15: 199-203.

Gupta RS, Kachhawa JB and Chaudhary R 2004. Antifertility effects of methanolic pod extract of *Albizia lebbeck* Benth. in male rats. *Asian Journal of Andrology*, 6(2):155-159.

Gupta S and Narayan R 2006. Species diversity in four contrasting sites in a peri-urban area in Indian dry tropics. *Tropical Ecology*, 47(2): 229-241.

Gupta V, Kumar P, Bansal P and Singh R. 2009. Anti-inflammatory and anti-nociceptive activity of *Mitragyna parvifolia*. *Asian Journal of Medical Sciences*. 1: 97-99.

Gurudeeban S, Satyavani K and Ramanathan T 2010. Bitter Apple (*Citrullus colocynthis*): An overview of chemical composition and biomedical potentials. *Asian Journal of Plant Sciences* 9(7): 394-401.

Gwalwanshi DR, Vyas D, Bishwas AJ, Salunkhe O and Tiwari P 2014. Ethnomedicinal and ecological studies on Fabaceae of Runj Forest Panna (MP), India. *International Journal of Recent Scientific Research Research*, 5(7): 1264-1269.

Hada L 2015. An analysis on Indian traditional herbal medicines used for the treatment of diabetes. *International Journal of Physical Education and Sport Sciences*, 8(15): 1-4.

Hamel K, Parsons J, Boule M and Feldman S 2001. An aquatic plant identification manual for Washington's freshwater plants. Washington State Department of Printing, p. 1-195.

Haq F 2012. The ethno-botanical uses of medicinal plants of Allai Valley, Western Himalaya, Pakistan. *International Journal of Plant Research*, 2(1): 21-34.

Harborne JB 1993. Introduction to Ecological Biochemistry, 4th ed. Harcourt Brace & Company, London.

Hartwell JL 1967-1971. Plants used against Cancer. A survey. Lloydia, p. 30-34.

Hasan S 2014. Pharmacological and medicinal uses of *Achyranthes aspera*. *International Journal of Science, Environmentand Technology*, 3(1): 123-129.

Hashim S, Bakht T, Marwat KB and Jan A 2014. Medicinal properties, phytochemistry and pharmacology of *Tribulus terrestris* L. (Zygophyllaceae). *Pakistan Journal of Botany*, 46(1): 399-404.

Hassan RA, Tawfik WA and Abou-Setta LM 2014. The flavonoid constitunts of *Leucaena leucocephala* growing in Egypt, and their biological activity. *African Journal of Traditional, Complementary, and Alternative Medicines*, 11(1): 67-72.

Hedimbi M and Chinsembu KC 2012. Ethnomedicinal study of plants used to manage HIV/AIDS-related disease conditions in the Ohangwena region, Namibia. *International Journal of Medicinal Plant Research*, 1(1): 4-11.

Heim E 2015. Flora and vegetation of Bali Indonesia: An Illustrated Field Guide, p. 224.

Heywood VH 1993. Flowering Plants of the World. Oxford University Press, New York.

Higgins SI and Richardson DM 1996. A review of models of alien plant spread. *Ecological Modelling*, 87(1-3): 249-265.

Hogge LR, Reed DW and Underhill EW 1988. HPLC separation of glucosinolates from leaves and seeds of *Arabidopsis thaliana* and their identification using thermospray liquid chramatography/mass spectrometry. *Journal of Chromatographic Science*, 26: 551-556.

Hong L, Guo Z, Huang K, Wei S, Liu B, Meng S and Long C 2015. Ethnobotanical study on medicinal plants used by Maonan people in China. *Journal of Ethnobiology and Ethnomedicine*, 11: 32. https://doi.org/10.1186/s13002-015-0019-1.

Horio T, Yoshida K, Kikuchi H, Kawabata J and Mizutani J 1993. A phenolic amide from roots of *Chenopodium album*. *Phytochemistry*, 33: 807-808.

Hsieh YJ, Tseng SP, Kuo YH, Cheng TL, Chiang CY, Tzeng YM and Tsai WC 2016. Ovatodiolide of *Anisomeles indica* exerts the anticancer potential on pancreatic cancer cell lines through STAT3 and NF-κB regulation. *Hindawi Publishing Corporation Evidence-Based Complementary and Alternative Medicine*, 2016: 1-10. http://dx.doi.org/10.1155/2016/8680372.

http://baliherbs.blogspot.in/ *Imperata cylindrica* (L.) Raeusch.

http://bioinfo.bisr.res.in/project/domap/. Database of Medicinal and Aromatic plants in Rajasthan.

http://blogs.sld.cu/marionod/2012/04/16/dominican-medicinal-plants-a-guide-for-health-care-providers-second-edition-web-text-version-without-illustrations/ Medicina Naturaly Tradicional/ Plantas medicinales etc.

http://bsienvis.nic.in/Database/bsi_3946.aspx. *Alternanthera pungens* Kunth

http://bsienvis.nic.in/Database/ImportantPlantswithChemicalConstituents_3946.aspx

http://entheology.com/category/plants/ *Datura innoxia* Mill.

http://folkmedsindh.com.pk/. Folk Medicine Sindh.

http://healingherbinfo.com. Healing Herb Info.

http://herbs.indianmedicinalplants.info/index.php/medicinal-plants-pictures-a-details. Encylopedia of Medicinal plants. Herbs-Medicinal plants usage and identification Data base.

http://herbs-treatandtaste.blogspot.in/. Herbs-Treat and Taste.

http://herbs-treatandtaste.blogspot.in/2011/07/. *Sorghum halepense* (L.) Pers.

http://medical-dictionary.thefreedictionary.com/. Medical Dictionary.

http://medicinalplantinindia.blogspot.in/. Medicinal Plants in India.

http://medicinalplants.us/. Medicinal Plants.

http://medicinalplants.us/scoparia-dulcis-l-sweet-broomweed. *Scoparia-dulcis* Linn.

http://medicinalplantsinsingapore.wikifoundry.com/page/Leucaena+leucocephala

http://medplants.blogspot.in/. Medicinal plants with usage, patents and their publications.

http://pankajinprintmedia.blogspot.in/2012/12/herbal-remedies-useful-with-crotalaria.html. Herbal remedies useful with *Crotalaria mysorensis* Roth. Null in old cases of Mirgi (Epilepsy) and associated troubles.

http://plants.jstor.org/stable/10.5555/al.ap.flora.sffa001370184100001. *Equisetum ramosissimum* Desf. *subsp. ramosissimum*

http://pzimedia.com/health/Health-Benefits-Of-Physalis-Minima--35841.html/. Health Benefits of *Physalis minima.*

http://shodhganga.inflibnet.ac.in/bitstream/10603/11731/5/chapter%203.pdf. *Amaranthus viridis* L.

http://shodhganga.inflibnet.ac.in/bitstream/10603/120517/3/ethnobotany..........pdf. *Hemarthria compressa* (L.f.) R.Br.

http://shodhganga.inflibnet.ac.in/bitstream/10603/120517/3/ethnobotany..........pdf *Brachiaria ramosa* (L.) Stapf.

http://shodhganga.inflibnet.ac.in/bitstream/10603/152758/13/13_chapter5.pdf. *Typha domingensis* Pers.

http://shodhganga.inflibnet.ac.in/bitstream/10603/50574/8/08_chapter%204.pdf. *Veronica anagallis-aquatica* L.

http://shodhganga.inflibnet.ac.in/bitstream/10603/75475/12/12_chapter%206.pdf. *Fimbristylis ferruginea* (L.) Vahl.

http://tropical.theferns.info/ Useful Tropical Plants Database.

http://uses.plantnet-project.org/en/Boerhavia_diffusa_(PROTA). *Boerhavia diffusa* L.

http://uses.plantnet-project.org/en/Cyperus_iria_(PROSEA). *Cyperus iria*

http://uses.plantnet-project.org/en/Equisetum_ramosissimum_(PROSEA). *Equisetum ramosissimum* Desf.

http://uses.plantnet-project.org/en/Ipomoea_obscura_(PROTA) *Ipomoea obscura* (L.) Ker Gawl.

http://uses.plantnet-project.org/en/Phragmites_karka_(PROTA) *Phragmites karka* (Retz.) Trin. ex Steud.

http://www.academia.edu/7422611/an_analysis_of_the_data_on_the_medicinal_plants_recorded_in_hortus_malabaricus. An analysis of the data on the medicinal plants.

http://www.airitilibrary.com/Publication/alDetailedMesh?docid=10004025-200512-25-12-2524-2528-a.

http://www.alwaysayurveda.com. Always ayurvada- a sister concern of planet Ayurvada.

http://www.ayurveda-florida.com/Ayurvedic_Materia_Medica_Articles/Table2.htm. Table of Ayurvedic Plants and Minerals with Sanskrit (and Synonyms), Common, and Botanical Names.

http://www.biodiversityofindia.org/index.php?title=Property:Medicinal_use_description *Equisetum ramosissimum* Desf. subsp. *ramosissimum.*

http://www.biodiversityofindia.org/index.php?title=Property:Modern_scientific_knowledge_description. *Equisetum ramosissimum* Desf.

https://www.botanical.com. Electronic version of "A Modern Herbal" by Maud Grieve, originally published in 1931.

http://www.bsienvis.nic.in/. ENVIS (BSI) 2016. ENVIS Centre on Floral Diversity.

http://www.bsienvis.nic.in/Database/MedicinalPlants_3939.aspx. ENVIS (BSI) 2017. ENVIS Centre on Floral Diversity. Medicinal Plants.

http://www.cabi.org/isc/datasheet/17494. *Cyperus compressus.*

http://www.cabi.org/isc/datasheet/30864. Invasive Species Compendium/ *Lindernia crustacea.*

http://www.efloraofgandhinagar.in/. E- Flora of Gandhinagar, Gujrat, India.

http://www.efloras.org/flora_page.aspx?flora_id=1. Flora of North America @ efloras.org.

http://www.efloras.org/flora_page.aspx?flora_id=2. Flora of China.

http://www.efloras.org/flora_page.aspx?flora_id=5. Flora of Pakistan @ efloras.org.

http://www.florakarnataka.ces.iisc.ac.in/hjcb2/. Herbarium JCB. Herbarium JCB – IISc/ Digital flora of Karnataka

http://www.flowersofindia.net/. Flowers of India.

http://www.frlhtenvis.nic.in/Database/ENVIS_Center_1613.aspx. Medicinal Plants Database - ENVIS Centre on Medicinal Plants.

http://www.henriettes-herb.com/. Henriette's Herbal Homepage.

http://www.herbwisdom.com/herblist.html. A-Z. Herb List-Herb Wisdom.

http://www.instituteofayurveda.org/plants/plants_detail.php?i=1043&s=Local_name, *Panicum antidotale* Retz.

http://www.missouribotanicalgarden.org/media/fact-pages/tropicos.aspx. Missouri Botanical Garden.

http://www.moef.gov.in/sites/default/files/Volume-II.pdf Ampelopteris prolifera (Retz.) Copel.

http://www.motherherbs.com/. Mother Herbs and Agro product.

http://www.nmpb-mpdb.nic.in/. National Medicinal Plants Board/ Database on Medicinal plants.

http://www.pfaf.org/user/Default.aspx. Plants for a Future.

http://www.planetayurveda.com/. Planet Ayurveda.

http://www.rain-tree.com. Tropical Plant Data Base.

http://www.saussurea-costus.com/list-of-medicinal-and-aromatic-plants.php. Nanda Medicinal Plants/ List of Medicinal and Aromatic Plants.

http://www.stuartxchange.org/Malatiki.html, *Salix tetrasperma.*

http://www.stuartxchange.org/Sabilau.html, *Cyanotis axillaris* (L.) D.Don ex Sweet.

http://www.uicnmed.org/nabp/web/documents/book/chapter3.pdf *Centaurium pulchellum* (Sw.) Druce

http://www.wildflower.org/ . Plant Database. Lady Bird Johnson Wildflower Center.

http://www.worldagroforestry.org/output/agroforestree-database. The Agroforestree Database. World Agroforestry Centre.

http://www.yourproduceman.com/news_august_22_05.html.

https://easyayurveda.com/2017/10/08/kans-grass-saccharum-spontaneum-kasha/ *Saccharum spontaneum* L.

https://florafaunaweb.nparks.gov.sg/special-pages/plant-detail.aspx?id=2472. Flora & Fauna Web- National parks board.

https://gobotany.newenglandwild.org/species/euphorbia/serpens/. *Euphorbia serpens* Kunth.

https://herbpathy.com/. Herbpathy.

https://herbpathy.com/Uses-and-Benefits-of-Desmostachya-Bipinnata-Cid1167 *Desmostachya bipinnata* (L.) Stapf

https://herbpathy.com/Uses-and-Benefits-of-Indigofera-Linifolia-Cid5796.

https://medical-dictionary.thefreedictionary.com/Panicum+capillare, *Panicum antidotale* Retz.

https://naturesaccess.com.

https://naturesaccess.com/2016/03/13/hairy-bittercress-cardamine-hirsuta/ Hairy Bittercress (*Cardamine hirsuta*)

https://phytochem.nal.usda.gov/phytochem/search. Dr. Duke's Phytochemical and Ethnobotanical Databases.

https://plants.usda.gov/plantguide/pdf/pg_mepo3.

https://thenaturopathicherbalist.com/herbal-actions/a/anti-catarrhal/. The Naturopathic Herbalist.

https://www.ayurleafherbals.com/billilotan. Ayurleaf Herbals.

https://www.bimbima.com/ayurveda/. Daily life experience of ayurvedic medicines, complementary therapies.

https://www.bimbima.com/ayurveda/medicinal-plant-wild-kulthi/457/

https://www.cabi.org/isc/datasheet/21634. *Eragrostis japonica* (Thunb.) Trin.

https://www.cabi.org/isc/datasheet/49775. *Setaria verticillata* (L.) P.Beauv.

https://www.drugs.com. Drugs.com. Know more. Be sure.

https://www.efloraofindia/ Database of Indian Plants- developed by the members of Efloraofindia Google Group.

https://www.gbif.org/species/113598580. *Citrullus colocynthis* (L.) Schrad. in Döring M (2015). English Wikipedia - Species Pages. Wikimedia Foundation. Checklist Dataset https://doi.org/10.15468/c3kkgh accessed via GBIF.org.

https://www.herbalstudies.net/_media/resources/library/Medicinal-Plants-of-Jamaica.pdf. Medicinal plants of Jamaica-part 1 & 2.

https://www.hort.purdue.edu/newcrop/duke_energy/Cannabis_sativa.html. *Cannabis sativa* L.

https://www.indiabiodiversity.org/. India Biodiversity Portal.

https://www.odishafdc.com/products_medicinal_plants.php. Odisha forest development corporation/List of Important Medicinal Plants.

https://www.prota4u.org.

https://www.prota4u.org/database/protav8.asp?g=pe&p=Euphorbia+prostrata+Aiton. *Euphorbia prostrata* Aiton

https://www.prota4u.org/database/protav8.asp?g=pe&p=Lepidium+didymum+L. *Lepidium didymum* L.

https://www.theplantlist.org/. The Plant List 2013. Version 1.1. Published on the Internet.

Huang Q, Zhang S, Huang R, Wei L, Chen Y, Lv S, Liang C, Tan S, Liang S, Zhuo L and Lin X 2013. Isolation and identification of an anti-hepatitis B virus compound from *Hydrocotyle sibthorpioides* Lam. *Journal of Ethnopharmacology*, 150(2): 568-575.

Huang QQ, Wu JM, Bai YY, Zhou L and Wang GX 2009. Identifying the most noxious invasive plants in China: role of geographical origin, life form and means of introduction. *Biodiversity and Conservation*, 18(2): 305-316.

Husain SZ, Malik RN, Javaid M and Bibi S 2008. Ethnobotanical properties and uses of medicinal plants of Morgah Biodiversity Park, Rawalpindi. *Pakistan Journal of Botany*, 40(5): 1897-1911.

Husin F, Chan YY, Siew Hua G, Sulaiman SA and Hanim SR 2015. The Effect of *Hydrocotyle sibthorpioides* Lam. extracts on in vitro dengue replication. Evidence-Based Complementary and Alternative Medicine. http://dx.doi.org/10.1155/2015/596109.

Hussain AI, Rathore H, Sattar MZA and Gilani AH 2014. *Citrullus colocynthis* (L.) Schrad (bitter apple fruit): A review of its phytochemistry, pharmacology, traditional uses and nutritional potential. *Journal of Ethnopharmacology*, 155(1): 54-66.

Hussain AZ and Kumaresan S 2014. Antimicrobial activity and heavy metal analysis of *Ipomoea obscura*. L. *Journal of Microbiology and Biotechnology Research*, 4(5):18-23.

Hussain HS and Deeni YY 1991. Plants in Kano ethnomedicine: Screening for antimicrobial activity and alkaloids. *International Journal of Pharmacognosy*, 29: 51-56.

Hussain I, Ullah R, Khan N, Ayaz S, Ahmad S, Shanzeb, Ahmed M, Pir TH and Khan FA 2011. Phytochemicals and inorganic profile of *Calendula officinale* and *Sonchus asper*. *African Journal of Pharmacy and Pharmacology*, 5(16): 1813-1818.

IABGC 2000. International agenda for botanic gardens in conservation. Botanic Gardens Conservation International, U.K.

Ibrahim J, Ajaegbu VC and Egharevba HO 2010. Pharmacognostic and phytochemical analysis of *Commelina benghalensis* L. *Ethnobotanical Leaflets*, 14: 610-15.

ICMR 1975. Manual of standards of quality of drinking water supplies, Indian Council of Medical Research, New Delhi.

Ignacimuthu S, Ayyanar M and Sivaraman S 2006. Ethnobotanical investigations among tribes in Madurai district of Tamil Nadu (India). *Journal of Ethnobiology and Ethnomedicine*, 2(25): 1-7.

Imam H, Zarnigar, Sofi G, Seikh A and Lone A 2013. The incredible benefits of Nagarmotha (*Cyperus rotundus*). *International Journal of Nutrition, Pharmacology, Neurological Diseases*, 4(1): 23-27.

Isaiah S, Kumar AC and Selvan NS 2016. Phytochemical screening, anti-microbial activity and GC-MS analysis of *Corchorus tridens* L. *International Journal of Pharmacological Research*, 6(12): 353-357.

Ishtiaq CM and Khan MA 2008. An ethnomedicinal inventory of plants used for family planning and sex diseases in Samahni Valley, Pakistan. *Indian Journal of Traditional Knowledge*, 7(2): 277-283.

Islam MM 2013. Biochemistry, medicinal and food values of jute (*Corchorus capsularis* L. and *C. olitorius* L.) leaf: A review. *International Journal of Enhanced Research in Science Technology & Engineering*, 2(11): 35-44.

Islam MN, Nyeem MAB, Taher MA and Awal A 2015. Analgesic and CNS depressant effect of the crude ethanolic extract of the *Operculina turpethum*. *Biosensors Journal*, 4(2): 1-4.

Islam S, Rahman MM, Alam MH, Uddin SJ, Bellah F, and Murshid GMM 2014. Anthelmintic and cytotoxic activity of the ethanolic extract of *Phoenix sylvestris* root. *World Journal of Pharmaceutical Science*, 2(8): 725-728.

IUCN 2001. Global invasive species programme. Global strategy on invasive alien species. IUCN, Gland, Switzerland and Cambridge.

Ivanova D, Gerova D, Chervenkov T and Yankova T 2005. Polyphenols and antioxidant capacity of Bulgarian medicinal plants. *Journal of Ethnopharmacology*, 96(1-2): 145-150.

Jadhav VM, Kamble SS and Kadam VJ 2009. Herbal medicine: *Syzygium cumini*: A review. *Journal of Pharmacy Research*, 2(8): 1212-1219.

Jagtap MD, Asabe AS, Telave AB, Mali BS, Chavan SJ and Kanade MB 2014. Antifungal potential of *Cuscuta reflexa* Roxb. *Central European Journal of Experimental Biology*, 3(3): 30-32.

Jagtap SD, Deokule SS and Bhosle SV 2006. Some unique ethnomedicinal uses of plants used by the Korku tribe of Amravati district of Maharashtra, India. *Journal of Ethnopharmacology*, 107: 463-469.

Jagtap SD, Deokule SS, Pawar PK and Harsulkar AM 2009. Traditional ethnomedicinal knowledge confined to the pawra tribe of Satpura hills, Maharashtra, India. *Ethnobotanical Leaflets*, 13: 98-115.

Jahan R, Al-Nahain A, Majumder S and Rahmatullah M 2014. Ethnopharmacological significance of *Eclipta alba* (L.) Hassk. (Asteraceae). *International Scholarly Research Notices*, 2014: 1-22.

Jain A, Choubey S, Singour PK, Rajak H and Pawar RS 2011. *Sida cordifolia* (Linn) – An overview. *Journal of Applied Pharmaceutical Science*, 1(2): 23-31.

Jain A, Jain D and Shrivastava S 2016. A short review on pharmacological activity of *Equisetum ramosissimum*. *Asian Journal of Pharmaceutical Education and Research,* 5(4): 1-8.

Jain A, Katewa SS, Chaudhary BL and Galav P 2004. Folk herbal medicines used in birth control and sexual diseases by tribals of southern Rajasthan, India. *Journal of Ethnopharmacology,* 90(1): 171-177.

Jain RA, Agarwal RC, Dubey D, Verma R and Jain R 2012. Evaluation of antibacterial and antioxidant activity of fruits extract of *Argemone mexicana* Linn. *International Journal of Research Article Pharmaceutical Innovations,* 2(1): 45-51.

Jain S, Bhatia G, Barik R, Kumar P, Jain A and Dixit VK 2010. Antidiabetic activity of *Paspalum scrobiculatum* Linn. in alloxan induced diabetic rats. *Journal of Ethnopharmacology,* 127(2): 325-328.

Jain SC and Singh B 1998. Bioeffeciency of *Heliotropium ellipticum* Ledeb.1. Antimicrobial screening. *Indian Journal of Pharmaceuticals Sciences,* 60(6): 394-395.

Jain SC, Jain R and Singh R 2009. Ethnobotanical survey of Sariska and Siliserh regions from Alwar district of Rajasthan, India. *Ethnobotanical Leaflets,* 13: 171-188.

Jain SK 1968. Medicinal plants. National Book Trust, New Delhi, p. 16-178.

Jain SK 1991. Dictionary of Indian folk medicine and ethnobotany. Deep Publications, New Delhi.

Jain SK and Srivastava S 2005. Traditional uses of some Indian plants among islanders of the Indian Ocean, *Indian Journal of Traditional Knowledge,* 4(4): 345-357.

Jain SK, and Tarafder CR 1970. Medicinal plant lore of the Santals. *Economic Botany,* 24: 214-78.

Jaiswal BS 2012. *Solanum torvum*: A review of its traditional uses, phytochemistry and pharmacology. *International Journal of Pharma and Bio Sciences,* 3(4): 104-111.

Jakhar S and Dahiya P 2017. Antimicrobial, antioxidant and phytochemical potential of *Alternanthera pungens* HB&K. *Journal of Pharmaceutical Sciences and Research,* 9(8): 1305-1311.

Jamal AK, Yaacob WA and Din LB 2008. A chemical study on *Phyllanthus reticulatus*. *Journal of Physical Science,* 19 (2): 45-50.

Jameel M, Ansari JA, Ali A, Ahamad J, Ali M and Tamboli E 2012. Pharmacological scientific evidence for the promise of *Tribulus terrestris*. *International Research Journal of Pharmacy,* 3(5): 403-406.

Jang E, Kim B-J, Lee K-T, Inn K-S and Lee J-H 2015. A survey of therapeutic effects of *Artemisia capillaris* in liver diseases. *Evidence-Based Complementary and Alternative Medicine,* http://dx.doi.org/10.1155/2015/728137.

Jani S, Shukla VJ and Harisha CR 2012. Phytochemical and pharmacognistical investigation on *Ammannia baccifera* Linn. (Stem and leaf). *International Journal of Pharmaceutical Archives,* 3(4): 884-887.

Jawonisi IO and Adoga GI 2013. Chemical constituents of essential oil of *Lantana camara* Linn. Leaves. *British Journal of Pharmacology and Toxicology,* 4(4): 155-157.

Jayaweera DMA 1980. Medicinal plants (indigenous and exotic) used in Ceylon, Vol. 1. National Science Council of Sri Lanka.

Jebastella J and Reginald AM 2015. Screening of antibacterial activity in medicinal grass (*Dactyloctenium aegyptium*) using two extracts. *International Journal of Recent Scientific Research,* 6(7): 5046-5048.

Jeeshna MV, Paulsamy S and Mallikadevi T 2011. Phytochemical constituents and antimicrobial studies of the exotic plant species, *Croton bonplandianum* Baill. *Life Science Journal,* 3(1): 23-27.

Jeeva S, Kiruba S, Mishra BP, Venugopal N, Dhas SSM, Regini GS, Kingston C, Kavitha A, Sukumaran S, Raj ADS and Laloo RC 2006. Weeds of Kanyakumari district and their value in rural life. *Indian Journal of Traditional Knowledge,* 5(4): 501-509.

Jena GSJP and Satapathy KB 2015. Weed diversity of rabi crops and their ethnomedicinal uses in Kendrapara District of Odisha, India. *International Research Journal of Biological Sciences,* 4(3): 33-38.

Jena J and Gupta AK 2012. *Ricinus communis* Linn: A phytopharmacological review. *International Journal of Pharmacy and Pharmaceutical Sciences,* 4(4): 25-29.

Jeyaprakash K, Ayyanar M, Geetha KN and Sekar T 2011. Traditional uses of medicinal plants among the tribal people in Theni District (Western Ghats), Southern India. *Asian Pacific Journal of Tropical Biomedicine,* 1(1): 20-25.

Jhuma S and Bhattacharya S 2011. *Cissampelos pareira*: A promising antifertility agent. *International Journal of Research in Ayurveda and Pharmacy,* 2(2): 439-442.

Joshi P 1982. An ethnobotanical study of Bhils- A preliminary survey. *Journal of Economic and Taxonomic Botany,* 3: 257-66.

Joshi Y, Joshi AK, Prasad N and Juyal D 2014. A review on *Ficus palmata* (Wild Himalayan Fig). *The Journal of Phytopharmacology,* 3(5): 374-377.

Joycharat N, Limsuwan S, Subhadhirasakul S, Voravuthikunchai SP, Pratumwan S, Madahin I, Nuankaew W and Promsawat A 2012. Anti-Streptococcus mutans efficacy of Thai herbal formula used as a remedy for dental caries. *Pharmaceutical Biology,* 50(8): 941-947.

Kage, Devendra N, Seetharam YN, Malashetty and Vijaykumar B 2009. In vitro antibacterial property and phytochemical profile of *Trichosanthes cucumerina* L. var. *cucumerina*. *Advances in Natural and Applied Sciences,* 3: 438-441.

Kainsa S and Singh R 2015. Chemical constituents, antihyperglycemic and antioxidant effects of *Nepeta hindostana* whole herb in alloxan and OGTT induced diabetes in rats. *Journal of Chemical and Pharmaceutical Research,* 7(9): 920-932.

Kainsa S, Kumar P and Rani P 2012. Medicinal plants of Asian origin having anticancer potential: Short review. *Asian Journal of Biomedical and Pharmaceutical Sciences,* 2(10): 1-7.

Kala CP 2005. Ethnomedicinal botany of the Apatani in the Eastern Himalayan region of India. *Journal of Ethnobiology and Ethnomedicine,* 1(11): 1-12.

Kala H, Dhyani PP and Sajwan BS 2006. Developing the medicinal plants sector in northern India: Challenges and opportunities. *Journal of Ethnobiology and Ethnomedicine,* 2(32): 1-15.

Kaladhar DSVGK, Harasreeramulu S, Duddukuri GR and Surekha CH 2011. In Vitro Regeneration of the Medicinal Herb, *Evolvulus nummularius* L. From Shoot Tip and Flower Explants. *Indian Journal of Fundamental and Applied Life Sciences,* 1(2): 81-89.

Kalita S, Kumar G, Karthik L and Rao KVB 2012. A review on medicinal properties of *Lantana camara* Linn. *Research Journal of Pharmacy and Technology,* 5(6): 711-715.

Kamali H, sani TA, Feyzi P and Mohammadi A 2015. Chemical composition and antioxidant activity from essential oil of *Capsella bursa-pastoris*. *International Journal of Pharm Tech Research,* 8(8): 01-04.

Kamaluddin, Rahman AHMM, and Islam AKMR 2014. Taxonomy and traditional medicine practices of Polygonaceae (Smartweed) Family at Rajshahi, Bangladesh. *International Journal of Advanced Research,* 2(11): 459-469.

Kamble SY, Patil1 SR, Sawant PS, Sawant S, Pawar SG and Singh EA 2010. Studies on plants used in traditional medicine by Bhilla tribe of Maharashtra, *Indian Journal of Traditional Knowledge,* 9(3): 591-598.

Kamboj A and Saluja AK 2008. *Ageratum conyzoides* L.: A review on its phytochemical and pharmacological profile. *International Journal of Green Pharmacy,* 2(2): 59-68.

Kamboj A and Saluja AK 2010. Phytopharmacological review of *Xanthium strumarium* L. (Cocklebur). *International Journal of Green Pharmacy,* 4(3): 129-139.

Kanjilal PC 1933. Forest flora of Pilibhit, Oudh, Gorakhpur and Bundelkhand. Allahabad, India: Narendra Publishing House. p. 427.

Kantamreddi VSS and Wright CW 2007. Investigation of Indian Diospyros Species for antiplasmodial properties. *Evidence-based Complementary and Alternative Medicine,* 5(2):187-90.

Kapoor BBS and Kumar S 2013. Ethnomedicinal plants of Barmer district, Rajasthan used in herbal and folk remedies. *Indian Journal of Pharmaceutical and Biological Research,* 1(3): 61-66.

Kapoor BBS and Lakhera S 2013. Ethnomedicinal plants of Jodhpur district, Rajasthan used in herbal and folk remedies. *Indian Journal of Pharmaceutical and Biological Research,* 1(4): 71-75.

Kapoor LD 2000. Handbook of ayurvedic medicinal plants. CRC Press, p. 424.

Kariyajjanavar P, Yargol S, Achar KGS and Parashurama TR 2016. Traditional healthcare knowledge of Sedum Taluk, Kalaburgi, Karnataka, South India. *International Journal of Current Microbiology and Applied Sciences,* 5(11): 321-328.

Karou SD, Nadembega WMC, Ilboudo DP, Ouermi D, Gbeassor M, Souza CD and Simpore J 2007. *Sida acuta* Burm. f.: A medicinal plant with numerous potencies. *African Journal of Biotechnology,* 6(25): 2953-2959.

Karthikeyan K, Dhanapal CK and Gopalakrishnan G 2014. A review on medicinal importance of *Alysicarpus monilifer*. *International Journal of Chemical and Pharmaceutical Sciences,* 5(1): 1-5.

Karthikeyan K, Dhanapal CK and Gopalakrishnan G 2015. Preliminary phytochemical investigation of whole plant of *Alysicarpus monilifer* (L.) DC. *International Journal of Pharma and Bio Sciences,* 6(2): 67-72.

Karthikeyan S, Sanjappa M and S. Moorthy 2009. Flowering plants of India: Dicotyledons, Vol. I. Acanthaceae – Avicenniaceae. Botanical Survey of India, Calcutta. p. 366.

Karthiyayini T, Sandu NR and Senthilkumar KL 2011. Anti-diabetic activity on the flowers of *Nymphaea pubescens* Willd. *Research Journal of Pharmaceutical, Biological and Chemical Sciences,* 2(1): 866-873.

Kasture VS, Chopade CT and Deshmukh VK 2000. Anticonvulsive activity of *Albizzia lebbeck, Hibiscus rosa sinesis* and *Butea monosperma* in experimental animals. *Journal of Ethnopharmacology,* 71(1-2): 65-75.

Kasture VS, Kasture SB and Pal SC 1996. Anticonvulsant activity of *Albizzia lebbeck* leaves. *Indian Journal of Experimental Biology,* 34: 78-80.

Katewa SS and Galav PK 2005. Traditional herbal medicines from Shekhawati region of Rajasthan. *Indian Journal of Traditional Knowledge,* 4(3): 237-245.

Katewa SS, Chaudhary BL, Jain A and Galav P 2003. Traditional uses of plant biodiversity from Aravalli hills of Rajasthan. *Indian Journal of Traditional Knowledge,* 2(1): 27-39.

Katewa SS, Guria BD and Jain A 2011. Ethnomedicinal and obnoxious grasses of Rajasthan, India. *Journal of Ethnopharmacology,* 76(2001): 293-297.

Katiri A, Mohamed B, Fouad M and Hassan B 2017. Ethnobotanical survey of medicinal plants used for the treatment of diabetes in the Tizi n' Test Region (Taroudant Province, Morocco). *Journal of Pharmacognosy and Natural Products*, 3(1): 1-10.

Kaul S and Dwivedi S 2010. Indigeneous Ayurvedic knowledge of some species in the treatment of human diseases and disorders. *International Journal of Pharmacy & Life Sciences*, 1(1): 44-49.

Kaur J, Pannu MS, Kaushal S, Wadhwa M and Bakshi MPS 2006. In vitro evaluation of *Phalaris minor* Seeds as Livestock Feed. *Asian Australasian Journal of Animal Sciences*, 19(3): 363-367.

Kaur J, Rani S, Singh G and Sood N 2015. Trends of diffusion pattern and denudation of traditional knowledge in two medicinally important plants- *Calotropis procera* and *Thevetia peruviana* in villages of Rohtak district, Haryana, India. *Journal of Medicinal Plants Studies*, 3(5): 14-19.

Kaur M and Arora R 2011. Antioxidant activity of *Cucumis melo* var. *agrestis* seeds for their therapeutic potential. *International Journal of Research in Ayurveda & Pharmacy*, 2(4): 1235-1238.

Kaur M and Kalia AN 2011. *Convolvulus arvensis* – A useful weed. *International Journal of Pharmacy and Pharmaceutical Sciences*, 4(1): 38-40.

Kaur R and Vashistha BD 2014. Ethnobotanical studies on Karnal district, Haryana, India. *International Research Journal of Biological Sciences*, 3(8): 46-55.

Kaur R, Singh N and Vashistha BD 2016. Flowering plant diversity of district Karnal, Haryana, India. *International Journal of Life Sciences*, 4(3): 361-371.

Kayed AM, Sayed E, Elghaly M and Hela AAE 2015. New epoxy megastigmane glucoside from *Dactyloctenium aegyptium* L.P.Beauv. Wild (Crowfoot grass). *Journal of Scientific and Innovative Research*, 4(6): 237-244.

Keat NB, Umar RU, Lajis NH, Chen TY, Li TY, Rahmani M and Sukari MA 2010. Chemical constituents from two weed species of *Spermacoce* (Rubiaceae). *The Malaysian Journal of Analytical Sciences*, 14(1): 6-11.

Khalid MS, Singh RK, Reddy N, Kumar SJ, Kumar BS, Kumar GNS and Rao KS 2011. Anti-inflammatory activity of aqueous extract of *Ipomoea carnea* Jacq. *Pharmacologyonline*, 1: 326-331.

Khan A, Mehmood S, Khan RK, and Khan SK 2013a. Study of wild medicinal xerophytes of district Bannu, Khyber Pakhtunkhwa, Pakistan. *Journal of Medicinal Plants Studies*, 1(4): 79-86.

Khan H, Jan SA, Javed M, Shaheen R, Khan Z, Ahmad A, Safi SZ and Imran M 2016. Nutritional composition, antioxidant and antimicrobial activities of selected wild edible plants. *Journal of Food Biochemistry*, 40: 61-70.

Khan I, AbdElsalam NM, Fouad H, Tariq A, Ullah R and Adnan M 2014a. Application of ethnobotanical indices on the use of traditional medicines against common diseases. Hindawi Publishing Corporation. *Evidence-Based Complementary and Alternative Medicine*, http://dx.doi.org/10.1155/2014/635371.

Khan J, Khan R and Qureshi RA 2013b. Ethnobotanical study of commonly used weeds of district Bannu, Khyber Pakhtunkhwa (Pakistan). *Journal of Medicinal Plants Studies*, 1(2): 1-6.

Khan JB and Singh GP 2012. Ethno-medicinal diversity of Churu district traditionally used by rural people. *Indian Journal of Fundamental and Applied Life Sciences*, 2(1): 109-114.

Khan M, Musharaf S, Ibrar M and Hussain F 2011a. Pharmacognostic evaluation of the *Amaranthus viridis* L. *Research in Pharmaceutical Biotechnology*, 3(1): 11-16.

Khan MA, Marwat KB, Bakhtiar, Wahid F, Khan H and Hashim S 2014b. *Pistia Stratiotes* L. (Araceae): Phytochemistry, use in medicines, phytoremediation, biogas and management options. *Pakistan Journal of Botany*, 46(3): 851-860.

Khan MA, Prakash R, Ali S, Aljarbou A and Khan MA 2011b. Comparative study of antibacterial activity and toxicity of certain plants used in Unani medicine. *Advances in Bioresearch*, 2(2): 10-13.

Khan R, Mehmood S, Khan S and Subhan M 2013c. Ethnobotanical study of common weed flora of sugarcane in district Bannu, Khyber Pakhtunkhawa, Pakistan. *Journal of Medicinal Plants Studies*, 1(4): 49-78.

Khan RA 2010a. Effects of *Launaea procumbens* on brain antioxidant enzymes and cognitive performance of rat. *BMC Complementary and Alternative Medicine*, 12(219):1-6.

Khan RA 2010b. Phytochemical screening and pharmacological evaluation of *Sonchus asper* (L.) Hill. and *Launaea procumbens* (Roxb.) Amin. Ph.D Thesis, Qauid-i-Azam University, Islamabad, Pakistan.

Khan RA 2012. Protective effect of *Launaea procumbens* (L.) on lungs against CCl4-induced pulmonary damages in rat. *BMC Complementary and Alternative Medicine*, 12(133): 1-7.

Khan RA, Khan MR, Sahreen S and Ahmed M 2012. Evaluation of phenolic contents and antioxidant activity of various solvent extracts of *Sonchus asper* (L.) Hill. *Chemistry Central Journal*, 6:12. doi: 10.1186/1752-153X-6-12.

Khan RA, Khan MR, Sahreen S and Bokhari J 2010a. Antimicrobial and phytotoxic screening of various fractions of *Sonchus asper. African Journal of Biotechnology*, 9(25): 3883-3887.

Khan S, Rawat R, Rawat AKS and Shirwaiker A 2010b. A report on the quality control parameters of aerial parts of *Pluchea lanceolata* (DC.) Oliv. & Hiern, Asteraceae. *Revista Brasileira de Farmacognosia Brazilian Journal of Pharmacognosy*, 20(4): 563-567.

Khan SM, Ud Din N, Ilyas M, Sohail, Ur Rahman I, Ijaz F, Iqbal Z and Ali Z 2015. Ethnobotanical study of some medicinal plants of Tehsil Kabal, District Swat, KP, Pakistan. *Medicinal & Aromatic Plants*, 4(3): 1-4.

Khan VA and Khan AA 2005. Herbal folklores for male sexual disorders and debilities in western Uttar Pradesh. *Indian Journal of Traditional Knowledge*, 4: 317-324.

Khan ZUH, Ali F, Khan SU and Ali I 2011c. Phytochemical study on the constituents from *Cirsium arvense*. *Mediterranean Journal of Chemistry*, 1(2): 64-69.

Khan ZUH, Khan S, Chen Y. Hassan SU and Wan P 2014c. Natural products – new antimicrobial constituents derived *from Cerium arvense*. *International Journal of Plant Research*, 4(1): 5-10.

Khanna KK and Mudgal V 1999. Dicotyledonous plants of Uttar Pradesh: A checklist. BSMPS, Botanical Survey of India, Calcutta.

Khare CP 2007. Indian medicinal plants: An illustrated dictionary. Springer-Verlag, New York, p. 900.

Khatri A, Rathore A and Patil UK 2010. *Prosopis cineraria* (L.) Druce: A boon plant of desert an overview. *International Journal of Biomedical and Advance Research*, 1(5): 141-149.

Khuroo AA, Rashid I, Reshi Z, Dar GH and Wafai BA 2007. The alien flora of Kashmir Himalaya. *Biological Invasions*, 9: 269-292.

Khuroo AA, Reshi ZA, Malik AH, Weber E, Rashid I, Dar GH 2012. Alien flora of India: taxonomic composition, invasion status and biogeographic affiliations. *Biological Invasions*, 14: 99-113.

Kightlinger LK, Seed JR and Kightlinger MB 1996. *Ascaris lumbricoides* aggregation in relation to child growth status, delayed cutaneous hypersensitivity, and plant anthelmintic use in Madagascar. *Journal of Parasitology*, 82(1):25-33.

Kirtikar KR and Basu BD 1975. Indian Medicinal Plants, Bishen Singh Mahendra Pal Singh, Dehradun.

Kishor K, Tripathi AM, Roy S and Chaudhary L B 2011. Assessment and preservation of tree diversity of Uttar Pradesh, India; p. 68–75. In: P. Singh *et al.* (eds.). Forest Biodiversity: Earth's Living Treasure. Lucknow: U.P. State Biodiversity Board. http://www.upsbdb.org/pdf/Souvenir2011/9.pdf.

Klimes L and Dickoré B 2005. A contribution to the vascular plant flora of Lower Ladakh (Jammu & Kashmir, India). *Willdenowia*, 35: 125-153.

Koetpanya C 1992. Chemical constituents and biological activities of *Sphenoclea zeylanica* Gaertn. (Chadaphon Koetpanya Chulalongkorn Univ., Bangkok (Thailand). Graduate School).

Koh HL, Tan CH and Chua TK 2009. A guide to medicinal plants: An illustrated, scientific and medicinal approach. World Scientific Publishing Co Pte. Ltd., Singapore.

Kohda H, Tanaka S, Yamaoka Y, Morinaga S and Ohhara Y 1994. Constituents of *Corchorus olitorius* L. *Natural Medicine*, 48: 213-214.

Komathi S and Durai MF 2015. Qualitative phytochemical screening of *Saccharum spontaneum* Linn. Leaf Extracts. *Journal of Academia and Industrial Research*, 4(6): 168-170.

Komorowski T 1982. Chemical constituents of *Veronica anagallis-aquatica* L. Food and Agriculture Organisation of the United Nation.

Komorowski T and Swiatek I 1982. Chemical constituents of *Veronica anagallis-aquatica* L. Food and Agriculture Organization of the United Nation. http://agris.fao.org/agris search/search.do?recordID=PL8400604.

Kor NM, Hajmohamadi S and Kor ZM 2013. Physiological and pharmaceutical effects of *Tribulus terrestris* as a multipurpose and valuable medicinal plant. *International journal of Advanced Biological and Biomedical Research*, 1(10):1289-1295.

Krishen P 2006. Trees of Delhi: A field Guide. Delhi; Penguin Books, p. 360.

Krishna MP, Varghese R, Mohan M and Hatha AAM 2014. Antibacterial activity of *Acmella paniculata* extract on human pathogenic bacteria. *International Journal of Herbal Medicine*, 2(1): 132-134.

Krishnamurthy N, Padmabai R, Natarajan CP, Kuppusnamy S 1975. Colour content of turmeric varieties and studies on its processing. *Journal of Food Science and Technology*, 12: 12-14.

Kshirsagar SR 2016. Ethnobotanical and ethnomedicinal plants of Kelzar and Tatani forest areas in Nasik district, Maharashtra. *International Journal of Science Info*, 1(2): 95-100.

Kumar A 2010. Medicinal plants, First Edition. International Scientific Publishing Academy, New Delhi, India.

Kumar A, Agarwal S, Singh A and Deepak D 2012a. Medico-botanical study of some weeds growing in Moradabad district of western Uttar Pradesh in India. *Indian journal of scientific research*, 3(1):107-111.

Kumar AS, Kavimani S and Jayaveera KN 2012b. Anti-diabetic and anti-hyperlipidemic effects of methanol extracts of *Chloris barbata* (SW.) in Streptozotocin-induced diabetic rats. *European Journal of Experimental Biology*, 2(4):1346-1353.

Kumar CAS, Varadharajan R, Muthumani P, Meera R, Devi P and Kameswari B 2010a. Psychopharmacological studies on the stem of *Saccharum spontaneum*. *International Journal of Pharm. Tech. Research*, 2(1): 319-321.

Kumar D, Gupta J, Kumar S, Arya R, Kumar T and Gupta A 2012c. Pharmacognostic evaluation of *Cayratia trifolia* (Linn.) leaf. *Asian Pacific Journal of Tropical Biomedicine*, 2(1): 6-10.

Kumar D, Kumar S, Gupta J, Arya R and Gupta A 2011a. A review on chemical and biological properties of *Cayratia trifolia* Linn. (Vitaceae). Pharmacognosy Reviews 5(10): 184-188.

Kumar DS, Banji D, Harani A, Rao AT, Rao SN, Kumar CP and Krupa DS 2013a. Screening of polyphenolic compounds in *Echinochloa crus galli* Roxb extracts by various analytical techniques. *Asian Journal of Chemistry*, 25(17): 9848-9852.

Kumar M and Singh M 2013. Study of plant diversity of Karnal district, Haryana, India. *International Journal of Pharmacy & Life Sciences*, 4(4): 2573-2582.

Kumar P and Kumar S 2015. Phytochemistry, traditional uses, pharmacology of Indian medicinal plant *Chenopodium album* (Linn). *World Journal of Pharmacy and Pharmaceutical Sciences*, 4(7): 404-421.

Kumar PGM and Shiddamallayya N 2016. Survey of wild medicinal plants of Hassan district, Karnataka. *Journal of Medicinal Plants Studies*, 4(1): 91-102.

Kumar PR 2014. Ethno medicinal plants used for oral health care in India. *International Journal of Herbal Medicine*, 2(1): 81-87.

Kumar PS, Suresh E and Kalavathy S 2013b. Review on a potential herb *Calotropis gigantea* (L.) R. Br. *Scholars Academic Journal of Pharmacy*, 2(2):135-143.

Kumar R 2013. Folk veterinary medicines in the Bareilly district of Uttar Pradesh, India. *Indian Journal of Traditional Knowledge*, 12(1): 40-46.

Kumar R and Saxena RS 2012. The sedges and grasses of district Saharanpur (U.P.), India. *Current Botany*, 3(5): 5-7.

Kumar RK and Mishra SS 2012. Use of Agnikuanri (*Ammannia baccifera* L.) in ringworm by some of the villagers of Birmaharajpur Block (Sonepur district). *Journal of Pharmaceutical and Scientific Innovation*, 1(4): 37-38.

Kumar RS, Rajkapoor B and Perumal P 2011b. Antitumor and cytotoxic activities of methanol extract of *Indigofera linnaei* Ali. *Asian Pacific Journal of Cancer Prevention*, 12: 613-618.

Kumar S and Chauhan AKS 2011. Diversity of flowering plants in Bharatpur Forest Division, Rajasthan. *The Journal of Indian Botanical Society*, 90(3&4): 229-238.

Kumar S and Dhankhar R 2015. Assessment of floristic and avian faunal diversity of Bhindawas wetland, Jhajjar (Haryana), India. *Plant Archives*, 15(2): 733-740.

Kumar S and Narain S 2010. Herbal remedies of wetlands macrophytes in India. *International Journal of Pharma and Bio Sciences*, 1(2): 1-12.

Kumar S and Satapathy MK 2011. Medicinal plants in an urban environment; herbaceous medicinal flora from the campus of Regional Institute of Education, Bhubaneswar, Odisha. *International Journal of Pharmacy and Life Sciences*, 2(11): 1206-1210.

Kumar S, Ahmad H, Sehgal S, Gupta R and Saraf SA 2011c. Detection and quantitation of ß-sitosterol in *Diospyros montana* Roxb. By HPTLC. *International Journal of Pharma and Bio Sciences*, 2(4): 519-527.

Kumar S, Bajwa B S, Singh K and Kalia AN 2013c. Anti – inflammatory activity of herbal plants: A review. *International Journal of advances in Pharmacy, Biology and Chemistry*, 2: 272-281.

Kumar S, Garg VK, Kumar N, Sharma PK, Chaudhary S and Upadhyay A 2011d. Pharmacognostical studies on the leaves of *Ziziphus nummularia* (Burm. F.). *European Journal of Experimental Biology*, 1(2): 77-83.

Kumar S, Malhotra R, and Kumar D 2010b. *Euphorbia hirta*: Its chemistry, traditional and medicinal uses, and pharmacological activities. *Pharmacognosy Reviews*, 4(7): 58-61.

Kumar S, Parveen F, Goyal S and Chauhan A 2008. Indigenous herbal coolants for combating heat stress in the hot Indian Arid zone. *Indian Journal of Traditional Knowledge*, 7(4): 679-682.

Kumar S, Saini M, Kumar V, Omprakash, Arya R, Rana M and Kumar D 2012d. Traditional medicinal plants curing diabetes: A promise for today and tomorrow. *Asian Journal of Traditional Medicines*, 7(4): 178-188.

Kumar SA, Gandhimathi R, Mohana LS, Rahul N, Kumar CKA 2009. Evaluation of the antinociceptive properties from *Indigofera tinctoria* leaves extracts. *Journal of Pharmaceutical Sciences and Research*, 1(2): 31-37.

Kumar V and Akhtar M 2013. Ethnomedicinal Solanaceous plants of eastern Uttar Pradesh. *Indian Journal of Life Sciences*, 2(2): 95-98.

Kumar VV, Kumar GJ and Kumar UP 2017. Pharmacological importance of *Cucumis melo* L.: An overview. *Asian Journal of Pharmaceutical and Clinical Research*, 10(3): 8-12.

Kumar Y, Haridasan K and Rao RR 1980. Ethnobotanical notes on certain medicinal plants among some Garo people around Balpakram Sanctuary in Meghalaya. *Bulletin of the Botanical Survey of India*, 22: 161-165.

Kumari R, Priyadarshi A and Mita Kotecha 2015. Plants used by the rural communities of village Eka, district Firozabad, Uttar Pradesh. *International Ayurvedic Medical Journal*, 3(2): 502-505.

Kunwar RM, Shrestha KP and Bussmann RW 2010. Traditional herbal medicine in Far-west Nepal: a pharmacological appraisal. *Journal of Ethnobiology and Ethnomedicine*, 6(35): 1-18.

Kuppuswamy KM, Jonnalagadda B and Arockiasamy S 2013. GC-MS analysis of chloroform extract of *Croton bonplandianum*. *International Journal of Pharma and Bio Sciences*, 4(4): 613-617.

Kurade NP, Jaitak V, Kaul VK and Sharma OP 2010. Chemical composition and antibacterial activity of essential oils of *Lantana camara, Ageratum houstonianum* and *Eupatorium adenophorum. Pharmaceutical Biology,* 48: 539-544.

Kurashov EA, Mitrukova GG, Krylovac JV, Aleshinad DG, Bataevae YV and Astafyeva OV 2016. Low-molecular weight metabolites in *Spirodela polyrhiza* (L.) Scheiden from northwest Russia in the middle of the growing season. *International scientific research journal,* 72(10): 10-22.

Kushwaha VB and Maurya S 2012. Biological utilities of *Parthenium hysterophorus. Journal of Applied and Natural Science,* 4(1): 137-143.

Kweon MH, Kwak JH, Ra KS, Sung HC and Yang HC 1996. Structural characterization of a flavonoid compound scavenging superoxide anion radical isolated from *Capsella bursapastoris. International Journal of Biochemistry and Molecular Biology,* 29: 423-428.

Ladewig D and Pletscher A 1993. Fifty Years of LSD: Current status and perspectives of hallucinogens (A symposium of the Swiss academy of medical sciences - Lugano-Agno. Parthenon Publishing, p. 238.

Laitonjam WS and Wangkheirakpam SD 2011. Comparative study of the major components of the Indigo dye obtained from *Strobilanthes flaccidifolius* Nees. and *Indigofera tinctoria* Linn. *International Journal of Plant Physiology and Biochemistry,* 3(7): 108-116.

Lan S 1996. Chemical components of *Lindernia ciliata* (Colsm) Pennell. *China journal of Chinese materia medica,* 21(1):38-9.

Lan Z and De-Zhi Z 2007. Identification of the chemical constituents from *Hydrocotyle sibthorpioides. Journal of Guangdong College of Pharmacy,* 2007-05

Larbie C, Torkornoo D and Dadson J 2014. Anti-diabetic and hypolipidaemic effect of botanicals: A review of medicinal weeds on KNUST Campus, Kumasi. *Journal of Applied Pharmaceutical Science,* 4(10): 97-104.

Lassak EV and McCarthy T 2001. Australian medicinal plants. New Holland Publishers.

Lavanya R and Pattar PV 2017. Traditional knowledge on folk medicine by rural women in Chikkanayakanahalli Taluk, Tumkur district, Karnataka. *International Journal of Herbal Medicine,* 5(2): 5-10.

Leishangthem S and Sharma LD 2014. Study of some important medicinal plants found in Imphal-East District, Manipur, India. *International Journal of Scientific and Research Publications,* 4(9): 1-5.

Levekar GS 2007. Database on medicinal plants used in Ayurveda and Siddha, Vol. VIII. CCRAS, New Delhi.

Levin DA 1976. Alkaloid-bearing plants: an ecogeographic perspective. *American Naturalist,* 110: 261-284.

Li CG, Qu GR, Niu HY, Dong CX and Zhu Y 2005. Chemical constituents of essential oil from *Sonchus brachyotus* DC. *Journal of Henan Normal University (Natural Science).*

Lim TK 2012. Edible medicinal and non-medicinal plants. Springer Netherlands.

Lin W-S, Chen J-Y, Wang J-C *et al.* 2014. The anti-aging effects of *Ludwigia octovalvis* on Drosophila melanogaster and SAMP8 mice. *Age,* 36(2):689-703. doi:10.1007/s11357-013-9606-z.

Liu J, Lin H and McIntosh H 2001. Genus *Phyllanthus* for chronic hepatitis B virus infection: a systematic review. *Journal of Viral Hepatitis,* 8(5): 358-366.

López V, Jäger AK, Akerreta S, Cavero RY, Calvo MI 2011. Pharmacological properties of *Anagallis arvensis* L. ("scarlet pimpernel") and *Anagallis foemina* Mill. ("blue pimpernel") traditionally used as wound healing remedies in Navarra (Spain). *Journal of Ethnopharmacology,* 134(3):1014-1047.

Lovleen and Saggoo MIS 2014. Male meiosis and meiotic abnormalities in two species of *Pouzolzia* Gaud. from western Himalayas, India. *Chromosome Botany,* 9: 1-5.

Luper S, 1998. A review of plants used in the treatment of liver disease: Part 1. *Alternative Medicine Review,* 3(6): 410-421.

Mack RN, Simberloff D, Lonsdale WM,Evans H, Clout M, and Bazzaz FA 2000. Biotic invasions: Causes, epidemiology, global consequences, and control. *Ecological Applications,* 10(3): 689-710.

Mahesh AR, Kumar H, Ranganath MK and Devkar RA 2012. Detail Study on *Boerhaavia diffusa* plant for its medicinal importance-A Review. *Research Journal of Pharmaceutical Sciences,* 1(1): 28-36.

Maheshwari JK 1963. The Flora of Delhi. Council of Scientific and Industrial Research, New Delhi, p. 1- 447.

Maheshwari JK 1966. Illustrations of the flora of Delhi. CSIR, New Delhi, India, p. 1-282.

Maheshwari JK and Paul SR 1975. The exotic flora of Ranchi. *Journal of Bombay Natural History Society,* 72(1): 158-188.

Maheswari PU, Reddy AM, Rambabu M and Basha SKM 2012. Traditional medicinal flora habituated in various regions of YSR (Kadapa) district, Andhra Pradesh, India. *Indian Journal of Fundamental and Applied Life Sciences,* 2(3): 162-175.

Mahmood A, Mahmood A, Hussain I and Kiyani WK 2011a. Indigenous medicinal knowledge of medicinal plants of Barnala area, district Bhimber, Pakistan. *International Journal of Medicinal and Aromatic Plants,* 1(3): 294-301.

Mahmood A, Mahmood A, Naveed I, Memon MM, Bux H, Majeed MY, Mujtaba G and Mumtaz MS 2011b. Indigenous medicinal knowledge of common plants used by local people of Hattian Bala district, Azad Jammu and Kashmir (AJK), Pakistan. *Journal of Medicinal Plants Research,* 5(23): 5517-5521.

Mahmood N, Piocente S, Burke A, Khan AI and Pizza C 1997. Constituents of *Cuscuta reflexa* are anti-HIV agents. *Antiviral Chemistry & Chemotherapy,* 8(1): 70-74.

Mahmoud T 2010. Desert plants of Egypt's Wadi El Gemal National Park. The American University in Cairo Press, p. 180.

Mahmoud T and Gairola S 2013. Traditional knowledge and use of medicinal plants in the eastern desert of Egypt: A case study from Wadi El-Gemal National Park. *Journal of Medicinal Plants Studies,* 1(6):10-17.

Majumdar K and Datta BK 2007. A study on ethnomedicinal usage of plants among the folklore herbalists and Tripuri medical practitioners: Part-II. *Natural Product Radiance,* 6(1): 66-73.

Makhija IK and Khamar D 2010. Anti-snake venom properties of medicinal plants. *Scholars Research Library,* 2 (5): 399-411.

Makhija IK, Sharma IP and Khamar D 2010. Phytochemistry and pharmacological properties of *Ficus religiosa:* An overview. *Annals of Biological Research,* 1(4):171-180.

Malakar C and Choudhury PPN 2015. Pharmacological potentiality and medicinal uses of *Ipomoea aquatica* Forsk: A review. *Asian Journal of Pharmaceutical and Clinical Research,* 8(2): 60-63.

Malhotra CL 1979. Flora of Pilibhit District (Uttar Pradesh), Vol. 2. Ph.D. Thesis, Guwahati University, Assam, India.

Mali MC and Harsh N 2015. Nutritional value estimation of the leaves and seeds of *Solanum surattense*. *Journal of Medicinal Plants Studies,* 3(1): 27-29.

Mali PY and Bhadane VV 2011. Ethno-medicinal wisdom of tribals of Aurangabad district (M. S.) India. *Indian Journal of Natural Products and Resources,* 2(1): 102-109.

Mali RG, Hundiwale JC Gavit RS, Patil DA and Patil KS 2006. Herbal abortifacients used in North Maharashtra. *Natural product Radiance,* 5(4): 315-318.

Malik AH, Khuroo AA, Dar GH and Khan ZS 2011. Ethnomedicinal uses of some plants in the Kashmir Himalaya. *Indian Journal of Traditional Knowledge,* 10(2): 362-366.

Malik S, Ahmad S, Sadiq A, Alam K, Wariss HM, Ahmad I, Hayat MQ, Anjum S and Mukhtar M 2015. A comparative ethno-botanical study of Cholistan (an arid area) and Pothwar (a semi-arid area) of Pakistan for traditional medicines. *Journal of Ethnobiology and Ethnomedicine,* 11(31): 1-20.

Malik V 2015a. Herbaceous flora of Muzaffarnagar district (U.P.). *Scholars Academic Journal of Biosciences,* 32B):182-196.

Malik V 2015b. A checklist of grasses (Poaceae) of Saharanpur Forest Division. *Indian Journal of Fundamental and Applied Life Sciences,* 5(2): 74-80.

Maliya SD 2012. Additions to the flora of Katarniyaghat Wildlife Sanctuary, Bahraich district, Uttar Pradesh. *Journal of Economic and Taxonomic Botany* 36(2): 419–426.

Maliya SD and Datt B 2010. A contribution to the flora of Katarniyaghat Wildlife Sanctuary, Bahraich District, Uttar Pradesh. *Journal of Economic and Taxonomic Botany* 34(1): 42–68.

Mall TP 2016. Nutrimental and therapeutic potential feed plants of Bahraich (U.P), India. *Agricultural Science Research Journal,* 6(9): 227-234.

Mall TP and Tripathi SC 2015. Nepeta Herb- A nutrimental potent ethnomedicinal plant from North Western Tarai Forests of U.P. India. *World Journal of Pharmaceutical Research,* 4(11): 1623-1630.

Mamadalieva NZ, Lafont R and Wink M 2014. Diversity of secondary metabolites in the genus *Silene* L. (Caryophyllaceae)—Structures, distribution, and biological properties. *Diversity — Open Access Biodiversity Journal,* 6: 415-499.

Mamidala E and Gujjeti RP 2013. Phytochemical and antimicrobial activity of *Acmella paniculata* plant extracts. *Journal of Bio Innovation,* 2(1):17-22.

Manandhar, NP 2002. Plants and people of Nepal. Portland, Oregon, USA: Timber Press, p. 220.

Manda RM and Seru G 2015. Hypoglycaemic activity of the leaves of *Mimosa rubicaulis* (Lam). *European Journal of Biomedical and Pharmaceutical sciences,* 2(1): 543-548.

Mandal A and Mandal S 2016. Eco-floristic survey of asansol industrial belt, West-Bengal with reference to their economic and medicinal value. *International Journal of Current Microbiology and Applied Sciences,* 5(6): 949-960.

Mandal FB 2011. The management of alien species in India. *International Journal of Biodiversity and Conservation,* 3: 467-473.

Manikandan PNA 2008. Ethno- medico - botanical studies of Badaga population in the Nilgiri district of Tamilnadu, South India. *Ancient Science of Life,* 17(3): 50-59.

Manikandaselvi S, Vadivel V and Brindha P 2016. Review on nutraceutical potential of *Cassia occidentalis* L. – An Indian traditional medicinal and food plant. *International Journal of Pharmaceutical Sciences Review and Research,* 37 (2): 141-146.

Mannan MM, Maridass M and Victor B 2008. A review on the potential uses of ferns. *Ethnobotanical Leaflets,* 12: 281-285.

Manoharan S 2014. An ethnomedical, pharmacological and phytochemical review of *Ammania baccifera* L. *World Journal of Pharmaceutical Research* 3(6):1771-1789.

Manral U, Raha A, Solanki R, Hussain SA, Babu MM, Mohan D, Veeraswami GG, Sivakumar K and Talukdar G 2013. Plant species of Okhla Bird Sanctuary: A wetland of Upper Gangetic Plains, India. *Check List,* 9 (2): 263–274.

Manvar D, Mishra M, Suriender Kumar S, and Pandey VN 2012. Identification and evaluation of anti-Hepatitis C Virus phytochemicals from *Eclipta alba.* J Ethnopharmacol. 144(3): 545–554.

Maoa Z, B, Fu H, Nan Z and Wan C 2015. Fatty acid, amino acid, and mineral composition of four common vetch seeds on Qinghai-Tibetan plateau. *Food Chemistry,* 171: 13-18.

Marandi RR and Britto SJ 2015. Medicinal properties of edible weeds of crop fields and wild plants eaten by Oraon tribals of Latehar district, Jharkhand. *International Journal of Life science and Pharma Research,* 5(2): 1-9

Maregesi SM, Ngassapa OJ, Pieters L, Vlietinck AJ 2007. Ethno pharmacological survey of the Bunda district, Tanzania: Plants used to treat infectious diseases. *Journal of Ethnopharmacology,* 113(3): 457-470.

Maridass M, Ghantikumar S and Raju G 2008. Preliminary phytochemical analysis of *Diospyros* species. *Ethnobotanical Leaflets,* 12: 868-872.

Maroyi A 2011. Ethnobotanical study of medicinal plants used by people in Nhema communal area, Zimbabwe. *Journal of Ethnopharmacology,* 136(2): 347-354.

Maroyi A 2013. Traditional use of medicinal plants in south-central Zimbabwe: Review and perspectives. *Journal of Ethnobiology and Ethnomedicine,* 9(31): 1-18.

Martinez GJ and Lujan MC 2011. Medicinal plants used for traditional veterinary in the Sierras de Cordoba (Argentina): An ethnobotanical comparison with human medicinal uses. *Journal of Ethnobiology and Ethnomedicine,* 7(23): 1-18.

Maruthupandian A, Mohan VR and Kottaimuthu R 2011. Ethnomedicinal plants used for the treatment of diabetes and jaundice by Palliyar tribals in Sirumalai hills, Western Ghats, Tamil Nadu, India. *Indian Journal of Natural Products and Resources,* 2(4): 493-497.

Marwat SK, Fazal-ur-Rehman and Khan IU 2015. Ethnobotanical importance and phytochemical constituents of Parthenium weed (*Parthenium hysterophorus* L.) – A Review. *Plant Science Today,* 2(2): 77-81.

Marwat SK, Khan MA, Ahmad M, Zafar M and Usman K 2012. Floristic account of the Asclepiadaceous species from the flora of Dera Ismail Khan District, KPK, Pakistan. *American Journal of Plant Sciences,* 3: 141-149.

Marwat SK, Khan MA, Fazalur-rehman, Ahmad M and Zafar M 2011a. Biodiversity and importance of floating weeds of Dara Ismail, khan district of KPK, Pakistan. *African Journal of Traditional, Complementary and Alternative medicines,* 8: 97-107.

Marwat SK, Rehman F, Khan MA, Ahmad M, Zafar M and Ghulam G 2011b. Medicinal folk recipes used as traditional phytotherapies in district Dera Ismail khan, KPK, Pakistan. *Pakistan Journal of Botany,* 43(3): 1453-1462.

Marzouk M, Hussein S, Ahmed E, Farid MM, Ibrahim LF and Abdel-Hameed ES 2016. Phenolic profiling of *Rorippa palustris* (L.) Besser (Brassicaceae) by LC-ESI-MS: Chemosystematic signifiance and cytotoxic activity. *Asian Pacific Journal of Tropical Disease,* 6: 633-637.

Mathad P and Mety SS 2009. Phytochemical and antimicrobial activity of *Digera muricata* (L.) Mart. *E-Journal of Chemistry,* 7(1): 275-280.

Mathur M, Nama KS and Choudhary K 2016. Economic and ethnomedicinal importance of the floral diversity on ancient walls of Kota District, Rajasthan. *International Journal of Pure & Applied Bioscience,* 4(4): 167-173.

Matthew KM 1995. An Excursion Flora of Central Tamil Nadu, India. CRC Press, p. 682.

Mazid M, Khan TA, Mohammad F 2012. Medicinal plants of rural India: A review of use by Indian Folks. *Indo Global Journal of Pharmaceutical Sciences,* 2(3): 286-304.

Mazumdar J and Choudhury MD 2015. Review on the ethnomedicobotany and phytochemistry of *Sonchus brachyotus* A.P. de Candolle (Asteraceae). *Pleione,* 9(1): 201-209.

Meena Devi VN, Ariharan VN and Nagendra Prasad P 2013. Nutritive value and potential uses of *Leucaena leucocephala* as biofuel – A mini review. *Research Journal of Pharmaceutical, Biological and Chemical Sciences,* 4(1):515-521.

Meena KL and Yadav BL 2010. Some ethnomedicinal plants of southern Rajasthan. *Indian Journal of Traditional Knowledge,* 9(3):169-172.

Mehrotra BN 1986. Research in pharmacognosy on Indian medicinal plants. Current Research on Medicinal Plants in India, INSA, New Delhi, p. 6.

Mei H, Zuo S, Ye L, Wang J and Ma S 2012. Review of the application of the traditional Chinese medicinal herb, *Ranunculus sceleratus* Linn. *Journal of Medicinal Plants Research,* 6(10): 1821-1826.

Meira M, da Silva EP, David JM and David JP 2012. Review of the genus *Ipomoea*: traditional uses, chemistry and biological activities. *Revista Brasileira de Farmacognosia,* 22(3): 662- 713.

Menut C, Lamaty G, Amvam Zollo PH, Kuiate J-R and Bessière J-MB 1993. Aromatic plants of tropical central Africa. Part X: Chemical composition of the essential oils of *Ageratum houstonianum* Mill. and *Ageratum conyzoides* L. from Cameroon. *Flavour and Fragrance Journal,* 8: 1-4.

Mgole MS, Luc P and David NO 2007. Ethno pharmacological survey of the Bunda district, Tanzania: Plants used to treat infectious diseases. *Journal of Ethnopharmacology,* 113(3): 457-470.

Mikawlrawng K, Kumar S and Vandana 2014. Current scenario of urolithiasis and the use of medicinal plants as antiurolithiatic agents in Manipur (North East India): A review. *International Journal of Herbal Medicine,* 2(1): 1-12.

Minu V, Harsh V, Ravikant T, Paridhi J and Noopur S 2012. Medicinal plants of Chhattisgarh with anti-snake venom property. *International Journal of Current Pharmaceutical Research,* 3: 1-10.

Mirjalili MH, Moyano E, Bonfill M, Cusido RM and Palazón J 2009. Steroidal lactones from *Withania somnifera,* an ancient plant for novel medicine. *Molecules,* 14: 2373-2393.

Mishra AK, Mir SA, Sharma MP and Singh H 2015b. Additions to the flora of Delhi: Part II. *Indian Journal of Plant Sciences,* 4(2): 1-6.

Mishra AK, Mir SA, Sharma MP and Singh HB 2014. Addition to the Flora of Delhi. *Indian Journal of Plant Sciences,* 3(3): 64-67.

Mishra AK, Mir SA, Sharma MP, Singh H 2015d. Alien plant species in Delhi flora. *International Journal of Geology, Earth & Environmental Sciences,* 5(2):128-140.

Mishra AK, Sharma MP and Khan MI 2014. Additions to the flora of Ghaziabad. *Trends in Biosciences,* 7(22): 3509-3512.

Mishra AK, Sharma MP and Singh H 2015f. Floristic diversity of Delhi, India: A checklist. *International Journal of Herbal Medicine,* 3(4): 8-18.

Mishra AK, Sharma MP and Singh H 2015a. Angiospermic flora of Delhi, India: An Updated Checklist. *Indian Journal of Plant Sciences,* 4(3): 58-84.

Mishra AK, Sharma MP and Singh H 2015c. Additions to the flora of Delhi: Part III. *Indian Journal of Plant Sciences,* 4(3): 38-40.

Mishra AK, Sharma MP and Singh H 2015e. Plant species of Delhi flora: a medicinal review. *Indian Journal of Plant Sciences,* 4(4): 73-111.

Mishra AK, Sharma MP, Khan MI and Singh H 2016. Diversity and status of Unani medicinal plants species of Delhi flora. *International Journal of Botany Studies,* 1(1):10-16.

Mishra KN and Pal M 2010. Tree wealth of eastern Uttar Pradesh, India. *Plant Archives* 10(2): 833–836. http://www.cabi.org/ISC/FullTextPDF/2012/20123401919.pdf

Mishra LC, Singh BB, Dagenais S 2000. Scientific basis for the therapeutic use of *Withania somnifera* (Ashwagandha): A review. *Alternative Medicine Review,* 5(4):334-346.

Mishra MR, Mishra A, Pradhan DK, Panda AK, Behera RK and Jha S 2013. Antidiabetic and antioxidant activity of *Scoparia dulcis* Linn. *Indian Journal of Pharmaceutical Sciences,* 75(5): 610–614.

Mishra S, Aeri V, Gaur PK, and Jachak SM 2014. Phytochemical, therapeutic, and ethnopharmacological overview for a traditionally important herb: *Boerhavia diffusa* Linn. *BioMed Research International,* 5: 1-19.

Misra VK 2015. Successional pattern and plant species diversity in the aquatic and wetland habitats of North-central Uttar Pradesh, India. Indian Forester, 141(1): 57-67.

Mithun NM, Shashidhara S and Kumar RV 2011. *Eclipta alba* (L.) A review on its phytochemical and pharmacological profile. *Pharmacologyonline,* 1: 345-357.

Mitra R and Jain SK 1991. Medicinal plant research in India. An overview. *Ethnobotany,* 3: 65-77.

Mitra S and Mukherjee SK 2009. Ethnobotany of some grasses of West Bengal, India. In Advances in Plant Biology (Debidas Bhattacharya Birth Centenary Commemorative Volume) (Eds. Sudhendu Mandal & Sukla Bhattacharya), pp. 221-253.

Mohammadi H, Sajjadi SE, Noroozi M and Mirhosseini M 2016. Collection and assessment of traditional medicinal plants used by the indigenous people of Dastena in Iran. *Journal of Herbmed Pharmacology,* 5(2): 54-60.

Mohanty S, Srivastava P, Maurya AK, Cheema HS, Shanker K, Dhawan S, Darokar MP, Dnyaneshwar Bawankule DU 2013. Antimalarial and safety evaluation of *Pluchea lanceolata* (DC.) Oliv. & Hiern: *In-vitro* and *in-vivo* study. *Journal of Ethnopharmacology,* 149(3):797-802.

Mohideen S, Sasikala E and Gopal V 2002. Pharmacognostic studies on *Sida acuta* Burm.f. *Ancient Science of Life,* 22(1): 57-66.

Mohite MS, Shelar PA, Raje VN, Babar SJ and Sapkal RK 2012. Review on pharmacological properties of *Abutilon indicum. Asian Journal of Pharmaceutical Research,* 2(4): 156-160.

Molla HA and Roy B 1985. Traditional uses of some medicinal plants by the Rabha tribals in Jalpaiguri District, West Bengal. *Journal of economic and taxonomic botany,* 7: 578-580.

Mondal P, Bhuyan N, Das S, Kumar M, Borah S and Mahato K 2013. Herbal medicines useful for the treatment of diabetes in north-east India: A review. *International Journal of Pharmacy and Biological Sciences,* 3 (1):575-589.

Mondal S and Bandyopadhyay A 2016. The wonders of a medicinal tree: *Holoptelea integrifolia* (Roxb.) Planch. *International Journal of Pharmacy and Pharmaceutical Sciences,* 8(8): 43-48.

Moran DQ 2009. Guide to Afro-Cuban Herbalism. Author House, p. 696.

Morebise O 2015. Medicinal plants of Dominica—Uses, chemical constituents, bioactivities and prospects. *Journal of Medicinal Plants Studies,* 3(5): 144-154.

Morshed MA, Azimuddin, Saifur R, Barua A and Haque A 2011. Evaluation of antimicrobial and cytotoxic properties of *Leucas aspera* and *Spilanthes paniculata*. *International Journal of Biosciences,* 1(2):7-16.

Moshi MJ, Mbwambo ZA, Nondo RSO, Masimbaa PJ, Kamuhabwa A, Kapingu MC, Thomas P and Richard M 2006. Evaluation of ethnomedical claims and brine shrimp toxicity of some plants used in Tanzania as traditional medicines. *African Journal of Traditional, Complementary and Alternative medicines,* 3(3): 48-58.

Mudgal V and Khanna KK 1999. Dicotyledonous plants of Uttar Pradesh: A Check List, Hardcover. Bishen Singh Mahendra Pal Singh, Dehradun, India.

Muhammad IC and Khan MA 2008. An ethnomedicinal inventory of plants used for family planning and sex diseases in Samahni valley, Pakistan. *Indian Journal of Traditional Knowledge,* 7(2): 277-283.

Mukhopadhyay MK, Banerjee P and Nath D 2012. Phytochemicals – biomolecules for prevention and treatment of human diseases- A review. *International Journal of* Scientific *& Engineering Research,* 3(7): 1-32.

Mulla SK and Swamy P 2010. Preliminary pharmacognostical and phytochemical evaluation of *Portulaca quadrifida* Linn. *International Journal of Pharm. Tech. Research,* 2(3): 1699-1702.

Mungole AJ, Awati R, Chaturvedi A, and Zanwar P 2010. Preliminary phytochemical screening of *Ipomoea obscura* (L) -A hepatoprotective medicinal plant. *International Journal of PharmTech Research,* 2(4):2307-2312.

Murad W, Ahmad A, Gilani SA and Khan MA 2011. Indigenous knowledge and folk use of medicinal plants by the tribal communities of Hazar Nao Forest, Malakand District, North Pakistan. *Journal of Medicinal Plants Research,* 5(7): 1072-1086.

Murthy EN 2012. Ethno-medicinal plants used by Gonds of Adilabad district, Andhra Pradesh. *International Journal of Pharmacy & Life Sciences,* 3(10): 2034-2043.

Murthy KSR, Reddy MC and Pullaiah T 2015. Ethnobotany, chemistry and pharmacology of an aromatic genus *Anisomeles* linn. In India. *International Journal of Life science and Pharma Research,* 5(4): 35-49.

Murugananthan G, Kumar SG, Chethan SP and Mohan S 2013. Anti-arthritic and anti-inflammatory constituents from medicinal plants. *Journal of Applied Pharmaceutical Science,* 3(4): 161-164.

Murugesan S, Senthilkumar N, Babu DS and Rajasugunasekar D 2016. Chemical constituents and toxicity assessment of the leaf oil of *Lantana camara* Linn from Tamilnadu regions. *Asian Journal of Plant Science and Research,* 6(3):32-42.

Mustapha AA 2013. Ethnomedicinal studies of medicinal plants with antifungal activities in Keffi local government, Nasarawa state, Nigeria. *Asian Journal of Plant Science and Research,* 3(4):109-115.

Mustapha AA, Owuna G and Uthman I 2013. Plant remedies practiced by Keffi people in the management of dermatosis. *Journal of Medicinal Plants Studies,* 1(5): 112-118.

Muthu C, Ayyanar M, Raja N, and Ignacimuthu S 2006. Medicinal plants used by traditional healers in Kancheepuram district of Tamil Nadu, India. *Journal of Ethnobiology and Ethnomedicine,* 243): 1-10.

Muthulingam M 2010. Antihepatotoxic efficacy of Nymphaea pubescens (willd.) on acetaminophen induced liver damage in male Wistar rats. *International Journal of Current Research,* 3: 12-16.

Naithani HB, Chandra S and Pal M 2006. Additions to the flora of Delhi. *Indian Forester,* 132(5): 589-600.

Naithani HB, Negi SS, Pal M, Chandra S and Khanduri VP 2007. Vegetational survey and inventorisation of species in the Ridge Forest of Delhi, Forest Research Institute (Indian Council of Forestry Research and Education) New Forest, Derhadun-248006) 1-624.

Najafabadi MP, Sichani MM, Kazemi MJ, Shirsalimian M and Tavakoli M 2016. Investigation of the chemical composition and different effects of a *Rumex dentatus* methanol extract against drug resistant *Pseudomonas aeruginosa* isolates. *Iranian Red Crescent Medical Journal,* 18(12): 1-8. DOI: http://dx.doi.org/10.5812/ircmj.27064.

Najeeb U, Saima S, Ali R, Hussain U, Farhan and Ijaz A 2013. Antibacterial and antifungal study of *Spergula arvensis*. *International Journal of Science Innovations and Discoveries,* 3(3): 348-352.

Namrata S, Kumar YU, Jana GK and Tripathi R 2014. Pharmacognostic, phytochemical and physiochemical study of *Evolvulus nammularius* (Convolvulaceae). *International Journal of Pharmacognosy and Phytochemical Research,* 6(4): 874-878.

Narayana VL and Narasimharao GM 2015. Plants used in ethnoveterinary medicine by tribals of Visakhapatnam and Vizianagarm districts, Andhra Pradesh, *International Journal of Pure & Applied Bioscience,* 3(2): 432-439.

Naskar K 1993. Plant wealth of the lower Ganga Delta: an eco-taxonomical approach, Vol. 1. Daya Pub. House, p. 810.

Nassar MI, Abdel-Razik AF, El-Khrisy ED, Dawidar AA, Bystrom A and Mabry TJ 2002. A benzoquinone and flavonoids from *Cyperus alopecuroides*. *Phytochemistry,* 60(4): 385-7.

Natrajan P, Elumalai A, Soundarajan C 2012. Study of antibacterial activity of *Chloris barbata* (sw) leaves. *International Research Journal of Pharmaceutical and Applied Sciences,* 2(1): 37-40.

Nayak P, Thirunavoukkarasu M 2016. A review of the plant *Boerhaavia diffusa*: its chemistry, pharmacology and therapeutical potential. *The Journal of Phytopharmacology,* 5(2): 83-92.

Nayar MP 1977. Changing patterns of the Indian flora. *Bulletin of the Botanical Survey of India,* 19(1-4): 145-155.

Naz I, Ahmad M and Hassan T 2014. Ethnobotanical investigation of medicinal flora used by indigenous people in district Attock, Pakistan. *Journal of advanced botany and zoology,* 1(4):1-7.

Naz R, Bano A 2013. Phytochemical screening, antioxidants and antimicrobial potential of *Lantana camara* in different solvents. *Asian Pacific Journal of Tropical Disease,* 3(6): 480-486.

Nazmul-Hasan AHM, Roy P, Bristy NJ, Paul SK, Wahed TB and Alam MN 2015. Evaluation of in vitro antioxidant and brine shrimp lethality bioassay of different extracts of Polygonum plebeium R. Br. *International Journal of Advanced Research,* 3(12): 97-107.

Neamsuvan O, Singdam P, Yingcharoen K and Sengnon N 2012. A survey of medicinal plants in mangrove and beach forests from Sating Phra Peninsula, Songkhla Province, Thailand. *Journal of Medicinal Plants Research,* 6(12): 2421-2437.

Neeraja PV and Margaret E 2012. *Eclipta alba* (L.) Hassk: A valuable medicinal herb. *International Journal of Current Pharmaceutical Review and Research,* 2(4): 188-197.

Neeta MP, Mukta N and Bilwa K 2015. Comparative qualitative phytochemical analysis of *Sesamum indicum* L. *International Journal of Current Microbiology and Applied Sciences,* 2: 172-181.

Neogi B, Prasad MNV and Rao RR 1989. Ethnobotany of some weeds of Khasi and Garo hills, Meghalaya, Northeastern India. *Economic Botany,* 43(4): 471-479.

Nguta JM and Mbaria JM 2013. Brine shrimp toxicity and antimalarial activity of some plants traditionally used in treatment of malaria in Msambweni district of Kenya. *Journal of Ethnopharmacology,* 148(3): 988-992.

Nibret E, Youns M, Krauth-Siegel RL, Wink M 2011. Biological activities of Xanthatin from *Xanthium strumarium* leaves. *Phytotherapy Research,* 25(12):1883-1890.

Nicollier GF, and Thompson AC 1982. Phytotoxic compounds from *Melilotus alba* (white sweet clover) and isolation and identification of two new flavonoids. *Journal of Agricultural and Food Chemistry,* 30(4): 760-764.

Nigam G, Babu GD and Maurya SK 2013. Folklore claims on some medicinal plants used in Jhansi District, Uttar Pradesh, India, by Rawat and Sahariya Tribes. Research and Reviews: *Journal of Pharmacognsoy and Phytochemistry,* 1(2): 1-4.

Nigussie Z and Alemayehu G 2013. *Sesbania sesban* (L.) Merrill: Potential uses of an underutilized multipurpose tree in Ethiopia. *African Journal of Plant Science,* 7(10): 468-475.

Nisar A, Shinwari ZK, Hussain J and Perveen R 2015. Phytochemicals, antibacterial and antioxidative investigations of *Alhagi maurorum* Medik. *Pakistan Journal of Botany,* 47(1): 121-124.

Nisar MF, Jaleel F, Waseem M, Ismail S, Toor Y, Haider SM and Zhong JL 2014. Ethno-medicinal uses of plants from District Bahawalpur, Pakistan. *Current Research Journal of Biological Sciences,* 6(5): 183-190.

Nivas RK and Boominathan M 2015. Antimicrobial evaluation of selected south Indian medicinal plants against *Streptococcus pneumoniae. International Journal of Current Microbiology and Applied Sciences,* 4(3): 835-840.

Njoroge GN and Bussmann RW 2007. Ethnotherapeautic management of skin diseases among the Kikuyus of Central Kenya. *Journal of Ethnopharmacology,* 111:303-307.

Noiarsa P, Ruchirawat S, Otsuka H, and Kanchanapoom T 2008. Chemical constituents from *Oldenlandia corymbosa* L. of Thai origin. *Journal of Natural Medicines,* 62(2): 249-250.

Nyananyo BL, Gijo AH and Ogamba EN 2007. The Physico-chemistry and distribution of water Hyacinth (*Eichhornia cressipes*) on the river Nun in the Niger Nelta. *Journal of Applied Sciences and Environmental Management,* 11(3):133-137.

O'Neill MJ and Lewis JA 1993. The renaissance of plant research in the pharmaceutical industry. *American Chemical Society,* 534: 48-55.

Odeja OO, Obi G, Ogwuche CE, Elemike EE and Oderinlo OO 2014. Phytochemical screening, antioxidant and antimicrobial activities of *Senna occidentalis* (L.) leaves. *International Journal of Herbal Medicine,* 2(4): 26-30.

Ogbobe O and Akano V 1993. The physico-chemical properties of the seed and seed oil of *Jatropha gossipifolia. Plant Foods for Human Nutrition,* 43(3):197-200.

Ogundajo AL, Akpome AS, Tijani NA and Ogunwande IA 2015. Chemical constituents of the leaf essential oil of *Physalis angulata* L. *Asian Journal of Applied Sciences,* 3(4): 652- 655.

Okeniyi SO, Adedoyin BJ and Garba S 2012. Phytochemical screening, cytotoxicity, antioxidant and antimicrobial activities of stem and leave extracts of *Euphorbia heterophylla. Bulletin of Environment, Pharmacology and Life Sciences,* 1(8): 87-91.

Okhale, Ehiabhi S, Amanabo, Omachonu M, Jegede, Adeola I, Egharevba, Omoregie H, Muazzam, Wudil I, Kunle and Folashade O 2010. Phytochemical and pharmacognostic investigation of antidiabetic *Scoparia dulcis* Linn Scrophulariaceae whole plant grown in Nigeria. *Researcher,* 2(6): 7-16.

Oleszek W, Price KR and Fenwick GR 1988. Triterpene saponins from the roots of *Medicago lupulina* L (black medick trefoil). *Journal of the Science of Food and Agriculture,* 43(4): 289-297.

Olutola DO, Onocha P, Ekundayo O and Ali M 2014. Isolation of aurantiamides from *Gomphrena celosioides* C. Mart. *Iranian journal of pharmaceutical research,* 13(1):143-147.

Oudhia P 2001. Parthenium hysterophorus: traditional medicinal uses. https://hort.purdue.edu/newcrop/CropFactSheets/parthenium.html

Packialakshmi N and Alwin 2014. Analysis of phytochemical and high performance liquid chromatography in *Desmostachya bipinnata. Biomedical Research,* 1(1): 1-6.

Padal SB, Chandrasekhar P and Vijayakumar Y 2013. Ethnomedicinal Uses Of Some Fabaceae Family Plants of Narsipatnam Division, Visakhapatnam District, Andhra Pradesh, India. *International Journal of Innovation in Research and Development,* 2(6): 808-822.

Padalia H and Chanda S 2015. Comparative phytochemical analysis of aerial parts of *A. procumbeans, F. dichotoma, S. sponteneum, S. nigra and T. angustifolia. Journal of Pharmacognosy and Phytochemistry*, 4(2): 11-16.

Padalia K 2015. Gewai saag: A folk medicine used by the tribal people of Central Himalayan region. *Indian Journal of Traditional Knowledge,* 14(1): 144-146.

Padma Y, Sarojinidevi N, Ratnam KV, Reddy GT and Raju PRV 2016. Herbal/Folk remedies used for snake bites by tribals/rural people of Rayalaseema region, Andhra Pradesh, India. *Journal of Medicinal Plants Studies,* 4(2): 52-56.

Pal DK and Nimse SB 2006. Little known uses of common aquatic plant, *Hydrilla verticillata* (Linn. f.) Royle. *Natural Product Radiance,* 5(2): 108-111.

Palaksha MN and Ravishankar K 2012. Phytochemical screening and evaluation of *in vitro* antibacterial and anthelmintic activities of *Sida acuta* leaf extracts. *Journal of chemical and pharmaceutical research,* 4(11):4757-4761.

Paliwal DNK 2017. *Eclipta alba* (Linn.) Hassk. – A review. *World Journal of Pharmaceutical and Life Sciences,* 3(1): 713-721.

Panchal MA, Murti K and Lambole V 2010. Pharmacological properties of *Verbascum Thapsus* - A review. *International Journal of Pharmaceutical Sciences Review and Research,* 5(2): 73-77.

Panda A and Mishra MK 2011. Ethnomedicinal survey of some wetland plants of South Orissa and their conservation. *Indian journal of traditional knowledge,* 10(2): 296-303.

Panda H 2002. Medicinal plants cultivation and their uses. Asia Pacific Business Press Inc., p. 598.

Panda SK, Rout SD, Mishra N and Panda T 2011. Phytotherapy and traditional knowledge of tribal communities of Mayurbhanj district, Orissa, India. *Journal of Pharmacognosy and Phytotherapy,* 3(7): 101-113.

Panda T and Mishra M 2016. Indigenous knowledge on animal health care practices in Kendrapara district of Odisha, India. *International Letters of Natural Sciences,* 53: 10-27.

Panda T, Mishra N, Pradhan BK and Mohanty R 2016. Some less known folk medicine in Bhadrak and Kendrapara districts of Odisha, India. *Micro Medicine,* 4(1): 8-20.

Pandaa T and Padhy RN 2008. Ethnomedicinal plants used by tribes of Kalahandi district, Orissa. *Indian Journal of Traditional Knowledge,* 7(2): 242-249.

Pandey AK, Mohan M, Singh P and Tripathi NN 2015. Chemical composition, antioxidant and antimicrobial activities of the essential oil of *Nepeta hindostana* (Roth) Haines from India. *Records of Natural Products,* 9(2): 224-233.

Pandey MM, Rastogi S, Rawat KS 2007. Indian herbal drug for general healthcare: An overview. *The Internet Journal of Alternative Medicine,* 6(1): 1-10.

Pandey RP and Parmar PJ 1994. The exotic flora of Rajasthan. *Journal of Economic and Taxonomic Botany,* 18: 105-121.

Pandya DJ and Anand IS 2011. Diuretic potential of various extracts of *Oxystelma esculentum* and its preliminary phytochemical screening. *Pharmacologyonline,* 1: 163-173.

Pandya PR 2009. A study of the weed flora of some cultivated fields of Bharuch district. PhD Thesis, Saurashtra University.

Panghal M, Arya V, Yadav S, Kumar S and Yadav JP 2010. Indigenous knowledge of medicinal plants used by Saperas community of Khetawas, Jhajjar district, Haryana, India. *Journal of Ethnobiology and Ethnomedicine,* 6(4): 1-11.

Panhwar AQ and Abro H 2007. Ethnobotanical studies of Mahal kohistan (khirthar National park). *Pakistan Journal of Botany,* 39(7): 2301-2315.

Pani M, Nahak G and Sahu RK 2015. Review on ethnomedicinal plants of Odisha for the treatment of malaria. *International Journal of Pharmacognosy and Phytochemical Research,* 7(1): 156-165.

Panigrahi G, Singh AN and Misra OP 1969. Contribution to the botany of the Tarai forests of the Bahraich district of Uttar Pradesh. *Bulletin of Botanical Survey of India* 11(1/2): 89–114.

Pappan A and Thomas B 2017. Contribution of invasive plants in herbal medicinal practices. *International Journal of Herbal Medicine,* 5(2): 73-77.

Parab N and Vaidya S 2016. Determination of some trace elements and macro minerals of *Sesbania bispinosa* (Jacq.) W. F. Wight. *International Journal of Pharmacy and Pharmaceutical Research,* 6(2): 383-401.

Pareek AK, Garg S, Kumar M, and Yadav SM 2015. *Prosopis cineraria*: A gift of nature for pharmacy. *International Journal of Pharma Sciences and Research,* 6(6): 958- 964.

Paridhavi, Sunil SJ, Nitin A, Patil MB, Chimkode R, Tripathi A. 2008. Antimicrobial and Wound healing activities of leaves of *Alternanthera sessilis* Linn. *International Journal of green pharmacy*, 2:141-144.

Park CJ, Park CB, Hong SS, Lee HS, Lee SY, Kim SC 2000. Characterization and cDNA cloning of two glycine- and histidine-rich antimicrobial peptides from the roots of shepherd's purse, *Capsella bursa-pastoris*. *Plant Molecular Biology*, 44(2): 187-197.

Parkavi V, Vignesh M, Selvakumar K, Mohamed JM and Ruby JJ 2012. Antibacterial activity of aerial parts of *Imperata cylindrica* (L) Beauv. *International Journal of Pharmaceutical Sciences and Drug Research*, 4(3): 209-212.

Patel DK, Prasad SK, Kumar R, and Hemalatha S 2012. An overview on antidiabetic medicinal plants having insulin mimetic property. *Asian Pacific Journal of Tropical Biomedicine*, 2(4): 320-330.

Patel JR and Dabgar YB 2016. Study on traditional therapy of rural people from selected talukas of Banaskantha district, North Gujarat. *Indian Journal of Fundamental and Applied Life Sciences*, 6(2): 62-68.

Patel R, Mahato AKR, Kumar VV and Asari RV 2013. Status of the medicinal plants in Tharawada-Gandher Reserve Forest of Kachchh, Gujarat and the ethnomedicinal practices of local community. *Journal of Medicinal Plants Studies*, 1(4): 1-10.

Patel RM, Dabgar YB and Joshi PN 2010. Some noteworthy ethno medicinal plants of Western Kachchh, Gujart. *Life Sciences Leaflets*, 9: 244-250.

Patel S 2011. Harmful and beneficial aspects of *Parthenium hysterophorus*: an update. *3 Biotech*, 1(1): 1-9.

Patel TD, Jain V and Dodia R 2014. *Oldenlandia corymbosa* L.: A phytopharmacological review. *International Journal of Phytopharmacy*, 4(3): 79-82.

Pathak GP, Sharma V, Singh G and Kour A 2013. Neem (*Azadirachta indica*): Traditional medicine for house-hold remedy against various human and animal ailments: review. *World Journal of Pharmacy and Pharmaceutical Sciences*, 2(5): 3854-3860.

Patil DA 2008. Herbal Cures: Traditional Approach. Pointer Publishers, Jaipur, Rajasthan.

Patil DA and Pawar S 2006. Ethnobotany of Jalgaon District, Maharashtra. Daya Publishing House, New Delhi, India.

Patil JU and Biradar SD 2011. Folkloric medicinal plants of Hingoli district, Maharashtra. *Indian journal of natural products and resources*, 2(1): 97-101.

Patil MV and Patil DA 2006. Ethnobotany of Nasik District Maharashtra. Daya publishing house, Delhi, p. 419.

Patil MV and Patil DA 2007a. Some herbal remedies used by the tribals of Nasik district, Maharashtra. *Natural Product Radiance*, 6(2): 152-157.

Patil SL and Patil DA 2007b. Ethnomedicinal plants of Dhule district, Maharashtra. *Natural Product Radiance*, 6(2): 148-151.

Patra B, Sahu D and Misra MK 2014. Ethno-medicobotanical studies of Mohana area of Gajapati district, Odisha, India. *International Journal of Herbal Medicine*, 2(4): 40-45.

Pattanaik C, Reddy CS and Dhal NK 2008. Phytochemical study of coastal sand dunes of Orissa. *Indian Journal of Traditional Knowledge*, 7(2): 263-268.

Patwari BN, Ahmed AB, Das T and Sengupta R 2014. Phytochemical screening and analgesic effects of ethanolic extract of plant *Murdania nudiflora* (l) brenan (commelinaceae) in albino mice using hot plate method. *International Journal of Pharmacy and Pharmaceutical Sciences*, 6(7): 512-515.

Pavithra PS, Sreevidya N and Verma RS 2009. Antibacterial and antioxidant activity of methanol extract of *Evolvulus nummularius*. *Indian Journal of Pharmacology*, 41(5): 233-236.

Payne A, Mukhopadhyay AK, Deka S, Saikia L and Nandi SP 2015. Anti-vibrio and antioxidant properties of two weeds: *Euphorbia serpens* and *Amaranthus viridis*. *Research Journal of Medicinal Plants*, 9(4): 170-178.

Perumal G 2010. Ethnomedicinal use of Pteridophyte from Kolli Hills, Namakkal district, Tamil Nadu, India. *Ethnobotanical Leaflets*, 14: 161-172.

Peter KV 2012. Handbook of Herbs and Spices, Vol. 1. Elsevier, p. 640.

Peter SL and Thomas B 2015. Some medicinal grasses in thalappilly taluk of Thrissur district, kerala, India. *American Journal of Biological and Pharmaceutical Research*, 2(3):135-138.

Pethakamsetty L, Seru G and Kandula L 2010. Phytochemical and biological examination of the aerial parts of *Tephrosia pumila*. *Journal of Pharmacy Research*, 3(1): 193-197.

Petrus AJA 2013. Ethnobotanical and pharmacological profile with propagation strategies of *Mukia maderaspatana* (L.) M. Roem.–A concise overview. *Indian Journal of Natural Products and Resources*, 4(1): 9-26.

Planning Commission 2000. Report of the task force on conservation and sustainable use of medicinal plants. Government of India.

Poongothai G and Sripathi SK 2013. A Review on insulinomimetic pinitol from plants. *International Journal of Pharma and Bio Sciences*, 4(2): 992-1009.

Poornima N, Umarajan KM and Babu K 2009. Studies on anatomical and phytochemical analysis of *Oxystelma esculentum* (L.f.) R.br. Ex Schltes. *Botany Research International*, 2(4): 239-243.

Porika R, Poojari S, Lunavath V and Mamidala E 2014. Preliminary phytochemical investigation and TLC analysis of *P. Angulata* fruit extract. *Journal of Pharmacy and Biological Sciences*, 9(2): 11-14.

Porwal A, Khobragade K and Jagtiani S 2015. *Euphorbia Prostrata* - a clinically proven drug in hemorrhoids – multiple pharmacological actions targeting pathological processes. *International Journal of Medical and Health Sciences*, 4(2): 269-273.

Prabha DS, Dahms HU and Malliga P 2014. Pharmacological potentials of phenolic compounds from *Prosopis* spp.-A review. *Journal of Coastal Life Medicine*, 2(11): 918-924.

Prabhavathi NB, Kowsalya B, Kumar SR, Sravani BJ, Sri GD, Sakila A and Jayachand P 2012. Analgesic activity of different solvent extract of *Operculina turpethum* by using Swiss Albino Mice. *Asian Journal of Pharmaceutical and Clinical Research*, 5(3): 215-218.

Prachi, Chauhan N, Kumar D and Kasana MS 2009. Medicinal plants of Muzaffarnagar district used in treatment of urinary tract and kidney stones. *Indian Journal of Traditional Knowledge*, 8(2): 191-195.

Pradhan V, Farooqui M, Khan TA, Khan PA and Khan JV 2017. Antimicrobial activity of *Caesulia axillaris* Roxb and *Psoralea corylifolia* L. *World Journal of Pharmaceutical Research*, 6(14): 352-358.

Pragada PM and Rao GMN 2012. Ethnoveterinary medicinal practices in tribal regions of Andhra Pradesh, India. *Bangladesh Journal of Plant Taxonomy*, 19(1): 7-16.

Prajapati N and Kumar U 2013. Agro's Dictionary of Medicinal Plants. Agrobios India, Jodhpur, India, p. 1-398.

Prajapati ND, Purohit SS, Sharma AK and Kumar T 2003. A Hand Book of Medicinal plants, Agrobios, Jodhpur, India, p. 373.

Prakash R 2014. Traditional uses of medicinal plants in Uttarakhand Himalayan Region. *Scholars Academic Journal of Biosciences*, 2(5): 345-353.

Praneetha P, Rani VP, Reddy YN and Kumar BR 2014. Hepatoprotective studies on methanolic extract of whole plant of *Lindernia ciliata*. *Bangladesh Journal of Pharmacology*, 9: 567-574.

Prasad KN, Shivamurthy GR and Aradhya SM 2008. *Ipomoea aquatica*, an underutilized green leafy vegetable: A review. *International Journal of Botany*, 4: 123-129.

Prasad KS and Savithramma N 2015. Tapping of aquatic plant *Nymphaea pubescens* Willd. for synthesis of silver nanoparticles and its antimicrobial evaluation. *International Journal of Pharmaceutical Sciences Review and Research*, 33(2): 63-66.

Prasad NP and Abraham Z 1984. Ethnobotany of Nayadis of North Kerala. *Journal of economic and taxonomic botany*, 5: 41-48.

Prasad VP, Mason D, Marburger JE, Ajith Kumar CR 1997. Illustrated Flora of Keoladeo National Park Bharatpur Rajasthan: A General Guide to the Wetland Flora of the Gangetic Plains. Bombay Natural History Society, p. 453.

Prashanth GK and Krishnaiah GM 2014. Chemical composition of the leaves of *Azadirachta indica* Linn. (Neem). *International Journal of Advancement in Engineering Technology, Management & Applied Science*, 1(5): 21-31.

Prashith KTR, Vinayaka KS and Raghavendra HL 2017. Ethnobotanical uses, phytochemistry and biological activities of *Salix tetrasperma* roxb. (Salicaceae) - A review. *Journal of Medicinal Plants Studies*, 5(5): 203-208.

Pratap B, Kumar R, Tiwari D, Yadav S, and Singh S 2010. Evaluation of antidiarrhoeal property of the hydroalcoholic extract of roots of *Calotropis gigantea* R.Br. on *Caster*-induced diarrhoea in rats. *Der Pharmacia Lettre*, 2(3): 309-314.

Preeti and Tripathi S 2014. *Ziziphus jujuba*: a phytopharmacological review. *International Journal of Research and Development in Pharmacy and Life Sciences*, 3(3): 959-966.

Preeti, Singh S and Chaudhary N 2014. *Ziziphus jujuba*: A phytopharmacological review. *International Journal for Pharmaceutical Research Scholars*, 3(1): 514-523.

Prince S, Roy RK, Anurag, Gupta D and Sharma VK 2013. *Hyptis suaveolens* (L.) Poit: A phyto-pharmacological review. *International Journal of Chemical and Pharmaceutical Sciences*, 4(1): 1-11.

Priya CL and Rao KVB 2012. Ethanobotanical and current ethanopharmacological aspects of *Argemone mexicana* Linn: An overview. *International Journal of Pharmaceutical Sciences and Research*, 3(7): 2143-2148.

Pulipati S, Babu SP, Devi BS, Devi GR and Bhanuja M 2015. Pharmacognostic Studies of *Alternanthera philoxeroides* (Mart.) Griseb. *Journal of Pharmacognosy and Phytochemistry*, 4(2): 202-204.

Pullaiah T 2006. Encyclopedia of world medicinal plants, Vol. 1-4. Regency Publications, New Delhi.

Pullaiah T 2014. Ethnobotany, phytochemistry and pharmacology of *Melochia corchorifolia*. *International Research Journal of Pharmacy*, 5(7): 543-545.

Pullaiah T and Naidu KC 2003. Antidiabetic plants in India and herbal based antidiabetic research. Regency Publications, p. 321.

Punjaji SA 2012. Some less known herbal remedies against wounds from Jamkhed Tahasil areas in Ahmednagar district (M.S.) India. *Journal of Pharmaceutical Research and Opinion*, 2(6): 58-62.

Punjaji SA 2016. Some less known traditional phyto-zootherapies in Kalsubai hills of Akole tahasil in Ahmednagar district (M.S.) India. *World Journal of Pharmacy and Pharmaceutical Sciences*, 5(3): 369-376.

Puri HS 2002. Rasayana: Ayurvedic herbs for longevity and rejuvenation, Vol. 1. of Traditional Herbal Medicines for Modern Times. CRC Press.

Puri HS 2003. Rasayana: Ayurvedic herbs for longevity and rejuvenation, Vol. 2. of Traditional Herbal Medicines for Modern Times. *The Journal of Alternative and Complementary Medicine,* 9(2): 331-332.

Purnomo Y, Soeatmadji DW, Sumitro SB and Widodo MA 2015. Anti-diabetic potential of *Urena lobata* leaf extract through inhibition of dipeptidylpeptidase IV activity. *Asian Pacific Journal of Tropical Biomedicine,* 5(8): 645-649.

Pushpagadan P 1994. Ethnobiology in India: A status report. Ministry of Environment and Forests. Govt. of India, New Delhi.

Putnam AR. 1985. Weed allelopathy. In: Duke, S.O, Ed, Weed Physiology, Vol. 1. Reproduction and Ecophysiology, CRC Press, Boca Raton, pp. 131-155.

Qamaruddin, Samiulla L, Singh VK and Jamil SS 2012. Phytochemical and pharmacological profile of *Withania somnifera* Dunal: A review. *Journal of Applied Pharmaceutical Science,* 2(1):170-175.

Quattrocchi U 2012. CRC World Dictionary of Medicinal and Poisonous Plants: Common Names, Scientific Names, Eponyms, Synonyms, and Etymology, Vol. 1-5. CRC Press, p. 3960.

Qureshi R 2013. Medicinal flora of Hingol National park, Baluchistan, Pakistan. *Pakistan Journal of Botany,* 44(2): 725-732.

Qureshi R, Bhatti GR and Memon RA 2010. Ethnomedicinal uses of herbs from northern part of Nara Desert, Pakistan. *Pakistan Journal of Botany,* 42(2): 839-851.

Rabia AA 2012. Ethno-botanic treatments for paralysis (Falij) in the Middle East. *Scientific Research Corporation,* 3: 157-166.

Rachchh RP and Galani VJ 2015. Evaluation of antinociceptive and antirheumatic activity of *Grangea maderaspatana* (L.) Poir. Using experimental models. *Ayu,* 36(4): 425-431.

Ragasa C, Austria JPM, Subosa AF and Shen CC 2015. Chemical constituents of *Amaranthus viridis*. *Chemistry of Natural Compounds,* 51(1): 146-147.

Rahman AHMM 2013a. An ethnobotanical investigation on Asteraceae family at Rajshahi, Bangladesh. *Academia Journal of Medicinal Plants,* 1(5): 92-100.

Rahman AHMM 2013b. Assessment of Angiosperm weeds of Rajshahi, Bangladesh with emphasis on medicinal plants. *Research in Plant Sciences,* 1(3): 62-67.

Rahman AHMM 2013c. Ethno-medico-botanical studies on cucurbits of Rajshahi Division, Bangladesh. *Journal of Medicinal Plants Studies,* 1(3): 118-125.

Rahmatullah M, Hasan MM, Ahmed M, Khan MW, Hossan MS, Rahman MM, Nasrin D, Miajee ZUMEU, Hossain MS, Jahan R and Khatun MA 2010. A survey of medicinal plants used by folk medicinal practitioners in Balidha village of Jessore District, Bangladesh. *American-Eurasian Journal of Sustainable Agriculture,* 4(2): 111-116.

Rahul J 2013. An ethnobotanical study of medicinal plants in Taindol Village, district Jhansi, region of Bundelkhand, Uttar Pradesh, India. *Journal of Medicinal Plants Studies,* 1(4): 59-71.

Rahuman AA, Venkatesan P, Gopalakrishnan G 2008. Mosquito larvicidal activity of oleic and linoleic acids isolated from *Citrullus colocynthis* (Linn.) Schrad. *Parasitology Research,* 103(6):1383-1390.

Rai M, Acharya D and Rios JL 2011. Ethnomedicinal Plants: Revitalizing of Traditional Knowledge of Herbs. Science Publishers.

Rai PK and Lalramnghinglova H 2010. Lesser known ethnomedicinal plants of Mizoram, North East India: An Indo-Burma hotspot region. *Journal of Medicinal Plants Research,* 4(13):1301-1307.

Raikwar H and Chowbey DA 2017. Use of *Leucaena leucocephala. International Journal of Green and*

Raj V, Sharma C, Rani S, Kumar B and Kumar A 2015. Plant *Opuntia dillenii*: A review on its traditional uses, phytochemical and pharmacological properties. *Pharmaceutical Science,* 1(1): 29-43.

Raja MMMK and Mishra SH 2010. Comprehensive review of *Clerodendrum phlomidis*: A traditionally used bitter. *Journal of Chinese Integrative Medicine,* 8(6): 510-524.

Rajakumar N and Srivanna MB 2012. Traditional veterinary healthcare practices in Shimoga district of Karnataka, India. *Indian Journal of Traditional Knowledge,* 11(2): 283-287.

Rajakumar N, Shivanna MB 2009. Ethno-medicinal application of plants in the eastern region of Shimoga district, Karnataka, India. *Journal of Ethnopharmacology,* 126: 64-73.

Rajalakshmi P, Vadivel V and Brindha P 2016. Review on medicinal plants recommended in Siddha literatures for the management of hypertension. *International Journal of Research in Pharmaceutical Sciences,* 7(1): 16-33.

Rajesh P, Latha S, Selvamani P and Kannan VR 2010. *Capparis sepiaria* Linn – Pharmacognostical standardization and toxicity profile with chemical compounds identification (GC-MS). *International Journal of Phytomedicine,* 2: 71-79.

Rajpoot K and Mishra RN 2011. *Boerhaavia diffusa* roots (Punarnava mool) – Review as rasayan (Rejuvenator / Antiaging). *International Journal of Research in Pharmaceutical and Biomedical Sciences,* 2(4): 1451-1460.

Rajput A, Pal SC and Bhagavan Patil 2011. Phytochemical screening, antibacterial activity and physicochemical evaluation of leaves of *Butea monosperma. International Journal of Pharmacy and Pharmaceutical Sciences,* 3(3): 189191.

Ram HNA, Shreedhara CS, Gajera FP and Zanwar SB 2009. Phamacognostical evaluation of *Phyllanthus reticulatus* Poir. *Research article, Pharmacognosy Magazine,* 5(18): 176-182.

Rama Shankar, Rawat MS, Deb S and Sharma BK 2012. Jaundice and its traditional cure in Arunachal Pradesh. *Journal of Pharmaceutical and Scientific Innovation,* 1(3): 93-97.

Ramakrishna N and Saidulu CH 2014. Medicinal plants used by ethnic people of Adilabad District, Andhra Pradesh, India. *International Journal of Pharmaceutical Research & Allied Sciences,* 3(2): 51-59.

Rambhau SS, Ravindra DP, Prabhakharrao PP and Chandrakant SV 2014. Phytochemical and pharmacological activity of Agnimantha (*Clerodendrum phlomidis* linn.f) - A review. *International journal of Ayurvedic & Herbal Medicine,* 4(5):1615-1621.

Rameshkumar A and Sivasudha T 2012. *In vitro* antioxidant and antibacterial activity of aqueous and methanolic extract of *Mollugo nudicaulis* Lam. leaves. *Asian Pacific Journal of Tropical Biomedicine,* 2(2): 895-900.

Rameshkumar S and Ramakritinan CM 2013. Floristic survey of traditional herbal medicinal plants for treatments of various diseases from coastal diversity in Pudhukkottai District, Tamil Nadu, India. *Journal of Coastal Life Medicine,* 1(3): 225-232.

Rana M, Dhamija H, Prashar B and Shivani Sharma 2012. *Ricinus communis* L. – A review. *International Journal of Pharm. Tech. Research,* 4(4): 1706-1711.

Rani S, Chahota RK and Sharma TR 2015. Ethnobotanical survey and traditional knowledge of some medicinal and aromatic plants of Himachal Pradesh, North-West Himalayas. *International Journal of Applied and Pure Science and Agriculture,* 1(12): 1-22.

Rani SL, Devi VK and Soris PT, Maruthupandian A. Mohan VR 2011. Ethnomedicinal plants used by Kanikkars of Agasthiarmalai Biosphere Reserve, Western Ghats. *Journal of Ecobiotechnology,* 3(7): 16-25.

Rani SS and Raju RRV 2014. Phytochemical analysis of *Phyllanthus maderaspatensis* and *Celosia argentea*. *Journal of Agriculture and Veterinary Science,* 7(3): 13-14.

Ranjit MR and John BS 2015. Medicinal properties of edible weeds of crop fields and wild plants eaten by Oraon tribals of Latehar district, Jharkhand. *International Journal of Life Science and Pharma Research,* 5(2): 9-20.

Rao AS and Verma DM 1982. Cyperaceae of North-East India. Published by Botanical Survey of India.

Rao AS, Merugu R and Atthapu T 2013. A new triterpene from *Ludwigia hyssopifolia* (G.Don) Exell. *International Journal of Chem. Tech. Research,* 5(1): 342-346.

Rao DS, Rao GMN and Murthy PP 2016a. Diversity and indigenous uses of some ethno medicinal plants in Papikondalu Wildlife Sanctuary, Eastern Ghats of Andhra Pradesh, India. *American Journal of Ethnomedicine,* 3(1): 6-25.

Rao KNV, Sunitha Ch, Banji D, Sandhya S and Mahesh V 2011. A study on the nutraceuticals from the genus *Rumex*. *Journal for drugs and medicines,* 3(1): 76-88.

Rao MRK, Saranya Y and Divya D 2016b. Preliminary phytochemical analysis of *Typha domingensis* rhizome aqueous extracts. *International Journal of Pharmaceutical Sciences Review and Research,* 37(1): 30-32.

Rao PP 2016. Ophthalmic uses of *Boerhaavia diffusa* L. (Punarnava): Review. *International Journal of Herbal Medicine,* 4(2): 5-9.

Rao RR and Jamir NS 1982a. Ethnobotanical studies in Nagaland-1. Medicinal plants. *Economic Botany,* 36:176-181.

Rao RR and Jamir NS 1982b. Ethnobotanical studies in Nagaland-2. 54 medicinal plants used by Nagas. *Journal of Economic and Taxonomic Botany,* 3(1):11-17.

Rao YK, Fang SH, Hsieh SC, Yeh TH and Tzeng YM 2009. The constituents of *Anisomeles indica* and their anti-inflammatory activities. *Journal of Ethnopharmacology,* 121(2): 292-296.

Rashid ANM, Hossain MS, Hassan N, Dash BK, Sapon MA, and Sen MK 2014. A review on medicinal plants with antidiabetic activity. *Journal of Pharmacognosy and Phytochemistry,* 3(4): 149-159.

Rastogi S, Kulshreshtha DK and Rawat AKS 2006. *Streblus asper* Lour. (Shakhotaka): A review of its chemical, pharmacological and ethnomedicinal properties. *Evidence-Based Complementary and Alternative Medicine,* 3(2): 217-222.

Rastogi S, Pandey MM and Rawat AK 2011. An ethnomedicinal, phytochemical and pharmacological profile of *Desmodium gangeticum* (L.) DC. and *Desmodium adscendens* (Sw.) DC. *Journal of Ethnopharmacology,* 136(2): 283-96.

Rathanavel C and Arasu PT 2013. Quantification of heavy metals and minerals in selected Indian medicinal plants using atomic absorption spectrophotometer. *International Journal of Pharma and Bio Sciences,* 4(3): 897-908.

Rathod MC and Dhale DA 2014. Pharmacognostic characterization and phytochemical screening of *Launaea procumbens*. *International Journal of Pharmaceutical Research Scholars,* 3: 41-50.

Ratsch C 1998. The Encyclopedia of Psychoactive Plants: Ethnopharmacology and its Applications. Rochester: Park Street Press.

Rattanata N, Daduang S, Phaetchanla S, Bunyatratchata W, Promraksa B, Tavichakorntrakool R, Uthaiwat P, Boonsiri P and Daduang J 2014. Antioxidant and antibacterial properties of selected Thai weed extracts. *Asian Pacific Journal of Tropical Biomedicine,* 4(11): 890-895.

Ravichandran N, Vajrai R, Raj CD and Brindha P 2012. Phytochemical analysis and *in vitro* cytotoxic effect of *Phyllanthus madraspatensis* L. *International Journal of Pharmacy and Pharmaceutical Sciences,* 4: 111-114.

Ravikanth K, Kanaujia A, Thakur D, Sharma A and Gautam B 2014. Nutritional Constituents of the Plants *Fumaria indica* and *Caesalpinia bonducella*. *International Journal of Advances in Pharmacy, Biology and Chemistry,* 3(3): 698-702.

Rawat DS and Kharwal AD 2014. Ethnobotanical studies of weed flora in Shivalik Hills, Himachal Pradesh (H.P.), India. *International Journal of Advanced Research,* 2(5): 218-226

Raza MA, Mukhtar F and Danish M 2015. *Cuscuta reflexa* and *Carthamus oxyacantha*: Potent sources of alternative and complimentary drug. *Springer Plus,* 4(76): 1-6.

Reddy CS 2008. Catalogue of invasive alien flora of India. *Life Science Journal,* 5(2): 84-89.

Reddy CS, Bagyanarayana G, Reddy KN and Raju VS 2008. Invasion alien fora of India. National Biological Information Infrastructure, USGS, USA.

Reddy CS, Reddy KN, Rao KT and Pattanaik C 2007. Rare and little known medicinal plants from Nallamalais of the Eastern Ghats, India. *Journal of Plant Sciences,* 2: 113-117.

Reddy KN, Subbaraju GV, Reddy CS, and Raju VS 2006. Ethnoveterinary medicine for treating livestock in Eastern Ghats of Andhra Pradesh. *Indian Journal of Traditional Knowledge,* 5(3): 368-372.

Reddy KN, Trimurthulu G and Reddy CS 2010a. Medicinal plants used by ethnic people of Medak district, Andhra Pradesh, India. *Indian journal of traditional knowledge,* 9(1):184-190.

Reddy KN, Trimurthulu G and Reddy CS 2010b. Plants used by the ethnic people of Krishna district, Andhra Pradesh. *Indian Journal of Traditional Knowledge,* 9(2): 313-317.

Reddy KVRM, Nagalakshmi D and Murty PP 2012. Some folk medicinal plants of Bhiravakona hills of Prakasam district, A. P. India. *Current Botany,* 3(5): 51-58.

Regina KMM, Adama H, Jeanne M and Odile N 2015. Ethnobotany and ethnopharmacognosy of Lamiaceae Species from Central Burkina Faso: *Leucas martinicensis* (Jacquin) R. Brown, *Hoslundia opposita* Vahl and *Orthosiphon pallidus* Royle Ex Benth. *American Journal of Ethnomedicine,* 2(4): 219-232.

Rehman SU, Faisal R, Shinwari, ZK, Ahmad N and Ahmad I 2017. Phytochemical screening and biological activities of *Trigonella incisa* and *Nonea edgeworthii. Pakistan Journal of Botany,* 49(3): 1161-1165.

Rekha D, Panneerselvam A and Thajuddin N 2014. Studies on medicinal plants of A.V.V.M. Sri Pushpam College Campus, Thanjavur district of Tamil Nadu, southern India. *World Journal of Pharmaceutical Research,* 3(5): 785-820.

Rengifo E and Arana GV 2013. *Physalis angulata* L. (Bolsa Mullaca): A review of its traditional uses, chemistry and pharmacology. *Boletín de Estudios Latinoamericanos y del Caribe,* 12(5): 431-445.

Riaz M, Zia-Ul-Haq M and Jaafar HZE 2013. Common mullein, pharmacological and chemical aspects. *Revista Brasileira de Farmacognosia,* 23(6): 948-959.

Rizk AM 1986a. Constituents of plants growing in Qatar. *Fitoterapia,* 57: 3-9.

Rizk AM 1986b. The Phytochemistry of Flora of Qatar. King Print of Richmond, London, p. 30-32.

Rodrigues JC, Ascensao L, Bonet MA and Valles J 2003. An ethnobotanical study of medicinal and aromatic plants in the Natural Park of "Serra de Sao Mamede" (Portugal). *Journal of Ethnopharmacology,* 89: 199-209.

Rosenthal J 1997. Integrating drug discovery, biodiversity conservation, and economic development: Early lessons from the international cooperative biodiversity groups. Biodiversity and Human Health. Island Press, Washington. DC, pp. 281-301.

Roy S, Roy VC, Roy RC. A review of *Mitragyna parvifolia* (Roxb) Korth. Pharmatutor-Art-1594.

Ruby GDR and Sara CS 2014. Bioactive compounds and its importance in Pteridophytes. *International Journal of Research in Engineering and Bioscience,* 2(6): 64-78.

Rungsung W, Dutta S, Das D and Hazra J 2013. A brief review on the botanical aspects and therapeutic potentials of important Indian medicinal plants. *International Journal of Herbal Medicine,* 1(3): 38-45.

Rustenbekova GB, Goryaev MI and Nizhinskaya GA 1974. Flavonoids of *Chenopodium botrys. Khimija Prirodnykh Soyedineniy,* 3: 406-406.

Saad B and Said O 2011. Greco-Arab and Islamic Herbal Medicine: Traditional System, Ethics, Safety, Efficacy, and Regulatory Issues. Wiley, p. 568

Safaa B, Chalak L, Dalleh H and Arnold N 2015. Ethnopharmacological survey of medicinal plants used in traditional medicine by the communities of Mount Hermon, Lebanon. Journal of Ethnopharmacology, 173: 139-156.

Saha A, Hoque A, Mallick SK and Panda S 2013. Medicinal Uses of Grasses by the Tribal People in West Bengal - An Overview. *International Journal of Basic and Applied Sciences,* 3(3): 63-70.

Saharia S and Sharma CS 2011. Ethno-medicinal studies on indigenous wetland plants in the tea garden tribes of Darrang and Udalguri district, Assam, India. *International Journal of Environment and Biodiversity (NeBIO),* 2(1): 27-33.

Saheb TS 2014. A study on medicinal climbers of Nallamalais, Andhra Pradesh. *International Journal of Multidisciplinary Research and Development,* 1(5): 172-176.

Sahoo HB, Mandal PK, Sagar R, Bhattamisra SK 2016. Evaluation of lactogenic activity of *Triumfetta rhomboidea* L. root: Validating its traditional usage. *Journal of Experimental and Integrative Medicine,* 6(1): 26-30.

Sahu CR, Nayak RK and Dhal NK 2013. Traditional herbal remedies for various diseases used by tribals of Boudh District, Odisha, India for sustainable development. *International Journal of Herbal Medicine,* 1(1): 12-20.

Sahu PK and Gupta S 2014. Medicinal plants of Morning glory: Convolvulaceae Juss. of Central India (Madhya Pradesh & Chhattishgarh). *International Quarterly Journal of Biology & Life Sciences,* 2(2):463-469.

Sahu TR 1982. An ethnobotanical survey of Madhya Pradesh I. Plants used against various disorders among tribal women. *Ancient Science of Life,* 1(1): 178-181.

Sahu TR, Sahu T and Dakwale RN 1983. Further contributions towards the ethnobotany of Madhya Pradesh, 2. Plants used against diarrhoea and dysentery. *Ancient Science of Life,* 2:169-70.

Saikia P and Khan ML 2011. Diversity of medicinal plants and their uses in homegardens of upper Assam, Northeast India. *Asian Journal of Pharmaceutical and Biological Research,* 1(3): 296-309.

Saikia R, Choudhury MD, Talukdar AD and Chetia P 2011. Antidiabetic activity of ethno medicinal plant *Scoparia dulcis* L. (Family: Scrophulariaceae): A review. *Assam University Journal of Science & Technology Biological and Environmental Sciences,* 7(1): 173-180.

Saini A, Gahlawat DK, Chauhan C, Gulia SK, Ganie SA, Archita and Yadav SS 2015a. Ethnomedicinal uses and phytochemistry of *Abutilon indicum* (Linn.) Sweet: An overview. *Journal of Pharmacognosy and Phytochemistry,* 3(5): 66-72.

Saini A, Soni HK and Gupta P 2016. A review on *Tridax procumbens*. *Imperial Journal of Interdisciplinary Research,* 2(8): 308-319.

Saini DC 2005. Flora of Bahraich district, Uttar Pradesh—I, II, III, IV, V, VI. *Journal of Economic and Taxonomic Botany* 29(3/4): 528–920.

Saini V, Mishra R, Mandloi S and Yadav N 2015b. Analysis of the phytochemical content of *Jatropha gossypifolia* L. *Chemical and Process Engineering Research,* 35: 99-104.

Saleem M, Karim M, Qadir MI, Ahmed B, Rafiq M and Ahmad B 2014. In vitro antibacterial activity and phytochemical analysis of hexane extract of *Vicia sativa. Bangladesh Journal of Pharmacology,* 9: 189-193.

Salem AZM, Mahemad ZM Salem, Manuel González Ronquillo and Cipriano M 2011. Major chemical constituents of *Leucaena leucocephala* and *Salix babylonica* leaf extracts. *Journal of Tropical Agriculture,* 49(1-2): 95-98.

Salgado ER and Arana GV 2013. *Physalis angulata* L. (*Bolsa Mullaca*): A review of its traditional uses, chemistry and pharmacology. *Boletín Latinoamericano y del Caribe de Plantas Medicinales y Aromáticas,* 12(5): 431-445.

Salihoglu EM, Akaydin G, Can EC and Akaydin SY 2013. Evalution of antioxidant activity of various herbal folk medicines. *Journal of Nutrition and Food Sciences,* 3(5): 1-9.

Salim KP and Paarakh PM 2009. A phyto-pharmacological review of *Syzygium cumini* (L.) Skeels. *Pharmacologyonline,* 2: 101-122.

Samant SS and Dhar U 1997. Diversity, endemism and economic potential of wild edible plants of Indian Himalaya. *International Journal of Sustainable Development and World Ecology,* 4: 179-191.

Samant SS, Dhar U and Palni LMS 1998. Medicinal plants of Indian Himalaya: Diversity, distribution, potential values. Nainital: Gyanodaya Prakashan.

Samarah NH and Ereifej K 2009. Chemical composition and mineral content of common vetch seeds during maturation. *Journal of Plant Nutrition,* 32(1): 177-186.

Samuel JK and Andrews B 2010. Traditional medicinal plant wealth of Pachalpur and Periyur hamlets Dindigul district, Tamil Nadu. *Indian Journal of Traditional Knowledge,* 9(2): 264-270.

Samvatsar S 2004. Plants used for the treatment of different types of fevers by Bhils and its sub-tribes in India. *Indian Journal of Traditional Knowledge,* 3(1): 96-100.

Sandhya S, Vinod KR, Sekhar JC, Aradhana R and Nath VS 2010. An updated review on *Tricosanthes cucumerina* L. *International Journal of Pharmaceutical Sciences Review and Research,* 1(2): 56-60.

Sandhyarani G, Ramesh A and Naik B 2014. Study on ethno-medico-botany of some plants of Tirupati district of Andhra Pradesh, India. *European Journal of Complementary and Alternative Medicine,* 1(1): 35-37.

Sandhyavali MS, Sivakamisundari P, Sharma P and Murugan V 2014. Phytochemical studies on the aerial parts of *Indigofera linnaei* Ali. *Pharmacophore,* 5(1): 94-97.

Sanghai DB, Kumar SV, Srinivasan KK, Aswatharam HN and Shreedhara CS 2013. Pharmacognostic and phytochemical investigation of the leaves of *Malvastrum coromandelianum* (L.) Garcke. *Ancient Science of Life,* 33(1): 39-44.

Sani KS, Ibrahim MH, Mustafa M and Rusea G 2015. Ethnobotanical survey of medicinal plants used for traditional maternal healthcare in Katsina state, Nigeria. *South African Journal of Botany,* 97:165-175.

Sanjay P 2015. A review on chemical and pharmacological profiles of Shakhotaka (*Streblus asper* Lour.). *Anveshana Ayurveda Medical Journal,* 1(4): 256-262.

Sankar JG and Satapathy KB 2015. Weed diversity of rabi crops and their ethnomedicinal uses in Kendrapara district of Odisha, India. *International Research Journal of Biological Sciences,* 4(3): 33-38.

Santhoshkumar B and Satyanarain S 2010. Herbal remedies of wetlands macrophytes in India. International Journal of Pharma and Bio Sciences, 2: 1-12.

Sanyal MN 1991. A handbook of excursion flora of the Gangetic Plains and adjoining hills. Mittal Publications, p. 179.

Sapkale AP, Thorat MS, Dighe DA, Munot NM and Singh MC 2013. Investigation of anti-inflammatory activity of whole flower extract of *Butea monosperma* by ex-vivo and in-vitro techniques and their correlation. *International Journal of Pharmacy & Pharmaceutical Sciences,* 5: 84-86.

Sara V, Franca T and Gelsomina F 2009. Traditional uses of medicinal plants in Valvestino (Italy). *Journal of Ethnopharmacology,* 121(1): 106-116.

Sarangzai AM, Ahmed A and Laghari SK 2013. Traditional uses of some useful medicinal plants of Ziarat district Balochistan, Pakistan. *FUUAST Journal of Biology,* 3(1): 101-107.

Sarma A, Sarmah P, Dolai DK and Medhi P 2015. *Oxalis debilis* var. *corymbosa* (DC.) Lourteig and *Oxalis corniculata* L: A comparative study of nutraceutical properties. *International Journal of Current Research,* 7(1): 11307-11310.

Sarwar AKMG 2015. Medicinal plant genetic resources of Bangladesh – genera represented by single species and their conservation needs. *Journal of Medicinal Plants Studies,* 3(2): 65-74.

Sarwar S, Sabir M, Raza SS, and Malik SA 2016. Analysis of antimicrobial activity of medicinal plant *Amaranthus viridis*. *International Journal of Innovation and Scientific Research,* 20(2): 494-499.

Sathyamurthi JBS 2015. *Vetiveria zizanioides*- a review. *International Journal of Scientific Research Today*. 1(1): 136-140.

Satynarayana NV, Lachake SH and Chauhan MG 1993. Pharmacognostical and phytochemical investigation of *Peristrophe bicalyculata*. *Ancient Science of Life,* 12(3-4): 420-427.

Savitha G and Balamurugan S 2014. Pharmacognostical studies on *Oxystelma esculentum* (L.F) R. Br. ex Schltes, a medicinal plant. *International Letters of Natural Sciences,* 17: 51-57.

Savithramma N, Yugandhar P and Suhrulatha D 2015. Traditional medicinal plants used by local people of Kailasakona- A sacred grove of Chittoor district, Andhra Pradesh, India. *International Journal of Pharmacy and Pharmaceutical Sciences,* 7(3): 407-411.

Sawant RS and Godghate AG 2013. Preliminary phytochemical analysis of leaves of *Tridax procumbens* Linn. *International Journal of Science, Environment and Technology,* 2(3): 388 –394.

Sawhney RS and Atal CK 1968. Phytochemical studies on the genus Crotalaria. VIII. Chemical constituents of *Crotalaria mysorensis*. *Journal of the Indian Chemical Society,* 45: 1052-1053.

Sawhney RS and Atal CK 1970. Genus Crotalaria. X. Methylethers of supinidine and heliotridine from *C. medicaginea* Lam. *Journal of the Indian Chemical Society,* 47: 741.

Saxena AP and Vyas KM 1983. Ethnobotany of Dhasan valley. *Journal of Economic and Taxonomic Botany,* 4: 121-128.

Saxena KG 1991. Biological invasion in the Indian subcontinent. Review of invasion by plants. In: Ecology of Bilogical Invasion in the Tropics, edited by Ramakrishnan PS (International Scientific Publications), New Delhi, India, p. 53-73.

Scherrer AM, Motti R, Weckerle CS 2005. Traditional plant use in the areas of Monte Vesole and Ascea, Cilento National Park (Campania, Southern Italy). *Journal of Ethnopharmacology,* 97: 129-143.

Schippmann U, Leaman JD and Cunningham AB 2002. Impact of cultivation and gathering of medicinal plants on biodiversity: Global trends and issues, Inter-Departmental Working Group on Biological Diversity for Food and Agriculture, Rome.

Schmelzer GH and Gurib-Fakim A (Editors) 2013. Plant resources of tropical Africa 11(2). Medicinal plants 2. PROTA Foundation, Wagenigen, Netherlands/CTA, pp. 384.

Sekar KC 2012. Invasive alien plants of Indian Himalayan Region- Diversity and implication. *American Journal of Plant Sciences,* 3: 177-184.

Selvam NT and Acharya MV 2015. Review of *Ipomoea pes-tigridis* L.: Traditional uses, botanical characteristics, chemistry and biological activities. *International Journal of Pharma Sciences and Research,* 6(12): 1443-1448.

Selvam V, Sumathy R and Kumuthakalavalli R 2014. Phytochemical constituents and *in vitro* anti - inflammatory activity of an Ivy gourd, *Coccinia grandis*. *World Journal of Pharmacy and Pharmaceutical Sciences,* 3(7): 1173-1179.

Semenya SS, Potgieter MJ and Tshisikhawe MP 2013. Use, conservation and present availability status of ethnomedicinal plants of Matebele-Village in the Limpopo Province, South Africa. *African Journal of Biotechnology,* 12(18): 2392 2405.

Sen S and Chakraborty R 2017. Revival, modernization and integration of Indian traditional herbal medicine in clinical practice: Importance, challenges and future *Journal of traditional and complementary medicine,* 7(2): 234–244.

Sen SA, Chakraborty RA and De BB 2011. Challenges and opportunities in the advancement of herbal medicine: India's position and role in a global context. *Journal of Herbal Medicine,* 1(3-4): 65-75.

Sengupta R, Sheorey SD and Hinge MA 2012. Analgesic and anti-inflammatory plants: An updated review. *International Journal of Pharmaceutical Sciences Review and Research,* 12(2):114-119.

Senthamizhselvi C 2013. Traditional knowledge of medicinal plants–A study in Thinnalore village, Kancheepuram district of Tamil Nadu, India. *Global research analysis*, 2(12): 230-233.

Senthil Kumar KK and Rajkapoor B 2010. Effect of *Oxalis corniculata* L. extracts on biogenic amines concentrations in rat brain after induction of seizure. *International Journal of Phytopharmacology*, 1(2): 87-91

Sergio Zarate 1999. Ethnobotany and domestication process of *Leucaena* in Mexico. *Joumal of Ethnobiology*, 19(1): 1-23.

Shah A, Marwat SK, Gohar F, Khan A, Bhatti KH, Amin M, Noor Ud Din, Ahmad M and Zafar M 2013a. Ethnobotanical study of medicinal plants of semi-tribal area of Makerwal & Gulla Khel (lying between Khyber Pakhtunkhwa and Punjab Provinces), Pakistan. *American Journal of Plant Sciences*, 4: 98-116.

Shah S, Sirajuddin, Rehmanullah and Zahir M 2013b. Pharmacognostic standardization and pharmacological study of *Sisymbrium irio* L. *American Journal of Research Communication*, 1(7): 241-253.

Shaheen H, Qureshi R, Akram A and Gulfraz M 2014. Inventory of medicinal flora from Thal Desert, Punjab, Pakistan. *African Journal of Traditional, Complementary and Alternative Medicine*, 11(3): 282-290.

Shahiladevi S, Jayanthi G and Jegadeesan M 2006. Preliminary phytochemical studies on *Solanum surattense* Burm. F. seeds. *Ancient Science of Life*, 26: 59-64.

Shahriar M, Chowdhury SA, Imam MZ and Datta BK 2015. Analgesic and neuropharmacological activities of methanol extract from the leaf of *Nicotiana plumbaginifolia* Viv. *Bangladesh Pharmaceutical Journal*, 18(2): 85-89.

Shajeela PS, Kalpanadevi V and Mohan VR 2012. Potential antidiabetic, hypolipidaemic and antioxidant effects of *Nymphaea pubescens* extract in alloxan induced diabetic rats. *Journal of Applied Pharmaceutical Science*, 2(2): 83-88.

Shalini CB, Pillai SC and Mohan VR 2014. Ethnomedicinal plants used by the Kanikkars of Southern Western Ghats. *International Journal of Pharmaceutical Sciences Review and Research*, 28(2): 101-107.

Shankar K, Singh SK, Kumar D, Varshney S, Gupta A, Rajan S, Srivastava A, Beg M, Srivastava AK, Kanojiya S, Mishra DK and Gaikwad AN 2015. *Cucumis melo* ssp. *agrestis* var. *agrestis* ameliorates high fat diet induced dyslipidemia in Syrian golden hamsters and inhibits adipogenesis in 3T3-L1 Adipocytes. *Pharmacognosy Magazine*, 11(Suppl 4): S501-S510. doi:10.4103/0973-1296.172945.

Shankar R, Rawat MS, Deb S, and Sharma BK 2012. Jaundice and its traditional cure in Arunachal Pradesh. *Journal of Pharmaceutical & Scientific Innovation*, 1(3): 93-97.

Shanmugam S, Rajendran K and Suresh K 2012. Traditional uses of medicinal plants among the rural people in Sivagangai district of Tamil Nadu, Southern India. *Asian Pacific Journal of Tropical Biomedicine*, 2(1): S429-S434.

Sharma A and Arora D 2012. Phytochemical and pharmacological potential of genus *Stellaria*: A review. *Journal of Pharmacy Research*, 5(7): 3591-3596.

Sharma AK and Dhakre JS 1995. Flora of Agra District. BSI, Calcutta, India.

Sharma BD and Panday DS 1984. Exotic Flora of Allahabad District. Botanical Survey of India, Howrah, India.

Sharma BM and Jamwal, PS 1989. *Emex australis* Steinh. A new record for India from Jammu. *Journal of Economic and Taxonomic Botany*, 11: 77–79.

Sharma D and Paul Y 2013. Preliminary and pharmacological profile of *Melia azedarach* L.: An overview. *Journal of Applied Pharmaceutical Science*, 3(12): 133-138.

Sharma GP, Singh JS and Raghubanshi AS 2005a. Plant invasions: Emerging trends and future implications. *Current Science*, 88(5): 726-734.

Sharma IP, Kanta C, Semwal SC and Goswami N 2017. Wild fruits of Uttarakhand (India): Ethnobotanical and medicinal uses. *International Journal of Complementary & Alternative Medicine*, 8(3): doi-10.15406/ijcam.2017.08.00260.

Sharma J and Singh V 2012. *Holoptelea Integrifolia*: an overview. *European Journal of Applied Sciences*, 4(1): 42-46.

Sharma J, Gairola S, Gaur RD, Painuli RM and Siddiqi TO 2013a. Ethnomedicinal plants used for treating epilepsy by indigenous communities of sub-Himalayan region of Uttarakhand, India. *Journal of Ethnopharmacology*, 150: 353-370.

Sharma J, Painuli RM and Gaur RD 2010. Plants used by the rural communities of district Shahjahanpur, Uttar Pradesh. *Indian Journal of Traditional Knowledge*, 9(4): 798-803.

Sharma JK 2016. Significance of Permanent Sample Plots (PSPs) established in different forest ecosystems in monitoring ecological attributes and conservation of biodiversity: A review. Pages 317-340. In: Plant Biodiversity: Monitoring, Assessment and Conservation. Eds. Ansari A, Gill SS, Abbas ZK and Naeem M, CABI, UK, December 2016. pp. 628.

Sharma L and Khandelwal S 2010. Traditional uses of plants as cooling agents by the tribal and traditional communities of Dang Region in Rajasthan, India. *Ethnobotanical Leaflets*, 14: 218- 224.

Sharma M and Kumar A 2013. Traditional medicinal plants of Rajasthan used in tribal medicine: A review. *International journal of life science and pharma research*, 3(2): 38-42.

Sharma M, Yusuf M, Hussain S and Hussain A 2012. Phytochemical constituents and pharmcological activities of *Eclipta alba* Linn. (Asteraceae): A review. *International Journal of Pharmaceutics*, 3(12): 5-53.

Sharma MP, Ahmad J, Hussain A and Khan S 1992. Folklore medicinal plants of Mewat (Gurgaon district), Haryana, India. *International Journal of Pharmacognosy*, 30(2): 129-134.

Sharma N and Vijayvergia R 2013. A review on *Digera muricata* (l.) Mart.-A great versatile medicinal plant. *International Journal of Pharmaceutical Sciences Review and Research*, 20(1): 114-119.

Sharma OP 2016. Plant taxonomy, Second Edition. Thirteenth reprint. McGraw Hill Education (India) Private Limited, New Delhi.

Sharma P, Rani S, Ojha SN, Sood SK and Rana JC 2014a. Indian herbal medicine as hepatoprotective and hepatocurative: A review of scientific evidence. *Life Sciences Leaflets*, 49: 61-115.

Sharma P, Sharma GN, Shrivastava B and Hemant R 2014b. Phytochemical screening, total phenol estimation, antioxidant activity of *Blainvillea acmella* leaf and stem successive extracts. *Malaysian Journal of Analytical Sciences*, 18(2): 351-359.

Sharma P, Shrivastava B, Sharma GN and Jadhav HR 2013b. Phytochemical and ethenomedicinal values of *Barleria prionitis* L.: An overview. *Journal of Harmonized Research in Pharmacy*, 2(3):190-199.

Sharma PC, Yelne MB, Dennis TJ 2001. Database on Medicinal Plants Used in Ayurveda, Vol. III. Central Council for Research in Ayurveda & Siddha, New Delhi.

Sharma R, Parihar JS and Vatsa VK 2005b. Flora of Yamuna ravines of district Mathura. *Vegetos- An International Journal of Plant Research*, 18(1&2): 87-88.

Sharma RA and Singh R 2013. A review on *Phyla nodiflora* Linn.- A wild wetland medicinal herb. *International Journal of Pharmaceutical Sciences Review and Research*, 20(1): 57-63.

Sharma RA, Singh B, Singh D and Chandrawat P 2009. Ethnomedicinal, pharmacological properties and chemistry of some medicinal plants of Boraginaceae in India. *Journal of Medicinal Plants Research*, 3(13): 1153-1175.

Sharma RK, Verma N, Jha KK, Singh NK and Kumar B 2014c. Phytochemistry, pharmacological activity, traditional and medicinal uses of erigeron species: A review. *International Journal of Advance Research and Innovation*, 2(2): 379-383.

Sharma S and Thokchom R 2014. A review on endangered medicinal plants of India and their conservation. *Journal of Crop and Weed*, 10(2): 205-218.

Sharmila S, Kalaichelvi K, Rajeshwari M and Anjanadevi N 2014. Studies on the folklore medicinal uses of some indigenous plants among the tribes of Thiashola, Manjoor, Nilgiris South Division, Western Ghats. *International Journal of Plant, Animal and Environmental Sciences*, 4(3): 14-22.

Shazia Bi 2013. Flora of Aligarh with special reference to Ethnobotany. Ph.D. Thesis, Department of Botany, A.M.U, Aligarh.

Shekhawat GS and Anand S 1981. An ethnobotanical profile of Indian desert. *Journal of Economic and Taxonomic Botany*, 5: 591-598.

Shekhawat MS, Manokari M, and Ravindran CP 2016. Micropropagation, micromorphological studies, and in vitro flowering in *rungia pectinata* L. *Hindawi Publishing Corporation Scientifica*, 2016: 1-7. http://dx.doi.org/10.1155/2016/5813851.

Shekhawat MS, Manokari M, and Ravindran CP 2016. Micropropagation, micromorphological studies, and in vitro flowering in *Rungia pectinata* L. *Scientifica*, 2016(2016): 1-7. http://dx.doi.org/10.1155/2016/581385.

Sher Z, Khan ZUD and Hussain F 2011. Ethnobotanical studies of some plants of Chagharzai valley, district Buner, Pakistan. *Pakistan Journal of Botany*, 43(3): 1445-1452.

Shiddamallayya N, Rama RV, Doddamani SH and Venkateshwarlu G 2016. A glimpse on forest flora and Indian system of medicine plants of Chitradurga district, Karnataka. *International Journal of Herbal Medicine*, 4(1): 25-33.

Shinwari ZK, Malik S, Karim AM, Faisal R and Qaiser M 2015. Biological activities of commonly used medicinal plants from Ghazi Brotha, Attock district. *Pakistan Journal of Botany*, 47(1): 113-120.

Shishodia J 2013. Study of flora of district Meerut in form of digital herbarium database. Ph.D. Thesis, Chaudhary Charan Singh University, Meerut (U.P.), India.

Shiva V 1996. Protecting our Biological and Intellectual Heritage in the Age of Biopiracy. The Research Foundation for Science, Technology and Natural Resources Policy, New Delhi, India.

Shivakoti C and Ramesh KRA 2015. Preliminary phytochemical screening of *Setaria verticillata*. *Indo American Journal of Pharmaceutical Research*, 5(6): 2425-2429.

Shivhare MK, Singour PK, Chaurasiya PK, and Pawar RS 2012. *Trianthema portulacastrum* Linn. (Bishkhapra). *Pharmacognosy Reviews*, 6(12): 132-140.

shodhganga.inflibnet.ac.in/bitstream/10603/7149/10/10_result%20part%20ii.pdf. *Imperata cylindrica* (L.) Raeusch.

Shrivastava R, Bhat SH, Malla MY and Mir MI 2013. Phytochemical investigation and in vitro antioxidant activity of *Argemone mexicana* Linn. *International Journal of Pharmacy and Biological Sciences*, 4(2): 960-965.

Shu YZ 1998. Recent natural products based drug development: a pharmaceutical industry perspective. *Journal of Natural Products*, 61: 1053–1071.

Shukla AN, Srivastava S and Rawat AKS 2010. An ethnobotanical study of medicinal plants of Rewa district, Madhya Pradesh. *Indian Journal of Traditional Knowledge*, 9(1): 191-202.

Shulka BK and Sinha BK 2004. Grasses of Bundelkhand (Uttar Pradesh)- Taxonomic studies. *Journal of Non- Timber Forest Products,* 11: 274-315.

Siddhuraju P, Vijayakumari K and Janardhanan K 1995. Studies on the underexploited legumes, *Indigofera linifolia* and *Sesbania bispinosa*: nutrient composition and antinutritional factors. *International Journal of Food Sciences and Nutrition,* 46(3): 195-203.

Siddique AA, Bhakuni RS, Misraa L, Gupta P and Darokar MP 2017. New triglycerides from antimicrobial extracts of *Nepeta hindostana* weed. *Indian Journal of Chemistry,* 56 B: 542-550.

Siddiqui MB and Hussain W 1991.Traditional treatment of diarrhoea and dysentery through herbal drugs in rural India. *Fitoterapia,* 62: 325-329.

Siddiqui MB and Hussain W 1992. Some aquatic and marshy land medicinal plants from Hardoi district of Uttar Pradesh. *Fitoterapia,* 63: 245-248.

Siddiqui S 1942. A note on the isolation of three new bitter principles from the nim oil. *Current Science,* 11: 278-279.

Sikarwar RLS and Kaushik JP 1992. Traditional medicine among the rural folk of Morena district Madhya Pradesh. *Ancient Science of Life,* 12: 274-279.

Silva FL, Fischer DCH, Tavares JF, Silva MS, De-Athayde-Filho PF, Barbosa-Filho JM 2011. Compilation of secondary metabolites from *Bidens pilosa* L. *Molecules,* 16(2): 1070-1102.

Silva TCDL, Filho JV, Amorim ELCD, Souza IAD, Albuquerque UPD and Araújo ECD 2012. Acute toxicity study of stone-breaker (*Phyllanthus tenellus* Roxb.). *Journal of Basic and Applied Pharmaceutical Sciences,* 33(2): 205-210.

Simberloff D, Parker IM and Windle PM 2005. Introduced species policy, management and future research needs. Frontiers *in Ecology and the Environment,* 3: 12-20.

Sindhu S, Uma G and Kumudha P 2012. Survey of medicinal plants in Chennimallai Hills, Erode Districts, Tamilnadu. *Asian Journal of Plant Science and Research,* 2(6): 712-717.

Sindhu T, Rajamanikandan S, Ragavendran P, Sophia D, Meenakshi P, PriyaDurga D and Gopalakrishnan VK 2010. Antidiabetic activity of *Mollugo nudicaulis* against alloxan induced diabetic rats. *International Journal of Applied Biology and Pharmaceutical Technology,* 1(3): 511-519.

Singh A 2015a. Folk medicinal uses of the plant fruits from Banaras Hindu University Main Campus, India. *International Journal of Research in Pharmacy and Biosciences,* 2(8): 18-22.

Singh A 2015b. Observations on the flora of Varanasi District in Uttar Pradesh State of India. *Global Journal of Environmental Sciences & Technology,* 3(10): 368-389.

Singh A 2016. An ethnobotanical study of medicinal plants in Bhiwani district of Haryana, India. *Journal of Medicinal Plants Studies,* 4(2): 212-215.

Singh A and Navneet A 2017. A review: Traditional, ethnomedicinal utilization, pharmacological properties and phytochemistry of *Barleria prionitis* Linn. *International Journal of Pharmaceutical Sciences Review and Research,* 44(2): 19-26.

Singh A and Sarin R 2010. *Artemisia scoparia* - A new source of artemisinin. *Bangladesh Journal of Pharmacology,* 5: 17-20.

Singh A, Satanker N, Kushwaha M, Disoriya R and Gupta AK 2013a. Ethno-botany and uses of non-graminaceous forage species of Chitrakoot region of Madhya Pradesh. *Indian Journal of Natural Products and Resources,* 4(4): 425–431.

Singh A, Singh MK and Singh R 2013b. Traditional medicinal flora of the district Buxar (Bihar, India). *Journal of Pharmacognosy and Phytochemistry,* 2(2): 41-49.

Singh B and Singh J 2014. Ethnobotanical uses of some plants from central Haryana, India. *Phytodiversity,* 1(1-2): 7-24.

Singh B, Singh VN, Phukan SJ, Sinha BK and Borthakur SK 2012. Contribution to the Pteridophytic flora of India: Nokrek Biosphere Reserve, Meghalaya. *Journal of Threatened Taxa,* 4(1): 2277-2294.

Singh D and Gupta S 2015. The study of plant people relation within the purview of traditional medicinal system in Agra. *International Journal of Research,* 2(8): 538-546.

Singh D and Misra BK 2014. Flora of Etawah and Auraiya districts, Uttar Pradesh State, India. Lambert Academic Publishing (LAP).

Singh D and Shukla CP 2015. Phytodiversity and traditional knowledge of Rae-Bareli District, Uttar Pradesh. *A Journal of Multidisciplinary Advance Research,* 4 No. 1: 4-9.

Singh H, Dhole PA, Saravanan R and Baske PK 2017. Ethnomedicinal plants used in sexual disorder in Balangir and Deogarh districts, Odisha, India. *International Journal of Current Science,* 20(3): 57-62.

Singh H, Mishra A and Mishra AK 2015a. *Cleome viscosa* Linn (Capparaceae): A review. *Pharmacognosy Journal,* 7(6): 326-329.

Singh JB 2009. Flora of Jhajjar district, Haryana. *Journal of Plant Development Sciences,* 1: 57-63.

Singh JS, Singh SP and Gupta SR 2006. Ecology Environment and Resource Conservation, Anamaya Publishers, New Delhi.

Singh K, Gupta S and Mathur PK 2010a. Investigation on ethnomedicinal plants of district Firozabad. *Journal of Advanced Laboratory Research in Biology,* 1(1): 64-66.

Singh KJ and Huidrom D 2013. *Ethnobotanical uses of medicinal plant, Justicia adhatoda L. by Meitei community of Manipur, India. Journal* of Coastal Life Medicine, 1(4): 322-325.

Singh KK 1997. Flora of Dudhwa National Park. Deheradun: Bishen Singh Mahendra pal Singh. p. 516.

Singh KP, Daboriya V, Kumar S and Singh S 2010b. Antibacterial activity and phytochemical investigations on *Nicotiana plumbaginifolia* viv. (Wild tobacco). *Romanian Journal of Biology - Plant Biology,* 55: 135–142.

Singh KP, Khanna KK and Sinha GP (Eds.) 2016. Flora of Uttar Pradesh, Vol. 1. Ranunculaceae – Apiaceae. Botanical Survey of India, Calcutta. p. 674.

Singh KP, Shukla AN and Singh JS 2010c. State level inventory of alien plants, their source regions and use potential. *Current Science,* 99: 107-114.

Singh M and Kumar M 2013. Study of plant diversity of Jind district, Haryana, India. Asian Journal of Plant Science and Research, 3(3): 44-53.

Singh NK, Ghosh A, Laloo D and Singh VP 2014a. Pharmacognostical and physicochemical evaluation of *Croton bonplandianum. International Journal of Pharmacy and Pharmaceutical Sciences,* 6(3): 286-290.

Singh NP 1969. Flora of Bulandshahr District. *Nelumbo,* Bulletin of the Botanical Survey of India, 11(1-2): 1-22.

Singh P, Sharma AK, Joshi KC, Jakupovic J and Bohlmann F 1985. Acanthospermolides and other constituents from *Blainvillea acmella. Phytochemistry,* 24(9): 2023-2028.

Singh PK, Kumar V, Singh RH, Tiwari RK, Sharma A, Rao CV and Singh RH 2010d. Medico-ethnobotany of 'Chatara' block of district Sonebhadra, Uttar Pradesh, India. Advances in Biological Research, 4(1): 65-80.

Singh RP, Rai N and Tiwari VK 2011. A study of polypetalous plant diversity of Moradabad district, Uttar Pradesh, India. *Webmed Central Ecology,* 2 (11): WMC001892.

Singh S and Singh R 2012. Ethnomedicinal use of Pteridophytes in reproductive health of tribal women of Pachmarhi biosphere reserve, Madhya pradesh, India. *International Journal of Pharmaceutical Sciences and Research,* 3(12): 4780-4790.

Singh S, Dixit RD and Sahu TR 2005. Ethnomedicinal use of Pteridophytes of Amarkantak, Madhya Pradesh. *Indian journal of traditional knowledge,* 4(4): 392-395.

Singh SB, Devi WR, Marina A, Devi WI, Swapana N and Singh CB 2013c. Ethnobotany, phytochemistry and pharmacology of *Ageratum conyzoides* Linn (Asteraceae). *Journal of Medicinal Plants Research,* 7(8): 371-385.

Singh SC and Srivastava GN 1999. Exotic medicinal plants of Lucknow district UP, India. *Journal Economic and Taxonomy Botany,* 23: 223-235.

Singh TT, Devi AR, Sharma HR and Sharma HM 2015b. Medicinal plants used in the treatment of various skin diseases by the scheduled caste community of Andro village in Imphal east district, Manipur (India). *DAMA International,* 2(2): 9-19.

Singh TT, Sharma HM, Devi AR and Sharma HR 2014b. Plants used in the treatment of piles by the scheduled caste community of Andro village in Imphal East District, Manipur (India). *Journal of Plant Sciences,* 2(3):113-119.

Singh U and Narain S 2010. Traditional treatment of leucoderma by Kol tribes of Vindhyan region of Uttar Pradesh. *Indian Journal of Traditional Knowledge,* 9(1): 173-174.

Singh Y 2016. Ethnobotanical survey of medicinal plants against some common diseases in Aligarh district, U.P., India. *International Journal of Advanced Research,* 4(12): 1548-1553.

Sinha P, Arora VK and Wahi SP 1984. Chemical investigation on *Alternanthera sessilis, Indian Drug,* 1: 139-40.

Sinhababu A and Banerjee A 2013. Ethno-botanical study of medicinal plants used by Tribals of Bankura district, West Bengal, India. *Journal of Medicinal Plants Studies,* 1(3): 98-104.

Sivakumar R and Sunmathi D 2016. Phytochemical screening and antimicrobial activity of ethanolic leaf extract of *Alternanthera sessilis* (L.) R.Br. Ex Dc and *Alternanthera philoxeroides* (Mart.) Griseb. *European Journal of Pharmaceutical and Medical Research,* 3(3): 409-412.

SivaKumar T, Srinivasan K, Rajavel R, Vasudevan M, Ganesh M, Kamalakannan K and Mallika P 2009. Isolation of chemical constituents from *Prosopis juliflora* bark and anti-inflammatory activity of its methanolic extracts. *Journal of Pharmacy Research,* 2(3): 551-556.

Sivaperumal R, Ramya S, Ravi AV, Rajasekaran C and Jayakumararaj R 2009. Herbal remedies practiced by Malayali's to treat skin diseases. *Environment & We - An International Journal of Science & Technology,* 4: 35-44.

Sivapraksam SSK, Karunakaran K, Subburaya U, Kuppusamy S, and Subashini TS 2014. A review on phytochemical and pharmacological profile of *Hedyotis corymbosa* Linn. *International Journal of Pharmaceutical Sciences Review and Research,* 26(1): 320-324.

Sivarajan VV and Balachandran I 1994. Ayurvedic drugs and their plant sources. Oxford and IBH Publishing Co. Pvt. Ltd., New Delhi, p. 570.

Snigdha M, Kumar SS, Sharmistha M, and Deepa C 2013. An overview on *Vetiveria zizanioides. Research Journal of Pharmaceutical, Biological and Chemical,* 4(3): 777-783.

Song K, Wang HQ, Liu C, Kang J, Li BM and Chen RY 2014. Chemical constituents from *Chenopodium ambrosioides*. *China Journal of Chinese Materia Medica*, 39(2): 254-7.

Soni P and Singh L 2012. *Marsilea quadrifolia* linn. - A valuable culinary and Remedial fern in Jaduguda, Jharkhand, India. *International Journal of Life science and Pharma Research*, 2(3): 99-104.

Sonju JJ, Islam MF, Sutradhar K and Akter T 2017. Analysis of phytochemical, antioxidant and microbial property of various extracts of the plant *Melilotus indica*. *World Journal of Pharmaceutical Research*, 6(4): 129-147.

Sood SK, Kumar S, Dogra KS and Sharma R 2011. Alien plants distribution and ecology in the temple-courtyards of Himachal Pradesh (N.W. Himalaya). *Himachal Pradesh University Journal*, 1-11.

Soosairaj S, Kala A, Raja P and Vijaya K 2013. Phytochemical screening and GC-MS analysis on *Spermacoce articularis* L.F. (Rubiaceae). *International Journal of ChemTech Research*, 5(6): 3070-3074.

Sorathia KD 2014. Ethnomedicinal explorations for certain convolvulaceae members of Anjar Taluka. *Life sciences leaflets*, 50: 28-33.

Sreeramulu N, Suthari S, Ragan A and Raju VS 2013. Ethno-botanico-medicine for common human ailments in Nalgonda and Warangal districts of Andhra Pradesh, India. *Annals of Plant Sciences*, 02(07): 220-229.

Sripathi SK and Sankari U 2010. Ethnobotanical documentation of a few medicinal plants in the Agasthiayamalai region of Tirunelveli district, India. *Ethnobotanical Leaflets*, 14: 173-81.

Srirama R, Deepak HB, Senthilkumar U, Ravikanth G, Gurumurthy BR, Shivanna MB, Chandrasekaran CV, Agarwal A and Shaanker RU 2012. Hepatoprotective activity of Indian *Phyllanthus*. *Pharmaceutical Biology*, 50(8): 948-953.

Srivastav S, Singh P, Mishra G, Jha KK and Khosa RL 2011. *Achyranthes aspera*-An important medicinal plant: A review. *Journal of Natural Product and Plant Resources*, 1(1): 1-14.

Srivastava AK 1993. Exotic weeds of Gorakhpur district UP. *Journal of Economic and Taxonomic Botany*, 17: 261-263.

Srivastava B, Sharma H, Dey YN, Wanjari MM and Jadhav AD 2014a. *Alhagi pseudalhagi*: A review of its phyto-chemistry, pharmacology, folklore claims and Ayurvedic studies. *International Journal of Herbal Medicine*, 2(2): 47-51.

Srivastava D 2017. Medicinal plants of genus *Ipomoea* found in Uttar-Pradesh, India. *Research Journal of Recent Sciences*, 6(12): 12-22.

Srivastava P and Shanker K 2012. *Pluchea lanceolata* (Rasana): Chemical and biological potential of Rasayana herb used in traditional system of medicine. *Fitoterapia*, 83(8): 1371-1385.

Srivastava RC, Singh RK and Mukherjee TK 2010. Indigenous biodiversity of Apatani plateau: Learning on biocultural knowledge of Apatani tribe of Arunachal Pradesh for sustainable livelihoods. *Indian Journal of Traditional Knowledge*, 9(3): 432-442.

Srivastava S, Dvivedi A, and Shukla RP 2014b. Invasive alien species of terrestrial vegetation of North-Eastern Uttar Pradesh. *International Journal of Forestry Research*, 2014: 1-9.

Srivastava TN 1976. Flora gorakhpurensis. New Delhi: Today & Tomorrow's Printer & Publisher, p. 411.

Ssegawa Paul and Kasenene John Massan 2007. Medicinal plant diversity and uses in the Sango bay area, Southern Uganda. *Journal of Ethnopharmacology*, 113: 521-540.

Stasi LCD, Oliveira GP, Carvalhaes MA, Queiroz-Junior M, Tien OS, Kakinami SH and Reis MS 2002. Medicinal plants popularly used in the Brazilian tropical Atlantic forest. *Fitoterapia*, 73: 69-91.

Steenkamp V 2003. Traditional herbal remedies used by South African women for gynaecological complaints. *Journal of Ethnopharmacology*, 86: 97-108.

Stefanović OD, Tešić JD and Ljiljana ČR 2015. *Melilotus albus* and *Dorycnium herbaceum* extracts as source of phenolic compounds and their antimicrobial, antibiofilm, and antioxidant potentials. *Journal of Food and Drug Analysis*, 23(3): 417-424.

Stepp JR 2004.The role of weeds as sources of pharmaceuticals. *Journal of Ethnopharmacology*, 92: 163-166.

Stepp JR and Moerman DE 2011. The importance of weeds in ethnopharmacology. *Journal of Ethnopharmacology*, 75 (1):19-23.

Subramaniam S and Sivasubramanian A 2015. Tradition to therapeutics: Sacrificial medicinal grasses *Desmostachya bipinnata* and *Imperata cylindrica* of India. *Boletín Latinoamericano y del Caribe de Plantas Medicinales y Aromáticas*, 14(3): 156 -170.

Sufian MA, Chowdhury TK, Mian MM, Mohiuddin M, Koly SF, Sahabuddin M and Sarwa MS 2017. In vitro study on thrombolytic and membrane stabilizing activities of *Alternanthera paronychioides* and *Vernonia patula* leaves and stems. *Annual Research & Review in Biology*, 15(3): 1-9.

Suksri S, Premcharoen S, Thawatphan C and Sangthongprow S 2005. Ethnobotany in Bung Khong Long non-hunting area, North-East Thailand. *Kasetsart Journal: Natural Science*, 39: 519-533.

Sultana R, Rahman MS, Bhuiyan MNI, Begum J and Anwar MN 2008. In vitro antibacterial and antifungal activity of *Borreria articularis*. *Bangladesh Journal of Microbiology*, 25(2): 95-98.

Sultana S, Khan MA, Ahmad M and Zafar M 2006. Indigenous knowledge of folk herbal medicines by the women of district Chakwal, Pakistan. *Ethnobotanical Leaflets*, 10: 243-253.

Sundarapandian M, Thekkekkara D and Nanjan MJ 2016. Nutritional assessment of the plant, *Spergula arvensis* L. *Indian Journal of Natural Products and Resources,* 7(2): 150-154.

Supe U, Dhote F and Roymon MG 2011. A review on micro propagation of *Withania somnifera* – A medicinal plant. *Journal of Agricultural Technology,* 7(6): 1475-1483.

Susanti RF, Kurnia K, Vania A and Reynaldo IJ 2015. Total phenol, flavanoid and antioxidant activity of *Physalis angulata* leaves extract by subcritical water extraction. *Modern Applied Science,* 9(7): 190-198.

Sushama B and Nishteswar K 2013. Ayurvedic and ethano–Botanical perspectives of Chirabilva (*Holoptelea integrifolia* Planch). *International Ayurvedic Medical Journal,* 1(3): 1-8.

Sutar AC, Kasture SB, Kalaichelvan Vk 2014. Phytochemical profile studies on the steroids of methanolic leaf extract of medicinally important plant *Holoptelea integrifolia* (roxb.) Planch using high-performance thin layer chromatography. *Asian Journal of Pharmaceutical and Clinical Research,* 7(4): 197-200.

Suthar P, Mathur K, Goyal M and Yadav SK 2016. Traditional uses, phytochemistry, pharmacological properties of plant *Alhagi maurorum* (Medik.): A review. *World Journal of Pharmacy and Pharmaceutical Sciences,* 5(4): 682-692.

Suthari S, Sreeramulu N, Omkar K, Reddy CS and Raju VS 2014. Intracultural cognizance of medicinal plants of Warangal north forest division, Northern Telangana, India. *Ethnobotany research and applications,* 12: 211-235.

Swain SR, Sinha BN and Murthy PN 2008. Antiinflammatory, diuretic and antimicrobial activities of *Rungia pectinata* Linn. and *Rungia repens* Nees. *Indian Journal of Pharmaceutical Sciences,* 70(5): 679-683.

Swami SB, Thakor NSJ, Patil MM and Haldankar PM 2012. Jamun (*Syzygium cumini* (L.): A review of its food and medicinal uses. *Food and Nutrition Sciences,* 3: 1100-1117.

Swapna MM, kumar PR, Anoop KP, Manju CN, and Rajith NP 2011. A review on the medicinal and edible aspects of aquatic and wetland plants of India. *Journal of Medicinal Plants Research,* 5(33): 7163-7176.

Swarnkar S and Katewa SS 2008. Ethanobotanical observation on tuberous plants from tribal areas of Rajasthan (India). *Ethanobotanical Leaflets* 12: 647-666.

Swathy B, Lakshmi SM and Kumar AS 2010. Evaluation of analgesic and anti-inflammatory properties of *Chloris barbata* (Sw.). *International Journal of Phytopharmacology,* 1(2): 92-96.

Tabaraki R, Yosefi Z and Gharneh HAA 2012. Chemical composition and antioxidant properties of *Malva sylvestris* L. *Journal of Research in Agricultural Science,* 8(1): 59-68.

Talukdar SN, Rahman MB and Paul S 2015. A review on *Barleria prionitis*: Its pharmacognosy, phytochemicals and traditional use. *Journal of Advances in Medical and Pharmaceutical Sciences,* 4(4): 1-13.

Tangjang S, Namsa ND, Aran C and Litin A 2011. An ethnobotanical survey of medicinal plants in the Eastern Himalayan zone of Arunachal Pradesh, India. *Journal of Ethnopharmacology,* 134(1): 18-25.

Tapadia G, Kale MA and Saboo S 2017. Phytochemical characterization, anti-cancer and antimicrobial activity of isolated fractions of *Alysicarpus vaginalis*. *Bangladesh Journal of Pharmacology,* 12: 77-83.

Tareen RB, Bibi T, Khan MA, Ahmad M and Zafar M 2010. Indigenous knowledge of folk medicine by the women of Kalat and Khuzdar regions of Balochistan, Pakistan. *Pakistan Journal of Botany,* 42(3): 1465-1485.

Tarnam YA, Ilyas MHM and Begum TN 2014. Biological potential and phytopharmacological screening of *Gomphrena* species. *International Journal of Pharma Research & Review,* 3(1): 58-66.

Tayade SK and Patil DA 2006. Ethnomedicinal wisdom of tribals of Nandurbar district (Maharashtra). *Natural Product Radiance,* 5(1): 64-69.

Tennyson S, Ravindran J, Eapen A and William J 2015. Ovicidal activity of *Ageratum houstonianum* Mill. (Asteraceae) leaf extracts against Anopheles stephensi, Aedes aegypti and Culex quinquefasciatus (Diptera: Culicidae). *Asian Pacific Journal of Tropical Disease,* 5(3): 199-203.

The Biological Diversity Act 2002 and Biological Diversity Rules 2004. National Biodiversity Authority, Chennai, India.

The Plant List 2013. Version 1.1. Published on the Internet; http://www.theplantlist.org.

Thomas CM, Wood RC 3rd, Wyatt JE, Pendleton MH, Torrenegra RD, Rodriguez OE, Harirforoosh S, Ballester M, Lightner J, Krishnan K and Ramsauer VP 2012. Anti-neoplastic activity of two flavone isomers derived from *Gnaphalium elegans* and *Achyrocline bogotensis*. *PLoS One,* 7(6): e39806. doi: 10.1371/journal.pone.0039806.

Thomas SCL 2006. Taiwanese native medicinal plants: Phytopharmacology and therapeutic values. CRC Press, p.400.

Thombre NA, Kshirsagar SJ, Kadam KM and Pingale S 2016. Roles of *Commelina benghalensis* in medicine – a review. *International Journal of Institutional Pharmacy and Life Sciences,* 6(3): 505-512.

Tirkey A 2006. Some ethnomedicinal plants of family Fabaceae of Chhattisgarh state. *Indian Journal of Traditional Knowledge,* 5(4): 551-553.

Tiwari JK, Ballabha R and Tiwari P 2010. Diversity and present status of medicinal plants in and around Srinagar Hydroelectric power project in Garhwal Himalaya, India: Needs for Conservation. *Researcher,* 2(2): 50-60.

Tiwari KP, Massod M, Rathore YKS and Minocha PK 1978. Study of anthocyanin from the flowers of some medicinal plants. *Vijana Parishad Anusandhan Patrika,* 21:277.

Tiwari R and Rana CS 2015. Phytomedicine for the diabetes: A traditional approach. *Annals of Phytomedicine,* 4(1): 108-110.

Tiwari VJ 2016. Evidence Based Ethnopharmacological uses of *Trichosanthes cucumerina* var. *cucumerina* (L.) Haines. (Cucurbitaceae) by Gond tribe of Gondia District of Maharashtra State, India. *Research Journal of Pharmacognosy and Phytochemistry,* 8(4): 209-216.

Tomar A 2009. Folk medicinal uses of plant roots from Meerut district, Uttar Pradesh. *Indian Journal of Traditional Knowledge,* 8(2): 298-301.

Tomar A, Singh H and Singh V 2008. Exotic elements in the flora of Baghpat district, Uttar Pradesh. *Indian Journal of Forestry,* 31(3): 463-471.

Tona L, Kambu K, Mesia K, Cimanga K, Apers S, de Bruyne T, Pieters L, Totte J and Vlietinck AJ 1999. Biological screening of traditional preparations from some medicinal plants used as antidiarrhoeal in Kinshasa, Congo. *Phytomedicine,* 6:59-66.

Tona L, Mesia K, Ngimbi NP, Chrimwami B, Okond'ahoka, Cimanga K, de Bruyne T, Apers S, Hermans N, Totte J, Pieters L and Vlietinck AJ 2001. *In-vivo* antimalarial activity of *Cassia occidentalis, Morinda morindoides* and *Phyllanthus niruri. Annals of Tropical Medicine and Parasitology,* 95(1): 47-57.

Tonny TS, Sultana S and Siddika F 2017. Study on medicinal uses of *Persicaria* and *Rumex* species of Polygonaceae family. *Journal of Pharmacognosy and Phytochemistry,* 6(6): 587-589.

Torkelson AR 1995. The cross name index to medicinal plants, Vol. 1-4. CRC Press, p. 1200.

Touqeer S, Saeed MA and Ajaib M 2013. A Review on the phytochemistry and pharmacology of genus Tephrosia. *Phytopharmacology,* 4(3): 598-637.

Tripathi P, Kumar R, Sharma AK, Mishra A, and Gupta R 2010. *Pistia stratiotes* (Jalkumbhi). *Pharmacognosy Reviews,* 4(8): 153-160.

Tripathi YC, Prabhu VV, Pal RS and Mishra RN 1996. Medicinal plants of Rajasthan in Indian System of Medicine. *Ancient science of life,* 15:190-212.

Tripti J, Kumar N and Kothiya P 2015. A review on *Cleome viscosa*: An endogenous Herb of Uttarakhand. *International Journal of Pharma Research & Review,* 4(7): 25-31.

Trivedi PC 2002. Ethnobotany. Jaipur: Aavinashkar Publishers and Distributors, p. 222.

Tulika T, Mala A 2015. Pharmaceutical potential of aquatic plant *Pistia stratiotes* (L.) and *Eichhornia crassipes. Journal of Plant Sciences,* 3(1-1):10-18.

Turker AU and Gurel E 2005. Common Mullein (*Verbascum thapsus* L.): Recent advances in research. *Phytotherapy Research,* 19: 733-739.

Ulhas AS 2015. Pharmacological review of Akashvel (*Cuscuta Reflexa* Roxb.). *International Journal of Pharmacology Research,* 5(3): 152-154.

Ullah R, Hussain Z, Iqbal Z, Hussain J, Khan FU, Khan N, Muhammad Z, Ayaz S, Ahmad S, Rehman NU and Hussain I 2010. Traditional uses of medicinal plants in Darra Adam Khel, NWFP, Pakistan. *Journal of Medicinal Plants Research,* 4(17): 1815-1821.

Ullah S, Khan MR, Shah NI, Shah SA, Majid M and Farooq MA 2014. Ethnomedicinal plant use value in the Lakki Marwat District of Pakistan. *Journal of Ethnopharmacology,* 158(2): 412-422.

Umadevi M, Rajeswari R, Rahale CS, Selvavenkadesh S, Pushpa R, Kumar KPS and Bhowmik D 2012. Traditional and medicinal uses of *Withania Somnifera. The pharma innovation,* 1(9): 102-110.

Une HD, Pal SC, Kasture VS and Kasture SB 2001. Phytochemical constituents and pharmacological profile of *Albizzia lebbeck. Journal of Natural Remedies,* 1(1):1-5.

Upadhayay N 2016. Phytosociological documentation of traditional knowledge of *Barleria acanthoides* Vahl. *International Journal of Science and Research,* 5(4):1188-1190.

Upadhyay H, Kumar A, Gupta MK, Sharma A and Rahal A 2013. Validation of medicinal values of traditionally used *Sonchus asper* (prickly sow thistle) leaves for the treatment of skin ailments. *Advancement in Medicinal Plant Research,* 1(1): 29-35.

Upasani SV, Beldar VG, Tatiyab AU, Upasani MS, Surana SJ and Patil DS 2017. Ethnomedicinal plants used for snakebite in India: a brief overview. *Integrative Medicine Research,* 6: 114-130.

Upton R, Graff A, Jolliffe G, Länger R and Williamson E 2011. American herbal pharmacopoeia: Botanical pharmacognosy - microscopic characterization of botanical medicines. CRC Press, p. 800.

Usman A Ashfaq and Sobia Idrees 2014. Medicinal plants against hepatitis C Virus. *World Journal of Gastroenterology,* 20(11): 2941-2947.

Usman LA, Hamid AA, Muhammad NO, Olawore NO, Edewor TI and Saliu BK 2010. Chemical constituents and anti-inflammatory activity of leaf essential oil of Nigerian grown *Chenopodium album* L. *Experimental and Clinical Sciences (EXCLI) Journal,* 9: 181-186.

Usman MRM, Patil SB, Patil SS and Patil RS 2013. *Sesbania sesban* Linn. : An overview. *International Journal of Pharmacy and Life Sciences*, 4(5): 2644-2648.

Vadhana R 2006. Floristic Plants of the World. Sarup & Sons publication, New Delhi.

Vadivel V and Janardhanan K 2002. Agrobotanical traits and chemical composition of *Cassia obtusifolia* L.: A lesser-known legume of the Western Ghats region of South India. *Plant Foods for Human Nutrition*, 57(2):151-164.

Vadivelan S, Sinha BN, Betanabhatla KS, Christina AJM and Pillai RN 2007. Anti-inflammatory activity of *Spermacoce articularis* Linn on carrageenan induced Paw edema in Westar male rats. *Pharmacologyonline*, 3: 478-484.

Vanila D, Ghanthikumar S and ManickamVS 2008. Ethnomedicinal uses of plants in the plains area of the Tirunelveli district, Tamil Nadu, India. *Ethnobotanical Leaflets*, 12: 1198-1205.

Vardhana R 2007. Flora of Ghaziabad District. Shree Publishers & Distibutors, New Delhi, p. 639.

Vardhana R 2008. Direct uses of medicinal plant and their identification. Sarup & Sons Publication, New Delhi, p. 423.

Vardhana R 2010. Aquatic plants of district Ghaziabad and adjacent areas. *Plant Archives*, 10 (2): 927-932.

Vardhana R 2016. Poisonous plants of district Ghaziabad (U.P.), India. *Plant Archives*, 16(1): 60-64.

Varsha V, Suresh SN and Prejeena V 2015. Phytochemical screening of *Cyanthellium cinereum* leaf extracts. *International Journal of Medicine and Pharmaceutical Research*, 3(6): 1238-1241.

Vashistha BD and Kaur M 2013. Floristic and ethno botanical survey of Ambala district, Haryana. *International Journal of Pharmacy and Biological Sciences*, 4(2): 353-360.

Vasudevan P, Manoharan S and Muthusamy J 2014. An ethnomedical, pharmacological and phytochemical review of *Ammannia baccifera* L. *World Journal of Pharmaceutical Research*, 3(6): 1771-1789.

Veale DJH, Furman KI and Oliver DW 1992. South African traditional herbal medicines used during pregnancy and childbirth. *Journal of Ethnopharmacology*, 36: 185-191.

Vedpal, Dhanabal SP, Dhamodaran P, Chaitnya MVNL, Duraiswamy B, Jayaram U and Srivastava N 2016. Ethnopharmacological and phytochemical profile of three potent *Desmodium* species: *Desmodium gangeticum* (L.) DC, *Desmodium triflorum* Linn and *Desmodium triquetrum* Linn. *Journal of Chemical and Pharmaceutical Research*, 8(7): 91-97.

Veena S, Chimmad BV, Naik RK and Shanthakumar G 2005. Physico-chemical and nutritional studies in Barnyard Millet. *Karnataka Journal of Agricultural Sciences*, 18(1):101-105.

Venkatachalapathi A, Sangeeth T, Ali MA, Tamilselvi SS and Paulsamy S 2016. Ethnomedicinal assessment of Irula tribes of Walayar valley of Southern Western Ghats, India. *Saudi Journal of Biological Sciences, https://doi.org/10.1016/j.sjbs.2016.10.011.*

Venkatraman S and Pari L 2003. Effect of *Coccinia indica* leaves on antioxidant status in streptozotocin-induced diabetic rats. *Journal of Ethnopharmacology*, 84:163-168.

Verma RK 2014. An ethnobotanical study of plants used for the treatment of livestock diseases in Tikamgarh District of Bundelkhand, Central India. *Asian Pacific Journal of Tropical Biomedicine*, 4 (1): S460-S467.

Verma RK and Verma D 2016. Traditional plant based therapy among rural communities of some villages of Pilibhit district of Rohilkhand division, India. *International Research Journal of Biological Sciences*, 5(5): 1-13.

Verma S 2016a. A review on ethnomedicinal plant *Acacia nilotica* (Linn.) Wild. *Journal of Pharmacognosy and Photochemistry*, 5(2): 241-242.

Verma S 2016b. A Review study on *Leucaena leucocephala*: A multipurpose tree. *International Journal of Scientific Research in Science, Engineering and Technology*, 2(2): 103-105.

Verma S, Sharma H and Garg M 2014. *Phyllanthus amarus*: A Review. *Journal of Pharmacognosy and Phytochemistry*, 3(2): 18-22.

Verma SM and Suresh KB 2002. Phytochemical investivations of *Indigofera tinctoria* Linn leaves. *Ancient Science of Life*, 21: 235-239.

Vethanarayanan P, Unnikannan P, Baskaran L and Sundaramoorthy P 2011. Survey on traditional medicinal plants used by the village peoples of Cuddalore district, Tamil Nadu, India. *Asian Journal of Biochemical and Pharmaceutical Research*, 1(3): 351-361.

Vijikumar S, Ramanathan.K and Devi BP 2011. *Cuscuta reflexa* ROXB.-A wonderful miracle plant in ethnomedicine. *Indian Journal of Natural Sciences*, 3(9): 676- 683.

Vishal R 2013. Protective role of Indian medicinal plants against liver damage. *The Journal of Phytopharmacology*, 2(3): 1-3.

Viswanathan MB, Thangadurai D, Vendan KT and Ramesh N 1999. Chemical analysis and nutritional assessment of *Teramnus labialis* (L.) Spreng. (Fabaceae). *Plant Foods for Human Nutrition*, 54(4):345-352.

Voeks RA 1996. Tropical forest healers and habitat preference. *Economic Botany*, 50(4): 381-400.

Volpato G, Kourkova P and Zeleny V 2012. Healing war wounds and perfuming exile: The use of vegetal, animal, and mineral products for perfumes, cosmetics, and skin healing among Sahrawi refugees of Western Sahara. *Journal of Ethnobiology and Ethnomedicine*, 8(49): 1-19.

Vulto AG and Smet PAGM 1988. In Dukes MNG (Ed.), Meyler's side effects of drugs, 11th Edition, Elsevier, Amsterdam, p. 999-1005.

Wagh VV 2016. Preliminary study on ethno-medicinal plants used for treating malarial fever in Pilibhit Tiger Reserve, Uttar Pradesh, India. *International Journal of Bioassays*, 5(7): 4672-4676.

Wang S, Zhang J, Zhang Z, Gao W, Yan Y, Li X, and Liu C 2014a. Identification of chemical constituents in the extract and rat serum from *Ziziphus jujuba* Mill by HPLC-PDA-ESI-MSn. *Iranian Journal of Pharmaceutical Research*, 13(3): 1055–1063.

Wang WQ, Song WB, Lan XJ, Huang M, Xuan LJ 2014b. Merremins A-G, resin glycosides from *Merremia hederacea* with multidrug resistance reversal activity. *Journal of Natural Products*, 77(10): 2234-2240.

Warrier PK, Nambiar VPK and Ramankutty C 2002. Indian medicinal plants compendium of 500 species. Orient Longman Pvt. Ltd. Chennai, p. 320-322.

Warrier PK, Nambiar VPK and Ramankutty C 2010. Indian medicinal plants: A compendium of 500 species (Arya Vaidya Sala), Vol. 1-5. Universities Press (Private Limited), Himayatnagar, Hyderabad, India.

Wasson E 2003. The complete encyclopedia of trees and shrubs: Descriptions, cultivation requirements, pruning, planting. *Thunder Bay Press*, San Diego, California.

Watanabe K, Shono Y, Kakimizu A, Okada A, Matsuo N, Satoh A and Nishimura H 1993. New mosquito repellent from *Eucalyptus camaldulensis*. *Journal of Agricultural and Food Chemistry*, 41: 2164-2166.

Watt JM and Breyer-Brandwijk MG 1962. The medicinal and poisonous plants of southern and eastern Africa, 2nd ed. Livingstone, London.

Weber E 2003. Invasive plant species of the world: A reference guide to environmental weeds (CABI Publishing) UK.

Wiart C 2006a. Ethnopharmacology of medicinal plants: Asia and the Pacific. *Humana Press*, Totowa, New Jersey.

Wiart C 2006b. Medicinal plants of the Asia-Pacific: Drugs for the future? World Scientific, p. 756.

Wiedenfeld H 2011. Plants containing pyrrolizidine alkaloids: toxicity and problems. *Food Additives and Contaminants*, 28(3): 282-292.

Wiersema JH and León B 2013. World Economic Plants: A Standard Reference, Second Edition. CRC Press, p. 1336.

Wu CH, Hsieh HT, Lin JA and Yen GC 2013. Alternanthera paronychioides protects pancreatic β-cells from glucotoxicity by its antioxidant, antiapoptotic and insulin secretagogue actions. *Food Chemistry*, 139(1-4): 362-370.

Wu SJ, Pang CS, Lin DL and Lean TN 2009. Supercritical carbon dioxide extract of *Physalis peruviana* induced cell cycle arrest and apoptosis in human lung cancer H661 cells. *Food and Chemical Toxicology*, 47(6):1132-1138.

Xian Q, Chen H, Zou H and Yin D 2006. Chemical composition of essential oils of two submerged macrophytes, *Ceratophyllum demersum* L. and *Vallisneria spiralis* L. *Flavour and Fragrance Journal*, 21(3): 524-526.

Xiao Y, Wang YL, Gao SX, Sun C and Zhou ZY 2007. Chemical composition of *Hydrilla verticillata* (L. f.) Royle in Taihu Lake. *Chinese Journal of Chemistry*, 25(5): 661-665.

Xiao-xiong W and Zhong-jian J 2005. Chemical constituents of *Equisetum ramosissimum*. *Acta Botanica Boreali-Occidentalia Sinica*, 25(12): http://en.cnki.com.cn/Article_en/CJFDTotal-DNYX200512026.htm

Yadav AS 2005. Supplement to the flora of north-east Rajasthan from Alwar district. *Journal of Phytological Research*, 18(1): 111-114.

Yadav J, Kumar S and Siwach P 2006. Folk medicine used in gynecological and other related problems by rural population of Haryana. *Indian Journal of Traditional Knowledge*, 5(3): 323-326.

Yadav JP, Arya V, Yadav S, Panghal M, Kumar S and Dhankhar S 2010a. *Cassia occidentalis:* A review on its ethnobotany, phytochemical and pharmacological profile. *Fitoterapia*, 81(4): 223-30.

Yadav N, Vasudeva N, Singh S and Sharma SK 2007. Medicinal properties of the genus *Chenopodium* Linn. *Natural Product Radiance*, 6(2): 131-134.

Yadav RH 2013. Medicinal plants in folk medicine system of Ethiopia. *Journal of Poisonous and Medicinal Plants Research*, 1(1): 7-11.

Yadav S, Arya V, Kumar S, Yadav M and Yadav JP 2012. Ethnomedicinal flora of Dosi Hills of Mahendergarh district (Haryana), India. *Annals of Biology*, 28(2): 152-157.

Yadav S, Yadav JP, Arya V and Panghal M 2010b. Sacred groves in conservation of plant biodiversity in Mahendragarh district of Haryana. *Indian Journal of Traditional Knowledge*, 9(4): 693-700.

Yadav SS and Bhandoria MS 2013. Ethnobotanical exploration in Mahendergarh district of Haryana (India). *Journal of Medicinal Plants Research*, 7(18): 1263-1271.

Yadav SS, Bhukal RK, Bhandoria MS, Ganie SA, Gulia SK and Raghav TBS 2014. Ethnoveterinary medicinal plants of Tosham block of district Bhiwani (Haryana) India. *Journal of Applied Pharmaceutical Science*, 4(6): 40-48.

Yakob HK, Sulaiman SF and Uyub AM 2012. Antioxidant and antibacterial activity of *Ludwigia octovalvis* on *Escherichia coli* O157:H7 and some pathogenic bacteria. *World Applied Sciences Journal*, 16(1): 22-29.

Yan J and Yang XW 2005. Studies on the chemical constituents in herb of *Ludwigia octovalvis*. *Zhongguo Zhong Yao Za Zhi*, 30(24):1923-26.

Yan X, Yin XJ, Wang GK, Hong J, Lin B and Qin MJ 2012. Chemical constituents of *Medicago polymorpha*. *Chinese Pharmaceutical Journal*, 2012-06.

Yang SH, Guo HZ, Guo DA and Zheng JH 2006. Studies on chemical constituents of hairy root *of Cassia obtusifolia*. *China Journal of Chinese materia medica,* 31(3):217-319.

Ye M, Yan YN, Qiao L and Ni XM 2002. Studies on chemical constituents of *Cuscuta chinensis*. *China Journal of Chinese Materia Medica,* 27(2):115-117.

Yellasubbaiah N, Nagasudha B, Kuamr SVS, Sowjanya B, Sultana MH, Lakshmi PH, Ismail S, Devi YY and Babu MS 2015. Evaluation of anti-oxidant and anthelminthic activity of *Rhynchosia minima* (Linn) DC. *Journal of Global Trends in Pharmaceutical Sciences,* 6(2): 2579-2588.

Yousafa Z, Wanga Y and Baydounc E 2013. Phytochemistry and pharmacological studies on *Solanum torvum* Swartz. *Journal of Applied Pharmaceutical Science,* 3(4): 152-160.

Youssef RSA 2013. Medicinal and non-medicinal uses of some plants found in the middle region of Saudi Arabia. *Journal of Medicinal Plants Research,* 7(34): 2501-2513.

Yuan NN, Huang WH, Li YL and Cen YZ 2009. Studies on the chemical constituents of *Blumea laciniata*. *Zhong Yao Cai = Zhongyaocai = Journal of Chinese Medicinal Materials,* 32(7): 1065-1066.

Yue-Zhong Shu 1998. Recent natural products based drug development: a pharmaceutical industry perspective. *Journal of Natural Products* 61: 1053-1071.

Yukes JE and Balick MJ 2010. Dominican medicinal plants: A guide for health care providers, Second Edition. The New York Botanical Garden.

Yumkham SD and Singh PK 2011. Less known ferns and fern allies of Manipur with ethnobotanic uses. *Indian journal of traditional knowledge,* 10(2): 287-291.

Yumnam JY and Tripathi OP 2012. Traditional knowledge of eating raw plants by the Meitei of Manipur as medicine/nutrient supplement in their diet. *Indian Journal of Traditional Knowledge,* 11(1): 45-50.

Zahoor A, Hussain H, Khan A, Ahmed I, Ahmad V and Krohn K 2012. Chemical Constituents from *Erigeron bonariensis* L. and their Chemotaxonomic Importance. *Records of Natural Products,* 6(4): 376-380.

Zakaria DM, Islam M, Anisuzzaman SMD, Kundu SK, Khan MS and Begum AA 2011. Ethnomedicinal survey of medicinal plants used by folk medical practitioners in four different villages of Gazipur district, Bangladesh. *Advances in Natural and Applied Sciences,* 5(4): 392-399.

Zenebe G, Zerihun M, and Solomon Z 2012. An ethnobotanical study of medicinal plants in Asgede Tsimbila district, Northwestern Tigray, Northern Ethiopia. *Ethnobotany Research & Applications,* 10:305-320.

Zeraati F, Zamani A, Goodarzi MT, Hashjin SMM and Razzaghi k 2010. *In Vitro* cytotoxic effects of *Cuscuta chinensis* whole extract on human acute Lymphoblastic Leukemia Cell Line. *Iranian Journal of Medical Sciences,* 35(4): 310-314.

Zereen A, Bokhari TZ and Khan Z 2013. Ethnobotanical usages of grasses in Central Punjab-Pakistan. *International Journal of Scientific & Engineering Research,* 4(9): 452-461.

Zhang YJ, Deng GF, Xu XR, Wu S, Li S and Li HB 2013. Chemical components and bioactivities of Cape gooseberry (*Physalis peruviana*). *International Journal of Food Nutrition and Safety,* 3(1): 15-24.

Zhasa NN, Hazarika P and Tripathi YC 2015. Indigenous knowledge on utilization of plant biodiversity for treatment and cure of diseases of human beings in Nagaland, India: A case study. *International Research Journal of Biological Sciences,* 4(4), 89-106.

Zheng R 2006. Studies on the chemical constituents of *Sonchus Brachyotus* DC. *Master's Thesis,* Jilin University, China.

Zheng X, Wang W, Piao H, Xu W, Shi H and Zhao C 2013. The Genus *Gnaphalium* L. (Compositae): Phytochemical and pharmacological characteristics. *Molecules,* 18: 8298-8318.

Zhou G, Yang X, Jiang Y and Tu P 2013. Flavanoids and xanthones isolated from *Arenaria serpyllifolia* L. *Journal of Chinese Pharmaceutical Sciences,* 22(3): 286-288.

Zhou Y-X, Xin H-L, Rahman K, Wang S-J, Peng C and Zhang H 2015. *Portulaca oleracea* L. A review of phytochemistry and pharmacological effects. *BioMed Research International,* 2015: 1-11. doi:10.1155/2015/925631.

Zimdahl RL1999. Fundamentals of Weed Science. 2nd ed. Academic Press, San Diego.

Zong XD, Fen YZ, Zhu ZY and Zou ZD 2011. Antioxidant and antibacterial activity of six edible wild plants (*Sonchus* spp.) in China. *Natural Product Research,* 25(20):1893-901.

Appendices

Appendix 1. Floristic and ethnobotanical studies undertaken in districts of Western Uttar Pradesh

SN	District in Western Uttar Pradesh	Area (km²)	Type of Study	No. of plant species recorded	No. of Families	No. of Genera	Plant Groups Represented			Reference
							Angiosperms		Pterid.	
							Dicot	Monocot		
1	Agra	4,027	Floristic Study	609	103	384	469	140	-	Sharma and Dhakre (1995)
			Ethnobotanical Study	50	-	-	-	-	-	Singh and Gupta (2015)
2	Aligarh	3,700.4	Floristic Study	751 (w+c)	134	525	609	142	-	Shazia (2013)
			Ethnobotanical Study	61	35	57	-	-	-	Singh (2016)
3	Amroha	2,321	Information not found							
4	Badaun	4,234	Information not found							
5	Baghpat (NCR)	1,321	Floristic Study	566	102	371	-	-	-	Ahamed and Gupta (2010)
				76	34	67	-	-	-	Tomar *et al.* (2008)
			Ethnobotanical Study	Information not found						
6	Bareilly	4,120	Floristic Study	151	45	91	-	-	-	Misra (2015)[#]
			Ethnobotanical Study	42	-	-	-	-	-	Kumar (2013)
7	Bijnor	4,049	Floristic Study	Information not found						
			Ethnobotanical Study	20	-	-	-	-	-	Chaudhary and Gupta (2011)
				40	28	37	-	-	-	Chaudhary (2011)
				40	25	40	-	-	-	Chaudhary and Kumar (2015)
8	Bulandshahr (NCR)	4,353	Floristic Study	376	-	-	-	-	-	Singh (1969)
				122	34	-	-	-	-	Gupta and Narayan (2006)
			Ethnobotanical Study	Information not found						
9	Etah	4,446	Information not found							
10	Etawah	2,311	Floristic Study	560	102	340	-	-	-	Singh and Misra (2014)[$]
			Ethnobotanical Study	Information not found						
11	Firozabad	2,362	Floristic Study	Information not found						
			Ethnobotanical Study	71	-	62	-	-	-	Singh *et al.* (2010)
				52	34	-	-	-	-	Kumari *et al.* (2015)
12	Gautam Buddha Nagar (NCR)	1,442	Floristic Study	257	65	214	196	58	3	Ansari *et al.* (2016)
			Ethnobotanical Study	Information not found						

SN	District in Western Uttar Pradesh	Area (km²)	Type of Study	No. of plant species recorded	No. of Families	No. of Genera	Plant Groups Represented			Reference
							Angiosperms		Pterid.	
							Dicot	Monocot		
13	Ghaziabad (NCR)	1,273	Floristic Study	1654 (w+c)	171	868	1312	342	-	Vardhana (2007)
				6	5	6	6	-	-	Mishra *et al.* (2014)
				262	-	-	-	-	-	Vardhana (2010)
			Ethnobotanical Study	50	32	44	-	-	-	Chaudhary and Kumar (2015)
				27	18	-	-	-	-	Vardhana (2016)
14	Hapur (NCR)	660	Information not found							
15	Hathras	1,840	Information not found							
16	Kasganj	1,993	Information not found							
17	Mainpuri	2,745	Information not found							
18	Mathura	3,329.40	Floristic Study	507	97	321	-	-	-	Sharma *et al.* (2005)
			Ethnobotanical Study	Information not found						
19	Meerut (NCR)	2,522	Floristic Study	862	118	521	668	194	-	Shishodia (2013)
			Ethnobotanical Study	39	28	39	-	-	-	Tomar (2009)
20	Moradabad	3,493	Floristic Study	153	60	-	-	-	-	Bhati *et al.* (2014)
				235	46	154	-	-	-	Singh *et al.* (2011)
			Ethnobotanical Study	25	18	-	-	-	-	Kumar *et al.* (2012)
21	Muzaffarnagar (NCR)	3,008	Floristic Study	577	77	321	427	150	-	Malik (2015a)
			Ethnobotanical Study	15	13	15	-	-	-	Prachi *et al.* (2009)
22	Pilibhit	3,504	Floristic Study	723 (w+c)	128	471	506	217	-	Malhotra (1979)
				151	45	91	-	-	-	Misra (2015)#
			Ethnobotanical Study	50	31	50	-	-	-	Verma and Verma (2016)
				23	16	-	-	-	-	Wagh (2016)
23	Rampur	2,367	Information not found							
24	Saharanpur	3,860	Floristic Study	142	-	74	-	-	-	Malik (2015b)
				110	-	64	-	-	-	Kumar and Saxena (2012)
			Ethnobotanical Study	74	42	69	-	-	-	Ahamed *et al.* (2010)
25	Sambhal	-	Information not found							
26	Shahjahanpur	4,575	Floristic Study	151	45	91	-	-	-	Misra (2015)#
			Ethnobotanical Study	70	32	-	-	-	-	Sharma *et al.* (2010)
27	Shamli	1,063	Information not found							

- Study included Bareilly, Lakhimpur-Kheri, Pilibhit and Shahjahanpur District; $ - Study also included Auraiya district of Uttar Pradesh; Pterid.-Pteridophytes; w+c - wild and cultivated

Appendix 2. Floristic and ethnobotanical studies undertaken in the districts under National Capital Region (NCR)

SN	District	Total Area (km^2)	Type of Study	No. of plant species recorded	No. of Families	No. of Genera	Plant Groups Represented			Reference
							Angiosperms		Pterid.	
							Dicot	Monocot		
National Capital Region -58,332 km²										
1	National Capital Territory (NCT) of Delhi	1,483	Floristic Study	604 (w+c)	92	399	-	-	-	Mishra *et al.* (2015a)
				8	7	8	7	1	-	Mishra *et al.* (2015b)
				3	-	-	3	-	-	Mishra *et al.* (2015c)
				102	33	69	-	-	-	Mishra *et al.* (2015d)
				192	46	-	-	-	-	Manral *et al.* (2013)
				531	-	-	-	-	-	Maheshwari (1963)
				278	66	206	-	-	-	Maheshwari (1966)
				6	4	6	6	-	-	Mishra *et al.* (2014)
				252	-	-	-	-	-	Krishen (2006)
				37	-	-	-	-	-	Naithani *et al.* (2006)
				460	85	326	393	67	-	Mishra *et al.* (2015e)
Uttar Pradesh - 243,290 km² (14,858 km²)										
2	Baghpat*	1,321								
3	Bulandshahr*	4,353								
4	Gautam Buddha Nagar*	1,442								
5	Ghaziabad*	1,273								
6	Hapur*	660								
7	Meerut*	2,522								
8	Muzaffarnagar*	3,008								
Haryana - 44,212 km² (28,545 km²)										
9	Bhiwani	5,140	Floristic Study	Information not found						
			Ethnobotanical Study	30	20	26	26	4	-	Singh (2016)
				54	37	-	-	-	-	Yadav *et al.* (2014)
10	Faridabad	2,151	Information not found							

SN	District	Total Area (km^2)	Type of Study	No. of plant species recorded	No. of Families	No. of Genera	Plant Groups Represented			Reference
							Angiosperms		Pterid.	
							Dicot	Monocot		
11	Gurgaon	1,253	Floristic Study	151	54		129	-	-	Debta and Srivastava (2015)
			Ethnobotanical Study	48	-	-	-	-	-	Sharma *et al.* (1992)
12	Jhajjar	1,834	Floristic Study	731	114	459	611	120	-	Singh (2009)
				84	41	-	-	-	-	Kumar and Dhankhar (2015)
			Ethnobotanical Study	57	35	51	-	-	-	Panghal *et al.* (2010)
13	Jind	2,702	Floristic Study	282	76	-	-	-	-	Singh and Kumar (2013)
			Ethnobotanical Study	72	35	-	-	-	-	Singh and Singh (2014)[#]
14	Karnal	2,520	Floristic Study	277	72	-	-	-	-	Kumar and Singh (2013)
				345	77	245	309	36	-	Kaur *et al.* (2016)
			Ethnobotanical Study	72	35	-	-	-	-	Singh and Singh (2014)[#]
				71	38	67	-	-	-	Kaur and Vashistha (2014)
15	Mahendragarh,	1,899	Floristic Study	50	-	-	-	-	-	Yadav *et al.* (2010)
				52	25	-	-	-	-	Yadav *et al.* (2012)
			Ethnobotanical Study	56	33	-	-	-	-	Yadav and Bhandoria (2013)
16	Nuh,	1,860	Information not found							
17	Palwal	1,359	Information not found							
18	Panipat	1,268	Floristic Study	Information not found						
			Ethnobotanical Study	72	35	-	-	-	-	Singh and Singh (2014)[#]
19	Riwari	1,594	Information not found							
20	Rohtak	1,668	Floristic Study	Information not found						
			Ethnobotanical Study	2	2	2	2	-	-	Kaur *et al.* (2015)

SN	District	Total Area (km²)	Type of Study	No. of plant species recorded	No. of Families	No. of Genera	Plant Groups Represented			Reference
							Angiosperms		Pterid.	
							Dicot	Monocot		
21	Sonipat	2,260	Floristic Study	Information not found						
			Ethnobotanical Study	72	35	-	-	-	-	Singh and Singh (2014)[#]
Rajasthan - 342,239 km² (13,446 km²)										
22	Alwar	8,380	Floristic Study	20	-		-	-	-	Yadav (2005)
				403	86	271	308	90	4	Dular (2015)
			Ethnobotanical Study	99	-	-	-	-	-	Dashahre *et al.* (2014)
				110	43	88	-	-	-	Jain *et al.* (2009)
23	Bharatpur	5,066	Floristic Study	323	77	237	-	-	-	Kumar and Chauhan (2011)
				372	-	-	-	-	-	Prasad *et al.* (1997)
			Ethnobotanical Study	Information not found						

* Information included under appendix 1 (Floristic and ethnobotanical studies undertaken in districts of Western Uttar Pradesh); # - Study included four districts of Haryana namely Karnal, Panipat, Sonipat and Jind; Pterid.-Pteridophytes; w+c - wild and cultivated

Appendix 3: Checklist of wild plants recorded in Chithara Village Panchayat including SNU Campus

S.N.	Name of plant species	Family	Types	Habit
1	*Abrus precatorius* L.*	Leguminosae	Dicot	CL/CR
2	*Abutilon indicum* (L.) Sweet*	Malvaceae	Dicot	SH
3	*Acacia nilotica* (L.) Delile	Leguminosae	Dicot	T
4	*Achyranthes aspera* L.*	Amaranthaceae	Dicot	H
5	*Acmella paniculata* (Wall. ex DC.) R.K.Jansen	Compositae	Dicot	H
6	*Ageratum conyzoides* (L.) L.*	Compositae	Dicot	H
7	*Ageratum houstonianum* Mill.*	Compositae	Dicot	H
8	*Albizia lebbeck* (L.) Benth.	Leguminosae	Dicot	T
9	*Alhagi maurorum* Medik.	Leguminosae	Dicot	US
10	*Alternanthera paronychioides* A.St.-Hil.*	Amaranthaceae	Dicot	H
11	*Alternanthera philoxeroides* (Mart.) Griseb.*	Amaranthaceae	Dicot	H
12	*Alternanthera pungens* Kunth*	Amaranthaceae	Dicot	H
13	*Alternanthera sessilis* (L.) R.Br. ex DC.	Amaranthaceae	Dicot	H
14	*Alysicarpus monilifer* (L.) DC.*	Leguminosae	Dicot	H
15	*Alysicarpus vaginalis* (L.) DC.*	Leguminosae	Dicot	H
16	*Amaranthus viridis* L.*	Amaranthaceae	Dicot	H
17	*Ammannia baccifera* L.	Lythraceae	Dicot	H
18	*Ampelopteris prolifera* (Retz.) Copel.	Thelypteridaceae	Pteridophyte	P
19	*Anagallis arvensis* L.*	Primulaceae	Dicot	H
20	*Anisomeles indica* (L.) Kuntze*	Lamiaceae	Dicot	US
21	*Arabidopsis thaliana* (L.) Heynh.*	Brassicaceae	Dicot	H
22	*Arenaria serpyllifolia* L.*	Caryophyllaceae	Dicot	H
23	*Argemone mexicana* L.*	Papaveraceae	Dicot	H
24	*Artemisia scoparia* Waldst. & Kitam.*	Compositae	Dicot	US
25	*Avena sterilis* L.*	Poaceae	Monocot	GR
26	*Azadirachta indica* A.Juss.*	Meliaceae	Dicot	T
27	*Bacopa monnieri* (L.) Wettst.*	Plantaginaceae	Dicot	H
28	*Barleria prionitis* L.	Acanthaceae	Dicot	US
29	*Bidens pilosa* L.*	Compositae	Dicot	H
30	*Blainvillea acmella* (L.) Philipson*	Compositae	Dicot	H
31	*Blumea obliqua* (L.) Druce*	Compositae	Dicot	H
32	*Blumea viscosa* (Mill.) V.M.Badillo*	Compositae	Dicot	H
33	*Blumea laciniata* (Wall. ex Roxb.) DC.	Compositae	Dicot	H
34	*Boerhavia diffusa* L.*	Nyctaginaceae	Dicot	H
35	*Brachiaria reptans* (L.) C.A.Gardner & C.E.Hubb.*	Poaceae	Monocot	GR
36	*Brachiaria ramosa* (L.) Stapf	Poaceae	Monocot	GR
37	*Butea monosperma* (Lam.) Taub.*	Leguminosae	Dicot	T
38	*Caesulia axillaris* Roxb.	Compositae	Dicot	H
39	*Cajanus scarabaeoides* (L.) Thouars*	Leguminosae	Dicot	CL/CR
40	*Calotropis gigantea* (L.) Dryand.*	Apocynaceae	Dicot	SH
41	*Calotropis procera* (Aiton) Dryand.*	Apocynaceae	Dicot	US
42	*Cannabis sativa* L.*	Cannabaceae	Dicot	H
43	*Capparis sepiaria* L.*	Capparaceae	Dicot	SH
44	*Capsella bursa-pastoris* (L.) Medik.*	Brassicaceae	Dicot	H

S.N.	Name of plant species	Family	Types	Habit
45	*Cardamine hirsuta* L.*	Brassicaceae	Dicot	H
46	*Carthamus oxyacantha* M.Bieb.	Compositae	Dicot	H
47	*Cayratia trifolia* (L.) Domin*	Vitaceae	Dicot	CL/CR
48	*Celosia argentea* L.*	Amaranthaceae	Dicot	H
49	*Cenchrus ciliaris* L.*	Poaceae	Monocot	GR
50	*Centaurium pulchellum* (Sw.) Druce*	Gentianaceae	Dicot	H
51	*Centella asiatica* (L.) Urb.*	Apiaceae	Dicot	H
52	*Ceratophyllum demersum* L.*	Ceratophyllaceae	Dicot	H
53	*Chenopodium album* L.*	Amaranthaceae	Dicot	H
54	*Chenopodium murale* L.*	Amaranthaceae	Dicot	H
55	*Chloris barbata* Sw.	Poaceae	Monocot	GR
56	*Chrysopogon zizanioides* (L.) Roberty*	Poaceae	Monocot	GR
57	*Cirsium arvense* (L.) Scop.*	Compositae	Dicot	H
58	*Cissampelos pareira* L.*	Menispermaceae	Dicot	CL/CR
59	*Citrullus colocynthis* (L.) Schrad.*	Cucurbitaceae	Dicot	CL/CR
60	*Cleome viscosa* L.*	Cleomaceae	Dicot	US
61	*Clerodendrum phlomidis* L.f.	Verbenaceae	Dicot	SH
62	*Coccinia grandis* (L.) Voigt*	Cucurbitaceae	Dicot	CL/CR
63	*Commelina benghalensis* L.*	Commelinaceae	Monocot	H
64	*Commelina forsskalii* Vahl*	Commelinaceae	Monocot	H
65	*Convolvulus arvensis* L.*	Convolvulaceae	Dicot	CL/CR
66	*Convolvulus prostratus* Forssk.*	Convolvulaceae	Dicot	H
67	*Corchorus aestuans* L.*	Malvaceae	Dicot	H
68	*Corchorus olitorius* L.*	Malvaceae	Dicot	H
69	*Corchorus tridens* L.*	Malvaceae	Dicot	H
70	*Cordia dichotoma* G.Forst.*	Boraginaceae	Dicot	T
71	*Crateva adansonii* subsp. *odora* (Buch.-Ham.) Jacobs*	Capparaceae	Dicot	T
72	*Crotalaria medicaginea* Lam.*	Leguminosae	Dicot	H
73	*Crotalaria mysorensis* Roth*	Leguminosae	Dicot	H
74	*Croton bonplandianus* Baill.*	Euphorbiaceae	Dicot	H
75	*Cucumis melo* L. var. *agrestis* Naudin*	Cucurbitaceae	Dicot	CL/CR
76	*Cuscuta chinensis* Lam.*	Convolvulaceae	Dicot	CL/CR
77	*Cuscuta reflexa* Roxb.	Convolvulaceae	Dicot	CL/CR
78	*Cyanotis axillaris* (L.) D.Don ex Sweet*	Commelinaceae	Monocot	H
79	*Cyanthillium cinereum* (L.) H.Rob.*	Compositae	Dicot	H
80	*Cynodon dactylon* (L.) Pers.*	Poaceae	Monocot	GR
81	*Cyperus alopecuroides* Rottb.*	Cyperaceae	Monocot	SE
82	*Cyperus compressus* L.*	Cyperaceae	Monocot	SE
83	*Cyperus difformis* L.*	Cyperaceae	Monocot	SE
84	*Cyperus iria* L.	Cyperaceae	Monocot	SE
85	*Cyperus rotundus* L.*	Cyperaceae	Monocot	SE
86	*Dactyloctenium aegyptium* (L.) Willd.*	Poaceae	Monocot	GR
87	*Dalbergia sissoo* DC.*	Leguminosae	Dicot	T
88	*Datura innoxia* Mill.*	Solanaceae	Dicot	H
89	*Desmodium gangeticum* (L.) DC.*	Leguminosae	Dicot	US
90	*Desmodium triflorum* (L.) DC.*	Leguminosae	Dicot	H

S.N.	Name of plant species	Family	Types	Habit
91	*Desmostachya bipinnata* (L.) Stapf*	Poaceae	Monocot	GR
92	*Dichanthium annulatum* (Forssk.) Stapf*	Poaceae	Monocot	GR
93	*Digera muricata* (L.) Mart.*	Amaranthaceae	Dicot	H
94	*Digitaria ciliaris* (Retz.) Koeler*	Poaceae	Monocot	GR
95	*Diospyros montana* Roxb.*	Ebenaceae	Dicot	T
96	*Dysphania ambrosioides* (L.) Mosyakin & Clemants*	Amaranthaceae	Dicot	US
97	*Echinochloa colona* (L.) Link*	Poaceae	Monocot	GR
98	*Echinochloa crus-galli* (L.) P.Beauv.	Poaceae	Monocot	GR
99	*Eclipta prostrata* (L.) L.*	Compositae	Dicot	H
100	*Eichhornia crassipes* (Mart.) Solms	Pontederiaceae	Monocot	H
101	*Eleusine indica* (L.) Gaertn.	Poaceae	Monocot	GR
102	*Emex australis* Steinh.*	Polygonaceae	Dicot	H
103	*Equisetum ramosissimum* Desf. *	Equisetaceae	Pteridophyte	P
104	*Equisetum ramosissimum* subsp. *debile* (Roxb. ex Vaucher) Hauke*	Equisetaceae	Pteridophyte	P
105	*Eragrostis amabilis* (L.) Wight & Arn.	Poaceae	Monocot	GR
106	*Eragrostis japonica* (Thunb.) Trin.	Poaceae	Monocot	GR
107	*Erigeron bonariensis* L.*	Compositae	Dicot	H
108	*Eriochloa procera* (Retz.) C.E.Hubb.*	Poaceae	Monocot	GR
109	*Euphorbia heterophylla* L.*	Euphorbiaceae	Dicot	H
110	*Euphorbia hirta* L.*	Euphorbiaceae	Dicot	H
111	*Euphorbia hypericifolia* L.*	Euphorbiaceae	Dicot	H
112	*Euphorbia prostrata* Aiton*	Euphorbiaceae	Dicot	H
113	*Euphorbia serpens* Kunth*	Euphorbiaceae	Dicot	H
114	*Evolvulus nummularius* (L.) L.*	Convolvulaceae	Dicot	H
115	*Ficus palmata* Forssk.	Moraceae	Dicot	SH
116	*Ficus religiosa* L.*	Moraceae	Dicot	T
117	*Fimbristylis dichotoma* (L.) Vahl*	Cyperaceae	Monocot	SE
118	*Fimbristylis ferruginea* (L.) Vahl	Cyperaceae	Monocot	SE
119	*Fumaria indica* (Hausskn.) Pugsley*	Papaveraceae	Dicot	H
120	*Gnaphalium purpureum* L.*	Compositae	Dicot	H
121	*Gomphrena celosioides* Mart.*	Amaranthaceae	Dicot	H
122	*Gonostegia pentandra* (Roxb.) Miq.	Urticaceae	Dicot	H
123	*Grangea maderaspatana* (L.) Poir.*	Compositae	Dicot	H
124	*Heliotropium ellipticum* Ledeb.*	Boraginaceae	Dicot	H
125	*Hemarthria compressa* (L.f.) R.Br.*	Poaceae	Monocot	GR
126	*Holoptelea integrifolia* Planch.*	Ulmaceae	Dicot	T
127	*Hydrilla verticillata* (L.f.) Royle*	Hydrocharitaceae	Monocot	H
128	*Hydrocotyle sibthorpioides* Lam.*	Araliaceae	Dicot	H
129	*Hyptis suaveolens* (L.) Poit.*	Lamiaceae	Dicot	US
130	*Imperata cylindrica* (L.) Raeusch.*	Poaceae	Monocot	GR
131	*Indigofera astragalina* DC.*	Leguminosae	Dicot	US
132	*Indigofera linifolia* (L.f.) Retz.*	Leguminosae	Dicot	H
133	*Indigofera linnaei* Ali	Leguminosae	Dicot	H
134	*Indigofera tinctoria* L.*	Leguminosae	Dicot	US
135	*Ipomoea aquatica* Forssk.*	Convolvulaceae	Dicot	CL/CR
136	*Ipomoea carnea* Jacq.*	Convolvulaceae	Dicot	SH

S.N.	Name of plant species	Family	Types	Habit
137	*Ipomoea coptica* (L.) Roth ex Roem. & Schult.*	Convolvulaceae	Dicot	CL/CR
138	*Ipomoea eriocarpa* R. Br.*	Convolvulaceae	Dicot	CL/CR
139	*Ipomoea obscura* (L.) Ker Gawl.*	Convolvulaceae	Dicot	CL/CR
140	*Ipomoea pes-tigridis* L.*	Convolvulaceae	Dicot	CL/CR
141	*Ipomoea triloba* L.*	Convolvulaceae	Dicot	CL/CR
142	*Jatropha gossypiifolia* L.	Euphorbiaceae	Dicot	SH
143	*Justicia adhatoda* L.*	Acanthaceae	Dicot	SH
144	*Justicia japonica* Thunb.*	Acanthaceae	Dicot	H
145	*Lantana camara* L.*	Verbenaceae	Dicot	SH
146	*Lathyrus aphaca* L.*	Leguminosae	Dicot	H
147	*Launaea procumbens* (Roxb.) Ramayya & Rajagopal*	Compositae	Dicot	H
148	*Lemna perpusilla* Torr.*	Araceae	Monocot	H
149	*Lepidium didymum* L.*	Brassicaceae	Dicot	H
150	*Leptochloa panicea* (Retz.) Ohwi	Poaceae	Monocot	GR
151	*Leucaena leucocephala* (Lam.) de Wit*	Leguminosae	Dicot	T
152	*Leucas cephalotes* (Roth) Spreng.*	Lamiaceae	Dicot	H
153	*Lindernia ciliata* (Colsm.) Pennell*	Linderniaceae	Dicot	H
154	*Lindernia crustacea* (L.) F.Muell.*	Linderniaceae	Dicot	H
155	*Ludwigia hyssopifolia* (G.Don) Exell	Onagraceae	Dicot	H
156	*Ludwigia octovalvis* (Jacq.) P.H.Raven*	Onagraceae	Dicot	H
157	*Malva parviflora* L.*	Malvaceae	Dicot	H
158	*Malvastrum coromandelianum* (L.) Garcke*	Malvaceae	Dicot	H
159	*Marsilea quadrifolia* L.*	Marsileaceae	Pteridophyte	P
160	*Mazus pumilus* (Burm.f.) Steenis*	Phrymaceae	Dicot	H
161	*Mecardonia procumbens* (Mill.) Small*	Plantaginaceae	Dicot	H
162	*Medicago monantha* (C.A.Mey.) Trautv.*	Leguminosae	Dicot	H
163	*Medicago polymorpha* L.*	Leguminosae	Dicot	H
164	*Medicago lupulina* L.	Leguminosae	Dicot	H
165	*Melia azedarach* L.	Meliaceae	Dicot	T
166	*Melilotus albus* Medik.*	Leguminosae	Dicot	H
167	*Melilotus indicus* (L.) All.*	Leguminosae	Dicot	H
168	*Melochia corchorifolia* L.*	Malvaceae	Dicot	H
169	*Merremia hederacea* (Burm. f.) Hallier f.	Convolvulaceae	Dicot	CL/CR
170	*Mimosa rubicaulis* Lam.	Leguminosae	Dicot	SH
171	*Mitragyna parvifolia* (Roxb.) Korth.*	Rubiaceae	Dicot	T
172	*Mollugo nudicaulis* Lam.*	Molluginaceae	Dicot	H
173	*Morus alba* L.*	Moraceae	Dicot	T
174	*Mukia maderaspatana* (L.) M.Roem.*	Cucurbitaceae	Dicot	CL/CR
175	*Murdannia nudiflora* (L.) Brenan*	Commelinaceae	Monocot	H
176	*Nepeta hindostana* (B.Heyne ex Roth) Haines	Lamiaceae	Dicot	H
177	*Nicotiana plumbaginifolia* Viv.*	Solanaceae	Dicot	H
178	*Nymphaea pubescens* Willd.*	Nymphaeaceae	Dicot	H
179	*Nymphoides cristata* (Roxb.) Kuntze*	Menyanthaceae	Dicot	H
180	*Oldenlandia corymbosa* L.*	Rubiaceae	Dicot	H
181	*Operculina turpethum* (L.) Silva Manso	Convolvulaceae	Dicot	CL/CR
182	*Oxalis corniculata* L.*	Oxalidaceae	Dicot	H

S.N.	Name of plant species	Family	Types	Habit
183	*Oxalis debilis* var. *corymbosa* (DC.) Lourteig*	Oxalidaceae	Dicot	H
184	*Oxystelma esculentum* (L. f.) Sm.*	Apocynaceae	Dicot	CL/CR
185	*Panicum antidotale* Retz.*	Poaceae	Monocot	GR
186	*Parthenium hysterophorus* L.*	Compositae	Dicot	H
187	*Paspalum distichum* L.	Poaceae	Monocot	GR
188	*Paspalum scrobiculatum* L.*	Poaceae	Monocot	GR
189	*Peristrophe bicalyculata* (Retz.) Nees*	Acanthaceae	Dicot	H
190	*Persicaria barbata* (L.) H.Hara*	Polygonaceae	Dicot	H
191	*Persicaria lanigera* (R.Br.) Soják*	Polygonaceae	Dicot	H
192	*Phalaris minor* Retz.*	Poaceae	Monocot	GR
193	*Phoenix sylvestris* (L.) Roxb.*	Arecaceae	Monocot	T
194	*Phragmites karka* (Retz.) Trin. ex Steud.*	Poaceae	Monocot	GR
195	*Phyla nodiflora* (L.) Greene*	Verbenaceae	Dicot	H
196	*Phyllanthus amarus* Schumach. & Thonn.*	Phyllanthaceae	Dicot	H
197	*Phyllanthus maderaspatensis* L.*	Phyllanthaceae	Dicot	H
198	*Phyllanthus reticulatus* Poir.*	Phyllanthaceae	Dicot	SH
199	*Phyllanthus tenellus* Roxb.	Phyllanthaceae	Dicot	H
200	*Physalis angulata* L.*	Solanaceae	Dicot	H
201	*Physalis peruviana* L.*	Solanaceae	Dicot	H
202	*Pistia stratiotes* L.*	Araceae	Monocot	H
203	*Pluchea lanceolata* (DC.) C.B.Clarke*	Compositae	Dicot	US
204	*Poa annua* L.*	Poaceae	Monocot	GR
205	*Polygonum plebeium* R.Br.*	Polygonaceae	Dicot	H
206	*Portulaca oleracea* L.*	Portulacaceae	Dicot	H
207	*Portulaca quadrifida* L.*	Portulacaceae	Dicot	H
208	*Potamogeton crispus* L.*	Potamogetonaceae	Monocot	H
209	*Prosopis cineraria* (L.) Druce*	Leguminosae	Dicot	T
210	*Prosopis juliflora* (Sw.) DC.*	Leguminosae	Dicot	T
211	*Ranunculus sceleratus* L.*	Ranunculaceae	Dicot	H
212	*Rhynchosia capitata* (Roth) DC.*	Leguminosae	Dicot	CL/CR
213	*Rhynchosia minima* (L.) DC.*	Leguminosae	Dicot	CL/CR
214	*Ricinus communis* L.*	Euphorbiaceae	Dicot	SH
215	*Rorippa palustris* (L.) Besser*	Brassicaceae	Dicot	H
216	*Rumex dentatus* L.*	Polygonaceae	Dicot	H
217	*Rungia pectinata* (L.) Nees	Acanthaceae	Dicot	H
218	*Saccharum bengalense* Retz.*	Poaceae	Monocot	GR
219	*Saccharum spontaneum* L.*	Poaceae	Monocot	GR
220	*Sagittaria guayanensis* Kunth*	Alismataceae	Monocot	H
221	*Salix tetrasperma* Roxb.	Salicaceae	Dicot	T
222	*Scoparia dulcis* L.*	Plantaginaceae	Dicot	US
223	*Senna obtusifolia* (L.) H.S.Irwin & Barneby	Leguminosae	Dicot	H
224	*Senna occidentalis* (L.) Link*	Leguminosae	Dicot	US
225	*Sesamum indicum* L.*	Pedaliaceae	Dicot	H
226	*Sesbania bispinosa* (Jacq.) W.Wight*	Leguminosae	Dicot	SH
227	*Sesbania sesban* (L.) Merr.*	Leguminosae	Dicot	T
228	*Setaria pumila* (Poir.) Roem. & Schult.*	Poaceae	Monocot	GR

S.N.	Name of plant species	Family	Types	Habit
229	*Setaria verticillata* (L.) P.Beauv.*	Poaceae	Monocot	GR
230	*Sida acuta* Burm.f.*	Malvaceae	Dicot	US
231	*Sida cordata* (Burm.f.) Borss.Waalk.*	Malvaceae	Dicot	H
232	*Sida cordifolia* L.*	Malvaceae	Dicot	US
233	*Silene conoidea* L.*	Caryophyllaceae	Dicot	H
234	*Sisymbrium irio* L.*	Brassicaceae	Dicot	H
235	*Solanum americanum* Mill.*	Solanaceae	Dicot	H
236	*Solanum surattense* Burm. f.*	Solanaceae	Dicot	H
237	*Solanum torvum* Sw.	Solanaceae	Dicot	SH
238	*Sonchus asper* (L.) Hill*	Compositae	Dicot	H
239	*Sonchus brachyotus* DC.*	Compositae	Dicot	H
240	*Sorghum halepense* (L.) Pers.*	Poaceae	Monocot	GR
241	*Spergula arvensis* L.*	Caryophyllaceae	Dicot	H
242	*Spermacoce articularis* L.f.*	Rubiaceae	Dicot	H
243	*Sphenoclea zeylanica* Gaertn.	Sphenocleaceae	Dicot	H
244	*Spirodela polyrrhiza* (L.) Schleid.*	Araceae	Monocot	H
245	*Stellaria media* (L.) Vill.*	Caryophyllaceae	Dicot	H
246	*Streblus asper* Lour.*	Moraceae	Dicot	T
247	*Stuckenia pectinata* (L.) Börner*	Potamogetonaceae	Monocot	H
248	*Syzygium cumini* (L.) Skeels*	Myrtaceae	Dicot	T
249	*Tamarix indica* Willd.*	Tamaricaceae	Dicot	SH
250	*Tephrosia pumila* (Lam.) Pers.*	Leguminosae	Dicot	H
251	*Tephrosia purpurea* (L.) Pers.*	Leguminosae	Dicot	US
252	*Tephrosia villosa* (L.) Pers.*	Leguminosae	Dicot	US
253	*Teramnus labialis* (L.f.) Spreng.	Leguminosae	Dicot	CL/CR
254	*Trianthema portulacastrum* L.*	Aizoaceae	Dicot	H
255	*Tribulus terrestris* L.*	Zygophyllaceae	Dicot	H
256	*Trichosanthes cucumerina* L.*	Cucurbitaceae	Dicot	CL/CR
257	*Tridax procumbens* (L.) L.*	Compositae	Dicot	H
258	*Trifolium tomentosum* L.*	Leguminosae	Dicot	H
259	*Triumfetta rhomboidea* Jacq.*	Malvaceae	Dicot	US
260	*Typha domingensis* Pers.*	Typhaceae	Monocot	H
261	*Urena lobata* L.*	Malvaceae	Dicot	SH
262	*Utricularia stellaris* L.f.	Lentibulariaceae	Dicot	H
263	*Vallisneria spiralis* L.	Hydrocharitaceae	Monocot	H
264	*Verbascum chinense* (L.) Santapau	Scrophulariaceae	Dicot	H
265	*Verbascum thapsus* L.	Scrophulariaceae	Dicot	H
266	*Veronica agrestis* L.*	Plantaginaceae	Dicot	H
267	*Veronica anagallis-aquatica* L.*	Plantaginaceae	Dicot	H
268	*Vicia sativa* L.*	Leguminosae	Dicot	H
269	*Withania somnifera* (L.) Dunal*	Solanaceae	Dicot	US
270	*Xanthium strumarium* L.*	Compositae	Dicot	H
271	*Ziziphus jujuba* Mill.*	Rhamnaceae	Dicot	T
272	*Ziziphus nummularia* (Burm.f.)Wight & Arn.*	Rhamnaceae	Dicot	SH

* also reported from SNU Campus

(H-Herb; SH-Shrub; US-Undershrub; T-Tree; CL/CR-Climber/Creeper; GR-Grass, SE-Sedge, P-Pteridophytic herb)

Appendix 4: Checklist of wild plants recorded in different seasons

SN	Name of plant species	Family	Seasonal vegetation				Permanent Vegetation (Shrubs and Trees)
			Winter	Summer	Rainy	Throughout year	
1	*Abrus precatorius* L.	Leguminosae					√
2	*Abutilon indicum* (L.) Sweet	Malvaceae				√	
3	*Acacia nilotica* (L.) Delile	Leguminosae					√
4	*Achyranthes aspera* L.	Amaranthaceae				√	
5	*Acmella paniculata* (Wall. ex DC.) R.K.Jansen	Compositae			√		
6	*Ageratum conyzoides* (L.) L.	Compositae	√				
7	*Ageratum houstonianum* Mill.	Compositae	√				
8	*Albizia lebbeck* (L.) Benth.	Leguminosae					√
9	*Alhagi maurorum* Medik.	Leguminosae		√			
10	*Alternanthera paronychioides* A.St.-Hil.	Amaranthaceae			√		
11	*Alternanthera philoxeroides* (Mart.) Griseb.	Amaranthaceae				√	
12	*Alternanthera pungens* Kunth	Amaranthaceae		√			
13	*Alternanthera sessilis* (L.) R.Br. ex DC.	Amaranthaceae				√	
14	*Alysicarpus monilifer* (L.) DC.	Leguminosae			√		
15	*Alysicarpus vaginalis* (L.) DC.	Leguminosae			√		
16	*Amaranthus viridis* L.	Amaranthaceae			√		
17	*Ammannia baccifera* L.	Lythraceae			√		
18	*Ampelopteris prolifera* (Retz.) Copel.	Thelypteridaceae			√		
19	*Anagallis arvensis* L.	Primulaceae	√				
20	*Anisomeles indica* (L.) Kuntze	Lamiaceae			√		
21	*Arabidopsis thaliana* (L.) Heynh.	Brassicaceae	√				
22	*Arenaria serpyllifolia* L.	Caryophyllaceae	√				
23	*Argemone mexicana* L.	Papaveraceae	√				
24	*Artemisia scoparia* Waldst. & Kitam.	Compositae		√	√		
25	*Avena sterilis* L.	Poaceae	√				
26	*Azadirachta indica* A.Juss.	Meliaceae					√
27	*Bacopa monnieri* (L.) Wettst.	Plantaginaceae				√	
28	*Barleria prionitis* L.	Acanthaceae				√	
29	*Bidens pilosa* L.	Compositae		√			
30	*Blainvillea acmella* (L.) Philipson	Compositae			√		
31	*Blumea obliqua* (L.) Druce	Compositae	√				
32	*Blumea viscosa* (Mill.) V.M.Badillo	Compositae	√				
33	*Blumea laciniata* (Wall. ex Roxb.) DC.	Compositae	√				
34	*Boerhavia diffusa* L.	Nyctaginaceae				√	
35	*Brachiaria reptans* (L.) C.A.Gardner & C.E.Hubb.	Poaceae			√		
36	*Brachiaria ramosa* (L.) Stapf	Poaceae			√		
37	*Butea monosperma* (Lam.) Taub.	Leguminosae					√
38	*Caesulia axillaris* Roxb.	Compositae			√		
39	*Cajanus scarabaeoides* (L.) Thouars	Leguminosae			√		
40	*Calotropis gigantea* (L.) Dryand.	Apocynaceae					√

SN	Name of plant species	Family	Seasonal vegetation				Permanent Vegetation (Shrubs and Trees)
			Winter	Summer	Rainy	Throughout year	
41	*Calotropis procera* (Aiton) Dryand.	Apocynaceae				√	
42	*Cannabis sativa* L.	Cannabinaceae				√	
43	*Capparis sepiaria* L.	Capparaceae					√
44	*Capsella bursa-pastoris* (L.) Medik.	Brassicaceae	√				
45	*Cardamine hirsuta* L.	Brassicaceae	√				
46	*Carthamus oxyacantha* M.Bieb.	Compositae		√			
47	*Cayratia trifolia* (L.) Domin	Vitaceae			√		
48	*Celosia argentea* L.	Amaranthaceae			√		
49	*Cenchrus ciliaris* L.	Poaceae			√		
50	*Centaurium pulchellum* (Sw.) Druce	Gentianaceae	√				
51	*Centella asiatica* (L.) Urb.	Apiaceae		√			
52	*Ceratophyllum demersum* L.	Ceratophyllaceae				√	
53	*Chenopodium album* L.	Amaranthaceae	√				
54	*Chenopodium murale* L.	Amaranthaceae	√				
55	*Chloris barbata* Sw.	Poaceae			√		
56	*Chrysopogon zizanioides* (L.) Roberty	Poaceae				√	
57	*Cirsium arvense* (L.) Scop.	Compositae	√				
58	*Cissampelos pareira* L.	Menispermaceae			√		
59	*Citrullus colocynthis* (L.) Schrad.	Cucurbitaceae		√			
60	*Cleome viscosa* L.	Cleomaceae			√		
61	*Clerodendrum phlomidis* L.f.	Verbenaceae					√
62	*Coccinia grandis* (L.) Voigt	Cucurbitaceae			√		
63	*Commelina benghalensis* L.	Commelinaceae			√		
64	*Commelina forsskalii* Vahl	Commelinaceae			√		
65	*Convolvulus arvensis* L.	Convolvulaceae			√		
66	*Convolvulus prostratus* Forssk.	Convolvulaceae			√		
67	*Corchorus aestuans* L.	Malvaceae			√		
68	*Corchorus olitorius* L.	Malvaceae			√		
69	*Corchorus tridens* L.	Malvaceae			√		
70	*Cordia dichotoma* G.Forst.	Boraginaceae					√
71	*Crateva adansonii* subsp. *odora* (Buch.-Ham.) Jacobs	Capparaceae					√
72	*Crotalaria medicaginea* Lam.	Leguminosae			√		
73	*Crotalaria mysorensis* Roth	Leguminosae			√		
74	*Croton bonplandianus* Baill.	Euphorbiaceae				√	
75	*Cucumis melo* var. *agrestis* Naudin	Cucurbitaceae			√		
76	*Cuscuta chinensis* Lam.	Convolvulaceae				√	
77	*Cuscuta reflexa* Roxb.	Convolvulaceae				√	
78	*Cyanotis axillaris* (L.) D.Don ex Sweet	Commelinaceae			√		
79	*Cyanthillium cinereum* (L.) H.Rob.	Compositae				√	
80	*Cynodon dactylon* (L.) Pers.	Poaceae				√	
81	*Cyperus alopecuroides* Rottb.	Cyperaceae			√		
82	*Cyperus compressus* L.	Cyperaceae			√		
83	*Cyperus difformis* L.	Cyperaceae			√		

SN	Name of plant species	Family	Seasonal vegetation				Permanent Vegetation (Shrubs and Trees)
			Winter	Summer	Rainy	Throughout year	
84	*Cyperus iria* L.	Cyperaceae			√		
85	*Cyperus rotundus* L.	Cyperaceae			√		
86	*Dactyloctenium aegyptium* (L.) Willd.	Poaceae			√		
87	*Dalbergia sissoo* DC.	Leguminosae					√
88	*Datura innoxia* Mill.	Solanaceae				√	
89	*Desmodium gangeticum* (L.) DC.	Leguminosae			√		
90	*Desmodium triflorum* (L.) DC.	Leguminosae			√		
91	*Desmostachya bipinnata* (L.) Stapf	Poaceae			√		
92	*Dichanthium annulatum* (Forssk.) Stapf	Poaceae			√		
93	*Digera muricata* (L.) Mart.	Amaranthaceae			√		
94	*Digitaria ciliaris* (Retz.) Koeler	Poaceae			√		
95	*Diospyros montana Roxb.*	Ebenaceae					√
96	*Dysphania ambrosioides* (L.) Mosyakin & Clemants	Amaranthaceae			√		
97	*Echinochloa colona* (L.) Link	Poaceae			√		
98	*Echinochloa crus-galli* (L.) P.Beauv.	Poaceae			√		
99	*Eclipta prostrata* (L.) L.	Compositae				√	
100	*Eichhornia crassipes* (Mart.) Solms	Pontederiaceae					√
101	*Eleusine indica* (L.) Gaertn.	Poaceae			√		
102	*Emex australis* Steinh.	Polygonaceae	√	√			
103	*Equisetum ramosissimum* Desf.	Equisetaceae				√	
104	*Equisetum ramosissimum* subsp. *debile* (Roxb. ex Vaucher) Hauke	Equisetaceae				√	
105	*Eragrostis amabilis* (L.) Wight & Arn.	Poaceae			√		
106	*Eragrostis japonica* (Thunb.) Trin.	Poaceae			√		
107	*Erigeron bonariensis* L.	Compositae		√			
108	*Eriochloa procera* (Retz.) C.E.Hubb.	Poaceae			√		
109	*Euphorbia heterophylla* L.	Euphorbiaceae			√		
110	*Euphorbia hirta* L.	Euphorbiaceae				√	
111	*Euphorbia hypericifolia* L.	Euphorbiaceae			√		
112	*Euphorbia prostrata* Aiton	Euphorbiaceae			√		
113	*Euphorbia serpens* Kunth	Euphorbiaceae			√		
114	*Evolvulus nummularius* (L.) L.	Convolvulaceae			√		
115	*Ficus palmata* Forssk.	Moraceae					√
116	*Ficus religiosa* L.	Moraceae					√
117	*Fimbristylis dichotoma* (L.) Vahl	Cyperaceae			√		
118	*Fimbristylis ferruginea* (L.) Vahl	Cyperaceae			√		
119	*Fumaria indica* (Hausskn.) Pugsley	Papaveraceae	√				
120	*Gnaphalium purpureum* L.	Compositae	√				
121	*Gomphrena celosioides* Mart.	Amaranthaceae			√		
122	*Gonostegia pentandra* (Roxb.) Miq.	Urticaceae			√		
123	*Grangea maderaspatana* (L.) Poir.	Compositae				√	
124	*Heliotropium ellipticum* Ledeb.	Boraginaceae		√			
125	*Hemarthria compressa* (L.f.) R.Br.	Poaceae			√		

SN	Name of plant species	Family	Seasonal vegetation				Permanent Vegetation (Shrubs and Trees)
			Winter	Summer	Rainy	Throughout year	
126	*Holoptelea integrifolia* Planch.	Ulmaceae					√
127	*Hydrilla verticillata* (L.f.) Royle	Hydrocharitaceae				√	
128	*Hydrocotyle sibthorpioides* Lam.	Araliaceae			√		
129	*Hyptis suaveolens* (L.) Poit.	Lamiaceae			√		
130	*Imperata cylindrica* (L.) Raeusch.	Poaceae			√		
131	*Indigofera astragalina* DC.	Leguminosae			√		
132	*Indigofera linifolia* (L.f.) Retz.	Leguminosae			√		
133	*Indigofera linnaei* Ali	Leguminosae		√			
134	*Indigofera tinctoria* L.	Leguminosae				√	
135	*Ipomoea aquatica* Forssk.	Convolvulaceae			√		
136	*Ipomoea carnea* Jacq.	Convolvulaceae					√
137	*Ipomoea coptica* (L.) Roth ex Roem. & Schult.	Convolvulaceae			√		
138	*Ipomoea eriocarpa* R. Br.	Convolvulaceae			√		
139	*Ipomoea obscura* (L.) Ker Gawl.	Convolvulaceae	√				
140	*Ipomoea pes-tigridis* L.	Convolvulaceae			√		
141	*Ipomoea triloba* L.	Convolvulaceae			√		
142	*Jatropha gossypiifolia* L.	Euphorbiaceae					√
143	*Justicia adhatoda* L.	Acanthaceae					√
144	*Justicia japonica* Thunb.	Acanthaceae			√		
145	*Lantana camara* L.	Verbenaceae					√
146	*Lathyrus aphaca* L.	Leguminosae	√				
147	*Launaea procumbens* (Roxb.) Ramayya & Rajagopal	Compositae	√				
148	*Lemna perpusilla* Torr.	Araceae				√	
149	*Lepidium didymum* L.	Brassicaceae	√				
150	*Leptochloa panicea* (Retz.) Ohwi	Poaceae			√		
151	*Leucaena leucocephala* (Lam.) de Wit	Leguminosae					√
152	*Leucas cephalotes* (Roth) Spreng.	Lamiaceae			√		
153	*Lindernia ciliata* (Colsm.) Pennell	Linderniaceae			√		
154	*Lindernia crustacea* (L.) F.Muell.	Linderniaceae			√		
155	*Ludwigia hyssopifolia* (G.Don) Exell	Onagraceae			√		
156	*Ludwigia octovalvis* (Jacq.) P.H.Raven	Onagraceae			√		
157	*Malva parviflora* L.	Malvaceae	√				
158	*Malvastrum coromandelianum* (L.) Garcke	Malvaceae			√		
159	*Marsilea quadrifolia* L.	Marsileaceae				√	
160	*Mazus pumilus* (Burm.f.) Steenis	Phrymaceae				√	
161	*Mecardonia procumbens* (Mill.) Small	Plantaginaceae			√		
162	*Medicago monantha* (C.A.Mey.) Trautv.	Leguminosae	√				
163	*Medicago polymorpha* L.	Leguminosae	√				
164	*Medicago lupulina* L.	Leguminosae	√				
165	*Melia azedarach* L.	Meliaceae					√
166	*Melilotus albus* Medik.	Leguminosae	√				
167	*Melilotus indicus* (L.) All.	Leguminosae	√				

SN	Name of plant species	Family	Seasonal vegetation				Permanent Vegetation (Shrubs and Trees)
			Winter	Summer	Rainy	Throughout year	
168	*Melochia corchorifolia* L.	Malvaceae			√		
169	*Merremia hederacea* (Burm. f.) Hallier f.	Convolvulaceae			√		
170	*Mimosa rubicaulis* Lam.	Leguminosae					√
171	*Mitragyna parvifolia* (Roxb.) Korth.	Rubiaceae					√
172	*Mollugo nudicaulis* Lam.	Molluginaceae			√		
173	*Morus alba* L.	Moraceae					√
174	*Mukia maderaspatana* (L.) M.Roem.	Cucurbitaceae			√		
175	*Murdannia nudiflora* (L.) Brenan	Commelinaceae			√		
176	*Nepeta hindostana* (B.Heyne ex Roth) Haines	Lamiaceae		√	√		
177	*Nicotiana plumbaginifolia* Viv.	Solanaceae				√	
178	*Nymphaea pubescens* Willd.	Nymphaeaceae			√		
179	*Nymphoides cristata* (Roxb.) Kuntze	Menyanthaceae			√		
180	*Oldenlandia corymbosa* L.	Rubiaceae			√		
181	*Operculina turpethum* (L.) Silva Manso	Convolvulaceae				√	
182	*Oxalis corniculata* L.	Oxalidaceae	√				
183	*Oxalis debilis* var. *corymbosa* (DC.) Lourteig	Oxalidaceae	√				
184	*Oxystelma esculentum* (L. f.) Sm.	Apocynaceae		√			
185	*Panicum antidotale* Retz.	Poaceae			√		
186	*Parthenium hysterophorus* L.	Compositae				√	
187	*Paspalum distichum* L.	Poaceae			√		
188	*Paspalum scrobiculatum* L.	Poaceae			√		
189	*Peristrophe bicalyculata* (Retz.) Nees	Acanthaceae			√		
190	*Persicaria barbata* (L.) H.Hara	Polygonaceae			√		
191	*Persicaria lanigera* (R.Br.) Soják	Polygonaceae			√		
192	*Phalaris minor* Retz.	Poaceae	√				
193	*Phoenix sylvestris* (L.) Roxb.	Arecaceae					√
194	*Phragmites karka* (Retz.) Trin. ex Steud.	Poaceae				√	
195	*Phyla nodiflora* (L.) Greene	Verbenaceae	√				
196	*Phyllanthus amarus* Schumach. & Thonn.	Phyllanthaceae			√		
197	*Phyllanthus maderaspatensis* L.	Phyllanthaceae			√		
198	*Phyllanthus reticulatus* Poir.	Phyllanthaceae					√
199	*Phyllanthus tenellus* Roxb.	Phyllanthaceae		√			
200	*Physalis angulata* L.	Solanaceae			√		
201	*Physalis peruviana* L.	Solanaceae			√		
202	*Pistia stratiotes L.*	Araceae		√	√		
203	*Pluchea lanceolata* (DC.) C.B.Clarke	Compositae				√	
204	*Poa annua* L.	Poaceae	√				
205	*Polygonum plebeium* R.Br.	Polygonaceae			√		
206	*Portulaca oleracea* L.	Portulacaceae		√	√		
207	*Portulaca quadrifida* L.	Portulacaceae		√	√		
208	*Potamogeton crispus* L.	Potamogetonaceae				√	

SN	Name of plant species	Family	Seasonal vegetation				Permanent Vegetation (Shrubs and Trees)
			Winter	Summer	Rainy	Throughout year	
209	*Prosopis cineraria* (L.) Druce	Leguminosae					√
210	*Prosopis juliflora* (Sw.) DC.	Leguminosae					√
211	*Ranunculus sceleratus* L.	Ranunculaceae	√				
212	*Rhynchosia capitata* (Roth) DC.	Leguminosae			√		
213	*Rhynchosia minima* (L.) DC.	Leguminosae			√		
214	*Ricinus communis* L.	Euphorbiaceae					√
215	*Rorippa palustris* (L.) Besser	Brassicaceae	√				
216	*Rumex dentatus* L.	Polygonaceae	√				
217	*Rungia pectinata* (L.) Nees	Acanthaceae			√		
218	*Saccharum bengalense* Retz.	Poaceae				√	
219	*Saccharum spontaneum* L.	Poaceae				√	
220	*Sagittaria guayanensis* Kunth	Alismataceae			√		
221	*Salix tetrasperma* Roxb.	Salicaceae					√
222	*Scoparia dulcis* L.	Plantaginaceae				√	
223	*Senna obtusifolia* (L.) H.S.Irwin & Barneby	Leguminosae			√		
224	*Senna occidentalis* (L.) Link	Leguminosae			√		
225	*Sesamum indicum* L.	Pedaliaceae			√		
226	*Sesbania bispinosa* (Jacq.) W.Wight	Leguminosae			√		
227	*Sesbania sesban* (L.) Merr.	Leguminosae					√
228	*Setaria pumila* (Poir.) Roem. & Schult.	Poaceae			√		
229	*Setaria verticillata* (L.) P.Beauv.	Poaceae			√		
230	*Sida acuta* Burm.f.	Malvaceae			√		
231	*Sida cordata* (Burm.f.) Borss.Waalk.	Malvaceae			√		
232	*Sida cordifolia* L.	Malvaceae				√	
233	*Silene conoidea* L.	Caryophyllaceae	√				
234	*Sisymbrium irio* L.	Brassicaceae	√				
235	*Solanum americanum* Mill.	Solanaceae				√	
236	*Solanum surattense* Burm. f.	Solanaceae			√		
237	*Solanum torvum* Sw.	Solanaceae					√
238	*Sonchus asper* (L.) Hill	Compositae	√				
239	*Sonchus brachyotus* DC.	Compositae	√				
240	*Sorghum halepense* (L.) Pers.	Poaceae			√		
241	*Spergula arvensis* L.	Caryophyllaceae	√				
242	*Spermacoce articularis* L.f.	Rubiaceae			√		
243	*Sphenoclea zeylanica* Gaertn.	Sphenocleaceae			√		
244	*Spirodela polyrrhiza* (L.) Schleid.	Araceae				√	
245	*Stellaria media* (L.) Vill.	Caryophyllaceae	√				
246	*Streblus asper* Lour.	Moraceae					√
247	*Stuckenia pectinata* (L.) Börner	Potamogetonaceae				√	
248	*Syzygium cumini* (L.) Skeels	Myrtaceae					√
249	*Tamarix indica* Willd.	Tamaricaceae					√
250	*Tephrosia pumila* (Lam.) Pers.	Leguminosae			√		
251	*Tephrosia purpurea* (L.) Pers.	Leguminosae		√			

SN	Name of plant species	Family	Seasonal vegetation				Permanent Vegetation (Shrubs and Trees)
			Winter	Summer	Rainy	Throughout year	
252	*Tephrosia villosa* (L.) Pers.	Leguminosae			√		
253	*Teramnus labialis* (L.f.) Spreng.	Leguminosae			√		
254	*Trianthema portulacastrum* L.	Aizoaceae			√		
255	*Tribulus terrestris* L.	Zygophyllaceae			√		
256	*Trichosanthes cucumerina* L.	Cucurbitaceae			√		
257	*Tridax procumbens* (L.) L.	Compositae		√			
258	*Trifolium tomentosum* L.	Leguminosae	√				
259	*Triumfetta rhomboidea* Jacq.	Malvaceae			√		
260	*Typha domingensis* Pers.	Typhaceae					√
261	*Urena lobata* L.	Malvaceae			√		
262	*Utricularia stellaris* L.f.	Lentibulariaceae				√	
263	*Vallisneria spiralis* L.	Hydrocharitaceae				√	
264	*Verbascum chinense* (L.) Santapau	Scrophulariaceae		√	√		
265	*Verbascum thapsus* L.	Scrophulariaceae		√	√		
266	*Veronica agrestis* L.	Plantaginaceae	√				
267	*Veronica anagallis-aquatica* L.	Plantaginaceae	√				
268	*Vicia sativa* L.	Leguminosae	√				
269	*Withania somnifera* (L.) Dunal	Solanaceae				√	
270	*Xanthium strumarium* L.	Compositae			√		
271	*Ziziphus jujuba* Mill.	Rhamnaceae					√
272	*Ziziphus nummularia* (Burm.f.)Wight & Arn.	Rhamnaceae					√

Appendix 5: Horticultural plants of the SNU Campus

SN	Name of the plant species	Family	Common Name	Hindi Name
	Trees (Small to Large-sized)			
1	*Acacia auriculiformis* Benth.	Leguminosae	Ear-Leaf Acacia, Australian Wattle	बंगाली बबूल
2	*Azadirachta indica* A.Juss.	Meliaceae	Neem	नीम
3	*Bauhinia purpurea* L.	Leguminosae	Purple Bauhinia, Purple Butterfly Tree	कचनार
4	*Bauhinia variegata* L.	Leguminosae	Kachnar, Orchid Tree, Varigated Bauhinia	कचनार
5	*Callistemon viminalis* (Sol. ex Gaertn.) G.Don	Myrtaceae	Weeping Bottle Brush	चील
6	*Carica papaya* L.	Caricaceae	Papaya	पपीता
7	*Cassia fistula* L.	Leguminosae	Golden Rain Tree	अमलतास
8	*Cassia javanica* subsp. *nodosa* (Roxb.) K.Larsen & S.S.Larsen	Leguminosae	Pink and White Shower, Java Cassia, Rainbow Shower Tree	...
9	*Casuarina equisetifolia* L.	Casuarinaceae	Coast Sheoak, Beach Casuarina	जंगली सुरू
10	*Ceiba speciosa* (A.St.-Hil.) Ravenna	Malvaceae	Silk Floss Tree	रेशम रूई
11	*Citrus* sp.	Rutaceae	Lemon	निम्बू
12	*Delonix regia* (Hook.) Raf.	Leguminosae	Flame Tree, Gold Mohar	गुलमोहर
13	*Erythrina variegata* L.	Leguminosae	Indian Coral Tree	पंगार
14	*Ficus* sp.4	Moraceae	Ficus	...
15	*Ficus virens* Aiton	Moraceae	White Fig	पाखंड़, पिलखन
16	*Ficus benghalensis* L.	Moraceae	Banyan Tree	बढ़
17	*Ficus benjamina* L.	Moraceae	Weeping Fig, Benjamin's Fig	पुकर
18	*Grevillea robusta* A.Cunn. ex R.Br.	Proteaceae	Silky/Silver Oak Tree	...
19	*Jacaranda mimosifolia* D.Don	Bignoniaceae	Jacaranda, Blue Jacaranda	नीली गुलमोहर
20	*Kigelia africana* (Lam.) Benth.	Bignoniaceae	African Sausage Tree	बलम खीरा
21	*Lagerstroemia speciosa* (L.) Pers.	Lythraceae	Giant Crape-Myrtle, Pride of India	जरूल
22	*Madhuca longifolia* var. *latifolia* (Roxb.) A.Chev.	Sapotaceae	Indian Butter Tree	महुआ
23	*Magnolia champaca* (L.) Baill. ex Pierre	Magnoliaceae	Champa, Golden Champa	चम्पा, सोन चम्पा
24	*Mangifera indica* L.	Anacardiaceae	Mango	आम
25	*Melaleuca bracteata* F.Muell.	Myrtaceae	Golden Bottle Brush	...
26	*Melaleuca* sp.2	Myrtaceae	Golden Bottle Brush	...
27	*Melia azedarach* L.	Meliaceae	Chinaberry Tree, Barbados Lilac	बकैन
28	*Millettia peguensis* Ali	Leguminosae	Moulmein Rosewood	करंज, सुखचैन
29	*Mimusops elengi* L.	Sapotaceae	Spanish Cherry	मौलसरी
30	*Neolamarckia cadamba* (Roxb.) Bosser	Rubiaceae	Kadam	कदम्ब
31	*Peltophorum pterocarpum* (DC.) K.Heyne	Leguminosae	Copperpod, Yellow Flametree	पीला गुलमोहर
32	*Polyalthia longifolia* (Sonn.) Thwaites	Annonaceae	Ashok	अशोक
33	*Psidium guajava* L.	Myrtaceae	Guava	अमरूद
34	*Putranjiva roxburghii* Wall.	Putranjivaceae	Putranjiva, Lucky Bean Tree	पुत्रजीव
35	*Schleichera oleosa* (Lour.) Merr.	Sapindaceae	Ceylon Oak, Lac Tree	कुसुम
36	*Senna siamea* (Lam.) H.S.Irwin & Barneby	Leguminosae	Kassod Tree, Yellow Cassia	कास्सोद, मंजरी
37	*Tabebuia aurea* (Silva Manso) Benth. & Hook.f. ex S.Moore	Bignoniaceae	Silver/Yellow Trumpet Tree	...

SN	Name of the plant species	Family	Common Name	Hindi Name
38	*Terminalia arjuna* (Roxb. ex DC.) Wight & Arn.	Combretaceae	Arjun	अर्जुन
	Shrubs			
39	*Acalypha wilkesiana* Müll.Arg.	Euphorbiaceae	Acalypha, Copper Leaf	…
40	*Bougainvillea spectabilis* Willd.	Nyctaginaceae	Bougainvillea	बूगनबेल
41	*Caesalpinia pulcherrima* (L.) Sw.	Leguminosae	Peacock Flower, Dwarf Poinciana	छोटा गुलमोहर
42	*Calliandra surinamensis* Benth.	Leguminosae	Pink Powderpuff	…
43	*Cascabela thevetia* (L.) Lippold	Apocynaceae	Yellow Oleander	पीली कनेर
44	*Euphorbia cotinifolia* L.	Euphorbiaceae	Caribbean Copper Bush	…
45	*Ficus* sp.3	Moraceae	Ficus Panda, Golden Ficus	पांडा फायकस
46	*Furcraea foetida* (L.) Haw.	Asparagaceae	Mauritius Hemp, Giant False Agave	…
47	*Jatropha integerrima* Jacq.	Euphorbiaceae	Peregrina, Spicy Jatropha	जंगली अरंडी
48	*Lantana camara* L.	Verbenaceae	Lantana	राई मुनिया
49	*Nerium oleander* L.	Apocynaceae	Oleander	कनेर
50	*Nyctanthes arbor-tristis* L.	Oleaceae	Night-Flowering Jasmine	हार सिंगार
51	*Plumeria alba* L.	Apocynaceae	Frangipani	गुलचीन, चम्पा
52	*Plumeria rubra* L.	Apocynaceae	Red Frangipani	लाल गुलचीन
53	*Rosa indica* L.	Rosaceae	Rose	गुलाब
54	*Senna* sp.	Leguminosae	…	…
55	*Tabernaemontana divaricata* (L.) R.Br. ex Roem. & Schult.	Apocynaceae	Pinwheel Flower, Crape Jasmine	चांदनी
56	*Tecoma stans* (L.) Juss. ex Kunth	Bignoniaceae	Yellow Trumpet Bush	पीलिया
57	*Yucca aloifolia* L.	Asparagaceae	Aloe Yucca, Dagger Plant	युक्का
	Shrubs (Hedge)			
58	*Bougainvillea* sp.2	Nyctaginaceae	Bougainvillea	बूगनबेल
59	*Duranta erecta* L.	Verbenaceae	Sky Flower, Golden Dew Drop	नीलकांता
60	*Hamelia patens* Jacq.	Rubiaceae	Hamelia, Firebush	…
61	*Hibiscus rosa-sinensis* L.	Malvaceae	China Rose	गुढ़ल
62	*Lagerstroemia indica* L.	Lythraceae	Crape Myrtle	सावनी
63	*Lawsonia inermis* L.	Lythraceae	Henna	मेहेंदी
64	*Murraya paniculata* (L.) Jack	Rutaceae	Orange Jasmine, Chinese Box	कामिनी
65	*Plumbago auriculata* Lam.	Plumbaginaceae	Blue Plumbago	नीला चित्रक
66	*Volkameria inermis* L.	Lamiaceae	Glory Bower, Indian Privet	संकुप्पी
	Climbers/Creepers			
67	*Allamanda cathartica* L.	Apocynaceae	Golden Trumpet Vine	…
68	*Asparagus racemosus* Willd.	Asparagaceae	Satawari	सतावरी
69	*Combretum indicum* (L.) DeFilipps	Combretaceae	Rangoon Creeper	मधु मालती
70	*Jasminum sambac* (L.) Aiton	Oleaceae	Arabian Jasmine	मोगरा
71	*Jasminum* sp.2	Oleaceae	Jasmine	मोगरा
72	*Passiflora coccinea* Aubl.	Passifloraceae	Red/Scarlet Passion Flower	झुमका लता
73	*Passiflora edulis* Sims	Passifloraceae	Passion Fruit	कृष्णा फल, झुमका लता
74	*Tarlmounia elliptica* (DC.) H.Rob., S.C.Keeley, Skvarla & R.Chan	Compositae	Curtain Creeper	पर्दा बेल

SN	Name of the plant species	Family	Common Name	Hindi Name
	Perennial herbs (Ground Cover Plants)			
75	*Alternanthera* sp.	Amaranthaceae	Red Leaf Alternanthera	लालसा
76	*Canna × generalis* L.H. Bailey & E.Z. Bailey	Cannaceae	Canna Lily	केली
77	*Catharanthus roseus* (L.) G.Don	Apocynaceae	Periwinkle, Vinca	सदाबहार
78	*Dracaena* sp.	Asparagaceae	Dracaena	…
79	*Euphorbia milii* Des Moul.	Euphorbiaceae	Crown of Thorns, Kiss Me Not	…
80	*Galphimia gracilis* Bartl.	Malpighiaceae	Galphimia, Gold Shower	…
81	*Ophiopogon jaburan* (Siebold) Lodd.	Asparagaceae	Mondo Grass, Giant/White Lily Turf	…
82	*Schefflera arboricola* (Hayata) Merr.	Araliaceae	Dwarf Schefflera, Dwarf Umbrella Tree	…
83	*Sphagneticola trilobata* (L.) Pruski	Compositae	Wedelia, Trailing Daisy	वेडेलिया
84	*Strobilanthes* sp.	Acanthaceae	…	…
	Palms			
85	*Dypsis lutescens* (H.Wendl.) Beentje & J.Dransf.	Arecaceae	Areca Palm	…
86	*Beaucarnea recurvata* Lem.	Asparagaceae	Nolina/Bottle Palm	…
87	*Bismarckia nobilis* Hildebr. & H.Wendl.	Arecaceae	Bismarckia Palm, Silver Blue Fan Palm	…
88	*Caryota mitis* Lour.	Arecaceae	Fish Tail Palm	…
89	*Hyophorbe lagenicaulis* (L.H.Bailey) H.E. Moore	Arecaceae	Bottle/Champagne Palm	…
90	*Phoenix roebelenii* O'Brien	Arecaceae	Pygmy/Miniature Date Palm	…
91	*Roystonea regia* (Kunth) O.F.Cook	Arecaceae	Royal Palm	…
92	*Washingtonia filifera* (Linden ex André) H.Wendl. ex de Bary	Arecaceae	Washington/Desert/California Palm	…
93	*Wodyetia bifurcata* A.K.Irvine	Arecaceae	Fox Tail Palm	…
	Banana			
94	*Musa x paradisica* L.	Musaceae	Banana	केला
	Bamboo			
95	*Bambusa vulgaris* Schrad.	Poaceae	Buddha Belly Bamboo	बांस
	Gymnosperms			
96	*Cycas revoluta* Thunb.	Cycadaceae	Japanese Sago Palm	…
97	*Pinus roxburghii* Sarg.	Pinaceae	Chir Pine	चीड़
98	*Thuja plicata* Donn ex D.Don	Cupressaceae	Western/Pacific Redcedar	मोर पंखी

Appendix 6: Plants in Thematic Botanic Garden

SN	Name of Plant Species	Family	Common Name	Hindi Name	Introduced in
	Trees (Small to Large-Sized)				
1	*Acacia auriculiformis* Benth.	Leguminosae	Ear-Leaf Acacia, Australian Wattle	बंगाली बबूल	OAG
2	*Acacia catechu* (L.f.) Willd.	Leguminosae	Cutch Tree	खैर	XG, OAG
3	*Acacia nilotica* (L.) Delile	Leguminosae	Gum Arabic Tree, Babul	बबूल	XG, OAG
4	*Acacia modesta* Wall.	Leguminosae	Amritsar Gum	फुलाई	XG
5	*Acacia senegal* (L.) Willd.	Leguminosae	Gum Acacia, Gum Arabic Tree	कुमुट	XG
6	*Adansonia digitata* L.	Malvaceae	Baobab, Kalpavriksha	गोरख ईमली	BP
7	*Adenanthera microsperma* Teijsm. & Binn.	Leguminosae	Bead Tree	रतन गुंज	BP
8	*Aegle marmelos* (L.) Corrêa	Rutaceae	Bel	बेल	FPG, BP
9	*Ailanthus excelsa* Roxb.	Simaroubaceae	Indian Tree of Heaven	महानिब, महारुख	OAG
10	*Alangium salviifolium* (L.f.) Wangerin	Cornaceae	Sage-Leaved Alangium	अंकोल	BP
11	*Albizia lebbeck* (L.) Benth.	Leguminosae	Siris Tree	सिरिस	XG, OAG
12	*Alstonia scholaris* (L.) R. Br.	Apocynaceae	Devil Tree	शैतान का झाड़, चितवन	BP
13	*Annona squamosa* L.	Annonaceae	Custard Apple	शरीफा	FPG
14	*Anogeissus pendula* Edgew.	Combretaceae	Buttontree	करधई	BP
15	*Artocarpus heterophyllus* Lam.	Moraceae	Jack Fruit, Kathal	कटहल	FPG
16	*Artocarpus lacucha* Buch.-Ham.	Moraceae	Monkey Fruit, Barhal	बरहल	FPG, BP
17	*Averrhoa carambola* L.	Oxalidaceae	Star Fruit	कमरख	FPG
18	*Azadirachta indica* A.Juss.	Meliaceae	Neem, Indian Lilac, Margosa	नीम	MPG, BP
19	*Barringtonia acutangula* (L.) Gaertn.	Lecythidaceae	Barringtonia	...	OAG
20	*Bauhinia* sp.4	Leguminosae	...	...	OAG
21	*Bauhinia purpurea* L.	Leguminosae	Purple Bauhinia, Purple Butterfly Tree	कचनार	OAG
22	*Bauhinia variegata* L.	Leguminosae	Kachnar, Orchid Tree, Varigated Bauhinia	कचनार	BP
23	*Bombax ceiba* L.	Malvaceae	Silk-Cotton Tree	सेमल	BP
24	*Boswellia serrata* Roxb. ex Colebr.	Burseraceae	Indian Frankincense	सालर, शल्लकी	XG
25	*Butea monosperma* (Lam.) Taub.	Leguminosae	Flame of the Forest, Bastard Teak	पलाश, ढाक, टेसू	MPG, BP, OAG
26	*Callistemon citrinus* (Curtis) Skeels	Myrtaceae	Bottle Brush	चील	OAG, BP
27	*Cassia fistula* L.	Leguminosae	Golden Rain Tree	अमलतास	BP, OAG

SN	Name of Plant Species	Family	Common Name	Hindi Name	Introduced in
28	*Castanopsis* sp.	Fagaceae	...	...	OAG
29	*Casuarina equisetifolia* L.	Casuarinaceae	Coast Sheoak, Beach Casuarina	जंगली सुरू	OAG
30	*Ceiba speciosa* (A.St.-Hil.) Ravenna[1]	Malvaceae	White Floss Silk Tree	...	OAG
31	*Ceiba speciosa* (A.St.-Hil.) Ravenna[2]	Malvaceae	Red Floss Silk Tree	...	OAG
32	*Ceiba speciosa* (A.St.-Hil.) Ravenna[3]	Malvaceae	Silk Floss Tree	रेशम रूई	BP
33	*Celtis* sp.	Cannabaceae	...	...	OAG
34	*Chukrasia tabularis* A.Juss.	Meliaceae	Indian Mahogany, Indian Redwood	चिकरासी	BP
35	*Cinnamomum camphora* (L.) J.Presl	Lauraceae	Camphor Tree	कपूर	MPG, OAG
36	*Cinnamomum tamala* (Buch.-Ham.) T.Nees & Eberm.	Lauraceae	Indian Bay Leaf	तेजपत्ता	MPG
37	*Citrus* × *aurantium* L.	Rutaceae	Bitter Orange	नारंगी	FPG
38	*Citrus aurantiifolia* (Christm.) Swingle	Rutaceae	Key Lime	कांगजी निम्बू	FPG
39	*Citrus japonica* Thunb.	Rutaceae	Chinese Orange, Nagami Kumquat	चायनीज संतरा	FPG
40	*Citrus limetta* Risso	Rutaceae	Sweet lemon	मौसंबी	FPG
41	*Citrus limon* (L.) Osbeck	Rutaceae	Lemon	निम्बू	FPG
42	*Citrus maxima* (Burm.) Merr.	Rutaceae	Pomelo	चकोतरा	FPG
43	*Citrus medica* L.	Rutaceae	Citron	बड़ा निम्बू	FPG
44	*Citrus* sp. (hybrid between 'King' (*C. nobilis*) × 'Willow Leaf' (*C.* × *deliciosa*)	Rutaceae	Mandarin, Kinnow	किन्नौ	FPG
45	*Citrus* × *sinensis* (hybrid between pomelo (*C. maxima*) & mandarin (*C. reticulata*)	Rutaceae	Sweet Orange	संतरा	FPG
46	*Colophospermum mopane* (Benth.) Leonard	Leguminosae	Mopane	...	XG
47	*Cordia dichotoma* G.Forst.	Boraginaceae	Indian Cherry, Glue Berry	लसौड़ा, टेंटी, डेला	FPG, XG
48	*Cordia sinensis* Lam.	Boraginaceae	Grey Leaved Saucerberry	गोंदी	XG
49	*Corymbia citriodora* (Hook.) K.D.Hill & L.A.S.Johnson	Myrtaceae	Lemon-Scented Eucalyptus	सफ़ैदा	BP
50	*Crateva adansonii* subsp. *odora* (Buch.-Ham.) Jacobs	Capparaceae	Garlic Pear Tree, Caper Tree, Three-Leaf Caper	बर्ना, बर्नी	BP, OAG
51	*Dalbergia sissoo* DC.	Leguminosae	Shisham, Indian Rosewood	शीशम	BP
52	*Delonix regia* (Hook.) Raf.	Leguminosae	Flame Tree, Gold Mohar	गुलमोहर	BP
53	*Dillenia indica* L.	Dilleniaceae	Elephant Apple	चलता, कमरबेल	OAG
54	*Dillenia pentagyna* Roxb.	Dilleniaceae	Dog Teak, Five-Carpelled Simpoh	करमल	OAG
55	*Diospyros malabarica* (Desr.) Kostel.	Ebenaceae	Gaub Tree, Malabar Ebony	तेंदू	MPG, BP
56	*Diospyros montana* Roxb.	Ebenaceae	Mountain Persimmon, Bombay Ebony	बिसतेंदु	MPG, BP
57	*Elaeocarpus serratus* L.	Elaeocarpaceae	Rudraksha	रुद्राक्ष	OAG

SN	Name of Plant Species	Family	Common Name	Hindi Name	Introduced in
58	*Eriobotrya japonica* (Thunb.) Lindl.	Rosaceae	Loquat	लोकाट	FPG
59	*Erythrina blakei* R.Parker	Leguminosae	Blake's Coral Tree	पंगार	BP
60	*Erythrina variegata* L.	Leguminosae	Indian Coral Tree	पंगार	BP
61	*Euphorbia tirucalli* L.	Euphorbiaceae	Pencil Tree, Indian Tree Spurge	...	XG
62	*Fernandoa adenophylla* (Wall. ex G.Don) Steenis	Bignoniaceae	Katsagon Tree	मरोड़फली	BP
63	*Ficus auriculata* Lour.	Moraceae	Fig	अंजीर	FPG
64	*Ficus benghalensis* L.	Moraceae	Banyan Tree	बढ़	FG, BP
65	*Ficus benjamina* L.	Moraceae	Weeping Fig, Benjamin's Fig	पुकर	FG
66	*Ficus binnendijkii* Miq.	Moraceae	Alii Fig	...	FG, BP
67	*Ficus carica* L.	Moraceae	Fig	अंजीर	FPG
68	*Ficus elastica* Roxb. ex Hornem.	Moraceae	Rubber Plant	...	FG
69	*Ficus hispida* L.f.	Moraceae	Devil Fig	गोबला	FG
70	*Ficus lyrata* Warb.	Moraceae	Fiddle Leaf Fig	...	FG
71	*Ficus natalensis* subsp. *leprieurii* (Miq.) C.C.Berg	Moraceae	Triangular leaf fig	...	FG
72	*Ficus racemosa* L.	Moraceae	Cluster Fig	गूलर	FG
73	*Ficus religiosa* L.	Moraceae	Peepal	पीपल	FG, BP
74	*Ficus retusa* L.	Moraceae	Laurel Fig	कामरूप	FG, BP
75	*Ficus* sp.11	Moraceae	Nuda Ficus	...	FG
76	*Ficus* sp.12	Moraceae	Ficus Long Island	...	FG
77	*Ficus virens* Aiton	Moraceae	White Fig	पाखड़, पिलखन	FG
78	*Gmelina arborea* Roxb.	Lamiaceae	White Teak, Gamari	गम्हड़	OAG
79	*Grevillea robusta* A.Cunn. ex R.Br.	Proteaceae	Silky/Silver Oak Tree	...	BP
80	*Holoptelea integrifolia* Planch.	Ulmaceae	Indian Elm	चिलबिल, काजू	BP
81	*Jacaranda mimosifolia* D.Don	Bignoniaceae	Jacaranda, Blue Jacaranda	नीली गूलमोहर	BP, OAG
82	*Juglans regia* L.	Juglandaceae	Walnut	अखरोट	FPG
83	*Kigelia africana* (Lam.) Benth.	Bignoniaceae	African Sausage Tree	बालम खीरा	MPG, BP
84	*Lagerstroemia speciosa* (L.) Pers.	Lythraceae	Giant Crape-Myrtle, Pride of India	जरुल	MPG, BP
85	*Limonia acidissima* Groff	Rutaceae	Wood Apple	कैथ	FPG
86	*Litchi chinensis* Sonn.	Sapindaceae	Lychee	लीची	FPG
87	*Madhuca longifolia* var. *latifolia* (Roxb.) A.Chev.	Sapotaceae	Indian Butter Tree	महुआ	OAG
88	*Magnolia grandiflora* L.	Magnoliaceae	Magnolia	हिम चम्पा	OAG

SN	Name of Plant Species	Family	Common Name	Hindi Name	Introduced in
89	*Magnolia champaca* (L.) Baill. ex Pierre	Magnoliaceae	Champa, Golden Champa	चम्पा, सोन चम्पा	BP
90	*Mallotus nudiflorus* (L.) Kulju & Welzen	Euphorbiaceae	False White Teak	पिण्डार	OAG
91	*Malus domestica* Borkh.	Rosaceae	Apple	सेब	FPG
92	*Mangifera indica* L.[1]	Anacardiaceae	Mango Amrapali	आम आम्रपाली	FPG
93	*Mangifera indica* L.[2]	Anacardiaceae	Mango Dasheri	आम दशहरी	FPG
94	*Manilkara hexandra* (Roxb.) Dubard	Sapotaceae	Ceylon Ironwood	खिरनी	FPG
95	*Manilkara zapota* (L.) P.Royen[1]	Sapotaceae	Sapota	चीकू	FPG
96	*Manilkara zapota* (L.) P.Royen[2]	Sapotaceae	Sapota Variegated	चीकू	FPG
97	*Markhamia lutea* (Benth.) K.Schum.	Bignoniaceae	Nile Trumpet, Siala Tree	…	BP
98	*Melaleuca bracteata* F.Muell.	Myrtaceae	Golden Bottle Brush	…	OAG
99	*Melia azedarach* L.	Meliaceae	Chinaberry Tree, Barbados lilac	बकैन	BP
100	*Mesua ferrea* L.	Calophyllaceae	Ironwood, Indian Rose Chestnut	नागकेसर, नाग चम्पा	OAG
101	*Millettia peguensis* Ali	Leguminosae	Moulmein Rosewood	करंज, सुखचैन	BP
102	*Mimusops elengi* L.	Sapotaceae	Spanish cherry	मौलसरी	MPG, BP
103	*Mitragyna parvifolia* (Roxb.) Korth.	Rubiaceae	Kaim, True Kadamb	कैम	BP
104	*Moringa oleifera* Lam.	Moringaceae	Drumstick Tree	सहजन	FPG, MPG, BP
105	*Morus alba* L.	Moraceae	Mulberry	शहतूत	FPG
106	*Murraya koenigii* (L.) Spreng.	Rutaceae	Curry Leaf Tree	करी पत्ता	MPG, BP
107	*Neolamarckia cadamba* (Roxb.) Bosser	Rubiaceae	Kadam	कदम्ब	BP, OAG
108	*Ochna obtusata* DC.	Ochnaceae	Golden Champak, Mickey Mouse Plant	रामधन चंपा	BP
109	*Operculicarya decaryi* H. Perrier	Anacardiaceae	Elephant Tree	…	XG
110	*Parkinsonia aculeata* L.	Leguminosae	Mexican Palo Verde, Jerusalem Thorn	विलायती कीकर, राम बबूल	XG
111	*Peltophorum pterocarpum* (DC.) K.Heyne	Leguminosae	Copperpod, Yellow Flametree	पीला गुलमोहर	BP
112	*Persea americana* Mill.	Lauraceae	Avocado, Alligator Pear	…	FPG
113	*Phyllanthus emblica* L.	Phyllanthaceae	Indian Gooseberry, Myrobalan	आमला	FPG, MPG, OAG
114	*Pimenta dioica* (L.) Merr.	Myrtaceae	Allspice	…	MPG
115	*Pithecellobium dulce* (Roxb.) Benth.	Leguminosae	Manila Tamarind, Madras Thorn	जंगल जलेबी	FPG
116	*Platanus orientalis* L.	Platanaceae	Oriental Plane, Chinar	चिनार	BP, OAG
117	*Polyalthia longifolia* var. *angustifolia*	Annonaceae	False Ashok	अशोक	BP

SN	Name of Plant Species	Family	Common Name	Hindi Name	Introduced in
118	*Polyalthia longifolia* (Sonn.) Thwaites	Annonaceae	Ashok	अशोक	BP
119	*Pongamia pinnata* (L.) Pierre	Leguminosae	Indian Beech, Pongam Oiltree	करंज	BP
120	*Populus deltoides* Marshall	Salicaceae	Eastern Cottonwood	पॉप्लर	OAG
121	*Prosopis cineraria* (L.) Druce	Leguminosae	Khejri Tree	खेजड़ी	XG, OAG
122	*Prosopis juliflora* (Sw.) DC.	Leguminosae	Algaroba, Mesquite	जंगली कीकर	XG, OAG
123	*Prunus armeniaca* L.	Rosaceae	Apricot	ख़ूबानी	FPG
124	*Prunus domestica* L.	Rosaceae	Plum	आलू बुख़ारा	FPG
125	*Prunus dulcis* (Mill.) D.A.Webb	Rosaceae	Almond	बादाम	FPG
126	*Prunus persica* (L.) Batsch	Rosaceae	Peach	आड़ू	FPG
127	*Psidium guajava* L.[1]	Myrtaceae	Guava	अमरूद	FPG
128	*Psidium guajava* L.[2]	Myrtaceae	Guava Allahabadi	अमरूद	FPG
129	*Psidium guajava* L.[3]	Myrtaceae	Guava L-49	अमरूद	FPG
130	*Pterocarpus santalinus* L.f.	Leguminosae	Red Sandalwood, Saunderswood	रक्तचंदन	OAG
131	*Pterospermum acerifolium* (L.) Willd.	Malvaceae	Karnikara Tree	कनक चम्पा	BP
132	*Pterygota alata* (Roxb.) R.Br.	Malvaceae	Buddha Coconut	...	OAG
133	*Putranjiva roxburghii* Wall.	Putranjivaceae	Putranjiva, Lucky Bean Tree	पुत्रजीव	MPG, BP
134	*Pyrus pyrifolia* (Burm.f.) Nakai	Rosaceae	Pear	नाशपाती	FPG
135	*Quercus griffithii* Hook.f. & Thomson ex Miq.	Fagaceae	Oak	...	OAG
136	*Salix* sp.4	Salicaceae	Willow	...	OAG
137	*Salix tetrasperma* Roxb.	Salicaceae	Indian Willow	लैला, बैंस	BP, OAG
138	*Salix alba* L.	Salicaceae	White willow	...	OAG
139	*Salix babylonica* L.	Salicaceae	English Willow	सेलिक्स, इंग्लिश विलो	BP
140	*Salvadora oleoides* Decne.	Salvadoraceae	......	पीलू	XG
141	*Salvadora persica* L.	Salvadoraceae	Toothbrush Tree, Miswak	मिस्वाक	XG
142	*Santalum album* L.	Santalaceae	Sandalwood	चंदन	MPG, OAG
143	*Sapindus mukorossi* Gaertn.	Sapindaceae	Chinese Soapberry	रीठा, रिष्ट	OAG
144	*Saraca asoca* (Roxb.) Willd.	Leguminosae	Sita Ashok	सीता अशोक	BP, OAG
145	*Schleichera oleosa* (Lour.) Merr.	Sapindaceae	Ceylon Oak, Lac Tree	कुसुम	BP
146	*Sesbania sesban* (L.) Merr.	Leguminosae	Common Sesban, Egyptian Rattlepod, Egyptian Riverhemp	जयंती, रेवासन	MPG

SN	Name of Plant Species	Family	Common Name	Hindi Name	Introduced in
147	*Shorea robusta* Gaertn.	Dipterocarpaceae	Sal	साल	OAG
148	*Spondias* sp.	Anacardiaceae	Hog Plums, Spanish Plums	अमरा	FPG
149	*Syzygium jambos* (L.) Alston	Myrtaceae	Malabar Plum, Malabar Plum	जामुन	FPG
150	*Syzygium samarangense* (Blume) Merr. & L.M.Perry	Myrtaceae	Rose Apple	जामुन	FPG
151	*Syzygium cumini* (L.) Skeels	Myrtaceae	Java Plum, Jamun, Black Plum	जामुन	FPG, BP
152	*Tabebuia aurea* (Silva Manso) Benth. & Hook.f. ex S.Moore	Bignoniaceae	Silver/Yellow Trumpet Tree	...	XG, BP
153	*Tabebuia rosea* (Bertol.) Bertero ex A.DC.	Bignoniaceae	Rosy Trumpet Tree	बसंत रानी	XG, BP
154	*Tamarindus indica* L.[1]	Leguminosae	Tamarind	इमली	FPG, XG, BP
155	*Tamarindus indica* L.[2]	Leguminosae	Sweet Tamarind	मीठी इमली	FPG
156	*Tamarix aphylla* (L.) H.Karst.	Tamaricaceae	Indian Tamarisk	लाल झाड़	BP
157	*Tecomella undulata* (Sm.) Seem.	Bignoniaceae	Roheda	...	XG
158	*Tectona grandis* L.f.	Lamiaceae	Yellow Trumpet	टीकोमा	BP
159	*Terminalia arjuna* (Roxb. ex DC.) Wight & Arn.	Combretaceae	Arjun	अर्जुन	MPG, BP
160	*Terminalia bellirica* (Gaertn.) Roxb.	Combretaceae	Beleric, Bastard Myrobalan	बहेड़ा, कर्षफल	MPG, BP
161	*Terminalia chebula* Retz.	Combretaceae	Black/Chebulic Myrobalan	हर्रा, हरड़	MPG, BP
162	*Terminalia mantaly* H.Perrier	Combretaceae	Madagascar Almond	...	OAG
163	*Toona ciliata* M.Roem.	Meliaceae	Indian Mahogany	तून	BP
164	*Ziziphus jujuba* Mill.	Rhamnaceae	Common Jujube	बेर	FPG
	Undershrubs/Shrubs/Scandent Shrubs/Succulent Shrubs/Cacti				
165	*Abutilon indicum* (L.) Sweet	Malvaceae	Indian Abutilon, Indian Mallow	कंघी, ककही	MPG
166	*Acacia concinna* (Willd.) DC.	Leguminosae	Soap Pod	शिकाकाई	XG, OAG
167	*Acalypha* sp.2	Euphorbiaceae	Acalypha	...	OAG
168	*Acalypha wilkesiana* Müll.Arg.	Euphorbiaceae	Acalypha, Copper Leaf	...	OAG
169	*Adenium obesum* (Forssk.) Roem. & Schult.	Apocynaceae	Desert Rose	...	IP-CP
170	*Agave americana* L.	Asparagaceae	Agave	कमल कैक्टस	XG
171	*Anisomeles indica* (L.) Kuntze	Lamiaceae	Indian/Malabar Catmint	काला भंग्रा	MPG
172	*Artemisia scoparia* Waldst. & Kitam.	Compositae	Redstem Wormwood	सीता–बानी	MPG
173	*Barleria prionitis* L.	Acanthaceae	Porcupine Flower	वज्रदंती	MPG
174	*Bauhinia tomentosa* L.	Leguminosae	Yellow Bauhinia	कचनार	OAG
175	*Bauhinia racemosa* Lam.	Leguminosae	Bidi Leaf Tree	कठमूली, झिंझेरी	BP

SN	Name of Plant Species	Family	Common Name	Hindi Name	Introduced in
176	*Bougainvillea* sp.	Nyctaginaceae	Bougainvillea	बूगनबेल	BP
177	*Buxus sempervirens* L.	Buxaceae	European Box, Boxwood	...	IP-CP
178	*Cactus* sp.1	Cactaceae	Cactus	...	XG
179	*Cactus* sp.2	Cactaceae	Cactus	...	XG
180	*Cactus* sp.3	Cactaceae	Star Cactus	...	XG
181	*Calliandra haematocephala* Hassk.	Leguminosae	Red Powder Puff	...	OAG
182	*Calliandra surinamensis* Benth.	Leguminosae	Pink Powderpuff	...	BP
183	*Calotropis gigantea* (L.) Dryand.	Apocynaceae	Giant Indian Milkweed, Crown Flower	सफ़ेद आक	MPG, BP
184	*Calotropis procera* (Aiton) Dryand.	Apocynaceae	Rubber Bush, Sodom Apple	आक, मंदार	MPG, XG
185	*Capparis decidua* (Forssk.) Edgew.	Capparaceae	Bare Caper	कैर, कीकर	XG
186	*Carissa carandas* L.	Apocynaceae	Bengal Currant, Karonda	करौंदा	FPG
187	*Cascabela thevetia* (L.) Lippold	Apocynaceae	Yellow Oleander	पीली कनेर	BP
188	*Cestrum diurnum* L.	Solanaceae	Day Jasmine	दिन का राजा	OAG
189	*Cestrum nocturnum* L.	Solanaceae	Night Blooming Jasmine	रात की रानी	BP
190	*Cleome viscosa* L.	Cleomaceae	Asian Spider Flower, Dog Mustard	बागरा, हुरहुर, पीली हुरहुर	MPG
191	*Clerodendrum phlomidis* L.f.	Lamiaceae	Arni	अर्नी	MPG
192	*Commiphora wightii* (Arn.) Bhandari	Burseraceae	Guggul, Indian Bdellium	गुग्गुल	XG
193	*Desmodium gangeticum* (L.) DC.	Leguminosae	Sal Leaved Desmodium	दीर्घमूली, शालपर्णी	MPG
194	*Dichrostachys cinerea* (L.) Wight & Arn.	Leguminosae	Sickle Bush	कुणाली	XG
195	*Duranta erecta* L.	Verbenaceae	Sky Flower, Golden Dew Drop	नीलकांता	IP-CP
196	*Dysphania ambrosioides* (L.) Mosyakin & Clemants	Amaranthaceae	Mexican Tea	सुगंध वास्तुक	MPG
197	*Euphorbia cotinifolia* L.	Euphorbiaceae	Caribbean Copper Bush	...	XG
198	*Euphorbia royleana* Boiss.	Euphorbiaceae	Royle's Spurge	थोर	XG
199	*Euphorbia* sp.	Euphorbiaceae	Euphorbia Cactus	...	XG
200	*Excoecaria cochinchinensis* Lour.	Euphorbiaceae	Chinese Croton	लैला-मजनू	OAG
201	*Ficus palmata* Forssk.	Moraceae	Wild Fig	जंगली अंजीर	FPG
202	*Ficus* sp.13	Moraceae	Ficus Panda, Golden Ficus	पांडा फयकस	FG
203	*Furcraea foetida* (L.) Haw.	Asparagaceae	Mauritius Hemp, Giant False Agave	...	XG
204	*Galphimia gracilis* Bartl.	Malpighiaceae	Galphimia, Gold Shower	...	IP-CP
205	*Gardenia jasminoides* J.Ellis	Rubiaceae	Gardenia, Cape Jasmine	गंधराज	BP

SN	Name of Plant Species	Family	Common Name	Hindi Name	Introduced in
206	*Gmelina philippensis* Cham.	Lamiaceae	Parrot's Beak	बधारा	OAG
207	*Grewia asiatica* L.	Malvaceae	Phalsa	फालसा	FPG
208	*Grewia hirsuta* Vahl	Malvaceae	Kukurbicha	नागबला	OAG
209	*Hamelia patens* Jacq.	Rubiaceae	Hamelia, Firebush	...	BP
210	*Hibiscus rosa-sinensis* L.	Malvaceae	China Rose	गुड़ल	BP
211	*Hibiscus rosa-sinensis* L.[1]	Malvaceae	China Rose (Double Variety)	गुड़ल	OAG
212	*Hibiscus rosa-sinensis* L.[2]	Malvaceae	China Rose (Pink)	गुड़ल	OAG
213	*Hibiscus rosa-sinensis* L.[3]	Malvaceae	China Rose (Red)	गुड़ल	OAG
214	*Hibiscus rosa-sinensis* L.[4]	Malvaceae	China Rose (Single Variety)	गुड़ल	OAG
215	*Hibiscus rosa-sinensis* L.[5]	Malvaceae	China Rose (Varigated Variety)	गुड़ल	OAG
216	*Hibiscus rosa-sinensis* L.[6]	Malvaceae	China Rose (White)	गुड़ल	OAG
217	*Hibiscus rosa-sinensis* L.[7]	Malvaceae	China Rose (Yellow)	गुड़ल	OAG
218	*Hylocereus undatus* (Haw.) Britton & Rose	Cactaceae	Dragon Fruit	ड्रैगन फल	FPG
219	*Hyptis suaveolens* (L.) Poit.	Lamiaceae	American Mint, Bush Mint	विलायती तुलसी	MPG
220	*Indigofera tinctoria* L.	Leguminosae	True Indigo, Common Indigo	नील	MPG
221	*Ipomoea carnea* Jacq.	Convolvulaceae	Bush Morning Glory	बेहया, बेशरम	MPG
222	*Ixora coccinea* L.	Rubiaceae	Ixora	...	OAG
223	*Jatropha curcas* L.	Euphorbiaceae	Barbados Nut	रतनजोती	XG, BP
224	*Jatropha gossypiifolia* L.	Euphorbiaceae	Bellyache Bush, Cotton-Leaf Physic Nut	रतनजोती	MPG
225	*Jatropha integerrima* Jacq.	Euphorbiaceae	Peregrina, Spicy Jatropha	जंगली अरंडी	BP
226	*Justicia adhatoda* L.	Acanthaceae	Malabar Nut, Vasaka, Adhatoda, Adulsa	अरुस, अडूसा	MPG
227	*Justicia* sp.	Acanthaceae	...	...	XG
228	*Lagerstroemia indica* L.	Lythraceae	Crape Myrtle	सावनी	BP
229	*Lantana camara* L.	Verbenaceae	Lantana	राईमुनिया	BP
230	*Lawsonia inermis* L.	Lythraceae	Henna	मेहेंदी	MPG, BP
231	*Malpighia emarginata* DC.	Malpighiaceae	West Indian Cherry	...	FPG
232	*Malpighia* sp.2	Malpighiaceae	Cherry	...	FPG
233	*Malvaviscus arboreus* var. *mexicanus* Schltdl.	Malvaceae	Mexican Turk's Cap	जुबा कुसुम	XG
234	*Manihot esculenta* Crantz	Euphorbiaceae	Variegated Tapioca, Cassava	साबूदाना, टेपीओका	OAG
235	*Murraya paniculata* (L.) Jack	Rutaceae	Orange Jasmine, Chinese Box	कामिनी	BP

SN	Name of Plant Species	Family	Common Name	Hindi Name	Introduced in
236	*Mussaenda erythrophylla* Schumach. & Thonn.	Rubiaceae	Red Flag Bush	बेदिना	OAG
237	*Mussaenda* sp.2	Rubiaceae	Musaaenda	...	OAG
238	*Nerium oleander* L.	Apocynaceae	Oleander	कनेर	BP
239	*Nyctanthes arbor-tristis* L.	Oleaceae	Night-Flowering Jasmine	हार सिंगार	MPG, BP
240	*Opuntia dillenii* (Ker Gawl.) Haw.	Cactaceae	Pricklypear	नागफनी	XG
241	*Osmanthus fragrans* Lour. (Kashmir)	Oleaceae	Sweet Osmanthus, Fragrant Olive	सिलंग	IP-CP
242	*Pachystachys lutea* Nees	Acanthaceae	Lollipop Plant, Golden Shrimp Plant	...	OAG
243	*Phyllanthus reticulatus* Poir.	Phyllanthaceae	Black Honey Shrub, Black Berried Featherfoil	काले मधु का पेड़	MPG
244	*Pluchea lanceolata* (DC.) C.B.Clarke	Compositae	Rasna	फार	MPG
245	*Plumbago auriculata* Lam.	Plumbaginaceae	Plumbago, Cape Leadwort	नीला चित्रक	MPG
246	*Plumeria alba* L.	Apocynaceae	Frangipani	गुलचीन चम्पा	OAG, BP
247	*Plumeria obtusa* L.	Apocynaceae	White Frangipani	लाल गुलचीन	OAG
248	*Plumeria pudica* Jacq.	Apocynaceae	Wild Plumeria, Fiddle Leaf Plumeria	...	OAG
249	*Plumeria rubra* L.	Apocynaceae	Red Frangipani	लाल गुलचीन	OAG, BP
250	*Plumeria* sp.	Apocynaceae	Frangipani	...	OAG
251	*Polyscias fruticosa* (L.) Harms	Araliaceae	Ming Aralia	...	OAG
252	*Psidium* sp.[4]	Myrtaceae	Cherry Guava	...	FPG
253	*Punica granatum* L.	Lythraceae	Pomegranate	अनार	FPG
254	*Ravenia spectabilis* Engl.	Rutaceae	Lemonia, Pink Ravenia	...	XG
255	*Ricinus communis* L.	Euphorbiaceae	Castor Bean, Castor Oil Plant	अंरडी, चिया	MPG
256	*Rosa indica* L.	Rosaceae	Rose	गुलाब	OAG
257	*Russelia equisetiformis* Schltdl. & Cham.	Plantaginaceae	Fountain Bush, Coral Plant	रसीली	XG
258	*Schefflera elegantissima* (Veitch ex Mast.) Lowry & Frodin	Araliaceae	False Aralia	...	OAG
259	*Scoparia dulcis* L.	Plantaginaceae	Sweet Broom Weed	मीठीपत्ती, घोड़ा तुलसी	MPG
260	*Senna alata* (L.) Roxb.	Leguminosae	Candle Bush, Ringworm Shrub	दादमुर्दन	OAG
261	*Senna occidentalis* (L.) Link	Leguminosae	Coffee Senna, Negro Coffee	कसुंडा, बड़ी कसौंदी	MPG
262	*Senna* sp.3	Leguminosae	...	...	MPG
263	*Senna* sp.4	Leguminosae	...	...	BP
264	*Sida acuta* Burm.f.	Malvaceae	Common Wireweed, Common Fanpetals	बरैरा	MPG
265	*Sida cordifolia* L.	Malvaceae	Heart-Leaf Sida, Country Mallow, Flannel Weed	खरींटा, खरैंटी	MPG

SN	Name of Plant Species	Family	Common Name	Hindi Name	Introduced in
266	*Solanum indicum* L.	Solanaceae	Brihati, Poison Berry	बड़ी कटेरी	MPG
267	*Tabernaemontana divaricata* (L.) R.Br. ex Roem. & Schult.	Apocynaceae	Pinwheel Flower, Crape Jasmine	चांदनी	BP
268	*Tamarix indica* Willd.	Tamaricaceae	Indian Tamarisk	झाऊ	XG
269	*Tecoma castanifolia* (D.Don) Melch.	Bignoniaceae	Gaudi Chaudi, Yellow Trumpet Bush	पीलिया	BP
270	*Tecoma stans* (L.) Juss. ex Kunth	Bignoniaceae	Yellow Trumpet Bush	पीलिया	BP
271	*Tephrosia purpurea* (L.) Pers.	Leguminosae	Wild Indigo, Fish Poison	सरफोंक, शरपुंखा	MPG
272	*Tephrosia villosa* (L.) Pers.	Leguminosae	Shaggy Wild Indigo, Hoary Tephrosia	...	MPG
273	*Triumfetta rhomboidea* Jacq.	Malvaceae	Diamond Burbark, Chinese Bur	चीकी	MPG
274	*Urena lobata* L.	Malvaceae	Caesarweed	बचिता, लपेटुआ	MPG
275	*Vitex trifolia* L.	Lamiaceae	Indian Wild Pepper, Simple-Leaf Chaste Tree	सम्भालू	BP
276	*Volkameria inermis* L.	Lamiaceae	Glory Bower, Indian Privet	संकुप्पी	OAG
277	*Withania somnifera* (L.) Dunal	Solanaceae	Winter Cherry, Indian Ginseng	अश्वगंधा, असगंध	MPG
278	*Yucca aloifolia* L.	Asparagaceae	Aloe Yucca, Dagger Plant	युक्का	XG
279	*Yucca glauca* Nutt.	Asparagaceae	Soapweed Yucca	युक्का	XG
280	*Ziziphus nummularia* (Burm.f.) Wight & Arn.	Rhamnaceae	Jhar Beri	झाड़ बेरी	FPG
	Climbers/Creepers				
281	*Abrus precatorius* L.	Leguminosae	Licorice, Coral Bead Vine, Crab's Eye	रत्ती, गुंची	MPG
282	*Allamanda cathartica* L.	Apocynaceae	Golden Trumpet Vine	...	OAG
283	*Asparagus racemosus* Willd.	Asparagaceae	Satawari	सतावरी	MPG
284	*Asparagus* sp. 2	Asparagaceae	Asparagus	...	OAG
285	*Bignonia capreolata* L.	Bignoniaceae	Crossvine	...	OAG
286	*Citrullus colocynthis* (L.) Schrad.	Cucurbitaceae	Colocynth, Desert Gourd	इंद्रायन	MPG
287	*Clematis gouriana* Roxb. ex DC.	Ranunculaceae	Indian Travelers's Joy	बेलकुम	OAG
288	*Clerodendrum laevifolium* Blume	Lamiaceae	Wallich's Glory Bower	...	OAG
289	*Clerodendrum* sp.3	Lamiaceae	...	...	OAG
290	*Clitoria ternatea* L.	Leguminosae	Butterfly Pea	अपराजिता	MPG
291	*Coccinia grandis* (L.) Voigt	Cucurbitaceae	Ivy Gourd, Scarlet Gourd	कुन्द्रू	MPG
292	*Combretum indicum* (L.) DeFilipps	Combretaceae	Rangoon Creeper	मधु मालती	OAG
293	*Convolvulus arvensis* L.	Convolvulaceae	Field Bindweed	हिरनपग	MPG
294	*Cucumis melo* var. *agrestis* Naudin	Cucurbitaceae	Wild Musk Melon	कचरिया	OAG

SN	Name of Plant Species	Family	Common Name	Hindi Name	Introduced in
295	*Epipremnum aureum* (Linden & André) G.S.Bunting	Araceae	Pothos, Money Plant, Silver Vine	Money Plant	OAG
296	*Hedera helix* L.	Araliaceae	Common Ivy, English Ivy	Ivy	OAG
297	*Ipomoea cairica* (L.) Sweet	Convolvulaceae	Railway Creeper	नीली बेलं	OAG
298	*Ipomoea coptica* (L.) Roth ex Roem. & Schult.	Convolvulaceae	Egyptian Morning Glory	…	MPG
299	*Ipomoea pes-tigridis* L.	Convolvulaceae	Tiger Foot Morning Glory	पंचपतिया	MPG
300	*Ipomoea* sp.3	Convolvulaceae	…	…	OAG
301	*Ipomoea batatas* (L.) Lam.	Convolvulaceae	Sweet Potato	शकरकंद	OAG
302	*Ipomoea* sp.4	Convolvulaceae	…	…	OAG
303	*Jasminum sambac* (L.) Aiton	Oleaceae	Arabian Jasmine	शकरकंद	OAG
304	*Jasminum* sp.2	Oleaceae	Jasmine	मोगरा	OAG
305	*Mansoa alliacea* (Lam.) A.H.Gentry	Bignoniaceae	Garlic Vine Plant	लहान बेल	OAG
306	*Monstera deliciosa* Liebm.	Araceae	Swiss Cheese Plant, Monstera	…	OAG
307	*Oxystelma esculentum* (L. f.) Sm.	Apocynaceae	Rosy Milkweed Vine	दुधिया लता	MPG
308	*Passiflora coccinea* Aubl.	Passifloraceae	Red/Scarlet Passion Flower	झुमका लता	FPG
309	*Passiflora edulis* Sims	Passifloraceae	Passion Fruit	कृष्ण फल, झुमका लता	FPG
310	*Piper betle* L.	Piperaceae	Betel	पान	MPG
311	*Sphagneticola trilobata* (L.) Pruski	Compositae	Wedelia, Trailing Daisy	वडेलिया	IP-CP
312	*Thunbergia mysorensis* (Wight) T.Anderson	Acanthaceae	Mysore Trumpetvine, Indian Clock Vine	…	OAG
313	*Tinospora sinensis* (Lour.) Merr.	Menispermaceae	Chinese Tinospora	गिलोय	MPG
314	*Vitis vinifera* L.	Vitaceae	Grapes	अंगूर	FPG
	Herbs including Grasses				
315	*Achyranthes aspera* L.	Amaranthaceae	Prickly Chaff Flower, Devil's Horsewhip	चिरचिटा, लटजीरा	MPG
316	*Ageratum houstonianum* Mill.	Compositae	Floss Flower, Blue Billygoat Weed	रक्तरोधी	MPG
317	*Allium tuberosum* Rottler ex Spreng.	Amaryllidaceae	Garlic Chives, Oriental Garlic, Asian/Chinese Chives	…	MPG
318	*Aloe vera* (L.) Burm.f.	Xanthorrhoeaceae	Aloe	घृत कुमारी	MPG, XG
319	*Alpinia galanga* (L.) Willd.	Zingiberaceae	Kulanjan, Thai Ginger	कुलजन	MPG
320	*Alternanthera paronychioides* A.St.-Hil.	Amaranthaceae	Smooth Chaff Flower	…	MPG
321	*Alternanthera* sp.2	Amaranthaceae	Red Leaf Alternanthera	लालसा	IP-CP
322	*Alternanthera* sp.3	Amaranthaceae	Green Leaf Alternanthera	…	IP-CP
323	*Amaranthus viridis* L.	Amaranthaceae	Slender/Green Amaranth	जंगली चौलाई	MPG

SN	Name of Plant Species	Family	Common Name	Hindi Name	Introduced in
324	*Aralia* sp.	Araliaceae	Aralia	...	IP-CP
325	*Argemone mexicana* L.	Papaveraceae	Mexican Prickly Poppy	सत्यानाशी, कटैला	MPG
326	*Bacopa monnieri* (L.) Wettst.	Plantaginaceae	Water Hyssop, Indian Pennywort	ब्राह्मी	MPG
327	*Blumea viscosa* (Mill.) V.M.Badillo	Compositae	Sticky Blumea, Clammy False Oxtongue	...	MPG
328	*Boerhavia diffusa* L.	Nyctaginaceae	Red Spiderling, Spreading Hog Weed	पुनर्नवा, गदहपुर्णा	MPG
329	*Bryophyllum pinnatum* (Lam.) Oken	Crassulaceae	Patharchatta	पत्थरचट्टा	MPG, XG
330	*Calendula officinalis* L.	Compositae	Pot Marigold, Calendula	गेंदा	MPG
331	*Canna indica* L.	Cannaceae	Wild Canna	केली	OAG
332	*Cannabis sativa* L.	Cannabaceae	Marijuana, Hemp	भांग, गाजा, चरस	MPG
333	*Carthamus oxyacantha* M.Bieb.	Compositae	Wild Safflower, Jeweled Distaff Thistle	कंडिअरी, पीली कटेरी	MPG
334	*Catharanthus roseus* (L.) G.Don	Apocynaceae	Periwinkle, Vinca	सदाबहार	MPG
335	*Centella asiatica* (L.) Urb.	Apiaceae	Indian Pennywort, Gotu kola	ब्राह्मी, ब्रह्मण्डूकी	MPG
336	*Ceratophyllum demersum* L.	Ceratophyllaceae	Hornwort, Coon's Tail	बृहतपुष्पी	APG
337	*Chenopodium album* L.	Amaranthaceae	Goosefoot, Wild Spinach, Lamb's Quarters	बथुआ	MPG
338	*Chenopodium murale* L.	Amaranthaceae	Nettle-leaved Goosefoot, Australian Spinach	जंगली बथुआ, खरतुआ	MPG
339	*Chlorophytum borivilianum* Santapau & R.R.Fern.	Asparagaceae	Safed Musli	सफ़ेद मुसली	MPG
340	*Chrysopogon zizanioides* (L.) Roberty	Poaceae	Vetiver, Khas-Khas Grass	खस–खस	MPG
341	*Cirsium arvense* (L.) Scop.	Compositae	Creeping/Californian Thistle	कंदई	MPG
342	*Convolvulus prostratus* Forssk.	Convolvulaceae	Prostrate Bindweed	शंखपुष्पी	MPG
343	*Crinum latifolium* L.	Amaryllidaceae	Pink-Striped Trumpet Lily	सुदर्शन	MPG
344	*Croton bonplandianus* Baill.	Euphorbiaceae	Croton, Rushfoil	बन तुलसी, काला भांगरा	MPG
345	*Curcuma longa* L.	Zingiberaceae	Turmeric	हल्दी	MPG
346	*Cymbopogon citratus* (DC.) Stapf	Poaceae	Lemon Grass	अघिनघास, गंधत्रिण	MPG
347	*Cynodon dactylon* (L.) Pers.	Poaceae	Bermuda Grass, Bahama Grass	दूब	MPG
348	*Datura metel* L.	Solanaceae	Thorn Apple	धतूरा	MPG
349	*Datura innoxia* Mill.	Solanaceae	Pricklyburr, Thorn Apple	सफ़ेद धतूरा	MPG
350	*Dieffenbachia bowmannii* Carrière	Araceae	Dumb Cane	...	IP-CP
351	*Dracaena fragrans* (L.) Ker Gawl.	Asparagaceae	Dracaena	...	IP-CP

SN	Name of Plant Species	Family	Common Name	Hindi Name	Introduced in
352	*Dracaena reflexa* var. *angustifolia* Baker	Asparagaceae	Dracaena	…	IP-CP
353	*Dracaena* sp.3	Asparagaceae	Dracaena	…	IP-CP
354	*Eclipta prostrata* (L.) L.	Compositae	False Daisy, Trailing Eclipta	भिंग्रराज, केशराज	MPG
355	*Eichhornia crassipes* (Mart.) Solms	Pontederiaceae	Water Hyacinth	जल कुम्भी	APG
356	*Eranthemum* sp.	Acanthaceae	Eranthemum	…	IP-CP
357	*Euphorbia hirta* L.	Euphorbiaceae	Asthma Weed, Asthma Plant	दुधी, बड़ा दुधी	MPG
358	*Euphorbia milii* Des Moul.	Euphorbiaceae	Crown of Thorns, Kiss Me Not	…	XG, IP-CP
359	*Euphorbia prostrata* Aiton	Euphorbiaceae	Prostrate Sandmat, Prostrate Spurge	छोटा दुधी	MPG
360	*Euphorbia tithymaloides* L.	Euphorbiaceae	Redbird Flower, Devil's-Backbone	विलायती शेर	IP-CP
361	*Evolvulus nummularius* (L.) L.	Convolvulaceae	Roundleaf Bindweed, Agracejo Rastrero	मुसाकर्णी	MPG
362	*Gomphrena celosioides* Mart.	Amaranthaceae	Prostrate Globe-Amaranth	…	MPG
363	*Heliconia* sp.	Heliconiaceae	Crab Claw	…	OAG
364	*Heliotropium ellipticum* Ledeb.	Boraginaceae	Heliotrope	हाथी सूंड़ा	MPG
365	*Hydrilla verticillata* (L.f.) Royle	Hydrocharitaceae	Waterthyme, Hydrilla	सेवार, झंगी	APG
366	*Imperata cylindrica* (L.) Raeusch.	Poaceae	Blady Grass, Cogon Grass, Kunai Grass, Japanese Bloodgrass	दाभ, उलू	MPG
367	*Jacobaea maritima* (L.) Pelser & Meijden	Compositae	Silver Ragwort	…	IP-CP
368	*Justicia gendarussa* Burm.f.	Acanthaceae	Warer Willo	नीली नारगंडी	IP-CP
369	*Kalanchoe blossfeldiana* Poelln.	Crassulaceae	Christmas Kalanchoe	…	IP-CP
370	*Launaea procumbens* (Roxb.) Ramayya & Rajagopal	Compositae	Creeping Launaea	पीली दुद्धी, जलगोभी	MPG
371	*Lemna perpusilla* Torr.	Araceae	Common Duckweed, Common Waterlens, Duck meat	…	APG
372	*Melilotus albus* Medik.	Leguminosae	White Sweet Clover, White Melilot	सफ़ेद बन मेथी	MPG
373	*Melilotus indicus* (L.) All.	Leguminosae	Indian Sweet Clover, Yellow Sweet Clover	बन मेथी मोरिला	MPG
374	*Melochia corchorifolia* L.	Malvaceae	Chocolate Weed	चित्रबीज	MPG
375	*Mentha arvensis* L.	Lamiaceae	Mint	पुदीना	MPG
376	*Mentha* sp.2	Lamiaceae	Ginger Mint	पुदीना	MPG
377	*Mimosa pudica* L.	Leguminosae	Touch-Me-Not	छुई-मुई	MPG
378	*Nicotiana plumbaginifolia* Viv.	Solanaceae	Tex-Mex Tobacco	जंगली तम्बाकू	MPG
379	*Nymphaea pubescens* Willd.	Nymphaeaceae	White Water Lily, White Lotus	को का कुमुद	APG
380	*Nymphoides cristata* (Roxb.) Kuntze	Menyanthaceae	Crested Floating Heart	कुमुदिनी	APG
381	*Ocimum* × *africanum* Lour.	Lamiaceae	Lemon Basil	तुलसी	MPG

SN	Name of Plant Species	Family	Common Name	Hindi Name	Introduced in
382	*Ocimum basilicum* L.	Lamiaceae	Sweet Basil	राम तुलसी	MPG
383	*Ocimum tenuiflorum* L.	Lamiaceae	Holy Basil	तुलसी	MPG
384	*Ophiopogon jaburan* (Siebold) Lodd.	Asparagaceae	Mondo Grass, Giant/White Lily Turf	...	IP-CP
385	*Origanum* sp.	Lamiaceae	Oregano	...	MPG
386	*Peristrophe bicalyculata* (Retz.) Nees	Acanthaceae	Panicled Foldwing, Kali Anghedi	अत्रिलाल	MPG
387	*Persicaria barbata* (L.) H.Hara	Polygonaceae	Bearded Knotweed	...	MPG
388	*Phyla nodiflora* (L.) Greene	Verbenaceae	Frog Fruit, Creeping Lip Plant	जलबूटी	MPG
389	*Phyllanthus amarus* Schumach. & Thonn.	Phyllanthaceae	Black Catnip, Stone Breaker	भूई आवला, हजारमणि	MPG
390	*Phyllanthus maderaspatensis* L.	Phyllanthaceae	Madras Leaf-Flower	हजारमणि	MPG
391	*Physalis angulata* L.	Solanaceae	Angular Winter Cherry, Cutleaf Groundcherry	रसभरी	MPG
392	*Physalis peruviana* L.	Solanaceae	Cape Gooseberry, Peruvian Ground Cherry, Poha Berry	रसभरी	MPG
393	*Pistia stratiotes* L.	Araceae	Water Cabbage, Water Lettuce	जलकुम्भी	APG
394	*Plectranthus amboinicus* (Lour.) Spreng.	Lamiaceae	Cuban Oregano, Indian Borage	पत्ता अजवाइन	MPG
395	*Portulaca oleracea* L.	Portulacaceae	Purslane	लूनिया, बड़ी नोनी	MPG
396	*Potamogeton crispus* L.	Potamogetonaceae	Curly-Leaf Pondweed	...	APG
397	*Pseuderanthemum carruthersii* (Seem.) Guillaumin	Acanthaceae	Purple False Eranthemum	...	IP-CP
398	*Rauvolfia serpentina* (L.) Benth. ex Kurz	Apocynaceae	Indian Snakeroot	सर्पगंधा	MPG
399	*Rumex dentatus* L.	Polygonaceae	Toothed Dock, Aegean Dock	जंगली पालक	MPG
400	*Sagittaria guayanensis* Kunth	Alismataceae	Guyanese Arrowhead	...	APG
401	*Sansevieria hyacinthoides* (L.) Druce	Asparagaceae	Snake Plant, Devil's Tongue	...	XG
402	*Sedum* sp.	Crassulaceae	Stonecrop	...	IP-CP
403	*Sida cordata* (Burm.f.) Borss.Waalk.	Malvaceae	Long-Stalk Sida	भूइनी, खरैंटी	MPG
404	*Solanum americanum* Mill.	Solanaceae	Black Nightshade	मकोय	MPG
405	*Solanum surattense* Burm. f.	Solanaceae	Thorny/Yellow Berried Nightshade	कटेली, ऊट कटेली	MPG
406	*Sonchus asper* (L.) Hill	Compositae	Prickly/Rough Sow-Thistle	दुधी	MPG
407	*Spirodela polyrrhiza* (L.) Schleid.	Araceae	Common Duck Meat, Duckweed	...	APG
408	*Strobilanthes* sp.	Acanthaceae	Goldfussia	...	IP-CP
409	*Stuckenia pectinata* (L.) Börner	Potamogetonaceae	Sago Pondweed, Fennel Pondweed	...	APG
410	*Syngonium podophyllum* Schott	Araceae	Arrowhead Vine	सिंगोनियम	IP-CP
411	*Thymus vulgaris* L.	Lamiaceae	Garden Thyme	थाइमस	MPG

SN	Name of Plant Species	Family	Common Name	Hindi Name	Introduced in
412	*Tradescantia pallida* (Rose) D.R.Hunt	Commelinaceae	...	...	IP-CP
413	*Tradescantia sillamontana* Matuda	Commelinaceae	White Velvet Tradescantia	...	IP-CP
414	*Trianthema portulacastrum* L.	Aizoaceae	Giant Pigweed, Horse-Purslane	विषखपरा, साबुनी	MPG
415	*Tribulus terrestris* L.	Zygophyllaceae	Puncture Vine, Caltrop	गोखरू	MPG
416	*Utricularia stellaris* L.f.	Lentibulariaceae	Bladderwort	...	APG
417	*Vallisneria spiralis* L.	Hydrocharitaceae	Tape Grass, Straight Vallisneria	सिवाल, शैवाल	APG
418	*Xanthium strumarium* L.	Compositae	Common Cocklebur, Clotbur	छोटा धतूरा, घाघरा, संखाहुली	MPG
	Palms				
419	*Areca catechu* L.	Arecaceae	Supari/Betelnut Palm	...	PG
420	*Beaucarnea recurvata* Lem.	Asparagaceae	Nolina/Bottle Palm	...	PG
421	*Bismarckia nobilis* Hildebr. & H.Wendl.	Arecaceae	Bismarckia/Silver Blue Fan Palm	...	PG
422	*Butia capitata* (Mart.) Becc.	Arecaceae	Jelly Palm	...	PG
423	*Caryota mitis* Lour.	Arecaceae	Fish Tail Palm	...	PG
424	*Chamaedorea metallica* O.F.Cook ex H.E. Moore	Arecaceae	Chaemaedorea Palm	...	PG
425	*Chamaerops humilis* L.	Arecaceae	Chamaerops Palm	...	PG
426	*Chamaerops humilis* var. *cerifera*	Arecaceae	Blue Mediterranean Fan Palm	...	PG
427	*Copernicia prunifera* (Mill.) H.E.Moore	Arecaceae	Copernicia Palm	...	PG
428	*Dypsis decaryi* (Jum.) Beentje & J.Dransf.	Arecaceae	Triangular Palm	...	PG
429	*Dypsis lutescens* (H.Wendl.) Beentje & J.Dransf.	Arecaceae	Areca Palm	...	PG
430	*Hyophorbe lagenicaulis* (L.H.Bailey) H.E. Moore	Arecaceae	Bottle/Champagne Palm	...	PG
431	*Latania lontaroides* (Gaertn.) H.E.Moore	Arecaceae	Red Latan Palm	...	PG
432	*Licuala grandis* H.Wendl.	Arecaceae	Ruffled Fan Palm	...	PG
433	*Livistona* sp.	Arecaceae	Livistona Palm	...	PG
434	*Pandanus utilis* Bory	Pandanaceae	Pandanus Palm	केवड़ा	PG
435	*Phoenix dactylifera* L.	Arecaceae	Date Palm	खजूर	PG
436	*Phoenix paludosa* Roxb.	Arecaceae	Mangrove Date Palm	...	PG
437	*Phoenix roebelenii* O'Brien	Arecaceae	Pygmy/Miniature Date Palm	...	PG
438	*Phoenix sylvestris* (L.) Roxb.	Arecaceae	Date Palm	खजूर	PG
439	*Ptychosperma elegans* (R.Br.) Blume	Arecaceae	Seaforthia Palm	...	PG
440	*Ravenala madagascariensis* Sonn.	Strelitziaceae	Travellers Palm	...	PG
441	*Ravenea* sp.	Arecaceae	Majesty Palm	...	PG
442	*Rhapis excelsa* (Thunb.) Henry	Arecaceae	Rhapis Palm	...	PG

SN	Name of Plant Species	Family	Common Name	Hindi Name	Introduced in
443	*Roystonea regia* (Kunth) O.F.Cook	Arecaceae	Royal Palm	...	PG
444	*Sabal palmetto* (Walter) Lodd. ex Schult. & Schult.f.	Arecaceae	Sabal or Cabbage Palm	...	PG
445	*Thrinax* sp.	Arecaceae	Thrinax Palm	...	PG
446	*Trachycarpus fortunei* (Hook.) H.Wendl.	Arecaceae	China Palm	...	PG
447	Unidentified1	Arecaceae	Nephtrox palm (?)	...	PG
448	Unidentified2 (Endemic to Andaman Islands)	Arecaceae	...	...	PG
449	*Washingtonia filifera* (Linden ex André) H.Wendl. ex de Bary	Arecaceae	Washington/Desert/California Palm	...	PG
450	*Wodyetia bifurcata* A.K.Irvine	Arecaceae	Fox Tail Palm	...	PG
	Papaya/Banana				
451	*Carica papaya* L.	Caricaceae	Papaya	पपीता	FPG
452	*Musa x paradisica* L.[1]	Musaceae	Banana Hyderabad	केला हैदराबादी	FPG
453	*Musa x paradisica* L.[2]	Musaceae	Banana Poona	केला पूना	FPG
	Bamboos				
454	*Bambusa heterostachya* (Munro) Holttum	Poaceae	Malay Dwarf Variegated Bamboo	बांस	BG
455	*Bambusa nana* Roxb.	Poaceae	Hedge Bamboo	बांस	BG
456	*Bambusa* sp. (Dehradun)	Poaceae	...	बांस	BG
457	*Bambusa sp.* (Variegated)	Poaceae	Male Bamboo, Solid Bamboo	बांस	BG
458	*Bambusa vulgaris* Schrad.[1]	Poaceae	Buddha Belly Bamboo	बांस	BG
459	*Bambusa vulgaris* Schrad.[2]	Poaceae	Wamin-Dwarf Buddha Belly Bamboo	बांस	BG
460	*Bambusa vulgaris* Schrad.[3]	Poaceae	Golden bamboo	बांस	BG
461	*Dendrocalamus asper* (Schult.) Backer	Poaceae	Giant Bamboo, Edible Bamboo	बांस	BG
462	*Dendrocalamus giganteus* Munro	Poaceae	Giant Bamboo	बांस	BG
463	*Dendrocalamus strictus* (Roxb.) Nees	Poaceae	Desi Bamboo	बांस	BG
	Gymnosperms including Cycads				
464	*Abies pindrow* (Royle ex D.Don) Royle	Pinaceae	West Himalayan Fir	झिल्ला	GSG
465	*Agathis robusta* (C.Moore ex F.Muell.) F.M.Bailey	Araucariaceae	Queensland Kauri Pine	...	GSG
466	*Araucaria bidwillii* Hook.	Araucariaceae	Bunya Pine	...	GSG
467	*Araucaria columnaris* (G.Forst.) Hook.	Araucariaceae	New Caledonia Pine, Cook Pine	...	GSG
468	*Araucaria cunninghamii* Mudie	Araucariaceae	Hoop Pine	...	GSG
469	*Cedrus deodara* (Roxb. ex D.Don) G.Don	Pinaceae	Himalayan Cedar, Devdar	देवदार	GSG
470	*Cryptomeria japonica* (Thunb. ex L.f.) D.Don	Cupressaceae	Sugi, Japanese Cedar	...	GSG

SN	Name of Plant Species	Family	Common Name	Hindi Name	Introduced in
471	*Cupressus sempervirens* L.	Cupressaceae	Pencil Pine, Mediterranean cypress	...	GSG
472	*Cupressus* sp.2	Cupressaceae	...	...	GSG
473	*Cycas circinalis* L.	Cycadaceae	Queen Sago	...	GSG
474	*Cycas revoluta* Thunb.	Cycadaceae	Japanese Sago Palm	...	GSG
475	*Dioon edule* Lindl.	Zamiaceae	Chestnut Dioon	...	GSG
476	*Dioon spinulosum* Dyer ex Eichl.	Zamiaceae	Giant Dioon	...	GSG
477	*Ginkgo biloba* L.	Ginkgoaceae	Ginkgo, Maidenhair Tree	...	GSG
478	*Juniperus indica* Bertol.	Cupressaceae	Black Juniper	...	GSG
479	*Juniperus* sp.[2]	Cupressaceae	Juniper Golden	...	GSG
480	*Juniperus* sp.[3]	Cupressaceae	Juniper var. Thailand	...	GSG
481	*Juniperus* sp.[4]	Cupressaceae	Juniper var. Thailand (variegated)	...	GSG
482	*Larix* sp.	Pinaceae	Larch	...	GSG
483	*Pinus roxburghii* Sarg.	Pinaceae	Chir Pine	चीड़	GSG
484	*Pinus strobus* var. *pendula* L.	Pinaceae	Pendula Eastern White Pine.	...	GSG
485	*Platycladus orientalis* (L.) Franco	Cupressaceae	Oriental Arbor-Vitae, Thuja	...	GSG
486	*Podocarpus neriifolius* D.Don	Podocarpaceae	Brown Pine	...	GSG
487	*Taxus baccata* L.	Taxaceae	English Yew, European Yew	...	GSG
488	*Thuja plicata* Donn ex D.Don	Cupressaceae	Western/Pacific Redcedar	मोर पंखी	GSG
489	*Tsuga canadensis* (L.) Carrière	Pinaceae	Canadian/Eastern Hemlock	...	GSG
490	*Zamia furfuracea* L.f. ex Aiton	Zamiaceae	Cardboard Palm, Mexican Cycad	...	GSG
491	*Zamia* sp.2	Zamiaceae	...	...	GSG
492	*Zamia variegata* Warsz.	Zamiaceae	Spanish-Camotillo	...	GSG
	Pteridophytes (Ferns and Fern Allies)				
493	*Adiantum* sp.	Pteridaceae	Fern	फ़र्न	IP-CP
494	*Ampelopteris prolifera* (Retz.) Copel.	Thelypteridaceae	Walking Fern	भैसाग	APG
495	*Equisetum ramosissimum* subsp. *debile* (Roxb. ex Vaucher) Hauke	Equisetaceae	Horsetail	...	MPG
496	*Equisetum ramosissimum* Desf. (Syn. *E. ramosissimum* subsp. *ramosissimum*)	Equisetaceae	Horsetail, Branched Horsetail	...	MPG
497	*Marsilea quadrifolia* L.	Marsileaceae	Four Leaf Clover, Water Shamrock, Common Water Clover	चौपतिया, सुस्नि	APG
	Algae				
498	*Chara* sp.	Characeae	Green Algae	...	APG

APG - Aquatic Plants Garden and Conservation Area; BG - Bamboo Garden (Bambusetum); BP - Boundary Plantation; FG - Ficus Garden; FPG - Fruit Plants Garden; GSG - Gymnosperms Garden; IP-CP - Indoor Plants-Cover Plants; MPG - Medicinal Plants Garden; OAG - Other Areas of Garden; PG - Palm Garden (Palmetum); XG - Xerophytes Garden (Arid & Semi-Arid Zone Plants)

Appendix 7: Plants under various themes in Thematic Botanic Garden

SN	Name of Plant Species	Family	Common Name	Hindi Name
	Fruit Plants Garden			
1	*Aegle marmelos* (L.) Corrêa	Rutaceae	Bel	बेल
2	*Annona squamosa* L.	Annonaceae	Custard Apple	शरीफा
3	*Artocarpus heterophyllus* Lam.	Moraceae	Jack Fruit, Kathal	कटहल
4	*Artocarpus lacucha* Buch.-Ham.	Moraceae	Monkey Fruit, Barhal	बरहल
5	*Averrhoa carambola* L.	Oxalidaceae	Star Fruit	कमरख
6	*Carica papaya* L.	Caricaceae	Papaya	पपीता
7	*Carissa carandas* L.	Apocynaceae	Bengal Currant, Karonda	करौंदा
8	*Citrus × aurantium* L.	Rutaceae	Bitter Orange	नारंगी
9	*Citrus aurantiifolia* (Christm.) Swingle	Rutaceae	Key Lime	कागजी निम्बू
10	*Citrus japonica* Thunb.	Rutaceae	Chinese Orange, Nagami Kumquat	चाइनीज संतरा
11	*Citrus limetta* Risso	Rutaceae	Sweet lemon	मौसंबी
12	*Citrus limon* (L.) Osbeck	Rutaceae	Lemon	निम्बू
13	*Citrus maxima* (Burm.) Merr.	Rutaceae	Pomelo	चकोतरा
14	*Citrus medica* L.	Rutaceae	Citron	बड़ा निम्बू
15	*Citrus* sp. (hybrid between 'King' (*C. nobilis*) × 'Willow Leaf' (*C. × deliciosa*)	Rutaceae	Mandarin, Kinnow	किन्नौ
16	*Citrus × sinensis* (hybrid between pomelo (*C. maxima*) & mandarin (*C. reticulata*)	Rutaceae	Sweet Orange	संतरा
17	*Cordia dichotoma* G.Forst.	Boraginaceae	Indian Cherry, Glue Berry	लसौड़ा, टेंटी, डेला
18	*Eriobotrya japonica* (Thunb.) Lindl.	Rosaceae	Loquat	लोकाट
19	*Ficus auriculata* Lour.	Moraceae	Fig	अंजीर
20	*Ficus carica* L.	Moraceae	Fig	अंजीर
21	*Ficus palmata* Forssk.	Moraceae	Wild Fig	जंगली अंजीर
22	*Grewia asiatica* L.	Malvaceae	Phalsa	फालसा
23	*Hylocereus undatus* (Haw.) Britton & Rose	Cactaceae	Dragon Fruit	ड्रैगन फल
24	*Juglans regia* L.	Juglandaceae	Walnut	अखरोट
25	*Limonia acidissima* Groff	Rutaceae	Wood Apple	कैथ
26	*Litchi chinensis* Sonn.	Sapindaceae	Lychee	लीची
27	*Malpighia emarginata* DC.	Malpighiaceae	West Indian Cherry	...

SN	Name of Plant Species	Family	Common Name	Hindi Name
28	*Malpighia* sp.2	Malpighiaceae	Cherry	…
29	*Malus domestica* Borkh.	Rosaceae	Apple	सेब
30	*Mangifera indica* L.[1]	Anacardiaceae	Mango Amrapali	आम आम्रपाली
31	*Mangifera indica* L.[2]	Anacardiaceae	Mango Dasheri	आम दशहरी
32	*Manilkara hexandra* (Roxb.) Dubard	Sapotaceae	Ceylon Ironwood	खिरनी
33	*Manilkara zapota* (L.) P.Royen[1]	Sapotaceae	Sapota	चीकू
34	*Manilkara zapota* (L.) P.Royen[2]	Sapotaceae	Sapota Variegated	चीकू
35	*Moringa oleifera* Lam.	Moringaceae	Drumstick Tree	सहजन
36	*Morus alba* L.	Moraceae	Mulberry	शहतूत
37	*Musa x paradisica* L.[1]	Musaceae	Banana Hyderabad	केला हैदराबादी
38	*Musa x paradisica* L.[2]	Musaceae	Banana Poona	केला पूना
39	*Passiflora coccinea* Aubl.	Passifloraceae	Red/Scarlet Passion Flower	झुमका लता
40	*Passiflora edulis* Sims	Passifloraceae	Passion Fruit	कृष्णा लता, झुमका लता
41	*Persea americana* Mill.	Lauraceae	Avocado, Alligator Pear	…
42	*Phyllanthus emblica* L.	Phyllanthaceae	Indian Gooseberry, Myrobalan	आमला
43	*Pithecellobium dulce* (Roxb.) Benth.	Leguminosae	Manila Tamarind, Madras Thorn	जंगल जलेबी
44	*Prunus armeniaca* L.	Rosaceae	Apricot	ख़ूबानी
45	*Prunus domestica* L.	Rosaceae	Plum	आलू बुखारा
46	*Prunus dulcis* (Mill.) D.A.Webb	Rosaceae	Almond	बादाम
47	*Prunus persica* (L.) Batsch	Rosaceae	Peach	आड़ू
48	*Psidium guajava* L.[1]	Myrtaceae	Guava	अमरुद
49	*Psidium guajava* L.[2]	Myrtaceae	Guava Allahabadi	अमरुद
50	*Psidium guajava* L.[3]	Myrtaceae	Guava L-49	अमरुद
51	*Psidium* sp.[4]	Myrtaceae	Cherry Guava	…
52	*Punica granatum* L.	Lythraceae	Pomegranate	अनार
53	*Pyrus pyrifolia* (Burm.f.) Nakai	Rosaceae	Pear	नाशपाती
54	*Spondias* sp.	Anacardiaceae	Hog Plums, Spanish Plums	अमरा
55	*Syzygium jambos* (L.) Alston	Myrtaceae	Malabar Plum, Malabar Plum	नाशपाती
56	*Syzygium samarangense* (Blume) Merr. & L.M.Perry	Myrtaceae	Rose Apple	अमरा
57	*Syzygium cumini* (L.) Skeels	Myrtaceae	Java Plum, Jamun, Black Plum	जामुन
58	*Tamarindus indica* L.[1]	Leguminosae	Tamarind	इमली

SN	Name of Plant Species	Family	Common Name	Hindi Name
59	*Tamarindus indica* L.[2]	Leguminosae	Sweet Tamarind	मीठी इमली
60	*Vitis vinifera* L.	Vitaceae	Grapes	अंगूर
61	*Ziziphus jujuba* Mill.	Rhamnaceae	Common Jujube	बेर
62	*Ziziphus nummularia* (Burm.f.) Wight & Arn.	Rhamnaceae	Jhar Beri	झाड़ बेर
	Palm Garden (Palmetum)			
1	*Areca catechu* L.	Arecaceae	Supari/Betelnut Palm	...
2	*Beaucarnea recurvata* Lem.	Asparagaceae	Nolina/Bottle Palm	...
3	*Bismarckia nobilis* Hildebr. & H.Wendl.	Arecaceae	Bismarckia/Silver Blue Fan Palm	...
4	*Butia capitata* (Mart.) Becc.	Arecaceae	Jelly Palm	...
5	*Caryota mitis* Lour.	Arecaceae	Fish Tail Palm	...
6	*Chamaedorea metallica* O.F.Cook ex H.E. Moore	Arecaceae	Chaemaedorea Palm	...
7	*Chamaerops humilis* L.	Arecaceae	Chamaerops Palm	...
8	*Chamaerops humilis* var. *cerifera*	Arecaceae	Blue Mediterranean Fan Palm	...
9	*Copernicia prunifera* (Mill.) H.E.Moore	Arecaceae	Copernicia Palm	...
10	*Dypsis decaryi* (Jum.) Beentje & J.Dransf.	Arecaceae	Triangular Palm	...
11	*Dypsis lutescens* (H.Wendl.) Beentje & J.Dransf.	Arecaceae	Areca Palm	...
12	*Hyophorbe lagenicaulis* (L.H.Bailey) H.E. Moore	Arecaceae	Bottle/Champagne Palm	...
13	*Latania lontaroides* (Gaertn.) H.E.Moore	Arecaceae	Red Latan Palm	...
14	*Licuala grandis* H.Wendl.	Arecaceae	Ruffled Fan Palm	...
15	*Livistona* sp.	Arecaceae	Livistona Palm	...
16	*Pandanus utilis* Bory	Pandanaceae	Pandanus Palm	केवड़ा
17	*Phoenix dactylifera* L.	Arecaceae	Date Palm	खजूर
18	*Phoenix paludosa* Roxb.	Arecaceae	Mangrove Date Palm	...
19	*Phoenix roebelenii* O'Brien	Arecaceae	Pygmy/Miniature Date Palm	...
20	*Phoenix sylvestris* (L.) Roxb.	Arecaceae	Date Palm	खजूर
21	*Ptychosperma elegans* (R.Br.) Blume	Arecaceae	Seaforthia Palm	...
22	*Ravenala madagascariensis* Sonn.	Strelitziaceae	Travellers Palm	...
23	*Ravenea* sp.	Arecaceae	Majesty Palm	...
24	*Rhapis excelsa* (Thunb.) Henry	Arecaceae	Rhapis Palm	...
25	*Roystonea regia* (Kunth) O.F.Cook	Arecaceae	Royal Palm	...
26	*Sabal palmetto* (Walter) Lodd. ex Schult. & Schult.f.	Arecaceae	Sabal or Cabbage Palm	...
27	*Thrinax* sp.	Arecaceae	Thrinax Palm	...
28	*Trachycarpus fortunei* (Hook.) H.Wendl.	Arecaceae	China Palm	...
29	Unidentified1	Arecaceae	Nephtrox Palm (?)	...
30	Unidentified2 (Endemic to Andaman Islands)	Arecaceae	...	...

SN	Name of Plant Species	Family	Common Name	Hindi Name
31	*Washingtonia filifera* (Linden ex André) H.Wendl. ex de Bary	Arecaceae	Washington/Desert/California Palm	...
32	*Wodyetia bifurcata* A.K.Irvine	Arecaceae	Fox Tail Palm	...
	Gymnosperms Garden			
	Cycads			
1	*Cycas circinalis* L.	Cycadaceae	Queen Sago	...
2	*Cycas revoluta* Thunb.	Cycadaceae	Japanese Sago Palm	...
3	*Dioon edule* Lindl.	Zamiaceae	Chestnut Dioon	...
4	*Dioon spinulosum* Dyer ex Eichl.	Zamiaceae	Giant Dioon	...
5	*Zamia furfuracea* L.f. ex Aiton	Zamiaceae	Cardboard Palm, Mexican Cycad	...
6	*Zamia* sp.2	Zamiaceae	...	...
7	*Zamia variegata* Warsz.	Zamiaceae	Spanish-Camotillo	...
	Other Gymnosperm species			
8	*Abies pindrow* (Royle ex D.Don) Royle	Pinaceae	West Himalayan Fir	झिल्ला
9	*Agathis robusta* (C.Moore ex F.Muell.) F.M.Bailey	Araucariaceae	Queensland Kauri Pine	...
10	*Araucaria bidwillii* Hook.	Araucariaceae	Bunya Pine	...
11	*Araucaria columnaris* (G.Forst.) Hook.	Araucariaceae	New Caledonia Pine, Cook Pine	...
12	*Araucaria cunninghamii* Mudie	Araucariaceae	Hoop Pine	...
13	*Cedrus deodara* (Roxb. ex D.Don) G.Don	Pinaceae	Himalayan Cedar, Devdar	देवदार
14	*Cryptomeria japonica* (Thunb. ex L.f.) D.Don	Cupressaceae	Sugi, Japanese Cedar	...
15	*Cupressus sempervirens* L.	Cupressaceae	Pencil Pine, Mediterranean cypress	...
16	*Cupressus* sp.2	Cupressaceae	...	...
17	*Ginkgo biloba* L.	Ginkgoaceae	Ginkgo, Maidenhair Tree	...
18	*Juniperus indica* Bertol.	Cupressaceae	Black Juniper	...
19	*Juniperus* sp.[2]	Cupressaceae	Juniper Golden	...
20	*Juniperus* sp.[3]	Cupressaceae	Juniper var. Thailand	...
21	*Juniperus* sp.[4]	Cupressaceae	Juniper var. Thailand (Variegated)	...
22	*Larix* sp.	Pinaceae	Larch	...
23	*Pinus roxburghii* Sarg.	Pinaceae	Chir Pine	चीड
24	*Pinus strobus* var. *pendula* L.	Pinaceae	Pendula Eastern White Pine	...
25	*Platycladus orientalis* (L.) Franco	Cupressaceae	Oriental Arbor-Vitae, Thuja	...
26	*Podocarpus neriifolius* D.Don	Podocarpaceae	Brown Pine	...
27	*Taxus baccata* L.	Taxaceae	English Yew, European Yew	...
28	*Thuja plicata* Donn ex D.Don	Cupressaceae	Western/Pacific Redcedar	मोर पंखी
29	*Tsuga canadensis* (L.) Carrière	Pinaceae	Canadian/Eastern Hemlock	...

SN	Name of Plant Species	Family	Common Name	Hindi Name
	Bamboo Garden (Bambusetum)			
1	*Bambusa heterostachya* (Munro) Holttum	Poaceae	Malay Dwarf Variegated Bamboo	बांस
2	*Bambusa nana* Roxb.	Poaceae	Hedge Bamboo	बांस
3	*Bambusa* sp. (Dehradun)	Poaceae	…	बांस
4	*Bambusa sp.* (Variegated)	Poaceae	Male Bamboo, Solid Bamboo	बांस
5	*Bambusa vulgaris* Schrad.[1]	Poaceae	Buddha Belly Bamboo	बांस
6	*Bambusa vulgaris* Schrad.[2]	Poaceae	Wamin-Dwarf Buddha Belly Bamboo	बांस
7	*Bambusa vulgaris* Schrad.[3]	Poaceae	Golden Bamboo	बांस
8	*Dendrocalamus asper* (Schult.) Backer	Poaceae	Giant Bamboo, Edible Bamboo	बांस
9	*Dendrocalamus giganteus* Munro	Poaceae	Giant Bamboo	बांस
10	*Dendrocalamus strictus* (Roxb.) Nees	Poaceae	Desi Bamboo	बांस
	Medicinal Plants Garden			
1	*Abrus precatorius* L.	Leguminosae	Licorice, Coral Bead Vine, Crab's Eye	रत्ती, गुंची
2	*Abutilon indicum* (L.) Sweet	Malvaceae	Indian Abutilon, Indian Mallow	कंघी, ककही
3	*Achyranthes aspera* L.	Amaranthaceae	Prickly Chaff Flower, Devil's Horsewhip	चिरचिटा, लटजीरा
4	*Ageratum houstonianum* Mill.	Compositae	Floss Flower, Blue Billygoat Weed	रक्त रोधी
5	*Allium tuberosum* Rottler ex Spreng.	Amaryllidaceae	Garlic Chives, Oriental Garlic, Asian/Chinese Chives	…
6	*Aloe vera* (L.) Burm.f.	Xanthorrhoeaceae	Aloe	घृतकुमारी
7	*Alpinia galanga* (L.) Willd.	Zingiberaceae	Kulanjan, Thai Ginger	कुलंजन
8	*Alternanthera paronychioides* A.St.-Hil.	Amaranthaceae	Smooth Chaff Flower	…
9	*Amaranthus viridis* L.	Amaranthaceae	Slender/Green Amaranth	जंगली चौलाई
10	*Anisomeles indica* (L.) Kuntze	Lamiaceae	Indian/Malabar Catmint	काला भंग्रा
11	*Argemone mexicana* L.	Papaveraceae	Mexican Prickly Poppy	सत्यानाशी कटैला
12	*Artemisia scoparia* Waldst. & Kitam.	Compositae	Redstem Wormwood	सीता-बानी
13	*Asparagus racemosus* Willd.	Asparagaceae	Satawari	सतावरी
14	*Azadirachta indica* A.Juss.	Meliaceae	Neem, Indian Lilac, Margosa	नीम
15	*Bacopa monnieri* (L.) Wettst.	Plantaginaceae	Water Hyssop, Indian Pennywort	ब्राह्मी
16	*Barleria prionitis* L.	Acanthaceae	Porcupine Flower	वज्रदंती
17	*Blumea viscosa* (Mill.) V.M.Badillo	Compositae	Sticky Blumea, Clammy False Oxtongue	…
18	*Boerhavia diffusa* L.	Nyctaginaceae	Red Spiderling, Spreading Hog Weed	पुनर्नवा, गदहपूर्णा
19	*Bryophyllum pinnatum* (Lam.) Oken	Crassulaceae	Patharchatta	पत्थरचट्टा

SN	Name of Plant Species	Family	Common Name	Hindi Name
20	*Butea monosperma* (Lam.) Taub.	Leguminosae	Flame of the Forest, Bastard Teak	पलाश, ढाक, टेसू
21	*Calendula officinalis* L.	Compositae	Pot Marigold, Calendula	गेंदा
22	*Calotropis gigantea* (L.) Dryand.	Apocynaceae	Crown Flower, Giant Indian Milkweed	सफ़ेद आक
23	*Calotropis procera* (Aiton) Dryand.	Apocynaceae	Rubber Bush, Sodom Apple	आक, मदार
24	*Cannabis sativa* L.	Cannabaceae	Marijuana, Hemp	भांग, गांजा, चरस
25	*Carthamus oxyacantha* M.Bieb.	Compositae	Wild Safflower, Jeweled Distaff Thistle	कडिअरी, पीली कटेरी
26	*Catharanthus roseus* (L.) G.Don	Apocynaceae	Periwinkle, Vinca	सदाबहार
27	*Centella asiatica* (L.) Urb.	Apiaceae	Indian Pennywort, Gotu kola	ब्राह्मी, ब्रह्मण डूकी
28	*Chenopodium album* L.	Amaranthaceae	Goosefoot, Wild Spinach, Lamb's Quarters	बथुआ
29	*Chenopodium murale* L.	Amaranthaceae	Nettle-leaved Goosefoot, Australian Spinach	जंगली बथुआ, खरतुआ
30	*Chlorophytum borivilianum* Santapau & R.R.Fern.	Asparagaceae	Safed Musli	सफ़ेद मुसली
31	*Chrysopogon zizanioides* (L.) Roberty	Poaceae	Vetiver, Khas-Khas Grass	खस-खस
32	*Cinnamomum camphora* (L.) J.Presl	Lauraceae	Camphor Tree	कपूर
33	*Cinnamomum tamala* (Buch.-Ham.) T.Nees & Eberm.	Lauraceae	Indian Bay Leaf	तेजपत्ता
34	*Cirsium arvense* (L.) Scop.	Compositae	Creeping/Californian Thistle	कंदई
35	*Citrullus colocynthis* (L.) Schrad.	Cucurbitaceae	Colocynth, Desert Gourd	इंद्रायन
36	*Cleome viscosa* L.	Cleomaceae	Asian Spider Flower, Dog Mustard	बागरा, हुरहुर, पीली हुरहुर
37	*Clerodendrum phlomidis* L.f.	Lamiaceae	Arni	अर्नी
38	*Clitoria ternatea* L.	Leguminosae	Butterfly Pea	अपराजिता
39	*Coccinia grandis* (L.) Voigt	Cucurbitaceae	Ivy Gourd, Scarlet Gourd	कुन्द्रू
40	*Convolvulus arvensis* L.	Convolvulaceae	Field Bindweed	हिरनपग
41	*Convolvulus prostratus* Forssk.	Convolvulaceae	Prostrate Bindweed	शखपुष्पी
42	*Crinum latifolium* L.	Amaryllidaceae	Pink-Striped Trumpet Lily	सुदर्शन
43	*Croton bonplandianus* Baill.	Euphorbiaceae	Croton, Rushfoil	बन तुलसी, काला भांगरा
44	*Curcuma longa* L.	Zingiberaceae	Turmeric	हल्दी
45	*Cymbopogon citratus* (DC.) Stapf	Poaceae	Lemon Grass	अघिनघास, गंधत्रिण
46	*Cynodon dactylon* (L.) Pers.	Poaceae	Bermuda Grass, Bahama Grass	दूब

SN	Name of Plant Species	Family	Common Name	Hindi Name
47	*Datura metel* L.	Solanaceae	Thorn Apple	धतूरा
48	*Datura innoxia* Mill.	Solanaceae	Pricklyburr, Thorn Apple	सफ़ेद धतूरा
49	*Desmodium gangeticum* (L.) DC.	Leguminosae	Sal Leaved Desmodium	दीर्घमूली, शालपर्णी
50	*Diospyros malabarica* (Desr.) Kostel.	Ebenaceae	Gaub Tree, Malabar Ebony	तेंदू
51	*Diospyros montana* Roxb.	Ebenaceae	Mountain Persimmon, Bombay Ebony	बिसतेंदू
52	*Dysphania ambrosioides* (L.) Mosyakin & Clemants	Amaranthaceae	Mexican Tea	सुगंध वास्तुका
53	*Eclipta prostrata* (L.) L.	Compositae	False Daisy, Trailing Eclipta	भृगराज, केशराज
54	*Equisetum ramosissimum* subsp. *debile* (Roxb. ex Vaucher) Hauke	Equisetaceae	Horsetail	...
55	*Equisetum ramosissimum* Desf. (Syn. *E. ramosissimum* subsp. *ramosissimum*)	Equisetaceae	Horsetail, Branched Horsetail	...
56	*Euphorbia hirta* L.	Euphorbiaceae	Asthma Weed, Asthma Plant	दुधी, बड़ा दुधी
57	*Euphorbia prostrata* Aiton	Euphorbiaceae	Prostrate Sandmat, Prostrate Spurge	छोटी दुधी
58	*Evolvulus nummularius* (L.) L.	Convolvulaceae	Roundleaf Bindweed, Agracejo Rastrero	मुसाकर्णी
59	*Gomphrena celosioides* Mart.	Amaranthaceae	Prostrate Globe-Amaranth	...
60	*Heliotropium ellipticum* Ledeb.	Boraginaceae	Heliotrope	हाथीसूंड़ा
61	*Hyptis suaveolens* (L.) Poit.	Lamiaceae	American Mint, Bush Mint	विलायती तुलसी
62	*Imperata cylindrica* (L.) Raeusch.	Poaceae	Blady Grass, Cogon Grass, Kunai Grass, Japanese Bloodgrass	दाभ, अलू
63	*Indigofera tinctoria* L.	Leguminosae	True Indigo, Common Indigo	नील
64	*Ipomoea carnea* Jacq.	Convolvulaceae	Bush Morning Glory	बेहया बेशरम
65	*Ipomoea coptica* (L.) Roth ex Roem. & Schult.	Convolvulaceae	Egyptian Morning Glory	...
66	*Ipomoea pes-tigridis* L.	Convolvulaceae	Tiger Foot Morning Glory	पंचपतिया
67	*Jatropha gossypiifolia* L.	Euphorbiaceae	Bellyache Bush, Cotton-Leaf Physic Nut	रतनजोती
68	*Justicia adhatoda* L.	Acanthaceae	Malabar Nut, Vasaka, Adhatoda, Adulsa	अरुस, अडूसा
69	*Kigelia africana* (Lam.) Benth.	Bignoniaceae	African Sausage Tree	बालम खीरा
70	*Lagerstroemia speciosa* (L.) Pers.	Lythraceae	Giant Crape-Myrtle, Pride of India	जरुल
71	*Launaea procumbens* (Roxb.) Ramayya & Rajagopal	Compositae	Creeping Launaea	पीली दुधी, जंगली गोभी
72	*Lawsonia inermis* L.	Lythraceae	Henna	मेहेंदी
73	*Melilotus albus* Medik.	Leguminosae	White Sweet Clover, White Melilot	सफ़ेद बन मेथी
74	*Melilotus indicus* (L.) All.	Leguminosae	Indian Sweet Clover, Yellow Sweet Clover	बन मेथी, मोरिला
75	*Melochia corchorifolia* L.	Malvaceae	Chocolate Weed	चित्रबीज

SN	Name of Plant Species	Family	Common Name	Hindi Name
76	*Mentha arvensis* L.	Lamiaceae	Mint	पुदीना
77	*Mentha* sp.2	Lamiaceae	Ginger Mint	पुदीना
78	*Mimosa pudica* L.	Leguminosae	Touch-Me-Not	छुई-मुई
79	*Mimusops elengi* L.	Sapotaceae	Spanish Cherry	मौलसरी
80	*Moringa oleifera* Lam.	Moringaceae	Drumstick Tree	सहजन
81	*Murraya koenigii* (L.) Spreng.	Rutaceae	Curry Leaf Tree	करी पत्ता
82	*Nicotiana plumbaginifolia* Viv.	Solanaceae	Tex-Mex Tobacco	जंगली तंबाकू
83	*Nyctanthes arbor-tristis* L.	Oleaceae	Night-Flowering Jasmine	हार सिंगार
84	*Ocimum* × *africanum* Lour.	Lamiaceae	Lemon Basil	तुलसी
85	*Ocimum basilicum* L.	Lamiaceae	Sweet Basil	राम तुलसी
86	*Ocimum tenuiflorum* L.	Lamiaceae	Holy Basil	तुलसी
87	*Origanum* sp.	Lamiaceae	Oregano	...
88	*Oxystelma esculentum* (L. f.) Sm.	Apocynaceae	Rosy Milkweed Vine	दुधिया तला, दुधी बेल
89	*Peristrophe bicalyculata* (Retz.) Nees	Acanthaceae	Panicled Foldwing, Kali Anghedi	अत्रिलाल
90	*Persicaria barbata* (L.) H.Hara	Polygonaceae	Bearded Knotweed	...
91	*Phyla nodiflora* (L.) Greene	Verbenaceae	Frog Fruit, Creeping Lip Plant	जल बूटी
92	*Phyllanthus amarus* Schumach. & Thonn.	Phyllanthaceae	Black Catnip, Stone Breaker	भूई आवला, हजारमणि
93	*Phyllanthus emblica* L.	Phyllanthaceae	Indian Gooseberry, Myrobalan	आमला
94	*Phyllanthus maderaspatensis* L.	Phyllanthaceae	Madras Leaf-Flower	हजारमणि
95	*Phyllanthus reticulatus* Poir.	Phyllanthaceae	Black Honey Shrub, Black Berried Featherfoil	रसभरी
96	*Physalis angulata* L.	Solanaceae	Angular Winter Cherry, Cutleaf Groundcherry	सरभरी
97	*Physalis peruviana* L.	Solanaceae	Cape Gooseberry, Peruvian Ground Cherry, Poha Berry	जलकुम्भी
98	*Pimenta dioica* (L.) Merr.	Myrtaceae	Allspice	...
99	*Piper betle* L.	Piperaceae	Betel	पान
100	*Plectranthus amboinicus* (Lour.) Spreng.	Lamiaceae	Cuban Oregano, Indian Borage	पत्ता अजवाईन
101	*Pluchea lanceolata* (DC.) C.B.Clarke	Compositae	Rasna	फार
102	*Plumbago auriculata* Lam.	Plumbaginaceae	Plumbago, Cape Leadwort	नीला चित्रक
103	*Portulaca oleracea* L.	Portulacaceae	Purslane	लूनिया, बड़ी नोनी
104	*Putranjiva roxburghii* Wall.	Putranjivaceae	Putranjiva, Lucky Bean Tree	पुत्रजीव

SN	Name of Plant Species	Family	Common Name	Hindi Name
105	*Rauvolfia serpentina* (L.) Benth. ex Kurz	Apocynaceae	Indian snakeroot	सर्पगंधा
106	*Ricinus communis* L.	Euphorbiaceae	Castor Bean, Castor Oil Plant	अंरडी, चिया
107	*Rumex dentatus* L.	Polygonaceae	Toothed Dock, Aegean Dock	जंगली पालक
108	*Santalum album* L.	Santalaceae	Sandalwood	चंदन
109	*Scoparia dulcis* L.	Plantaginaceae	Sweet Broom Weed	मीठीपत्ती, घोड़ा तुलसी
110	*Senna occidentalis* (L.) Link	Leguminosae	Coffee Senna, Negro Coffee	कसुंडा, बड़ी कसौंदी
111	*Senna* sp.	Leguminosae	…	…
112	*Sesbania sesban* (L.) Merr.	Leguminosae	Common Sesban, Egyptian Rattlepod, Egyptian Riverhemp	जयंती, रेवासन
113	*Sida acuta* Burm.f.	Malvaceae	Common Wireweed, Common Fanpetals	बरैरा
114	*Sida cordata* (Burm.f.) Borss.Waalk.	Malvaceae	Long-Stalk Sida	भूइनी, खरैंटी
115	*Sida cordifolia* L.	Malvaceae	Heart-Leaf Sida, Country Mallow, Flannel Weed	खरींटा, खरैंटी
116	*Solanum americanum* Mill.	Solanaceae	Black Nightshade	मकोय
117	*Solanum surattense* Burm. f.	Solanaceae	Thorny/Yellow Berried Nightshade	कटेली, ऊंट कटेली
118	*Solanum indicum* L.	Solanaceae	Brihati, Poison Berry	बड़ी कटेली
119	*Sonchus asper* (L.) Hill	Compositae	Prickly/Rough Sow-Thistle	दुधी
120	*Tephrosia purpurea* (L.) Pers.	Leguminosae	Wild Indigo, Fish Poison	सरफोंक, शरपुंखा
121	*Tephrosia villosa* (L.) Pers.	Leguminosae	Shaggy Wild Indigo, Hoary Tephrosia	…
122	*Terminalia arjuna* (Roxb. ex DC.) Wight & Arn.	Combretaceae	Arjun	अर्जुन
123	*Terminalia bellirica* (Gaertn.) Roxb.	Combretaceae	Beleric, Bastard Myrobalan	बहेड़ा, कर्षफल
124	*Terminalia chebula* Retz.	Combretaceae	Black/Chebulic Myrobalan	हर्रा, हरड़
125	*Thymus vulgaris* L.	Lamiaceae	Garden Thyme	थाइमस
126	*Tinospora sinensis* (Lour.) Merr.	Menispermaceae	Chinese Tinospora	गिलोय
127	*Trianthema portulacastrum* L.	Aizoaceae	Giant pigweed, Horse-Purslane	विषखपरा, साबुनी
128	*Tribulus terrestris* L.	Zygophyllaceae	Puncture Vine, Caltrop	गोखरू
129	*Triumfetta rhomboidea* Jacq.	Malvaceae	Diamond Burbark, Chinese Bur	चीकी
130	*Urena lobata* L.	Malvaceae	Caesarweed	बचिता, लपेटुआ
131	*Withania somnifera* (L.) Dunal	Solanaceae	Winter Cherry, Indian Ginseng	अश्वगंधा, असगंध

SN	Name of Plant Species	Family	Common Name	Hindi Name
132	*Xanthium strumarium* L.	Compositae	Common Cocklebur, Clotbur	छोटा धतूरा, घाघरा, संखहुली
	Xerophytes Garden (Arid & Semi-Arid Zone Plants)			
1	*Acacia catechu* (L.f.) Willd.	Leguminosae	Cutch Tree	खैर
2	*Acacia concinna* (Willd.) DC.	Leguminosae	Soap Pod	शिकाकाई
3	*Acacia nilotica* (L.) Delile	Leguminosae	Gum Arabic Tree, Babul	बबूल
4	*Acacia modesta* Wall.	Leguminosae	Amritsar Gum	फुलाई
5	*Acacia senegal* (L.) Willd.	Leguminosae	Gum Acacia, Gum Arabic Tree	कुमुट
6	*Agave americana* L.	Asparagaceae	Agave	कमल कैक्टस
7	*Albizia lebbeck* (L.) Benth.	Leguminosae	Siris Tree	सिरिस
8	*Aloe vera* (L.) Burm.f.	Xanthorrhoeaceae	Aloe	घृतकुमारी
9	*Boswellia serrata* Roxb. ex Colebr.	Burseraceae	Indian Frankincense	सालर, शल्लकी
10	*Bryophyllum pinnatum* (Lam.) Oken	Crassulaceae	Patharchatta	पत्थरचट्टा
11	*Cactus* sp.1	Cactaceae	Cactus	…
12	*Cactus* sp.2	Cactaceae	Cactus	…
13	*Cactus* sp.3	Cactaceae	Star Cactus	…
14	*Calotropis procera* (Aiton) Dryand.	Apocynaceae	Rubber Bush, Sodom Apple	आक, मदार
15	*Capparis decidua* (Forssk.) Edgew.	Capparaceae	Bare Caper	कैर, करीर
16	*Colophospermum mopane* (Benth.) Leonard	Leguminosae	Mopane	…
17	*Commiphora wightii* (Arn.) Bhandari	Burseraceae	Guggul, Indian Bdellium	गुगुल
18	*Cordia dichotoma* G.Forst.	Boraginaceae	Indian Cherry, Glue Berry	लसौड़ा, टेंटी, डेला
19	*Cordia sinensis* Lam.	Boraginaceae	Grey Leaved Saucerberry	गोंदी
20	*Dichrostachys cinerea* (L.) Wight & Arn.	Leguminosae	Sickle Bush	कुणाली
21	*Euphorbia cotinifolia* L.	Euphorbiaceae	Caribbean Copper Bush	…
22	*Euphorbia milii* Des Moul.	Euphorbiaceae	Crown of Thorns, Kiss Me Not	…
23	*Euphorbia tirucalli* L.	Euphorbiaceae	Pencil Tree, Indian Tree Spurge	…
24	*Euphorbia royleana* Boiss.	Euphorbiaceae	Royle's Spurge	थोर
25	*Euphorbia* sp.	Euphorbiaceae	Euphorbia Cactus	…
26	*Furcraea foetida* (L.) Haw.	Asparagaceae	Mauritius Hemp, Giant False Agave	…
27	*Jatropha curcas* L.	Euphorbiaceae	Barbados Nut	रत्नजोत
28	*Justicia* sp.	Acanthaceae	…	…
29	*Malvaviscus arboreus* var. *mexicanus* Schltdl.	Malvaceae	Mexican Turk's Cap	जुबा कुसुम

SN	Name of Plant Species	Family	Common Name	Hindi Name
30	*Operculicarya decaryi* H. Perrier	Anacardiaceae	Elephant Tree	...
31	*Opuntia dillenii* (Ker Gawl.) Haw.	Cactaceae	Pricklypear	नागफनी
32	*Parkinsonia aculeata* L.	Leguminosae	Mexican Palo Verde, Jerusalem Thorn	विलायती कीकर, राम बबूल
33	*Prosopis cineraria* (L.) Druce	Leguminosae	Khejri Tree	खेजड़ी
34	*Prosopis juliflora* (Sw.) DC.	Leguminosae	Algaroba, Mesquite	जगली कीकर
35	*Ravenia spectabilis* Engl.	Rutaceae	Lemonia, Pink Ravenia	...
36	*Russelia equisetiformis* Schltdl. & Cham.	Plantaginaceae	Fountain Bush, Coral Plant	रसीली
37	*Salvadora oleoides* Decne.	Salvadoraceae	Pilu	पीलू
38	*Salvadora persica* L.	Salvadoraceae	Toothbrush Tree, Miswak	मिस्वाक
39	*Sansevieria hyacinthoides* (L.) Druce	Asparagaceae	Snake plant, Devil's tongue	...
40	*Tabebuia aurea* (Silva Manso) Benth. & Hook.f. ex S.Moore	Bignoniaceae	Silver/Yellow Trumpet Tree	...
41	*Tabebuia rosea* (Bertol.) Bertero ex A.DC.	Bignoniaceae	Rosy Trumpet Tree	बंसत रानी
42	*Tamarindus indica* L.[1]	Leguminosae	Tamarind	इमली
43	*Tamarix indica* Willd.	Tamaricaceae	Indian Tamarisk	झाऊ
44	*Tecomella undulata* (Sm.) Seem.	Bignoniaceae	Roheda	...
45	*Yucca aloifolia* L.	Asparagaceae	Aloe Yucca, Dagger Plant	युक्का
46	*Yucca glauca* Nutt.	Asparagaceae	Soapweed Yucca	युक्का
	Aquatic Plants Garden and Conservation Area			
1	*Ampelopteris prolifera* (Retz.) Copel.	Thelypteridaceae	Walking Fern	भैसाग
2	*Ceratophyllum demersum* L.	Ceratophyllaceae	Hornwort, Coon's Tail	बृहतपुष्पी
3	*Chara* sp.	Characeae	Green Algae	...
4	*Eichhornia crassipes* (Mart.) Solms	Pontederiaceae	Water Hyacinth	जलकुम्भी
5	*Hydrilla verticillata* (L.f.) Royle	Hydrocharitaceae	Waterthyme, Hydrilla	सेवार, झंगी
6	*Lemna perpusilla* Torr.	Araceae	Common Duckweed, Common Waterlens, Duck Meat	...
7	*Marsilea quadrifolia* L.	Marsileaceae	Four Leaf Clover, Water Shamrock, Common Water Clover	चौपतिया, सुस्नि
8	*Nymphaea pubescens* Willd.	Nymphaeaceae	White Water Lily, White Lotus	को का, कुमुद
9	*Nymphoides cristata* (Roxb.) Kuntze	Menyanthaceae	Crested Floating Heart	कुमुदिनी
10	*Pistia stratiotes* L.	Araceae	Water Cabbage, Water Lettuce	जलकुम्भी
11	*Potamogeton crispus* L.	Potamogetonaceae	Curly-Leaf Pondweed	...
12	*Sagittaria guayanensis* Kunth	Alismataceae	Guyanese Arrowhead	...
13	*Spirodela polyrrhiza* (L.) Schleid.	Araceae	Common Duck Meat, Duckweed	...

SN	Name of Plant Species	Family	Common Name	Hindi Name
14	*Stuckenia pectinata* (L.) Börner	Potamogetonaceae	Sago Pondweed, Fennel Pondweed	...
15	*Utricularia stellaris* L.f.	Lentibulariaceae	Bladderwort	...
16	*Vallisneria spiralis* L.	Hydrocharitaceae	Tape Grass, Straight Vallisneria	सिवाल, शैवाल
	Ficus Garden			
1	*Ficus benghalensis* L.	Moraceae	Banyan Tree	बढ़
2	*Ficus benjamina* L.	Moraceae	Weeping Fig, Benjamin's Fig	पुकर
3	*Ficus binnendijkii* Miq.	Moraceae	Alii Fig	...
4	*Ficus elastica* Roxb. ex Hornem.	Moraceae	Rubber Plant	...
5	*Ficus hispida* L.f.	Moraceae	Devil Fig	गोबला
6	*Ficus lyrata* Warb.	Moraceae	Fiddle Leaf Fig	...
7	*Ficus natalensis* subsp. *leprieurii* (Miq.) C.C.Berg	Moraceae	Triangular Leaf Fig	...
8	*Ficus racemosa* L.	Moraceae	Cluster Fig	गूलर
9	*Ficus religiosa* L.	Moraceae	Peepal	पीपल
10	*Ficus retusa* L.	Moraceae	Laurel Fig	कामरूप
11	*Ficus* sp.11	Moraceae	Nuda Ficus	...
12	*Ficus* sp.12	Moraceae	Ficus Long Island	...
13	*Ficus* sp.13	Moraceae	Ficus Panda, Golden Ficus	पांडा फायकस
14	*Ficus virens* Aiton	Moraceae	White Fig	पाखड़, पिलखन

Appendix 8: Medicinal properties of plants and their uses in curing different ailments/diseases

SN	Name of Disease/Medicinal properties	Name of plant species	No. of plants
1	**Abdominal pain, Abdominal disorders, Stomach pain, Stomachache, Abdominal colic pain, Colic**	*Abrus precatorius, Ageratum conyzoides, Alhagi maurorum, Alternanthera philoxeroides, Alysicarpus monilifer, A. vaginalis, Ampelopteris prolifera, Anisomeles indica, Argemone mexicana, Boerhavia diffusa, Cajanus scarabaeoides, Calotropis procera, Cannabis sativa, Capsella bursa-pastoris, Celosia argentea, Cleome viscosa, Clerodendrum phlomidis, Centaurium pulchellum, Centella asiatica, Chenopodium album, Cissampelos pareira, Citrullus colocynthis, Cordia dichotoma, Convolvulus arvensis, Corchorus aestuans, Cyanthillium cinereum, Cyperus rotundus, Dalbergia sissoo, Desmodium triflorum, Diospyros montana, Dysphania ambrosioides, Euphorbia heterophylla, E. hirta, E. hypericifolia, Holoptelea integrifolia, Hyptis suaveolens, Ipomoea obscura, I. triloba, Lantana camara, Leucas cephalotes, Lindernia crustacea, Malva parviflora, Malvastrum coromandelianum, Melilotus indicus, Mitragyna parvifolia, Mukia maderaspatana, Nepeta hindostana, Oxalis corniculata, O. debilis* var. *corymbosa, Persicaria barbata, Physalis peruviana, Polygonum plebeium, Phyla nodiflora, Phyllanthus amarus, P. maderaspatensis, Physalis angulata, Portulaca oleracea, P. quadrifida, Saccharum bengalense, Scoparia dulcis, Senna obtusifolia, Sida cordata, S. cordifolia, Solanum americanum, S. surattense, Sonchus brachyotus, Trichosanthes cucumerina, Triumfetta rhomboidea, Typha domingensis, Urena lobata, Verbascum thapsus, Withania somnifera, Ziziphus jujuba*	73
2	**Abortion, Abortifacient**	*Abrus precatorius, Achyranthes aspera, Butea monosperma, Calotropis gigantea, Calotropis procera, Cannabis sativa, Chenopodium murale, Citrullus colocynthis, Cyanotis axillaris, Cyanthillium cinereum, Cynodon dactylon, Dalbergia sissoo, Diospyros montana, Dysphania ambrosioides, Eclipta prostrata, Euphorbia hirta, Hydrocotyle sibthorpioides, Justicia adhatoda, Leucaena leucocephala, Physalis angulata, Rhynchosia minima, Ricinus communis, Saccharum bengalense, Salix tetrasperma, Trianthema portulacastrum, Tribulus terrestris, Trichosanthes cucumerina , Urena lobata, Withania somnifera*	29
3	**Abscesses, Whitlow, Whitlow lesions**	*Ammannia baccifera, Celosia argentea, Centella asiatica, Cissampelos pareira, Crateva adansonii* subsp. *odora, Croton bonplandianus, Cynodon dactylon, Desmodium triflorum, Euphorbia heterophylla, E. hirta, E. prostrata, Gonostegia pentandra, Hydrocotyle sibthorpioides, Hyptis suaveolens, Hydrilla verticillata, Ipomoea aquatica, Marsilea quadrifolia, Phyllanthus amarus, P. reticulatus, Physalis angulata, Portulaca oleracea, P. quadrifida, Ranunculus sceleratus, Ricinus communis, Senna occidentalis, Sesbania sesban, Sida acuta, Solanum americanum, Spermacoce articularis, Stellaria media, Tribulus terrestris, Typha domingensis, Xanthium strumarium, Veronica anagallis-aquatica*	34
4	**Absorbent**	*Eclipta prostrata, Ficus religiosa*	2
5	**Acne, Acne vulgaris**	*Cissampelos pareira, Euphorbia hirta, Oldenlandia corymbosa*	3
6	**Acrid**	*Achyranthes aspera, Ageratum conyzoides, Alternanthera sessilis, Argemone mexicana, Artemisia scoparia, Azadirachta indica, Butea monosperma, Calotropis gigantea, Centella asiatica, Chenopodium album, C. murale, Cleome viscosa, Cordia dichotoma, Cyperus rotundus, Dalbergia sissoo, Desmodium triflorum, Eclipta prostrata, Holoptelea integrifolia, Marsilea quadrifolia, Mitragyna parvifolia, Operculina turpethum, Phyla nodiflora, Phyllanthus amarus, Prosopis cineraria, Ranunculus sceleratus, Ricinus communis, Sesamum indicum, Sida cordata, Solanum americanum, S. surattense, Syzygium cumini, Tephrosia purpurea, Ziziphus jujuba*	33
7	**Adaptogenic, Adaptogen**	*Sida cordifolia, Withania somnifera*	2
8	**Adenopathy**	*Coccinia grandis*	1
9	**Agalactia, Failure of lactation, Hypogalactia, Agalactorrhea, Lactation disorders**	*Alternanthera sessilis, Cissampelos pareira, Coccinia grandis, Saccharum spontaneum*	4
10	**Ageing, Antiageing, Lifespan, Rejuvenating, Vitality**	*Abutilon indicum, Bacopa monnieri, Barleria prionitis, Centella asiatica, Convolvulus prostratus, Cordia dichotoma, Diospyros montana, Ficus religiosa, Hydrilla verticillata, Ludwigia octovalvis, Sida cordifolia, Sesbania sesban, Solanum americanum, S. surattense, Withania somnifera*	15
11	**Agglutinating**	*Rhynchosia minima*	1
12	**Aggravate**	*Rumex dentatus*	1

SN	Name of Disease/Medicinal properties	Name of plant species	No. of plants
13	**Ague**	*Ageratum conyzoides, Euphorbia heterophylla*	2
14	**Albuminuria**	*Cyanthillium cinereum, Portulaca oleracea*	2
15	**Alcohol deaddiction**	*Blainvillea acmella*	1
16	**Alcohol withdrawal**	*Cannabis sativa*	1
17	**Aldose-reductase-inhibitor**	*Euphorbia hirta*	1
18	**Alexiteric**	*Arenaria serpyllifolia, Coccinia grandis, Cynodon dactylon, Phyla nodiflora, Pluchea lanceolata, Portulaca oleracea, Tephrosia purpurea, T. villosa, Trichosanthes cucumerina, Xanthium strumarium*	10
19	**Allelopathic**	*Lantana camara*	1
20	**Allergy, Allergic rhinitis, Hay fever, Rhinitis, Rhinopathy, Antiallergic**	*Abrus precatorius, Ageratum conyzoides, Albizia lebbeck, Azadirachta indica, Bidens pilosa, Calotropis gigantea, Cynodon dactylon, Euphorbia hirta, Lepidium didymum, Prosopis juliflora, Setaria pumila, Solanum americanum, S. surattense, Streblus asper, Xanthium strumarium*	15
21	**Alopecia**	*Euphorbia hirta, Holoptelea integrifolia, Sesamum indicum, Trichosanthes cucumerina*	4
22	**Alterative**	*Ampelopteris prolifera, Capparis sepiaria, Clerodendrum phlomidis, Convolvulus prostratus, Croton bonplandianus, Cuscuta reflexa, Cyanthillium cinereum, Cynodon dactylon, Dalbergia sissoo, Ficus religiosa, Merremia hederacea, Phyllanthus reticulatus, Portulaca quadrifida, Solanum americanum, Trichosanthes cucumerina, Veronica agrestis, Veronica anagallis-aquatica, Withania somnifera*	18
23	**Alternative medicine**	*Cyperus rotundus, Stellaria media, Verbascum thapsus*	3
24	**Amebicide**	*Coccinia grandis, Euphorbia hirta*	2
25	**Amenorrhea**	*Azadirachta indica, Cannabis sativa, Chenopodium murale, Chrysopogon zizanioides, Citrullus colocynthis, Cyperus iria, Dysphania ambrosioides, Euphorbia hirta, Indigofera linifolia, Leucas cephalotes, Lindernia crustacea, Melia azedarach, Nepeta hindostana, Oxalis corniculata, Solanum surattense, Trianthema portulacastrum*	16
26	**Amentia**	*Bacopa monnieri, Chrysopogon zizanioides, Centella asiatica*	3
27	**Anabolic**	*Cynodon dactylon, Withania somnifera*	2
28	**Anaemia, Anaemic, Antianaemic**	*Achyranthes aspera,* Amaranthus viridis, *Alternanthera philoxeroides, Bacopa monnieri, Barleria prionitis, Boerhavia diffusa, Brachiaria reptans, Cajanus scarabaeoides, Cayratia trifolia, Chenopodium album, Coccinia grandis, Fumaria indica, Jatropha gossypiifolia Melia azedarach, Morus alba, Tephrosia purpurea, Trianthema portulacastrum, Tribulus terrestris, Phyllanthus amarus, P. reticulatus, Physalis angulata, Pistia stratiotes, Portulaca quadrifida, Scoparia dulcis, Operculina turpethum, Oxalis corniculata, Solanum torvum, Stellaria media, Ziziphus jujuba*	29
29	**Analgesic, Anodyne, Pain relief**	*Abrus precatorius, Abutilon indicum, Achyranthes aspera, Acmella paniculata, Ageratum conyzoides, Amaranthus viridis, Anagallis arvensis, Anisomeles indica, Argemone mexicana, Bacopa monnieri, Boerhavia diffusa, Cajanus scarabaeoides, Calotropis procera, Capparis sepiaria, Cannabis sativa, Carthamus oxyacantha, Cenchrus ciliaris, Cissampelos pareira, Chloris barbata, Cleome viscosa, Clerodendrum phlomidis, Corchorus olitorius, Cordia dichotoma, Cyanthillium cinereum, Cyperus rotundus, Datura innoxia, Desmodium gangeticum, Dysphania ambrosioides, Euphorbia hirta, E. prostrata, Ficus religiosa, Grangea maderaspatana, Gomphrena celosioides, Gonostegia pentandra, Hydrocotyle sibthorpioides, Ipomoea obscura, I. triloba, Jatropha gossypiifolia, Ludwigia hyssopifolia, Malvastrum coromandelianum, Marsilea quadrifolia, Melilotus indicus, Melia azedarach, Mitragyna parvifolia, Mukia maderaspatana, Oldenlandia corymbosa, Operculina turpethum, Oxalis corniculata, Parthenium hysterophorus, Peristrophe bicalyculata, Phyllanthus amarus, P. tenellus, Physalis angulata, Pistia stratiotes, Pluchea lanceolata, Ranunculus sceleratus, Ricinus communis, Salix tetrasperma, Scoparia dulcis, Senna obtusifolia, S. occidentalis, Sesbania sesban, Sida cordata, Sisymbrium irio, Solanum americanum, S. surattense, S. torvum, Sonchus brachyotus, Tephrosia purpurea, Trianthema portulacastrum, Tribulus terrestris, Trichosanthes cucumerina, Tridax procumbens, Triumfetta rhomboidea, Verbascum thapsus, Withania somnifera, Ziziphus jujuba, Z. nummularia*	80

SN	Name of Disease/Medicinal properties	Name of plant species	No. of plants
30	**Anasarca**	*Calotropis gigantea, Calotropis procera, Cynodon dactylon*	3
31	**Angina pectoris**	*Tephrosia purpurea*	1
32	**Antacid**	*Azadirachta indica*	1
33	**Anthrax**	*Capparis sepiaria, Cannabis sativa, Crateva adansonii* subsp. *odora*	3
34	**Antiadipogenic**	*Cucumis melo* var. *agrestis*	1
35	**Anti-amoebic, Amoebicidal**	*Dysphania ambrosioides Parthenium hysterophorus, Urena lobata*	3
36	**Antianaphylactic**	*Coccinia grandis*	1
37	**Antiangiogenic**	*Bidens pilosa*	1
38	**Antibacterial, Antibiotic, Bacterial infection, Bactericide**	*Acacia nilotica, Acmella paniculata, Ageratum conyzoides, Ageratum houstonianum, Albizia lebbeck, Alternanthera pungens, Ammannia baccifera, Ampelopteris prolifera, Artemisia scoparia, Azadirachta indica, Barleria prionitis, Bidens pilosa, Blumea viscosa, Caesulia axillaris, Calotropis procera, Cayratia trifolia, Chenopodium album, Cirsium arvense, Citrullus colocynthis, Coccinia grandis, Commelina benghalensis, Convolvulus arvensis, Crateva adansonii* subsp. *odora, Cuscuta reflexa, Cyperus rotundus, Dalbergia sissoo, Diospyros montana, Eleusine indica, Euphorbia hirta, Evolvulus nummularius, Ficus religiosa, Heliotropium ellipticum, Hyptis suaveolens, Imperata cylindrica, Indigofera linifolia, Ipomoea triloba, Jatropha gossypiifolia, Ludwigia octovalvis, Ludwigia hyssopifolia, Malvastrum coromandelianum, Medicago monantha, M. lupulina, Melilotus indicus, Melochia corchorifolia, Morus alba, Oldenlandia corymbosa, Operculina turpethum, Oxalis corniculata, Panicum antidotale, Phyllanthus amarus, P. tenellus, Physalis angulata, Polygonum plebeium, Portulaca oleracea, Prosopis cineraria, Ranunculus sceleratus, Rhynchosia minima, Rumex dentatus, Sesbania sesban, Sida acuta, S. cordifolia, Spergula arvensis, Syzygium cumini, Tephrosia pumila, Trichosanthes cucumerina, Triumfetta rhomboidea, Urena lobata, Verbascum thapsus, Withania somnifera, Xanthium strumarium, Ziziphus nummularia*	69
39	**Antibiofilm**	*Melilotus albus*	1
40	**Anticoagulant, Anticlotting**	*Ageratum conyzoides, Melilotus albus, M. indicus, Tridax procumbens*	4
41	**Anticomplement**	*Azadirachta indica*	1
42	**Anticonvulsant**	*Abrus precatorius, Bacopa monnieri, Calotropis procera, Centella asiatica, Cleome viscosa, Cuscuta reflexa, Cyperus rotundus, Euphorbia hirta*	8
43	**Antidental decay**	*Barleria prionitis*	1
44	**Antidyslipidemic, Hypolipidemic**	*Coccinia grandis, Cucumis melo* var. *agrestis, Echinochloa crus-galli, Ficus palmata, Indigofera linnaei, Peristrophe bicalyculata*	6
45	**Antifertility**	*Abrus precatorius, Barleria prionitis, Butea monosperma, Capsella bursa-pastoris, Cissampelos pareira, Eclipta prostrata, Euphorbia hirta, Hyptis suaveolens, Syzygium cumini, Trianthema portulacastrum, Trichosanthes cucumerina*	11
46	**Antifibrotic**	*Artemisia scoparia, Cleome viscosa*	2
47	**Antifungal, Antimycotic, Fungal infection, Fungicidal**	*Acacia nilotica, Ageratum conyzoides, Argemone mexicana, Azadirachta indica, Barleria prionitis, Blainvillea acmella, Blumea obliqua, Butea monosperma, Caesulia axillaris, Cajanus scarabaeoides, Calotropis procera, Chrysopogon zizanioides, Chenopodium album, Cirsium arvense, Clerodendrum phlomidis, Coccinia grandis, Commelina benghalensis, Convolvulus arvensis, Crateva adansonii* subsp. *odora, Cucumis melo* var. *agrestis, Dysphania ambrosioides, Equisetum ramosissimum, Euphorbia heterophylla, E. hirta, E. serpens, Ficus palmata, Ficus religiosa, Heliotropium ellipticum, Hydrocotyle sibthorpioides, Hyptis suaveolens, Ipomoea carnea, I. triloba, Justicia adhatoda, Lantana camara, Medicago monantha, Pistia stratiotes, Phyla nodiflora, Phyllanthus tenellus, Portulaca oleracea, P. quadrifida, Prosopis juliflora, Ranunculus sceleratus, Senna obtusifolia, S. occidentalis, Sida cordifolia, Spergula arvensis, Syzygium cumini, Tephrosia pumila, Tridax procumbens, Verbascum thapsus, Xanthium strumarium*	51
48	**Antihepatotoxic**	*Butea monosperma, Phyllanthus amarus, P. tenellus, Tridax procumbens*	4

SN	Name of Disease/Medicinal properties	Name of plant species	No. of plants
49	**Antihistaminic**	*Euphorbia hirta*	1
50	**Antihyperalgesia**	*Blumea viscosa*	1
51	**Antihyperglycemic**	*Bidens pilosa, Cucumis melo* var. *agrestis, Cyperus rotundus, Phyllanthus amarus, Prosopis cineraria*	5
52	**Antihyperlipidemic**	*Amaranthus viridis, Azadirachta indica, Prosopis cineraria*	3
53	**Antihypersensitive**	*Peristrophe bicalyculata*	1
54	**Anti-hyperuricaemia**	*Coccinia grandis*	1
55	**Anti-inflammatory, Inflammation, Painful Inflammation, Antiphlogistic, Demulcent**	*Abrus precatorius, Abutilon indicum, Achyranthes aspera, Acmella paniculata, Ageratum conyzoides, Albizia lebbeck, Alysicarpus monilifer, Amaranthus viridis, Anagallis arvensis, Anisomeles indica, Argemone mexicana, Artemisia scoparia, Azadirachta indica, Barleria prionitis, Bacopa monnieri, Bidens pilosa, Blainvillea acmella, Blumea viscosa, Boerhavia diffusa, Butea monosperma, Calotropis gigantea, C. procera, Cannabis sativa, Capparis sepiaria, Capsella bursa-pastoris, Cannabis sativa, Cayratia trifolia, Celosia argentea, Centaurium pulchellum, Chenopodium album, Chloris barbata, Chrysopogon zizanioides, Cirsium arvense, Cissampelos pareira, Citrullus colocynthis, Cleome viscosa, Clerodendrum phlomidis, Coccinia grandis, Commelina benghalensis, Convolvulus arvensis, Corchorus olitorius, Cordia dichotoma, Crateva adansonii* subsp. *odora, Cuscuta chinensis, Cuscuta reflexa, Cynodon dactylon, Cyanotis axillaris, Cyanthillium cinereum, Cyperus rotundus, Dactyloctenium aegyptium, Dalbergia sissoo, Datura innoxia, Desmodium gangeticum, D. triflorum, Euphorbia heterophylla, E. hirta, Eichhornia crassipes, Eclipta prostrata, Ficus palmata, F. religiosa, Gomphrena celosioides, Gonostegia pentandra, Grangea maderaspatana, Heliotropium ellipticum, Holoptelea integrifolia, Hyptis suaveolens, Imperata cylindrica, Indigofera astragalina, I. linnaei, Imperata cylindrica, Ipomoea aquatica, I. carnea, I. eriocarpa, Jatropha gossypiifolia, Justicia adhatoda, Lantana camara, Leucas cephalotes, Launaea procumbens, Ludwigia hyssopifolia, Malva parviflora, Malvastrum coromandelianum, Marsilea quadrifolia, Medicago monantha, Melia azedarach, Mitragyna parvifolia, Morus alba, Mukia maderaspatana, Nepeta hindostana, Oldenlandia corymbosa, Operculina turpethum, Oxalis corniculata, Peristrophe bicalyculata, Phoenix sylvestris, Phyllanthus maderaspatensis, P. tenellus, Physalis angulata, P. peruviana, Pistia stratiotes, Pluchea lanceolata, Polygonum plebeium, Portulaca oleracea, P. quadrifida, Prosopis cineraria, P. juliflora, Ricinus communis, Rumex dentatus, Rungia pectinata, Saccharum bengalense, Salix tetrasperma, Scoparia dulcis, Senna obtusifolia, S. occidentalis, Sesbania bispinosa, S. sesban, Sida acuta, S. cordata, S. cordifolia, Solanum americanum, S. surattense, S. torvum, Sonchus asper, S. brachyotus, Sorghum halepense, Spermacoce articularis, Stellaria media, Syzygium cumini, Teramnus labialis, Tephrosia purpurea, Trianthema portulacastrum, Tribulus terrestris, Trichosanthes cucumerina, Tridax procumbens, Triumfetta rhomboidea, Urena lobata, Vallisneria spiralis, Verbascum thapsus, Vicia sativa, Withania somnifera, Xanthium strumarium, Ziziphus nummularia*	140
56	**Antimicrobial**	*Abrus precatorius, Acmella paniculata, Alysicarpus vaginalis, Anagallis arvensis, Anisomeles indica, Azadirachta indica, Barleria prionitis, Blumea viscosa, Carthamus oxyacantha, Calotropis procera, Capsella bursa-pastoris, Centella asiatica, Chloris barbata, Cleome viscosa, Commelina benghalensis, Corchorus olitorius, Cordia dichotoma, Croton bonplandianus, Cyperus rotundus, Euphorbia heterophylla, Eichhornia crassipes, Echinochloa crus-galli, Gomphrena celosioides, Heliotropium ellipticum, Indigofera astragalina, I. linnaei, Ipomoea aquatica, I. carnea, Jatropha gossypiifolia, Lantana camara, Malva parviflora, Melia azedarach, Melilotus albus, Mollugo nudicaulis, Nicotiana plumbaginifolia, Peristrophe bicalyculata, Phyllanthus maderaspatensis, Pistia stratiotes, Rungia pectinata, Senna occidentalis, Sida cordata, Sisymbrium irio, Solanum torvum, Sonchus brachyotus, Spermacoce articularis, Sphenoclea zeylanica, Syzygium cumini, Tamarix indica*	48
57	**Antimitotic**	*Xanthium strumarium*	1
58	**Antimutagenic**	*Acacia nilotica, Cyperus rotundus, Phyllanthus tenellus, Senna occidentalis*	4
59	**Antimycobacterial**	*Physalis peruviana*	1
60	**Antinematocidal**	*Ageratum conyzoides*	1
61	**Antinociceptive**	*Amaranthus viridis, Blumea viscosa, Calotropis procera, Chenopodium album, Indigofera linnaei, Grangea maderaspatana, Hyptis suaveolens, Ipomoea eriocarpa, Tamarix indica, Xanthium strumarium*	10

SN	Name of Disease/Medicinal properties	Name of plant species	No. of plants
62	**Antioxidant**	*Abrus precatorius, Acacia nilotica, Alternanthera paronychioides, A. pungens, Anagallis arvensis, Anisomeles indica, Artemisia scoparia, Azadirachta indica, Bacopa monnieri, Barleria prionitis, Blumea viscosa, Boerhavia diffusa, Cardamine hirsuta, Carthamus oxyacantha, Calotropis procera, Chrysopogon zizanioides, Capsella bursa-pastoris, Centella asiatica, Chenopodium album, Coccinia grandis, Crateva adansonii* subsp. *odora, Croton bonplandianus, Cuscuta reflexa, Cucumis melo* var. *agrestis, Cyperus rotundus, Dactyloctenium aegyptium, Digera muricata, Eleusine indica, Equisetum ramosissimum, Euphorbia prostrata, E. serpens, Echinochloa crus-galli, Eichhornia crassipes, Ficus religiosa, Hydrilla verticillata, Indigofera linnaei, Ipomoea aquatica, I. carnea, I. eriocarpa, Lantana camara, Launaea procumbens, Ludwigia octovalvis, Malva parviflora, Medicago monantha, Melia azedarach, Melilotus albus, Melochia corchorifolia, Mitragyna parvifolia, Mollugo nudicaulis, Morus alba, Oldenlandia corymbosa, Oxalis debilis* var. *corymbosa, Physalis peruviana, Prosopis cineraria, Polygonum plebeium, Portulaca oleracea, Rhynchosia capitata, Streblus asper, Scoparia dulcis, Senna occidentalis, Sisymbrium irio, Sida cordifolia, Solanum torvum, Sonchus asper, S. brachyotus, Spergula arvensis, Spermacoce articularis, Syzygium cumini, Tamarix indica, Trianthema portulacastrum, Tridax procumbens, Verbascum thapsus, Withania somnifera, Xanthium strumarium, Ziziphus jujuba*	75
63	**Antiperiodic**	*Azadirachta indica, Ceratophyllum demersum, Cissampelos pareira, Crateva adansonii* subsp. *odora, Justicia adhatoda, Melia azedarach*	6
64	**Antiplatelet, Antiplatelet aggregation, Antiaggregant**	*Acacia nilotica, Euphorbia hirta, E. prostrata, Solanum torvum*	4
65	**Antiproliferative**	*Leucaena leucocephala*	1
66	**Antiprostaglandin**	*Euphorbia hirta*	1
67	**Anti-respiratory syncytial virus**	*Barleria prionitis*	1
68	**Antiscorbutic**	*Capsella bursa-pastoris, Cleome viscosa, Indigofera linifolia, I. linnaei, Lepidium didymum, Portulaca oleracea, P. quadrifida, Rorippa palustris, Rumex dentatus*	9
69	**Antisecretory**	*Ipomoea eriocarpa*	1
70	**Antiseptic**	*Ageratum conyzoides, A. houstonianum, Albizia lebbeck, Alhagi maurorum, Ampelopteris prolifera, Artemisia scoparia, Blumea laciniata, Calotropis gigantea, Chrysopogon zizanioides, Cissampelos pareira, Cleome viscosa, Coccinia grandis, Croton bonplandianus, Cynodon dactylon, Dalbergia sissoo, Desmodium triflorum, Dysphania ambrosioides, Eclipta prostrata, Euphorbia hirta, Ficus religiosa, Heliotropium ellipticum, Hyptis suaveolens, Justicia adhatoda, Lantana camara, Leucaena leucocephala, Leucas cephalotes, Melia azedarach, Oxalis corniculata, Oxystelma esculentum, Paspalum scrobiculatum, Phyllanthus amarus, P. maderaspatensis, Physalis peruviana, Pistia stratiotes, Portulaca oleracea, Prosopis juliflora, Ranunculus sceleratus, Setaria verticillata, Solanum americanum, Spergula arvensis, Verbascum thapsus, Vicia sativa*	42
71	**Antisoporific, Soporific**	*Centella asiatica, Hyptis suaveolens, Ziziphus jujuba*	3
72	**Antispasmodic**	*Acacia nilotica, Bacopa monnieri, Calotropis gigantea, Chrysopogon zizanioides, Cannabis sativa, Chenopodium murale, Cissampelos pareira, Coccinia grandis, Datura innoxia, Desmodium triflorum, Dysphania ambrosioides, Euphorbia hirta, Grangea maderaspatana, Hyptis suaveolens, Justicia adhatoda, Melilotus indicus, Nepeta hindostana, Nicotiana plumbaginifolia, Phyllanthus tenellus, Physalis angulata, P. peruviana, Solanum americanum, S. torvum, Verbascum thapsus*	24
73	**Antispermatogenic**	*Abrus precatorius*	1
74	**Antisteatotic**	*Artemisia scoparia*	1
75	**Antithrombin, Antithrombotic**	*Alternanthera paronychioides, Capsella bursa-pastoris, Euphorbia prostrata*	3
76	**Antitoxic**	*Abrus precatorius, Commelina benghalensis, Indigofera tinctoria, Peristrophe bicalyculata*	4
77	**Antitussive**	*Anagallis arvensis, Arenaria serpyllifolia, Coccinia grandis, Cucumis melo* var. *agrestis, Cyperus rotundus, Equisetum ramosissimum, Phoenix sylvestris, Verbascum thapsus, Xanthium strumarium*	9

SN	Name of Disease/Medicinal properties	Name of plant species	No. of plants
78	**Antivirus, Antiviral**	*Abrus precatorius, Ampelopteris prolifera, Amaranthus viridis, Anagallis arvensis, Argemone mexicana, Artemisia scoparia, Azadirachta indica, Bidens pilosa, Blumea viscosa, Cayratia trifolia, Cuscuta reflexa, Cyanthillium cinereum, Eclipta prostrata, Eichhornia crassipes, Equisetum ramosissimum, Euphorbia hirta, Ficus religiosa, Imperata cylindrica, Lantana camara, Medicago monantha, Melia azedarach, Panicum antidotale, Phyllanthus tenellus, Ranunculus sceleratus, Senna obtusifolia, Solanum torvum, Spergula arvensis, Verbascum thapsus*	28
79	**Apéritif**	*Tamarix indica*	1
80	**Aphrodisiac**	*Abrus precatorius, Abutilon indicum, Ammannia baccifera, Argemone mexicana, Albizia lebbeck, Bacopa monnieri, Boerhavia diffusa, Butea monosperma, Capparis sepiaria, Cuscuta chinensis, Cannabis sativa, Cayratia trifolia, Celosia argentea, Centella asiatica, Chenopodium album, C. murale, Coccinia grandis, Cordia dichotoma, Cynodon dactylon, Cyperus rotundus, Dalbergia sissoo, Datura innoxia, Desmodium gangeticum, Eclipta prostrata, Euphorbia hirta, Ficus religiosa, Ipomoea carnea, Malvastrum coromandelianum, Marsilea quadrifolia, Melia azedarach, Morus alba, Nepeta hindostana, Nymphaea pubescens, Oxystelma esculentum, Phyla nodiflora, Phyllanthus reticulatus, Ricinus communis, Rumex dentatus, Saccharum bengalense, S. spontaneum, Sesamum indicum, Sida cordata, S. cordifolia, Solanum americanum, S. surattense, Teramnus labialis, Tribulus terrestris, Triumfetta rhomboidea, Verbascum thapsus, Vicia sativa, Withania somnifera, Ziziphus jujuba, Z. nummularia*	53
81	**Aphtha/Aphthae**	*Ipomoea obscura, Urena lobata*	2
82	**Appendicitis, Appendix**	*Bidens pilosa, Oldenlandia corymbosa, Phyllanthus amarus*	3
83	**Appetite loss, Appetite stimulant, Appetizer, Anorexia, Improves appetite**	*Ageratum conyzoides, Amaranthus viridis, Ammannia baccifera, Azadirachta indica, Butea monosperma, Caesulia axillaris, Calotropis gigantea, C. procera, Cannabis sativa, Capparis sepiaria, Centaurium pulchellum, Cissampelos pareira, Convolvulus prostratus, Corchorus olitorius, Crateva adansonii* subsp. *odora, Datura innoxia, Desmodium gangeticum, Eclipta prostrata, Gnaphalium purpureum, Grangea maderaspatana, Hyptis suaveolens, Imperata cylindrica, Lindernia ciliata, Medicago monantha, Morus alba, Operculina turpethum, Oxalis debilis* var. **corymbosa**, *Leucas cephalotes, Phoenix sylvestris, Phyla nodiflora, Phyllanthus maderaspatensis, Ranunculus sceleratus, Spermacoce articularis, Solanum americanum, S. surattense, Streblus asper, Syzygium cumini, Tephrosia purpurea, Tribulus terrestris, Trichosanthes cucumerina, Vallisneria spiralis, Veronica anagallis-aquatica, Vicia sativa, Xanthium strumarium, Ziziphus jujuba, Ziziphus nummularia*	46
84	**Arrow poison**	*Euphorbia heterophylla*	1
85	**Ascites**	*Bacopa monnieri, Calotropis gigantea, C. procera, Citrullus colocynthis, Corchorus olitorius, Cyanotis axillaris, Phyllanthus maderaspatensis, Ricinus communis, Trianthema portulacastrum*	9
86	**Asthenia, Weakness**	*Bacopa monnieri, Chenopodium album, C. murale, Eichhornia crassipes, Eclipta prostrata, Ficus religiosa, Holoptelea integrifolia, Ipomoea carnea, Morus alba, Phoenix sylvestris, Sorghum halepense, Tribulus terrestris, Trichosanthes cucumerina*	13
87	**Asthma, Bronchial asthma, Allergic asthma, Antiasthmatic**	*Abrus precatorius, Acacia nilotica, Achyranthes aspera, Albizia lebbeck, Alhagi maurorum, Alternanthera philoxeroides, Amaranthus viridis, Argemone mexicana, Barleria prionitis, Blumea obliqua, Boerhavia diffusa, Calotropis gigantea, C. procera, Cannabis sativa, Centella asiatica, Chenopodium murale, Chrysopogon zizanioides, Cissampelos pareira, Clerodendrum phlomidis, Coccinia grandis, Convolvulus arvensis, C. prostratus, Cynodon dactylon, Desmostachya bipinnata, Datura innoxia, Desmodium gangeticum, Dysphania ambrosioides, Euphorbia heterophylla, E. hirta, E. hypericifolia, Evolvulus nummularius, Ficus religiosa, Hyptis suaveolens, Indigofera linnaei, I. tinctoria, Ipomoea aquatica, I. carnea, Lantana camara, Justicia adhatoda, Leucas cephalotes, Lindernia ciliata, L. crustacea, Melia azedarach, Mukia maderaspatana, Murdannia nudiflora, Nepeta hindostana, Phoenix sylvestris, Phyla nodiflora, Phyllanthus amarus, P. reticulatus, Physalis angulata, P. peruviana, Pluchea lanceolata, Portulaca oleracea, P. quadrifida, Pistia stratiotes, Prosopis cineraria, P. juliflora, Rhynchosia minima, Senna obtusifolia, S. occidentalis, Sesamum indicum, Sisymbrium irio, Sida acuta, Solanum americanum, S. surattense, S. torvum, Sonchus asper, Stellaria media, Tephrosia purpurea, T. villosa, Trianthema portulacastrum, Tribulus terrestris, Verbascum thapsus, Vicia sativa, Withania somnifera , Ziziphus jujuba, Z. nummularia*	78

SN	Name of Disease/Medicinal properties	Name of plant species	No. of plants
88	**Astringent**	*Abutilon indicum, Acacia nilotica, Albizia lebbeck, Alternanthera sessilis, Amaranthus viridis, Anisomeles indica, Azadirachta indica, Bacopa monnieri, Butea monosperma, Blumea obliqua, Boerhavia diffusa, Calotropis gigantea, C. procera, Cannabis sativa, Capsella bursa-pastoris, Cayratia trifolia, Cirsium arvense, Cissampelos pareira, Coccinia grandis, Commelina benghalensis, Cordia dichotoma, Cynodon dactylon, Cyperus iria, Cyperus rotundus, Dalbergia sissoo, Dactyloctenium aegyptium, Desmostachya bipinnata, Desmodium gangeticum, D. triflorum, Digera muricata, Diospyros montana, Erigeron bonariensis, Euphorbia hirta, E. hypericifolia, E. prostrata, Equisetum ramosissimum, E. ramosissimum* subsp. *debile, Ficus religiosa, Fumaria indica, Grangea maderaspatana, Holoptelea integrifolia, Hydrocotyle sibthorpioides, Imperata cylindrica, Justicia adhatoda, Ludwigia hyssopifolia, Marsilea quadrifolia, Melia azedarach, Melilotus indicus, Morus alba, Murdannia nudiflora, Nymphaea pubescens, Oxalis corniculata, Persicaria barbata, Phyla nodiflora, Phyllanthus amarus, P. maderaspatensis, P. reticulatus, Prosopis cineraria, P. juliflora, Ranunculus sceleratus, Ricinus communis, Ceratophyllum demersum, Rumex dentatus, Saccharum spontaneum, Sesamum indicum, Sesbania bispinosa, S. sesban, Sida acuta, S. cordata, S. cordifolia, Solanum americanum, Stellaria media, Stuckenia pectinata, Syzygium cumini, Tamarix indica, Teramnus labialis, Tephrosia purpurea, Tribulus terrestris, Trichosanthes cucumerina, Tridax procumbens, Triumfetta rhomboidea, Typha domingensis, Verbascum thapsus, Veronica anagallis-aquatica, Xanthium strumarium, Ziziphus jujuba, Z. nummularia*	87
89	**Atherosclerosis**	*Cyperus rotundus*	1
90	**Athlete's foot**	*Euphorbia hirta, Hyptis suaveolens*	2
91	**Baldness, Hair growth, Prevent Hair fall, Hair loss**	*Azadirachta indica, Cajanus scarabaeoides, Caesulia axillaris, Calotropis procera, Citrullus colocynthis, Convolvulus prostratus, Cuscuta reflexa, Datura innoxia, Eclipta prostrata, Equisetum ramosissimum, Holoptelea integrifolia, Indigofera tinctoria, Lepidium didymum, Leucaena leucocephala, Phyllanthus amarus, Sesamum indicum, Sida acuta, Sphenoclea zeylanica, Tephrosia pumila, Tridax procumbens, Ziziphus jujuba, Ziziphus nummularia*	22
92	**Beri-beri**	*Trianthema portulacastrum*	1
93	**Bile (Pitta) disorders**	*Alhagi maurorum, Ammannia baccifera, Azadirachta indica, Convolvulus arvensis, Eclipta prostrata, Lindernia crustacea, Saccharum spontaneum, Sida acuta*	8
94	**Biliary afflictions, Bilious complaints, Biliousness**	*Ammannia baccifera, Azadirachta indica, Bacopa monnieri, Ceratophyllum demersum, Convolvulus prostratus, Chenopodium album, Coccinia grandis, Cuscuta reflexa, Dactyloctenium aegyptium, Desmodium gangeticum, Desmodium triflorum, Digera muricata, Lantana camara, Lindernia crustacea, Mukia maderaspatana, Oxalis corniculata, Phragmites karka, Teramnus labialis, Trichosanthes cucumerina, Xanthium strumarium, Ziziphus nummularia*	21
95	**Bites/Stings - Scorpion, Snake, Insects, Spider, Bug, Bed bug, Caterpillar; Antidote, Anti-snake/scorpion venom, Antidote for poisoning, Alexipharmic,** **Antipoison, Antivenom**	*Abrus precatorius, Abutilon indicum, Achyranthes aspera, Ageratum conyzoides, Albizia lebbeck, Alysicarpus monilifer, Amaranthus viridis, Ammannia baccifera, Anagallis arvensis, Anisomeles indica, Argemone mexicana, Artemisia scoparia, Azadirachta indica, Barleria prionitis, Bidens pilosa, Blumea laciniata, Butea monosperma, Calotropis gigantea, Cannabis sativa, Capparis sepiaria, Cayratia trifolia, Celosia argentea, Centaurium pulchellum, Ceratophyllum demersum, Chenopodium album, Chrysopogon zizanioides, Cissampelos pareira, Citrullus colocynthis, Coccinia grandis, Convolvulus arvensis, Croton bonplandianus, Cyanthillium cinereum, Cynodon dactylon, Cyperus rotundus, Datura innoxia, Diospyros montana, Eclipta prostrata, Euphorbia heterophylla, E. hirta, Evolvulus nummularius, Ficus religiosa, Heliotropium ellipticum, Holoptelea integrifolia, Hyptis suaveolens, Indigofera linnaei, I. tinctoria, Ipomoea aquatica, I. pes-tigridis, Justicia adhatoda, Leucaena leucocephala, Leucas cephalotes, Marsilea quadrifolia, Melia azedarach, Melochia corchorifolia, Mimosa rubicaulis, Mitragyna parvifolia, Oxalis corniculata, O. debilis* var. *corymbosa, Peristrophe bicalyculata, Phyllanthus amarus, P. maderaspatensis, Portulaca oleracea, Prosopis juliflora, Ranunculus sceleratus, Rhynchosia minima, Salix tetrasperma, Scoparia dulcis, Senna obtusifolia, Sesbania sesban, Solanum surattense, Sonchus brachyotus, Sphenoclea zeylanica, Streblus asper, Trianthema portulacastrum, Triumfetta rhomboidea, Urena lobata, Verbascum thapsus, Vicia sativa, Xanthium strumarium, Ziziphus jujuba*	80

SN	Name of Disease/Medicinal properties	Name of plant species	No. of plants
96	**Bitter**	*Achyranthes aspera, Ageratum conyzoides, Alternanthera sessilis, Argemone mexicana, Artemisia scoparia, Azadirachta indica, Bacopa monnieri, Blumea laciniata, Boerhavia diffusa, Butea monosperma, Calotropis gigantea, C. procera, Centella asiatica, Ceratophyllum demersum, Cissampelos pareira, Citrullus colocynthis, Clerodendrum phlomidis, Coccinia grandis, Commelina benghalensis, Convolvulus prostratus, Cordia dichotoma, Cyperus rotundus, Dalbergia sissoo, Dactyloctenium aegyptium, Desmodium gangeticum, Diospyros montana, Dysphania ambrosioides, Eclipta prostrata, Equisetum ramosissimum, Holoptelea integrifolia, Indigofera tinctoria, Lantana camara, Mitragyna parvifolia, Mukia maderaspatana, Nymphaea pubescens, Operculina turpethum, Oxystelma esculentum, Phyllanthus amarus, Pistia stratiotes, Portulaca oleracea, P. quadrifida, Prosopis cineraria, Ranunculus sceleratus, Ricinus communis, Rhynchosia minima, Senna occidentalis, Sesamum indicum Sida acuta, S. cordata, Solanum americanum, S. surattense, Tamarix indica, Teramnus labialis, Tephrosia purpurea, Tridax procumbens, Xanthium strumarium, Ziziphus jujuba*	56
97	**Bleeding, Bleeding from lungs, kidney and uterus, Bleeding diathesis, Bleeding of injuries, Uncontrolled bleeding**	*Abutilon indicum, Blumea obliqua, Butea monosperma, Capsella bursa-pastoris, Cayratia trifolia, Celosia argentea, Cirsium arvense, Commelina benghalensis, Euphorbia heterophylla, E. hirta, E. hypericifolia, E. prostrata, Fimbristylis dichotoma, Holoptelea integrifolia, Imperata cylindrica, Jatropha gossypiifolia, Justicia adhatoda, Leucas cephalotes, Nymphaea pubescens, Oldenlandia corymbosa, Phoenix sylvestris, Phyllanthus reticulatus, Portulaca oleracea, Ricinus communis, Sida cordata, S. cordifolia, Solanum torvum, Teramnus labialis, Tribulus terrestris, Tridax procumbens, Typha domingensis, Vallisneria spiralis, Verbascum thapsus, Veronica agrestis, Withania somnifera, Ziziphus nummularia*	36
98	**Blindness**	*Achyranthes aspera, Butea monosperma, Phyllanthus maderaspatensis, Solanum americanum, Spermacoce articularis, Trianthema portulacastrum*	6
99	**Bloating**	*Argemone mexicana*	1
100	**Blood circulation**	*Argemone mexicana, Carthamus oxyacantha, Calotropis gigantea, Hyptis suaveolens, Justicia adhatoda, Melilotus indicus, Oldenlandia corymbosa, Ranunculus sceleratus*	8
101	**Blood morbidity, Blood disease, Blood disorders**	*Ammannia baccifera, Azadirachta indica, Blumea laciniata, Cissampelos pareira, Cuscuta reflexa, Dalbergia sissoo, Eichhornia crassipes, Sida acuta, Tephrosia villosa, Trianthema portulacastrum*	10
102	**Blood pressure**	*Argemone mexicana, Centella asiatica, Croton bonplandianus, Cyperus rotundus, Imperata cylindrica, Ipomoea aquatica, Morus alba, Senna occidentalis, Solanum americanum, Tridax procumbens*	10
103	**Blood purifier, Cleansing the blood**	*Amaranthus viridis, Ampelopteris prolifera, Arenaria serpyllifolia, Bacopa monnieri, Capparis sepiaria, Centella asiatica, Chenopodium album, Cissampelos pareira, Citrullus colocynthis, Clerodendrum phlomidis, Cuscuta reflexa, Cyanthillium cinereum, Evolvulus nummularius, Fumaria indica, Gomphrena celosioides, Hyptis suaveolens, Imperata cylindrica, Leucas cephalotes, Nymphaea pubescens, Prosopis cineraria, Scoparia dulcis, Senna obtusifolia, Sonchus asper, Tridax procumbens, Triumfetta rhomboidea, Veronica anagallis-aquatica, Xanthium strumarium, Ziziphus jujuba, Z. nummularia*	29
104	**Body cleansing**	*Leucaena leucocephala*	1
105	**Body pain, Back pain, Aches, Pains, Shoulder pain, Lumbar pain, Chest pain/complaints, pain in pelvic region**	*Abutilon indicum, Alysicarpus vaginalis, Capparis sepiaria, Commelina benghalensis, Corchorus tridens, Cordia dichotoma, Cuscuta reflexa, Eclipta prostrata, Equisetum ramosissimum* subsp. *debile, Euphorbia heterophylla, Fumaria indica, Grangea maderaspatana, Hydrocotyle sibthorpioides, Hyptis suaveolens, Indigofera astragalina, Justicia adhatoda, Ludwigia octovalvis, Malva parviflora, Malvastrum coromandelianum, Marsilea quadrifolia, Melia azedarach, Mollugo nudicaulis, Oxalis corniculata, Phyllanthus amarus, P. maderaspatensis, Physalis angulata, Ricinus communis, Silene conoidea, Sisymbrium irio, Solanum surattense, Spergula arvensis, Withania somnifera*	32
106	**Body resistance scurf**	*Stellaria media*	1
107	**Boils**	*Albizia lebbeck, Ageratum conyzoides, Alternanthera sessilis, Alysicarpus monilifer, Barleria prionitis, Cayratia trifolia, Chrysopogon zizanioides, Citrullus colocynthis, Cordia dichotoma, Crotalaria medicaginea, Cyanotis axillaris, Cyanthillium cinereum, Dalbergia sissoo, Diospyros montana, Euphorbia hirta, Heliotropium ellipticum, Holoptelea integrifolia, Hydrilla verticillata, Hydrocotyle sibthorpioides, Hyptis suaveolens, Ipomoea aquatica, I. pes-tigridis, Lemna perpusilla, Lindernia crustacea, Ludwigia hyssopifolia, L. octovalvis, Malva parviflora, Merremia hederacea, Murdannia nudiflora, Oldenlandia corymbosa, Oxalis corniculata, Physalis angulata, Prosopis cineraria, Ricinus communis, Rhynchosia minima, Sonchus asper, Sesamum indicum, Stellaria media, Streblus asper, Tephrosia purpurea, Trichosanthes cucumerina, Triumfetta rhomboidea, Verbascum thapsus, Withania somnifera, Xanthium strumarium, Ziziphus jujuba, Ziziphus nummularia*	47

SN	Name of Disease/Medicinal properties	Name of plant species	No. of plants
108	**Bone cancer**	*Scoparia dulcis*	1
109	**Bowel complaints, Bowel movement, Bowels of costiveness,** **Constipation, Constipating**	*Acmella paniculata, Ageratum conyzoides, Alternanthera philoxeroides, A. sessilis, Amaranthus viridis, Ammannia baccifera, Anisomeles indica, Argemone mexicana, Azadirachta indica, Bacopa monnieri, Barleria prionitis, Blumea viscosa, Boerhavia diffusa, Calotropis gigantea, Citrullus colocynthis, Cleome viscosa, Coccinia grandis, Cordia dichotoma, Croton bonplandianus, Cuscuta reflexa, Cynodon dactylon, Cyperus rotundus, Dalbergia sissoo, Digera muricata, Echinochloa colona, Emex australis, Euphorbia heterophylla, E. hirta, E. hypericifolia, Ficus palmata, Fimbristylis dichotoma, Fumaria indica, Grangea maderaspatana, Hyptis suaveolens, Indigofera tinctoria, Ipomoea aquatica, Justicia adhatoda, Lantana camara, Lindernia ciliata, Marsilea quadrifolia, Medicago lupulina, M. polymorpha, Melia azedarach, Melilotus indicus, Mimosa rubicaulis, Morus alba, Nicotiana plumbaginifolia, Nymphaea pubescens, Operculina turpethum, Oxalis corniculata, Phyla nodiflora, Phyllanthus amarus, P. maderaspatensis, Physalis peruviana, Pluchea lanceolata, Polygonum plebeium, Ricinus communis, Rumex dentatus, Senna obtusifolia, S. occidentalis, Sida acuta, Solanum surattense, Spermacoce articularis, Stellaria media, Syzygium cumini, Teramnus labialis, Trichosanthes cucumerina, Urena lobata, Verbascum thapsus, Xanthium strumarium, Ziziphus jujuba, Ziziphus nummularia*	72
110	**Breast cancer, Breast disorders**	*Cissampelos pareira, Citrullus colocynthis, Diospyros montana, Equisetum ramosissimum* subsp. *debile*	4
111	**Bright's disease**	*Tribulus terrestris*	1
112	**Bronchitis, Bronchial disorder, Bronchial spasm, Bronchial catarrh**	*Abrus precatorius, Achyranthes aspera, Albizia lebbeck, Amaranthus viridis, Azadirachta indica, Bacopa monnieri, Bidens pilosa, Blumea laciniata, B. obliqua, Boerhavia diffusa, Calotropis gigantea, C. procera, Convolvulus prostratus, Cannabis sativa, Carthamus oxyacantha, Centella asiatica, Cissampelos pareira, Cleome viscosa, Coccinia grandis, Cordia dichotoma, Cynodon dactylon, Cyperus rotundus, Datura innoxia, Desmodium gangeticum, D. triflorum, Eclipta prostrata, Erigeron bonariensis, Euphorbia hirta, E. hypericifolia, Indigofera tinctoria, Ipomoea pes-tigridis, Lantana camara, Justicia adhatoda, Leucas cephalotes, Lindernia ciliata, Malva parviflora, Marsilea quadrifolia, Melilotus indicus, Mimosa rubicaulis, Operculina turpethum, Oxystelma esculentum, Phoenix sylvestris, Phyla nodiflora, Phyllanthus amarus, Phyllanthus maderaspatensis, Pluchea lanceolata, Prosopis cineraria, Ranunculus sceleratus, Ricinus communis, Scoparia dulcis, Senna occidentalis, Sida acuta, Solanum americanum, S. surattense, Sonchus asper, S. brachyotus, Stellaria media, Teramnus labialis, Tephrosia purpurea, Trichosanthes cucumerina, Tridax procumbens, Urena lobata, Verbascum thapsus, Vicia sativa, Ziziphus nummularia*	65
113	**Bronchodilator, Bronchorelaxant**	*Bacopa monnieri, Euphorbia hirta, Justicia adhatoda*	3
114	**Bruises, contusion**	*Blumea viscosa, Equisetum ramosissimum, Euphorbia hirta, Hydrocotyle sibthorpioides, Malva parviflora, Phyllanthus amarus, Portulaca oleracea, Tephrosia purpurea, Tridax procumbens, Verbascum thapsus*	10
115	**Burn wounds, Burns**	*Ageratum conyzoides, Amaranthus viridis, Artemisia scoparia, Brachiaria ramosa, Cannabis sativa, Centella asiatica, Chenopodium album, Cissampelos pareira, Commelina benghalensis, Cucumis melo* var. *agrestis, Equisetum ramosissimum* subsp. *debile, Evolvulus nummularius, Ficus religiosa, Indigofera linnaei, Leucas cephalotes, Ludwigia hyssopifolia, Mimosa rubicaulis, Murdannia nudiflora, Oxalis corniculata, Phyla nodiflora, Phyllanthus maderaspatensis, Phyllanthus reticulatus, Portulaca oleracea, Ricinus communis, Scoparia dulcis, Sesamum indicum, Solanum torvum, Stellaria media, Veronica anagallis-aquatica, Verbascum thapsus*	30
116	**Burning, Burning sensation**	*Acacia nilotica, Alternanthera sessilis, Azadirachta indica, Blumea laciniata, Calotropis procera, Cissampelos pareira, Coccinia grandis, Cordia dichotoma, Crateva adansonii* subsp. *odora, Cynodon dactylon, Dalbergia sissoo, Desmodium gangeticum, D. triflorum, Fumaria indica, Hyptis suaveolens, Mitragyna parvifolia, Morus alba, Mukia maderaspatana, Oldenlandia corymbosa, Oxalis corniculata,, Phoenix sylvestris, Phyla nodiflora, Phyllanthus reticulatus, Portulaca oleracea, Saccharum bengalense, S. spontaneum, Sida cordata, Syzygium cumini, Teramnus labialis, Trichosanthes cucumerina, Ziziphus jujuba, Z. nummularia*	32

SN	Name of Disease/Medicinal properties	Name of plant species	No. of plants
117	**Cancer, Anticancer, Anticarcinogenic**	*Abrus precatorius, Acacia nilotica, Alysicarpus vaginalis, Anisomeles indica, Amaranthus viridis, Azadirachta indica, Bacopa monnieri, Barleria prionitis, Bidens pilosa, Calotropis procera, Cannabis sativa, Capparis sepiaria, Capsella bursa-pastoris, Cayratia trifolia, Chenopodium album, Cirsium arvense, Convolvulus arvensis, Corchorus aestuans, Cuscuta chinensis, Cyanthillium cinereum, Cynodon dactylon, Cyperus rotundus, Dactyloctenium aegyptium, Diospyros montana, Echinochloa crus-galli, Eichhornia crassipes, Equisetum ramosissimum* subsp. *debile, Euphorbia heterophylla, E. hirta, E. hypericifolia, Ficus religiosa, Gnaphalium purpureum, Gonostegia pentandra, Hyptis suaveolens, Imperata cylindrica, Ipomoea pes-tigridis, Jatropha gossypiifolia, Justicia adhatoda, Lantana camara, Launaea procumbens, Lepidium didymum, Leucaena leucocephala, Melia azedarach, Morus alba, Mukia maderaspatana, Oldenlandia corymbosa, Operculina turpethum, Oxystelma esculentum, Physalis angulata, P. peruviana, Ricinus communis, Senna occidentalis, Solanum americanum, Streblus asper, Sonchus asper, S. brachyotus, Trianthema portulacastrum, Tribulus terrestris, Triumfetta rhomboidea, Verbascum thapsus, Xanthium strumarium, Ziziphus jujuba*	62
118	**Carbuncles**	*Cayratia trifolia, Crateva adansonii* subsp. *odora, Cynodon dactylon, Echinochloa crus-galli, Ipomoea pes-tigridis, Paspalum scrobiculatum, Verbascum thapsus, Withania somnifera*	8
119	**Cardiac disorders, Cardiac pain, Cardiovascular ailments**	*Achyranthes aspera, Boerhavia diffusa, Capsella bursa-pastoris, Chenopodium album, C. murale, Cissampelos pareira, Cleome viscosa, Corchorus olitorius, Datura innoxia, Euphorbia hirta, Ipomoea carnea, Launaea procumbens, Morus alba, Nepeta hindostana, Peristrophe bicalyculata, Phoenix sylvestris, Sida cordifolia, Solanum americanum, S. surattense, Sonchus asper*	20
120	**Cardiac tonic, Cardiotonic, Cardiac weakness, Cardiac stimulant, Cardiac debility, Cardio-protective**	*Bacopa monnieri, Blumea laciniata, Boerhavia diffusa, Centella asiatica, Chenopodium album, Cleome viscosa, Cuscuta reflexa, Desmodium gangeticum, Ficus religiosa, Justicia adhatoda, Melia azedarach, Nepeta hindostana, Nymphaea pubescens, Salix tetrasperma, Senna occidentalis, Sida cordifolia, Spirodela polyrrhiza, Streblus asper, Tribulus terrestris*	19
121	**Cardiodepressant**	*Euphorbia hirta, Justicia adhatoda, Sesbania sesban, Verbascum thapsus*	4
122	**Cardiopathy**	*Desmodium gangeticum, Indigofera tinctoria, Nepeta hindostana, Oxalis corniculata, Phyla nodiflora, Solanum americanum, Tribulus terrestris*	7
123	**Carminative**	*Achyranthes aspera, Alternanthera paronychioides, Anisomeles indica, Bacopa monnieri, Boerhavia diffusa, Calotropis procera, Centella asiatica, Ceratophyllum demersum, Chenopodium album, C. murale, Chrysopogon zizanioides, Cissampelos pareira, Citrullus colocynthis, Cleome viscosa, Corchorus aestuans, Cuscuta reflexa, Cynodon dactylon, Cyperus rotundus, Desmodium gangeticum, Dysphania ambrosioides, Grangea maderaspatana, Holoptelea integrifolia, Hyptis suaveolens, Leucas cephalotes, Ludwigia hyssopifolia, Melilotus albus, Mukia maderaspatana, Nepeta hindostana, Operculina turpethum, Oxalis corniculata, Parthenium hysterophorus, Persicaria barbata, Phalaris minor, Phyllanthus maderaspatensis, Polygonum plebeium, Ricinus communis, Solanum surattense, Spirodela polyrrhiza, Stellaria media, Streblus asper, Syzygium cumini, Trichosanthes cucumerina, Tridax procumbens*	43
124	**Catarrh, Anticatarrhal**	*Albizia lebbeck, Anisomeles indica, Barleria prionitis, Calotropis gigantea, Calotropis procera, Cannabis sativa, Chenopodium murale, Coccinia grandis, Cynodon dactylon, Datura innoxia, Desmodium gangeticum, Dysphania ambrosioides, Euphorbia hirta, Eclipta prostrata, Hyptis suaveolens, Justicia adhatoda, Lantana camara, Ludwigia hyssopifolia, Prosopis juliflora, Solanum americanum, S. surattense, Teramnus labialis, Tridax procumbens, Veronica agrestis, Verbascum thapsus*	25
125	**Cathartic**	*Citrullus colocynthis, Euphorbia heterophylla, Ipomoea carnea, Lindernia crustacea, Melia azedarach, Ricinus communis, Solanum americanum, Trianthema portulacastrum*	8
126	**Cellulitis**	*Melia azedarach*	1
127	**Cephalalgia**	*Cynodon dactylon, Eclipta prostrata, Euphorbia hypericifolia, Indigofera tinctoria, Oxalis corniculata, Portulaca oleracea, Ziziphus jujuba*	7
128	**Cerebral affections, Cerebral thrombosis, Cerebral weakness**	*Anagallis arvensis, Carthamus oxyacantha, Ipomoea carnea*	3
129	**Cervical spondylosis**	*Sida cordifolia*	1
130	**Chemopreventive**	*Azadirachta indica, Leucaena leucocephala, Syzygium cumini, Withania somnifera*	4
131	**Chest complaints in children**	*Ipomoea coptica*	1

SN	Name of Disease/Medicinal properties	Name of plant species	No. of plants
132	**Chickenpox**	*Lantana camara, Sida cordifolia*	2
133	**Chikungemya Virus**	*Azadirachta indica*	1
134	**Cholagogue**	*Alternanthera sessilis, Anagallis arvensis, Artemisia scoparia, Convolvulus arvensis, Heliotropium ellipticum, Portulaca oleracea, Rumex dentatus, Vicia sativa*	8
135	**Cholera**	*Cajanus scarabaeoides, Cannabis sativa, Clerodendrum phlomidis, Croton bonplandianus, Mimosa rubicaulis, Persicaria barbata, Sesamum indicum, Sida cordata, Triumfetta rhomboidea*	9
136	**Cholesterol level, Cholesterol absorption, Anticholesterolemic, Antihypercholesterolemic**	*Argemone mexicana, Artemisia scoparia, Coccinia grandis, Convolvulus prostratus, Diospyros montana, Physalis peruviana, Sida cordifolia*	7
137	**Chyluria**	*Capsella bursa-pastoris*	1
138	**Cicatrizant, Cicatrizing**	*Blumea viscosa, Euphorbia hirta, Hydrilla verticillata*	3
139	**Cirrhosis**	*Hydrocotyle sibthorpioides, Solanum americanum*	2
140	**Cognitive**	*Withania somnifera*	1
141	**Cold**	*Ageratum conyzoides, Ammannia baccifera, Anisomeles indica, Bacopa monnieri, Bidens pilosa, Caesulia axillaris, Calotropis gigantea, Cannabis sativa, Cissampelos pareira, Citrullus colocynthis, Convolvulus prostratus, Cyanthillium cinereum, Datura innoxia, Desmodium gangeticum, Euphorbia hirta, Fumaria indica, Hydrocotyle sibthorpioides, Hyptis suaveolens, Imperata cylindrica, Justicia adhatoda, Leucas cephalotes, Melilotus indicus, Mollugo nudicaulis, Mukia maderaspatana, Phoenix sylvestris, Phyla nodiflora, Phyllanthus amarus, P. reticulatus, P. tenellus, Prosopis cineraria, Ranunculus sceleratus, Senna occidentalis, Sida acuta, Solanum americanum, S. torvum, Spirodela polyrrhiza, Stellaria media, Syzygium cumini, Urena lobata, Ziziphus nummularia*	40
142	**Collyrium**	*Crateva adansonii* subsp. *odora, Portulaca quadrifida*	2
143	**Colporrhagia**	*Cynodon dactylon*	1
144	**Complexion**	*Convolvulus prostratus*	1
145	**Condiment**	*Dysphania ambrosioides, Solanum americanum*	2
146	**Conjunctivitis**	*Abrus precatorius, Acacia nilotica, Argemone mexicana, Boerhavia diffusa, Cissampelos pareira, Cyanthillium cinereum, Cynodon dactylon, Dysphania ambrosioides, Eclipta prostrata, Equisetum ramosissimum, Euphorbia hirta, E. hypericifolia, Merremia hederacea, Oxalis corniculata, Senna obtusifolia, Solanum americanum, Spermacoce articularis, Withania somnifera*	18
147	**Consumption**	*Coccinia grandis, Tribulus terrestris, Ziziphus jujuba*	3
148	**Contraceptive, Male contraceptive**	*Abrus precatorius, Azadirachta indica, Butea monosperma, Chenopodium album, Crateva adansonii* subsp. *odora, Leucaena leucocephala, Melia azedarach, Sesbania sesban*	8
149	**Convulsion**	*Cannabis sativa, Coccinia grandis, Datura innoxia, Desmodium gangeticum, D. triflorum, Evolvulus nummularius, Hyptis suaveolens, Oxalis corniculata, Senna occidentalis, Solanum americanum, Verbascum thapsus*	11
150	**Cooling**	*Alternanthera sessilis, Argemone mexicana, Bacopa monnieri, Boerhavia diffusa, Butea monosperma, Chrysopogon zizanioides, Centella asiatica, Ceratophyllum demersum, Coccinia grandis, Commelina benghalensis, Cordia dichotoma, Cucumis melo* var. *agrestis, Cynodon dactylon, Cyperus rotundus, Desmostachya bipinnata, Desmodium triflorum, Digera muricata, Eclipta prostrata, Equisetum ramosissimum* subsp. *debile, Ficus religiosa, Fumaria indica, Lantana camara, Morus alba, Nymphaea pubescens, Oxalis corniculata, Persicaria barbata, Phoenix sylvestris, Phyla nodiflora, Phyllanthus amarus, P. reticulatus, Pistia stratiotes, Portulaca oleracea, Prosopis cineraria, Rungia pectinata, Rumex dentatus, Saccharum bengalense, Sonchus asper, Stellaria media, Syzygium cumini, Teramnus labialis, Tribulus terrestris, Trichosanthes cucumerina, Triumfetta rhomboidea, Xanthium strumarium, Ziziphus jujuba, Z. nummularia*	46
151	**Cornea opacity**	*Achyranthes aspera, Oxalis corniculata*	2

SN	Name of Disease/Medicinal properties	Name of plant species	No. of plants
152	**Corneal ulcers**	*Trianthema portulacastrum*	1
153	**Coryza**	*Cissampelos pareira, Clerodendrum phlomidis, Euphorbia hirta*	3
154	**Cough, Expectorant**	*Abrus precatorius, Abutilon indicum, Achyranthes aspera, Acmella paniculata, Ageratum conyzoides, Albizia lebbeck, Alhagi maurorum, Alternanthera sessilis, Alysicarpus vaginalis, Ammannia baccifera, Ampelopteris prolifera, Anagallis arvensis, Argemone mexicana, Azadirachta indica, Bacopa monnieri, Barleria prionitis, Bidens pilosa, Boerhavia diffusa, Caesulia axillaris, Calotropis gigantea, C. procera, Cannabis sativa, Capparis sepiaria, Chrysopogon zizanioides, Cirsium arvense, Cissampelos pareira, Citrullus colocynthis, Cleome viscosa, Clerodendrum phlomidis, Coccinia grandis, Cordia dichotoma, Cucumis melo* var. *agrestis, Cuscuta reflexa, Cynodon dactylon, Cyanthillium cinereum, Dalbergia sissoo, Datura innoxia, Desmodium gangeticum, D. triflorum, Eclipta prostrata, Euphorbia hirta, Ficus religiosa, Hydrocotyle sibthorpioides, Hyptis suaveolens, Indigofera astragalina, I. tinctoria, Ipomoea aquatica, Justicia adhatoda, J. japonica, Lantana camara, Lepidium didymum, Leucaena leucocephala, Leucas cephalotes, Lindernia ciliata, Malva parviflora, Malvastrum coromandelianum, Marsilea quadrifolia, Medicago lupulina, Melia azedarach, Mimosa rubicaulis, Mitragyna parvifolia, Morus alba, Mukia maderaspatana, Nicotiana plumbaginifolia, Operculina turpethum, Oxystelma esculentum, Panicum antidotale, Peristrophe bicalyculata, Phoenix sylvestris, Phyllanthus maderaspatensis, P. tenellus, Physalis angulata, Portulaca oleracea, P. quadrifida, Pistia stratiotes, Prosopis cineraria, Ranunculus sceleratus, Ricinus communis, Senna obtusifolia, S. occidentalis, Sesamum indicum, Sisymbrium irio, Solanum americanum, S. surattense, S. torvum, Sonchus brachyotus, Stellaria media, Syzygium cumini, Tephrosia pumila, T. purpurea, Tribulus terrestris, Triumfetta rhomboidea, Urena lobata, Veronica agrestis, Verbascum thapsus, Vicia sativa, Withania somnifera, Ziziphus jujuba, Ziziphus nummularia*	99
155	**Cracks in hands and feet**	*Merremia hederacea, Solanum torvum*	2
156	**Cramp, Muscle cramp**	*Capsella bursa-pastoris, Cissampelos pareira, Coccinia grandis, Datura innoxia, Evolvulus nummularius, Hyptis suaveolens, Solanum americanum, Verbascum thapsus*	8
157	**Curare**	*Desmodium triflorum, Euphorbia hirta*	2
158	**Cutaneous affections, Cutaneous diseases**	*Argemone mexicana, Calotropis gigantea, Holoptelea integrifolia, Hyptis suaveolens, Paspalum scrobiculatum, Rumex dentatus*	6
159	**Cuts, Cuts and wounds**	*Ageratum houstonianum, Barleria prionitis, Blumea laciniata, B. viscosa, Caesulia axillaris, Croton bonplandianus, Cynodon dactylon, Evolvulus nummularius, Hemarthria compressa, Hydrilla verticillata, Hyptis suaveolens, Indigofera linifolia, Ludwigia hyssopifolia, Melilotus indicus, Nicotiana plumbaginifolia, Physalis angulata, Sida cordata, Stellaria media, Verbascum thapsus, Veronica anagallis-aquatica, Urena lobata, Ziziphus jujuba*	22
160	**Cyanogenetic**	*Cynodon dactylon*	1
161	**Cystitis**	*Abutilon indicum, Acacia nilotica, Arenaria serpyllifolia, Boerhavia diffusa, Cannabis sativa, Cissampelos pareira, Corchorus olitorius, Cynodon dactylon, Cyperus rotundus, Phyllanthus amarus, Portulaca oleracea, Sida cordata, S. cordifolia, Solanum torvum*	14
162	**Cytopathogenic**	*Ranunculus sceleratus*	1
163	**Cytoprotective**	*Barleria prionitis*	1
164	**Cytotoxic, Toxic, Poisonous**	*Abrus precatorius, Acacia nilotica, Anagallis arvensis, Anisomeles indica, Argemone mexicana, Calotropis gigantea, C. procera, Centella asiatica, Datura innoxia, Diospyros montana, Equisetum ramosissimum, Euphorbia heterophylla, E. hirta, Heliotropium ellipticum, Malva parviflora, Oldenlandia corymbosa, Operculina turpethum, Physalis angulata, Ranunculus sceleratus, Rhynchosia minima, Ricinus communis, Rumex dentatus, Scoparia dulcis, Senna occidentalis, Solanum americanum, S. torvum, Tamarix indica, Verbascum thapsus*	28
165	**Dandruff**	*Ageratum conyzoides, Cuscuta chinensis, Cayratia trifolia, Datura innoxia, Malva parviflora, Phyla nodiflora, Ricinus communis, Sida acuta, Vicia sativa*	9
166	**Deafness**	*Cajanus scarabaeoides, Dysphania ambrosioides*	2
167	**Decongestant**	*Hyptis suaveolens*	1

SN	Name of Disease/Medicinal properties	Name of plant species	No. of plants
168	**Delirium**	*Butea monosperma, Cannabis sativa, Datura innoxia, Diospyros montana, Sida cordifolia*	5
169	**Dengue fever, Dengue virus**	*Cissampelos pareira, Euphorbia hirta, Hydrocotyle sibthorpioides, Physalis angulata*	4
170	**Dental caries**	*Argemone mexicana, Solanum surattense*	2
171	**Deobstruent**	*Corchorus olitorius, Grangea maderaspatana, Melia azedarach, Merremia hederacea, Phyllanthus maderaspatensis, Ranunculus sceleratus, Tephrosia purpurea, Withania somnifera*	8
172	**Depression, Anxiety, Restlessness, Melancholia, Anxiolytic, Antidepressant**	*Abrus precatorius, Achyranthes aspera, Anagallis arvensis, Bacopa monnieri, Convolvulus arvensis, C. prostratus, Diospyros montana, Dysphania ambrosioides, Equisetum ramosissimum, Euphorbia hirta, Ipomoea carnea, I. pes-tigridis, Leucaena leucocephala, Morus alba, Nepeta hindostana, Oldenlandia corymbosa, Senna occidentalis, Withania somnifera*	18
173	**Depurative**	*Achyranthes aspera, Alternanthera sessilis, Arenaria serpyllifolia, Argemone mexicana, Azadirachta indica, Bacopa monnieri, Calotropis gigantea, C. procera, Cissampelos pareira, Coccinia grandis, Cordia dichotoma, Cynodon dactylon, Cyperus rotundus, Dalbergia sissoo, Eclipta prostrata, Eleusine indica, Holoptelea integrifolia, Hydrocotyle sibthorpioides, Justicia adhatoda, Leucas cephalotes, Marsilea quadrifolia, Mitragyna parvifolia, Oldenlandia corymbosa, Oxalis corniculata, Oxystelma esculentum, Ricinus communis, Saccharum bengalense, Solanum americanum, S. surattense, Stellaria media, Streblus asper, Tephrosia purpurea, Tribulus terrestris, Ziziphus jujuba*	34
174	**Dermatitis, Dermatosis, Dermatopathy, Antidermatitis**	*Albizia lebbeck, Alternanthera paronychioides, Bacopa monnieri, Coccinia grandis, Cynodon dactylon, Dysphania ambrosioides, Euphorbia hirta, Hyptis suaveolens, Ludwigia octovalvis, Malva parviflora, Nymphaea pubescens, Physalis angulata, P. peruviana, Pistia stratiotes, Rumex dentatus, Sesamum indicum, Solanum americanum, Stellaria media, Syzygium cumini, Trichosanthes cucumerina*	20
175	**Desiccant**	*Typha domingensis*	1
176	**Detersive**	*Argemone mexicana*	1
177	**Detoxifier, Detoxification, Detoxication, Detoxifying**	*Cajanus scarabaeoides, Centella asiatica, Cyperus rotundus, Hydrilla verticillata, Hydrocotyle sibthorpioides, Senna occidentalis, Sonchus brachyotus, Sisymbrium irio*	8
178	**Diabetes, Diabetes mellitus, Antidiabetic**	*Abrus precatorius, Acacia nilotica, Alternanthera paronychioides, Amaranthus viridis, Ampelopteris prolifera, Avena sterilis, Azadirachta indica, Barleria prionitis, Bacopa monnieri, Boerhavia diffusa, Butea monosperma, Cajanus scarabaeoides, Capsella bursa-pastoris, Cayratia trifolia, Centaurium pulchellum, Celosia argentea, Chloris barbata, Cirsium arvense, Citrullus colocynthis, Coccinia grandis, Convolvulus arvensis, C. prostratus, Cordia dichotoma, Crotalaria medicaginea, Cuscuta reflexa, Cyperus rotundus, Digera muricata, Echinochloa crus-galli, Euphorbia hirta, E. prostrata, Equisetum ramosissimum, Ficus palmata, F. religiosa, Holoptelea integrifolia, Ipomoea aquatica, I. carnea, I. eriocarpa, Jatropha gossypiifolia, Justicia adhatoda, Leucaena leucocephala, Lindernia crustacea, Ludwigia octovalvis, Malvastrum coromandelianum, Marsilea quadrifolia, Melia azedarach, Mimosa rubicaulis, Mollugo nudicaulis, Morus alba, Mukia maderaspatana, Nymphaea pubescens, Paspalum scrobiculatum, Phragmites karka, Peristrophe bicalyculata, Phyla nodiflora, Phyllanthus amarus, P. tenellus, P. reticulatus, Physalis angulata, P. peruviana, Pistia stratiotes, Portulaca oleracea, Prosopis juliflora, Scoparia dulcis, Senna occidentalis, Sida cordifolia, Solanum americanum, S. torvum, Sonchus asper, Syzygium cumini, Trichosanthes cucumerina, Tridax procumbens, Urena lobata, Verbascum thapsus*	73
179	**Diarrhoea, Antidiarrhoeal**	*Acacia nilotica, Ageratum conyzoides, Albizia lebbeck, Alhagi maurorum, Alternanthera sessilis, Alysicarpus monilifer, Argemone mexicana, Bacopa monnieri, Barleria prionitis, Blumea viscosa, Butea monosperma, Cajanus scarabaeoides, Calotropis procera, Cannabis sativa, Capsella bursa-pastoris, Celosia argentea, Ceratophyllum demersum, Chenopodium album, Chloris barbata, Cissampelos pareira, Cleome viscosa, Clerodendrum phlomidis, Commelina benghalensis, Corchorus aestuans, Cordia dichotoma, Cuscuta reflexa, Cyanthillium cinereum, Cynodon dactylon, Cyperus rotundus, Dalbergia sissoo, Desmostachya bipinnata, Datura innoxia, Desmodium gangeticum, D. triflorum, Diospyros montana, Eichhornia crassipes, Equisetum ramosissimum, Erigeron bonariensis, Euphorbia hirta, E. serpens, Ficus religiosa, Fumaria indica, Gnaphalium purpureum, Hydrocotyle sibthorpioides, Hyptis suaveolens, Imperata cylindrica, Indigofera astragalina, I. tinctoria, Justicia adhatoda, Lindernia crustacea, Ludwigia hyssopifolia, L. octovalvis, Malvastrum coromandelianum, Marsilea quadrifolia, Melilotus indicus, Morus alba, Nymphaea pubescens, Oxalis corniculata, Oxalis debilis* var. *corymbosa, Parthenium hysterophorus, Phoenix sylvestris, Phyla nodiflora, Phyllanthus amarus, P. maderaspatensis, P. reticulatus, Physalis angulata, P. peruviana, Polygonum plebeium, Portulaca quadrifida, Prosopis juliflora, Rhynchosia minima, Scoparia dulcis, Senna obtusifolia, S. occidentalis, Sesbania sesban, Setaria verticillata, Sida cordata, Sida acuta, Solanum americanum, Sonchus brachyotus, Sorghum halepense, Spermacoce articularis, Streblus asper, Syzygium cumini, Tamarix indica, Tephrosia pumila, T. purpurea, Trichosanthes cucumerina, Tridax procumbens, Triumfetta rhomboidea, Urena lobata, Verbascum thapsus, Veronica anagallis-aquatica, Withania somnifera, Xanthium strumarium, Ziziphus jujuba, Z. nummularia*	97

SN	Name of Disease/Medicinal properties	Name of plant species	No. of plants
180	**Digestive system disorders, Digestion, Digestive, Digesting agent**	*Achyranthes aspera, Ageratum conyzoides, Alternanthera pungens, A. sessilis, Amaranthus viridis, Ammannia baccifera, Argemone mexicana, Bacopa monnieri, Bidens pilosa, Butea monosperma, Calotropis gigantea, C. procera, Chenopodium album, C. murale, Cissampelos pareira, Cleome viscosa, Cordia dichotoma, Cucumis melo* var. *agrestis, Cyperus rotundus, Dalbergia sissoo, Digera muricata, Echinochloa colona, Euphorbia hirta, E. hypericifolia, Ficus religiosa, Holoptelea integrifolia, Hydrocotyle sibthorpioides, Imperata cylindrica, Justicia adhatoda, Lepidium didymum, Leucas cephalotes, Malva parviflora, Medicago polymorpha, Melilotus indicus, Morus alba, Oldenlandia corymbosa, Oxalis corniculata, Phoenix sylvestris, Phyla nodiflora, Physalis angulata, Pluchea lanceolata, Portulaca oleracea, Prosopis juliflora, Ranunculus sceleratus, Ricinus communis, Sesamum indicum, Sida acuta, Solanum americanum, S. surattense, Syzygium cumini, Tephrosia purpurea, Tribulus terrestris, Trichosanthes cucumerina, Vicia sativa, Xanthium strumarium*	55
181	**Diphtheria**	*Justicia adhatoda*	1
182	**Dipsia**	*Ceratophyllum demersum, Desmostachya bipinnata, Mukia maderaspatana, Nymphaea pubescens, Portulaca oleracea, Saccharum bengalense, Solanum americanum, Ziziphus jujuba*	8
183	**Discutient**	*Melilotus indicus*	1
184	**Disinfectant, Vaginal disinfectant**	*Ficus religiosa, Justicia adhatoda, Panicum antidotale, Physalis angulata, Streblus asper*	5
185	**Diuretic**	*Abrus precatorius, Abutilon indicum, Achyranthes aspera, Ageratum conyzoides, Alhagi maurorum, Alternanthera pungens, Alysicarpus monilifer, A. vaginalis, Amaranthus viridis, Anagallis arvensis, Arenaria serpyllifolia, Argemone mexicana, Artemisia scoparia, Barleria prionitis, Bacopa monnieri, Blainvillea acmella, Blumea viscosa, Boerhavia diffusa, Butea monosperma, Cannabis sativa, Capsella bursa-pastoris, Cardamine hirsuta, Carthamus oxyacantha, Cayratia trifolia, Cenchrus ciliaris, Centella asiatica, Chenopodium album, C. murale, Chrysopogon zizanioides, Cissampelos pareira, Commelina benghalensis, Convolvulus arvensis, Corchorus olitorius, Cordia dichotoma, Crateva adansonii* subsp. *odora, Cucumis melo* var. *agrestis, Cuscuta reflexa, Cyanthillium cinereum, Cyperus iria, Dalbergia sissoo, Desmodium gangeticum, D. triflorum, Desmostachya bipinnata, Digera muricata, Dysphania ambrosioides, Eclipta prostrata, Eleusine indica, Erigeron bonariensis, Euphorbia hirta, E. hypericifolia, Equisetum ramosissimum* subsp. *debile, E. ramosissimum, Fimbristylis dichotoma, Fumaria indica, Gonostegia pentandra, Grangea maderaspatana, Hydrocotyle sibthorpioides, Indigofera linifolia, I. linnaei, I. tinctoria, Ipomoea aquatica, Justicia adhatoda, Launaea procumbens, Ludwigia hyssopifolia, Marsilea quadrifolia, Melia azedarach, Merremia hederacea, Morus alba, Mukia maderaspatana, Nicotiana plumbaginifolia, Nymphaea pubescens, Oldenlandia corymbosa, Oxalis corniculata, Oxystelma esculentum, Persicaria barbata, Phoenix sylvestris, Phyla nodiflora, Phyllanthus amarus, P. maderaspatensis, P. reticulatus, P. tenellus, Phragmites karka, Physalis angulata, P. peruviana, Pistia stratiotes, Portulaca oleracea, P. quadrifida, Ricinus communis, Rorippa palustris, Rungia pectinata, Rumex dentatus, Saccharum bengalense, S. spontaneum, Senna occidentalis, Sida acuta, S. cordata, S. cordifolia, Solanum americanum, S. surattense, S. torvum, Sonchus asper, Sorghum halepense, Spergula arvensis, Spirodela polyrrhiza, Stellaria media, Tamarix indica, Tephrosia purpurea, Trianthema portulacastrum, Tribulus terrestris, Trichosanthes cucumerina, Triumfetta rhomboidea, Typha domingensis, Urena lobata, Verbascum thapsus, Veronica agrestis, V. anagallis-aquatica, Vicia sativa, Withania somnifera, Xanthium strumarium, Ziziphus jujuba*	120
186	**Dizziness**	*Dysphania ambrosioides, Ipomoea coptica, Morus alba, Senna obtusifolia*	4
187	**Dog bite**	*Blumea laciniata, Heliotropium ellipticum, Indigofera tinctoria, Ipomoea carnea, I. pes-tigridis, Senna occidentalis*	6
188	**Dropsy, Oedema, Abdominal dropsy, Dropsical affections, Edema, Antiedemic**	*Achyranthes aspera, Anagallis arvensis, Argemone mexicana, Barleria prionitis, Capsella bursa-pastoris, Cayratia trifolia, Centella asiatica, Cissampelos pareira, Crateva adansonii* subsp. *odora, Croton bonplandianus, Cyanthillium cinereum, Desmodium gangeticum, Eclipta prostrata, Euphorbia hirta, Hydrocotyle sibthorpioides, Imperata cylindrica, Ipomoea pes-tigridis, Mitragyna parvifolia, Morus alba, Operculina turpethum, Phyllanthus amarus, Pistia stratiotes, Senna occidentalis, Solanum americanum, S. surattense, S. torvum, Spermacoce articularis, Spirodela polyrrhiza, Trianthema portulacastrum, Tribulus terrestris*	30

SN	Name of Disease/Medicinal properties	Name of plant species	No. of plants
189	**Dysentery, Blood dysentery, Amoebic dysentery, Antidysenteric**	*Acacia nilotica, Acmella paniculata, Ageratum conyzoides, Albizia lebbeck, Alternanthera philoxeroides, Alysicarpus vaginalis, Amaranthus viridis, Ampelopteris prolifera, Arenaria serpyllifolia, Argemone mexicana, Blumea viscosa, Butea monosperma, Caesulia axillaris, Cajanus scarabaeoides, Calotropis gigantea, C. procera, Cannabis sativa, Capparis sepiaria, Capsella bursa-pastoris, Cardamine hirsuta, Celosia argentea, Centella asiatica, Ceratophyllum demersum, Chenopodium album, Cirsium arvense, Cissampelos pareira, Clerodendrum phlomidis, Cordia dichotoma, Cynodon dactylon, C. rotundus, Dalbergia sissoo, Dactyloctenium aegyptium, Desmostachya bipinnata, Dichanthium annulatum, Desmodium gangeticum, D. triflorum, Dysphania ambrosioides, Erigeron bonariensis, Evolvulus nummularius, Eclipta prostrata, Equisetum ramosissimum, Euphorbia hirta, E. hypericifolia, Ficus religiosa, Hyptis suaveolens, Imperata cylindrica, Ipomoea obscura, Justicia adhatoda, Lindernia crustacea, Ludwigia hyssopifolia, L. octovalvis, Malvastrum coromandelianum, Medicago polymorpha, Melochia corchorifolia, Nymphaea pubescens, Oxalis corniculata, Parthenium hysterophorus, Persicaria barbata, Phyllanthus amarus, Pistia stratiotes, Portulaca oleracea, P. quadrifida, Prosopis cineraria, Rhynchosia minima, Saccharum spontaneum, Scoparia dulcis, Senna obtusifolia, S. occidentalis, Sesamum indicum, Sisymbrium irio, Sida acuta, S. cordata, S. cordifolia, Solanum americanum, Sonchus brachyotus, Spermacoce articularis, Streblus asper, Tamarix indica, Tridax procumbens, Triumfetta rhomboidea, Urena lobata, Verbascum thapsus, Ziziphus jujuba, Z. nummularia*	84
190	**Dysgeusia**	*Coccinia grandis*	1
191	**Dyspepsia**	*Achyranthes aspera, Ageratum conyzoides, Alternanthera sessilis, Anisomeles indica, Azadirachta indica, Bacopa monnieri, Bidens pilosa, Blumea viscosa, Boerhavia diffusa, Butea monosperma, Calotropis gigantea, C. procera, Chenopodium album, C. murale, Cissampelos pareira, Cleome viscosa, Clerodendrum phlomidis, Cordia dichotoma, Dalbergia sissoo, Desmodium gangeticum, Fumaria indica, Hemarthria compressa, Holoptelea integrifolia, Hyptis suaveolens, Imperata cylindrica, Ipomoea carnea, Marsilea quadrifolia, Mimosa rubicaulis, Mitragyna parvifolia, Morus alba, Nymphaea pubescens, Oxalis corniculata, Phyla nodiflora, Phyllanthus amarus, Ricinus communis, Saccharum spontaneum, Sida cordifolia, Solanum americanum, S. surattense, Tephrosia purpurea, Tribulus terrestris, Trichosanthes cucumerina, Ziziphus jujuba*	43
192	**Dyspnoea**	*Cissampelos pareira, Euphorbia hirta, Morus alba*	3
193	**Dysuria, Difficulty in urinating/urination, Painful urination, Urodynia, Odynuria**	*Capsella bursa-pastoris, Cissampelos pareira, Clerodendrum phlomidis, Coccinia grandis, Corchorus olitorius, Cucumis melo* var. *agrestis, Cyanthillium cinereum, Desmostachya bipinnata, Diospyros montana, Euphorbia hypericifolia, Hyptis suaveolens, Imperata cylindrica, Launaea procumbens, Portulaca oleracea, Ricinus communis, Saccharum bengalense, Sesbania sesban, Sida cordifolia, Solanum surattense, Trianthema portulacastrum, Tribulus terrestris, Typha domingensis*	22
194	**Ear troubles, Earache, Otalgia, Otopathy, Remove insect from ear**	*Alysicarpus vaginalis, Artemisia scoparia, Azadirachta indica, Cannabis sativa, Cleome viscosa, Coccinia grandis, Cuscuta reflexa, Datura innoxia, Desmodium triflorum, Eclipta prostrata, Grangea maderaspatana, Heliotropium ellipticum, Hydrocotyle sibthorpioides, Justicia adhatoda, Lindernia crustacea, Melochia corchorifolia, Phyllanthus maderaspatensis, Physalis angulata, Portulaca oleracea, Scoparia dulcis, Solanum americanum, S. surattense, Spermacoce articularis, Tridax procumbens, Verbascum thapsus*	25
195	**Ecbolic**	*Rhynchosia minima*	1
196	**Eczema**	*Abrus precatorius, Acacia nilotica, Ageratum conyzoides, Amaranthus viridis, Ampelopteris prolifera, Anisomeles indica, Azadirachta indica, Calotropis gigantea, C. procera, Cayratia trifolia, Celosia argentea, Centella asiatica, Cyanthillium cinereum, Erigeron bonariensis, Euphorbia hirta, Holoptelea integrifolia, Hyptis suaveolens, Jatropha gossypiifolia, Justicia adhatoda, Parthenium hysterophorus, Pistia stratiotes, Polygonum plebeium, Ranunculus sceleratus, Senna obtusifolia, S. occidentalis, Solanum americanum, Stellaria media, Verbascum thapsus*	27
197	**Elephantiasis**	*Bacopa monnieri, Calotropis gigantea, C. procera, Centella asiatica, Eclipta prostrata, Sida acuta, S. cordifolia, Streblus asper, Tephrosia purpurea*	9
198	**Emaciation**	*Eichhornia crassipes, Sida cordifolia, Withania somnifera*	3
199	**Embrocation**	*Anisomeles indica, Urena lobata*	2
200	**Emetic, Antiemetic**	*Abrus precatorius, Achyranthes aspera, Argemone mexicana, Bacopa monnieri, Boerhavia diffusa, Calotropis gigantea, C. procera, Cannabis sativa, Chrysopogon zizanioides, Coccinia grandis, Convolvulus arvensis, C. prostratus, Cucumis melo* var. *agrestis, Cynodon dactylon, Cyperus rotundus, Datura innoxia, Desmodium gangeticum, Euphorbia heterophylla, E. hirta, Heliotropium ellipticum, Ipomoea aquatica, Lindernia crustacea, Melia azedarach, Mimosa rubicaulis, Persicaria barbata, Phragmites karka, Portulaca quadrifida, Sesbania bispinosa, Solanum americanum, Trichosanthes cucumerina*	30

SN	Name of Disease/Medicinal properties	Name of plant species	No. of plants
201	**Emmenagogue**	*Bacopa monnieri, Centella asiatica, Chrysopogon zizanioides, Capsella bursa-pastoris, Chenopodium murale, Cissampelos pareira, Cynodon dactylon, Cyperus rotundus, Dalbergia sissoo, Dysphania ambrosioides, Euphorbia hirta, E. hypericifolia, Gonostegia pentandra, Grangea maderaspatana, Hyptis suaveolens, Leucas cephalotes, Mazus pumilus, Melia azedarach, Nepeta hindostana, Oxalis corniculata, Parthenium hysterophorus, Ranunculus sceleratus, Sesbania sesban, Solanum surattense, Stellaria media, Trianthema portulacastrum, Trichosanthes cucumerina, Urena lobata, Withania somnifera*	29
202	**Emollient**	*Amaranthus viridis, Butea monosperma, Cenchrus ciliaris, Chenopodium murale, Commelina benghalensis, Cordia dichotoma, Desmostachya bipinnata, Ficus palmata, Imperata cylindrica, Leucaena leucocephala, Malvastrum coromandelianum, Marsilea quadrifolia, Malva parviflora, Melilotus albus, M. indicus, Morus alba, Pistia stratiotes, Portulaca oleracea, Ricinus communis, Saccharum spontaneum, Sesamum indicum, Silene conoidea, Solanum americanum, Sonchus asper, Tribulus terrestris, Triumfetta rhomboidea, Urena lobata, Verbascum thapsus, Xanthium strumarium, Ziziphus jujuba*	30
203	**Encephalitis**	*Ampelopteris prolifera*	1
204	**Enlarged scrotum**	*Barleria prionitis, Centella asiatica*	2
205	**Enlargement of glands, gland swelling**	*Centella asiatica, Crateva adansonii* subsp. *odora, Eclipta prostrata*	3
206	**Enteritis**	*Erigeron bonariensis, Equisetum ramosissimum, Euphorbia hirta, Ludwigia hyssopifolia, Malvastrum coromandelianum, Oxalis corniculata, Portulaca oleracea, Urena lobata*	8
207	**Enterosis**	*Coccinia grandis, Solanum torvum*	2
208	**Eosinophilia**	*Albizia lebbeck*	1
209	**Epilepsy, Epileptic attacks, Antiepileptic**	*Abrus precatorius, Ageratum conyzoides, Anagallis arvensis, Bacopa monnieri, Chrysopogon zizanioides, Convolvulus prostratus, Cannabis sativa, Centella asiatica, Crateva adansonii* subsp. *odora, Crotalaria mysorensis, Cynodon dactylon, Cyperus rotundus, Datura innoxia, Evolvulus nummularius, Ficus religiosa, Indigofera linnaei, Ipomoea carnea, I. eriocarpa, I. tinctoria, Salix tetrasperma, Solanum surattense, Withania somnifera, Xanthium strumarium*	23
210	**Erysipelas**	*Bacopa monnieri, Cynodon dactylon, Cyperus rotundus, Desmodium triflorum, Mitragyna parvifolia, Phyla nodiflora, Portulaca quadrifida, Saccharum bengalense, Solanum americanum, Stellaria media*	10
211	**Excoriation**	*Dalbergia sissoo*	1
212	**Excrescences, fleshy excrescences**	*Cynodon dactylon, Euphorbia hirta, Oxalis corniculata*	3
213	**Exhaustion, Fatigue**	*Azadirachta indica, Convolvulus prostratus, Centella asiatica, Cirsium arvense, Morus alba, Phoenix sylvestris, Senna obtusifolia, Teramnus labialis, Withania somnifera, Ziziphus jujuba*	10
214	**Expelling placenta**	*Argemone mexicana, Cyanthillium cinereum*	2
215	**Expelling spine from skin**	*Barleria prionitis*	1
216	**External afflictions**	*Anisomeles indica*	1
217	**Eye diseases, Eye ailments, Eye inflammation, Eye treatments, Eye troubles/ problems, Eye sight, Vision improvement, Eyewashes, Epiphora, Sore eyes, Cataract, Glaucoma, Trachoma, Lacrimation, Corneal opacity, Watery eyes, Yellowness of eyes**	*Abrus precatorius, Acacia nilotica,, Ageratum conyzoides, Albizia lebbeck, Alhagi maurorum, Alternanthera philoxeroides, A. sessilis, Alysicarpus vaginalis, Amaranthus viridis, Argemone mexicana, Boerhavia diffusa, Capsella bursa-pastoris, Cardamine hirsuta, Celosia argentea, Chenopodium album, Cirsium arvense, Cissampelos pareira, Commelina benghalensis, Crateva adansonii* subsp. *odora, Cuscuta reflexa, Cyanthillium cinereum, Cynodon dactylon, Dalbergia sissoo, Datura innoxia, Eclipta prostrata, Equisetum ramosissimum, Euphorbia hirta, E. serpens, Ficus religiosa, Grangea maderaspatana, Indigofera tinctoria, Ipomoea obscura, Leucas cephalotes, Marsilea quadrifolia, Melilotus albus, Morus alba, Nymphaea pubescens, Oxalis corniculata, Oldenlandia corymbosa, Phoenix sylvestris Physalis peruviana, Portulaca quadrifida, Prosopis cineraria, Prosopis juliflora, Rorippa palustris, Senna obtusifolia, S. occidentalis, Sida cordifolia, Silene conoidea, Sisymbrium irio, Solanum americanum, S. surattense, Spermacoce articularis, Stellaria media, Tamarix indica, Tribulus terrestris, Tridax procumbens, Ziziphus nummularia*	57

SN	Name of Disease/Medicinal properties	Name of plant species	No. of plants
218	**Fattening**	*Xanthium strumarium*	1
219	**Febrifuge**	*Ageratum conyzoides, Alternanthera sessilis, Ammannia baccifera, Anisomeles indica, Arenaria serpyllifolia, Bacopa monnieri, Boerhavia diffusa, Calotropis gigantea, C. procera, Capparis sepiaria, Convolvulus prostratus, Centella asiatica, Cissampelos pareira, Cleome viscosa, Commelina benghalensis, Corchorus aestuans, C. olitorius, Cordia dichotoma, Crateva adansonii* subsp. *odora, Cucumis melo* var. *agrestis, Cuscuta reflexa, Cyanthillium cinereum, Cyperus iria, Cyperus rotundus, Datura innoxia, Desmodium gangeticum, Diospyros montana, Eclipta prostrata, Eleusine indica, Euphorbia hirta, Evolvulus nummularius, Fumaria indica, Heliotropium ellipticum, Hydrocotyle sibthorpioides, Hyptis suaveolens, Imperata cylindrica, Lantana camara, Ludwigia hyssopifolia, Justicia adhatoda, Lepidium didymum, Lindernia crustacea, Marsilea quadrifolia, Mazus pumilus, Melia azedarach, Mitragyna parvifolia, Oldenlandia corymbosa, Oxalis corniculata, Peristrophe bicalyculata, Phyla nodiflora, Phyllanthus amarus, P. maderaspatensis, Physalis angulata, Pluchea lanceolata, Ricinus communis, Rungia pectinata, Salix tetrasperma, Senna obtusifolia, S. occidentalis, Sesbania bispinosa, Sisymbrium irio, Sida acuta, S. cordata, S. cordifolia, Solanum americanum, S. surattense, Sonchus asper, Syzygium cumini, Trichosanthes cucumerina, Ziziphus jujuba*	69
220	**Febrile, Febrile eruption, Antifebrile**	*Artemisia scoparia, Imperata cylindrica, Indigofera linifolia, Lantana camara, Mazus pumilus, Merremia hederacea*	6
221	**Female sex disorders, infertility, Female reproductive system, Increase fertility in women**	*Argemone mexicana, Cissampelos pareira, Coccinia grandis, Digera muricata, Equisetum ramosissimum* subsp. *debile, Euphorbia hirta, Physalis angulata, Solanum surattense*	8
222	**Fever, High fever, Chronic fever, Intermittent fever, High body temperature, Antipyretic**	*Abrus precatorius, Abutilon indicum, Acacia nilotica, Acmella paniculata, Ageratum conyzoides, Alternanthera philoxeroides, A. sessilis, Alysicarpus monilifer, Ammannia baccifera, Amaranthus viridis, Anisomeles indica, Arenaria serpyllifolia, Argemone mexicana, Artemisia scoparia, Azadirachta indica, Bacopa monnieri, Barleria prionitis, Bidens pilosa, Blainvillea acmella, Blumea laciniata, B. obliqua, Butea monosperma, Cajanus scarabaeoides, Calotropis gigantea, C. procera, Cannabis sativa, Capparis sepiaria, Capsella bursa-pastoris, Cayratia trifolia, Celosia argentea, Centaurium pulchellum, Centella asiatica, Ceratophyllum demersum, Chenopodium murale, Chloris barbata, Chrysopogon zizanioides, Cissampelos pareira, Citrullus colocynthis, Cleome viscosa, Clerodendrum phlomidis, Coccinia grandis, Commelina benghalensis, Convolvulus arvensis, C. prostratus, Corchorus olitorius, Cordia dichotoma, Crateva adansonii* subsp. *odora, Cyanthillium cinereum, Cynodon dactylon, Cyperus rotundus, Dalbergia sissoo, Dactyloctenium aegyptium, Dalbergia sissoo, Datura innoxia, Desmodium gangeticum, D. triflorum, Desmostachya bipinnata, Diospyros montana, Eclipta prostrata, Eichhornia crassipes, Equisetum ramosissimum, Euphorbia hirta, E. hypericifolia, Ficus religiosa, Fimbristylis dichotoma, Fumaria indica, Gonostegia pentandra, Holoptelea integrifolia, Hydrocotyle sibthorpioides, Hyptis suaveolens, Imperata cylindrica, Ipomoea aquatica, I. eriocarpa, I. pes-tigridis, Jatropha gossypiifolia, Justicia adhatoda, Lantana camara, Launaea procumbens, Lepidium didymum, Leucas cephalotes, Lindernia ciliata, L. crustacea, Ludwigia octovalvis, Malva parviflora, Malvastrum coromandelianum, Marsilea quadrifolia, Melia azedarach, Merremia hederacea, Mimosa rubicaulis, Mitragyna parvifolia, Mollugo nudicaulis, Morus alba, Nepeta hindostana, Oldenlandia corymbosa, Operculina turpethum, Oxalis corniculata, O. debilis* var. *corymbosa, Panicum antidotale, Parthenium hysterophorus, Peristrophe bicalyculata, Phyla nodiflora, Phyllanthus amarus, P. maderaspatensis, Physalis angulata, P. peruviana, Pluchea lanceolata, Portulaca quadrifida, Prosopis juliflora, Rhynchosia minima, Ricinus communis, Salix tetrasperma, Scoparia dulcis, Senna obtusifolia, S. occidentalis, Sida acuta, S. cordata, S. cordifolia, Sisymbrium irio, Solanum americanum, S. surattense, S. torvum, Sonchus brachyotus, Spermacoce articularis, Spirodela polyrrhiza, Streblus asper, Syzygium cumini, Teramnus labialis, Tephrosia pumila, T. purpurea, T. villosa, Trichosanthes cucumerina, Urena lobata, Verbascum thapsus, Vicia sativa, Xanthium strumarium, Ziziphus jujuba*	136
223	**Fibromyalgia**	*Ricinus communis*	1
224	**Filariasis/Filaria**	*Ageratum conyzoides, Bacopa monnieri, Butea monosperma, Streblus asper, Urena lobata*	5
225	**Fish poison**	*Acmella paniculata, Diospyros montana, Euphorbia heterophylla, Persicaria barbata, Senna obtusifolia, Spergula arvensis, Tephrosia purpurea, Verbascum thapsus*	8
226	**Fistula**	*Cissampelos pareira*	1

SN	Name of Disease/Medicinal properties	Name of plant species	No. of plants
227	**Flu, Influenza**	*Albizia lebbeck, Bidens pilosa, Blumea obliqua, Commelina benghalensis, Euphorbia hirta, Fumaria indica, Hydrocotyle sibthorpioides, Hyptis suaveolens, Lantana camara, Oxalis corniculata, Phyllanthus amarus, P. tenellus*	12
228	**Fomentation**	*Abutilon indicum, Ageratum conyzoides, Anisomeles indica, Fumaria indica, Solanum americanum, Stellaria media*	6
229	**Foulness of breath**	*Cynodon dactylon*	1
230	**Fractures, Bone fracture, Limb fracture, Strengthens bone**	*Abrus precatorius, Alysicarpus vaginalis, Butea monosperma, Cayratia trifolia, Cuscuta chinensis, Equisetum ramosissimum* subsp. *debile, E. ramosissimum, Euphorbia hirta, Malva parviflora, Peristrophe bicalyculata, Phragmites karka, Physalis peruviana, Sesamum indicum, Stellaria media, Urena lobata, Withania somnifera*	16
231	**Freckles**	*Chenopodium album*	1
232	**Galactagogue, Lactagogue, Increase milk production in breast-feeding, Improve secretion of milk, Induce lactation, Lactogenic**	*Alternanthera sessilis, Cissampelos pareira, Coccinia grandis, Corchorus olitorius, Crateva adansonii* subsp. *odora, Cuscuta chinensis, Cyperus rotundus, Desmostachya bipinnata, Desmodium gangeticum, D. triflorum, Euphorbia heterophylla, E. hirta, E. serpens, Ficus palmata, Grangea maderaspatana, Hyptis suaveolens, Imperata cylindrica, Medicago lupulina, Oxystelma esculentum, Phyllanthus amarus, Ranunculus sceleratus, Ricinus communis, Sesamum indicum, Sonchus asper, Stellaria media, Streblus asper, Teramnus labialis, Triumfetta rhomboidea*	28
233	**Gall stones**	*Chrysopogon zizanioides, Phyllanthus amarus, Spermacoce articularis*	3
234	**Gangrene**	*Lepidium didymum*	1
235	**Gangrenous rectitis**	*Withania somnifera*	1
236	**Gas troubles, Flatulence, Antiflatulent**	*Achyranthes aspera, Argemone mexicana, Bacopa monnieri, Chenopodium album, C. murale, Cissampelos pareira, Cleome viscosa, Cuscuta reflexa, Cyanthillium cinereum,C. rotundus, Desmodium triflorum, Dysphania ambrosioides, Grangea maderaspatana, Hyptis suaveolens, Ipomoea aquatica, Ludwigia hyssopifolia, L. octovalvis, Melia azedarach, Mitragyna parvifolia, Mukia maderaspatana, Operculina turpethum, Oxystelma esculentum, Phyla nodiflora, Solanum americanum, S. surattense, Tephrosia purpurea, Tribulus terrestris, Trichosanthes cucumerina, Ziziphus jujuba*	29
237	**Gastralgia**	*Cyperus rotundus, Melochia corchorifolia, Solanum torvum, Typha domingensis*	4
238	**Gastric dysfunction/ailments/ problems**	*Anisomeles indica, Centaurium pulchellum, Eclipta prostrata, Euphorbia hirta, Ficus religiosa, Gonostegia pentandra, Phyla nodiflora, Sonchus brachyotus*	8
239	**Gastroenteritis**	*Bidens pilosa, Butea monosperma, Celosia argentea*	3
240	**Gastrohelcosis**	*Ficus religiosa, Solanum americanum*	2
241	**Gastrointestinal disorders/ diseases/infections**	*Anisomeles indica, Barleria prionitis, Chenopodium murale, Cleome viscosa, Crateva adansonii* subsp. *odora, Cyanthillium cinereum, Dysphania ambrosioides, Erigeron bonariensis, Euphorbia hirta, E. hypericifolia, Ficus palmata, Gomphrena celosioides, Hyptis suaveolens, Ludwigia octovalvis, Melilotus indicus, Phyla nodiflora, Phyllanthus amarus, P. maderaspatensis, Sonchus asper, Syzygium cumini, Verbascum thapsus*	21
242	**Gastrointestinal irritant**	*Trichosanthes cucumerina*	1
243	**Gastropathy**	*Alhagi maurorum, Dalbergia sissoo, Eclipta prostrata, I. tinctoria, Mitragyna parvifolia, Phyllanthus reticulatus, Pluchea lanceolata, Portulaca oleracea, Ricinus communis, Syzygium cumini*	10
244	**General debility**	*Achyranthes aspera, Bacopa monnieri, Boerhavia diffusa, Chenopodium album, C. murale, Clerodendrum phlomidis, Cynodon dactylon, Cyperus rotundus, Desmodium gangeticum, Eclipta prostrata, Ipomoea aquatica, Nymphaea pubescens, Panicum antidotale, Phoenix sylvestris, Portulaca quadrifida, Rhynchosia minima, Saccharum bengalense, S. spontaneum, Sida cordifolia, Solanum americanum, Withania somnifera, Xanthium strumarium, Ziziphus jujuba*	23
245	**Genito-urinary tract disorders, Urino-genitory tract, Urogenital disorders, Genito-urinary diseases**	*Centella asiatica, Cynodon dactylon, Euphorbia hirta, Phoenix sylvestris, Phyllanthus amarus, Tribulus terrestris*	6

SN	Name of Disease/Medicinal properties	Name of plant species	No. of plants
246	**Genotoxicity**	*Cyperus rotundus*	1
247	**Giddiness**	*Murdannia nudiflora, Solanum americanum*	2
248	**Gingivitis**	*Albizia lebbeck, Senna obtusifolia, Streblus asper, Tamarix indica, Tephrosia purpurea, Urena lobata, Ziziphus jujuba*	7
249	**Glandular swellings**	*Barleria prionitis*	1
250	**Gleet**	*Abutilon indicum, Tribulus terrestris*	2
251	**Glossitis**	*Coccinia grandis*	1
252	**Glycosuria**	*Clerodendrum phlomidis, Coccinia grandis*	2
253	**Goitre**	*Caesulia axillaris, Eichhornia crassipes, Nymphaea pubescens, Urena lobata*	4
254	**Gonorrhoea, Blennorrhagia**	*Abutilon indicum, Alternanthera pungens, Argemone mexicana, Cannabis sativa, Celosia argentea, Cirsium arvense, Cissampelos pareira, Clerodendrum phlomidis, Coccinia grandis, Corchorus aestuans, Corchorus olitorius, Cordia dichotoma, Dalbergia sissoo, Euphorbia heterophylla, E. hirta, Equisetum ramosissimum* subsp. *debile, E. ramosissimum, Ficus religiosa, Imperata cylindrica, Indigofera tinctoria, Justicia adhatoda, Lantana camara,, Launaea procumbens, Lindernia ciliata, Ludwigia hyssopifolia, Paspalum distichum, Phoenix sylvestris, Phyllanthus amarus, P. maderaspatensis, P. reticulatus, Physalis angulata, Portulaca quadrifida, Scoparia dulcis, Senna occidentalis, Sesbania sesban, Sida cordata, S. cordifolia, Solanum surattense, S. torvum, Tephrosia purpurea, Tribulus terrestris, Triumfetta rhomboidea, Urena lobata*	43
255	**Gout**	*Anagallis arvensis, Barleria prionitis, Cannabis sativa, Coccinia grandis, Dalbergia sissoo, Desmodium gangeticum, Heliotropium ellipticum, Justicia adhatoda, Pluchea lanceolata, Ricinus communis, Senna obtusifolia, Solanum americanum, Teramnus labialis, Verbascum chinense*	14
256	**Griping**	*Grangea maderaspatana, Phyllanthus maderaspatensis*	2
257	**Grippe in children**	*Cynodon dactylon*	1
258	**Gum troubles, Bleeding gums, Tender gums, Treating gum**	*Abutilon indicum, Eclipta prostrata, Syzygium cumini, Tridax procumbens, Verbascum thapsus, Ziziphus nummularia*	6
259	**Gynaecological disorders, Obstetrics and Gynaecology: induction of labour, Labour during childbirth, Labour Pain, Childbirth, Easy delivery, Postpartum pain**	*Abrus precatorius, Achyranthes aspera, Amaranthus viridis, Calotropis gigantea, Cannabis sativa, Cissampelos pareira, Dactyloctenium aegyptium, Datura innoxia, Diospyros montana, Lindernia ciliata, L. crustacea, Ludwigia hyssopifolia, Mitragyna parvifolia, Oldenlandia corymbosa, Physalis angulata, Saccharum spontaneum, Triumfetta rhomboidea, Urena lobata*	18
260	**Haematuria, Hematuria**	*Capsella bursa-pastoris, Cissampelos pareira, Cynodon dactylon, Erigeron bonariensis, Ficus religiosa, Hyptis suaveolens, Imperata cylindrica, Merremia hederacea, Pistia stratiotes, Portulaca oleracea, Sida cordifolia, Typha domingensis*	12
261	**Haemoglobinuria**	*Abrus precatorius*	1
262	**Haemoptysis**	*Cynodon dactylon, Lantana camara, Teramnus labialis, Tribulus terrestris*	4
263	**Haemorrhage, Haemostatic, Antihaemorrhagic, Antibleeding**	*Ageratum conyzoides, Butea monosperma, Capsella bursa-pastoris, Chrysopogon zizanioides, Convolvulus arvensis, Cynodon dactylon, Desmodium gangeticum, Dysphania ambrosioides, Echinochloa crus-galli, Equisetum ramosissimum, Erigeron bonariensis, Euphorbia heterophylla, E. hirta, E. prostrata, Ficus religiosa, Imperata cylindrica, Jatropha gossypiifolia, Mitragyna parvifolia, Oxalis corniculata, Physalis angulata, Saccharum spontaneum, Solanum torvum, Tribulus terrestris, Triumfetta rhomboidea, Typha domingensis, Veronica agrestis, Ziziphus jujuba*	27

SN	Name of Disease/Medicinal properties	Name of plant species	No. of plants
264	**Haemorrhoids, Haemorrhoidal bleeding, Bleeding haemorrhoids, Bleeding piles**	*Abutilon indicum, Ageratum conyzoides, Alhagi maurorum, Alternanthera sessilis, Ammannia baccifera, Anagallis arvensis, Azadirachta indica, Bidens pilosa, Capsella bursa-pastoris, Chenopodium album, C. murale, Cirsium arvense, Clerodendrum phlomidis, Cyanthillium cinereum, Dalbergia sissoo, Desmodium gangeticum, Equisetum ramosissimum, Euphorbia prostrata, Ficus religiosa, Holoptelea integrifolia, Hyptis suaveolens, Indigofera tinctoria, Ipomoea aquatica, I. obscura, Justicia adhatoda, Lepidium didymum, Leucas cephalotes, Marsilea quadrifolia, Melilotus indicus, Nicotiana plumbaginifolia, Nymphaea pubescens, Operculina turpethum, Oxalis corniculata, Phyla nodiflora, Portulaca oleracea, P. quadrifida, Ranunculus sceleratus, Ricinus communis, Rhynchosia minima, Saccharum spontaneum, Scoparia dulcis, Sesamum indicum, Sida cordifolia, Solanum americanum, S. surattense, Spermacoce articularis, Stellaria media, Tephrosia purpurea, Verbascum thapsus, Withania somnifera*	50
265	**Haemostyptic**	*Capsella bursa-pastoris*	1
266	**Halitosis**	*Coccinia grandis*	1
267	**Hallucinogenic**	*Cannabis sativa, Datura innoxia, Paspalum distichum*	3
268	**Headache, Headache with retching, Head itch, Cephalgia, Anticephalalgic**	*Ageratum conyzoides, Blainvillea acmella, Cannabis sativa, Capsella bursa-pastoris, Cayratia trifolia, Centella asiatica, Celosia argentea, Cleome viscosa, Corchorus aestuans, Cordia dichotoma, Crateva adansonii* subsp. *odora,, Datura innoxia, Desmodium gangeticum, Dysphania ambrosioides, Eclipta prostrata, Euphorbia hirta, Grangea maderaspatana, Hyptis suaveolens, Ipomoea pes-tigridis, I. eriocarpa, I. triloba, Justicia adhatoda, Lantana camara, Lepidium didymum, Leucas cephalotes, Lindernia ciliata, Malva parviflora, Melia azedarach, Melilotus albus, Melochia corchorifolia, Morus alba, Murdannia nudiflora, Phyllanthus maderaspatensis, P. reticulatus, Pistia stratiotes, Portulaca oleracea, Ricinus communis, Senna obtusifolia, Scoparia dulcis, Sida acuta, Solanum americanum, S. torvum, Spermacoce articularis, Tribulus terrestris, Trichosanthes cucumerina, Triumfetta rhomboidea, Urena lobata, Verbascum thapsus*	49
269	**Heart ailments, Heart burn, Heart palpitation, Heart diseases, Heart irregularity**	*Capsella bursa-pastoris, Carthamus oxyacantha, Cissampelos pareira, Cuscuta reflexa, Ficus religiosa, Phragmites karka, Rhynchosia minima, Scoparia dulcis, Sida cordifolia, Solanum surattense, Tridax procumbens*	11
270	**Hemafecia**	*Capsella bursa-pastoris*	1
271	**Hematospermia**	*Vallisneria spiralis*	1
272	**Hemoptosis, Hemoptysis, Haematemesis, Spitting of blood**	*Ceratophyllum demersum, Justicia adhatoda, Ludwigia hyssopifolia, Phoenix sylvestris, Peristrophe bicalyculata, Typha domingensis*	6
273	**Hepatic disorder Hepatoprotective, Hepatoprotectant, Hepatosis, Hepatopathy, Hepatic obstruction**	*Anagallis arvensis, Azadirachta indica, Barleria prionitis, Bidens pilosa, Blumea viscosa, Boerhavia diffusa, Butea monosperma, Calotropis procera, Capparis sepiaria, Capsella bursa-pastoris, Chenopodium album, Cleome viscosa, Coccinia grandis, Cordia dichotoma, Crateva adansonii* subsp. *odora, Digera muricata, Eclipta prostrata, Indigofera linifolia, Ipomoea carnea, Melochia corchorifolia, Morus alba, Mukia maderaspatana, Oldenlandia corymbosa, Operculina turpethum, Oxalis corniculata, Peristrophe bicalyculata, Phyllanthus maderaspatensis, Physalis peruviana, Scoparia dulcis, Senna occidentalis, Sida cordata, Trianthema portulacastrum, Ziziphus jujuba*	33
274	**Hepatitis, Hepatitis virus A, B, C; Infective hepatitis, Anti-hepatitis**	*Abrus precatorius, Acacia nilotica, Ageratum conyzoides, Artemisia scoparia, Bidens pilosa, Butea monosperma, Desmodium triflorum, Eclipta prostrata, Erigeron bonariensis, Equisetum ramosissimum* subsp. *debile, Hydrocotyle sibthorpioides, Indigofera tinctoria, Lindernia crustacea, Malvastrum coromandelianum, Morus alba, Oldenlandia corymbosa, Operculina turpethum, Phyllanthus amarus, P. tenellus, Physalis angulata, P. peruviana, Ranunculus sceleratus, Salix tetrasperma, Solanum americanum, Sonchus brachyotus, Tribulus terrestris*	26
275	**Hepatocellular carcinoma, Hepatoma**	*Artemisia scoparia, Hydrocotyle sibthorpioides*	2
276	**Hepatomegaly, Hepatosplenomegaly**	*Eclipta prostrata, Solanum americanum, Tephrosia purpurea*	3
277	**Herbicidal**	*Parthenium hysterophorus*	1

SN	Name of Disease/Medicinal properties	Name of plant species	No. of plants
278	**Herpes, Herpes infection, Herpes zoster**	*Ammannia baccifera, Holoptelea integrifolia, Lemna perpusilla, Nicotiana plumbaginifolia, Parthenium hysterophorus, Senna obtusifolia, Xanthium strumarium*	7
279	**Hiccough, Hiccups**	*Centella asiatica, Ficus religiosa, Solanum americanum, S. surattense*	4
280	**High blood pressure**	*Croton bonplandianus, Cyperus rotundus, Imperata cylindrica, Ipomoea aquatica, Phyllanthus tenellus, Physalis peruviana, Solanum americanum, Tridax procumbens*	8
281	**Hirsutism**	*Cyperus rotundus*	1
282	**Hoarseness**	*Bacopa monnieri, Centella asiatica, Morus alba, Solanum americanum*	4
283	**Hormonal imbalance**	*Cissampelos pareira, Launaea procumbens*	2
284	**Horn Cancer**	*Ipomoea obscura*	1
285	**Hot**	*Blumea laciniata, Citrullus colocynthis, Oxystelma esculentum, Portulaca quadrifida, Solanum surattense*	5
286	**Human Immunodeficiency Virus (HIV)**	*Acacia nilotica, Euphorbia heterophylla, Syzygium cumini*	3
287	**Hydragogue**	*Euphorbia hirta, Operculina turpethum, Solanum americanum*	3
288	**Hydrocele**	*Butea monosperma, Calotropis gigantea, Centella asiatica, Nicotiana plumbaginifolia*	4
289	**Hydrophobia**	*Achyranthes aspera, Anagallis arvensis, Datura innoxia, Indigofera linnaei, I. tinctoria, Solanum americanum*	6
290	**Hydrops articulation of the knee**	*Ranunculus sceleratus*	1
291	**Hygroma in cattle**	*Mukia maderaspatana*	1
292	**Hyperdipsia**	*Azadirachta indica, Cordia dichotoma, Cyperus rotundus, Ziziphus jujuba*	4
293	**Hyperdiuresis**	*Sida cordata*	1
294	**Hypertension, Antihypertensive**	*Acacia nilotica, Anisomeles indica, Argemone mexicana, Artemisia scoparia, Bidens pilosa, Cannabis sativa, Capsella bursa-pastoris, Centaurium pulchellum, Centella asiatica, Chenopodium murale, Crateva adansonii* subsp. *odora, Cuscuta reflexa, Cynodon dactylon, Cyperus rotundus, Eclipta prostrata, Equisetum ramosissimum, Jatropha gossypiifolia, Marsilea quadrifolia, Morus alba, Mukia maderaspatana, Peristrophe bicalyculata, Phyllanthus amarus, Scoparia dulcis, Solanum surattense, S. torvum, Spermacoce articularis, Tribulus terrestris, Triumfetta rhomboidea, Withania somnifera*	29
295	**Hypertonia**	*Ziziphus jujuba*	1
296	**Hypnotic**	*Argemone mexicana, Cannabis sativa, Datura innoxia, Ipomoea pes-tigridis, Marsilea quadrifolia, Verbascum thapsus, Withania somnifera, Ziziphus jujuba*	8
297	**Hypocholesteremic**	*Nepeta hindostana*	1
298	**Hypochondriac condition**	*Cyperus rotundus, Nepeta hindostana*	2
299	**Hypoglycaemic, Hypoglycaemia**	*Azadirachta indica, Cayratia trifolia, Chenopodium album, Clerodendrum phlomidis, Coccinia grandis, Euphorbia hirta, Ficus palmata, Malvastrum coromandelianum, Phyllanthus tenellus, Trichosanthes cucumerina, Xanthium strumarium*	11
300	**Hypotensive**	*Anagallis arvensis, Capsella bursa-pastoris, Centella asiatica, Chenopodium album, Cyperus rotundus, Desmodium gangeticum, Eclipta prostrata, Euphorbia hirta, Justicia adhatoda, Nepeta hindostana, Phyllanthus tenellus*	11
301	**Hypothermic**	*Argemone mexicana*	1

SN	Name of Disease/Medicinal properties	Name of plant species	No. of plants
302	**Hysteria**	*Bacopa monnieri, Chrysopogon zizanioides, Cynodon dactylon, Dysphania ambrosioides, Evolvulus nummularius, Grangea maderaspatana, Melia azedarach, Nepeta hindostana, Ziziphus jujuba*	9
303	**Immune system, Immune system deficiencies, Immune system modulator, immune-modulating, Immunity booster, immune-enhancing, Immunomodulatory**	*Abrus precatorius, Anisomeles indica, Artemisia scoparia, Azadirachta indica, Bidens pilosa, Clerodendrum phlomidis, Coccinia grandis, Convolvulus arvensis, Diospyros montana, Ipomoea carnea, Phyllanthus tenellus, Physalis angulata, P. peruviana, Portulaca oleracea, Senna occidentalis, Tridax procumbens, Withania somnifera*	17
304	**Immunosuppressive**	*Bidens pilosa, Euphorbia hirta*	2
305	**Impetigo**	*Ludwigia octovalvis*	1
306	**Impotence, Impotency, Erectile dysfunction, Penile dysfunction, Sexual dysfunction, Genital organs (Male) disorders, Night fall, Male infertility, Increase fertility in male**	*Acacia nilotica, Cannabis sativa, Carthamus oxyacantha, Cuscuta chinensis, C. reflexa, Datura innoxia, Euphorbia hirta,, Justicia adhatoda, Melilotus indicus, Phyllanthus reticulatus, Ranunculus sceleratus, Sida acuta, Sonchus asper, Tribulus terrestris, Triumfetta rhomboidea, Withania somnifera*	16
307	**Improving animal immunity**	*Phalaris minor*	1
308	**Incipient**	*Phyllanthus maderaspatensis*	1
309	**Incontinence**	*Morus alba*	1
310	**Increase concentration of mind**	*Bacopa monnieri*	1
311	**Indigestion**	*Bacopa monnieri, Barleria prionitis, Blumea viscosa, Boerhavia diffusa, Butea monosperma, Caesulia axillaris, Cajanus scarabaeoides, Calotropis gigantea, Cardamine hirsuta, Centaurium pulchellum, Cannabis sativa, Centella asiatica, Cissampelos pareira, Commelina benghalensis, Cynodon dactylon, Desmostachya bipinnata, Echinochloa crus-galli, Erigeron bonariensis, Hemarthria compressa, Hyptis suaveolens, Imperata cylindrica, Medicago polymorpha, Melilotus albus, Melia azedarach, Operculina turpethum, Oxalis corniculata, Polygonum plebeium, Phyla nodiflora, Physalis angulata, Solanum surattense, Triumfetta rhomboidea, Urena lobata, Ziziphus nummularia*	33
312	**Indolent**	*Azadirachta indica, Desmodium triflorum, Phyla nodiflora*	3
313	**Inducing mobility**	*Indigofera tinctoria*	1
314	**Infantile diarrhoea**	*Melilotus indicus*	1
315	**Infected fingernails**	*Lindernia crustacea*	1
316	**Infections of uterus**	*Hyptis suaveolens*	1
317	**Infectious diseases**	*Bidens pilosa, Ficus religiosa, Gomphrena celosioides*	3
318	**Inflammation of the gall bladder/ eardrum/lymph glands**	*Artemisia scoparia, Cyanotis axillaris, Holoptelea integrifolia*	3
319	**Insanity**	*Bacopa monnieri, Cannabis sativa, Datura innoxia, Ipomoea obscura, Withania somnifera*	5

SN	Name of Disease/Medicinal properties	Name of plant species	No. of plants
320	**Insecticidal, Insect repellent, Repel flies, Insectifuge**	*Abrus precatorius, Ageratum houstonianum, Azadirachta indica, Chrysopogon zizanioides, Erigeron bonariensis, Euphorbia heterophylla, E. hirta, Hyptis suaveolens, Ipomoea eriocarpa, Jatropha gossypiifolia, Justicia adhatoda, Lantana camara, Leucas cephalotes, Melilotus indicus, Parthenium hysterophorus, Ranunculus sceleratus, Senna occidentalis, Solanum americanum, Streblus asper, Tridax procumbens, Xanthium strumarium*	21
321	**Insomnia, Sleeplessness, Sleep disorders**	*Centella asiatica, Chrysopogon zizanioides, Ludwigia hyssopifolia, Marsilea quadrifolia, Peristrophe bicalyculata, Withania somnifera, Ziziphus jujuba*	7
322	**Intestinal cancer, Stomach cancer**	*Holoptelea integrifolia, Equisetum ramosissimum* subsp. *debile, Indigofera tinctoria*	3
323	**Intestinal complaints, Bellyache Intestinal colic, Intestinal bleeding, Intestinal affections**	*Argemone mexicana, Cyanthillium cinereum, Dactyloctenium aegyptium, Dysphania ambrosioides, Emex australis, Gnaphalium purpureum, Ipomoea aquatica, Malva parviflora, Phyllanthus maderaspatensis, Prosopis juliflora, Syzygium cumini, Tribulus terrestris, Typha domingensis*	13
324	**Intestinal haemorrhage**	*Phyllanthus reticulatus*	1
325	**Intoxication, Intoxicant**	*Cannabis sativa, Datura innoxia, Ipomoea coptica, Operculina turpethum, Paspalum scrobiculatum*	5
326	**Invigorate health**	*Azadirachta indica, Hyptis suaveolens, Ipomoea aquatica*	3
327	**Invigorates sex**	*Bacopa monnieri, Eclipta prostrata*	2
328	**Invigorating insanity**	*Bacopa monnieri, Sida cordifolia, Ziziphus jujuba*	3
329	**Itches, Itching, Itchiness**	*Alternanthera sessilis, Ammannia baccifera, Ampelopteris prolifera, Azadirachta indica, Carthamus oxyacantha, Celosia argentea, Cissampelos pareira, Cyanotis axillaris, Dalbergia sissoo, Desmodium triflorum, Eclipta prostrata, Euphorbia hirta, Lantana camara, L. procumbens, Lindernia crustacea, Melia azedarach, Oxystelma esculentum, Persicaria barbata, Phyllanthus amarus, Senna occidentalis, Stellaria media, Trianthema portulacastrum, Triumfetta rhomboidea*	23
330	**Jaundice, Antihepatic**	*Abrus precatorius, Acmella paniculata, Alternanthera pungens, Alysicarpus monilifer, Ammannia baccifera, Ampelopteris prolifera, Argemone mexicana, Artemisia scoparia, Barleria prionitis, Boerhavia diffusa, Carthamus oxyacantha, Calotropis procera, Capparis sepiaria, Cannabis sativa, Celosia argentea, Chenopodium album, Citrullus colocynthis, Coccinia grandis, Convolvulus arvensis, C. prostratus, Corchorus tridens, Crateva adansonii* subsp. *odora, Croton bonplandianus, Cuscuta reflexa, Dalbergia sissoo, Desmostachya bipinnata, Diospyros montana, Eclipta prostrata, Equisetum ramosissimum, Euphorbia hirta, Equisetum ramosissimum* subsp. *debile, Ficus religiosa, Fimbristylis dichotoma, Fumaria indica, Hydrocotyle sibthorpioides, Indigofera linnaei, Imperata cylindrica, Ipomoea obscura, Justicia adhatoda, Launaea procumbens, Leucas cephalotes, Lindernia ciliata, L. crustacea, Malvastrum coromandelianum, Melia azedarach, Mimosa rubicaulis, Mollugo nudicaulis, Nymphaea pubescens, Oxalis corniculata, Oxystelma esculentum, Panicum antidotale, Phyla nodiflora, Phyllanthus amarus, P. maderaspatensis, Physalis angulata, P. peruviana, Portulaca oleracea, Scoparia dulcis, Sesbania sesban, Solanum americanum, S. surattense, Spermacoce articularis, Streblus asper, Trianthema portulacastrum, Trichosanthes cucumerina, Veronica agrestis*	66
331	**Kidney disorders/troubles/ obstructions, Kidney pain, Anuria, Nephropathy**	*Argemone mexicana, Cenchrus ciliaris, Centella asiatica, Chenopodium murale, Cuscuta chinensis, Cyanthillium cinereum, Eleusine indica, Emex australis, Equisetum ramosissimum, Erigeron bonariensis, Fimbristylis dichotoma, Imperata cylindrica, Phyllanthus amarus, Pistia stratiotes, Portulaca oleracea, Ranunculus sceleratus, Saccharum bengalense, Sesamum indicum, Solanum americanum, Sonchus asper, Stellaria media, Syzygium cumini, Tephrosia purpurea, Tridax procumbens*	24
332	**Lactation stimulant in animals**	*Eragrostis japonica*	1
333	**Larvicidal, Pupicidal**	*Blainvillea acmella, Calotropis procera, Capsella bursa-pastoris, Coccinia grandis*	4

SN	Name of Disease/Medicinal properties	Name of plant species	No. of plants
334	**Laryngitis**	*Euphorbia hirta, Merremia hederacea*	2
335	**Laxative, Purgative, Aperient**	*Abrus precatorius, Abutilon indicum, Achyranthes aspera, Amaranthus viridis, Alhagi maurorum, Alternanthera philoxeroides, Alysicarpus vaginalis, Ammannia baccifera, Ampelopteris prolifera, Anagallis arvensis, Argemone mexicana, Artemisia scoparia, Bacopa monnieri, Boerhavia diffusa, Carthamus oxyacantha, Calotropis gigantea, C. procera, Chenopodium album, C. murale, Cissampelos pareira, Citrullus colocynthis, Coccinia grandis, Commelina benghalensis, Convolvulus arvensis, Corchorus olitorius, Cordia dichotoma, Crateva adansonii* subsp. *odora, Cuscuta reflexa, Cynodon dactylon, Desmodium gangeticum, D. triflorum, Digera muricata, Dysphania ambrosioides, Eleusine indica, Erigeron bonariensis, Euphorbia heterophylla, E. hirta, E. hypericifolia, Emex australis, Ficus palmata, F. religiosa, Fumaria indica, Holoptelea integrifolia, I. tinctoria, Ipomoea aquatica, I. carnea, I. pes-tigridis, Justicia adhatoda, Lindernia crustacea, Ludwigia hyssopifolia, Malva parviflora, Mazus pumilus, Medicago polymorpha, Medicago lupulina, Melia azedarach, Melilotus indicus, Merremia hederacea, Morus alba, Mukia maderaspatana, Operculina turpethum, Oxystelma esculentum, Persicaria barbata, Phoenix sylvestris, Phyllanthus maderaspatensis, Physalis peruviana, Pluchea lanceolata, Portulaca oleracea, P. quadrifida, Rhynchosia minima, Pistia stratiotes, Ricinus communis, Rumex dentatus, Rungia pectinata, Saccharum spontaneum, Senna obtusifolia, S. occidentalis, Sesamum indicum, Sesbania sesban, Sida cordata, Solanum americanum, S. surattense, Sonchus asper, Spergula arvensis, Stellaria media, Streblus asper, Tephrosia purpurea, Trianthema portulacastrum, Tribulus terrestris, Trichosanthes cucumerina, Vallisneria spiralis, Vicia sativa, Xanthium strumarium, Ziziphus jujuba*	93
336	**Leishmaniasis, Antileishmanial**	*Diospyros montana, Syzygium cumini*	2
337	**Leprosy, Antileprotic**	*Achyranthes aspera, Ageratum conyzoides, Albizia lebbeck, Alternanthera sessilis, Alysicarpus vaginalis, Amaranthus viridis, Anagallis arvensis, Argemone mexicana, Azadirachta indica, Bacopa monnieri, Calotropis procera, Centella asiatica, Cissampelos pareira, Coccinia grandis, Commelina benghalensis, Cordia dichotoma, Crateva adansonii* subsp. *odora, Cyanthillium cinereum, Cynodon dactylon, Cyperus rotundus, Dalbergia sissoo, Datura innoxia, Evolvulus nummularius, Eclipta prostrata, Holoptelea integrifolia, Ipomoea eriocarpa, Justicia adhatoda, Lantana camara, Marsilea quadrifolia, Melia azedarach, Mitragyna parvifolia, Murdannia nudiflora, Phyllanthus amarus, Pistia stratiotes, Prosopis cineraria, Ricinus communis, Solanum americanum, S. surattense, Streblus asper, Tephrosia purpurea, T. villosa, Tribulus terrestris, Triumfetta rhomboidea, Xanthium strumarium, Ziziphus jujuba*	45
338	**Leucoderma**	*Abrus precatorius, Albizia lebbeck, Alysicarpus monilifer, Azadirachta indica, Bacopa monnieri, Barleria prionitis, Cuscuta chinensis, Cyanthillium cinereum, Cynodon dactylon, Dalbergia sissoo, Eclipta prostrata, Fumaria indica, Holoptelea integrifolia, Ipomoea carnea, Melia azedarach, Oxystelma esculentum, Prosopis cineraria, Sesamum indicum, Streblus asper, Tamarix indica, Xanthium strumarium*	21
339	**Leucorrhoea, Genital organ (Female) disorder (Leucorrhea), White discharge, Vaginal discharge**	*Acacia nilotica, Blumea laciniata, Boerhavia diffusa, Cayratia trifolia, Celosia argentea, Centella asiatica, Cirsium arvense, Cissampelos pareira, Citrullus colocynthis, Crotalaria medicaginea, Cuscuta reflexa, Desmostachya bipinnata, Desmodium gangeticum, Euphorbia hypericifolia, Ficus religiosa, Gomphrena celosioides, Justicia adhatoda, Lantana camara, Lindernia crustacea, Ludwigia hyssopifolia, Merremia hederacea, Panicum antidotale, Peristrophe bicalyculata, Phyllanthus amarus, Portulaca oleracea, Saccharum spontaneum, Sida cordata, S. cordifolia, Solanum americanum, Syzygium cumini, Vallisneria spiralis, Withania somnifera, Ziziphus jujuba*	33
340	**Leukaemia, Antileukaemic**	*Abrus precatorius, Bidens pilosa, Cirsium arvense, Corchorus aestuans, Euphorbia hirta, Physalis peruviana*	6
341	**Libido, Promote libido**	*Boerhavia diffusa, Tribulus terrestris*	2
342	**Lice**	*Cyanthillium cinereum, Datura innoxia, Eclipta prostrata, Melia azedarach*	4
343	**Liver Complaints, Liver Diseases, Liver congestion, Liver lesions, Liver and Gall bladder disorders, Liver inflammation**	*Acmella paniculata, Ageratum conyzoides, Alhagi maurorum, Anagallis arvensis, Artemisia scoparia, Azadirachta indica, Barleria prionitis, Boerhavia diffusa, Carthamus oxyacantha, Celosia argentea, Citrullus colocynthis, Cleome viscosa, Commelina benghalensis, Cuscuta chinensis, Coccinia grandis, Cuscuta reflexa, Croton bonplandianus, Cyperus rotundus, Eclipta prostrata, Eleusine indica, Equisetum ramosissimum* subsp. *debile, E. ramosissimum, Fumaria indica, Grangea maderaspatana, Indigofera linnaei, Ipomoea aquatica, Ipomoea pes-tigridis, Launaea procumbens, Lindernia ciliata, Malvastrum coromandelianum, Morus alba, Nymphoides cristata, Oxalis corniculata, Panicum antidotale, Phyllanthus amarus, Physalis angulata, Pistia stratiotes, Ranunculus sceleratus, Ricinus communis, Scoparia dulcis, Senna occidentalis, Sisymbrium irio, Solanum americanum, S. torvum, Sonchus asper, Stuckenia pectinata, Tephrosia pumila, T. villosa, Trianthema portulacastrum, Tribulus terrestris, Trichosanthes cucumerina, Tridax procumbens, Ziziphus jujuba*	54

SN	Name of Disease/Medicinal properties	Name of plant species	No. of plants
344	**Liver regeneration**	*Sida cordifolia*	1
345	**Liver tonic**	*Artemisia scoparia, Azadirachta indica, Chenopodium murale, Oxalis corniculata, Phyllanthus maderaspatensis, Phyllanthus amarus, Withania somnifera, Ziziphus jujuba*	8
346	**Lockjaw**	*Cannabis sativa*	1
347	**Loose motion**	*Desmostachya bipinnata*	1
348	**Low blood pressure**	*Capsella bursa-pastoris*	1
349	**Lumbago**	*Azadirachta indica, Boerhavia diffusa, Datura innoxia, Desmodium gangeticum, Melia azedarach, Morus alba, Ricinus communis, Solanum surattense, Urena lobata, Xanthium strumarium*	10
350	**Lung disease**	*Ficus palmata, Grangea maderaspatana, Malvastrum coromandelianum, Morus alba, Sesamum indicum, Tephrosia purpurea*	6
351	**Lymphadenitis**	*Phyla nodiflora, Streblus asper*	2
352	**Malaria, Malarial fever, Antimalarial, Plasmodicide, Antiplasmodial**	*Abrus precatorius, Acacia nilotica, Achyranthes aspera, Ammannia baccifera, Argemone mexicana, Artemisia scoparia, Azadirachta indica, Barleria prionitis, Bidens pilosa, Blumea obliqua, Caesulia axillaris, Calotropis procera, C. gigantea, Cannabis sativa, Cissampelos pareira, Cleome viscosa, Clerodendrum phlomidis, Crateva adansonii* subsp. *odora, Croton bonplandianus, Cyanthillium cinereum, Cyperus rotundus, Datura innoxia, Dysphania ambrosioides, Eclipta prostrata, Equisetum ramosissimum* subsp. *debile, Euphorbia hirta, Holoptelea integrifolia, Hyptis suaveolens, Justicia adhatoda, Lantana camara, Leucas cephalotes, Melia azedarach, Nepeta hindostana, Parthenium hysterophorus, Phyla nodiflora, Phyllanthus reticulatus, Physalis angulata, P. peruviana, Pluchea lanceolata, Ranunculus sceleratus, Rhynchosia minima, Scoparia dulcis, Senna occidentalis, Solanum americanum, Sonchus asper, Streblus asper, Tamarix indica, Trichosanthes cucumerina, Tridax procumbens, Urena lobata, Xanthium strumarium*	51
353	**Malignant brain tumour**	*Scoparia dulcis*	1
354	**Mania**	*Anagallis arvensis, Cannabis sativa*	2
355	**Marasmus**	*Euphorbia hypericifolia*	1
356	**Mast cell stabilizing**	*Coccinia grandis*	1
357	**Mastitis**	*Portulaca oleracea, Tribulus terrestris, Urena lobata*	3
358	**Mastogenic**	*Euphorbia hirta*	1
359	**Measles**	*Azadirachta indica, Clerodendrum phlomidis, Crateva adansonii* subsp. *odora, Cynodon dactylon, Desmodium triflorum, Euphorbia hypericifolia, Hyptis suaveolens, Lantana camara, Leucaena leucocephala, Melia azedarach, Mimosa rubicaulis, Sida cordifolia, Spirodela polyrrhiza*	13
360	**Medicated oil**	*Centella asiatica, Sida cordifolia*	2
361	**Memory, Memory aid, Memory capacity, Memory enhance /improving, Intellect promoting, Brain tonic, Brain and heart tonic, Sharpen dull memory**	*Abrus precatorius, Bacopa monnieri, Centella asiatica, Convolvulus prostratus, Cynodon dactylon, Cyperus rotundus, Eclipta prostrata, Evolvulus nummularius, Holoptelea integrifolia, Ipomoea carnea, Hyptis suaveolens, Morus alba, Nepeta hindostana, Physalis peruviana, Withania somnifera, Xanthium strumarium*	16
362	**Meningitis**	*Ampelopteris prolifera*	1

SN	Name of Disease/Medicinal properties	Name of plant species	No. of plants
363	**Menstrual and Premenstrual problems, Menstrual bleeding, Menses, Irregular menses, Menstrual pain, Menstruation (Excessive), Menorrhagia, Menoxenia, Antimenorrhagial, Dysmenorrhoea, Premenstrual syndrome**	*Abrus precatorius, Amaranthus viridis, Ampelopteris prolifera, Bacopa monnieri, Boerhavia diffusa, Butea monosperma, Cannabis sativa, Capsella bursa-pastoris, Chenopodium album, Cissampelos pareira, Citrullus colocynthis, Coccinia grandis, Commelina benghalensis, Convolvulus arvensis, Cyanthillium cinereum, Cynodon dactylon, Cyperus iria, Cyperus rotundus, Dalbergia sissoo, Desmostachya bipinnata, Dichanthium annulatum, Diospyros montana, Dysphania ambrosioides, Erigeron bonariensis, Eragrostis amabilis, Euphorbia heterophylla, E. hirta, E. serpens, E. hypericifolia, Ficus religiosa, Gonostegia pentandra, Grangea maderaspatana, Holoptelea integrifolia, Hyptis suaveolens, Ipomoea aquatica, I. eriocarpa, Justicia adhatoda, Leucaena leucocephala, Leucas cephalotes, Lindernia ciliata, Melia azedarach, Oxalis corniculata, Oxystelma esculentum, Persicaria barbata, Phyla nodiflora, Phyllanthus amarus, P. maderaspatensis, P. reticulatus, Saccharum bengalense, S. spontaneum, Salix tetrasperma, Scoparia dulcis, Sesbania sesban, Solanum americanum, S. surattense, S. torvum, Stellaria media, Tephrosia purpurea, Tribulus terrestris, Trichosanthes cucumerina, Typha domingensis, Urena lobata, Veronica agrestis, V. anagallis-aquatica*	64
364	**Mental disorders, Mental confusion, Mental problems**	*Anagallis arvensis, Argemone mexicana, Centella asiatica, Cleome viscosa*	4
365	**Metabolic syndrome**	*Bidens pilosa*	1
366	**Metrorrhagia**	*Cirsium arvense, Cynodon dactylon*	2
367	**Migraine**	*Abrus precatorius, Barleria prionitis, Calotropis gigantea, Cannabis sativa, Chenopodium murale, Cissampelos pareira, Eclipta prostrata, Euphorbia heterophylla, Ficus religiosa, Hyptis suaveolens, Marsilea quadrifolia, Sesamum indicum, Verbascum thapsus*	13
368	**Miscarriage**	*Cissampelos pareira, Eclipta prostrata, Prosopis cineraria, Withania somnifera*	4
369	**Moles**	*Ficus palmata*	1
370	**Molluscicide, Molluscicidal**	*Acacia nilotica, Anagallis arvensis, Calotropis procera, Euphorbia hirta, Jatropha gossypiifolia, Physalis angulata*	6
371	**Morphine withdrawal**	*Cannabis sativa*	1
372	**Mosquitoes, Mosquito repellent, Mosquitocidal, Larvae of mosquito, Antimosquito**	*Ageratum houstonianum, Anisomeles indica, Capsella bursa-pastoris, Cayratia trifolia, Croton bonplandianus, Chrysopogon zizanioides, Cyperus rotundus, Eichhornia crassipes, Erigeron bonariensis, Hyptis suaveolens, Peristrophe bicalyculata, Sesbania bispinosa, Tridax procumbens*	13
373	**Mouth diseases, Mouth ulcers, Mouth sores, Mouthwash**	*Arabidopsis thaliana, Barleria prionitis, Blumea laciniata, Chrysopogon zizanioides, Cirsium arvense, Celosia argentea, Cordia dichotoma, Euphorbia hirta, Grangea maderaspatana, Leucas cephalotes, Nepeta hindostana, Oxystelma esculentum, Prosopis cineraria, Solanum americanum, Spermacoce articularis, Verbascum thapsus*	16
374	**Mumps**	*Datura innoxia, Lantana camara*	2
375	**Muscle relaxation, Myorelaxant, Muscle relaxant**	*Ageratum conyzoides, Bacopa monnieri, Cuscuta reflexa, Cyperus rotundus, Euphorbia hirta, Withania somnifera*	6
376	**Muscular pain, Myalgia**	*Boerhavia diffusa, Cayratia trifolia, Equisetum ramosissimum, Grangea maderaspatana, Imperata cylindrica, Mitragyna parvifolia, Tridax procumbens*	7
377	**Muscular stiffness, Stiffness of limbs**	*Barleria prionitis, Physalis angulata*	2
378	**Muscular weakness, Muscular swelling, Improve muscular strength**	*Convolvulus arvensis, Oxalis corniculata, Ziziphus jujuba*	3
379	**Mycosis**	*Coccinia grandis*	1
380	**Myocardial depressant**	*Argemone mexicana, Eclipta prostrata*	2

SN	Name of Disease/Medicinal properties	Name of plant species	No. of plants
381	**Narcotic**	*Cannabis sativa, Datura innoxia, Euphorbia hirta, E. hypericifolia, Lathyrus aphaca, Melia azedarach, Melilotus indicus, Paspalum scrobiculatum, Solanum americanum, Verbascum thapsus, Withania somnifera, Ziziphus jujuba*	12
382	**Nasal polyps**	*Alhagi maurorum, Equisetum ramosissimum* subsp. *debile*	2
383	**Nematicidal, Nematicide**	*Calotropis procera, Clerodendrum phlomidis, Eclipta prostrata, Indigofera tinctoria, Lantana camara*	5
384	**Nephritis**	*Anagallis arvensis, Boerhavia diffusa, Capsella bursa-pastoris, Cynodon dactylon, Lathyrus aphaca, Physalis angulata, Urena lobata, Ziziphus jujuba*	8
385	**Nervine tonic**	*Abutilon indicum, Bacopa monnieri, Centella asiatica, Cyperus rotundus, Malva parviflora, Pluchea lanceolata, Salix tetrasperma, Sida cordifolia, Teramnus labialis, Withania somnifera*	10
386	**Nervous system relaxant, neuroprotective, neuromuscular**	*Abrus precatorius, Centella asiatica, Ipomoea carnea, Launaea procumbens*	4
387	**Neurasthenia**	*Cyperus rotundus*	1
388	**Neuritis**	*Cannabis sativa, Sida cordifolia*	2
389	**Neurological disorders, Nerve diseases, Neuropathy, Chorea, Neuralgia, Nervous debility/ weakness/ breakdown, Nervousness**	*Albizia lebbeck, Anagallis arvensis, Bacopa monnieri, Calotropis procera, Cannabis sativa, Cayratia trifolia, Centella asiatica, Clerodendrum phlomidis, Convolvulus arvensis, C. prostratus, Datura innoxia, Diospyros montana, Dysphania ambrosioides, Eclipta prostrata, Ficus religiosa, Imperata cylindrica,, Indigofera linnaei, I. tinctoria, Ipomoea carnea, Marsilea quadrifolia, Morus alba, Mukia maderaspatana, Oldenlandia corymbosa, Parthenium hysterophorus, Phoenix sylvestris, Physalis peruviana, Pistia stratiotes, Pluchea lanceolata, Solanum americanum, Sida cordifolia, Teramnus labialis, Withania somnifera, Ziziphus jujuba*	33
390	**Neurosis**	*Bacopa monnieri*	1
391	**Neuroticism**	*Convolvulus prostratus*	1
392	**Neutraceutical**	*Oxalis debilis* var. *corymbosa*	1
393	**Night blindness**	*Achyranthes aspera, Albizia lebbeck, Boerhavia diffusa, Trianthema portulacastrum*	4
394	**Night sweats**	*Imperata cylindrica, Ziziphus jujuba*	2
395	**Nociception**	*Cyperus rotundus*	1
396	**Nocturnal emission**	*Anisomeles indica*	1
397	**Nootropic**	*Bacopa monnieri*	1
398	**Nose troubles, Nosebleed, Nasal congestion, Chronic nasal catarrh, Sinusitis, Epistaxis**	*Ageratum conyzoides, Capsella bursa-pastoris, Celosia argentea, Coccinia grandis, Cynodon dactylon, Euphorbia hirta, Fumaria indica, Hydrocotyle sibthorpioides, Grangea maderaspatana, Ipomoea aquatica, Morus alba, Solanum americanum, Streblus asper, Tribulus terrestris, Typha domingensis, Xanthium strumarium*	16
399	**Nyctalopia**	*Ricinus communis*	1
400	**Obesity**	*Echinochloa crus-galli, Phyllanthus reticulatus, Sesamum indicum, Stellaria media, Ziziphus jujuba*	5
401	**Odontalgia, Odontalgic**	*Chenopodium album, Eclipta prostrata, Mukia maderaspatana, Portulaca oleracea*	4
402	**Oestrogenic/Estrogenic**	*Anagallis arvensis, Calotropis procera, Verbascum thapsus*	3
403	**Oestrous cycle**	*Chenopodium album*	1
404	**Oleaginous**	*Chenopodium album*	1
405	**Oliguria**	*Merremia hederacea*	1

SN	Name of Disease/Medicinal properties	Name of plant species	No. of plants
406	**Oogenesis**	*Ziziphus nummularia*	1
407	**Ophthalmia, Ophthalmic problems, Ophthalmopathy**	*Achyranthes aspera, Ageratum conyzoides, Albizia lebbeck, Argemone mexicana, Boerhavia diffusa, Chenopodium murale, Coccinia grandis, Dalbergia sissoo, Euphorbia hypericifolia, Justicia adhatoda, J. japonica, Marsilea quadrifolia, Oxalis corniculata, Paspalum scrobiculatum, Phyla nodiflora, Phyllanthus amarus, P. maderaspatensis, Physalis angulata, Saccharum bengalense, Sesamum indicum, Sida cordifolia, Silene conoidea, Solanum americanum, Spermacoce articularis, Tribulus terrestris, Tridax procumbens, Ziziphus jujuba*	27
408	**Ophthalmodynia**	*Cordia dichotoma, Datura innoxia, Phyllanthus reticulatus*	3
409	**Orchitis**	*Ludwigia hyssopifolia*	1
410	**Otorrhoea**	*Datura innoxia, Peristrophe bicalyculata*	2
411	**Ovarian cancer**	*Indigofera tinctoria*	1
412	**Ovicidal**	*Blainvillea acmella, Cyperus rotundus, Eclipta prostrata*	3
413	**Oxytocic**	*Capsella bursa-pastoris, Euphorbia hirta, Justicia adhatoda*	3
414	**Palliative**	*Tribulus terrestris*	1
415	**Paralysis, Palsy, Facial paralysis, Hemiplegia**	*Abutilon indicum, Calotropis gigantea, C. procera, Cannabis sativa, Cuscuta reflexa, Datura innoxia, Ficus religiosa, Grangea maderaspatana, Holoptelea integrifolia, Phyllanthus reticulatus, Sida cordifolia, Teramnus labialis*	12
416	**Parasitic cutaneous diseases**	*Hyptis suaveolens*	1
417	**Parasitic infection, Antiparasitic, Parasitosis, Parasiticide, Antiprotozoal, Antigiardial, Antitrypanosomal, Verminosis**	*Albizia lebbeck, Artemisia scoparia, Cayratia trifolia, Coccinia grandis, Euphorbia hirta, Ficus religiosa, Hyptis suaveolens, Lantana camara, Leucaena leucocephala, Nymphaea pubescens, Physalis angulata, Ricinus communis, Sida acuta, Spergula arvensis, Tephrosia pumila, T. purpurea, Tribulus terrestris, Trichosanthes cucumerina, Tridax procumbens, Xanthium strumarium*	20
418	**Parkinson's disease**	*Tribulus terrestris, Withania somnifera*	2
419	**Parturition, Parturient**	*Argemone mexicana, Coccinia grandis, Justicia adhatoda, Ricinus communis*	4
420	**Pectoral**	*Albizia lebbeck, Azadirachta indica, Cordia dichotoma, Dysphania ambrosioides, Euphorbia heterophylla, E. hirta, Lantana camara, Malva parviflora, Oldenlandia corymbosa, Prosopis cineraria, Sida cordata, Tephrosia purpurea, Urena lobata, Verbascum thapsus, Ziziphus jujuba*	15
421	**Peptic ulcer**	*Chenopodium album, C. murale, Cirsium arvense*	3
422	**Perineal damage after child birth**	*Ranunculus sceleratus*	1
423	**Periodontitis**	*Gonostegia pentandra*	1
424	**Peritonitis**	*Streblus asper*	1
425	**perspiration (inducing), Stimulate secretion of sweat, Diaphoretic**	*Alhagi maurorum, Anagallis arvensis, Anisomeles indica, Argemone mexicana, Boerhavia diffusa, Calotropis gigantea, C. procera, Cayratia trifolia, Chenopodium murale, Chrysopogon zizanioides, Coccinia grandis, Cuscuta reflexa, Cyanthillium cinereum, Euphorbia hirta, E. hypericifolia, Fumaria indica, Lantana camara, Leucas cephalotes, Morus alba, Nepeta hindostana, Oldenlandia corymbosa, Phragmites karka, Phyllanthus maderaspatensis, Senna occidentalis, Sida acuta, Solanum americanum, Spirodela polyrrhiza, Tamarix indica, Xanthium strumarium*	29
426	**Perspiration (lessening)**	*Grangea maderaspatana, Ziziphus jujuba*	2
427	**Petechial haemorrhage**	*Sida cordifolia*	1

SN	Name of Disease/Medicinal properties	Name of plant species	No. of plants
428	**Pharyngodynia, Pharyngopathy**	*Ageratum conyzoides, Chenopodium album, C. murale, Cordia dichotoma, Morus alba*	5
429	**Phlegm**	*Ammannia baccifera, Eclipta prostrata*	2
430	**Photophobia**	*Celosia argentea*	1
431	**Phthisis**	*Anagallis arvensis, Azadirachta indica, Justicia adhatoda, Saccharum spontaneum*	4
432	**Phytopesticide**	*Oxystelma esculentum*	1
433	**Piles**	*Acacia nilotica, Ageratum conyzoides, Albizia lebbeck, Amaranthus viridis, Blumea laciniata, Butea monosperma, Calotropis gigantea, C. procera, Cannabis sativa, Chenopodium album, Cirsium arvense, Citrullus colocynthis, Commelina benghalensis, Corchorus olitorius, Cyanthillium cinereum, Cynodon dactylon, Datura innoxia, Desmodium triflorum, Eclipta prostrata, Erigeron bonariensis, Euphorbia hirta, Grangea maderaspatana, Imperata cylindrica, Ipomoea aquatica, I. carnea, Melia azedarach, Mukia maderaspatana, Murdannia nudiflora, Nicotiana plumbaginifolia, Oxalis debilis* var. *corymbosa, Pistia stratiotes, Prosopis cineraria, Ricinus communis, Rhynchosia minima, Ceratophyllum demersum, Saccharum spontaneum, Salix tetrasperma, Sida cordata, S. cordifolia, Solanum americanum, S. surattense, Streblus asper, Tephrosia purpurea, Tridax procumbens, Verbascum chinense, V. thapsus, Vicia sativa*	47
434	**Pimples**	*Butea monosperma, Dalbergia sissoo, Euphorbia heterophylla, E. hirta, Ipomoea aquatica, I. pes-tigridis, Ludwigia hyssopifolia, L. octovalvis, Oxalis corniculata, Silene conoidea, Solanum torvum, Tephrosia purpurea, Triumfetta rhomboidea*	13
435	**Pityriasis versicolor**	*Pluchea lanceolata*	1
436	**Pleurisy**	*Diospyros montana, Mimosa rubicaulis*	2
437	**Pneumonia**	*Ageratum conyzoides, Azadirachta indica, Commelina benghalensis, Corchorus aestuans, Diospyros montana, Equisetum ramosissimum, Justicia adhatoda, Operculina turpethum, Polygonum plebeium, Urena lobata, Triumfetta rhomboidea*	11
438	**Poison (Veterinary)**	*Celosia argentea, Cirsium arvense, Cyanthillium cinereum, Euphorbia heterophylla, Peristrophe bicalyculata, Solanum americanum*	6
439	**Poisoning (All types)**	*Ageratum conyzoides, Argemone mexicana, Centaurium pulchellum, Cannabis sativa, Cissampelos pareira, Datura innoxia, Dysphania ambrosioides, Ipomoea aquatica, I. carnea, Melia azedarach, Mitragyna parvifolia, Oxalis corniculata, Phyllanthus reticulatus, Ricinus communis, Tephrosia purpurea, Trianthema portulacastrum, Tridax procumbens, Peristrophe bicalyculata*	18
440	**Polyuria**	*Eleusine indica, Holoptelea integrifolia, Justicia adhatoda, Sida cordifolia, Streblus asper*	5
441	**Postnatal complaints**	*Euphorbia hirta*	1
442	**Postpartum uterine haemorrhage, Postpartum haemorrhage, Cessation of postpartum bleeding**	*Achyranthes aspera, Ageratum conyzoides, Justicia adhatoda, Physalis angulata, Tribulus terrestris*	5
443	**Poultice**	*Ageratum conyzoides, Alternanthera paronychioides, Anagallis arvensis, Cuscuta chinensis, Cayratia trifolia, Celosia argentea, Centella asiatica, Chenopodium album, Cissampelos pareira, Commelina benghalensis, Crateva adansonii* subsp. *odora, Cucumis melo* var. *agrestis, Cyanthillium cinereum, Datura innoxia, Desmodium gangeticum, D. triflorum, Eichhornia crassipes, Ficus palmata, Holoptelea integrifolia, Hyptis suaveolens, Ipomoea aquatica, I. triloba, Lindernia crustacea, Ludwigia hyssopifolia, L. octovalvis, Malva parviflora, Malvastrum coromandelianum, Melia azedarach, Merremia hederacea, Melilotus albus, Melochia corchorifolia, Oxalis corniculata, Phyla nodiflora, Phyllanthus amarus, Pluchea lanceolata, Portulaca oleracea, P. quadrifida, Physalis peruviana, Prosopis juliflora, Ricinus communis, Sesbania bispinosa, S. sesban, Sida acuta, Solanum americanum, S. torvum, Spermacoce articularis, Stellaria media, Sisymbrium irio, Urena lobata, Veronica anagallis-aquatica, Withania somnifera, Ziziphus jujuba*	52
444	**Pregnancy**	*Abutilon indicum, Equisetum ramosissimum, Holoptelea integrifolia, Nymphaea pubescens, Prosopis cineraria, Sesbania sesban, Sida cordata, Triumfetta rhomboidea*	8
445	**Pregnancy (inhibit)**	*Azadirachta indica*	1

SN	Name of Disease/Medicinal properties	Name of plant species	No. of plants
446	**Premature ejaculation**	*Acacia nilotica, Cannabis sativa, Clerodendrum phlomidis, Cuscuta reflexa, Datura innoxia, Euphorbia hirta, Malvastrum coromandelianum*	7
447	**Prevent graying of hair**	*Barleria prionitis, Eclipta prostrata*	2
448	**Prevent growth of hair**	*Leucaena leucocephala*	1
449	**Proctitis**	*Withania somnifera*	1
450	**Prolapsed Ani, Prolapsus**	*Cirsium arvense*	1
451	**Prophylactic**	*Calotropis gigantea, Indigofera tinctoria, Tamarix indica*	3
452	**Prostate cancer**	*Cuscuta chinensis, Diospyros montana*	2
453	**Prostatitis**	*Cyperus rotundus, Phyllanthus amarus*	2
454	**Proteolytic**	*Anagallis arvensis, Calotropis procera*	2
455	**Protisticide**	*Coccinia grandis*	1
456	**Psoriasis**	*Amaranthus viridis, Azadirachta indica, Coccinia grandis, Cyanthillium cinereum, Hydrocotyle sibthorpioides, Leucas cephalotes, Setaria verticillata, Solanum americanum, Stellaria media, Tribulus terrestris*	10
457	**Psychopathy, Psychosis**	*Marsilea quadrifolia, Peristrophe bicalyculata*	2
458	**Psychopharmacological, Psychostimulant**	*Cleome viscosa, Convolvulus prostratus*	2
459	**Pulmonary troubles/ailments, COPD (Chronic Obstructive Pulmonary Disease)**	*Alysicarpus vaginalis, Citrullus colocynthis, Hydrocotyle sibthorpioides, Solanum americanum, Stellaria media, Tephrosia pumila, Verbascum thapsus, Ziziphus jujuba*	8
460	**Pungent**	*Blumea laciniata, Convolvulus prostratus, Dysphania ambrosioides, Erigeron bonariensis, Oxystelma esculentum, Pistia stratiotes, Solanum surattense*	7
461	**Purify breast milk secretion**	*Cissampelos pareira*	1
462	**Pus, Pus formation, Pus in ear, Pyogenic**	*Barleria prionitis, Dalbergia sissoo, Digera muricata, Hyptis suaveolens, Melia azedarach, Senna occidentalis, Sida acuta*	7
463	**Pyorrhea**	*Barleria prionitis, Eclipta prostrata*	2
464	**Pyrexia**	*Fumaria indica, Phoenix sylvestris*	2
465	**Rabies**	*Abrus precatorius, Anagallis arvensis, Datura innoxia, Indigofera tinctoria, Ipomoea pes-tigridis, Lantana camara*	6
466	**Radioprotective**	*Syzygium cumini*	1
467	**Rat bite**	*Solanum americanum*	1
468	**Rat poisoning**	*Ranunculus sceleratus, Tephrosia purpurea*	2
469	**Rectal diseases**	*Butea monosperma*	1
470	**Refrigerant**	*Arenaria serpyllifolia, Azadirachta indica, Chrysopogon zizanioides, Cordia dichotoma, Dalbergia sissoo, Digera muricata, Ficus religiosa, Lantana camara, Justicia adhatoda, Marsilea quadrifolia, Melia azedarach, Mukia maderaspatana, Rungia pectinata, Saccharum spontaneum, Sida cordata, Stellaria media, Syzygium cumini, Trichosanthes cucumerina, Urena lobata, Vallisneria spiralis, Vicia sativa, Ziziphus jujuba*	22
471	**Renoprotective**	*Senna occidentalis*	1
472	**Reproductive disorder, Diseases of reproductive organs**	*Capsella bursa-pastoris, Launaea procumbens, Sida cordifolia, Tephrosia pumila*	4
473	**Resolutive, Resolvent**	*Argemone mexicana, Malvastrum coromandelianum, Lathyrus aphaca, Melia azedarach*	4
474	**Respiratory depressant, Respirodepressant**	*Tridax procumbens, Euphorbia hirta*	2

SN	Name of Disease/Medicinal properties	Name of plant species	No. of plants
475	**Respiratory diseases/ disorders, Respiratory infections, Respiratory tract infection**	*Achyranthes aspera, Amaranthus viridis, Azadirachta indica, Commelina benghalensis, Cyanthillium cinereum, Euphorbia heterophylla, E. hirta, Hyptis suaveolens, Justicia adhatoda, Morus alba, Phyla nodiflora, Pluchea lanceolata, Saccharum spontaneum, Solanum surattense, Urena lobata*	15
476	**Respirostimulant**	*Euphorbia hirta*	1
477	**Restorative**	*Imperata cylindrica, Nymphaea pubescens, Phoenix sylvestris, Sisymbrium irio, Withania somnifera*	5
478	**Revulsive**	*Achyranthes aspera*	1
479	**Rheumatic diseases, Rheumatic-arthritis, Rheumatic swellings, Rheumatism, Rheumatic fever, Arthritis, Rheumarthritis, Rheumatoid arthritis, Arthralgia, Joint pain, Knee joint pain, Swollen joints/feet, Rheumatalgia, Osteoarthritis, Antirheumatic, Antiarthritic**	*Abrus precatorius, Ageratum conyzoides, Alhagi maurorum, Ammannia baccifera, Amaranthus viridis, Anagallis arvensis, Anisomeles indica, Argemone mexicana, Avena sterilis, Azadirachta indica, Barleria prionitis, Bidens pilosa, Blainvillea acmella, Blumea viscosa, Boerhavia diffusa, Cannabis sativa, Capsella bursa-pastoris, Carthamus oxyacantha, Cajanus scarabaeoides, Calotropis gigantea, C. procera, Centella asiatica, Centaurium pulchellum, Chenopodium album, Chloris barbata, Chrysopogon zizanioides, Cissampelos pareira, Citrullus colocynthis, Cleome viscosa, Clerodendrum phlomidis, Coccinia grandis, Commelina benghalensis, Crateva adansonii* subsp. *odora, Cuscuta reflexa, Cyperus iria, C .rotundus, Datura innoxia, Equisetum ramosissimum, E. ramosissimum* subsp. *debile, Euphorbia hirta, Evolvulus nummularius, Fumaria indica, Grangea maderaspatana, Heliotropium ellipticum, Holoptelea integrifolia, Hydrilla verticillata, Hydrocotyle sibthorpioides, Hyptis suaveolens, Imperata cylindrica, Indigofera linnaei, Ipomoea eriocarpa, Jatropha gossypiifolia, Justicia adhatoda, Lantana camara, Launaea procumbens, Lindernia crustacea, Malvastrum coromandelianum, Medicago polymorpha, Melia azedarach, Mitragyna parvifolia, Morus alba, Mukia maderaspatana, Nepeta hindostana, Nicotiana plumbaginifolia, Oldenlandia corymbosa, Operculina turpethum, Parthenium hysterophorus, Phyla nodiflora, Phyllanthus maderaspatensis, Physalis angulata, P. peruviana, Phragmites karka, Pluchea lanceolata, Portulaca oleracea, Prosopis cineraria, Ranunculus sceleratus, Ricinus communis, Salix tetrasperma, Senna obtusifolia, Sesbania sesban, Setaria pumila, S. verticillata, Sisymbrium irio, Sida cordata, S. cordifolia, Solanum americanum, S. surattense, S. torvum, Sonchus brachyotus, Stellaria media, Tamarix indica, Tephrosia pumila, T. purpurea, Teramnus labialis, Trianthema portulacastrum, Tribulus terrestris, Urena lobata, Verbascum chinense, V. thapsus, Vicia sativa, Withania somnifera, Xanthium strumarium*	102
480	**Rickets**	*Holoptelea integrifolia*	1
481	**Ringworm**	*Ammannia baccifera, Azadirachta indica, Calotropis gigantea, Cleome viscosa, Coccinia grandis, Cordia dichotoma, Croton bonplandianus, Erigeron bonariensis, Euphorbia hirta, E. hypericifolia, Holoptelea integrifolia, Imperata cylindrica, Ipomoea aquatica, Lindernia crustacea, Ludwigia hyssopifolia, Malvastrum coromandelianum, Mukia maderaspatana, Phyllanthus amarus, Pistia stratiotes, Senna occidentalis, Sesbania bispinosa, Solanum americanum, S. surattense, Syzygium cumini, Withania somnifera*	25
482	**Rubefacient**	*Ammannia baccifera, Barleria prionitis, Cayratia trifolia, Cissampelos pareira, Cleome viscosa, Crateva adansonii* subsp. *odora, Euphorbia heterophylla, Ranunculus sceleratus*	8
483	**Salivation (reducing)**	*Carthamus oxyacantha*	1
484	**Salivation, Saliva (inducing)**	*Xanthium strumarium*	1
485	**Scabies**	*Ageratum conyzoides, Ammannia baccifera, Argemone mexicana, Azadirachta indica, Bacopa monnieri, Barleria prionitis, Boerhavia diffusa, Carthamus oxyacantha, Cissampelos pareira, Coccinia grandis, Cyperus rotundus, Dalbergia sissoo, Datura innoxia, Equisetum ramosissimum* subsp. *debile, Evolvulus nummularius, Ficus religiosa, Fumaria indica, Holoptelea integrifolia, Jatropha gossypiifolia, Lantana camara, Leucas cephalotes, Morus alba, Mukia maderaspatana, Persicaria barbata, Phyllanthus amarus, P. maderaspatensis, Physalis angulata, Senna occidentalis, Solanum americanum, Spermacoce articularis, Tribulus terrestris, Tridax procumbens, Xanthium strumarium, Ziziphus nummularia*	35
486	**Scalds, Burns**	*Cannabis sativa, Portulaca oleracea, Stellaria media, Artemisia scoparia*	4
487	**Schistosomiasis**	*Physalis angulata*	1

SN	Name of Disease/Medicinal properties	Name of plant species	No. of plants
488	**Sciatica**	*Barleria prionitis, Centaurium pulchellum, Citrullus colocynthis, Datura innoxia, Melia azedarach, Ricinus communis, Senna obtusifolia, Sida cordifolia*	8
489	**Scrofula, Scrofulous skin**	*Crateva adansonii* subsp. *odora, Fumaria indica, Melia azedarach, Ranunculus sceleratus, Streblus asper, Tephrosia purpurea, Typha domingensis, Xanthium strumarium*	8
490	**Scurf**	*Leucaena leucocephala*	1
491	**Scurvy**	*Acmella paniculata, Cayratia trifolia, Cleome viscosa, Melia azedarach, Oxalis corniculata, Oxalis debilis* var. *corymbosa, Portulaca oleracea, Stellaria media, Veronica anagallis-aquatica*	9
492	**Secretory**	*Azadirachta indica*	1
493	**Sedative**	*Argemone mexicana, Azadirachta indica, Cannabis sativa, Capsella bursa-pastoris, Centella asiatica, Chenopodium album, Datura innoxia, Desmodium gangeticum, Evolvulus nummularius, Euphorbia hirta, Grangea maderaspatana, Justicia adhatoda, Physalis angulata, Salix tetrasperma, Solanum americanum, S. torvum, Verbascum thapsus, Withania somnifera, Xanthium strumarium, Ziziphus jujuba*	20
494	**Seminal weakness, Semen quality improvement, Semen discharge**	*Abutilon indicum, Bacopa monnieri, Boerhavia diffusa, Chenopodium album, C. murale, Cuscuta chinensis, Ludwigia hyssopifolia, Withania somnifera, Ziziphus nummularia*	9
495	**Sex stimulant, sex-enhancing**	*Butea monosperma, Withania somnifera*	2
496	**Sexual disorders, Sexual debility, Sexual weakness, Sexual vigour**	*Chrysopogon zizanioides, Corchorus tridens, Eclipta prostrata, Euphorbia hirta, Ficus religiosa, Ipomoea carnea, Justicia adhatoda, Launaea procumbens, Phoenix sylvestris, Saccharum spontaneum, Sida cordifolia, Teramnus labialis, Tribulus terrestris, Withania somnifera*	14
497	**Shingles**	*Leucaena leucocephala*	1
498	**Sialagogue, Sialogenic**	*Imperata cylindrica, Xanthium strumarium*	2
499	**Sickle-cell disease**	*Scoparia dulcis*	1
500	**Skin diseases, Skin diseases like itching, allergy, lesions and rashes; Skin fungal infections, Skin disorders/ ailments, Skin eruptions, Eruptions, Chronic eruptions, Skin sores, Skin plagues, Skin ulcers, Blisters, Skin tumour, Pustules, Papillomas, Pruritus, Antipruritic**	*Achyranthes aspera, Acmella paniculata, Ageratum conyzoides, Albizia lebbeck, Alternanthera sessilis, Alysicarpus monilifer, Alysicarpus vaginalis, Amaranthus viridis, Ammannia baccifera, Ampelopteris prolifera, Anagallis arvensis, Anisomeles indica, Argemone mexicana, Avena sterilis, Azadirachta indica, Bacopa monnieri, Barleria prionitis, Bidens pilosa, Boerhavia diffusa, Butea monosperma, Chloris barbata, Calotropis gigantea, C. procera, Celosia argentea, Centella asiatica, Chenopodium album, Chrysopogon zizanioides, Cirsium arvense, Cissampelos pareira, Citrullus colocynthis, Cleome viscosa, Clerodendrum phlomidis, Coccinia grandis, Cuscuta chinensis, Commelina benghalensis, Convolvulus arvensis, Croton bonplandianus, Cyanthillium cinereum, Cynodon dactylon, Cyperus difformis, Dalbergia sissoo, Desmodium triflorum, Desmostachya bipinnata, Dysphania ambrosioides, Eclipta prostrata, Erigeron bonariensis, Euphorbia heterophylla, E. hirta, E. hypericifolia, E. serpens, Ficus religiosa, Fumaria indica, Gonostegia pentandra, Holoptelea integrifolia, Hyptis suaveolens, Indigofera tinctoria, Ipomoea obscura, Jatropha gossypiifolia, Justicia adhatoda, Lathyrus aphaca, Leucas cephalotes, Lindernia ciliata, Malvastrum coromandelianum, Marsilea quadrifolia, Medicago polymorpha, Melia azedarach, Melilotus indicus, Mitragyna parvifolia, Mukia maderaspatana, Oldenlandia corymbosa, Operculina turpethum, Oxalis corniculata, O. debilis* var. *corymbosa, Oxystelma esculentum, Parthenium hysterophorus, Peristrophe bicalyculata, Phyllanthus amarus, P. reticulatus, Physalis angulata, Portulaca oleracea, P. quadrifida, Prosopis cineraria, P. juliflora, Ranunculus sceleratus, Ricinus communis, Rumex dentatus, Saccharum bengalense, Sagittaria guayanensis, Senna obtusifolia, S. occidentalis, Sesamum indicum, Sesbania bispinosa, S. sesban, Setaria pumila, S. verticillata, Sida acuta, Solanum americanum, S. surattense, S. torvum, Sorghum halepense, Spermacoce articularis, Spirodela polyrrhiza, Stellaria media, Streblus asper, Syzygium cumini, Tephrosia purpurea, Tribulus terrestris, Trichosanthes cucumerina, Tridax procumbens, Urena lobata, Verbascum thapsus, Veronica agrestis, Vicia sativa, Xanthium strumarium, Ziziphus jujuba, Z. nummularia*	116
501	**Sleeping sickness**	*Physalis angulata*	1
502	**Smallpox**	*Coccinia grandis, Dactyloctenium aegyptium, Melochia corchorifolia, Mimosa rubicaulis, Panicum antidotale, Phyllanthus reticulatus, Physalis angulata, Rungia pectinata, Xanthium strumarium*	9

SN	Name of Disease/Medicinal properties	Name of plant species	No. of plants
503	**Sores**	*Abrus precatorius, Ageratum conyzoides, Ammannia baccifera, Cajanus scarabaeoides, Cannabis sativa, Cayratia trifolia, Cenchrus ciliaris, Celosia argentea, Chrysopogon zizanioides, Cirsium arvense, Cissampelos pareira, Coccinia grandis, Desmodium triflorum, Echinochloa crus-galli, Euphorbia hirta, Hyptis suaveolens, Indigofera tinctoria, Ipomoea aquatica, I. obscura, I. pes-tigridis, Jatropha gossypiifolia, Lindernia crustacea, Ludwigia hyssopifolia, Malva parviflora, Malvastrum coromandelianum, Melochia corchorifolia, Murdannia nudiflora, Nicotiana plumbaginifolia, Paspalum scrobiculatum, Phyllanthus amarus, P. reticulatus, Portulaca oleracea, Prosopis cineraria, Salix tetrasperma, Senna obtusifolia, Spermacoce articularis, Setaria pumila, Sida cordifolia, Solanum americanum, S. torvum, Trianthema portulacastrum, Verbascum thapsus, Veronica anagallis-aquatica , Xanthium strumarium*	42
504	**Sour**	*Oxalis corniculata, Morus alba, Portulaca oleracea, P. quadrifida, Sida cordata, Syzygium cumini*	6
505	**Spasmogenic**	*Acacia nilotica, Calotropis procera*	2
506	**Spasms, Spasmolytic**	*Ageratum conyzoides, Chenopodium album, Cyanthillium cinereum, Diospyros montana, Eclipta prostrata, Hyptis suaveolens, Justicia adhatoda, Phyllanthus amarus, P. maderaspatensis, Senna occidentalis*	10
507	**Spermatopoietic**	*Sida cordifolia*	1
508	**Spermatorrhea, Spermicidal**	*Ampelopteris prolifera, Hyptis suaveolens, Salix tetrasperma, Sida cordifolia, Tribulus terrestris*	5
509	**Spinal disease**	*Ammannia baccifera*	1
510	**Spleen diseases, Splenopathy, Spleen enlargement, Splenomegaly**	*Alternanthera sessilis, Bacopa monnieri, Calotropis procera, Cayratia trifolia, Chenopodium album, C. murale, Citrullus colocynthis, Cordia dichotoma, Cynodon dactylon, Echinochloa crus-galli, Eclipta prostrata, Grangea maderaspatana, Indigofera tinctoria, Lindernia ciliata, Morus alba, Panicum antidotale, Phyllanthus amarus, Portulaca oleracea, Sesbania sesban, Solanum americanum, S. torvum, Sisymbrium irio, Spermacoce articularis, Syzygium cumini, Tamarix indica, Tephrosia purpurea, T. villosa*	27
511	**Sprain, Torn muscles, Muscle disease**	*Albizia lebbeck, Capparis sepiaria, Chrysopogon zizanioides, Cyanthillium cinereum, Hyptis suaveolens, Ipomoea carnea, Jatropha gossypiifolia, Justicia adhatoda, Oxalis corniculata, Peristrophe bicalyculata, Portulaca oleracea*	11
512	**Sprue**	*Fumaria indica, Ludwigia hyssopifolia*	2
513	**Sterility**	*Bacopa monnieri, Crateva adansonii* subsp. *odora, Physalis angulata, Withania somnifera*	4
514	**Stimulant**	*Ageratum conyzoides, Anagallis arvensis, Cardamine hirsuta, Chrysopogon zizanioides, Capsella bursa-pastoris, Chenopodium album, Cleome viscosa, Cyperus iria, Dalbergia sissoo, Euphorbia hirta, Grangea maderaspatana, Hyptis suaveolens, Indigofera linifolia, Lantana camara, Justicia adhatoda, Lepidium didymum, Leucas cephalotes, Nepeta hindostana, Operculina turpethum, Persicaria barbata, Phoenix sylvestris, Rorippa palustris, Sida cordifolia, Sisymbrium irio, Solanum americanum, S. surattense, Withania somnifera*	27
515	**Stomach disorders, Gastritis**	*Anisomeles indica, Bacopa monnieri, Blumea viscosa, Calotropis gigantea, Capparis sepiaria, Clerodendrum phlomidis, Crateva adansonii* subsp. *odora, Cyanthillium cinereum, Emex australis, Equisetum ramosissimum, Fimbristylis ferruginea, Grangea maderaspatana, Holoptelea integrifolia, Hyptis suaveolens, Leucaena leucocephala, Malva parviflora, Melilotus albus, Murdannia nudiflora, Oxalis corniculata, Phyllanthus amarus, Physalis angulata, P. peruviana, Pistia stratiotes, Portulaca quadrifida, Prosopis juliflora, Scoparia dulcis, Sisymbrium irio*	27
516	**Stomach/Duodenal/Intestine ulcers, Ulcerative colitis**	*Centella asiatica, Chenopodium album, Cordia dichotoma, Morus alba, Syzygium cumini*	5
517	**Stomachalgia**	*Syzygium cumini, Tephrosia purpurea*	2

SN	Name of Disease/Medicinal properties	Name of plant species	No. of plants
518	**Stomachic**	*Achyranthes aspera, Ammannia baccifera, Anisomeles indica, Argemone mexicana, Calotropis gigantea, C. procera, Capparis sepiaria, Chrysopogon zizanioides, Cannabis sativa, Cayratia trifolia, Centella asiatica, Chenopodium murale, Cissampelos pareira, Clerodendrum phlomidis, Cucumis melo* var. *agrestis, Cuscuta reflexa, Cyperus iria, C. rotundus, Dysphania ambrosioides, Grangea maderaspatana, Hyptis suaveolens, Melia azedarach, Mitragyna parvifolia, Mukia maderaspatana, Nymphaea pubescens, Oldenlandia corymbosa, Oxalis corniculata, Parthenium hysterophorus, Phyla nodiflora, Phyllanthus amarus, P. maderaspatensis, P. reticulatus, Portulaca oleracea, Sida acuta, S. cordata, Senna obtusifolia, Solanum surattense, Spermacoce articularis, Syzygium cumini, Teramnus labialis, Trianthema portulacastrum, Tribulus terrestris, Trichosanthes cucumerina, Tridax procumbens, Vallisneria spiralis, Ziziphus nummularia*	46
519	**Stomatalgia, Stomatosis, Stomatitis**	*Abrus precatorius, Cannabis sativa, Coccinia grandis, Euphorbia hypericifolia, Tribulus terrestris*	5
520	**Stones, Kidney stone/ gravel, Renal colic, Renal disorders, Renal lithiasis, Renal insufficiency-diuresis, Calculus, Calculi (renal and vesical), Urinary stones, Lithiasis, Urolithiasis, Vesicular calculus, Antilithic, Antilithiatic, Litholytic, Lithotriptic, Lithontriptic**	*Achyranthes aspera, Ageratum conyzoides, Alhagi maurorum, Alysicarpus monilifer, A. vaginalis, Anagallis arvensis, Anisomeles indica, Arenaria serpyllifolia, Argemone mexicana, Avena sterilis, Boerhavia diffusa, Capsella bursa-pastoris, Chrysopogon zizanioides, Centaurium pulchellum, Cissampelos pareira, Coccinia grandis, Crateva adansonii* subsp. *odora, Cynodon dactylon, Cyperus rotundus, Desmodium gangeticum, D. bipinnata, Digera muricata, Equisetum ramosissimum, E. ramosissimum* subsp. *debile, Euphorbia hirta, Gomphrena celosioides, Hydrocotyle sibthorpioides, Imperata cylindrica, Justicia adhatoda, Melia azedarach, Oxalis corniculata, Phyllanthus amarus, P. tenellus, Physalis angulata, Portulaca oleracea, Saccharum spontaneum, Salix tetrasperma, Senna obtusifolia, Scoparia dulcis, Solanum surattense, Spermacoce articularis, Trianthema portulacastrum, Tribulus terrestris, Veronica agrestis*	44
521	**Strangulated hernia**	*Nicotiana plumbaginifolia*	1
522	**Strangury, Urination, Promote urination, Micturition, Urine-inducing**	*Abutilon indicum, Achyranthes aspera, Ammannia baccifera, Arenaria serpyllifolia, Bacopa monnieri, Boerhavia diffusa, Centella asiatica, Chenopodium album, Clerodendrum phlomidis, Convolvulus arvensis, Cordia dichotoma, Crateva adansonii* subsp. *odora, Cuscuta reflexa, Cyanthillium cinereum, Cynodon dactylon, Dalbergia sissoo, Desmodium gangeticum, Desmostachya bipinnata, Euphorbia hirta, Launaea procumbens, Marsilea quadrifolia, Morus alba, Oxalis corniculata, Phyla nodiflora, Phyllanthus maderaspatensis, P. reticulatus, Portulaca oleracea, Ricinus communis, Sida cordifolia, Solanum surattense, Spirodela polyrrhiza, Syzygium cumini, Tephrosia purpurea, Trianthema portulacastrum, Tribulus terrestris, Veronica anagallis-aquatica*	36
523	**Strength, Strengthens body**	*Acacia nilotica, Boerhavia diffusa, Corchorus olitorius, Eclipta prostrata, Ludwigia hyssopifolia, Solanum surattense*	6
524	**Stress, Antistress**	*Bacopa monnieri, Centella asiatica, Convolvulus arvensis, C. prostratus, Ficus religiosa, Withania somnifera*	6
525	**Stroke**	*Tephrosia pumila*	1
526	**Styptic**	*Ageratum conyzoides, Ceratophyllum demersum, Dalbergia sissoo, Imperata cylindrica, Justicia adhatoda, Mitragyna parvifolia, Oxalis corniculata, Tephrosia purpurea, Tribulus terrestris, Triumfetta rhomboidea, Urena lobata, Ziziphus jujuba*	12
527	**Substitute for ipecacuanha**	*Calotropis gigantea*	1
528	**Sudorific**	*Achyranthes aspera, Bacopa monnieri, Centella asiatica, Chrysopogon zizanioides, Cleome viscosa, Hyptis suaveolens, Justicia adhatoda, Leucas cephalotes, Mukia maderaspatana, Portulaca oleracea, Ricinus communis, Solanum americanum, S. surattense*	13
529	**Sunburn, Sun protective**	*Amaranthus viridis, Chenopodium album, Prosopis juliflora, Senna occidentalis*	4
530	**Sunstroke**	*Chenopodium album, Merremia hederacea, Mimosa rubicaulis, Rumex dentatus*	4
531	**Suppuration**	*Phyllanthus reticulatus, Sesbania sesban, Sida acuta*	3
532	**Sweet**	*Centella asiatica, Ceratophyllum demersum, Chenopodium album, Cynodon dactylon, Desmostachya bipinnata, Equisetum ramosissimum, Ficus religiosa, Melia azedarach, Saccharum spontaneum, Sesamum indicum*	10

SN	Name of Disease/Medicinal properties	Name of plant species	No. of plants
533	**Swelling, Body swelling**	*Abrus precatorius, Ageratum conyzoides, Ammannia baccifera, Anisomeles indica, Barleria prionitis, Cajanus scarabaeoides, Cannabis sativa, Cayratia trifolia, Clerodendrum phlomidis, Convolvulus arvensis, Crateva adansonii* subsp. *odora, Croton bonplandianus, Cyanotis axillaris, Desmodium gangeticum, Diospyros montana, Eclipta prostrata, Ficus religiosa, Hyptis suaveolens, Ipomoea carnea, I. obscura, I. pes-tigridis, Launaea procumbens, Malva parviflora, Melilotus indicus, Melochia corchorifolia, Phyllanthus amarus, Polygonum plebeium, Portulaca quadrifida, Ranunculus sceleratus, Ricinus communis, Rungia pectinata, Salix tetrasperma, Sida acuta, S. cordifolia, Sisymbrium irio, Solanum americanum, Sonchus asper, Stellaria media, Urena lobata, Withania somnifera*	40
534	**Swollen cervical glands, Neck glands**	*Phyla nodiflora, Ludwigia hyssopifolia*	2
535	**Syndrome-X**	*Coccinia grandis*	1
536	**Syphilis, Syphilitic sores, Syphilitic ulcers, Antisyphilitic**	*Azadirachta indica, Bacopa monnieri, Calotropis gigantea, C. procera, Centella asiatica, Cirsium arvense, Clerodendrum phlomidis, Coccinia grandis, Corchorus tridens, Crateva adansonii* subsp. *odora, Dalbergia sissoo, Euphorbia heterophylla, E. hirta, Hydrocotyle sibthorpioides, Ludwigia hyssopifolia, Mimosa rubicaulis, Nicotiana plumbaginifolia, Phyllanthus reticulatus, Pistia stratiotes, Sesbania sesban, Tephrosia purpurea, Trichosanthes cucumerina, Withania somnifera, Ziziphus jujuba*	24
537	**T helper cell modulator**	*Bidens pilosa*	1
538	**Teething pain**	*Solanum americanum*	1
539	**Tenesmus**	*Sida cordifolia*	1
540	**Teratogen**	*Ipomoea carnea*	1
541	**Testosterone booster**	*Tribulus terrestris*	1
542	**Tetanus, Tetany**	*Abrus precatorius, Ageratum conyzoides, Cannabis sativa, Eclipta prostrata*	4
543	**Thermogenic**	*Achyranthes aspera, Ageratum conyzoides, Butea monosperma, Cleome viscosa, Dalbergia sissoo, Desmodium gangeticum, Eclipta prostrata, Holoptelea integrifolia, Indigofera tinctoria, Leucas cephalotes, Melia azedarach, Operculina turpethum, Oxalis corniculata, Pluchea lanceolata, Portulaca oleracea, Ricinus communis, Sesamum indicum, Sida cordata, Solanum americanum, S. surattense, Tephrosia purpurea*	21
544	**Throat infection/irritation, Sore throat, Pharyngitis**	*Bidens pilosa, Coccinia grandis, Commelina benghalensis, Dalbergia sissoo, Hydrocotyle sibthorpioides, Justicia adhatoda, Lindernia crustacea, Malva parviflora, Malvastrum coromandelianum, Merremia hederacea, Morus alba, Oxystelma esculentum, Panicum antidotale, Prosopis juliflora, Sesbania sesban, Solanum americanum, S. surattense, Sisymbrium irio, Syzygium cumini, Urena lobata, Verbascum thapsus, Ziziphus nummularia*	22
545	**Thrush**	*Commelina benghalensis, Euphorbia hirta*	2
546	**Tinnitus**	*Morus alba*	1
547	**Tonic**	*Abrus precatorius, Acacia nilotica, Acmella paniculata, Ageratum conyzoides, Albizia lebbeck, Ampelopteris prolifera, Anisomeles indica, Barleria prionitis, Boerhavia diffusa, Butea monosperma, Cardamine hirsuta, Carthamus oxyacantha, Cajanus scarabaeoides, Calotropis procera, Capparis sepiaria, Chenopodium album, Chrysopogon zizanioides, Convolvulus prostratus, Corchorus aestuans, C. olitorius, Cordia dichotoma, Crateva adansonii* subsp. *odora, Cuscuta chinensis, C. reflexa, Cyperus iria, Desmodium triflorum, Eichhornia crassipes, Echinochloa crus-galli, Erigeron bonariensis, Euphorbia hirta, Ficus religiosa, Imperata cylindrica, Indigofera tinctoria, Ipomoea aquatica, Lantana camara, Justicia adhatoda, Lepidium didymum, Mazus pumilus, Medicago polymorpha, Merremia hederacea, Melia azedarach, Melilotus indicus, Mukia maderaspatana, Nepeta hindostana, Nymphaea pubescens, Oldenlandia corymbosa, Persicaria barbata, Phoenix sylvestris, Phyla nodiflora, Rumex dentatus, Rhynchosia minima, Saccharum spontaneum, Senna occidentalis, Sida acuta, S. cordifolia, Solanum americanum, S. torvum, Sonchus asper, Stellaria media, Streblus asper, Stuckenia pectinata, Syzygium cumini, Trichosanthes cucumerina, Tephrosia purpurea, Triumfetta rhomboidea, Veronica agrestis, Vicia sativa, Withania somnifera, Xanthium strumarium, Ziziphus jujuba, Ziziphus nummularia*	71
548	**Tonsillitis**	*Acacia nilotica, Barleria prionitis, Evolvulus nummularius, Merremia hederacea, Solanum americanum, Urena lobata*	6

SN	Name of Disease/Medicinal properties	Name of plant species	No. of plants
549	**Toothache, Teeth problem, Carious teeth, Antidontalgic**	*Abutilon indicum, Achyranthes aspera, Acmella paniculata, Anisomeles indica, Argemone mexicana, Barleria prionitis, Calotropis gigantea, C. procera, Cannabis sativa, Chenopodium album, Cynodon dactylon, Desmodium gangeticum, D. triflorum, Eclipta prostrata, Euphorbia hirta, E. hypericifolia, Equisetum ramosissimum, Ficus palmata, Ficus religiosa, Gonostegia pentandra, Hyptis suaveolens, Indigofera astragalina, Lantana camara, Mukia maderaspatana, Nepeta hindostana, Phoenix sylvestris, Physalis angulata, Scoparia dulcis, Sesamum indicum, Solanum surattense, S. torvum, Spermacoce articularis, Streblus asper, Syzygium cumini, Tridax procumbens, Urena lobata, Verbascum thapsus*	37
550	**Toxemia**	*Capparis sepiaria, Ludwigia octovalvis*	2
551	**Tracheitis**	*Verbascum thapsus*	1
552	**Tranquilizer**	*Convolvulus prostratus, Cannabis sativa, Centella asiatica, Chenopodium murale, Convolvulus arvensis*	5
553	**Trauma, Traumatic injury**	*Carthamus oxyacantha, Oxalis corniculata, Oxystelma esculentum*	3
554	**Travel sickness**	*Nicotiana plumbaginifolia*	1
555	**Tremor**	*Prosopis cineraria*	1
556	**Tubercular glands**	*Azadirachta indica*	1
557	**Tuberculosis**	*Abrus precatorius, Arenaria serpyllifolia, Butea monosperma, Cirsium arvense, Coccinia grandis, Diospyros montana, Ficus religiosa, Lantana camara, Justicia adhatoda, Leucas cephalotes, Mimosa rubicaulis, Mukia maderaspatana, Operculina turpethum, Peristrophe bicalyculata, Phyllanthus amarus, Solanum americanum, S. surattense, S. torvum, Spergula arvensis, Teramnus labialis, Trichosanthes cucumerina, Verbascum thapsus, Xanthium strumarium*	23
558	**Tumours, Antitumour**	*Abrus precatorius, Ageratum conyzoides, Albizia lebbeck, Alhagi maurorum, Azadirachta indica, Bacopa monnieri, Barleria prionitis, Bidens pilosa, Butea monosperma, Calotropis gigantea, C. procera, Cannabis sativa, Capparis sepiaria, Cayratia trifolia, Centella asiatica, Citrullus colocynthis, Cleome viscosa, Convolvulus arvensis, Corchorus olitorius, Cordia dichotoma, Croton bonplandianus, Cuscuta chinensis, Cynodon dactylon, Datura innoxia, Eichhornia crassipes, Euphorbia heterophylla, E. hypericifolia, E. serpens, Ficus palmata, Hyptis suaveolens, Indigofera linnaei, Ipomoea carnea, Ludwigia hyssopifolia, Melilotus indicus, Operculina turpethum, Portulaca oleracea, Ricinus communis, Rumex dentatus, Scoparia dulcis, Sesbania sesban, Solanum americanum, Sonchus brachyotus, Tephrosia purpurea, Triumfetta rhomboidea, Verbascum thapsus, Withania somnifera, Xanthium strumarium*	46
559	**Tympanites**	*Cyanotis axillaris, Oxalis corniculata, Tephrosia purpurea*	3
560	**Typhoid, Typhoid fever**	*Ageratum conyzoides, Ammannia baccifera, Mazus pumilus, Melia azedarach, Senna occidentalis, Ziziphus jujuba*	6
561	**Ulcerative stomatitis**	*Portulaca oleracea, Tribulus terrestris*	2
562	**Ulcers, Antiulcerogenic, Antiulcer**	*Abutilon indicum, Ammannia baccifera, Amaranthus viridis, Albizia lebbeck, Anagallis arvensis, Argemone mexicana, Azadirachta indica, Bacopa monnieri, Barleria prionitis, Bidens pilosa, Butea monosperma, Carthamus oxyacantha, Calotropis gigantea, Cannabis sativa, Cayratia trifolia, Celosia argentea,Centella asiatica, Ceratophyllum demersum, Cissampelos pareira, Coccinia grandis, Corchorus tridens, Cordia dichotoma, Cyperus rotundus, Dalbergia sissoo, Dactyloctenium aegyptium, Datura innoxia, Desmodium gangeticum, D. triflorum, Dysphania ambrosioides, Eclipta prostrata, Equisetum ramosissimum, Euphorbia hirta, E. hypericifolia, Equisetum ramosissimum* subsp. *debile, Ficus palmata, Ficus religiosa, Heliotropium ellipticum, Hydrocotyle sibthorpioides, Indigofera linnaei, I. tinctoria, Ipomoea obscura, I. eriocarpa, Lantana camara, Leucas cephalotes, Ludwigia octovalvis, Malva parviflora, Malvastrum coromandelianum, Melia azedarach, Medicago lupulina, Melilotus albus, Mollugo nudicaulis, Mukia maderaspatana, Operculina turpethum, Oxalis corniculata, Oxystelma esculentum, Persicaria barbata, Peristrophe bicalyculata, Phyla nodiflora, Phyllanthus amarus, P. maderaspatensis, Portulaca oleracea, P. quadrifida, Pistia stratiotes, Prosopis cineraria, Ranunculus sceleratus, Senna obtusifolia, Sesamum indicum, Sida acuta, S. cordifolia, Solanum americanum, S. torvum, Spermacoce articularis, Sphenoclea zeylanica, Stellaria media, Streblus asper, Syzygium cumini, Tephrosia purpurea, T. villosa, Trianthema portulacastrum, Tridax procumbens, Triumfetta rhomboidea, Verbascum thapsus, Veronica anagallis-aquatica, Withania somnifera, Xanthium strumarium, Ziziphus jujuba*	86
563	**Ulemorrhagia**	*Phyllanthus reticulatus*	1
564	**Ulitis**	*Tribulus terrestris*	1
565	**Ulocace**	*Tribulus terrestris*	1

SN	Name of Disease/Medicinal properties	Name of plant species	No. of plants
566	**Umbilical infections**	*Lindernia crustacea*	1
567	**Urethritis, Urethral pain, Urethrorrhea, Urethralgia, Urethral discharge**	*Abutilon indicum, Centella asiatica, Cordia dichotoma, Euphorbia hirta, Melia azedarach, Solanum torvum, Syzygium cumini*	7
568	**Urinary Tract Infections, Urinary affections, Uterine infections, Urine complaints, Urinary disorders, Urinary bladder complaints, Cystalgia, Bladder infections, Bladder inflammation, Scalding of urine, Blood in urine, Uropathy**	*Acacia nilotica, Alhagi maurorum, Amaranthus viridis, Anisomeles indica, Arenaria serpyllifolia, Barleria prionitis, Cajanus scarabaeoides, Capsella bursa-pastoris, Centella asiatica, Chenopodium album, Cissampelos pareira, Citrullus colocynthis, Clerodendrum phlomidis, Convolvulus prostratus, Cordia dichotoma, Crateva adansonii* subsp. *odora, Cyanthillium cinereum, Dactyloctenium aegyptium, Dalbergia sissoo, Desmostachya bipinnata, Digera muricata, Eclipta prostrata, Equisetum ramosissimum* subsp. *debile, Erigeron bonariensis, Euphorbia hirta, Ficus palmata, Gomphrena celosioides, Holoptelea integrifolia, Hyptis suaveolens, Imperata cylindrica, Indigofera tinctoria, Ipomoea carnea, Lindernia ciliata, Melia azedarach, Melochia corchorifolia, Morus alba, Oxalis corniculata, Oxalis debilis* var. *corymbosa, Panicum antidotale, Parthenium hysterophorus, Phoenix sylvestris, Phragmites karka, Phyllanthus amarus, Portulaca oleracea, P. quadrifida, Ricinus communis, Saccharum spontaneum, Senna obtusifolia, S. occidentalis, Sesamum indicum, Sida acuta, S. cordata, S. cordifolia, Syzygium cumini, Vicia sativa*	55
569	**Urorrhea**	*Azadirachta indica*	1
570	**Urticaria**	*Abrus precatorius, Artemisia scoparia, Clerodendrum phlomidis, Xanthium strumarium*	4
571	**Uterine contractions**	*Calotropis gigantea*	1
572	**Uterine discharges**	*Coccinia grandis*	1
573	**Uterine haemorrhage**	*Ageratum conyzoides, Capsella bursa-pastoris, Cissampelos pareira, Eclipta prostrata, Lindernia crustacea*	5
574	**Uterine prolapse**	*Cannabis sativa*	1
575	**Vaccinia Virus**	*Azadirachta indica*	1
576	**Vaginitis**	*Phyllanthus amarus*	1
577	**Vasoconstrictor**	*Acacia nilotica, Capsella bursa-pastoris*	2
578	**Vasodilator, Vasodilatory**	*Artemisia scoparia, Centella asiatica*	2
579	**Vasoprotective actions**	*Euphorbia prostrata*	1
580	**Venereal diseases, Venereal sores, Sexually transmitted diseases**	*Amaranthus viridis, Ammannia baccifera, Blumea obliqua, Cajanus scarabaeoides, Centella asiatica, Cissampelos pareira, Croton bonplandianus, Euphorbia hirta, Imperata cylindrica, Phyllanthus amarus, P. reticulatus, Salix tetrasperma, Senna occidentalis, Sida cordifolia, Sonchus asper*	15
581	**Venous insufficiency**	*Centella asiatica*	1
582	**Vermifuge, Prevent worms, Roundworms, Threadworms, Tapeworms, Hookworms, Pinworms, Worm infestation, Helminthiasis, Vermicidal, Abdominal worms in children, Intestinal worms/ Intestinal worm infestation, Anthelmintic**	*Abrus precatorius, Achyranthes aspera, Acmella paniculata, Ageratum conyzoides, Amaranthus viridis, Ampelopteris prolifera, Argemone mexicana, Artemisia scoparia, Azadirachta indica, Barleria prionitis, Boerhavia diffusa, Blumea obliqua, Butea monosperma, Carthamus oxyacantha, Cajanus scarabaeoides, Calotropis gigantea, C. procera, Centaurium pulchellum, Chenopodium album, Chenopodium murale, Cirsium arvense, Cissampelos pareira, Citrullus colocynthis, Cleome viscosa, Commelina benghalensis, Convolvulus arvensis, Cordia dichotoma, Crateva adansonii* subsp. *odora, Cuscuta reflexa, Cucumis melo* var. *agrestis, Cyanthillium cinereum, Cynodon dactylon, Cyperus rotundus, Dactyloctenium aegyptium, Datura innoxia, Desmodium gangeticum, Dysphania ambrosioides, Eclipta prostrata, Eleusine indica, Euphorbia heterophylla, E. hirta, Evolvulus nummularius, Fumaria indica, Grangea maderaspatana, Holoptelea integrifolia, Hydrocotyle sibthorpioides, Hyptis suaveolens, Imperata cylindrica, Indigofera linifolia, I. tinctoria, Ipomoea eriocarpa, Jatropha gossypiifolia, Justicia adhatoda, Leucaena leucocephala, Leucas cephalotes, Ludwigia hyssopifolia, Melia azedarach, Melilotus albus, Melochia corchorifolia, Mitragyna parvifolia, Morus alba, Oldenlandia corymbosa, Operculina turpethum, Oxalis corniculata, Oxystelma esculentum, Parthenium hysterophorus, Peristrophe bicalyculata, Persicaria barbata, Phalaris minor, Phoenix sylvestris, Phyla nodiflora, Phyllanthus reticulatus, Physalis angulata, P. peruviana, Portulaca quadrifida, Prosopis cineraria, Ranunculus sceleratus, Rhynchosia minima, Ricinus communis, Rungia pectinata, Senna obtusifolia, S. occidentalis, Sesbania sesban, Setaria pumila, Sida acuta, S. cordifolia, Solanum surattense, S. torvum, Syzygium cumini, Tamarix indica, Tephrosia purpurea, T. villosa, Trianthema portulacastrum, Tribulus terrestris, Trichosanthes cucumerina, Triumfetta rhomboidea, Urena lobata, Xanthium strumarium, Ziziphus jujuba*	97
583	**Vertigo**	*Calotropis procera, Desmodium triflorum, Dysphania ambrosioides, Euphorbia hypericifolia, Fumaria indica, Mukia maderaspatana, Tribulus terrestris*	7

SN	Name of Disease/Medicinal properties	Name of plant species	No. of plants
584	**Vesicant**	*Ammannia baccifera, Calotropis gigantea, Ranunculus sceleratus*	3
585	**Veterinary medicine, Urinary problems in cattle, Udder infection of ruminants, Sucking insect infesting livestock, Poultry- insect and fungal infestation, Guinea-worm infestation, Germicide**	*Argemone mexicana, Cajanus scarabaeoides, Capsella bursa-pastoris, Dalbergia sissoo, Eclipta prostrata, Eichhornia crassipes, Euphorbia heterophylla, Fumaria indica, Hyptis suaveolens, Ipomoea eriocarpa, Medicago lupulina, Mitragyna parvifolia, Nicotiana plumbaginifolia, Persicaria barbata, Prosopis cineraria, Saccharum bengalense, Sesbania bispinosa, S. sesban, Setaria pumila, Solanum americanum, Sorghum halepense*	21
586	**Viral infections**	*Oldenlandia corymbosa, Phyllanthus maderaspatensis*	2
587	**Virility, Restore virility, Virility tonic, Vigour, Stamina increasing**	*Alternanthera sessilis, Barleria prionitis, Solanum americanum, Withania somnifera, Ziziphus jujuba*	5
588	**Vitiligo**	*Cuscuta chinensis, Euphorbia heterophylla, Tribulus terrestris*	3
589	**Voice disorders**	*Sisymbrium irio*	1
590	**Vomiting, Nausea**	*Achyranthes aspera, Ageratum conyzoides, Alternanthera sessilis, Azadirachta indica, Calotropis gigantea, Convolvulus prostratus, Cannabis sativa, Capsella bursa-pastoris, Ceratophyllum demersum, Cissampelos pareira, Coccinia grandis, Cuscuta reflexa, Cynodon dactylon, Cyperus rotundus, Dalbergia sissoo, Desmodium gangeticum, Dysphania ambrosioides, Diospyros montana, Ficus religiosa, Fumaria indica, Holoptelea integrifolia, Justicia adhatoda, Lindernia crustacea, Melia azedarach, Melochia corchorifolia, Nicotiana plumbaginifolia, Polygonum plebeium, Physalis angulata, Portulaca oleracea, Senna obtusifolia, Solanum americanum, S. torvum, Syzygium cumini, Tephrosia purpurea, Withania somnifera, Ziziphus jujuba*	36
591	**Warts, Corns**	*Anagallis arvensis, Cannabis sativa, Citrullus colocynthis, Euphorbia heterophylla, E. hirta, E. hypericifolia, Oxalis corniculata, Salix tetrasperma, Scoparia dulcis, Verbascum thapsus*	9
592	**Wasting diseases**	*Fumaria indica, Withania somnifera*	2
593	**Weight deficiency in children**	*Prosopis juliflora*	1
594	**Wheezing**	*Citrullus colocynthis*	1
595	**Whooping cough, Pertussis**	*Argemone mexicana, Barleria prionitis, Cannabis sativa, Justicia adhatoda, Mollugo nudicaulis, Portulaca oleracea, Salix tetrasperma, Senna occidentalis, Sesbania bispinosa*	9
596	**Wounds, Wound healing, Vulnerary, Fresh wounds, Wounds and ulcers**	*Acacia nilotica, Ageratum conyzoides, A. houstonianum, Albizia lebbeck, Alysicarpus vaginalis, Ammannia baccifera, Ampelopteris prolifera, Anagallis arvensis, Argemone mexicana, Azadirachta indica, Barleria prionitis, Bidens pilosa, Butea monosperma, Calotropis gigantea, Capparis sepiaria, Capsella bursa-pastoris, Cayratia trifolia, Cenchrus ciliaris, Centaurium pulchellum, Celosia argentea, Centella asiatica, Ceratophyllum demersum, Chenopodium album, Cissampelos pareira, Cleome viscosa, Convolvulus arvensis, Corchorus tridens, Cordia dichotoma, Crotalaria medicaginea, Croton bonplandianus, Cuscuta chinensis, Cyanthillium cinereum, Cynodon dactylon, Cyperus rotundus, Dactyloctenium aegyptium, Dalbergia sissoo, Datura innoxia, Desmodium gangeticum, D. triflorum, Diospyros montana, Dysphania ambrosioides, Echinochloa crus-galli, Eichhornia crassipes,Equisetum ramosissimum, Evolvulus nummularius, Eclipta prostrata, Euphorbia heterophylla, E. hirta, Ficus religiosa, Heliotropium ellipticum, Holoptelea integrifolia, Hydrilla verticillata, Hydrocotyle sibthorpioides, Hyptis suaveolens, Imperata cylindrica, Indigofera tinctoria, Ipomoea carnea, I. pes-tigridis, I. triloba, Jatropha gossypiifolia, Justicia adhatoda, Lantana camara, Launaea procumbens, Lepidium didymum, Lindernia crustacea, Ludwigia hyssopifolia, Malva parviflora, Malvastrum coromandelianum, Medicago polymorpha, M. lupulina, Melia azedarach, Melilotus indicus, Melochia corchorifolia, Mitragyna parvifolia, Mollugo nudicaulis, Mukia maderaspatana, Nicotiana plumbaginifolia, Operculina turpethum, Oxalis corniculata, Panicum antidotale, Phalaris minor, Phoenix sylvestris, Phyla nodiflora, Phyllanthus amarus, Physalis angulata, Portulaca oleracea, Prosopis juliflora, Ricinus communis, Rhynchosia minima, Scoparia dulcis, Senna occidentalis, Setaria verticillata, Sida acuta, S. cordata, S. cordifolia, Sisymbrium irio, Silene conoidea, Solanum americanum, S. torvum, Sonchus asper, Spergula arvensis, Spermacoce articularis, Stellaria media, Streblus asper, Syzygium cumini, Tephrosia purpurea, Tridax procumbens, Typha domingensis, Urena lobata, Verbascum thapsus, Veronica anagallis-aquatica, Ziziphus jujuba*	112
597	**Yaws**	*Urena lobata*	1
598	**Yellow Fever**	*Crateva adansonii* subsp. *odora*	1

Appendix 9: Glossary of medical/technical terms (Source: Medical Dictionary and Google)

SN	Name of Ailments/ Diseases/ Medical/Technical Terms	Meaning/Definition
1	Abdominal colic pain	Severe abdominal pain, often caused by spasm, obstruction, or distention of any of the hollow viscera, such as the intestines. Colicky pain is described as pain which comes and goes.
2	Abdominal disorders/pain	Abdominal pain, also known as stomach pain, is a common symptom associated with both temporary, non-serious disorders and more serious conditions. Common causes of pain in the abdomen include gastroenteritis and irritable bowel syndrome.
3	Abdominal dropsy	Abnormal accumulation of fluid in the abdomen.
4	Abdominal worms in children	Parasitic roundworm infection in intestines which usually does not have any symptoms in more than 85% of cases. This may be followed by symptoms of abdominal swelling, abdominal pain, and diarrhoea.
5	Abortifacient	An agent (as a drug) that induces abortion.
6	Abortion	The deliberate termination of a human pregnancy, most often performed during the first 28 weeks.
7	Abscesses	A swollen area within body tissue, containing an accumulation of pus.
8	Absorbent	A substance that soaks up liquid easily.
9	Aches	A continuous prolonged dull pain in a part of one's body that is unpleasant but may not be very strong.
10	Acne vulgaris	Acne vulgaris (or simply acne) is a long-term skin disease that occurs when hair follicles become clogged with dead skin cells and oil from the skin.
11	Acrid	An acrid smell or taste is strong and bitter and causes a burning feeling in the throat.
12	Acute appendicitis	Acute appendicitis is inflammation of the appendix, the narrow, finger-shaped organ that branches off the first part of the large intestine on the right side of the abdomen.
13	Acute constipation	Acute constipation is usually defined by a slowing of intestinal transit generating a decrease in bowel movements and the appearance of dehydration. The person will have difficulty defecating or may not be able to at all.
14	Adaptogenic	Generating a substance that balances the body, particularly when the body is under stress, by either stimulating or relaxing.
15	Adenopathy	Large or swollen lymph nodes. Lymph nodes can become enlarged as a result of inflammatory diseases, infection, or cancer. Synonymous with lymphadenopathy.
16	Affections of uterus	Uterine affections are intimately connected to rheumatoid gout, dysmenorrhoea, and acute / chronic leucorrhea, chronic inflammation of cervical uteri and host of other chronic ailments.
17	Affections of stomach and bowels	Derangement of the stomach due to too much food intake and its quality is followed by a long train of unpleasant and distressing such as violent headaches, intolerable dullness, great vertigo, general lassitude, pain in the abdomen, constipation, diarrohoea, and a host of diseases.
18	Agalactia	The failure of the secretion of milk from any cause other than the normal ending of the lactation period.
19	Ageing	The process of growing old.
20	Agglunating	A reaction in which particles (such as red blood cells or bacteria) suspended in a liquid collect into clumps and which occurs especially as a serological response to a specific antibody.
21	Aggravate	To make worse, more serious, or more severe.
22	Ague	A fever (as malaria) marked by paroxysms of chills, fever, and sweating that recur at regular intervals; A fit of shivering: Chill.
23	Albuminuria	the presence of albumin in the urine that is usually a symptom of disease of the kidneys but sometimes a response to other diseases or physiological disturbances of benign nature
24	Alexipharmic	Expelling or counteracting poison: antidote.
25	Alexiteric	(i) A preservative against infectious diseases. (ii) A preservative against the effects of poison.
26	Allelopathic	The suppression of growth of one plant species by another due to the release of toxic substances.
27	Allergic rhinitis	More commonly referred to as hay fever, it is an inflammation of the nasal passages caused by allergic reaction to airborne substances.

SN	Name of Ailments/ Diseases/ Medical/Technical Terms	Meaning/Definition
28	Allergy	Hypersensitivity caused by exposure to a particular antigen (allergen) resulting in a marked increase in reactivity to that antigen on subsequent exposure, sometimes resulting in harmful immunologic consequences.
29	Alopecia	Alopecia simply means hair loss (baldness).
30	Alterative	Alteratives are herbs that gradually restore the proper function of the body and increase health and vitality.
31	Alternative medicine	Any of various systems of healing or treating disease (as homeopathy, chiropractic, naturopathy, Ayurveda, or faith healing).
32	Amebicide	A substance used to kill or capable of killing amoebas and especially parasitic amoebas.
33	Amenorrhea	Amenorrhea is the absence of menstruation — one or more missed menstrual periods.
34	Amentia	(i) Profound mental retardation. (ii) Severe intellectual disability.
35	Amoebicidal	A substance/ agent used to kill or capable of killing amoebas and especially parasitic amoebas of medical importance.
36	Anabolic	Related to the synthetic phase of metabolism in which simple substances are synthesized into the complex materials of living tissue.
37	Anaemia, Anemic	The condition of having less than the normal number of red blood cells or less than the normal quantity of haemoglobin in the blood. The oxygen-carrying capacity of the blood is, therefore, decreased.
38	Analgesic	Relieving pain; pertaining to analgesia; an agent that relieves pain without causing loss of consciousness.
39	Anasarca	Anasarca, or extreme generalized oedema, is a medical condition characterized by widespread swelling of the skin due to effusion of fluid into the extracellular space.
40	Angina pectoris	Commonly known as angina, is the sensation of chest pain, pressure, or squeezing, often due to ischemia of the heart muscle from obstruction or spasm of the coronary arteries.
41	Anodyne	Non-steroidal anti-inflammatory drug/agent that relieves pain without causing loss of consciousness.
42	Anorexia	Lack or loss of appetite.
43	Antacid	An agent that counteracts acidity.
44	Anthelmintic	Expelling or destroying parasitic worms (as tapeworms) especially of the intestine.
45	Anti-ageing	Delaying or lessening the effects of aging, especially on the skin and overall vigour
46	Antiaggregant	An antiplatelet drug that decreases platelet aggregation and inhibit thrombus formation. They are effective in the arterial circulation, where anticoagulants have little effect.
47	Anti-allergic	Relating to any agent or measure that prevents, inhibits, or alleviates an allergic reaction.
48	Anti-amoebic	Able to destroy or suppress amoebae of medical importance.
49	Antianaphylactic	Hypersensitivity induced by preliminary exposure to a substance and usually producing a contraction of smooth muscle and a dilation of blood vessels.
50	Antiangiogenic, Antiangiogenesis	Naturally occurring substance, drug, or other compound that can destroy or interfere with the fine network of blood vessels needed by tumours to grow and metastasize.
51	Antiasthmatic	(i) Tending to relieve or prevent asthma. (ii) An agent that prevents or halts an asthmatic attack.
52	Antibacterial	Anything that destroys bacteria or suppresses their growth or their ability to reproduce.
53	Antibiofilm	Agents that inhibit bacterial/fungal biofilm formation (over 65% of nosocomial (hospital-acquired) infections are caused by biofilms).
54	Antibiotic	A drug /plant effective against bacterial infections by inhibiting the growth of bacteria.
55	Antibleeding	Slowing, stopping, or preventing bleeding.
56	Anticancer	Used or effective against cancer causing diseases.
57	Anticarcinogenic	Tending to inhibit or prevent the activity of a carcinogen or the development of carcinoma/ cancer.
58	Anticatarrhal	Prevent / cure inflammation of the mucous membranes in one of the airways or cavities of the body, usually with reference to the throat and paranasal sinuses and remove/eliminate excess mucus.

SN	Name of Ailments/ Diseases/ Medical/Technical Terms	Meaning/Definition
59	Anticephalalgic	Relieving or preventing headaches.
60	Anticholesterolemic	Promoting a reduction of serum cholesterol levels in the blood.
61	Anticoagulant	An agent/substance that prevents the formation of blood clots, especially coagulation of blood.
62	Anticonvulsant	Anticonvulsant are medicines used to prevent or treat convulsions (seizures) in people with epilepsy.
63	Antidepressant	Antidepressant are medicines that relieve symptoms of depressive disorders or prevent or treat depression.
64	Antidiuretic	An antidiuretic is an agent or drug that, when administered to an organism, helps control body water balance by reducing urination, opposing diuresis.
65	Antidontalgic	A substance that has the ability to relieve toothache.
66	Antidote	A medicine taken or given to counteract a particular poison.
67	Antifibrotic	A drug that inhibits or reduces fibrosis.
68	Antihemorrhagic	An antihemorrhagic (antihæmorrhagic) agent is a substance that promotes hemostasis (stops bleeding).
69	Antihypercholesterolemic	A drug that prevents or controls an increase of cholesterol in the blood.
70	Antihyperglycemic	Drugs used in diabetes treat diabetes mellitus by lowering glucose levels in the blood.
71	Antihyperlipidemic	A drug used in the treatment of high levels of fats (lipids), such as cholesterol, in the blood (hyperlipidemia).
72	Antihypertensive	A drug used to lower high blood pressure.
73	Anti-hyperuricaemia	An abnormal level of uric acid in the blood, as a result of serum urate levels that exceed the limit of solubility. This results in the deposition of crystals of monosodium urate monohydrate in joints and tendons, causing gout.
74	Antileprotic	A leprostatic agent is a drug that interferes with proliferation of the bacterium that causes leprosy.
75	Antilithic	Preventing the formation of calculi or promoting their dissolution.
76	Antimitotic	Having an arresting action on mitosis. A drug having such an effect; a folic acid antagonist that is used in leukemia to inhibit the multiplication of white blood cells.
77	Antinociceptive	The action or process of blocking the detection of a painful or injurious stimulus by sensory neurons.
78	Antiperiodic	Preventing regular recurrence of the symptoms of a disease, as in malaria.
79	Antiphlogistic	An agent that reduces inflammation.
80	Antiplasmodial	An antiparasitic drug which is effective against parasites in the genus *Plasmodium*, responsible for causing malaria.
81	Antiplatelet	A drugs which eliminates or reduces the risk of blood clots. They're often called blood thinners
82	Antiproliferative	Inhibiting or preventing the growth or development of malignant cells in the body.
83	Antiprostaglandin	A drug or a substance (such as aspirin or ibuprofen) that blocks the production of prostaglandin or inhibits its action in the body.
84	Antipruritic	A drug used to relieve itching.
85	Antipyretic	A drug used to prevent or reduce fever.
86	Antiscorbutic	A drug having the effect of preventing or curing scurvy.
87	Antisecretary	Any drug that inhibits or decreases secretion, especially gastric secretion.
88	Antisoporific	A medicine or drug causes sleep or sleepiness, sleepy, drowsy.
89	Antispasmodic	A drug used to relieve spasm of involuntary muscle.
90	Antisteatotic	Inhibiting steatosis (process describing the abnormal retention of lipids (fat) within a cell).
91	Antisudorific	Any agent that prevents sweating.
92	Antitussive	A drug used to prevent or relieve a cough.
93	Antiulcerative	Antibacterial agent used to treat ulcers in the stomach and the upper part of the small intestine.
94	Anxiolytic	A drug used to reduce anxiety.

SN	Name of Ailments/ Diseases/ Medical/Technical Terms	Meaning/Definition
95	Aperient	A drug used to relieve constipation.
96	Apéritifs	An alcoholic drink taken before a meal to stimulate the appetite.
97	Aphrodisiac	A drug that arouses sexual desire.
98	Aphtha	A small ulcer occurring in groups in the mouth or on the tongue.
99	Aphthosis	A condition marked by presence of aphthae.
100	Arthralgia	Pain in a joint.
101	Arthrosis	Arthrosis refers to a joint of bones or an area where two bones are fixed for the purpose of allowing movement of body parts.
102	Articular rheumatism	A chronic articular disease of the joints, developing insidiously, or following the acute or subacute forms. It usually affects two or more joints, and is most likely dependent on the same conditions which give rise to the acute forms.
103	Ascites	The accumulation of fluid in the peritoneal cavity, causing abdominal swelling.
104	Asthenia	Abnormal physical weakness or lack of energy.
105	Astringent	Astringents cause constriction or contraction of mucous membranes or exposed tissues and are often used internally to check discharge of blood serum or mucous secretions. This can happen with a sore throat, hemorrhages, diarrohoea, or with peptic ulcers.
106	Atherosclerosis	A disease of the arteries characterized by the deposition of fatty material on their inner walls.
107	Atonic ulcers	A chronic ulcer with little tendency to heal.
108	Atrophy	A wasting away or progressive decline of a body part or tissue (arrested development or loss of a part or organ incidental to the normal development or life of an animal or plant)
109	Bilious	Affected by or associated with nausea or vomiting.
110	Blennorrhagia	Blennorrhagia is an excess of such discharge, often specifically referring to that seen in gonorrhea.
111	Bloating	Bloating is any abnormal general swelling, or increase in diameter of the abdominal area. As a symptom, the patient feels a full and tight abdomen, which may cause abdominal pain.
112	Blood morbidity	Sick/unhealthy blood.
113	Bloody flux	Diarrohoea in which blood is mixed with the intestinal discharge. 2: swine dysentery.
114	Bowels of costiveness	A change in normal bowel habits with decreased frequency of defecation or passage of hard dry stool.
115	Bright's disease	Chronic inflammation of the blood vessels in the kidney with protein, specifically albumin, in the urine. There are a number of disorders that lead to Bright's disease.
116	Bronchial catarrh	Acute bronchitis. An acute inflammatory disorder, involving the mucous membranes lining the bronchial tubes.
117	Bronchodilator	A drug that causes widening of the bronchi, for example any of those taken by inhalation for the alleviation of asthma.
118	Burning Micturition	Painful urination (dysuria) with discomfort or burning sensation.
119	Carbuncles	A severe abscess or multiple boil in the skin, typically infected with staphylococcus bacteria.
120	Carcinogenic	Having the potential to cause cancer.
121	Cardiac insufficiencies	Insufficiency of the pulmonary valve, permitting blood to flow into the right ventricle of the heart.
122	Cardiac oedema	Cardiac edema is a manifestation of congestive heart failure, due to increased venous and capillary pressures and often associated with renal sodium retention.
123	Cardiopathy	A disease or disorder of the heart.
124	Carminative	A drug that relieves flatulence.
125	Catarrh	Excessive discharge or build-up of mucus in the nose or throat, associated with inflammation of the mucous membrane.
126	Cathartic	A purgative drug.
127	Cellulitis	Spreading bacterial infection just below the skin surface. It is most commonly caused by *Streptococcus pyogenes* or *Staphylococcus aureus.*

SN	Name of Ailments/ Diseases/ Medical/Technical Terms	Meaning/Definition
128	Cephalalgia	Pain in the head caused by dilation of cerebral arteries or muscle contractions or a reaction to drugs- headache, ache, aching - a dull persistent (usually moderately intense) pain.
129	Cerebral affections	An abnormal condition of body or mind.
130	Cervical spondylosis	Cervical spondylosis is a general term for age-related wear and tear affecting the spinal disks in the neck. As the disks dehydrate and shrink, signs of osteoarthritis develop, including bony projections along the edges of bones (bone spurs). It is very common and worsens with age.
131	Cestocidal,	A drug/ agent that destroys/kills tapeworms.
132	Cheickenpox	A highly infectious viral disease also known medically as varicella that causes a blister-like rash, itching, fatigue and fever.
133	Chickengunya virus	A febrile disease that resembles dengue, and is caused by a togavirus (species Chikungunya virus of the genus *Alphavirus*) transmitted especially by aedes mosquitoes.
134	Cholagogue	A cholagogue is a medicinal agent which promotes the discharge/flow of bile from the gall bladder downward into the duodenum.
135	Cholecystolithiasis	The occurrence of gallstones within the gallbladder.
136	Choleretic	A drug which stimulates the production of bile by the liver.
137	Chorea	A neurological disorder characterized by jerky involuntary movements affecting especially the shoulders, hips, and face.
138	Chyluria	Chyluria, also called chylous urine, is a medical condition involving the presence of chyle in the urine stream, which results in urine appearing milky white. It is a condition that is more prevalent among people of Africa and the Indian subcontinent.
139	Cicatrizant,	Promoting the healing of a wound or the formation of a cicatrix - the scar of a healed wound.
140	Cirrhosis	A chronic disease of the liver marked by degeneration of cells, inflammation, and fibrous thickening of tissue. It is typically a result of alcoholism or hepatitis.
141	Cognitive	Generic term embracing the mental activities associated with thinking, learning, and memory.
142	Collyrium	Collyrium is an antique term for a lotion or liquid wash used as a cleanser for the eyes, particularly in diseases of the eye.
143	Colporrhagia	A vaginal hemorrhage.
144	Contusions	A region of injured tissue or skin in which blood capillaries have been ruptured; a bruise.
145	Cooling	Herbs are classified according to whether they have a warming or cooling effect on the body. Usually this is related to taste of herb. Generally, sweet herbs tonify, sour herbs are astringent, bitter herbs dry damp and clear heat, acrid herbs disperse cold and stagnation, and salty herbs have a softening, purging effect.
146	Coryza	Catarrhal inflammation of the mucous membrane in the nose, caused especially by a cold or by hay fever.
147	Cough	A rapid expulsion of air from the lungs, typically in order to clear the lung airways of fluids, mucus, or other material
148	Curare	A bitter resinous substance obtained from the bark and stems of some plants. It paralyses the motor nerves and is traditionally used by tribal people to poison their arrows and blowpipe darts.
149	Cyanogenetic	Chemical compounds contained in foods that release hydrogen cyanide when chewed or digested.
150	Cystalgia	Pain in a bladder, especially the urinary bladder.
151	Cystitis	Inflammation of the urinary bladder. It is often caused by bacterial infection and is usually accompanied by frequent painful urination.
152	Cytopathogenic	Pertaining to an agent or substance that causes a diseased condition in cells, in contrast to histologic changes; or producing pathological changes in cells.
153	Demulcent	A substance that relieves irritation or inflammation of the mucous membranes in the mouth by forming a protective film.
154	Dengue fever	An acute mosquito-borne viral illness of sudden onset with headache, fever, prostration, severe joint and muscle pain, swollen glands (lymphadenopathy), and rash. It is caused by a flavivirus (species Dengue virus of the genus *Flavivirus*), transmitted by aedes mosquitoes.
155	Dental caries	Dental caries is the scientific term for tooth decay or cavities.

SN	Name of Ailments/ Diseases/ Medical/Technical Terms	Meaning/Definition
156	Deobstruent	A drug which removes obstructions in the body by aiding the opening of ducts.
157	Depurative	Ability of a substance to decontaminate or purify.
158	Detersive	A cleansing agent: detergent.
159	Diaphoretic	A drug inducing perspiration. A person sweating heavily.
160	Diphtheria	An acute infectious upper respiratory tract disease that affects the throat. It is caused by the bacteria *Corynebacterium diphtheriae.*
161	Dipsia	Condition of thirst.
162	Discutient,	A discutient agent or drug or process causing dispersal or disappearance of a tumour or lesion.
163	Diuretic	Diuretics are medicines that help reduce the amount of water in the body. These drugs act on the kidneys to increase urine output. This reduces the amount of fluid in the bloodstream, which in turn lowers blood pressure.
164	Dropsy	An old term for the swelling of soft tissues or now called Edema due to the accumulation of excess water.
165	Dysgeusia	Dysgeusia or parageusia is a distortion of the sense of taste. It is also often associated with ageusia, which is the complete lack of taste, and hypogeusia, which is a decrease in taste sensitivity.
166	Dysmenorrhoea	Painful menstruation, typically involving abdominal cramps.
167	Dyspepsia	Dyspepsia can be defined as painful, difficult, or disturbed digestion, which may be accompanied by symptoms such as nausea and vomiting, heartburn, bloating, and stomach discomfort.
168	Dyspnoea	Difficult or labored breathing; shortness of breath.
169	Dysuria	Pain during urination, or difficulty urinating. Dysuria is usually caused by inflammation of the urethra, frequently as a result of infection.
170	Emaciation	Emaciation is defined as extreme weight loss and unnatural thinness due to a loss of subcutaneous fat (the fatty, or adipose tissue beneath the skin) and muscle throughout the body. The state of being abnormally thin or weak.
171	Embrocation	A liquid used for rubbing on the body to relieve pain from sprains and strains.
172	Emetic	A medicine or other substance which causes vomiting.
173	Emmenagogue	A substance that stimulates or increases menstrual flow.
174	Emollient	A drug/ medicine having the quality of softening or soothing the skin.
175	Encephalitis	Encephalitis is an inflammation of the brain, usually caused by a direct viral infection or a hyper-sensitivity reaction to a virus or foreign protein.
176	Encephalopathy	A disease in which the functioning of the brain is affected by some agent or condition (such as viral infection or toxins in the blood).
177	Enteritis	Inflammation of the intestine, especially the small intestine, usually accompanied by diarrohoea.
178	Eosinophilia	An increase in the number of eosinophils in the blood, occurring in response to some allergens, drugs, and parasites, and in some types of leukaemia.
179	Epilepsy	A pattern of repeated seizures is referred to as epilepsy. When nerve cells in the brain fire electrical impulses at a rate of up to four times higher than normal, this causes a sort of electrical storm in the brain, known as a seizure.
180	Ertussis	Bleeding from the nose.
181	Erysipelas	An acute, sometimes recurrent disease caused by a bacterial infection, characterized by large raised red patches on the skin.
182	Excoriation	A place on body where skin is scraped or abraded or wearing off.
183	Expectorant	A medication that helps bring up mucus and other material from the lungs, bronchi, and trachea.
184	Febrifuge	A medicine used to reduce fever - an antipyretic.
185	Fibromyalgia	A chronic disorder characterized by widespread pain, tenderness, and stiffness of muscles and associated connective tissue structures that is typically accompanied by fatigue, headache, and sleep disturbances - called also fibromyalgia syndrome, fibromyositis

SN	Name of Ailments/ Diseases/ Medical/Technical Terms	Meaning/Definition
186	Flu/Influenza	Flu is a colloquialism for influenza. The flu is an acute respiratory illness caused by influenza viruses A or B
187	Fungicidal	A chemical substance/drug that destroys or inhibits the growth of fungi.
188	Furuncle	It is also known as a "boil." A boil is a tender, dome-shaped skin lesion caused by an infection around a hair follicle with *Staphylococcus aureus.*
189	Galactagogue	A food or drug that promotes or increases the flow of a mother's milk.
190	Gamboge	A gum resin produced by various East Asian trees, used as a yellow pigment and in medicine as a purgative.
191	Gangrenous rectitis	Decay or death of mucous membrane of the rectum caused by a lack of blood supply.
192	Gastralgia	Pain in the stomach or epigastrium especially of a neuralgic type.
193	Gastritis	Commonly refers to inflammation of the lining of the stomach, but the term is often used to cover a variety of symptoms resulting from stomach lining inflammation and symptoms of burning or discomfort.
194	Gastrohelcosis	Ulceration of the stomach.
195	Gastropathy	A generic term for any disease of the stomach.
196	Gingivitis	Inflammation of the gums.
197	Gleet	A watery discharge from the urethra caused by gonorrheal infection.
198	Glossitis	Inflammation of the tongue.
199	Glycosuria	A condition characterized by an excess of sugar in the urine, typically associated with diabetes or kidney disease.
200	Gravel	Aggregations of crystals formed in the urinary tract.
201	Griping	Suffering from gastric or intestinal pain.
202	Haematuria	The presence of blood in urine.
203	Haemoglobinuria	It is a condition in which the oxygen transport protein hemoglobin is found in abnormally high concentrations in the urine.
204	Haemoptysis	Spitting or coughing of blood derived from the lungs or bronchial tubes as a result of pulmonary or bronchial hemorrhage.
205	Haemostatic/ Hemostatic	A drug arresting hemorrhage; styptic; .a drug or agent that retards or stops bleeding.
206	Halitosis	Technical term for chronic bad breath.
207	Hemafecia	A drug used as an anti-inflammatory agent to treat hepatitis, boils and piles, and as a hemostatic agent for the treatment of hematemesis, epistaxis and hemafecia.
208	Hematospermia	Presence of blood in semen possibly due to granulomatous prostatitis.
209	Hemiplegia	Total or partial paralysis of one side of the body that results from disease of or injury to the motor centers of the brain.
210	Hemoptosis	Coughing up of blood.
211	Hepatoma	A cancer of the cells of the liver.
212	Hepatomegaly	Abnormal enlargement of the liver.
213	Hepatopathy	Liver dysfunction due to venous congestion, usually due to congestive heart failure.
214	Hepatoprotective/ Hepatoprotectant	The ability to prevent damage to the liver.
215	Herbicidal	A substance used to destroy or inhibit the growth of plants, especially weeds.
216	Hiccough/Hiccups	Hiccups happen when the diaphragm, the muscle that separates the chest from the abdomen, suddenly contracts. When that happens, air is taken in, and then the voice box closes shut, which is what causes the "hic" sound.
217	Hirsutism	Abnormal growth of hair on a woman's face and body.
218	Human immunodeficiency virus (HIV)	Human immunodeficiency virus (HIV) causes potentially life-threatening chronic condition Acquired immunodeficiency syndrome (AIDS).
219	Hydragogue	A cathartic that causes evacuation of watery bowel movement.
220	Hydrocoele	Accumulation of fluid in the coat around the testis.

SN	Name of Ailments/ Diseases/ Medical/Technical Terms	Meaning/Definition
221	Hyperdipsia	Intense thirst that is relatively temporary.
222	Hyperdiuresis	Excessive secretion of urine.
223	Hypertonia	Hypertonia is a condition marked by an abnormal increase in muscle tension and a reduced ability of a muscle to stretch. It is caused by injury to motor pathways in the central nervous system, which carry information from the central nervous system to the muscles and control posture, muscle tone, and reflexes.
224	Hyperuricemia	Hyperuricemia is an abnormally high level of uric acid in the blood. In the pH conditions of body fluid, uric acid exists largely as urate, the ion form.
225	Hypotensive	Relating to or suffering from abnormally low blood pressure.
226	Impetigo	A contagious bacterial skin infection forming pustules and yellow crusty sores.
227	Incontinence	Lack of voluntary control over urination or defecation.
228	Indolent	A disease or condition causing little or no pain.
229	Insectifuge	A substance that repels, but does not kill, insects.
230	Insomnia, Sleeplessness	Inability to obtain an adequate amount or quality of sleep.
231	Intestinal parasitosis	A disease of intestines resulting from parasitic infestation.
232	Intoxication	A condition that follows the administration of a psychoactive substance, drug, alcohol and results in disturbances in the level of consciousness, cognition, perception, judgement, affect, or behaviour, or other psychophysiological functions and responses.
233	Lacrimation	Secretion of tears, especially in excess.
234	Lactogenic	Inducing or stimulating the production of milk.
235	Laryngitis	An inflammation of the larynx most often caused by viral infections resulting in sore throat, cough, difficulty in swallowing, and fever.
236	Leishmaniasis	Infection with or disease (as kala-azar or oriental sore) caused by *Leishmanias*, a genus of kinetoplastid flagellate protozoans of the family Trypanosomatidae that are parasitic in the tissues of vertebrates, are transmitted by sand flies.
237	Leprosy	Leprosy is a chronic, mildly contagious disease of tropical and subtropical regions, caused by the bacterium *Mycobacterium leprae.*
238	Libido	Sexual drive.
239	Litholytic	Ability of an agent/drug to dissolve/break calculi (Kidney stones).
240	Lithotriptic, Lithontriptic	An agent/drug that effects the dissolution of a calculus.
241	Loose motion/Diarrhea	It is medically termed as diarrhea which means an increased frequency or decreased consistency of bowel movements.
242	Lumbago	Pain in the muscles and joints of the lower back.
243	Lymphadenitis	Inflammation of the lymph nodes.
244	Malignant brain tumour	A malignant brain tumour is a fast-growing cancer that spreads to other areas of the brain and spine.
245	Malignant melanoma	A skin cancer that begins in cells called melanocytes.
246	Marasmus	Chronic wasting of body tissues, especially in young children, commonly due to prolonged dietary deficiency of protein and calories. Also called athrepsia.
247	Mastitis	Inflammation of one or more mammary glands within the breast, usually in a lactating woman caused by an infection in the breast or by a plugged milk duct.
248	Mastogenic	Drugs/herbs having property of enlarging mammary glands in females.
249	Measles	An acute highly contagious viral disease, Rubella, with fever, runny nose, cough, red eyes, and a spreading reddish colour skin rash.
250	Melancholia	A mental condition characterized by great depression of spirits and gloomy forebodings.
251	Melanoma	A tumour of melanin-forming cells, especially a malignant tumour associated with skin cancer.
252	Meningitis	Inflammation of the meninges, the three membranes that envelop the brain and the spinal cord caused by infection by bacteria, viruses, and protozoa.
253	Menorrhagia	Abnormally heavy bleeding at menstruation.
254	Menoxenia	Any abnormality relating to menstruation.

SN	Name of Ailments/ Diseases/ Medical/Technical Terms	Meaning/Definition
255	Metabolic syndrome	A constellation of conditions that place people at high risk for coronary artery disease. These conditions include type 2 diabetes, obesity, high blood pressure, and a poor lipid profile with elevated LDL ("bad") cholesterol, low HDL ("good") cholesterol, elevated triglycerides.
256	Metrorrhagia	Abnormal bleeding from the uterus.
257	Micturition	Urination; the discharge of urine from the bladder.
258	Moles	A pigmented spot on the skin. The medical term for mole is melanocytic nevus (a nevus composed of melanocytes).
259	Molluscicidal/ Molluscicide	Molluscicides, also known as snail baits and snail pellets, are pesticides against molluscs, which are usually used in agriculture or gardening, in order to control gastropod pests specifically slugs and snails which damage crops or other valued plants by feeding on them.
260	Mumps	An acute contagious virus disease caused by a paramyxovirus of the genus Rubulavirus (species Mumps virus) and marked by fever and by swelling especially of the parotid gland -called also epidemic parotitis.
261	Mycosis	A type of non-Hodgkin's lymphoma that first appears on the skin. Also known as chronic or rapidly progressive form of cutaneous T-cell lymphoma.
262	Myorelaxant	A muscle relaxant is a drug that affects skeletal muscle function and decreases the muscle tone. It may be used to alleviate symptoms such as muscle spasms, pain, and hyperreflexia.
263	Narcotic	An addictive drug affecting mood or behavior, especially an illegal one.
264	Nasal polyps	Benign growths that originate in the mucous membranes lining the nasal passages or paranasal sinuses. Polyps are believed to arise in the nasal mucosa as a result of chronic inflammation.
265	Nausea (Vomiting)	A stomach distress with distaste for food and an urge to vomit.
266	Nematicidal	A substance or agent having property to kill nematodes.
267	Nephritis	Inflammation of the kidneys.
268	Neuralgia	Intense, typically intermittent pain along the course of a nerve, especially in the head or face.
269	Neurasthenia	A group of symptoms, including chronic physical and mental fatigue, weakness, and generalized aches and pains, formerly thought to result from exhaustion of the nervous system and now usually considered a psychological disorder. The term is no longer in clinical use in many parts of the world.
270	Neuritis	Inflammation of a peripheral nerve or nerves, usually causing pain and loss of function.
271	Neuroprotective	Serving to protect nerve cells against damage, degeneration, or impairment of function.
272	Neurosis	A relatively mild mental illness that is not caused by organic disease, involving symptoms of stress (depression, anxiety, obsessive behavior, hypochondria) but not a radical loss of touch with reality.
273	Neutraceutical	A food or part of a food (as a fortified food or a dietary supplement) that allegedly provides medicinal or health benefits, including the prevention and treatment of disease, called also functional food.
274	Nociception	Nociception is the sensory nervous system's response to certain harmful or potentially harmful stimuli.
275	Nocturnal emission	An involuntary ejaculation of semen during sleep.
276	Nootropic	A drug) used to enhance memory or other cognitive functions.
277	Nyctalopia	A condition characterized by an abnormal inability to see in dim light or at night, typically caused by vitamin A deficiency.
278	Obesity	An abnormal accumulation of body fat, usually 20% or more over an individual's ideal body weight.
279	Odontalgia	Toothache.
280	Odontalgic	Pain felt in the area from which a tooth has been extracted.
281	Odynuria	Pain while urinating.
282	Oedema	The swelling of soft tissues as a result of excess fluid accumulation.
283	Oestrogenic/Estrogenic	1. Promoting estrus; 2. Relating to, caused by, or being an estrogen.
284	Oestrous cycle	The recurring periods of estrus in adult females of most mammalian species and the correlated changes in the reproductive tract from one period to another.
285	Oleaginous	Resembling or having the properties of oil or containing or producing oil.

SN	Name of Ailments/ Diseases/ Medical/Technical Terms	Meaning/Definition
286	Oliguria	Production of abnormally small amounts of urine.
287	Oogenesis	The production of egg cells (ova) in the ovaries and their preparation for release and fertilization.
288	Ophthalmia	Acute inflammation of the eyes of a newborn from infection during passage through the birth canal.
289	Ophthalmodynia	A pain in the eye.
290	Opthalmopathy	Ophthalmopathy (also known as thyroid eye disease (TED), dysthyroid/thyroid-associated orbitopathy (TAO), Graves' orbitopathy) is an autoimmune inflammatory disorder affecting the orbit around the eye, characterized by upper eyelid retraction, lid lag, swelling, redness (erythema), conjunctivitis, and bulging.
291	Orchitis	Inflammation of the testis caused by mumps and other infections and injury.
292	Otalgia	Earache.
293	Otopathy	Nonspecific term for any disease of the ear; while this term fills a lexical void, plain English alternatives are increasingly preferred—e.g., ear disease or ear pathology, with modifiers as to the location (as in middle or inner ear disease).
294	Otorrhoea	A discharge from the external ear.
295	Oxytocic	Hastening or facilitating childbirth, especially by stimulating contractions of the uterus.
296	Palliative	Specialized medical care for people with serious illnesses. It is focused on providing people with relief from the symptoms, pain and stress of a serious illness.
297	Palsy	Paralysis, especially that which is accompanied by involuntary tremors.
298	Papillomas	A small solid benign tumour with a clear-cut border that projects above the surrounding tissue. A raised wart is an example of a papilloma.
299	Paralysis	Loss of voluntary movement (motor function) that affects only one muscle or limb is partial paralysis, also known as palsy; paralysis of all muscles is total paralysis, as may occur in cases of botulism.
300	Parkinson's disease	A slowly progressive neurologic disease that is characterized by a fixed inexpressive face, tremor at rest, slowing of voluntary movements, gait with short accelerating steps, peculiar posture and muscle weakness.
301	Paroxysm	A sudden recurrence or attack of a disease.
302	Parturient	A woman or female mammal) about to give birth; in labour.
303	Pectoral	Relating to the breast or chest.
304	Pediculi/Pediculus	A genus of parasitic lice (family Pediculidae) that live in the hair/body and feed periodically on blood.
305	Penile dysfunction	A disorder that impairs the nerve supply or the blood flow to the penis. Medical definition of erectile dysfunction which is a form of impotence.
306	Peritonitis	Inflammation of the peritoneum (Peritoneum is the tissue layer of cells lining the inner wall of the abdomen and pelvis).
307	Pharyngitis	Inflammation of the pharynx, also called sore throat.
308	Pharyngodynia	Pain in the pharynx.
309	Pharyngopathy	Any disorder of the pharynx.
310	Phlebitis	Inflammation of the walls of a vein.
311	Phthisis	Pulmonary tuberculosis or a similar progressive wasting disease.
312	Piles	Also known as Haemorrhoids: Dilated (enlarged) veins in the walls of the anus and sometimes around the rectum, usually caused by untreated constipation but occasionally associated with chronic diarrhea.
313	Pityriasis versicolor	Pityriasis versicolor is a common skin complaint in which flaky discoloured patches appear mainly on the chest and back. It is sometimes called tinea versicolor, although the term 'tinea' should strictly refer to infection with a dermatophyte fungus.
314	Pneumonia	An acute disease that is marked by inflammation of lung tissue accompanied by infiltration of alveoli and often bronchioles with white blood cells (as neutrophils) and fibrinous exudate, is characterized by fever, chills, cough, difficulty in breathing, fatigue, chest pain, and reduced lung expansion.

SN	Name of Ailments/ Diseases/ Medical/Technical Terms	Meaning/Definition
315	Polyuria	Polyuria is a condition usually defined as excessive or abnormally large production or passage of urine (greater than 2.5 or 3 L over 24 hours in adults). Frequent urination is sometimes included by definition but is nonetheless usually an accompanying symptom.
316	Poultice	A soft, moist mass of material, typically consisting of bran, flour, herbs, etc., applied to the body to relieve soreness and inflammation and kept in place with a cloth.
317	Prolapsed Ani	A part or organ of the body) slip forward or down, especially of rectum
318	Prolapsus	The falling down or slipping out of place of an organ or part, such as the uterus.
319	Prophylactic	Treatment given or action taken to prevent disease.
320	Prostatitis	Inflammation of the prostate gland.
321	Prosthetic hyperplasia	Adenomatous hyperplasia of the periurethral part of the prostate gland that occurs especially in men over 50 years old and that tends to obstruct urination by constricting the urethra - abbreviation BPH; called also benign prostatic hypertrophy.
322	Proteolytic	Hydrolysis of proteins or peptides with formation of simpler and soluble products (as in digestion)
323	Proteolytic,	The breaking down of proteins into simpler compounds, as in digestion.
324	Protisticide	Protisticide is an agent that kills any member of the kingdom Protista. Protista, composed of "organisms which are unicellular or unicellular-colonial and which form no tissues.
325	Pruritis	Severe itching of the skin, as a symptom of various ailments.
326	Psoriasis	A skin disease marked by red, itchy, scaly patches.
327	Psychopathy	Mental illness or disorder.
328	Psychopharmacological	Psychopharmacology is the scientific study of the effects drugs have on mood, sensation, thinking, and behavior.
329	Pungent	Causing a sharp or irritating/ acrid sensation affecting the organs of taste or smell.
330	Purgative	Strongly laxative in effect.
331	Pustules	A pustule is a small collection of pus in the top layer of skin (epidermis) or beneath it in the dermis.
332	Pyoderma	A skin infection with formation of pus.
333	Pyogenic	Producing pus due to pyogenic bacteria.
334	Pyorrhea	Pyorrhea is an older term that used to refer to any kind of gum disease or periodontal disease. An advanced form of chronic periodontitis marked especially by the discharge of pus from the alveoli - called also Riggs' disease.
335	Pyrexia	Raised body temperature; fever.
336	Rabies	An acute virus disease of the nervous system of mammals that is caused by a rhabdovirus (species Rabies virus of the genus *Lyssavirus*) usually transmitted through the bite of a rabid animal.
337	Radioprotective	An agent that provides protection against the toxic effects of ionizing radiation.
338	Renal lithiasis	Kidney stones (renal lithiasis, nephrolithiasis) are small, hard mineral deposits that form inside your kidneys. The stones are made of mineral and acid salts.
339	Renoprotective,	Protecting the kidney against harmful effects such as of a drug or other chemical.
340	Resolutive	Having the power or ability to dissolve or dispel something.
341	Restorative	A thing that restores health, strength, or well-being, especially a medicine or drink.
342	Revulsive	One of the physiologic principles used in hydrotherapy to modify blood circulation. The alternating applications of heat and cold increase the blood flow through a body part.
343	Rhinitis	Inflammation of the mucous membrane of the nose, caused by a virus infection (e.g. the common cold) or by an allergic reaction (e.g. hay fever).
344	Rhinopathy	Any disease or malformation of the nose.
345	Rickets	A deficiency disease that affects the young during the period of skeletal growth, is characterized especially by soft and deformed bones, and is caused by failure to assimilate and use calcium and phosphorus normally due to inadequate sunlight or vitamin D.
346	Rubefacient	A rubefacient is a substance for topical application that produces redness of the skin e.g. by causing dilation of the capillaries and an increase in blood circulation.

SN	Name of Ailments/ Diseases/ Medical/Technical Terms	Meaning/Definition
347	Scalds	A burn or other injury caused by hot liquid or steam.
348	Schistosomiasis	Disease, also known as snail fever and bilharzia, of liver, gastrointestinal tract and bladder caused by parasitic flatworms called schistosomes. Infection is usually from infested water.
349	Sciatica	Pain affecting the back, hip, and outer side of the leg, caused by compression of a spinal nerve root in the lower back, often owing to degeneration of an intervertebral disc.
350	Scrofula	A disease with glandular swellings, probably a form of tuberculosis.
351	Scurvy	A disease caused by a deficiency of vitamin C, characterized by swollen bleeding gums and the opening of previously healed wounds, which particularly affected poorly nourished sailors until the end of the 18th century.
352	Secretory	Relating to secretion or the secretions.
353	Sedurific	A drug that induces sweating.
354	Sialagogue, Sialogenic	A drug that promotes the secretion of saliva.
355	Sickle cell disease	A group of inherited blood disorders characterized by chronic anemia, painful events, and various complications due to associated tissue and organ damage.
356	Sinusitis	Inflammation of a nasal sinus.
357	Sleeping sickness	Also called African trypanosomiasis, a systemic disease caused by the parasite *Trypanosoma brucei* that is transmitted by the bite of the tsetse fly, a gray-brown insect about the size of a honeybee.
358	Smallpox	An acute contagious febrile disease of humans that is caused by a poxvirus and characterized by skin eruption with pustules, sloughing, and scar formation.
359	Spasmogenic	Inducing spasm
360	Spasmolytic	A drug or treatment able to relieve spasm of smooth muscle.
361	Spermatopoietic	Promoting the formation and secretion of semen.
362	Spermatorrhea	Spermatorrhea is a condition of excessive, involuntary ejaculation. It is a recognized disorder in Traditional Chinese Medicine, in which certain patterns of involuntary ejaculation reflect problems with kidney. In Ayurvedic Medicine, Ashwagandha and Bala are used to treat this *vata* ailment.
363	Splenitis	Inflammation of the spleen.
364	Splenomegaly	Abnormal enlargement of the spleen.
365	Splenopathy	Disease of the spleen.
366	Sprue	1. A chronic form of malabsorption syndrome, occurring in both tropical and nontropical forms.
		2. In dentistry, the hole through which metal or other material is poured or forced into a mold.
367	Stimulant	A substance that raises levels of physiological or nervous activity in the body.
368	Stomachalgia	Pain in the stomach; called also gastralgia and stomachalgia.
369	Stomachic	A stomachic medicine or tonic promoting the appetite or assisting digestion.
370	Stomatalgia	Pain in the mouth; also called stomatodynia.
371	Stomatosis, Stomatitis	Inflammation of the mucous membrane of the mouth.
372	Strangury	A condition caused by blockage or irritation at the base of the bladder, resulting in severe pain and a strong desire to urinate.
373	Stroke	The sudden death of brain cells due to lack of oxygen, caused by blockage of blood flow or rupture of an artery to the brain. Sudden loss of speech, weakness, or paralysis of one side of the body can be symptoms.
374	Styptic	A substance) capable of causing bleeding to stop when it is applied to a wound.
375	Substitute for ipecacuanha	Ipecac is a medicine commonly extracted from the dried root of Uragoga (*Carapichea*) ipecacuanha having expectorant, emetic, and antidysenteric properties. It is commonly used to induce vomiting in cases of accidental poisoning.
376	Sudorific	A drug that induces sweating.
377	Sunstroke	A serious condition that happens when someone has been in the sun too long and that causes a person to stop sweating, have a very high body temperature, and become exhausted or unconscious.

SN	Name of Ailments/ Diseases/ Medical/Technical Terms	Meaning/Definition
378	Suppuration	Suppuration is the process of pus forming. Pus is the white substance that sometimes oozes from a sore or a pimple.
379	Tenesmus	A continual or recurrent inclination to evacuate the bowels, caused by disorder of the rectum or other illness.
380	Teratogen	An agent or factor which causes malformation of an embryo.
381	Tetanus	An acute infectious disease characterized by tonic spasm of voluntary muscles and especially of the muscles of the jaw and caused by an exotoxin produced by a bacterium of the genus *Clostridium* (*C. tetani*) which is usually introduced through a wound.
382	Tetany	A condition that is due usually to low blood calcium (hypocalcemia) and is characterized by spasms of the hands and feet, cramps, spasm of the voice box (larynx), and overactive neurological reflexes.
383	Thermogenic	Tending to produce heat; and the term is commonly applied to drugs which increase heat through metabolic stimulation
384	Thrombosis	The formation or presence of a blood clot in a blood vessel.
385	Tinea capitis	An infection of the scalp caused by fungi of the genera *Microsporum* or *Trichophyton*, characterized by itchy, scaly patches and sometimes hair loss; ringworm of the scalp.
386	Tinnitus	Ringing or buzzing in the ears.
387	Tonic	A medicinal substance/drug taken to give a feeling of vigour or well-being.
388	Tonsillitis	Inflammation of the tonsils and especially the palatine tonsils typically due to viral or bacterial infection and marked by red enlarged tonsils usually with sore throat, fever, difficult swallowing, hoarseness or loss of voice, and tender or swollen lymph nodes.
389	Toxemia	A condition in pregnancy, also known as pre-eclampsia (or preeclampsia) characterized by abrupt hypertension (a sharp rise in blood pressure), albuminuria (leakage of large amounts of the protein albumin into the urine) and edema (swelling) of the hands, feet, and face.
390	Tracheitis	Inflammation/infection of the trachea.
391	Tranquilizer	A medicinal drug taken to reduce tension or anxiety.
392	Trauma	A disordered psychic or behavioral state resulting from severe mental or emotional stress or physical injury.
393	Travel sickness	Also called as motion sickness is due to irritation of a portion of the inner ear called the labyrinth. The symptoms of motion sickness include nausea, vomiting, and vertigo.
394	Tympanites	Swelling of the abdomen with air or gas.
395	Typhoid	An acute illness characterized by fever caused by infection with the bacterium *Salmonella typhi.*
396	Ulcer	Ulcer is the breach of the continuity of skin, epithelium or mucous membrane caused by sloughing out of inflamed necrotic tissue.
397	Ulemorrhagia	Bleeding of the gums; the flow of blood from a ruptured blood vessel.
398	Ulitis/ Gingivitis	Inflammation of the gums. Bleeding is a primary symptom, and other symptoms include swelling, redness, pain, and difficulty in chewing.
399	Ulocace	Ulceration of the gums.
400	Urethritis	Urethritis is inflammation of the urethra. Pain with urination is the main symptom of urethritis caused due to infection by bacteria.
401	Urethrorrhea	Abnormal discharge from the urethra.
402	Urinary tract infection	An infection of the kidney, ureter, bladder, or urethra; abbreviated as UTI. Common symptoms include a frequent urge to urinate and pain or burning when urinating.
403	Urodynia	Pain during urination.
404	Urogenitory /Genito-urinary tract disorders	Disorders relating to the genital and urinary organs or their functions.
405	Urolithiasis	The formation of stony concretions in the bladder or urinary tract.
406	Uropathy	A disease of the urinary or urogenital organs.
407	Urorrhea	Involuntary flow of urine.
408	Urticaria	A rash of round, red weal on the skin which itch intensely, sometimes with dangerous swelling, caused by an allergic reaction, typically to specific foods.

SN	Name of Ailments/ Diseases/ Medical/Technical Terms	Meaning/Definition
409	Uterine Prolapse	Uterine prolapse occurs when pelvic floor muscles and ligaments stretch and weaken, providing inadequate support for the uterus. The uterus then slips down into or protrudes out of the vagina.
410	Vaginitis	Inflammation of the vagina (as from bacterial or fungal infection, allergic reaction, or hormone deficiency) that may be marked by irritation and vaginal discharge.
411	Vasoconstrictor	Vasoconstriction is the narrowing of the blood vessels resulting from contraction of the muscular wall of the vessels, in particular the large arteries and small arterioles. The process is the opposite of vasodilation, the widening of blood vessels.
412	Vasodilator	A drug, agent, or nerve that can cause dilatation (vasodilatation) of the walls of blood vessels.
413	Vasoprotective	Vasoprotective is a medication which acts to alleviate certain conditions of the blood vessels. For example agents used in the treatment of hemorrhoids and varicose veins and hormones such as estrogen can have a vasoprotective effect.
414	Venous insufficiency	Venous insufficiency is a condition where the flow of blood through the veins is inadequate, causing blood to pool in the legs. It can be caused by several different vein disorders, but it's most often caused by either blood clots or varicose veins.
415	Vermicidal	A substance or agent used to kill worms, especially a drug used to kill parasitic intestinal worms.
416	Vermifuge	Any drug or agent able to destroy or expel intestinal worms; an anthelmintic medicine.
417	Verminosis	Infestation with or disease caused by parasitic worms.
418	Vertigo	A sensation of whirling and loss of balance, associated particularly with looking down from a great height, or caused by disease affecting the inner ear or the vestibular nerve; giddiness.
419	Vesicant	An agent that causes blistering.
420	Vitiligo	A condition in which the pigment is lost from areas of the skin, causing whitish patches, often with no clear cause.
421	Vulnerary	A medicine used in the healing of wounds.
422	Warts	A small, hard, benign growth on the skin, caused by a virus.
423	Whitlow	An infection of the finger and especially the fingertip that is typically caused by infection of a virus (as the herpes simplex virus) and is marked by redness, tenderness, and fluid-filled blisters.
424	Whooping cough, Pertussis	A highly contagious, acute respiratory illness characterized by fits of coughing and caused by the bacterium *Bordetella pertussis*.
425	Yaws	A contagious disease of tropical countries, caused by a bacterium that enters skin abrasions and gives rise to small crusted lesions which may develop into deep ulcers.

Appendix 10: Glossary of taxonomic terms

Terms	Description
Habit and Habitat	
Annual	Plant that complete their life cycle in one year.
Aquatic	Plants growing in water.
Biennial	Plant that complete their life cycle in two years.
Climber	A weak-stemmed plant which climb up another plants and other structure to support itself with the help of its tendrils. They generally twining around the support.
Creeper	A plant which spread along the ground.
Habit	The growth form of a plant such as herb, shrub, tree.
Habitat	Habitat is a place where a particular plant found.
Herb	A small, non-woody plant (usually annual).
Liana	A long-stemmed, generally woody climber that are rooted in the soil and often climbing on trees as well as other means for support.
Parasitic plant	A plant which obtains all or part of its nutritional requirements from another living plant (host).
Perennial	Plant that live more than two years.
Shrub	A much-branched, woody, perennial plant; usually less than 5 m high and multi-stemmed at ground level.
Tree	A woody, perennial plant with an elongated stem or trunk, which usually support branches.
Vine	A plant which climb to the support or creeps along the ground.
Plant	
Dioecious	If the plants bear male and female flowers on separate plants then it is called dioecious plants.
Monoecious	If the plants bear separate male and female flowers on the same plant then it is called monoecious plant.
Stem	
Acaulescent	Without a stem or bearing inconspicuous stem.
Aerial	Stem above the ground.
Angular	Many angles in cross section.
Ascending	Becoming erect from a curved or slanted base.
Branched	With numerous branches.
Bulb	An underground, thick storage organ, comprising of a short stem and fleshy leaf bases.
Cladode	A compressed leaf-like photosynthetic stem.
Culm	Flowering and fruiting stem of sedges and grasses.
Decumbent	Lying along the ground but apex growing upwards.
Dichotomous	Stem dividing into two equal branches.
Diffuse	Stem branches spreading in all direction.
Erect	Growing upright.
Fistular	Tube-shaped; hollow stem.
Internode	Region of a stem between two nodes.
Laticiferous	Bearing latex.
Lenticel	Lens-shaped or wart-like pore in the bark.
Node	The part of a stem, where leaves or branches originate.
Prickle	A sharp pointed extension on the bark or epidermis of a plant.
Procumbent	Spreading along the ground but not rooting at the nodes.
Prostrate	Growing flat on the ground.
Rhizome	A creeping underground stem, generally growing horizontally.
Scandent	Stem having a climbing habit.
Scape	A long, leafless flowering stalk arising from an underwater or underground stem.
Semiterete	Rounded on one side, but flat on the other.

Terms	Description
Spine	A stiff, sharp outgrowth.
Stolon	A creeping or horizontal stem rooting at the nodes.
Stoloniferous	Having stolons.
Terete	Circular in cross-section; more or less cylindrical without grooves or ridges.
Thorn	A sharp pointed, woody, usually a modified stem.
Trailing	Stem spreading along the ground; usually with the help of adventitious roots.
Tuber	An underground stem, swollen as a storage organ.
Tuberous	Producing tubers.
Twinning	Long stem, climbing above the ground by twisting their stem around a support.
Phyllotaxy	
Alternate	One leaf at each node.
Basal	Leaves at the base of the stem or plant.
Cauline	Leaves borne usually on the upper part of a stem.
Opposite	Pair of leaves occur at each node on opposite sides
Opposite-decussate	Opposite, with two successive pairs of leaves borne at right angles to each other.
Radical	Leaves found at the base of a plant.
Whorled	When three or more leaves occur at each node.
Types of Stipule	
Exstipulate	Without stipule.
Foliaceous	Stipule resembling a leaf.
Interpetiolar	Present in between the petioles of opposite leaves.
Ochreate	Existing in the form of a tubular sheath around the internode (characteristic of family Polygonaceae).
Leaf Margin	
Ciliate	Fringed with hairs (hairs on margin).
Crenate	Blunt-toothed; dentate with rounded teeth.
Crispate	Curly.
Cuspidate	With a sharp, elongated, rigid tip; tipped with a cusp.
Dentate	Sharp marginal teeth pointing outward
Denticulate	Finely toothed
Dissected	Divided into many segments
Entire	Smooth margin throughout (without toothing).
Involute	Margin rolled towards upper (adaxial) surface.
Repand	With wavy margin
Revolute	Margin rolling downwards (towards the abaxial surface)
Runcinate	Irregularly saw-toothed or pinnately incised with the lobes or teeth directed downwards.
Serrate	Saw-toothed; with marginal teeth pointing towards the apex.
Serrulate	Finely serrate; margin with very minute teeth.
Sinuate	Margin with deep, wave-like depressions.
Undulate	Slightly wave-like margin.
Leaf Apex	
Acuminate	Tapering to a long, narrow point.
Acute	The leaf apex ending into a short, sharp angle (less than 90°).
Apiculate	Leaf blade terminating in a short, sharp, slender tip.
Asymmetrical	With the blade shape different on each side of the midrib.
Caudate	Narrow tail-like at the tip.
Emarginate	Leaf apex with deep or shallow notch.

Terms	Description
Incised	Bearing sharp and irregular incisions.
Laciniate	Margin irregularly divided into deep narrow segments
Lobed	Being divided by many lobes
Mucronate	The tip ending in a short, sharp point.
Obcordate	Leaf tip deeply notched.
Obtuse	A blunt or rounded tip.
Pinnatified	In pinnate way; divided almost to the midrib.
Retuse	Slightly notched in a round apex.
Rounded	The leaf blade is rounded at the tip.
Subobtuse	Slightly blunted.
Truncate	The leaf blade is square at the apex.
Leaf Base	
Amplexicaul	Clasping the stem.
Attenuate	Leaf blade gradually tapers to a long narrow base.
Auriculate	Bearing ear-shaped appendages near the petiole.
Cordate	Base is broad and notched; heart-shaped.
Cuneate	Leaf blade tapers to a narrow wedge-shaped base.
Oblique	Each side of the base being unequal-sized.
Rounded	Leaf blade has a rounded base.
Truncate	Somewhat square at the base.
Leaf Shape	
Cordate	Heart-shaped in outline with broad and notched base.
Elliptic	Ellipse-shaped; widest in the middle and more or less rounded at both ends.
Hastate	Like an arrow head. Lobes near base directed outwards.
Lanceolate	Much longer than broad or shaped like a lance-head; broadest in the lower half and tapering towards the apex.
Linear	Long and very narrow with almost parallel sides; length is usually more than four times longer than the width.
Oblanceolate	Inversely lanceolate; broadest portion near the apex.
Oblong	Longer than wide with slightly parallel sides.
Obovate	Reversed ovate; broadest portion towards the apex.
Orbicular	Circular.
Ovate	Usually egg-shaped; narrower towards the apex and broadest portion near the base.
Palmate	A compound leaf with leaflets attaching to a common point.
Reniform	Shaped like a kidney.
Rhomboid	Rhombic or diamond-shaped.
Sagittate	Arrowhead-shaped. Lobes near base directed downwards.
Spathulate	Spoon shaped; broad at the tip and tapered towards the base.
Leaf Surface	
Glabrescent	Becoming glabrous
Glabrous	The surface is smooth, without hairs.
Glandular	Glands cover the surface.
Glaucous	Covered with a bluish-white waxy coating.
Hairy	Covered with hairs.
Pubescent	Short, soft hairs cover the surface.
Scabrous	Covered with short, prickly hairs.
Tomentose	Covered with wool-like hair.

Terms	Description
Villous	Long and soft hairs cover the surface.
Types of Inflorescence	
Catkin	A spike, usually with long pendulous axis which bear unisexual flowers.
Compound corymb	Branched corymb.
Corymb (Corymbose)	A type of racemose inflorescence in which lower stalks are longer than the upper ones. In this type; lower and upper stalks reaching about the same height, thus the inflorescence give flat-topped appearance.
Cyathium	A special type of inflorescence in which a large female flower are surrounded by several male flowers in a cup shaped involucre that have nectar secreting glands. (Common in Euphorbia spp.)
Cyme (Cymose)	A type of inflorescence in which the main axis and lateral axis terminated in a flower. In such type; terminal flower is oldest and younger flowers, originate below the terminal one.
Dichasial cyme	A cyme inflorescence in which the lateral branches are opposite and have equal length.
Head or Capitulum	It is characteristic inflorescence in the family Compositae where dense cluster of sessile or sub-sessile flowers (florets) are arranged on the receptacle. The flowers (florets) are usually consists of marginal ray florets and centrally disc florets.
Panicle	Compound (branched) raceme.
Raceme (Racemose)	An indeterminate, unbranched inflorescence in which the main axis does not terminate in a flower but it keeps on growing continuously and produces a series of pedicellate flowers laterally. In such type, older flowers are present at the base and younger flower.
Solitary	Single flowered.
Spadix	A spike-like inflorescence with very small flowers often surrounded by a large sheathing bract or spathe.
Spike	Unbranched, indeterminate inflorescence with sessile flowers (flowers without pedicel) along the axis.
Spikelet	A small spike; characteristic inflorescence of the family Cyperaceae and Poaceae with one to many florets (flowers). Each floret is enclosed by two bracts, lemma and palea.
Umbel	An inflorescence in which all the pedicels are usually of equal length and originate from the common point.
Flower	
Actinomorphic	The flower which have radial symmetry.
Bisexual/hermaphrodite	Flower having both male and female reproductive organs.
Bracteate	Flower with bract.
Complete	Flower containing sepals, petals, stamens, and pistil.
Ebracteate	Flowers without bracts.
Incomplete	Flower lacking sepals, petals, stamens, and/or pistils.
Pedicellate	Flower with pedicel.
Pentamerous	Flowers in which various floral parts are arranged in groups of five.
Pistillate	Flowers containing only female parts.
Sessile	Without a stalk.
Staminate	Flowers containing only male parts.
Unisexual	Flower with one sex only; either male (stamen) or female (pistil).
Zygomorphic	The flower which have bilateral symmetry.
Flower's parts	
Anther	Terminal end of filament, which bears pollen.
Bract	A leaf-like structure (modified leaf) which develop in the axil of flower.
Bracteole	Small bract borne on the pedicel of the flower.
Calyx	The outer whorl of a flower (whorl of sepals), usually green.
Calyx tube	A tube formed by the fusion of the sepals (calyx), at least at the base.
Carpel	The basic unit of gynoecium; or floral organ that bears ovules.
Corolla	Collective term for the petals of a flower; whorl of petals.
Epicalyx	A whorl of bracts, just below the true calyx.
Filament	The slender stalk of a stamen below the anther.

Terms	Description
Ovary	Swollen lower portion of a carpel or many fused carpels.
Pedicel	The stalk of an individual flower.
Perianth	A collective term for calyx and corolla. Term generally used when calyx and corolla jointed and undifferentiated from each other.
Petal	Individual unit of corolla.
Pistil/Gynoecium	The female part of a flower; comprising of one or more carpels.
Sepal	Individual unit of calyx.
Stamen	A stamen is a male reproductive organ of a flower, consisting of anther and filament.
Stigma	Sticky end of style, which receives the pollen.
Style	Long, slender part of a carpel that connects stigma and ovary.
Tepal	Individual unit of perianth.
Types of fruit	
Achene	The achene is 1-seeded, dry, indehiscent fruit formed from a monocarpellary superior ovary.
Berry	A fleshy, indehiscent fruit with seed or seeds embedded in the pulp.
Capsule	A dry, dehiscent fruit developing from two to more united carpels.
Caryopsis	A dry indehiscent fruit of family Poaceae formed from monocarpellary ovary.
Drupe	It is one-seeded indehiscent fruit with a stony endocarp, fleshy mesocarp and thin epicarp.
Follicle	A dry, dehiscent fruit formed from a single carpel and splitting only along one suture.
Hypanthodium	It is a special type of inflorescence found in some members of family Moraceae (eg. Ficus) where receptacle form a pear-shaped hollow cavity with an opening (ostiole) at the apex. Both male and female flower are arranged on the inner surface of the hollow cavity
Pod	A dry 1 to many seeded dehiscent fruit with marginal placentation, dehisces along two sutures. The term is usually applied to the fruits of Leguminosae family.
Samara	Samara is a simple, indehiscent, dry and winged fruit, eg. *Holoptelea integrifolia.*
Silicle	The silicle is identical to siliqua except it is much shorter (as broad as or even broader than long).
Siliquae	An elongated, slender, dry, dehiscent fruit derived from two fused carpels, with parietal placentation and separated into two loculi by a false septum. It is characteristic fruits of the mustard family (Brassicaceae).
Sorosis	A fleshy multiple fruit formed from the ovaries of many flowers, eg. Mulberry.
Utricle	The utricle is a very small, bladder-like, dry, indehiscent, one seeded fruit.

Index – A. Wild and Horticultural Flora

Bold font denotes part of the flora

C

D

Index – B. Vernacular and Common Names of Plants

SN	Hindi/Vernacular Name	Sanskrit Name	Common Name	Botanical Name	Page No.
15	**Arni** (अर्नी)	Agnimantha (अग्निमंथा), Sriparni (श्रीपर्णी)	Arni	*Clerodendrum phlomidis* L.f.	252
16	**Arus** (अरुस), Arusa (अडूसा)	Vasaka (वसाका), Vasuka (वसुका), Atarusa (अतरुसा), Pra-Madya (प्रामाद्य)	Malabar Nut, Vasaka, Adhatoda, Adulsa	*Justicia adhatoda* L.	242
17	**Aswagandha** (अश्वगंधा), Asgandh (असंगध)	Aswagandha (अश्वगंधा), Ashwagandhi (अश्वगंधी)	Indian Ginseng, Winter Cherry	*Withania somnifera* (L.) Dunal	223
18	**Atrilal** (अत्रीलाल)	Nadikanta (नादीकांता), Kakajangha (काकजंघा), Sulomasha (सुलोमशा)	Panicled Foldwing, Kali Anghedi	*Peristrophe bicalyculata* (Retz.) Nees	245
19	**Babool** (बबूल), Kikar (कीकर)	Babbula (बब्बूला)	Gum Arabic Tree, Babool	*Acacia nilotica* (L.) Delile	89
20	…….	…….	**Baby Jump Up**	*Mecardonia procumbens (Mill.) Small*	234
21	**Bachita** (बचीता), Lapetua (लपेटुआ)	…….	Caesar weed	*Urena lobata* L.	72
22	**Bagra** (बागरा), Hurhur (हुरहुर), Pili Hulhul (पीली हुलहुल)	…….	Asian Spider Flower, Dog Mustard	*Cleome viscosa* L.	45
23	**Bakain** (बकैन), Mahaneem (महानीम)	Mahanimbah (महानिंब)	Chinaberry Tree, Persian Lilac, Bead Tree	*Melia azedarach* L.	81
24	**Ban Dhania** (बन धनिया)	…….	Corn Spurry	*Spergula arvensis* L.	52
25	**Ban long** (बन लौंग)	Bhu lavangah (भूलवंग)	Mexican Primrose-Willow, Swamp Primrose, Water Primrose	*Ludwigia octovalvis* (Jacq.) P.H.Raven	135
26	**Ban long** (बन लौंग), Bishkatali (बिष्टकाटली)	Bhu lavangah (भूलवंग)	Linear Leaf Water Primrose, Primrose Willow, Swamp Primrose, Water Primrose	*Ludwigia hyssopifolia* (G.Don) Exell	134
27	**Ban Methi** (बन मेथी), Morila (मोरिला)	…….	Indian Sweet Clover, Yellow Sweet Clover	*Melilotus indicus* (L.) All.	113
28	**Ban Tulsi** (बन तुलसी), Kala Bhangra (काला भांगरा)	…….	Croton, Rushfoil	*Croton bonplandianus* Baill.	280
29	**Bandra** (बंदरा), Bandri (बंदरी), Ban Kauni (बन कौनी), Siun (सीऊं)	…….	Yellow Fox-Tail Grass, Yellow Bristle Grass	*Setaria pumila* (Poir.) Roem. & Schult.	366
30	**Bara Nagar Motha** (बड़ा नागर मोथा), Motha (मोथा)	Mustaka (मुस्तका), Varida (वरिदा), Chakranksha (च्रकांक्षा), Charukesara (चारुकेसरा)	Nut Sedge, Common Nut Sedge, Nut Grass, Jawa Grass, Coco Grass	*Cyperus rotundus* L.	335
31	**Baraira** (बरैरा)	Bala (बाला).	Common Wireweed, Common Fanpetals	*Sida acuta* Burm.f.	67
32	**Barik Chirayata** (बारीक चिरायता), Khet Chirayata (खेत चिरायता)	…….	Pink Centaury, Field Centaury	*Centaurium pulchellum* (Sw.) Druce	191
33	**Barna** (बर्ना), Barni (बर्नी)	…….	Garlic Pear Tree, Caper Tree, Three-Leaf Caper	*Crateva adansonii* subsp. *odora* (Buch.-Ham.) Jacobs	48
34	**Bathua** (बथुआ)	…….	Goosefoot, Wild Spinach, Lamb's-Quarters	*Chenopodium album* L.	269
35	…….	…….	**Bearded Knotweed**	*Persicaria barbata* (L.) H.Hara	276
36	**Behaya** (बेहया), Besharam (बेशरम)	…….	Bush Morning Glory	*Ipomoea carnea* Jacq.	203

SN	Hindi/Vernacular Name	Sanskrit Name	Common Name	Botanical Name	Page No.
60	**Chakwar** (चकवड़)	Cakramarda (चक्रमर्दा), Prapunnata (प्रपुण्णता)	Sicklepod, Chinese Senna	*Senna obtusifolia* (L.) H.S.Irwin & Barneby	119
61	**Chamsuru** (चम्सुरु)	…….	Yellow Marsh Cress, Bog Yellow-cress	*Rorippa palustris* (L.) Besser	42
62	**Chatri** (छतरी), Atri (अत्रि), Matri (मात्री)	…….	Vetch, Garden Vetch, Black-Pod Vetch, Common Vetch	*Vicia sativa* L.	129
63	**Chaupatiya** (चौपतिया), Susni (सुस्नि)	Chatuh Patri (चतु:पत्री), Chatuspatri (चतुष्पत्री), Svastika (स्वस्तिका)	Four Leaf Clover, Water Shamrock, Common Water Clover	*Marsilea quadrifolia* L.	375
64	**Chiki** (चिकी)	Jhinjharita (झिझारिता), Gippit (गिपपित)	Diamond Burbark, Chinese Bur	*Triumfetta rhomboidea* Jacq.	71
65	**Chilbil** (चिलबिल), Papri (पापड़ी)	Chirivilva (चिरिविलवा), Pootikaranja (पूतिकरंजा)	Indian Elm, Indian Beech Tree, Jungle Cork Tree	*Holoptelea integrifolia* Planch.	296
66	**Chirchita** (चिरचिटा), Latjira (लटजीरा)	Apamarga (अपामार्ग)	Prickly Chaff Flower, Devil's Horsewhip	*Achyranthes aspera* L.	261
67	**Chitrabeej** (चित्रबीज)	…….	Chocolate Weed	*Melochia corchorifolia* L.	66
68	Chota Dhatura (छोटा धतूरा), Ghaghra (घाघरा), Sankhahuli (संखाहुली)	Arishta (अरिष्ट), Medhya (मेध्य), Sarpakshi (सर्पक्षी)	Common Cocklebur, Clotbur	*Xanthium strumarium* L.	180
69	**Choti Dudhi** (छोटी दुधी)	…….	Prostrate Sandmat, Prostrate Sandmat	*Euphorbia prostrata* Aiton	285
70	…….	…….	**Comb Rungia**	*Rungia pectinata* (L.) Nees	246
71	…….	…….	**Common Duckweed,** Common Waterlens, Duck meat.	*Lemna perpusilla* Torr.	325
72	…….	…….	**Common Duck Meat**, Duckweed	*Spirodela polyrrhiza* (L.) Schleid.	327
73	…….	…….	**Common Fringe Sedge**, Tall Fringe Rush, Forked Fimbry	*Fimbristylis dichotoma* (L.) Vahl	337
74	…….	…….	**Common Small Justicia**	*Justicia japonica* Thunb.	244
75	…….	…….	**Cone Catchfly**, Weed Campion, Weed Silene	*Silene conoidea* L.	51
76	…….	…….	**Curly-Leaf Pondweed**	*Potamogeton crispus* L.	330
77	**…….**	…….	**Cutleaf Blumea,** Cutleaf False Oxtongue.	*Blumea laciniata* (Wall. ex Roxb.) DC.	162
78	**Dabh** (दाभ), Davoli (दावोली)	…….	Daabh, Salt Reed-Grass, Halfa Grass	*Desmostachya bipinnata* (L.) Stapf	347
79	**Dabh** (दाभ), Uloo (उलू)	Darbha (दरभा).	Blady Grass, Cogon Grass, Kunai Grass, Japanese Blood Grass	*Imperata cylindrica* (L.) Raeusch.	356
80	**Dagadia** (दगडिया)	…….	Silky Indigo	*Indigofera astragalina* DC.	103
81	**Dahia** (दहिया), Sihora (सीहोरा)	Shakotakah (शाखोटक:)	Sand Paper Tree, Toothbrush tree	*Streblus asper* Lour.	304
82	**Daman Pappar** (दामन पप्पड़), Pitpapra (पित्तपापड़ा)	Parpatah (पर्पट:)	Diamond Flower, Corymbose Hedyotis	*Oldenlandia corymbosa* L.	152
83	Dhaincha (ढैंचा), Ikad (इकड़), Dadon (दादों), Daden (ददेन)	…….	Prickly Sesban	*Sesbania bispinosa* (Jacq.) W.Wight	122
84	**Dila** (दिला), Motha (मोथा)	…….	Small-Flowered Nut Sedge, Variable Flat Sedge, One-Arm Sedge, Rice Sedge, Small-Flower Umbrella-Sedge	*Cyperus difformis* L.	333

SN	Hindi/Vernacular Name	Sanskrit Name	Common Name	Botanical Name	Page No.
112	**Indrayan** (इन्द्रायन)	Gavakshi (गावकशी), Chitra (चित्रा), Indravaruni (इंद्रवारुणी)	Colocynth, Desert Gourd	*Citrullus colocynthis* (L.) Schrad.	137
113	....	....	**Ivy Woodrose**	*Merremia hederacea* (Burm. f.) Hallier f.	209
114	**Jal Buti** (जल बूटी)	....	Frog Fruit, Creeping Lip Plant	*Phyla nodiflora* (L.) Greene	249
115	**Jal Kumbhi** (जल कुम्भी)	Jalakumbhi (जल कुम्भी), Variparni (वरिपर्णी)	Water Hyacinth	*Eichhornia crassipes* (Mart.) Solms	315
116	**Jaldhaniya** (जलधनिया)	Kandakatuka (कंदकतुका), Kandira (कंदिरा)	Cursed Buttercup, Poisonous Buttercup, Celery-Leaved Buttercup, Blister Buttercup	*Ranunculus sceleratus* L.	29
117	**Jalkumbhi** (जलकुम्भी)	....	Water Cabbage, Water Lettuce	*Pistia stratiotes* L.	326
118	**Jamun** (जामुन)	Jambula (जंबूल)	Java Plum, Jamun, Black Plum.	*Syzygium cumini* (L.) Skeels	130
119	**Jangal Tor** (जंगल तोर), Ban Kulthi (बन कुलथी), Ban Kultha (बन कुलथा)	....	Showy Pigeonpea, Peanut Grass, Tropical Clover	*Cajanus scarabaeoides* (L.) Thouars	96
120	**Jangali Chachinda** (जंगली चचिंडा), Jangali Chichonda (जंगली चिचौंडा)	....	Wild Snake Gourd	*Trichosanthes cucumerina* L.	141
121	**Jangali Jai** (जंगली जई)	....	Sterile Oat, Winter Wild Oat	*Avena sterilis* L.	339
122	**Jangali Marua** (जंगली मरुआ)	Nandimukhi (नन्दीमुखी)	Goosegrass, Indian Goosegrass, Crowfoot Grass	*Eleusine indica* (L.) Gaertn.	352
123	**Jangli Anjir** (जंगली अंजीर)	....	Indian Fig	*Ficus palmata* Forssk.	300
124	Jangli Bathua (जंगली बथुआ), Khartua (खरतुआ)	....	Nettle-Leaved Goosefoot, Australian Spinach	*Chenopodium murale* L.	270
125	**Jangli Chaulai** (जंगली चौलाई)	....	Slender Amaranth, Green Amaranth	*Amaranthus viridis* L.	266
126	**Jangli Hala** (जंगली हाला)	....	Bitter Cress, Lesser Swine-cress, Wart Cress	*Lepidium didymum* L.	41
127	**Jangli Jhangora** (जंगली झंगोरा), Jharwa (झरवा), Shama (शामा)	....	Jungle Rice, Shama Millet, Small Barnyard Grass	*Echinochloa colona* (L.) Link	350
128	**Jangli Jowar** (जंगली-जोवार), Baru (बारू), Chinna (चिन्ना)	....	Johnson Grass, Aleppo Grass, Aleppo Millet Grass	*Sorghum halepense* (L.) Pers.	368
129	**Jangli jute**, Chonch (चोंच), Hade-ka-khet (हाडे-का-खेत)	Chunchuh (चुनचुह)	East Indian Mallow, Wild Jute, West African Mallow	*Corchorus aestuans* L.	61
130	**Jangli Mattar** (जंगली मटर)	....	Yellow Pea, Yellow Vetchling	*Lathyrus aphaca* L.	107
131	**Jangli Palak** (जंगली पालक)	....	Toothed Dock, Aegean Dock	*Rumex dentatus* L.	279
132	**Jangli Pudina** (जंगली पुदीना), Gha Buti (घा बूटी), Bhakumbar (भकुंवर)	....	Billygoat-Weed, Chick Weed, Goatweed	*Ageratum conyzoides* (L.) L.	156
133	**Jangli Tambakoo** (जंगली तम्बाकू)	....	Tex-Mex Tobacco	*Nicotiana plumbaginifolia* Viv.	213

SN	Hindi/Vernacular Name	Sanskrit Name	Common Name	Botanical Name	Page No.
156	**Karad** (कराड़), Delhi Grass (डेल्ही ग्रास), Sheda Grass (षेड ग्रास), Jargu Grass (जरगू ग्रास)	....	Marvel Grass, Ringed Dichanthium	*Dichanthium annulatum* (Forssk.) Stapf	348
157	**Kasunda** (कसुंडा), Bari Kasondi (बड़ी कसौंदी)	Kasamarda (कसमरदा), Vimarda (विमर्दा), Arimarda (अरिमर्दा)	Coffee Senna, Negro Coffee	*Senna occidentalis* (L.) Link	120
158	**Kateli** (कटेली), Oonth Kateli (ऊंट कटेली)	Kantakari (कंटकारी), Vyaghri (व्याघ्री)	Thorny Nightshade, Yellow Berried Nightshade	*Solanum surattense* Burm. f.	220
159	**Khajoor** (खजूर)	....	Wild Date Palm, Silver Date Palm	*Phoenix sylvestris* (L.) Roxb.	321
160	**Khaki-Weed** (खाकी-वीड), Kante Wali Santhi (कांटेवाली साठी), Isharmul (ईशरमूल)	....	Chaff-Flower	*Alternanthera pungens* Kunth	264
161	**Khal-muriya** (खाल-मुरिया), Tal-muriya (ताल-मुरिया), Ghamra (घमरा)	Jayanti Veda (जयंती वेदा)	Tridax Daisy, Coat Buttons, Mexican Daisy	*Tridax procumbens* (L.) L.	179
162	**Kharenti** (खरैंटी)	....	False Mallow, Broom Weed, Clock Plant, Prickly Malvastrum	*Malvastrum coromandelianum* (L.) Garcke	65
163	**Kharinta** (खरींटा), Kharenti (खरैंटी)	Balaa (बाला) (yellow-flowered var.), Sumanganaa (सुमन गाना), Bhadrabalaa (भद्रबाला), Bhadraudani (भदरौदानी)	Heart-Leaf Sida, Bala, Country Mallow, Flannel Weed	*Sida cordifolia* L.	69
164	**Khas-Khas Ghas** (खस-खस घास)	....	Vetiver, Khas-Khas Grass	*Chrysopogon zizanioides* (L.) Roberty	344
165	**Khatti-mithi** (खट्टी-मीठी)	....	Large-Flowered Pink Sorrel, Lilac Oxalis, Pink Wood Sorrel	*Oxalis debilis var. corymbosa* (DC.) Lourteig	77
166	**Khejri** (खेजड़ी), Sami (सामी)	....	Khejri Tree	*Prosopis cineraria* (L.) Druce	115
167	**Khubkalan** (खूबकलाँ), Asalio (असलियो)	Khakasi (खाकसी), Khubakala (खूबकला)	London Rocket	*Sisymbrium irio* L.	43
168	**Ko Kaa** (को का)	Kumud (कुमुद)	White Water Lily, White Lotus	*Nymphaea pubescens* Willd.	33
169	**Kodo** (कोदो)	....	Kodo Millet, Scrobic Paspalum	*Paspalum scrobiculatum* L.	360
170	**Kokhima** (कोखिमा), Gadar-tambaku (गदर-तम्बाकू), Kalhar (कल्हार)	Bhutakeshi (भूतकेशि)	Chinese Mullein	*Verbascum chinense* (L.) *Santapau*	226
171	**Kulata** (कुलटा), Kulthi (कुलथी)	....	Papra	*Rhynchosia capitata* (Roth) DC.	117
172	**Kulata** (कुलटा), Kulthi (कुलथी)	....	Burn-Mouth-Vine, Least Snout-Bean, Rhynchosia	*Rhynchosia minima* (L.) DC.	118
173	**Kumra** (कुमरा), Kumur (कुमुर)	....	Black-Jack, Beggar Tick, Cobbler's Pegs, Spanish Needle	*Bidens pilosa* L.	160
174	**Kumudini** (कुमुदिनी)	....	Crested Floating Heart	*Nymphoides cristata* (Roxb.) Kuntze	192
175	**Kundru** (कुंदरू)	....	Ivy Gourd, Scarlet Gourd	*Coccinia grandis* (L.) Voigt	138

SN	Hindi/Vernacular Name	Sanskrit Name	Common Name	Botanical Name	Page No.
200	**Mumiri** (मुमीरी)	…….	Shepherd's Purse, Blind weed	*Capsella bursa-pastoris* (L.) Medik	39
201	**Musakarni** (मुसाकर्णी)	Vishnukrantha (विष्णुक्रान्त:)	Roundleaf Bindweed, Agracejo Rastrero	*Evolvulus nummularius* (L.) L.	201
202	**Mustaru** (मुस्तरु), Bhediachim (भेदियाछिम)	…….	Madras Carpet	*Grangea maderaspatana* (L.) Poir	173
203	…….	…….	**Naked-Stem Carpetweed**, Daisy-leaved Chickweed	*Mollugo nudicaulis* Lam.	144
204	**Narkul** (नरकुल), Nal (नाल), Doka-Ghas (डोका-घास)	…….	Tall Reed	*Phragmites karka* (Retz.) Trin. ex Steud.	362
205	**Neel** (नील), Dharti Dhak (धरती धक्), Krishnaneel (कृष्णनील)	…….	Blue Pimpernel	*Anagallis arvensis* L.	183
206	**Neelini** (नीलिनी), Neel (नील)	Neelika (नीलिका), Neelapushpa (नीलपुष्पा)	True Indigo, Common Indigo	*Indigofera tinctoria* L.	105
207	**Neem** (नीम)	Arishtha (अरिष्ठा)	Neem, Indian Lilac, Margosa	*Azadirachta indica* A. Juss.	79
208	**Nisoth** (निसोथ), Panila (पनिला), Pithori (पिठोरी);	Triputa (त्रिपुता), Nishotra (निशोत्रा)	Transparent Wood Rose, White day glory, Indian Jalap	*Operculina turpethum* (L.) Silva Manso	210
209	…….	…….	**Nut Grass,** Yellow Nut Sedge. Chufa Sedge	*Cyperus alopecuroides* Rottb.	332
210	**Palash** (पलाश), Dhak (डाक), Tesu (टेसू)	Palasha (पलाशा)	Flame of the Forest, Bastard Teak	*Butea monosperma* (Lam.) Taub.	94
211	**Pan bel** (पान बेल)	Vachagandha (वचगंधा), Lakshmana (लक्ष्मना)	Obscure Morning Glory	*Ipomoea obscura* (L.) Ker Gawl.	206
212	**Panchpatia** (पंचपतिया)	…….	Tiger Foot Morning Glory	*Ipomoea pes-tigridis* L.	207
213	**Panghas** (पानघास)	…….	Pond Lovegrass	*Eragrostis japonica* (Thunb.) Trin.	353
214	**Panirak** (पनिरक)	…….	Cheeseweed Mallow, Small-Flowered Mallow	*Malva parviflora* L.	64
215	**Papara** (पापड़ा), Pittpapra (पित्तपापड़ा)	Parpata (परपता)	Indian Fumitory	*Fumaria indica* (Hausskn.) Pugsley	36
216	**Para Ghas** (पारा घास)	…….	Running Grass, Creeping Panic Grass, Sprawling Signal Grass	*Brachiaria reptans* (L.) C.A.Gardner & C.E.Hubb.	340
217	**Patera** (पटेरा), Pater (पटेर)	…….	Elephant Grass, Reed, Lesser Indian Reed Mace, Small Bulrush, Narrow leaf cattail	*Typha domingensis* Pers.	323
218	**Pat-Sag** (पात-साग), Mitha-Pat (मीठा-पात)	Mahachanchu (महाचंचु)	Nalta Jute, Tossa Jute	*Corchorus olitorius* L.	62
219	**Paviri** (पविरी)	…….	Chicken Weed, Wild Purslane	*Portulaca quadrifida* L.	56
220	**Pedda-Sama** (पेड्डा-समा), Popti (पोपटी)	…….	Browntop Millet	*Brachiaria ramosa* (L.) Stapf	340
221	**Peeli Duddhi** (पीली दुद्धी, Jangali gobhi (जंगली गोभी)	…….	Creeping Launaea	*Launaea procumbens* (Roxb.) Ramayya & Rajagopal	174
222	**Peeli kateri** (पीली कटेरी), Kandiari (कंडिअरी)	…….	Wild Safflower, Jewelled Distaff Thistle	*Carthamus oxyacantha* M.Bieb.	166
223	**Peepal** (पीपल)	Plaksha (प्लक्ष), Bodhivriksha (बोधिवृक्ष), Ashwattha (अश्वथामा)	Peepal, Bo Tree, Bodhi Tree, Holy Fig Tree, Sacred Fig Tree	*Ficus religiosa* L.	301

SN	Hindi/Vernacular Name	Sanskrit Name	Common Name	Botanical Name	Page No.
246	**Sakarkandi** (सकरकंदी), Sankooli (संकूली)	....	Common Floss Flower	*Blumea obliqua* (L.) Druce	163
247	**Sanwak** (संवक), Kayada (कायदा)	Varuka (वारुका)	Barnyard Grass, Japanese Millet, Jungle Rice, Cockspur Grass, Prickly Grass	*Echinochloa crus-galli* (L.) P.Beauv.	351
248	**Sarphonk** (सरफोंक), Sharpunkha (शरपुंखा)	....	Wild Indigo, Fish Poison	*Tephrosia purpurea* (L.) Pers.	125
249	**Satyanashi** (सत्यानाशी), Kataila (कटैला)	....	Mexican Prickly Poppy	*Argemone mexicana* L.	35
250	**Seeta-Bani** (सीता-बानी)	....	Redstem Wormwood	*Artemisia scoparia* Waldst. & Kitam.	159
251	**Sevar** (सेवार), Jhangi (झंगी)	....	Waterthyme, Hydrilla.	*Hydrilla verticillata* (L.f.) Royle	312
252	....	....	**Shaggy Wild Indigo,** Hoary Tephrosia	*Tephrosia villosa* (L.) Pers.	126
253	**Shahtoot** (शहतूत), Kattha (कट्ठा)	....	White Mulberry, Silkworm Mulberry	*Morus alba* L.	302
254	**Shankhpushpi** (शंखपुष्पी)	....	Prostrate Bindweed	*Convolvulus prostratus Forssk.*	197
255	**Shisham** (शीशम), Sisu (सिसु)	....	Indian Rosewood, North Indian Rosewood, Sisu	*Dalbergia sissoo* DC.	99
256	**Siris** (सिरिस)	Sirisah (सिरिसः)	Lebbeck Tree, Woman's Tongues Tree	*Albizia lebbeck* (L.) Benth.	90
257	**Sival** (सिवाल), Shaival (शैवाल)	....	Tape Grass, Straight Vallisneria	*Vallisneria spiralis* L.	313
258	....	....	**Smooth Chaff Flower**	*Alternanthera paronychioides* A.St.-Hil.	262
259	....	....	**Sticky Blumea**, Clammy False Oxtongue	*Blumea viscosa* (Mill.) V.M.Badillo	164
260	**Subabul** (सुबबूल)	....	Leucaena, Ipil-Ipil, White Babool, Subabul	*Leucaena leucocephala* (Lam.) de Wit	107
261	**Sugandha Vastook** (सुगंध वास्तुक)	Sugandha-Vastooka (सुगंध वास्तुका), Kshetra-Vaastuuka (क्षेत्र-वास्तुका)	Mexican Tea	*Dysphania ambrosioides* (L.) Mosyakin & Clemants	272
262	....	....	**Thale Cress**, Mouse-Ear Cress	*Arabidopsis thaliana* (L.) Heynh	38
263	....	....	**Thyme-Leaved Sandwort**	*Arenaria serpyllifolia* L.	50
264	**Tipatiya** (तिपतिया), Kudaliya (कुदालिया)	Hamsapadi (हंसपदी), Tripadi (त्रिपदी)	Creeping Tick Trefoil, Tropical Trefoil, Three-Flower Beggarweed	*Desmodium triflorum* (L.) DC.	101
265	**Tipatiya Ghaas** (तिपतिया घास)	....	Woolly Clover, Cottonball Clover	*Trifolium tomentosum* L.	128
266	....	....	**Three Corner Jack,** Devil's Thorn, Double Gee	*Emex australis* Steinh.	275
267	....	....	**Tropical Cupgrass,** Cup Grass, Spring Grass.	*Eriochloa procera* (Retz.) C.E.Hubb.	354
268	**Vajradanti** (वज्रदंती)	Artagalah (आर्तगल), Dasi Kurantakah (दासी कुरण्टक)	Porcupine Flower	*Barleria prionitis* L.	241
269	**Vilaiti Tulsi** (विलायती तुलसी)	Bhustrna (भुस्त्रैण)	American Mint, Bush Mint	*Hyptis suaveolens* (L.) Poit.	253
270	**Vishakhapara** (विषखपरा), Sabuni (साबुनी)	Chiratika (चिराटिका), Dhanapatra (धानपत्रा)	Giant pigweed, Horse-Purslane	*Trianthema portulacastrum* L.	145
271	....	....	**Wild Crabgrass,** Bamboo Grass, Summer Grass, Tropical Finger Grass	*Digitaria ciliaris* (Retz.) Koeler	349
272	....	....	**Wild Poinsettia,** Painted Spurge, Milkweed, Lesser Green Poinsettia	*Euphorbia heterophylla* L.	281

ABOUT THE AUTHORS

Dr. Jyoti Kumar Sharma

Dr. Jyoti Kumar Sharma, Former Director, Kerala Forest Research Institute (KFRI), Kerala, India, a Fellow of Phytopathological Society of India, has over 40 years' of research experience in the field of Forestry. He also held the post of the Member Secretary, Kerala State Biodiversity Board during 2005-07. He was the founding Professor and Dean, School of Environment and Natural Resources, Doon University, Dehradun. Currently, he is Professor & Head, Center for Environmental Sciences & Engineering, Shiv Nadar University. He has published more than 110 research papers. He served as an FAO/IPGRI Expert Member to Frame Guidelines for safe movement of Eucalypt germplasm in Bangkok, Thailand and Pine germplasm in Opocne, Prague, Czech Republic. He has worked as an International Forestry Consultant, FAO (UN) in Vietnam, Myanmar, Nepal, Bhutan, East Timor, and Thailand solving various forestry problems. He served as Chairman/ Coordinator, IUFRO Working Party on Diseases of Tropical Plantations, 1990-2000 and organized and chaired WP Symposium during XX World IUFRO Congress in Finland and during XXI World IUFRO Congress, Malaysia. He has been the recipient of several research fellowships awarded by British Council, Commonwealth Foundation, NORAD, JICA, and ITTO, Japan. He has served as a member of several expert committees of MoEF & CC and DBT, Govt. of India. Currently, he is a member of Research Coordination Committee, Central Silk Board, Ministry of Textiles, GOI.

Dr. Amit Kumar Tripathi

Dr. Amit Kumar Tripathi worked on "Quantitative assessment of floristic diversity and species populations in hill forests of Meghalaya" for his PhD degree in Botany awarded by North-Eastern Hill University (NEHU), Shillong, Meghalaya. Currently, he is Post-Doctoral Research Associate in the Center for Environmental Sciences & Engineering, Shiv Nadar University, working on Biodiversity documentation of Chithara Village Panchayat. Earlier, he worked in projects on "Mapping and quantitative assessment of geographic distribution and population status of plant resources of Eastern Himalayan Region" in the Department of Botany, NEHU, Shillong and conducted biodiversity surveys in National Parks, Wildlife Sanctuaries, Sacred Groves, Community Forests as well as Reserve Forests of Meghalaya. He has wide experience in Plant Taxonomy, especially field collection of wild plants and their identification. He has published research papers in reputed journals and attended several National Conferences and presented research papers.

Mohd Ahmad

Mohd Ahmad is M.Sc. in Botany and Post Graduate Diploma in Horticulture from Aligarh Muslim University, Aligarh, Uttar Pradesh. Currently, he is working in the Center for Environmental Sciences & Engineering, Shiv Nadar University, in research projects on Biodiversity documentation of Chithara Village Panchayat and Establishment of Thematic Botanic Garden in SNU. He is deeply involved with the collection of variety of plants and planting them in different themes in SNU Botanic Garden. He has keen research interest in development of new gardens and their landscaping, medicinal plants, biodiversity and Plant Taxonomy.